Springer Monographs *in* Mathematics

For further volumes:
http://www.springer.com/series/3733

Winfried Bruns · Joseph Gubeladze

Polytopes, Rings, and K-Theory

Springer

Winfried Bruns
FB Mathematik/Informatik
Universität Osnabrück
Osnabrück
Germany
wbruns@uos.de

Joseph Gubeladze
Department of Mathematics
San Francisco State University
San Francisco, CA 94132
USA
soso@math.sfsu.edu

ISBN 978-0-387-76355-2 e-ISBN 978-0-387-76356-9
DOI 10.1007/b105283
Springer Dordrecht Heidelberg London New York

Library of Congress Control Number: 2009926997

Mathematics Subject Classification (2000): 05E99, 11D04 11H06 13B21 13B26

Printed on acid-free paper

Springer is part of Springer Science+Business Media (www.springer.com)

To our children and grandchildren

Annika, Nils, Julia, Linus, Charlotte, and Sophie

Nino, Ana, Jimsher, Tamar, and Lazare

Preface

For every mathematician, ring theory and K-theory are intimately connected: algebraic K-theory is largely the K-theory of rings. At first sight, polytopes, by their very nature, must appear alien to surveyors of this heartland of algebra.

But in the presence of a discrete structure, polytopes define affine monoids, and, in their turn, affine monoids give rise to monoid algebras. Their spectra are the building blocks of toric varieties, an area that has developed rapidly in the last four decades.

From a purely systematic viewpoint, "monoids" should therefore replace "polytopes" in the title of the book. However, such a change would conceal the geometric flavor that we have tried to preserve through all chapters.

Before delving into a description of the contents we would like to mention three general features of the book: (1) the exhibiting of interactions of convex geometry, ring theory, and K-theory is not the only goal; we present some of the central results in each of these fields; (2) the exposition is of constructive (i. e., algorithmic) nature at many places throughout the text—there is no doubt that one of the driving forces behind the current popularity of combinatorial geometry is the quest for visualization and computation; (3) despite the large amount of information from various fields, we have strived to keep the polytopal perspective as the major organizational principle.

We start with polyhedra and triangulations (Chapter 1), which is the minimum of the structures, considered in the book. Important aspects of this chapter are the duality of cones and regular subdivisions of polytopes.

The interaction of convex bodies with the lattice of integer points quickly leads to normal affine monoids and Hilbert bases of rational cones, dealt with in Chapter 2. In the same chapter we discuss naive geometric characterizations of Hilbert bases that had been an open problem until a counterexample was found in 1998. Nevertheless there remain difficult problems in this area.

Chapter 3 presents one of the pearls in the theory of lattice polytopes—the Knudsen–Mumford theorem on unimodular triangulations of high multiples of lattice polytopes. It dates back to the origins of toric geometry in the early 1970s. Being conceived in the context of semistable reductions, it remained largely unnoticed by the combinatorial world. The situation changed when Sturmfels found the link between (unimodular) triangulations and Gröbner bases. In an attempt to develop an effective variant of the Knudsen–Mumford result, we derive dimensionally uniform polynomial bounds for unimodularly covered multiples of lattice polytopes, together with a companion result for rational cones.

Part II puts the affine monoids in their natural habitat of monoid algebras, where they have settled since the days of Gordan and Hilbert. Chapter 4 develops the basic theory. Already here one encounters divisor class groups and Picard groups—algebraic invariants outside the category of bare monoids. The discussion of seminormal monoid rings, prepared on the monoid level in Chapter 2, points forward to K-theory in Part III.

In Chapter 5 it is shown that an affine monoid ring remembers its combinatorial genesis: the underlying monoid is uniquely determined by the algebra structure of its monoid ring. In the same chapter we compute the group of graded automorphism of a normal affine monoid ring. Informally, these linear groups relate to the general linear groups in the same way that arbitrary lattice polytopes relate to unimodular simplices. The basic tools in this chapter come from the theory of linear algebraic groups, and Borel's theorem on maximal tori is used in a crucial way.

The origin of combinatorial commutative algebra is Stanley's very successful attempt to base the enumerative theory of linear diophantine systems and the Ehrhart theory of lattice polytopes on Hilbert functions of graded rings. A key theorem is Hochster's result that normal affine monoid algebras are Cohen–Macaulay. Our treatment in Chapter 6 is based on a minimum of homological algebra. For lack of space we have omitted local cohomology; in view of the extensive treatment in the books by Bruns and Herzog [72] and Miller and Sturmfels [261] this omission seems acceptable.

Chapter 7 develops Gröbner basis theory of toric ideals. It moves regular subdivisions of lattice polytopes center stage. Via degrees of (unimodular) triangulations and initial ideals it is related to the Koszul property, another major theme of the chapter. We do not go into finite free resolutions of toric rings, as this topic is extensively covered in [261].

Part III is devoted to K-theory. While the previous chapters consider affine monoid rings as graded objects, we now turn to projective modules that are invisible in the essentially local structure imposed by the grading—should nontrivial such modules exist at all. By a theorem of the second author, nontrivial (nongraded) projective modules do indeed not exist over affine monoid rings, provided the weakest necessary condition, namely seminormality, is satisfied. The proof of this result, based on the Quillen–Suslin theorem, is the main subject of Chapter 8. It is accompanied by a discussion of several related results, in particular the K_0-regularity of affine monoid rings.

In sharp contrast to the behavior of K_0, the K_1-groups of affine monoid rings, are much larger than the K_1-groups of their rings of coefficients, at least in a very large class. This is proved in Chapter 9, together with several accompanying topics, like the action of the general linear group on unimodular rows, and a survey of the nilpotence of higher K-groups over affine monoid rings. All the K-theoretical background for Chapters 8 and 9, except several short digressions into higher K-theory, is found in the classical books by Bass [19] and Milnor [262].

In the book there are two places of paradigmatic shift: from polytopes to rings in Chapter 4 and from rings to nonaffine varieties in Chapter 10. In the latter we have restricted ourselves to invariants reminiscent of those studied in the preceding chapters.

Namely, in the first part we survey Grothendieck groups, Chow groups, and intersection theory, and their rich interaction with each other via Grothendieck–Riemann–Roch and with the combinatorial structure of toric varieties. In full treatment, these topics would constitute a book of its own. Our survey, apart from original papers, draws from such well-known sources as Fulton [133], [134], and Oda [275].

In the second part of Chapter 10 two topics that extend the K-theory of Chapters 8 and 9 directly are highlighted by a full discussion: the existence of simplicial projective toric varieties with "huge" K_0-groups, and the triviality of equivariant vector bundles on a representation space of an abelian group.

Each chapter is accompanied by exercises. Their degrees of difficulty vary considerably, ranging from mere tests on the basic notions to guided tours into research. Some of the exercises will be used in the text. For such we have included extensive hints or pointers to the literature.

This book has many facets and draws from many sources. A priori it had been clear to us that a completely self-contained treatment was impossible. The theory of polyhedra and affine monoids is developed from scratch, while for ring theory and K-theory, let alone algebraic geometry, some prerequisites are necessary. For ring theory we expect some fluency, acquired by a study of an introductory text like Atiyah and Macdonald [10]. For projective modules (over the type of rings of our interest) we recommend Lam's book [237]. Sources for the background material in K-theory, toric varieties, and intersection theory have been mentioned above. At many places we have inserted small introductions to the concepts needed, citing the main theorems on which we build. This is also true for Chapter 10, but the reader needs a certain acquaintance with algebraic geometry, in the context of the first three chapters of Hartshorne [178].

The idea of writing this book came up in 2001 during the preparation of notes for the lecture series given by the first author at the summer school on toric geometry in Grenoble 2000 [64]. The actual implementation started in Spring 2003 when both authors were visiting Mathematical Sciences Research Institute in Berkeley for a semester-long program in commutative algebra. Over the years work on the book was supported by grants from DFG, DAAD, NSF and by the RiP program of the Mathematisches Forschungsinstitut Oberwolfach. The Universität Osnabrück and San Francisco State University hosted the respective authors several times. We thank all these institutions for their generous support.

We are grateful to all our friends, colleagues, and students who have helped us by providing valuable information or pointing out errors in preliminary versions, among them D. Cox, T. Römer, H. Spindler, and the members of the Osnabrück algebra seminar. J. Moyano, M. von Thaden and J. Uliczka read parts of the book and pointed out several inaccuracies. Our special thanks go to Bogdan Ichim for his very precise study of several chapters, and to Gesa Kämpf, Peyman Nasehpour, Peter Schenzel and Kirsten Schmitz for their support in the final production phase. We are indebted to Karl-Heinz Fieseler and Sam Payne for their most helpful suggestions for Chapter 10.

Osnabrück and San Francisco
October 2008

Winfried Bruns
Joseph Gubeladze

Contents

Part I Cones, monoids, and triangulations

Part II Affine monoid algebras

Part III K-theory

Part I

Cones, monoids, and triangulations

1

Polytopes, cones, and complexes

The algebraic objects discussed in this book are determined by two types of structures: continuous ones related to convexity and discrete ones determined by lattices. In this chapter we develop notions of convex geometry and of combinatorial topology related to convexity. The basic convex objects are polyhedra, polytopes, and cones, and the related combinatorial constructions are polyhedral complexes, triangulations, and fans. In the last section we unite convexity and lattice structures.

1.A Polyhedra and their faces

We use the symbols $\mathbb{Z}$, $\mathbb{Q}$, $\mathbb{R}$, and $\mathbb{C}$ to denote the integral, rational, real, and complex numbers respectively. The subsets of nonnegative elements in $\mathbb{Z}$, $\mathbb{Q}$, $\mathbb{R}$ will be referred to by $\mathbb{Z}_+$, $\mathbb{Q}_+$, $\mathbb{R}_+$. The symbol $\mathbb{N}$ stands for the set of positive integers.

Polyhedra are subsets of finite-dimensional $\mathbb{R}$-vector spaces V, but to some extent only the affine structure of V is essential. We recall that an *affine subspace* of V (of *dimension* d) is a subset of the form $v + U$ for some $v \in V$ and a vector subspace U (of dimension d); the empty set is also an affine subspace (of dimension -1). Equivalently, an affine subspace is a subset closed under all *affine linear combinations*

$$\sum_{i=1}^{n} a_i x_i, \qquad a_i \in \mathbb{R}, \quad \sum_{i=1}^{n} a_i = 1$$

(Exercise 1.1). For a subset X of V we let $\operatorname{aff}(X)$ denote the *affine hull* of X, i. e., the smallest affine subspace of V containing X. For $x_0, \dots, x_n \in V$ one has $\dim \operatorname{aff}(x_0, \dots, x_n) \le n$. If $\dim \operatorname{aff}(x_0, \dots, x_n) = n$, then $x_0, \dots, x_n$ are *affinely independent*; in other words, if $x = a_0 x_0 + \dots + a_n x_n$ is an affine combination of $x_0, \dots, x_n$, then $a_0, \dots, a_n$ are uniquely determined. They are called the *barycentric coordinates* of x with respect to $x_0, \dots, x_n$.

Affine subspaces $v + U$ and $v' + U'$ are *parallel* if $U \subset U'$ or $U' \subset U$. Subsets of V are parallel if their affine hulls are parallel.

A subset X of V is *convex* if it contains the *line segments*

W. Bruns, J. Gubeladze, *Polytopes, Rings, and K-Thoery*, Springer Monographs in Mathematics, DOI 10.1007/b105283_1, © Springer Science+Business Media, LLC 2009

$$[x, y] = \{ax + (1 - a)y : 0 \le a \le 1\}$$

for all points $x, y \in X$. (Open or halfopen line segments (x, y) and $(x, y]$ are defined respectively.) The intersection of convex sets is obviously convex. Therefore one may define the *convex hull* $\operatorname{conv}(X)$ of an arbitrary subset X of V to be the smallest convex subset of V containing X. If $X = \{x_1, \dots, x_n\}$ is finite, then

$$\operatorname{conv}(X) = \left\{ \sum_{i=1}^{n} a_i x_i \ : \ 0 \le a_i \le 1,\ i = 1, \dots, n,\ \sum_{i=1}^{n} a_i = 1 \right\},$$

and for arbitrary X the convex hull of X is just the union of the convex hulls of the finite subsets of X. Linear combinations as in the formula for $\operatorname{conv}(X)$ are called *convex linear combinations*.

The functions considered in affine geometry are the affine forms $\alpha : v \to \mathbb{R}$. An *affine form* is given by

$$\alpha(x) = \lambda(x) + a_0, \qquad a_0 = \alpha(0),$$

with a unique linear form λ on V. The space of linear forms on V is denoted by V^*. More generally, a map f from V into an $\mathbb{R}$-vector space W is *affine* if $f(v) = g(v) + w_0$ for some linear map $g : V \to W$ and some $w_o \in W$.

Affine forms are continuous with respect to the natural topologies on $V \cong \mathbb{R}^d$ and $\mathbb{R}$. So the *open halfspaces*

$$H_\alpha^> = \{x \in V : \alpha(x) > 0\},$$

associated with the nonconstant forms α are open sets. With such a form we associate also the *hyperplane*

$$H_\alpha = \{x \in V : \alpha(x) = 0\}$$

and the *closed halfspace*

$$H_\alpha^+ = \{x \in V : \alpha(x) \ge 0\}.$$

For flexibility of notation we set

$$H_\alpha^< = H_{-\alpha}^> \qquad \text{and} \qquad H_\alpha^- = H_{-\alpha}^+.$$

Every hyperplane H in V bounds exactly two open halfspaces that we denote by $H^>$ and $H^<$ (depending on the choice of α with $H = H_\alpha$). The corresponding closed halfspaces are denoted H^+ and H^-.

Topological notions are defined with respect to the standard topology. In dealing with polyhedra P we will usually have to consider the *relative interior* $\operatorname{int}(P)$ and the *relative boundary* ∂P of P that are taken within the affine subspace $\operatorname{aff}(P)$. Consequently, we drop the attribute "relative" from now on. The closure of P is denoted by $\overline{P}$. (There is no need for a relative closure.) Note that $\operatorname{int}(P) = P$ and $\overline{P} = P$ if $P = \{x\}$ consists of a single point.

Every neighborhood of 0 contains a basis of the vector space V. It follows that V is the affine hull of each of its open subsets.

The interior of a convex set X in V is also convex: if $x, y \in \operatorname{int}(X)$, then there exists an open neighborhood U of 0 such that $x + U, y + U \subset X$, and for $z \in [x, y]$ the neighborhood $z + U$ is contained in X.

We now introduce the central objects of this section.

Definition 1.1. A subset $P \subset V$ is called a *polyhedron* if it is the intersection of finitely many closed halfspaces. The *dimension* of P is given by $\dim \operatorname{aff}(P)$. A d-polyhedron has dimension d.

A *polytope* is a bounded polyhedron. A 2-polytope is called a *polygon.*

A *morphism of polyhedra* P and Q is a map $\varphi : P \to Q$ that can be extended to an affine map $\tilde{\varphi} : \operatorname{aff}(P) \to \operatorname{aff}(Q)$.

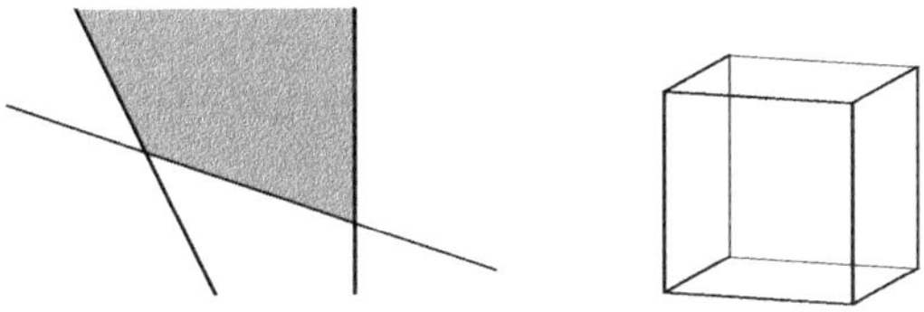

Fig. 1.1. A polyhedron and a polytope

Equivalently, P is a polyhedron if it is the set of solutions of a linear system of inequalities. Since halfspaces (open or closed) are convex, polyhedra are convex subsets of V, and the interior of a polyhedron is again convex (since the interior of a convex set is convex, as observed above). It will be shown in Section 1.C that a subset P of V is a polytope if and only if it is the convex hull of a finite subset of V.

In Proposition 1.29 we will see that $\varphi(P)$ is a polyhedron if φ is a morphism of polyhedra. For simplicity we will always consider a morphism $\varphi : P \to Q$ to be defined on $\operatorname{aff}(P)$ so that we need not distinguish between φ and $\tilde{\varphi}$.

It is convenient to always orient the halfspaces cutting out a polyhedron P in such a way that $P = H^+_{\alpha_1} \cap \dots \cap H^+_{\alpha_m}$. Often we simply write H^+_i for $H^+_{\alpha_i}$, assuming that a suitable affine form α_i has been chosen to define the halfspace.

From the representation of P as an intersection of halfspaces we can easily determine the affine hull $\operatorname{aff}(P)$.

Proposition 1.2. *Let $P = H^+_1 \cap \dots \cap H^+_m$ be a polyhedron. Then $\operatorname{aff}(P)$ is the intersection of those hyperplanes H_i, $i = 1, \dots, m$, that contain P.*

Proof. Let A be the intersection of the hyperplanes H_i containing P. Then P is contained in the affine space A. We replace V by A and all the halfspaces and hyperplanes by their intersections with A. Then the hyperplanes containing A and the corresponding halfspaces drop out. The proof of the proposition is thus reduced to showing that $\operatorname{aff}(P) = V$ if $P \not\subset H_i$ for $i = 1, \dots, m$. In this case P contains a point $x_i \in H^>_i$ for $i = 1, \dots, m$. It follows that

$$\frac{1}{m}(x_1 + \dots + x_m)$$

belongs to $H_i^>$ for $i = 1, \dots, m$. Therefore $P \supset H_1^> \cap \dots \cap H_m^> \neq \emptyset$ contains a nonempty open subset of V, and $V = \operatorname{aff}(P)$. □

Support hyperplanes, facets, and faces. If we visualize the notion of polytope in $\mathbb{R}^3$, then we see a solid body whose boundary is composed of polygons. These facets (unless they are affine subspaces) are bounded by line segments, called edges, and each edge has two endpoints. All these endpoints form the vertices of P. It is our first goal to describe the face structure for polyhedra of arbitrary dimension.

Definition 1.3. A hyperplane H is called a *support hyperplane* of the polyhedron P if P is contained in one of the two closed halfspaces bounded by H and $H \cap P \neq \emptyset$.

The intersection $F = H \cap P$ is a *face* of P, and H is called a *support hyperplane associated with* F. A *facet* of P is a face of dimension $\dim P - 1$. The polyhedron P itself and $\emptyset$ are the *improper faces* of P.

Faces of dimension 0 are called *vertices*, and $\operatorname{vert}(P)$ is the set of vertices of P. An *edge* is a face of dimension 1.

Below we will often use that the faces of P only depend on P in the following sense: if $V' \subset V$ is an affine subspace and $P \subset V'$, then the faces of P in V are its faces in V', and conversely. That a face in V is also a face in V' is trivial, and the converse follows immediately from the fact that a halfspace G^+ associated with a hyperplane G in V' can be extended to a halfspace H^+ of V such that $G^+ = H^+ \cap V'$, as observed above.

Another easy observation: if F is a face of P, then $F = \operatorname{aff}(F) \cap P$. In fact, $\operatorname{aff}(F)$ is contained in each support hyperplane associated with F.

Proposition 1.4. *Let $P \subset V$ be a polyhedron, and H a hyperplane such that $H \cap P \neq \emptyset$. Then H is a support hyperplane associated with a proper face of P if and only if $H \cap P \subset \partial P$.*

Proof. We may assume that $V = \operatorname{aff}(P)$. Suppose first that H is a support hyperplane. If $H \cap \operatorname{int}(P) \neq \emptyset$, then both inclusions $P \subset H^+$ or $P \subset H^-$ are impossible since none of the closed halfspaces contains a neighborhood of $x \in H \cap \operatorname{int}(P)$. Thus $H \cap P \subset \partial P$.

As to the converse implication, if H intersects P, then one of the open halfspaces, say $H^>$, must contain an interior point x of P since $\operatorname{int}(P) \subset H$ is impossible. If H is not a support hyperplane, then the other open halfspace $H^<$ must also contain a point $y \in P$. But all points on the half open line segment $[x, y)$ belong to $\operatorname{int}(P)$, as is easily checked. Exactly one of these points is in H so that $H \cap \operatorname{int}(P) \neq \emptyset$. □

In the representation

$$P = H_1^+ \cap \dots \cap H_n^+$$

the halfspace H_i^+ can be omitted if it contains the intersection of the remaining halfspaces. Therefore P has an irredundant representation as an intersection of halfspaces. It is evident from one-dimensional polyhedra in $\mathbb{R}^2$ that we can achieve the uniqueness of these halfspaces only if $\dim P = \dim V$. But this is not an essential restriction, since we can always consider P as a polyhedron in $\operatorname{aff}(P)$. First a lemma:

Lemma 1.5. *Suppose $P = H_1^+ \cap \cdots \cap H_n^+$ is a polyhedron. Then a convex set $X \subset \partial P$ is contained in H_i for some i.*

Proof. On the contrary we assume that $X \not\subset H_i$ for all i and choose $x_i \in X \setminus H_i$. Then

$$x = \frac{1}{n}(x_1 + \cdots + x_n) \in X \subset P,$$

but $x \in H_i^>$ for all i (as one sees immediately upon choosing an affine form defining H_i^+). The intersection $H_1^> \cap \cdots \cap H_n^>$ is open and contained in P. Thus $x \in \operatorname{int}(P)$, a contradiction. □

Theorem 1.6. *Let $P \subset V$ be a polyhedron such that $d = \dim P = \dim V$. Then the halfspaces $H_1^+, \dots, H_n^+$ in an irredundant representation $P = H_1^+ \cap \cdots \cap H_n^+$ are uniquely determined. In fact, the sets $F_i = P \cap H_i$, $i = 1, \dots, n$, are the facets of P.*

Proof. We choose an irredundant representation $P = H_1^+ \cap \cdots \cap H_n^+$.

By Proposition 1.4 and Lemma 1.5, every facet F of P is contained in one of the sets F_i. Let H be the support hyperplane defining F. Then $H = \operatorname{aff}(P \cap H) = \operatorname{aff}(F)$ since F is a facet. But H is contained in H_i, and so $H = H_i$ and $F = P \cap H = P \cap H_i = F_i$.

It remains to show that each F_i is a facet of P. It is enough to consider F_n. Let $P' = H_1^+ \cap \cdots \cap H_{n-1}^+$ and set $H = H_n$. Then $H \cap P' \neq \emptyset$, but H is not a support hyperplane of P'. By Proposition 1.4 H intersects the interior of P'. Therefore $H \cap P' = H \cap P$ contains a nonempty subset X that is open in H. But then $\operatorname{aff}(P \cap H) \supset \operatorname{aff}(X) = H$, and so $\dim P \cap H = \dim H = \dim P - 1$. □

The face structure of a polyhedron. We note an immediate corollary of Theorem 1.6:

Corollary 1.7. *Let P be a polyhedron.*
(a) Then ∂P is the union of the facets of P.
(b) Each proper face of P is contained in a facet.

Part (b) of the corollary is very useful for inductive arguments, provided the relation "F is a face of P" is transitive, and this is indeed true.

Proposition 1.8. *Let F be a face of the polyhedron P and $G \subset F$. Then G is a face of P if and only if it is a face of F.*

Proof. Evidently a face G of P is a face of every face $F \supset G$.

For the converse implication there is nothing to show if $F = P$, and this includes the case in which $\dim P = 0$. Suppose $\dim P > 0$. Then the proper face F is contained in a facet F', as we have seen in the corollary above, and, by induction, G is a face of F'. Therefore we may assume that F itself is a facet.

For simplicity we may further assume that $V = \operatorname{aff}(P)$. Let H' be a support hyperplane in $\operatorname{aff}(F)$ for the face G of F, and H the (uniquely determined) support hyperplane of P through F. There exists a point $x \in H^<$ that is contained in the interior of the polyhedron P' defined as the intersection of the halfspaces associated with the remaining facets of P. Set $\tilde{H} = \operatorname{aff}(H' \cup \{x\})$. Then $\tilde{H}$ is a hyperplane in V that intersects P exactly in G. We leave the verification of this simple fact to the reader. □

As we have seen, the maximal elements (with respect to inclusion) among the proper faces are the facets. Now we turn to the minimal faces:

Proposition 1.9. *Let P be a polyhedron and let U be the vector subspace given by the intersection of the hyperplanes through 0 parallel to the facets of P.*

(a) The sets $x + U$, $x \in P$, are the maximal affine spaces contained in P.
(b) Every minimal face F of P is a maximal affine subspace of P.
(c) In particular, if P has a bounded face, then its minimal nonempty faces are its vertices.

Proof. Clearly $x + U \subset P$ since the affine forms defining P are constant on $x + U$. On the other hand, if A is an affine subspace contained in P, then these forms must be constant on A, and so $A \subset x + U$ for $x \in A$. This proves (a).

For (b) there is nothing to prove if P has no proper face. Otherwise every minimal face of P is contained in a facet, and it follows by induction on $\dim P$ that the minimal faces F are affine subspaces of V. So $F \subset x + U \subset P$ for every $x \in F$. If H is a support hyperplane associated with F, then $U + x \subset F$, as is easily seen (for an arbitrary face F).

(c) follows immediately from (b). □

We can now prove several important properties of the set of faces of P:

Theorem 1.10. *Let P be a polyhedron and A a maximal affine subspace of P. Then the following hold:*

(a) The intersection $F \cap G$ of faces F, G is a face of F and G.
(b) Every face F is the intersection of facets. In particular, the number of faces is finite.
(c) Let $F_0 \subset \cdots \subset F_{m-1} \subset F_m = P$ be a strictly ascending maximal chain of nonempty faces of P. Then $\dim F_0 = \dim A$ and $\dim F_{i+1} = \dim F_i + 1$ for all $i = 0, \dots, m-1$.
(d) For each $x \in P$ there exists exactly one face F such that $x \in \operatorname{int}(F)$. It is the unique minimal element in the set of faces of P containing x.

Proof. (a) If $F \cap G = \emptyset$, then $F \cap G$ is a face by definition. Otherwise, a support hyperplane of P associated with the face F is also a support hyperplane of the polyhedron G that intersects G exactly in $F \cap G$. Thus $F \cap G$ is a face of G, and therefore of P.

(b) If F is a facet, then it is certainly an intersection of facets. Otherwise F is strictly contained in a facet F', and, by induction, the intersection of facets of F'. But the facets of F' arise as intersections of facets of P with F', as follows immediately from Theorem 1.6.

(c) Follows immediately by induction from Corollary 1.7 and Proposition 1.9.

(d) By (a) and (b) the intersection of all faces of P containing x is a face F of P. By Corollary 1.7 $x \in \operatorname{int}(F)$. Every other face G of P with $x \in G$ has F as a proper face, and so $x \in \partial G$. □

It follows from Theorem 1.10 that for all faces F, G of P there are a unique maximal face contained in $F \cap G$ (namely $F \cap G$ itself) and a unique minimal face containing

$F \cup G$. Therefore the set of faces, partially ordered by inclusion, forms a lattice, the *face lattice of* P.[1]

In general a polyhedron need not have a bounded face, and as we see from Proposition 1.9, it has a bounded face if and only if its minimal nonempty faces are vertices. As the next proposition shows, the structure of polyhedra in general is essentially determined by those with vertices.

For each affine subspace A of V we can find a *complementary subspace* A' characterized by the conditions that $A \cap A'$ consists of a single point x_0 and $V = A + (A' - x_0) = (A - x_0) + A'$. The map $A \times A' \to V$, $(x, x') \mapsto x + x' - x_0$, is then a bijection, and we may write $V = A \times A'$.

Proposition 1.11. *Let $P \subset V$ be a nonempty polyhedron, A a maximal affine subspace of P, A' a complementary subspace, and $P' = A' \cap P$. Then $P = A \times P'$ under the identification $V = A \times A'$, and P' has vertices.*

Proof. Let x_0 be the intersection point of A and A'. We can assume that $x_0 = 0$. For a point $y = x + x' \in P$, $x \in A$, $x' \in A'$, one has $y - x \in P$, since $y + A \subset P$ for all $y \in P$. But $y - x = x' \in P'$. Conversely, if $x' \in P' \subset P$, then $x' + x \in P$ for all $x \in A$. This shows $P = A \times P'$. Since $A \times B \subset P$ for each affine subspace B of P', the largest affine subspaces contained in P' are points. □

There meet exactly two edges in a vertex of a polyhedron in $\mathbb{R}^2$ and two facets in an edge of a 3-dimensional polyhedron in $\mathbb{R}^3$. This fact can be generalized to all dimensions:

Proposition 1.12. *Let F be a nonempty face of the polyhedron P. Then* $\dim F = \dim P - 2$ *if and only if there exist exactly two facets F' and G' containing F. In this case one has $F = F' \cap G'$.*

Proof. If there exist exactly two facets F' and G' containing F, then $\dim F = \dim P - 2$ as follows immediately from Theorem 1.10, and clearly $F = F' \cap G'$.

For the converse implication we can assume that $V = \operatorname{aff}(P)$. Let $\dim F = \dim P - 2$, and let F' be a facet containing F. Then F is a facet of F'. By Theorem 1.10 the facets of F' are of the form $F' \cap G'$ where G' is another facet of P. It is enough to show that G' is uniquely determined. Suppose that $F = F' \cap G''$ for a third facet G'' of P.

We choose affine-linear forms α, β, γ with nonnegative values on P such that $F' = P \cap H_\alpha$, $G' = P \cap H_\beta$, and $G'' = P \cap H_\gamma$. The affine linear forms vanishing on F, and therefore on $\operatorname{aff}(F)$, form a two-dimensional vector subspace of the space of all affine-linear forms. Thus there is a linear relation

$$a\alpha + b\beta + c\gamma = 0.$$

Since α, β, γ define pairwise different facets, none of a, b, c can be zero. We can assume that two of a, b, c are positive, say a and b. If $c < 0$, then H_γ^+ contains $H_\alpha^+ \cap H_\beta^+$—an

[1] A partially ordered set M is called a *lattice* if each two elements $x, y \in M$ have a unique infimum and a unique supremum. Most often we will use the term "lattice" in a completely different meaning; see Section 1.G.

impossibility, since the facets define an irreducible representation of P. But $c > 0$ is also impossible, since α, β, γ are simultaneously positive on an interior point of P. □

Finally we want to give another, internal, characterization of faces. We call a subset X of P *extreme* if it is convex and if it contains every line segment $[x, y]$ for which $x, y \in P$ and $(x, y) \cap X \neq \varnothing$.

Proposition 1.13. *Let P be a polyhedron. Then the faces of P are exactly its extreme subsets.*

Proof. Every face is an extreme subset for trivial reasons. Conversely, let X be an extreme subset. If $X \cap \operatorname{int}(P) = \varnothing$, then X is contained in a facet by Lemma 1.5, and we are done by induction on $\dim P$. Otherwise X contains an interior point x of P. Let $y \in P, x \neq y$. Then the line through x and y contains a point $z \in P$ such that x is strictly between y and z. Hence $[y, z] \subset X$, and in particular $y \in X$, which shows $X = P$. □

The notion of face and extreme subset make sense for arbitrary convex sets. However, while every face of a convex set is an extreme subset, the converse is false in general (the reader should find an example).

Convex and strictly convex functions. Let $X \subset V$ be a convex set. (As above, V is a finite-dimensional $\mathbb{R}$-vector space.) Then a function $f : X \to \mathbb{R}$ is *convex* if $f(tx+(1-t)y) \le tf(x) + (1-t)f(y)$ for all $x, y \in X$ and $t \in [0,1]$. In other words: f is convex if the graph of $f|[x, y]$ is below the line segment $[(x, f(x)), (y, f(y))] \subset V \times \mathbb{R}$. (We use $|$ to denote the restriction of functions or mappings to subsets.)

One can immediately prove *Jensen's inequality*: $f(t_1x_1 + \cdots + t_nx_n) \le t_1f(x_1) + \cdots + t_nf(x_n)$ for all convex combinations of points in X.

The function f is *strictly convex* if $f(tx + (1-t)y) < tf(x) + (1-t)f(y)$ for all $x, y \in X, x \neq y$, and $t \in (0,1)$.

Concave and *strictly concave* functions are defined in the same manner, the inequalities being reversed.

1.B Finite generation of cones

We have introduced polyhedra as the set of solutions of finite linear systems of inequalities. In this subsection we study the special case in which the inequalities are homogeneous, or, equivalently, the affine forms are linear forms on the finite-dimensional vector space V (over the field $\mathbb{R}$, just as in Section 1.A). The main advantage of the linear case is the duality between a vector space V and the space $V^* = \operatorname{Hom}_{\mathbb{R}}(V, \mathbb{R})$, for which there is no affine analogue.

Definition 1.14. A *cone* in V is the intersection of finitely many linear closed halfspaces. (We say that a halfspace is *linear* if it is defined by a linear form on V.)

A *morphism* of cones C and C' is a map $\varphi : C \to C'$ that extends to a linear map $\tilde{\varphi} : \mathbb{R}C \to \mathbb{R}C'$.

A subset X of V is *conical* if it is closed under nonnegative linear combinations: $a_1x_1 + \cdots + a_nx_n \in C$ for all $x_1, \dots, x_n \in C$ and all $a_1, \dots, a_n \in \mathbb{R}_+$, $n \in \mathbb{Z}_+$.

A cone C is evidently conical. Often this property is used to define the notion of cone. However, all our conical sets are polyhedra, and therefore the term "cone" will always include the attribute "polyhedral."

The faces of a cone are themselves cones since each support hyperplane H of a cone C must contain 0: the ray $\mathbb{R}_+x$, $x \in C \cap H$, $x \neq 0$, cannot leave C, and $0 \in H^>$ is impossible.

The intersection of conical sets is certainly conical. Therefore we may define the *conical hull* $\mathbb{R}_+X$ of a subset X of V as the smallest conical set containing X. In analogy with the convex hull, one has

$$\mathbb{R}_+X = \{a_1x_1 + \cdots + a_nx_n : n \in \mathbb{Z}_+,\ x_1, \dots, x_n \in X,\ a_i \in \mathbb{R}_+,\ i = 1, \dots, n\}$$

if $X = \{x_1, \dots, x_n\}$ is a finite set, and if X is infinite, then $\mathbb{R}_+X$ is the union of the sets $\mathbb{R}_+X'$, where $X' \subset X$ is finite. We say that a conical set is *finitely generated* if it is the conical hull of a finite set.

The central theorem in this subsection shows that finitely generated conical sets are just cones, and conversely. In addition to the external representation of cones as intersections of halfspaces, one has an internal description by finitely many elements generating C as a conical set.

Theorem 1.15. *Let $C \subset V$ be a conical set. Then the following are equivalent:*

(a) C is finitely generated;
(b) C is a cone.

Proof. Suppose that $C = \mathbb{R}_+x_1 + \cdots + \mathbb{R}_+x_m$. For (a) $\Longrightarrow$ (b) we have to find linear forms $\lambda_1, \dots, \lambda_n$ such that $C = H^+_{\lambda_1} \cap \cdots \cap H^+_{\lambda_n}$. We use induction on m, starting with the trivial case $m = 0$ for which $C = \{0\}$ is certainly polyhedral.

Suppose that

$$C' = \mathbb{R}_+x_1 + \cdots + \mathbb{R}_+x_{m-1} = H^+_{\kappa_1} \cap \cdots \cap H^+_{\kappa_r}$$

for linear forms $\kappa_1, \dots, \kappa_r$. The process by which we now construct $\lambda_1, \dots, \lambda_n$ is known as (the dual version of) *Fourier–Motzkin elimination*. (Instead of solving a system of inequalities we construct such a system for a given set of solutions.)

Set $x = x_m$. We may assume that

$$\kappa_i(x) \begin{cases} = 0 & i = 1, \dots, p, \\ > 0 & i = p+1, \dots, q, \\ < 0 & i = q+1, \dots, r. \end{cases}$$

Set

$$\mu_{ij} = \kappa_i(x)\kappa_j - \kappa_j(x)\kappa_i, \qquad i = p+1, \dots, q, \quad j = q+1, \dots, r.$$

Then $\mu_{ij}(x_u) \geq 0$ for $u = 1, \dots, m-1$ and $\mu_{ij}(x) = 0$. Let

$$\{\lambda_1, \dots, \lambda_n\} = \{\kappa_1, \dots, \kappa_q\} \cup \{\mu_{ij} : i = p+1, \dots, q,\ j = q+1, \dots, r\}.$$

Evidently

$$C \subset D = H^+_{\lambda_1} \cap \cdots \cap H^+_{\lambda_n}.$$

In fact, $x_1, \dots, x_m$ are contained in D, and so this holds for each of their nonnegative linear combinations.

It remains to prove the converse inclusion $D \subset C$. Choose $y \in D$. We must find $t \in \mathbb{R}_+$ and $z \in C'$ for which $y = tx + z$. So we consider the ray $y - tx$, $t \in \mathbb{R}_+$, and have to show that it meets C'.

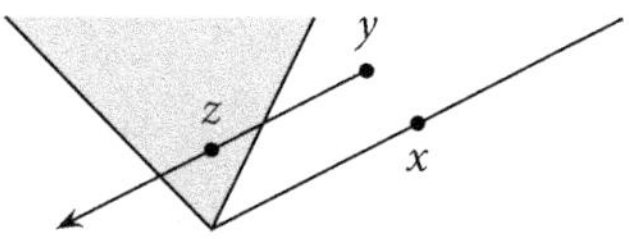

Fig. 1.2. Construction of z

We have to find the values of $t \in \mathbb{R}$ for which $y - tx \in C'$. Equivalently, we have to find out when $y - tx \in H^+_{\kappa_i}$ for $i = 1, \dots, r$. We distinguish three cases:

(i) $i \le p$. Then $\kappa_i(x) = 0$, $\kappa_i \in \{\lambda_1, \dots, \lambda_n\}$, and so $\kappa_i(y) \ge 0$. Clearly $y - tx \in H^+_{\kappa_i}$ for all $t \in \mathbb{R}$.

(ii) $p + 1 \le i \le q$. Then $\kappa_i(x) > 0$, $\kappa_i \in \{\lambda_1, \dots, \lambda_n\}$ and $\kappa_i(y) \ge 0$. Evidently $y - tx \in H^+_{\kappa_i}$ if and only if $t \le \kappa_i(y)/\kappa_i(x)$.

(iii) $q + 1 \le i \le r$. Then $\kappa_i(x) < 0$, and $y - tx \in H^+_{\kappa_i}$ if and only if $t \ge \kappa_i(y)/\kappa_i(x)$.

Thus the range of values of $t \in \mathbb{R}$ for which $y - tx \in C'$ is an intersection of intervals of type $(-\infty, a_i]$, $i = p + 1, \dots, q$, and $[b_j, \infty)$, $j = q + 1, \dots, r$. The intersection of all these intervals and $\mathbb{R}_+$ is nonempty if and only if $a_i \ge 0$ and $a_i \ge b_j$ for all values of i and j. But $a_i = \kappa_i(y)/\kappa_i(x) \ge 0$ since $\kappa_i(y) \ge 0$ and $\kappa_i(x) > 0$, and $a_i \ge b_j$, as follows immediately from the inequality $\mu_{ij}(y) \ge 0$ (observe that $\kappa_i(x)\kappa_j(x) < 0$).

The implication (a) $\Longrightarrow$ (b) has been proved, and we will see below that (b) $\Longrightarrow$ (a) follows from it (and is in fact equivalent to it). □

We are now justified in speaking of the *cone C generated by* $x_1, \dots, x_m$. Clearly, each face F of C is generated by the intersection $F \cap \{x_1, \dots, x_m\}$.

For the efficiency of Fourier–Motzkin elimination one should always produce a shortest possible list $\lambda_1, \dots, \lambda_n$ of linear forms in each step.

Duality. The concept by which we will complete the proof of Theorem 1.15 is that of *duality* of cones. For a subset X of V we set

$$X^* = \{\lambda \in V^* : \lambda(x) \ge 0 \text{ for all } x \in X\}$$

and call (the evidently conical set) X^* the *dual conical set* of X. Since $\dim V < \infty$, the natural map $h : V \to V^{**}$, defined by

$$(h(v))(\lambda) = \lambda(v),$$

is an isomorphism of vector spaces. This shows that X^* is not only conical, but also an intersection of linear halfspaces. (The symbol * has two different meanings; applied to V it always denotes the space of linear forms.)

Furthermore, the identification $V \cong V^{**}$ allows us to consider the bidual conical set X^{**} as a subset of V, and, by definition,

$$X \subset X^{**}.$$

Since X^{**} is always conical and an intersection of halfspaces, the inclusion $X \subset X^{**}$ is strict in general.

Using the implication (a) $\Longrightarrow$ (b) of Theorem 1.15 we now derive the *duality theorem for cones*:

Theorem 1.16. *Let $C \subset V$. Then the following hold:*

(a) C is the intersection of (possibly infinitely many) halfspaces if and only if $C = C^{**}$.
(b) In particular, $C = C^{**}$ if C is a cone.
(c) Suppose C is a cone, and let $\lambda_1, \dots, \lambda_n \in V^*$. Then $C = H^+_{\lambda_1} \cap \dots \cap H^+_{\lambda_n}$ if and only if $C^* = \mathbb{R}_+\lambda_1 + \dots + \mathbb{R}_+\lambda_n$.

Proof. (a) By definition, C^{**} is the intersection of linear halfspaces, namely of the halfspaces H^+_λ, $\lambda \in C^*$. On the other hand, if $C = \bigcap_{\lambda \in \Lambda} H^+_\lambda$ for a subset Λ of V^*, then $\Lambda \subset C^*$, and

$$C^{**} = \bigcap_{\lambda \in C^*} H^+_\lambda \subset \bigcap_{\lambda \in \Lambda} H^+_\lambda = C.$$

Together with the inclusion $C \subset C^{**}$, this shows that $C = C^{**}$.

(b) follows immediately from (a).

(c) Suppose that $C = H^+_{\lambda_1} \cap \dots \cap H^+_{\lambda_n} \subset V^*$ and set $C' = \mathbb{R}_+\lambda_1 + \dots + \mathbb{R}_+\lambda_n$. Then C' is a finitely generated conical set, $C' \subset C^*$, and $C = (C')^*$. By the implication (a) $\Longrightarrow$ (b) of Theorem 1.15, C' is a cone, and so (b) implies

$$C' = (C')^{**} = ((C')^*)^* = C^*.$$

Conversely, if $C^* = \mathbb{R}_+\lambda_1 + \dots + \mathbb{R}_+\lambda_n$, then $(C^*)^* = \{\lambda_1, \dots, \lambda_n\}^*$, and so $C = C^{**} = \{\lambda_1, \dots, \lambda_n\}^*$, as desired. □

As a first application we complete the proof of Theorem 1.15. Let C be a cone. We have to show that C is finitely generated as a conical set. By Theorem 1.16(c) the dual conical set C^* is finitely generated. So we can apply Theorem 1.15(a) $\Longrightarrow$ (b) and obtain that C^* is a cone. It follows that C^{**} is finitely generated, and since $C = C^{**}$, we are done.

Let F be a face of the cone C. Then

$$F^*_C = \{\lambda \in C^* : \lambda(F) = 0\}$$

is a conical set. Moreover, if $a\lambda + b\mu$, $\lambda, \mu \in C^*$, $a, b > 0$, vanishes on F, then $\lambda, \mu \in F^*_C$. So F^*_C is an extreme subset of C^*, and by Proposition 1.13 it is a face of C^*, the *face dual* to F (the subscript C is necessary since F^*_C depends on C in an essential way).

Theorem 1.17. *Let C be a cone in the vector space V. Then the assignment $F \mapsto F^*_C$ defines an order reversing bijection of the face lattices of C and C^*. Moreover, one has* $\dim F + \dim F^*_C = \dim V$.

Proof. Let us first show that the assignment is injective. Choose a support hyperplane H of C associated with F and a linear form λ such that $H = H_\lambda$ and $C \subset H^+_\lambda$. Then λ is contained in G^*_C for a face G of C if and only if $G \subset F$. Thus $F^*_C = G^*_C$ implies $F = G$.

The assignment is evidently order reversing, and if we apply it to C^* and $C^{**} = C$, then the composition of the two maps is an injective order preserving map of the face lattice of C to itself. But the face lattice is a finite set, and so the composition must be the identity.

Since we can exchange the roles of C and C^*, the first statement of the theorem has been proved. By Theorem 1.10, which describes the chain structure of the face lattice, we must have $\dim F^*_C = \dim G^*_C + 1$ if $\dim G = \dim F + 1$. Therefore it is enough for the second statement to show $\dim C + \dim C^*_C = \dim V$. This however is evident: all linear forms vanishing on C belong to C^*, and a linear form vanishes on C if and only if it vanishes on the vector subspace generated by C. □

Pointed cones. A cone C is *pointed* if 0 is a face of C. Equivalently we can require that

$$x, -x \in C \Longrightarrow x = 0,$$

and every cone decomposes into the direct sum of a pointed one and a vector subspace. This follows immediately from the following:

Proposition 1.18. *Let C be a cone and $C_0 = \{x \in C : -x \in C\}$.*

(a) C_0 is the maximal vector subspace of C and the unique minimal face of C.
(b) Let U be a vector subspace complement of C_0. Then $C = C_0 \oplus (C \cap U)$, and $C \cap U$ is a pointed cone.

The proposition is just a special case of Proposition 1.11. By $C = C_0 \oplus C'$ we indicate that each element of C has a unique decomposition $x = x_0 + x'$ with $x_0 \in C_0$ and $x' \in C'$.

Pointed cones can be characterized in terms of their dual cones:

Proposition 1.19. *A cone C in the vector space V is pointed if and only if* $\dim C^* = \dim V$.

Proof. A vector subspace U of V is contained in C if and only if C^* is contained in the dual vector space U^*. □

The proposition shows that the class of full-dimensional pointed cones is closed under duality. Very often we will use that a pointed cone has an essentially unique minimal system of generators:

Proposition 1.20. *Let C be a pointed cone and choose an element $x_i \neq 0$ from each 1-dimensional face $R_1, \dots, R_n$ of C. Then $x_1, \dots, x_n$ is, up to order and positive scalar factors, the unique minimal system of generators of C.*

Proof. By Proposition 1.19 one has $\dim C^* = \dim V = \dim V^*$. By Theorem 1.6 C^* has a unique description as an irredundant intersection of halfspaces $H_1^+, \dots, H_n^+$. So Theorem 1.16 implies that C has a unique minimal system of generators, up to positive scalar factors, since the elements x_i of $V = V^{**}$ with $H_{x_i}^+ = H_i^+$ are uniquely determined up to positive scalar factors. Moreover, the faces of C that are dual to the facets of C^* are exactly the 1-dimensional faces of C, and the face dual to $C^* \cap H_i$ contains x_i. □

The 1-dimensional faces of a pointed cone C are called its *extreme rays*. As a consequence of Proposition 1.20 we obtain that pointed cones have polytopes as cross-sections:

Proposition 1.21. *Let C be a pointed cone, $\lambda \in \operatorname{int}(C^*)$, and $a \in \mathbb{R}$, $a > 0$. Then the hyperplane $H = \{x : \lambda(x) = a\}$ intersects C in a polytope P, and one has $\mathbb{R}_+ P = C$.*

Proof. Since $\lambda \in \operatorname{int}(C^*)$, the hyperplane $H_0 = \{x : \lambda(x) = 0\}$ intersects C only in 0, as follows from Theorem 1.17 and Proposition 1.19. Therefore $a\lambda(x)^{-1}x \in P$ for all $x \in C$, $x \neq 0$, and $C = \mathbb{R}_+ P$.

It remains to be verified that P is bounded. Let R be an extreme ray of C. It suffices to show that R intersects H in exactly one point. Then Proposition 1.20 implies that P is the convex hull of the finite set of the intersection points of H with the extreme rays of C.

Let G be the facet of C^* dual to R, and U the vector subspace generated by G. Then the linear forms vanishing on R are exactly the elements of U (note that $\dim U = \dim C - 1$). Since $\lambda \notin U$, the hyperplane H intersects R in exactly one point. □

In the situation of Proposition 1.21 we will say that H *defines a cross-section* of C.

1.C Finite generation of polyhedra

We have introduced polyhedra as the set of solutions of linear systems of inequalities. It is an elementary fact from linear algebra that the set S of solutions of a linear system of equations is an affine subspace of the form $S = x + S_0$ where S_0 is the vector space of solutions of the associated homogeneous system of equations. As we will see, polyhedra have a similar description that in fact characterizes them.

Recession cone and projectivization. In the same way as one associates a homogeneous system of equations with an inhomogeneous one, we associate a cone with a polyhedron:

Definition 1.22. Let $P = H_1^+ \cap \dots \cap H_n^+$ be a polyhedron. Then the *recession cone* of P is

$$\operatorname{rec}(P) = \tilde{H}_1^+ \cap \dots \cap \tilde{H}_n^+,$$

where $\tilde{H}_i^+$ is the vector halfspace parallel to the affine halfspace H_i^+, $i = 1, \dots, n$ (so $\tilde{H}_i^+ = H_i^+ - x$ for some $x \in H_i^+$). (Halfspaces are *parallel* if their bounding hyperplanes are parallel.)

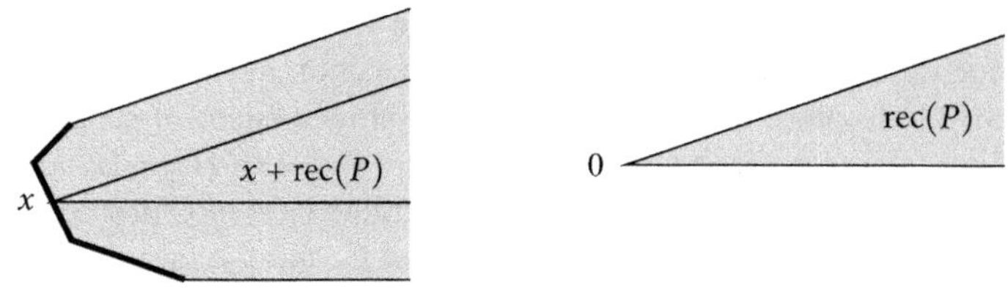

Fig. 1.3. A polyhedron and its recession cone

It is easy to see and left to the reader that the recession cone can also be described as the set of infinite directions in P:

Proposition 1.23. *Let P be a polyhedron, $x \in P$, and $v \in V$. Then $v \in \operatorname{rec}(P)$ if and only if $x + av \in P$ for all $a \in \mathbb{R}_+$.*

We have constructed the recession cone by passing from an inhomogeneous system of linear equations to the associated homogeneous system. There is a second construction leading from an inhomogeneous system to a homogeneous one, namely *projectivization* (or *homogenization*): we introduce a new variable and view the constant as its coefficient. In other words, with an affine form

$$\alpha = \lambda + \beta, \qquad \beta = \alpha(0),$$

we associate the linear form

$$\overline{\alpha} : V \oplus \mathbb{R} \to \mathbb{R}, \qquad \overline{\alpha}(v, h) = \lambda(v) + \beta h.$$

The affine hyperplane and the affine halfspaces associated with α are then extended to the linear hyperplane and the linear halfspaces associated with $\overline{\alpha}$. We indicate the extension by $\overline{}$.

We have chosen the letter h for the auxiliary variable since we interpret h as the height of the point (v, h) over V. In the following we have to embed subsets of V into $\overline{V} = V \oplus \mathbb{R}$ at a specific height. Thus we set

$$(X, b) = X \times \{b\} \subset \overline{V}$$

for $X \subset V$.

Definition 1.24. Let $P = H_1^+ \cap \cdots \cap H_n^+ \subset V$ be a polyhedron. Then we define the *cone over P* by

$$C(P) = \overline{H}_1^+ \cap \cdots \cap \overline{H}_n^+ \cap H_\chi^+$$

where $\chi : \overline{V} \to \mathbb{R}$ is given by $\chi(v, h) = h$.

Clearly $(P,1) = \{(v, h) \in C(P) : h = 1\}$. Thus $C(P)$ is just the cone with cross-section P at height 1, a statement made precise by the following:

Proposition 1.25. *Let C be the conical hull of $(P,1)$ in $\overline{V}$. Then*

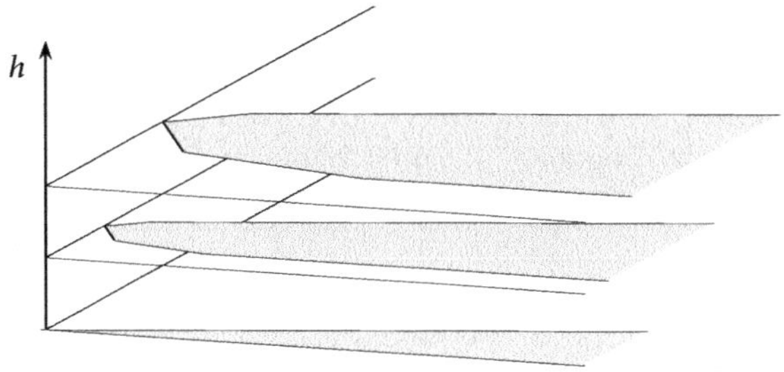

Fig. 1.4. Cross-sections of $C(P)$

(a) $C(P) \cap (V,0) = (\text{rec}(P), 0)$;
(b) $C(P) \cap (V,a) = C \cap (V,a)$ *for* $a \neq 0$;
(c) $C(P)$ *is the closure of* C *in* $\overline{V}$.

Proof. (a) is trivial and, moreover, $C(P) = C \cup (\text{rec}(P), 0)$, as is easily checked. This proves (b). For (c) it is therefore enough to show that $(\text{rec}(P), 0)$ is contained in the closure of C.

Pick $v \in \text{rec}(P)$ and choose $y \in P$. Then $y + tv \in P$ for all $t \in \mathbb{R}_+$, and $W = \{u(y + tv, 1) : t, u \in \mathbb{R}_+\} \subset C$. But W contains a point from every neighborhood of $(v, 0)$ in $\overline{V}$. Therefore $(v, 0)$ is contained in the closure of W, and, a fortiori, in the closure of C. □

Figure 1.4 illustrates the construction of $C(P)$, showing the recession cone at $h = 0$, $(P, 1)$ and $(2P, 2)$.

The theorems of Minkowski and Motzkin. According to Theorem 1.15 $C(P)$ is finitely generated, say $C(P) = \mathbb{R}_+ y_1 + \cdots + \mathbb{R}_+ y_m$. If y_i has height $h > 0$, then we can replace it by $h^{-1} y_i$. After this operation, $\{y_1, \dots, y_m\}$ decomposes into a subset Y_0 of elements of height 0 and a subset Y_1 of elements of height 1. Let $x_i = \pi(y_i)$ where π is the projection $V \oplus \mathbb{R} \to V$.

Choose $x \in P$. Then $z = (x, 1)$ can be written as a linear combination

$$z = \sum_{y \in Y_0} a_y y + \sum_{y \in Y_1} a_y y, \qquad a_y \in \mathbb{R}_+, \ \sum_{y \in Y_1} a_y = 1. \tag{1.1}$$

A comparison of the last components shows that indeed $\sum_{y \in Y_1} a_y = 1$. Furthermore $\pi(y) \in P$ for $y \in Y_1$ and $\pi(y) \in \text{rec}(P)$ for $y \in Y_0$.

A first consequence of these observations is *Minkowski's theorem*:

Theorem 1.26. *Let* $P \subset V$. *Then the following are equivalent:*

(a) *P is a polytope;*
(b) *P is a polyhedron and* $P = \text{conv}(\text{vert}(P))$;
(c) *P is the convex hull of a finite subset of* V.

Proof. For the implication (a) $\Longrightarrow$ (b) one notes that $\text{rec}(P) = 0$ if P is a polytope. So

Proposition 1.25 implies that $C(P)$ is a pointed cone with $C(P) \cap (V,0) = 0$, and thus we can apply Proposition 1.20: the elements of height 1 in the extreme rays of $C(P)$ form a minimal system of generators of $C(P)$, and they are evidently the vertices of P. We have $Y_0 = \emptyset$ in equation (1.1), and it follows immediately that $P = \operatorname{conv}(\operatorname{vert}(P))$.

The implication (b) $\Longrightarrow$ (c) is trivial since every polyhedron has only finitely many vertices.

As to (c) $\Longrightarrow$ (a), if P is the convex hull of a finite set, then it is certainly bounded, and the conical hull C of $(P,1)$ is finitely generated. So Theorem 1.15 implies that C is a cone. But then $C \cap (V,1)$ is a polyhedron that can be identified with P. To sum up: P is a polytope, and the proof of Minkowski's theorem is complete. □

If we give up the restriction $\operatorname{rec}(P) = 0$, then we obtain *Motzkin's theorem*:

Theorem 1.27. *Let $P \subset V$. Then the following are equivalent:*

(a) P is a polyhedron;
(b) there exist a polytope Q and a cone C such that $P = Q + C$.

Proof. For the implication (a) $\Longrightarrow$ (b) we choose $Q = \operatorname{conv}(\pi(Y_1))$ and let C be the cone generated by $\pi(Y_0)$. The notation is as for equation (1.1), and this equation proves the claim.

For the converse we choose a finite system X of generators of C. Then the conical hull of $(X,0) \cup \operatorname{vert}(Q,1)$ is a cone. Its cross-section at height 1 is not only equal to $(Q + C, 1)$, but it is also a polyhedron. □

Since a polytope is the convex hull of its finite vertex set, and since a cone is finitely generated, Theorem 1.27 allows us to say that a polyhedron is finitely generated.

If P has vertices, then we call the union of the bounded faces the *bottom* of P. (In Figure 1.3 the bottom is indicated by a thick line.) With this notion we can give a more precise statement about the generation of polyhedra.

Proposition 1.28. *Let P be a polyhedron, A a maximal affine subspace contained in P, A' a complementary affine subspace, and $P' = P \cap A'$. Furthermore, let $\pi : V \to A'$ denote the parallel projection along A, and B' the bottom of P'.*

Then $P = B' + \operatorname{rec}(P)$. Moreover, if $P = B + C$ for a bounded set B and a cone C, then $B' \subset \pi(B)$ and $C = \operatorname{rec}(P)$.

Proof. Since $P = P' + \operatorname{rec}(A)$ by Proposition 1.11, the equation $P = B' + \operatorname{rec}(P)$ follows from the analogous equation for P'. In proving the first statement we may therefore assume that $P = P'$, or, in other words, that P has vertices.

Let $x \in P$ and choose $y \in \operatorname{rec}(P)$. Then $x + ty \in P$ for all $t \in \mathbb{R}_+$, but the line $x + \mathbb{R}y$ cannot be contained in P; see Proposition 1.9. So there exists $t \geq 0$ for which $x' = x - ty$ belongs to a facet of P. But F has vertices, too, and we can apply induction since the proposition is obviously true in dimension 0: there exists x'' in the bottom of F and $y' \in \operatorname{rec}(F)$ such that $x' = x'' + y'$. But x'' belongs to the bottom of P and $y' \in \operatorname{rec}(P)$. So $x = x'' + (y' + ty)$ with $y' + ty \in \operatorname{rec}(P)$.

We leave the proof of the second assertion to the reader. □

First of all, note that the proof of the theorem contains a new demonstration of the implications (a) $\Longrightarrow$ (b) of Minkowski's and Motzkin's theorems.

Second, the proposition has the following interpretation in terminology that will be introduced in Definition 2.5, but is essentially self-explanatory: B' is a minimal system of generators of P as a $\operatorname{rec}(P)$-module, and a subset B of P is a minimal system of generators if and only if it differs from B' by units of $\operatorname{rec}(P)$, namely elements of the maximal subgroup contained in $\operatorname{rec}(P)$. In particular, if P has vertices, then the bottom of P is the unique minimal system of generators of P over $\operatorname{rec}(B)$. Exercise 1.4 compares the face lattices of P, $\operatorname{rec}(P)$ and $C(P)$.

Minkowski sums and joins. Minkowski and Motzkin's theorems help us to understand the behavior of polyhedra under affine maps.

Proposition 1.29. *Let V, V' be vector spaces over $\mathbb{R}$, $\varphi : V \to V'$ an affine map, P a polyhedron in V, and P' a polyhedron in V'. Then*

(a) $\varphi(P)$ and $\varphi^{-1}(P')$ are polyhedra;
(b) $\varphi(P)$ is a polytope if P is a polytope;
(c) if φ is linear and P, P' are cones, then $\varphi(P)$ and $\varphi^{-1}(P')$ are cones.

Proof. Since $\varphi = \psi + v'$ for a linear map $\psi : V \to V'$ and some $v' \in V'$, we may assume that φ is linear.

For the preimage we start from a description $P = H_1^+ \cap \cdots \cap H_n^+$. The preimage of a (linear) halfspace under a linear map is a (linear) halfspace, and the preimage of the intersection is the intersection of the preimages.

The image $\varphi(C)$ of a finitely generated conical set $C \subset V$ is certainly conical and finitely generated. Therefore, if P is a cone, then so is $\varphi(P)$. For (b) we argue similarly, using Minkowski's theorem and convex sets.

In order to show that the image of a polyhedron is a polyhedron, we write it in the form $Q + C$ where Q is a polytope and C is a cone. Then $\varphi(P) = \varphi(Q) + \varphi(C)$ is a polyhedron, as follows from (b), (c) and Motzkin's theorem. □

As a consequence of Proposition 1.29 the *Minkowski sum*

$$P + P' = \{x + x' : x \in P,\ x' \in P'\}$$

of polyhedra (polytopes) is a polyhedron (polytope):

Theorem 1.30. *Let P, P' be polyhedra in V. Then*

(a) $P + P'$ is a polyhedron;
(b) $P + P'$ is a polytope if (and only if) P and P' are polytopes;
(c) if P and P' are cones, then $P + P'$ is a cone.

Proof. Consider $P \times P'$ in $V \oplus V$. Together with P and P' it is a polyhedron, polytope or cone, respectively. Now we can apply Proposition 1.29 to the linear map $V \oplus V' \to V$, $(v, v') \mapsto v + v'$. □

Another natural construction associated with polytopes P and P' is the convex hull $Q = \operatorname{conv}(P \cup P')$ of their union. Evidently Q is the convex hull of the finite set $\operatorname{vert}(P) \cup \operatorname{vert}(P')$, and thus it is a polytope by Minkowski's theorem. It is likewise easy to see that

$$D = \operatorname{conv}(C \cup C') = C + C'$$

is a cone if C and C' are cones. However, $\operatorname{conv}(P \cup P')$ need not be a polyhedron if P and P' are just polyhedra. For example if P is a line in $\mathbb{R}^2$ and P' consists of a single point x not in P, then the smallest polyhedron containing P and P' is the strip bounded by P and its parallel through x. Of all the points on the parallel, only x is in $\operatorname{conv}(P \cup P')$. But $\operatorname{conv}(P \cup P')$ may fail to be a polyhedron only because it need not be closed. For this reason we introduce the *closed convex hull*

$$\overline{\operatorname{conv}}(X) = \overline{\operatorname{conv}(X)}$$

as an additional construction.

Proposition 1.31. *Let P and P' be polyhedra. Then $\overline{\operatorname{conv}}(P \cup P')$ is a polyhedron.*

Proof. We write $P = Q + \operatorname{rec}(P)$ and $P' = Q' + \operatorname{rec}(Q)$ with polytopes Q and Q'. Then

$$R = \operatorname{conv}(Q \cup Q') + \operatorname{rec}(P) + \operatorname{rec}(Q)$$

is a polyhedron by Motzkin's theorem and it contains $\operatorname{conv}(P \cup P')$ as is easily verified. On the other hand, $\operatorname{rec}(T) \supset \operatorname{rec}(P) + \operatorname{rec}(P')$ for every polyhedron $T \supset P \cup P'$, as follows immediately from Proposition 1.23. Therefore R is the smallest polyhedron containing $P \cup P'$. Clearly $R \supset \overline{\operatorname{conv}}(P \cup P')$.

It remains to show that $R \subset \overline{\operatorname{conv}}(P \cup P')$. Pick $s \in R$,

$$s = ax + (1-a)y + w + z, \qquad x \in Q,\ y \in Q',\ a \in [0,1], w \in \operatorname{rec}(P),\ z \in \operatorname{rec}(P').$$

If $a \neq 0, 1$, then $s \in \operatorname{conv}(P \cup P')$, and if $a = 0$, then each neighborhood of s contains a point $s' = a'x + (1-a')y + w + z$, $a' > 0$, with $s' \in \operatorname{conv}(P \cup P')$. For $a = 1$ one argues similarly. $\square$

We leave it to the reader to find the exact conditions under which $\overline{\operatorname{conv}}(P \cup P') = \operatorname{conv}(P \cup P')$.

As a last construction principle for polyhedra we introduce the *join*. Roughly speaking, it is the "free convex hull" that we obtain by considering polyhedra in positions independent of each other. Let $P \subset V$ and $Q \subset W$ be polyhedra. Then we set $V' = V \oplus W \oplus \mathbb{R}$,

$$P' = \{(x, 0, 0) : x \in P\}, \qquad Q' = \{(0, y, 1) : y \in Q\},$$

and

$$\operatorname{join}(P, Q) = \overline{\operatorname{conv}}(P' \cup Q').$$

If P and Q are polyhedra in $\mathbb{R}^d$ such that $\operatorname{aff}(P)$ and $\operatorname{aff}(Q)$ are disjoint and there are no parallel lines $l \subset \operatorname{aff}(P)$ and $l' \subset \operatorname{aff}(Q)$, then $\overline{\operatorname{conv}}(P \cup Q)$ will also be called the *join* of P and Q. (Prove that this polyhedron is isomorphic to $\overline{\operatorname{conv}}(P' \cup Q')$.) By Proposition 1.31 the join of P and Q is a polyhedron. If P and Q are polytopes, then $\operatorname{join}(P, Q)$ is a polytope.

Separation of polyhedra. A characteristic feature of convexity are separation theorems: convex sets that are disjoint or just touch each other can be separated by a hyperplane.

Theorem 1.32. *Let P and Q be polyhedra in the vector space V. Let F be the smallest face of P containing $P \cap Q$, and define the face G of Q analogously ($F = G = \varnothing$ if $P \cap Q = \varnothing$). Then there exists a hyperplane H such that $P \setminus F \subset H^{>}$ and $Q \setminus G \subset H^{<}$.*

Proof. Replacing P and Q by $C = C(P)$ and $D = C(Q)$ reduces the theorem to the case of cones P and Q.

Assume first that $P + (-Q) = V$, and choose $v \in \operatorname{int}(P)$. Then $-v = v' - w$ for some $v' \in P$, $w \in Q$. Hence $w = v + v' \in \operatorname{int}(P)$, and so $F = P$. But $P + (-Q) = V$ implies $-P + Q = V$ as well, and we conclude that $G = Q$. In this case there is nothing to prove.

It remains the case in which $P + (-Q) \neq V$. It is a cone by Theorem 1.15 and has a support hyperplane H through its minimal face, namely the vector space of its invertible elements. Suppose that $v \in P \cap H$. Then $-v \in H \cap (P + (-Q))$, $-v = v' - w'$ with $v' \in P$, $w' \in Q$. It follows that $v + v' = w' \in P \cap Q$. Hence $v + v' \in F$, and so $v \in F$: H intersects P in F, and analogously it intersects Q in G. It follows immediately that H is the desired hyperplane. □

Remark 1.33. Separation theorems like 1.32 often appear as *Farkas' lemma* (see Ziegler [381]). A far reaching generalization is the *Hahn–Banach separation theorem*. The most general version of Minkowski's theorem is the theorem of *Krein–Milman* (see Jarchow [211]).

1.D Polyhedral complexes

It is evident (in the true sense of the word) that a polygon, i. e., a 2-dimensional polytope can be triangulated, as illustrated by Figure 1.5(a):

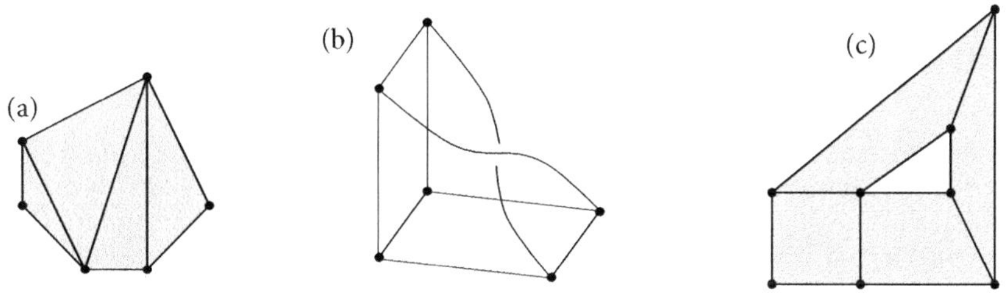

Fig. 1.5. Polyhedral complexes

In higher dimensions triangles are generalized by simplices:

Definition 1.34. A d-polytope with exactly $d+1$ vertices is called a *d-simplex*. In other words, a polytope is a simplex if its vertex set is affinely independent. The conical analogue is a *simplicial cone*, generated by a linearly independent set of vectors.

In the triangulation in Figure 1.5(a) the intersection of two triangles P_1 and P_2 and, more generally, of faces $F_1 \subset P_1$, $F_2 \subset P_2$, is a face of P_1 and of P_2. This compatibility condition defines a polyhedral complex. In the example above the polygon P and all triangles are contained in a single affine space. This condition is too strict in general. We weaken it as follows:

Definition 1.35. A *polyhedral complex* consists of (i) a finite family Π of sets, (ii) a family P_p, $p \in \Pi$, of polyhedra, and (iii) a family $\pi_p : P_p \to p$ of bijections satisfying the following conditions:

(a) For each face F of P_p, $p \in \Pi$, there exists $f \in \Pi$ with $\pi_p(F) = f$.
(b) For all $p, q \in \Pi$ there exist faces F of P_p and G of P_q such that $p \cap q = \pi_p(F) = \pi_q(G)$ and, furthermore, the restriction of $\pi_q^{-1} \circ \pi_p$ to F is an isomorphism of the polyhedra F and G.

We speak of a *polytopal complex* if the polyhedra P_p are polytopes, and of a *conical complex* if they are cones. In the latter case we require that the restriction of $\pi_q^{-1} \circ \pi_p$ is an isomorphism of cones.

If all the polyhedra P are simplices, then the complex is *simplicial*. A *simplicial conical complex* consists of simplicial cones.

We denote a polyhedral complex simply by Π, assuming that the polyhedra P_p and the maps π_p, $p \in \Pi$, are given implicitly. Moreover, $|\Pi|$ stands for the *support* $\bigcup_{p \in \Pi} p$ of Π.

A *subcomplex* of Π is a subset Π' that is itself a polyhedral complex (with respect to the families P_p and π_p, $p \in \Pi'$).

An *isomorphism* of polyhedral complexes Π, Π' is a bijective map $\varphi : |\Pi| \to |\Pi'|$ such that $\Pi' = \{\varphi(p) : p \in \Pi\}$ and for each $p \in \Pi$ the map $(\pi'_{p'})^{-1} \circ \varphi \circ \pi_p$ is an isomorphism of the polyhedra P_p and $P'_{p'}$, $p' = \varphi(p)$. In the case of conical complexes one requires that $(\pi'_{p'})^{-1} \circ \varphi \circ \pi_p$ is an isomorphism of cones.

Example 1.36. We choose the following 6 points in $\mathbb{R}^3$:

$$x_1 = (0,0,0), \quad x_2 = (1,0,0), \quad x_3 = (1,1,0),$$
$$x_4 = (0,1,0), \quad x_5 = (0,0,1), \quad x_6 = (0,1,1).$$

Let M be the subset of $\mathbb{R}^3$ formed by the union of $Q_1 = \operatorname{conv}(x_1, x_2, x_3, x_4)$, $Q_2 = \operatorname{conv}(x_1, x_4, x_5, x_6)$ and the set S that is obtained by mapping the unit square $Q_3 = \operatorname{conv}((0,0),(1,0),(1,1),(0,1))$ in $\mathbb{R}^2$ so that the image is a scroll turning by 180 degrees, identifying the left edge $E_1 = [(0,0),(0,1)]$ with $[x_2, x_3]$ and the right edge $E_2 = [(1,0),(1,1)]$ with $[x_6, x_5]$, reversing the orientation. (The image is a topological realization of the Möbius strip; see Figure 1.5(b).) Then the family consisting of

(i) the squares Q_1, Q_2, the set S, the 7 edges of Q_1 and Q_2, the images of $[(0,0),(1,0)]$ and $[(0,1),(1,1)]$, the vertices $x_1, \dots, x_6$, the improper face $\varnothing$; and

(ii) the identity mappings for all the polytopes except Q_3, E_1, E_2, for which we choose the (restrictions of) the scroll map

is a polyhedral complex.

Example 1.37. Let Π be a polyhedral complex and $c > 0$. Then we obtain an isomorphic complex $c\Pi$ as follows. The sets $p \in \Pi$ remain unchanged; however, each polytope P_p is replaced by cP_p, and π_p is replaced by $\pi_p \circ c^{-1}$, where c^{-1} (as a map) denotes multiplication by $c^{-1} \in \mathbb{R}$. While this example is rather irrelevant at this stage, it will be very important later, where an additional discrete structure will break the isomorphism.

An important class of polyhedral complexes is given by those with a global embedding:

Proposition 1.38. *Let Π be a finite family of polyhedra $P \subset \mathbb{R}^d$ satisfying the following conditions:*

(a) if F is a face of $P \in \Pi$, then $F \in \Pi$;
(b) $P \cap Q$ is a face of both P and Q for all $P, Q \in \Pi$.

Then Π, together with the family $\pi_P = \mathbf{1}_P$, $P \in \Pi$, of bijections, is a polyhedral complex.

In fact, the conditions for a polyhedral complex are obviously satisfied.

Definition 1.39. The polyhedral complexes given by Proposition 1.38 are called *embedded.* An embedded conical complex is called a *fan.*

The face lattice of a polyhedron P is a polyhedral complex as we have seen in Section 1.A. The set of all faces $F \neq P$ of P is also a polyhedral complex, the *boundary complex* of P. More generally, if X is the union of faces of P, then the family of all faces $F \subset X$ forms an embedded polyhedral complex. We call such complexes *boundary subcomplexes.* If the boundary complex of a polytope P is simplicial, then it is called a *simplicial polytope.* Evidently a polytope is simplicial if and only if all its facets are simplices. (The dual notion is that of *simple polytope*: P is simple if and only if each vertex is *simple*, i. e., lies in exactly d facets, $d = \dim P$.)

Let us now show that for a polyhedral complex Π the operation $(p, q) \mapsto p \cap q$ has properties that we have found already in the face lattice of a polyhedron:

Proposition 1.40. *Let Π be a polyhedral complex, and $p, q, f \in \Pi$ with $f \subset p$ and $p \neq q$. Then*

(a) there exists a (unique) face F of P_p such that $\pi_p(F) = f$, and $\pi_p \circ \pi_f^{-1}$ maps F isomorphically onto P_f;
(b) $\pi_p(\operatorname{int}(P_p)) \cap \pi_q(\operatorname{int}(P_q)) = \emptyset$;
(c) $p \cap q \in \Pi$.

Proof. (a) is just a specialization of 1.35(b) to the case in which $q \subset p$.

(b) We can assume that $p \not\subset q$. Then $p \cap q$ is a proper subset of p and the face F of p with $\pi_p(F) = p \cap q$ is a proper face of P. We have $\pi_p(\operatorname{int}(P_p)) \cap \pi_q(\operatorname{int}(P_q)) \subset \pi_p(\operatorname{int}(P_p)) \cap \pi_p(F) = \emptyset$— again we have used the injectivity of π_p.

(c) There exists a face F of P_p with $\pi_p(F) = p \cap q$, and $\pi_p(F) \in \Pi$ by 1.35(a). □

Example 1.41. Let $H_1, \dots, H_n$ be affine hyperplanes in $\mathbb{R}^d$. Such a set of hyperplanes defines an (embedded) polyhedral complex Π consisting of $\varnothing$ and all those polyhedra $H_{i_1}^+ \cap \dots \cap H_{i_r}^+ \cap H_{j_1}^- \cap \dots \cap H_{j_s}^-$, $0 \le r \le n$, $0 \le j \le n$, such that $\{i_1, \dots, i_r, j_1, \dots, j_s\} = \{1, \dots, n\}$. Such polyhedral complexes are called *hyperplane dissections*. The support of the complex is $\mathbb{R}^d$: with each $x \in \mathbb{R}^d$ we associate its sign vector $\in \{+1, -1, 0\}^n$ in an obvious way, and the sign vector determines the face of Π containing x.

Hyperplane arrangements are closely related to matroids; see Björner et al. [33].

Remark 1.42. In an embedded polyhedral complex the sets $p \in \Pi$ coincide with the polytopes P_p. But also in the abstract setting this identification is allowed after we have realized the polyhedra P_p, $p \in \Pi$ in disjoint affine spaces (vector spaces in the case of a conical complex): we then identify each $x \in P_p$ with $\pi_p(x) \in p$. Under this identification, each face F of P_p is identified with the set $f \in \Pi$ for which $\pi_p(F) = f$. If $p \cap q = \pi_p(F) = \pi_q(G)$ for faces F of P_p and G of P_q, then, under the identification of P with P_p and Q with P_q, $P \cap Q = p \cap q$. Henceforth we identify p and P_p.

Thus we find ourselves in a situation which is almost identical to that of an embedded complex, except that taking convex (or conical) hulls, or linear combinations or intersections with halfspaces does not make sense globally. However, such combinations can be formed locally. We return to the formal framework of Definition 1.35 for the argument to be given. Let $x_1, \dots, x_n \in |\Pi|$ and suppose that there exist $p, q \in \Pi$ with $x_1, \dots, x_n \in p \cap q$. We have to show that

$$\pi_p(a_1\pi_p^{-1}(x_1) + \dots + a_n\pi_p^{-1}(x_n)) = \pi_q(a_1\pi_q^{-1}(x_1) + \dots + a_n\pi_q^{-1}(x_n))$$

for all $a_1, \dots, a_n \in \mathbb{R}_+$ with $\sum_i a_i = 1$. But the equation follows immediately upon the application of π_q^{-1} to both sides, since $\pi_q^{-1} \circ \pi_p$ (where defined) respects affine linear combination by definition (this is the first time we use this property). In the case of conical complexes $\pi_q^{-1} \circ \pi_p$ respects linear combinations.

It is a useful observation that line segments $[x, y]$ do not ramify: suppose that $[x, y] \subset P \in \Pi$ and $[x, y'] \subset Q \in \Pi$; if there exists $z \in (x, y] \cap (x, y']$, then $y, y' \in P \cap Q$ and $[x, y] \subset [x, y']$ or vice versa. In fact, we can replace P and Q by their smallest faces containing $[x, y]$ and $[x, y']$, respectively. Then $(x, z) \subset \operatorname{int}(P) \cap \operatorname{int}(Q)$. It follows that $P = Q$ (by the disjoint decomposition of $|\Pi|$ into relative interiors) so that we can argue in $\operatorname{aff}(P)$ where the claim is obvious. (The reader should generalize this observation from line segments to subpolytopes of higher dimension.)

The preceding discussion shows that it is justified to extend the notions introduced for single polyhedra to polyhedral complexes. The polytopes $P \in \Pi$ are the *faces* of Π. The 0-dimensional faces are called *vertices*, and the 1-dimensional faces are the *edges*. The edges of a conical complex are also called its *rays*. The maximal faces are the *facets* of Π— so the facets of a polyhedron Q are the facets of its boundary complex. By $\max \Pi$ we denote the set of facets of Π. The *dimension* of Π is the maximum of the dimensions of its faces.

Often it is useful to work with the skeletons of a polyhedral complex, obtained as follows: the *e-skeleton* $\Pi^{(e)}$ consists of all polyhedra in Π that have dimension $\le e$. The 1-skeleton is a *graph* (undirected, without multiple edges and loops).

Example 1.43. In a simplicial complex Σ there are no affine dependencies between the vertices of the simplices, and such a complex can always be embedded. Let $x_1, \dots, x_n$ be the vertices of Σ. Then we choose affinely independent points $v_1, \dots, v_n$ in $\mathbb{R}^{n-1}$, and let Σ' be the collection of simplices $\operatorname{conv}(v_{i_1}, \dots, v_{i_k})$ for which $x_{i_1}, \dots, x_{i_k}$ span a simplex in Σ. Clearly $\Sigma \cong \Sigma'$.

This consideration shows that a simplicial complex is determined up to isomorphism by the sets $\{v_{i_1}, \dots, v_{i_k}\}$ that appear as vertex sets of its simplices, and the collection of these sets is called an *abstract simplicial complex.* An abstract simplicial complex Δ is nothing but a collection of subsets of a finite set V such that $F \in \Delta$ implies $F' \in \Delta$ for all subsets F' of F, and a simplicial complex Σ that gives rise to the abstract simplicial complex Δ is called a *geometric realization* of Δ.

Similarly one sees that every simplicial conical complex can be realized as a fan.

Let now Π be a polyhedral complex. Then a function $\alpha : |\Pi| \to \mathbb{R}^d$ is *affine, (strictly) convex* or *(strictly) concave*, respectively, if this attribute applies to $\alpha|P$ for each $P \in |\Pi|$. Note that such a function need not be extendable to an affine, convex, or concave function on $\operatorname{conv}(|\Pi|)$ if Π is an embedded polyhedral complex: the property is only required to hold piecewise.

Example 1.44. Let us return to Example 1.36. Suppose α is an affine function on Π, and set $y_i = \alpha(x_i)$. Then $y_1, \dots, y_6$ must satisfy the following system of equations resulting from the fact that the two diagonals in a quadrangle meet at their common midpoint:

$$y_1 + y_3 = y_2 + y_4, \qquad y_1 + y_6 = y_4 + y_5, \qquad y_2 + y_6 = y_3 + y_5.$$

It follows that $y_1 = y_4$, $y_2 = y_3$, $y_5 = y_6$.

This shows that Π is not embeddable: otherwise the coordinate functions in the surrounding space would separate y_1 and y_4, and as just seen, this is impossible.

1.E Subdivisions and triangulations

We have motivated the notion of polyhedral complex by the example of a triangulation in the plane (see Figure 1.5). Now we introduce the notion of triangulation and, more generally, that of subdivision, formally.

Definition 1.45. Let Π, Π' be polyhedral complexes. We say that Π' is a *subdivision* of Π if $|\Pi| = |\Pi'|$ and each face of Π is the union of faces $P' \in \Pi'$. (Equivalently, we could require that every face of Π' is contained in a face of Π.)

If Q is a polyhedron, then a *subdivision* of Q is a subdivision of its face lattice.

If Π and Π' are polytopal (or conical) complexes and Π' is simplicial, then it is called a *triangulation* of Π.

Let Π be a polyhedral complex and Π' a subdivision. Since the decomposition $|\Pi| = \bigcup_{P' \in \Pi'} \operatorname{int}(P')$ is disjoint, it follows immediately that each nonempty $P' \in \Pi'$ is contained in (the uniquely determined) $P \in \Pi$ for which $\operatorname{int}(P) \cap \operatorname{int}(P') \neq \emptyset$.

We want to show that every polytopal complex Π has a triangulation whose vertices can be chosen freely, provided the vertices of Π are taken into account. In order to construct it inductively we have to know the local structure of Π at a subset $X \subset |\Pi|$. It is determined by two subcomplexes of Π (and their difference):

Definition 1.46. The *star* of X is the subcomplex

$$\operatorname{star}_\Pi(X) = \{F : F \text{ is a face of some } G \in \Pi \text{ with } X \subset G\}.$$

The *link* of X is the subcomplex

$$\operatorname{link}_\Pi(X) = \{F \in \operatorname{star}_\Pi(X) : X \cap F = \emptyset\},$$

and the *open star neighborhood* is the subset

$$\operatorname{openstar}_\Pi(X) = \bigcup_{\substack{G \in \Pi \\ X \subset G}} \operatorname{int} G.$$

of Π. If there is no ambiguity about the complex Π, then we suppress the reference to it.

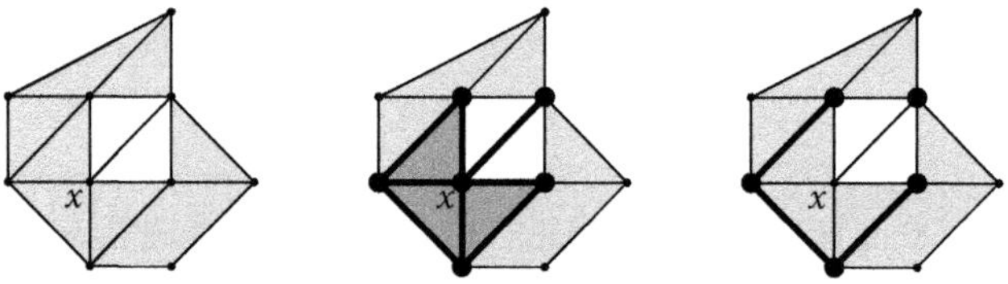

Fig. 1.6. A polytopal complex, star, and link of a vertex

As we have seen in Section 1.D the notion $\operatorname{conv}(X)$ is well-defined if there exists $P \in \Pi$ with $X \subset P$. Similarly we are justified in writing $x \in \operatorname{aff}(X)$ or $x \notin \operatorname{aff}(X)$ if $x \in P$ and $X \subset P$ for some $P \in \Pi$.

Proposition 1.47. *Let Π be a polytopal complex and $x \in |\Pi|$. Suppose that $F, G \in \operatorname{link}(x)$ and $P \in \Pi$ such that $x \in P$. Then the following hold:*

(a) $x \notin \operatorname{aff}(F)$;
(b) *if F is the union of the sets $M_i \subset F$, then* $\operatorname{conv}(F, x) = \bigcup_i \operatorname{conv}(M_i, x)$;
(c) $P = \bigcup\{\operatorname{conv}(F', x) : F' \in \operatorname{link}(x),\ F' \subset P\}$;
(d) $|\operatorname{star}(x)| = \bigcup\{\operatorname{conv}(F', x) : F' \in \operatorname{link}(x)\}$;
(e) $\operatorname{conv}(M, x) \cap |\operatorname{link}(x)| = M$ *for every convex subset M of F;*
(f) $\operatorname{conv}(M, x) \cap \operatorname{conv}(N, x) = \operatorname{conv}(M \cap N, x)$ *for all convex sets $M \subset F$ and $N \subset G$ (where* $\operatorname{conv}(M \cap N, x)$ *is formed within a face R of Π containing x and $F \cap G$).*

Proof. (a) Let Q be a face of Π such that $x \in Q$ and F is a face of $\operatorname{link}(x)$. Then $x \notin \operatorname{aff}(F)$ since $F = Q \cap \operatorname{aff}(F)$.

(b) Since F is convex, $\operatorname{conv}(F, x)$ is the union of the line segments $[x, y]$ with $y \in F$. This observation immediately implies the claim.

(c) The containment $\supset$ follows from the convexity of P. For the converse let $y \in P$, $y \neq x$. Passing to a face of P if necessary, we can assume that no proper face of P contains x and y. The ray (in $\operatorname{aff}(P)$) from x through y leaves P at a point $z \in P$. The facet Q of P containing z cannot contain x—otherwise it would contain y as well. So $Q \in \operatorname{link}(x)$, and $y \in \operatorname{conv}(Q, x)$.

(d) follows immediately from (c).

(e) We choose $Q \in \operatorname{star}(x)$ as in (a) and work in $\operatorname{aff}(Q)$. The line through x and a point $z \in M \subset F$ meets F only in z since $x \notin \operatorname{aff}(F)$.

(f) Note that we can find a suitable $R \in \Pi$ since $F \cap G \in \operatorname{link}(x)$. Let $y \in \operatorname{conv}(M, x) \cap \operatorname{conv}(N, x)$. If $y = x$, then certainly $y \in \operatorname{conv}(M \cap N, x)$. Otherwise there exist line segments $(x, z]$, $z \in M$, and $(x, z']$, $z' \in N$, both containing x. According to Remark 1.42 one has $z \in (x, z']$, or $z' \in (x, z]$. We can assume that $z \in (x, z']$. It follows that $z \in \operatorname{conv}(N, x) \cap |\operatorname{link}(x)|$, and so $z \in N$ by (e) (and $z = z'$). But this implies $z \in M \cap N$ and $y \in \operatorname{conv}(M \cap N, x)$. □

Let P be a polytope and $x \notin \operatorname{aff}(P)$. Then one calls $\operatorname{conv}(P, x)$ the *pyramid* over (the *base*) P with *apex* x. The faces of such a pyramid are easily described, as the reader may check:

Proposition 1.48. *Let Q be the pyramid over P with apex x. Then the faces of Q are given by the faces of P and the pyramids over the faces of P with apex x.*

Very often the notion of visibility is useful: we say that a point $y \in \mathbb{R}^d$ is *visible* from $x \in \mathbb{R}^d$ with respect to the obstacle $M \subset \mathbb{R}^d$ if the semi-open line segment $[x, y)$ does not intersect M. For an interpretation of Proposition 1.47 in terms of visibility imagine an embedded complex $|\Pi|$ as a subset of the surrounding space, $|\Pi|$ filled with some opaque material. Suppose $|\operatorname{star}_\Pi(x)|$ is covered by d-dimensional polytopes. Then $|\operatorname{link}_\Pi(x)|$ is the set of all those points of Π that are invisible from x, but become visible if we remove from $|\Pi|$ the interior of all faces containing x except x itself.

It is easily checked that a subdivision Π' of Π can be restricted to a subcomplex Π'' of Π: the polytopes $P \in \Pi'$ with $P \subset |\Pi''|$ form a subdivision of Π''. One often constructs subdivisions by going the other way around. Then one has to patch subdivisions of subcomplexes:

Proposition 1.49. *Let Π_1 and Π_2 be subcomplexes of Π, and let Π'_i be a subdivision of Π_i, $i = 1, 2$. If the subdivisions of $\Pi_1 \cap \Pi_2$ induced by Π'_1 and Π'_2 coincide, then $\Pi'_1 \cup \Pi'_2$ is a subdivision of $\Pi_1 \cup \Pi_2$.*

The straightforward proof is also left to the reader. We are mainly interested in polytopal and conical complexes, and will not try to extend the results proved below to arbitrary polyhedral complexes. While we have derived assertions about polytopes from their conical counterparts in the previous section, we will take the opposite direction now.

An important technique for the construction of subdivisions is *stellar subdivision* of a polytopal complex Π with respect to some point $x \in |\Pi|$ as introduced in the following:

Lemma 1.50. *Let Π be a polytopal complex and $x \in |\Pi|$. Suppose that Π' is a subdivision of $\{F \in \Pi : x \notin F\}$. Let Λ be the set of polytopes $F \in \Pi'$ that subdivide* $\operatorname{link}_\Pi(x)$. *Then*

$$\Pi'' = \Pi' \cup \{\operatorname{conv}(F,x) : F \in \Lambda\}$$

is a subdivision of Π with $\operatorname{vert}(\Pi'') = \operatorname{vert}(\Pi') \cup \{x\}$.

If Π' is a triangulation, then so is Π''.

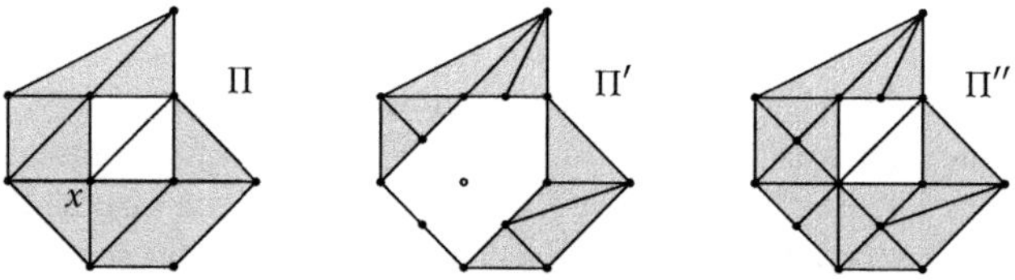

Fig. 1.7. Stellar subdivision

Proof. Let us first deal with the very last assertion: if F is a simplex, then the pyramid $\operatorname{conv}(F,x)$ is a simplex as well.

In view of the patching principle 1.49 it is enough for the main part of the lemma to show that $\Sigma = \{\operatorname{conv}(F,x) : F \in \Lambda\} \cup \Lambda$ is a subdivision of $\operatorname{star}_\Pi(x)$ containing x as a vertex, since (i) $\operatorname{star}_\Pi(x) \cap \{F \in \Pi : x \notin F\} = \operatorname{link}_\Pi(x)$, and (ii) both Σ and Π' restrict to the same subdivision on $\operatorname{link}_\Pi(x)$. Fact (ii) follows from Proposition 1.47(e): $\operatorname{conv}(F,x) \cap |\operatorname{link}_\Pi(x)| = F$ for all $F \in \Lambda$ (and F is a face of $\operatorname{conv}(F,x)$).

After this preparation one notes that $\{x\}$ is a face of Σ, obtained by taking $F = \varnothing$. Therefore $x \in \operatorname{vert}(\Sigma)$. Next it follows immediately from 1.47(a) and 1.48 that Σ contains all the faces of its polytopes.

We have to check that the intersection of two polytopes $P, Q \in \Sigma$ is a face of P as well as of Q. This is clear if $x \notin P \cup Q$, for both P and Q are members of Λ in this case.

So assume that $x \in P$. Then $P = \operatorname{conv}(F,x)$ with $F \in \Lambda$. If $x \in Q$, too, then $Q = \operatorname{conv}(G,x)$ with $G \in \Lambda$, and $P \cap Q = \operatorname{conv}(F \cap G, x)$ by Proposition 1.47(f). Now $F \cap G$ is a face of both F and G, and the pyramid $\operatorname{conv}(F \cap G, x)$ is a face of both pyramids P and Q with apex x.

Let $x \notin Q$. Then $Q \subset |\operatorname{link}_\Pi(x)|$. We claim that $P \cap Q = F \cap Q$. This suffices since $F \cap Q$ is certainly a face of Q and of F, and therefore of P. That $P \cap Q = F \cap Q$ follows from 1.47(e): $P \cap Q = P \cap |\operatorname{link}_\Pi(x)| \cap Q = F \cap Q$.

To sum up: Σ is a polytopal complex. It remains to verify that every $P \in \operatorname{star}_\Pi(x)$ is the union of those $Q \in \Sigma$ that are contained in P. This is clear if $x \notin P$, and follows from 1.47(b) and (c) if $x \in P$. □

Lemma 1.50 is a very powerful tool, as the following theorem shows:

Theorem 1.51. *Let* Π *be a polytopal complex and* V *a finite set of points with* $\operatorname{vert}(\Pi) \subset V \subset |\Pi|$. *Then there exists a triangulation* Σ *of* Π *with* $\operatorname{vert}(\Sigma) = V$.

Proof. It is enough to prove the theorem for $V = \operatorname{vert}(\Pi)$ since all extra vertices can be added by stellar subdivision as the lemma has shown.

However, the lemma solves the problem also for $V = \operatorname{vert}(\Pi)$. We use induction on V starting with the trivial case $V = \emptyset$. For $V \neq \emptyset$ we choose $x \in V$ and set $V' = V \setminus \{x\}$. The polytopal complex $\{F \in \Pi : x \notin F\}$ in the lemma has V' as its set of vertices. By induction it has a triangulation Σ', and the lemma tells us how to extend it to a triangulation of Π. □

Remark 1.52. (a) Let us consider a single polytope P of dimension d. Then the triangulation constructed in the proof of Theorem 1.51 does not include a d-simplex before the very last vertex of P has been inserted. The emergence of this triangulation is illustrated in Figure 1.8.

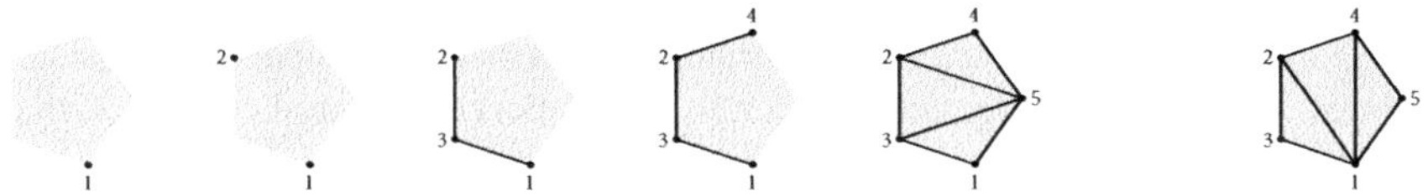

Fig. 1.8. Two triangulations of a pentagon

Suppose that the elements of $V \supset \operatorname{vert}(P)$ are given in some order, $V = \{x_1, \dots, x_m\}$. Then it is perhaps more natural to successively build up a triangulation of P by extending a triangulation of $P_{n-1} = \operatorname{conv}(x_1, \dots, x_{n-1})$ to a triangulation of $P'_n = \operatorname{conv}(x_1, \dots, x_n)$, $n = 1, \dots, m$. This construction is illustrated by the rightmost triangulation in Figure 1.8.

Simplifying notation we set $P' = \operatorname{conv}(x_1, \dots, x_{n-1})$ and $P = \operatorname{conv}(x_1, \dots, x_n)$. Let Σ be a triangulation of P'. If x_n in P', we use stellar subdivision, refining the triangulation of $P' = P$. (Or we omit x_n if we are only interested in finding a triangulation of P.) If $x_n \notin P'$ we must distinguish the cases (i) $x_n \notin \operatorname{aff}(P')$ and (ii) $x_n \in \operatorname{aff}(P')$.

In case (i) we add to Σ all the simplices $\operatorname{conv}(F, x)$, $F \in \Sigma$; that we obtain a triangulation in this way follows from Lemma 1.50. In case (ii) we augment Σ by all simplices $P = \operatorname{conv}(F, x)$ where $F \in \Sigma$ is contained in a support hyperplane H of P' such that $x_n \in H^-$. Now Lemma 1.50 does not cover our claim. It is however not too difficult to extend it accordingly. One should prepare the extension by a suitable variant of Proposition 1.47, in which $\operatorname{link}(x)$ is replaced by the set of all the faces F of Σ just mentioned. (See also Exercise 1.17.)

(b) Obviously the triangulation constructed via Lemma 1.50 or the procedure outlined in (a) (applied to $V \cap P$ for each polytope in the complex) depends not only on the set V, but also on the order in which the points of V are inserted as vertices. On the other hand, if V is given, then it depends only on this order, and more precisely, the triangulation of each $P \in \Pi$ depends only on the order in which the elements of $V \cap P$ are inserted. This shows that we could have carried out the construction in each polytope separately (after fixing an order on the elements of V), patching together the individual triangulations via 1.49. In particular, part (a) can be extended to polytopal complexes.

As an immediate consequence of Theorem 1.51 we obtain a classical result, *Carathéodory's theorem:*

Theorem 1.53. *Let $X \subset \mathbb{R}^d$. Then for every $x \in \operatorname{conv}(X)$ there exists an affinely independent subset X' of X such that $x \in \operatorname{conv}(X')$.*

Proof. There exists a finite subset X'' of X such that $x \in \operatorname{conv}(X'')$. We may assume $X = X''$. Then $\operatorname{conv}(X)$ is a polytope. It has a triangulation Σ with vertex set X. We now choose $X' = \operatorname{vert}(F)$ where $F \in \Sigma$ contains x. □

There also exists a very elementary proof of Carathéodory's theorem, not using the existence of triangulations; see [381, Prop. 1.15].

Now we transfer the theorem on triangulations from polytopal to conical complexes. Note that there is a unique point in the support of a conical complex Γ that is identified with $0 \in \mathbb{R}C$ for each $C \in \Gamma$. We denote it by 0 as well.

Theorem 1.54. *Let Γ be a conical complex, and $V \subset |\Gamma|$ be a finite set of vectors $v \neq 0$ such that $V \cap C$ generates C for each $C \in \Gamma$. Then there exists a triangulation Σ of Γ such that $\{\mathbb{R}_+ v : v \in V\}$ is the set of 1-dimensional faces of Γ.*

Proof. For each cone $C \in \Gamma$ we consider the convex hull $P_C = \operatorname{conv}(V \cap C, 0)$. These polytopes (together with all their faces) form a polytopal complex Π. The essential point in proving this claim is that one has $P_C \cap F = P_F$ if F is a face of C.

Set $\Lambda = \operatorname{link}_\Pi(0)$. Let $y \in |\Gamma|$, $y \neq 0$, and choose a face C of Γ containing y. The ray $\mathbb{R}_+ y$ leaves P_C at a point y' which belongs to $|\Lambda|$, as we have seen in the proof of 1.47. We can replace each $v \in V$ by v'; in other words, we can assume that $V \subset |\Lambda|$. By Theorem 1.51 we find a triangulation Σ' of Λ such that $\operatorname{vert}(\Sigma') = V$.

Lemma 1.50 tells us how to extend Σ' to a triangulation Σ'' of Π. That Σ'' is a triangulation is evidently equivalent to the fact that the cones $\mathbb{R}_+ D'$, $D \in \Sigma'$, form a conical complex Σ, and since each D is a simplex, Σ is a complex of simplicial cones. Its 1-dimensional faces are exactly the rays $\mathbb{R}_+ v$, $v \in V$, and that it covers $|\Gamma|$ follows since $|\Lambda|$ consists exactly of the points y', $y \in |\Gamma|$, $y \neq 0$. We finish the construction by choosing Σ as the collection of cones $\mathbb{R}_+ P$, $P \in \Sigma'$. □

As in the case of a polytopal complex we have a Carathéodory theorem:

Theorem 1.55. *Let $X \subset \mathbb{R}^d$. Then for every $x \in \mathbb{R}_+ X$ there exists an linearly independent subset X' of X such that $x \in \mathbb{R}_+ X'$.*

1.F Regular subdivisions

There is a distinguished class of subdivisions of polyhedral complexes which will be important in our study of monoid rings and toric varieties—the class of *regular subdivisions*.

Let P be a polyhedron and consider a function $f : P \to \mathbb{R}$. We say that a connected subset $W \subset P$ is a *domain of linearity* of f if the restriction $f|W$ can be extended to an affine function on $\operatorname{aff}(P)$ and there is no connected subset $V \subset P$ containing W strictly for which $f|V$ extends to an affine mapping. If Π is a polyhedral complex and $f : |\Pi| \to \mathbb{R}$ is a function, then a subset W of $|\Pi|$ is a domain of linearity of f if there exists a facet $P \in \Pi$ such that $W \subset P$ and W is a domain of linearity of $f|P$.

Definition 1.56. A subdivision Π' of a polyhedral complex Π is called *regular* if there is a convex function $f : |\Pi| \to \mathbb{R}$ whose domains of linearity are the facets of Π'. Such a function f is called a *support function* for the subdivision Π'. The set of all support functions will be denoted by $\operatorname{SF}(\Pi, \Pi')$.

To give a trivial example: the constant 0 is a support function of the trivial subdivision Π of Π, and so Π is a regular subdivision of itself. Several constructions of nontrivial regular subdivisions will be discussed below. An example of a nonregular subdivision will be given in Remark 1.63.

Remark 1.57. Regular subdivisions are called *projective* in Kempf et al. [221] (perhaps the most descriptive name in view of the discussion below). One also encounters the name *coherent* (for instance, in Gelfand, Kapranov, and Zelevinsky [140]). Our choice "regular" is common in the literature dealing with applications to Gröbner bases (Sturmfels [338]) and (computational) geometry (De Loera, Rambau, and Santos [106], Ziegler [381]).

Remark 1.58. Let $P \subset \mathbb{R}^d$ and $Q \subset \mathbb{R}^d \times \mathbb{R}_- \subset \mathbb{R}^{d+1}$ be polytopes such that P is a proper face of Q and, moreover, the projection $\mathbb{R}^{d+1} \to \mathbb{R}^d$, $(z_1, \dots, z_d, z_{d+1}) \mapsto (z_1, \dots, z_d)$ maps Q onto P. (Here we identify $\mathbb{R}^d$ and $(\mathbb{R}^d, 0)$.) Then the faces of Q that consist of lower endpoints of maximal line segments $[(z, 0), (z, h)]$, $h \le 0$, in Q, form the *bottom of Q below P*. (In this section the term "bottom" is used in this sense. It is reconciled with the notion of bottom in Section 1.C if we replace Q by the union of all vertical rays emerging from points in Q.) These faces project up to P to certain

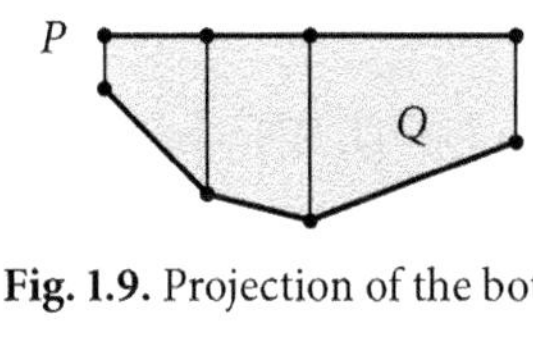

Fig. 1.9. Projection of the bottom

polytopes inside P, defining a subdivision of P. This subdivision is regular. It is supported by the function $f : P \to \mathbb{R}$, which assigns to a point $z \in P$ the height (i. e., the

$(d+1)$th coordinate) of the point in the bottom of Q right below z. In other words, the bottom is the graph of a support function. Conversely, all regular subdivisions of P are obtained this way.

Although we have defined regular subdivisions for arbitrary polyhedral complexes, in practice we will only encounter two special cases—those of polytopal complexes and those of conical complexes consisting of pointed cones. Essentially everything that will be said on polytopal complexes admits a natural conical version, usually not mentioned explicitly.

For a polytopal complex Π the set of functions $f : |\Pi| \to \mathbb{R}$ that are affine on the faces of Π is denoted by $\mathrm{PA}(\Pi)$ —the set of *piecewise affine functions*. Clearly, $\mathrm{PA}(\Pi)$ is a finite-dimensional real vector space, and if Π' is a subdivision of Π, then $\mathrm{PA}(\Pi)$ is a subspace of $\mathrm{PA}(\Pi')$.

The convexity of a function $f \in \mathrm{PA}(\Pi')$ can be tested as follows. For each facet $P' \in \Pi'$ we let $f_{P'}$ be the affine function (on $\operatorname{aff}(P)$ for the facet P of Π with $P' \subset P$) such that $f|P' = f_{P'}|P'$. Then f is convex if and only if

$$f(x) = \max\{f_{P'}(x) : P' \subset P\}$$

for all $x \in |\Pi|$ and all facets P of Π containing x. Moreover, f is a support function if and only if for each facet P of Π all the functions $f_{P'}$, $P' \subset P$, are pairwise different. These conditions can be tested by the application of suitable linear forms on the space of all functions, as we will see now,

Let Π' be an arbitrary subdivision of a polytopal complex Π. A finite *nonempty* system $\mathcal{D}$ of real-valued linear functionals on the space of all functions $\mathbb{R}^{|\Pi|}$ is a *regularity test system* for the pair (Π, Π') if the following equivalence holds for all functions $f \in \mathrm{PA}(\Pi')$:

$$\delta(f) \geq 0 \text{ for all } \delta \in \mathcal{D} \quad \Longleftrightarrow \quad f \text{ is convex on } \Pi.$$

That regularity test systems exist and do in fact test regularity (and not only convexity of functions) is shown in the following:

Lemma 1.59. *Let Π' be a subdivision of a polytopal complex Π. Then*

(a) there exists a regularity test system for (Π, Π');

(b) the set of all functions in $\mathrm{PA}(\Pi')$ that are convex with respect to Π form a cone $C \subset \mathrm{PA}(\Pi')$;

(c) $\mathrm{SF}(\Pi, \Pi')$ is the absolute interior of C in $\mathrm{PA}(\Pi')$, and so $\mathrm{SF}(\Pi, \Pi') \neq \emptyset$ if and only if C is full-dimensional;

(d) if $\mathcal{D}$ is a regularity test system for (Π, Π') and $f \in \mathrm{PA}(\Pi')$, then $f \in \mathrm{SF}(\Pi, \Pi')$ if and only if $\delta(f) > 0$ for all $\delta \in \mathcal{D}$.

Proof. $P \in \Pi$ is a facet and $P' \in \Pi'$ a face of dimension $\dim P - 1$ such that $\operatorname{int}(P') \subset \operatorname{int}(P)$. There are exactly two facets $P_1, P_2 \in \Pi'$ such that $P_1, P_2 \subset P$ and $P_1 \cap P_2 = P'$. Fix points $z_1 \in \operatorname{int}(P_1)$ and $z_2 \in \operatorname{int}(P_2)$ in such a way that the line segment $[z_1, z_2]$ intersects $\operatorname{int}(P')$, say in a point z. We have the linear map

$$\delta_{P1,P_2}:\mathbb{R}^{|\Pi|}\to\mathbb{R},\qquad \delta_{P1,P_2}(f)=g(z)-f(z),$$

where g is the affine map from the segment $[z_1,z_2]$ to $\mathbb{R}$ such that $g(z_1)=f(z_1)$ and $g(z_2)=f(z_2)$. It is easily checked that $g(z)$ depends linearly on f.

By running P through the facets of Π and P' through the faces of Π', so that $\operatorname{int}(P')\subset\operatorname{int}(P)$ and $\dim P'=\dim P-1$, and fixing for each P' two points $z_1,z_2\in|\Pi|$ on both sides of P' as above, we get a finite system $\mathscr{D}$ of linear functionals $\mathbb{R}^{|\Pi|}\to\mathbb{R}$.

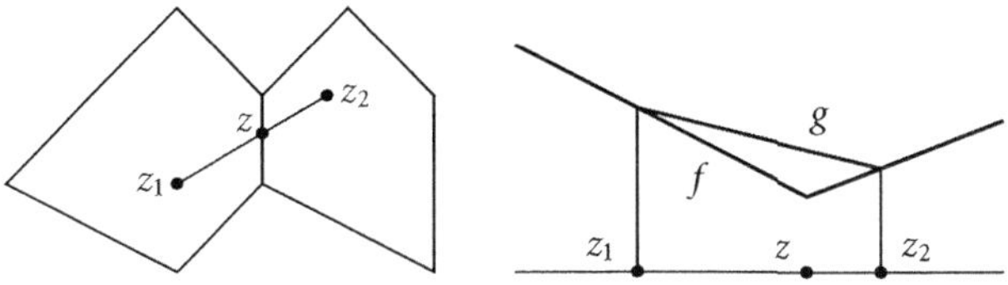

Fig. 1.10. Construction of the regularity test system

In order to check that $\mathscr{D}$ is a regularity test system assume first that $\delta_{P_1,P_2}(f)<0$ for some $f\in\mathbb{R}^{|\Pi|}$. Then f is not convex along the line segment chosen for the definition of δ_{P_1,P_2}. Conversely suppose that $\delta_{P_1,P_2}(f)\ge 0$ for some $f\in\operatorname{PA}(\Pi')$ and all functionals $\delta_{P_1,P_2}\in\mathscr{D}$. Let L be a line segment in a facet P of Π, and let $Q_1,\dots,Q_t$ be the facets of Π' traversed by L. In checking the convexity of f we are allowed to assume that $Q_i\cap Q_{i+1}$ is a face of dimension $\dim P-1$ for $i=1,\dots,t-1$; otherwise we perturb the endpoints of L, and use a continuity argument. Finally it is enough to consider the case $t=2$. We can assume that f is of constant value 0 on Q_1 and must show that it has nonnegative values on Q_2. This follows immediately from $\delta_{Q_1,Q_2}(f)\ge 0$.

This shows (a), and (b) follows immediately from (a). Moreover, in view of (b), every regularity test system $\mathscr{D}'$ defines the same cone, and its absolute interior is given by $\{f:\delta(f)>0\text{ for all }\delta\in\mathscr{D}'\}$. For (c) and (d) it is therefore enough to consider the system $\mathscr{D}$ constructed above. Now one observes that the graph of f is broken along the common boundary of P_1 and P_2 if and only if $\delta_{P1,P_2}(f)>0$, and so f is a support function for Π' if the strict inequality holds for all pairs P_1, P_2 that are contained in a facet P of Π' and meet along a common face of dimension $\dim P-1$. □

For a polytopal complex Π the evaluation of the functions at the vertices of Π gives rise to a linear embedding $\operatorname{PA}(\Pi)\to\mathbb{R}^n$, $n=\operatorname{vert}(\Pi)$. In the special case when Π is a simplicial complex this is an isomorphism of real vector spaces.

The existence of regularity test systems implies the following:

Corollary 1.60. *Let Π be a polytopal complex and Π' a triangulation of Π. If Π' is regular, then* $\operatorname{SF}(\Pi,\Pi')$ *is the interior of an n-dimensional cone in $\mathbb{R}^n\cong\operatorname{PA}(\Pi')$, $n=\#\operatorname{vert}(\Pi')$.*

In particular, the corollary says that regular triangulations Π' are stable with respect to small perturbations of the values $f(x)$, $x\in\operatorname{vert}(\Pi')$.

Next we give a quick overview of several basic regular triangulations. More specific triangulations will be considered in the next chapters.

Intersections and dissections. Let $\Pi_1, \dots, \Pi_k$ be a system of subdivisions of a polyhedral complex Π. Then all possible intersections $P_1 \cap \dots \cap P_k$ of faces $P_i \subset \Pi_i$, $i \in [1, k]$ form a subdivision of Π which we denote by $\Pi_1 \cap \dots \cap \Pi_k$ and call the *intersection* of the Π_i (by abuse of notation and terminology). The intersection of regular subdivisions is regular because if $f_i \in \mathrm{SF}(\Pi, \Pi_i)$, $i = 1, \dots, k$, then $f_1 + \dots + f_k \in \mathrm{SF}(\Pi, \Pi_1 \cap \dots \cap \Pi_k)$.

One special case of this construction deserves special consideration. Let $P \subset \mathbb{R}^d$ be a polytope and $H \subset \mathbb{R}^d$ a hyperplane that cuts P into two parts of the same dimension, say $P = Q \cup R$. Then the union of the face lattices of Q and R forms a regular subdivision of P. In fact, if H is given by an equation $a_1 X_1 + \dots + a_d X_d = b$ with $a_1, \dots, a_d, b \in \mathbb{R}$, then the function $f : P \to \mathbb{R}$, $f(z) = |a_1 z_1 + \dots + a_d z_d - b|$, $z = (z_1, \dots, z_d)$, is a support function of the subdivision under consideration. Taking intersections of such subdivisions one concludes that an arbitrary finite system of hyperplanes $H_1, \dots, H_k \subset \mathbb{R}^d$ cuts P into smaller polytopes that define a regular subdivision of P. Thus the hyperplane dissections introduced in Example 1.41 are regular.

Composition and patching. If Π' is a subdivision of Π and Π'' is one of Π', then Π'' is a subdivision of Π in a natural way. In this situation Π'' can be viewed as a *composite* of two successive subdivisions. Compositions enjoy the following transitivity of regularity.

Proposition 1.61. *If Π' is a regular subdivision of a polytopal complex Π and Π'' is one of Π', then Π'' is a regular subdivision of Π.*

Proof. We choose a regularity test system $\mathscr{D}''$ for the pair (Π, Π'') as in the proof of Lemma 1.59. For each pair of facets of Π'' that are contained in the same facet P of Π and share a common face of dimension $\dim P - 1$, $\mathscr{D}''$ contains a function δ_{P_1,P_2}. We split $\mathscr{D}''$ into two subsets, namely the set $\mathscr{D}'$ of those δ_{P_1,P_2} for which P_1 and P_2 are contained in the same facet of Π' and its complement $\mathscr{D} = \mathscr{D}'' \setminus \mathscr{D}'$.

Then $\mathscr{D}'$ is a regularity test system for (Π', Π'') of the type constructed in the proof of Lemma 1.59. Moreover, it is easily seen that $\mathscr{D}$ is a regularity test system for (Π, Π'). In fact the line segment used in the construction of $\delta_{P_1,P_2} \in \mathscr{D}$ connects interior points in different facets Q_1, Q_2 of Π' and meets the interior of the face $Q_1 \cap Q_2$, $\dim Q_1 \cap Q_2 = \dim Q_1 - 1 = \dim Q_2 - 1$.

Now choose $f \in \mathrm{SF}(\Pi, \Pi')$ and $g \in \mathrm{SF}(\Pi', \Pi'')$. Then $\delta(f) > 0$ for all $\delta \in \mathscr{D}$ and $\delta'(g) > 0$ for all $\delta' \in \mathscr{D}'$. Moreover, $\delta(f) = 0$ for all $\delta \in \mathscr{D}'$. Therefore, if the real number $\varepsilon > 0$ is small enough, then $\delta(f + \varepsilon g) > 0$ for all $\delta \in \mathscr{D}''$. □

The definition of regular subdivision contains the following patching principle.

Corollary 1.62. *Let Π be a polytopal complex with regular subdivision Π'. Furthermore let Π''_P be a regular subdivision of each facet $P \in \Pi'$. Then the subdivisions Π''_P can be patched to a regular subdivision Π'' of Π if and only if there are functions $f_P \in \mathrm{SF}(P, \Pi''_P)$ (we simply write P for the face complex of P) such that $(f_P)|P \cap Q = f_Q|P \cap Q$ for all facets $P, Q \in \Pi'$.*

Proof. In view of Proposition 1.61 we can assume that $\Pi = \Pi'$. If the Π''_P define a global regular subdivision with support function f, then we can take $f_P = f|P$ for each facet

$P \in \Pi$, and the condition $(f_P)|P \cap Q = f_Q|P \cap Q$ is obviously satisfied for all facets $P, Q \in \Pi$. Conversely, if the support functions f_P satisfy this condition, then (i) we can patch them up to a function on $|\Pi|$, and (ii) we can patch up the subdivisions to a global subdivision Π'' of Π, which is automatically regular. (Regularity is decided on each facet $P \in \Pi$ separately.) □

Remark 1.63. If the subdivisions Π''_P can merely be patched up to a global subdivision Π' of Π, then Π' need not be regular. The following example (taken from [221]) in dimension 2 shows this. Let Π' be the (visibly) regular subdivision of the equilateral

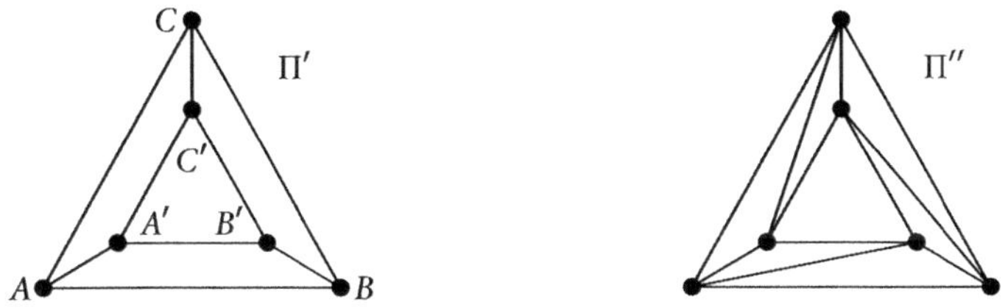

Fig. 1.11. A regular subdivision and a nonregular refinement

triangle T with face lattice Π (see Figure 1.11) into 3 isosceles trapezoids and one triangle. We refine it by cutting each of the quadrangles along a line so that each of them is regularly subdivided. However, the resulting triangulation Π'' of Π is not regular.

To this end we can assume that the support function f vanishes on the entire inner triangle. Let D be the intersection point of the line segments $[A, B']$ and $[A', B]$. Then $D = \lambda A + \mu B' = \lambda B + \mu A'$ for some $\lambda, \mu > 0$, $\lambda + \mu = 1$. Since $f(D) = \lambda f(A) + \mu f(B') < \lambda f(B) + \mu f(A')$, one obtains $f(A) < f(B)$. Running this argument around the triangle, we arrive at the contradiction $f(A) < f(A)$.

The nonregularity of Π'' follows also from a criterion of Gelfand, Kapranov and Zelevinsky [140, Ch. 7, Th. 1.7]. Let P be a polytope and A a finite subset of Π containing $\operatorname{vert}(\Pi)$. Each triangulation Π of P with vertex set A defines a function $\chi_P : A \to \mathbb{R}$ whose value at $a \in A$ is the sum of the volumes of the facets of Π adjacent to a. *If Π and Π' are different regular triangulations with vertex set A, then* $\chi_\Pi \neq \chi_{\Pi'}$. The proof of the criterion requires the theory of the secondary polytope developed in [140].

In our example, Π'' and its mirror image with respect to the vertical axis have the same vertex set and the same characteristic function, so they cannot be regular, since they are different.

Triangulations with a given set of vertices. Let Π be a polytopal complex and $V \supset \operatorname{vert}(\Pi)$ a finite subset of $|\Pi|$. By Theorem 1.51 there exists a triangulation Σ of Π such that $\operatorname{vert}(\Sigma) = V$. Now we derive the regular version of this fact.

There are two ways of doing so: (i) by exhibiting a fast process, based on the existence of a global strictly convex function, and (ii) by constructing a triangulation step by step, preserving the regularity. The first method is particularly well suited for a single polytope but becomes a bit complicated for general complexes. The second method is relatively slow but makes no difference from the complexity point of view between single polytopes and polytopal complexes. As we will see, it uses the regularity of stellar triangulations.

Let us first discuss the construction via a strictly convex function.

Proposition 1.64. *Let* Π *be a polytopal complex and* $f : \Pi \to \mathbb{R}$ *be a strictly convex function with* $f(x) \le 0$ *for all* $x \in |\Pi|$. *Let* V *be a subset of* Π *containing* $\operatorname{vert}(\Pi)$. *For each face* $P \in \Pi$ *let* P' *be the convex hull of the set* $(P, 0) \cup \{(v, f(v)) : v \in V \cap P\} \subset \operatorname{aff}(P) \times \mathbb{R}$. *Then the projection of the bottom of* P' *onto* P, $P \in \Pi$, *defines a regular subdivision* Π' *of* Π *such that* $V = \operatorname{vert}(\Pi')$.

Proof. In view of Remark 1.58 and Corollary 1.62 the only critical point is the claim that $V = \operatorname{vert}(\Pi')$. In checking it, we can assume that Π is the face lattice of a single polytope P. Then we have to check that $(v, f(v))$ is a vertex of P' for each $v \in V$. Since P' is a polytope, it is enough that $(v, f(v))$ is an extreme point of P'; see Proposition 1.13. This follows immediately from the definition of strict convexity. □

First we consider the case of a single polytope P and a finite subset $V \subset P$ containing $\operatorname{vert}(P)$. The existence of a strictly convex function f on P is shown by the following construction. Choose a large real number $r \gg 0$ such that the d-ball $\{x \in \mathbb{R}^d \mid \|x\| \le r\}$ contains P. Then the subset

$$\left\{z = (z_1, \dots, z_{d+1}) \in \mathbb{R}^{d+1} \;:\; (z_1, \dots, z_d) \in P,\ z_{d+1} \le 0,\ \|z\| = r\right\}$$

of the standard d-sphere $\{z \in \mathbb{R}^{d+1} : \|z\| = r\}$ of radius r is a graph of a strictly convex function f on P. (A similar construction is the Delaunay subdivision; see Exercise 1.19.)

The subdivision defined by Proposition 1.64 need not be a triangulation. However, by a small perturbation of the values $f(v)$, $v \in V$, that replaces $(v, f(v))$ by $(v, f(v) + \varepsilon_v)$, it can be refined to a regular triangulation. This is made precise in Exercise 1.20.

The case of a general polytopal complex makes no difference once the existence of a strictly convex function on such a complex has been established. To this end we prove the following claim: *for every polytope* P *and a function* $\varphi : \partial P \to \mathbb{R}$, *whose restriction to any face* $F \subset P$ *is strictly convex, there exists a strictly convex function* $f : P \to \mathbb{R}$ *such that* $f|\partial P = \varphi$. Then a strictly convex function on Π can be constructed inductively via the skeleton filtration

$$\Pi^{(0)} \subset \Pi^{(1)} \subset \cdots \subset \Pi^{(d)}, \quad d = \dim \Pi.$$

On the 0-skeleton every function is strictly convex.

In its turn the claim above follows from the existence of a function $g : P \to \mathbb{R}$ with $g|\partial P = 0$ that is convex on P and strictly convex on $\operatorname{int}(P)$. In fact, we can extend φ to a convex function on P by considering the convex hull R of the set

$$(\operatorname{int}(P), 0) \cup \{(x, \varphi(x)) : x \in \partial P\} \subset \mathbb{R}^{d+1}.$$

The bottom of R is a graph of a convex function $h : P \to \mathbb{R}$, and $(g + h) : P \to \mathbb{R}$ is a strictly convex function on P extending φ.

We construct the function $g : P \to \mathbb{R}$ with the desired properties as follows. Fix d linearly independent vectors $v_1, \dots, v_d \in \mathbb{R}^d$, none of them parallel to a facet of P. For each index i we consider all line segments $L_x = (x + \mathbb{R}v_i) \cap P$, $x \in P$. They are parallel to v_i (some of them may be degenerated into a point). For each of these segments $[u, v]$ we consider the semicircle γ with diameter $[u, v]$ passing through $\big((u+v)/2, -\|u-v\|/2\big) \in \mathbb{R}^{d+1}$. (A semicircle is a figure congruent to $\{(a, b) : a^2+b^2 = r^2,\ b \le 0\} \subset \mathbb{R}^2$ for some $r \ge 0$.) Again, some of these semicircles may be degenerated into a point. The union of these semicircles (for fixed i) defines a graph of a function $g_i : P \to \mathbb{R}$. We claim that the function $g = g_1 + \dots + g_d$ has the desired property.

The vanishing of each g_i on ∂P follows from its very construction. The convexity of g_i is not hard to see. We must only show its convexity on line segments $[x, y] \subset P$. If $[x, y]$ is parallel to v_i, then the convexity follows from the convexity of the semicircle. In the other case we consider the plane through $[x, y]$ parallel to v_i, and in it the convex hull Q of the endpoints of the line segments L_x and L_y. One or both of these segments can be a single point. Correspondingly, Q can be either a trapezoid, a triangle, or a line segment. Let $\tilde{g}$ be the function obtained by applying the same construction to Q (with respect to v_i). Then $\tilde{g}(x) = g_i(x)$, $\tilde{g}(y) = g_i(y)$, and $\tilde{g}$ is a convex function, as the reader may easily check— its graph is a piece of a conical (or cylindrical) surface that projects onto Q. Moreover, $g_i(z) \le \tilde{g}(z)$ since the semicircle under the segment $(z + \mathbb{R}v_i) \cap Q$, $z \in Q$ is contained in the semicircles under L_z.

It follows that g vanishes on ∂P and is convex on P, as convexity of g is to be checked only along line segments, too. But since each g_i is convex and at least one of them is strictly convex along a given line segment, their sum g is strictly convex.

Now we discuss the regularity of stellar triangulations. The proof of Theorem 1.51 gives an explicit construction of a triangulation Σ of Π such that $\operatorname{vert}(\Sigma) = V$. It is an inductive process consisting of successive stellar subdivisions, described in Lemma 1.50. We have seen that the composite of regular triangulations is regular. Therefore, the regularity of the triangulation constructed in the proof of Theorem 1.51 follows from the following:

Lemma 1.65. *Let Π be a polytopal complex and $x \in |\Pi|$. Then its subdivision*

$$\Pi(x) = (\Pi \setminus \operatorname{star}_\Pi(x)) \cup \operatorname{link}_\Pi(x) \cup \{\operatorname{conv}(F, x) : F \in \operatorname{link}_\Pi(x)\}$$

is regular.

Proof. Let $G \in \operatorname{star}_\Pi(x) \setminus \operatorname{link}_\Pi(x)$, say $G \subset \mathbb{R}^d$. Consider the pyramid $P_G = \operatorname{conv}(G, (x, -1)) \subset \mathbb{R}^{d+1}$ with base G and apex x. Then the bottom of P_G, viewed as a graph, defines a function $f_G : G \to \mathbb{R}$. When G runs through $\operatorname{star}_\Pi(x) \setminus \operatorname{link}_\Pi(x)$, these functions can be patched up to a function $f : |\operatorname{star}_\Pi(x)| \to \mathbb{R}$. Extending it by 0 to the whole set $|\Pi(x)| = |\Pi|$ we get a function $f \in \operatorname{SF}(\Pi, \Pi(x))$. □

Corollary 1.66.

(a) *Let Π be a polytopal complex and V a finite set of points with* $\operatorname{vert}(\Pi) \subset V \subset |\Pi|$. *Then there exists a regular triangulation Σ of Π with* $\operatorname{vert}(\Sigma) = V$.

(b) *Let Γ be a conical complex, and $V \subset |\Gamma|$ be a finite set of vectors $v \neq 0$ such that $V \cap C$ generates C for each $C \in \Gamma$. Then there exists a regular triangulation Σ of Γ such that $\{\mathbb{R}_+ v : v \in V\}$ is the set of 1-dimensional faces of Γ.*

Part (b), which is Theorem 1.54 with regularity added, follows from its polytopal analogue in the same way as Theorem 1.54 follows Theorem 1.51. The details are left to the reader (Exercise 1.21).

Projective fans. Recall that a fan $\mathscr{F}$ is an embedded conical complex. Consider the situation when $\dim \mathscr{F} = d$ and $\mathbb{R}^d$ is the ambient space. If $|\mathscr{F}| = \mathbb{R}^d$, we say $\mathscr{F}$ is a *complete fan*. A regular subdivision of $\mathbb{R}^d$ into pointed cones is called a *projective fan*. A subcomplex of a projective fan is a *quasiprojective fan*. Thus a quasiprojective fan is projective if and only if it is complete. This terminology is explained by the link to algebraic geometry via toric varieties, see Section 10.B.

For fans we will simply write $\operatorname{SF}(\mathscr{F})$ instead of $\operatorname{SF}(\mathscr{F}, \mathbb{R}^d)$.

Suppose Q is a d-polytope in the dual vector space $(\mathbb{R}^d)^* = \operatorname{Hom}_{\mathbb{R}}(\mathbb{R}^d, \mathbb{R})$. For each vertex $v \in Q$ consider the cone $C_v = \mathbb{R}_+(-v + Q) \subset (\mathbb{R}^d)^*$, i. e., C_v is the cone $\mathbb{R}_+(-v + Q)$ spanned by the shifted polytope $-v + Q$. Then we form the collection of cones $\mathscr{N}(Q)$ consisting of the $(C_v)^*$, $v \in \operatorname{vert}(Q)$ and all their faces (in the original vector space $\mathbb{R}^d$).

Proposition 1.67. *Let $\mathscr{F}$ be a d-dimensional fan in $\mathbb{R}^d$. Then the following conditions are equivalent:*

(a) *$\mathscr{F}$ is projective;*

(b) *there exists a d-polytope $Q \subset \mathbb{R}^d$ such that $0 \in \operatorname{int}(Q)$ and the cones in $\mathscr{F}$ are exactly the cones over the faces of Q;*

(c) *$\mathscr{F} = \mathscr{N}(P)$ for some d-polytope $P \subset (\mathbb{R}^d)^*$.*

In particular, $\mathscr{N}(P)$ is a fan, called the *normal fan of P*. Thus two d-polytopes

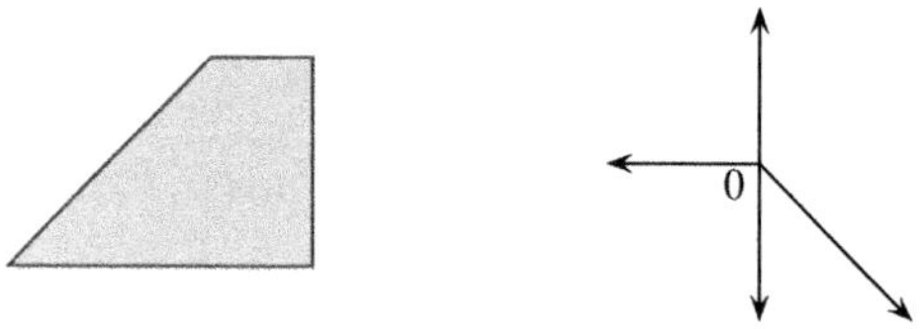

Fig. 1.12. A polygon and its normal fan

$P, Q \subset \mathbb{R}^d$ have the same normal fans if they have the same combinatorial type and the corresponding pairs of facets span parallel affine hulls. Clearly, for every projective fan there are infinitely many polytopes that support it.

Proof. (a) $\implies$ (b) For each maximal cone (i. e., facet) $C \in \max \mathscr{F}$ there exists an element $\varphi \in \operatorname{SF}(\mathscr{F})$ such that $\varphi^{-1}(0) = C$. In fact, choose a function $\psi \in \operatorname{SF}(\mathscr{F})$ with $\psi(0) = 0$ and let ψ^C denote the linear extension of $\psi|C$ to the whole space $\mathbb{R}^d$. Then the function $\psi - \psi^C$ is an element of $\operatorname{SF}(\mathscr{F})$ with the desired property.

For each $C \in \max \mathscr{F}$ we choose an element $\varphi_C \in \mathrm{SF}(\mathscr{F})$ such that $\varphi_C^{-1}(0) = C$ and set $\chi = \sum_{C \in \max \mathscr{F}} \varphi_C$. Then the polytope

$$Q = \chi^{-1}([0,1]) \subset \mathbb{R}^d$$

has the property as in (b).

(b) $\Longrightarrow$ (a) Consider the polytope $(Q,1) = \{(z,1) : z \in Q\} \subset \mathbb{R}^{d+1}$ and the cone $C(Q) = \mathbb{R}_+(Q,1)$ over it. For each point $x \in |\mathscr{F}|$ there is a unique point $z(x) \in \partial C(Q)$ that projects down to x. Let $h(z(x))$ denote the height of $z(x)$ above $\mathbb{R}^d$. Then the mapping $x \mapsto h(z(x))$, $x \in \mathbb{R}^d$, defines an element of $\mathrm{SF}(\mathscr{F})$.

(a) $\Longleftrightarrow$ (c) The following correspondences between $\mathrm{SF}(\mathscr{F})$ and the set of d-polytopes $P \subset (\mathbb{R}^d)^*$ for which $\mathscr{N}(P) = \mathscr{F}$ are mutually inverse:

$$\varphi \mapsto P := \{h \in (\mathbb{R}^d)^* : h(x) \le \varphi(x) \text{ for all } x \in \mathbb{R}^d\},$$
$$P \mapsto \varphi := \max_{g \in P} g(-) : |\mathscr{F}| \to \mathbb{R}. \qquad \square$$

The support function constructed in the proof of (b) $\Longrightarrow$ (a) is nonnegative on $\mathbb{R}^d$ and has value 0 at the origin.

1.G Rationality and integrality

The polyhedra we are mainly interested in later on are defined over the field of rational numbers, and especially the polytopes usually have their vertices in the integral lattice. In the following we introduce these more special objects.

Rational polyhedra. Let V be an $\mathbb{R}$-vector space. A *rational structure* on V is a $\mathbb{Q}$-subspace Q such that $V = \mathbb{R}Q$ and $\dim_{\mathbb{Q}} Q = \dim_{\mathbb{R}} V$. In other words, the embedding $Q \to V$ induces an isomorphism $\mathbb{R} \otimes_{\mathbb{Q}} Q \cong V$. We now define affine rational structures:

Definition 1.68. Let V be a vector space with rational structure Q, and $x \in V$. Then $Q' = x + Q$ is called an *affine rational structure* on V.

Let V and W be vector spaces with affine rational structures $Q' = x + Q$ and $R' = y + R$. An affine map $\alpha : V \to W$ is *rational* if $\alpha(Q') \subset R'$. An affine subspace A of V is called *rational* if $A = \mathrm{aff}(A \cap Q')$.

A halfspace is *rational* if its bounding hyperplane is rational, and a polyhedron is *rational* if it is the intersection of rational closed halfspaces. A polytope is *rational* if it is rational as a polyhedron. A cone is *rational* if it is the intersection of linear rational halfspaces. Later on, when we speak about rational polyhedra, it is always understood that a rational structure has been introduced previously.

A *morphism* $\varphi : P \to P'$ *of rational polyhedra* P and P' is a map $P \to P'$ that can be extended to a rational affine map $\mathrm{aff}(P) \to \mathrm{aff}(P')$. (We will see below that $\mathrm{aff}(P)$ is rational if P is rational.)

A polyhedral complex Π is *rational* if all its member polyhedra P_p are rational and the maps $\pi_q^{-1} \circ \pi_p$ (where defined) are morphisms of rational polyhedra. In particular, a fan is *rational* if it consists of rational cones.

A piecewise affine function is *rational* if its domains of linearity are rational polyhedra and the restrictions of f to them are rational (with respect to the fixed rational structure $\mathbb{Q}$ on $\mathbb{R}$).

Let V be a vector space with an affine rational structure $Q' = x + Q$. Then the triple (V, x, Q') can be identified with the triple $(\mathbb{R}^d, 0, \mathbb{Q}^d)$ where $d = \dim V$. After this identification, Gaussian elimination shows that an affine subspace U of V is rational if and only if it is of the form $U = \operatorname{Ker} \varphi$ where $\varphi : V \to \mathbb{R}^m$ is a suitable rational affine map. (We can choose $m = \dim V - \dim U$.) In particular, a rational halfspace is of the form $\{x \in V : \alpha(x) \geq 0\}$ with a rational affine form $\alpha : V \to \mathbb{R}$. It follows that a polyhedron P is rational if and only if it is given by affine rational inequalities $\alpha_i(x) \geq 0$.

Let Π be a rational polyhedral complex. Then the subspace of all rational functions in $\mathrm{PA}(\Pi)$ defines a rational structure on this vector space, because the values $f(v)$, $v \in \operatorname{vert}(\Pi)$, completely determine f and are subject only to a rational system of equations resulting from the affine dependence relations of the vertices of the faces $P \in \Pi$.

If we speak of a rational object in $\mathbb{R}^d$, then we always refer to the (affine) rational structure $0 + \mathbb{Q}^d \subset \mathbb{R}^d$, unless stated otherwise.

In order not to overload the text of the next proposition, we leave some details to the reader, for example the definition of the rational structures considered in part (d).

Proposition 1.69. *Let V be a vector space with an affine rational structure. Then the following hold:*

(a) The affine hull of a rational polyhedron in V is rational.
(b) The intersection of rational polyhedra is rational.
(c) Each face of a rational polyhedron P is rational.
(d) The recession cone $\operatorname{rec}(P)$ *and the projectivization* $C(P)$ *of a rational polyhedron P are rational cones.*
(e) A cone $C \subset V$ is rational if and only if the dual cone C^* is rational (with respect to the associated rational structure on V^*).
(f) A cone C is rational if and only if it is generated by rational vectors.
(g) A polytope $P \subset V$ is rational if and only if its vertices are rational points.
(h) A polyhedron P is rational if and only if it is the Minkowski sum of a rational polytope and a rational cone.
(i) Let Π' be a regular rational subdivision of a rational polytopal (or conical) complex. Then Π' has a rational support function.

Proof. (a) follows immediately from Proposition 1.2: $\operatorname{aff}(P)$ is the intersection of rational hyperplanes. (b) follows immediately from the definition. Thus the facets of P are rational, and so is each face, being an intersection of facets. This proves (c). Part (d) is again obvious.

For (e) and (f) we use Fourier–Motzkin elimination. By definition and by Theorem 1.16, a cone is rational if the dual cone is generated by rational linear forms. Then Fourier–Motzkin elimination yields a system of rational generators for $C^{**} = C$.

(g) and (h) follow from (f) via projectivization, in the same way as Theorem 1.27 has been proved.

(i) follows from Proposition 1.59: $\mathrm{SF}(\Pi, \Pi')$ is the interior of a full-dimensional cone and therefore its intersection with the set of rational piecewise affine functions is dense in it. □

For a subset X of $\mathbb{Q}^d$ we can form the convex or conical hull within $\mathbb{Q}^d$, using only rational coefficients in convex or conical linear combinations. However, these notions yield nothing new:

Proposition 1.70. *Let $X \subset \mathbb{Q}^d$. Then $\operatorname{conv}_{\mathbb{Q}}(X) = \operatorname{conv}_{\mathbb{R}}(X) \cap \mathbb{Q}^d$ and $\mathbb{Q}_+ X = \mathbb{R}_+ X \cap \mathbb{Q}^d$.*

Proof. Let $x \in \operatorname{conv}_{\mathbb{Q}}(X)$. In view of Carathéodory's theorem we can assume that X is affinely independent. In that case the presentation of $x \in \operatorname{conv}_{\mathbb{R}}(X) \cap \mathbb{Q}^d$ as a (convex) linear combination of X is uniquely determined, and therefore only possible with rational coefficients. The same argument works for conical hulls. □

The proposition can be transferred immediately to vector spaces with (affine) rational structures.

Lattices and lattice polytopes. In analogy to a rational structure on an $\mathbb{R}$-vector space V one defines a *lattice structure*: it is a finitely generated subgroup L with $V = \mathbb{R}L$ and $\operatorname{rank} L = \dim V$; in other words, the embedding $L \to V$ induces an isomorphism $\mathbb{R} \otimes_{\mathbb{Z}} L \cong V$.

If we speak of a lattice L in V, then we do not insist that $\operatorname{rank} L = \dim V$. We only require that L be a subgroup generated by $\mathbb{R}$-linearly independent vectors. Clearly, L is a lattice structure on the subspace $\mathbb{R}L$ of V. The standard lattice structure on $\mathbb{R}^d$ is defined by $\mathbb{Z}^d$.

Definition 1.71. An *affine lattice structure* on V is a subset of type $L' = x + L$ where $x \in V$ and L is a lattice structure on V.

Let V and W be vector spaces with affine lattice structures $L' = x + L$ and $M' = y + M$. Then an affine map $\varphi : V \to W$ is called *integral* if $\varphi(L') \subset M'$. In the case $M' = \mathbb{Z}$ we say that φ is *primitive* if $\varphi(L') = \mathbb{Z}$.

In particular, a *primitive integral form* on $\mathbb{R}^d$ is a linear form such that $\varphi(\mathbb{Z}^d) = \mathbb{Z}$.

Similarly as in the case of vector spaces, an *affine lattice* in V is a subset $x+L$ where L is a lattice in V.

A *morphism* of affine lattices L' and M' is a mapping $\varphi : L' \to M'$ such that $\varphi(\sum_i a_i x_i) = \sum_i a_i \varphi(x_i)$ for all affine combinations (i.e., $x_i \in L'$, $a_i \in \mathbb{Z}$, $\sum_i a_i = 1$).

Remark 1.72. Let $L \subset V$ be an affine lattice and H a hyperplane of $A = \operatorname{aff}(L)$ such that $H = \operatorname{aff}(H \cap L)$. We choose one of the two halfspaces of A with respect to H as H^+ and an affine form α on A with $\alpha(H) = 0$ and $\alpha(H^+) \subset \mathbb{R}_+$.

We write $L = x_0 + L_0$ with a lattice $L_0 \subset V$ and $x_0 \in L_0 \cap H$. In $H^>$ we find a point $x \in L$ such that the residue class of $x - x_0$ generates L_0/U where $U = (H \cap L) - x_0$. In fact, $L_0/U \cong \mathbb{Z}$. After division by $\alpha(x)$ we can assume $\alpha(x) = 1$. Then α satisfies the following conditions: (i) $\alpha(L \cap H^+) = \mathbb{Z}_+$ and (ii) $\alpha(H) = \{0\}$. We call it the *L-height above H* (with respect to the choice of H^+),

$$\operatorname{ht}_{L,H}(y) = \alpha(y), \qquad y \in A.$$

If $y \in L$, then the absolute value of the height counts the number of proper parallels to H through points of L that lie between y and H.

If $L' = x + L$ is an affine lattice structure on V, then we can identify the triple (V, x, L') with $(\mathbb{R}^d, 0, \mathbb{Z}^d)$, $d = \dim V$, after having chosen a basis $e_1, \dots, e_d$ of V in L. Evidently every lattice structure $x + L$ on V defines a rational structure, namely $x + \mathbb{Q}L$.

Definition 1.73. Let L be an affine lattice in V. Then the convex hull $P = \operatorname{conv}(x_1, \dots, x_m)$ of finitely many points $x_i \in L$ is called an *L-polytope* or, less precisely, a *lattice polytope*. The set $P \cap L$ of *L-points* of P will also be denoted by $\operatorname{lat}(P)$ if there is no ambiguity in regard to L.

Let P be an L-polytope and P' be an L'-polytope. Then a *morphism of lattice polytopes* $\varphi : P \to P'$ is a morphism of polytopes with the additional condition $\varphi(\operatorname{aff}(P) \cap L) \subset L'$.

Sometimes it will be necessary to restrict the lattice L to $\operatorname{aff}(P)$. Therefore we introduce the notation

$$L_P = L \cap \operatorname{aff}(P).$$

The smallest lattice that admits P as a lattice polytope is

$$\mathscr{L}(P) = v_0 + \sum_{v \in \operatorname{vert}(P)} \mathbb{Z}(v - v_0)$$

where v_0 is an arbitrary vertex of P.

Figure 1.13 shows a lattice polygon with lattice $\mathbb{Z}^2$. The lattice points of P are marked by solid circles.

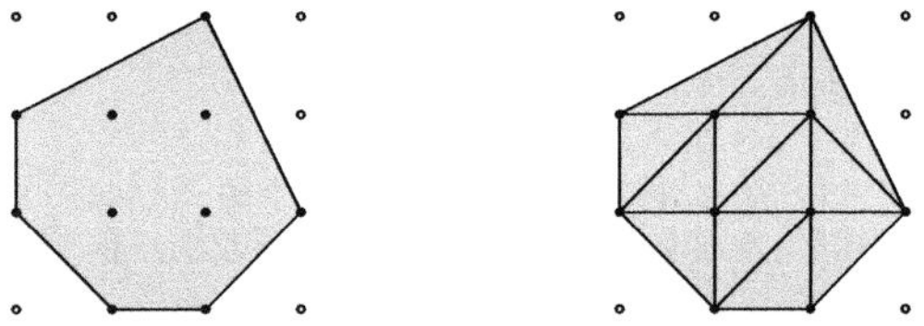

Fig. 1.13. A lattice polygon and one of its lattice triangulations

Lattice polytopal complexes. The definition of a lattice polytopal complex needs some care. In order to simplify it as much as possible, we identify the sets $p \in \Pi$ with the associated polytopes P_p and simply write $P \in \Pi$. Correspondingly the map π_P, $P \in \Pi$, is just the identity map on P, and we can consider $\operatorname{aff}(P \cap Q)$ as an affine subspace of $\operatorname{aff}(P)$ and of $\operatorname{aff}(Q)$.

Definition 1.74. A *lattice polytopal complex* is a polytopal complex Π augmented by a family $\Lambda_P \subset \operatorname{aff}(P)$, $P \in \Pi$, of affine lattices such that the following conditions are satisfied:

(a) every face $P \in \Pi$ is a Λ_P-polytope;
(b) $\Lambda_P \cap \operatorname{aff}(P \cap Q) = \Lambda_Q \cap \operatorname{aff}(P \cap Q)$ for all faces P, Q of Π.

In an *embedded lattice polytopal complex* consisting of polytopes $P \in \mathbb{R}^d$ all the affine lattices Λ_P, $P \in \Pi$, coincide with a fixed affine lattice L in $\mathbb{R}^d$ (so that condition (b) is automatically satisfied).

The system $\Lambda = \{\Lambda_P : P \in \Pi\}$ is called a *lattice structure* on Π. If we want to emphasize the lattice structure being considered we will write (Π, Λ) instead of just Π. The collection of all the lattice points in the faces $P \in \Pi$ is the set of *lattice points* of Π.

The simplest examples of lattice polytopal complexes are given by the face lattices of lattice polytopes: the system

$$\{(L_F) : f \text{ a face of } P\}$$

is a lattice structure on the face lattice of P.

In a lattice subdivision we require that the associated lattice not change if we pass from a polytope to one of its subdividing parts:

Definition 1.75. A *lattice subdivision* of a lattice polytopal complex (Π, Λ) is a lattice polytopal complex (Π', Λ') such that the following conditions are satisfied:

(a) Π' is a subdivision of Π;
(b) $\operatorname{vert}(\Pi')$ consists of lattice points in Π;
(c) for all faces $P \in \Pi$ and $P' \in \Pi'$ with $P' \subset P$ we have $\Lambda_{P'} = \Lambda_P \cap \operatorname{aff}(P')$.

An example of a lattice subdivision is given in Figure 1.13.

Remark 1.76. Condition (b) of Definition 1.74 does in general not follow from (a). The reason is that for an L-polytope one cannot always reconstruct $L \cap \operatorname{aff}(P)$ from $L \cap P$: the smallest affine lattice containing $L \cap P$ may be a proper subset of $L \cap \operatorname{aff}(P)$, as the following example shows. Let $P = \operatorname{conv}\big((0,0,0), (0,1,0), (1,0,0), (1,1,2)\big) \subset \mathbb{R}^3$. Then the vertices are the only lattice points of the $\mathbb{Z}^3$-polytope P, and the smallest lattice containing $\operatorname{lat}(P)$ is $\mathbb{Z}^2 \oplus 2\mathbb{Z}$.

Now we identify $\mathbb{R}^3$ with $(\mathbb{R}^3, 0) \subset \mathbb{R}^4$ and set $Q = \operatorname{conv}(P, (0,0,0,1))$. The polytopal complex Π consisting of Q and all its faces is not a lattice polytopal complex if we choose $\Lambda_Q = \mathbb{Z}^4$ and Λ_F as the smallest affine lattice in $\mathbb{R}^4$ containing F for all proper faces F of Q, despite the fact that condition (a) is satisfied for this collection of data.

Remark 1.77. Let L be an affine lattice and P, Q be L-polytopes. Then $P + Q$ is again a lattice polytope, namely with respect to the affine lattice $v + L$ where $v \in \operatorname{vert}(P)$. Similarly cP is a lattice polytope with respect to $(c-1)v + L$.

If Π is a lattice polytopal complex, then we can endow $c\Pi$ with a lattice structure by applying the construction above to cP for every $P \in \Pi$. Without the lattice structure, $c\Pi$ is isomorphic to Π (see Example 1.37), but as lattice polytopal complexes Π and $c\Pi$ are isomorphic only in trivial cases.

Exercises

1.1. Let V, W be $\mathbb{R}$-vector spaces. Prove:

(a) A subset A of V is an affine subspace if and only if it is closed under affine linear combinations.

(b) A map $f : V \to W$ is affine if and only if it respects affine linear combinations.

1.2. Let $X \subset \mathbb{R}^d$ and H be a hyperplane. Prove: if $X \subset H^+$, then $\operatorname{conv}(X) \cap H = \operatorname{conv}(X \cap H)$.

1.3. Let V be a d-dimensional vector space.

(a) Suppose that the cone C is generated by a basis $x_1, \dots, x_d$ of V, let H_i be the linear hyperplane generated by the elements x_j, $j \neq i$, and choose H_i^+ such that $x_i \in H_i^+$. Show that $C = H_1^+ \cap \dots \cap H_d^+$ is the unique irredundant representation of C as an intersection of halfspaces.

(b) Suppose that $C = H_{\kappa_1}^+ \cap \dots \cap H_{\kappa_r}^+$ is a cone of dimension d, given as an intersection of halfspaces, and let G be a system of generators of C. Show that H_{κ_j} intersects C in a facet if and only if $H_{\kappa_j} \cap G$ is a maximal element (with respect to inclusion) in the collection of intersections $H_{\kappa_i} \cap G$, $i = 1, \dots, r$.

1.4. Let P be a polyhedron, and F a nonempty face of P. Prove:

(a) Then $\operatorname{rec}(F)$ is face of $\operatorname{rec}(P)$. Moreover, F is unbounded if and only if $\operatorname{rec}(F) \neq 0$.

(b) $C(F)$ is a face of $C(P)$ that is not contained in $(V, 0)$.

(c) Every face of $\operatorname{rec}(P)$ is of the form $\operatorname{rec}(G)$ for a face G of P.

(d) The faces of $C(P) \cap (V, 0)$ are the faces of $(\operatorname{rec}(P), 0)$.

(e) The faces of $C(P)$ not contained in $(V, 0)$ are of the form $C(G)$ for a face G of P.

(f) The following are equivalent:

(i) P has vertices;
(ii) $\operatorname{rec}(P)$ is pointed;
(iii) $C(P)$ is pointed.

1.5. Let P and Q be polyhedra, $x \in P$, $y \in Q$. Show that $x + y$ is a vertex of $P + Q$ if and only if x is a vertex of P, y is a vertex of Q, and there exists an affine form α that attains its maximum on P exactly in x and its maximum on Q exactly in y.

1.6. Let C be a pointed cone in $\mathbb{R}^d$ and $f : \mathbb{R}^d \to \mathbb{R}^e$ a linear map with kernel U. Show that $f(C)$ is a pointed cone if and only if the following holds for each face of C: if $F \cap U \neq \emptyset$, then $F \subset U$.

1.7. Prove that the boundaries of polyhedra are *connected in codimension* 1 (apart from an obvious exceptional case). More precisely, let P be a polyhedron with at least two nonparallel facets. Then for each pair F, G of facets of P there exists a chain

$$F = F_0, F_1, \ldots, F_n = G$$

of facets of P such that $F_i \cap F_{i+1}$ is a facet of F_i and F_{i+1} for $i = 0, \ldots, n-1$ (so the intersection $F_i \cap F_{i+1}$ has codimension 1 with respect to F_i and F_{i+1}). In particular, ∂P is connected.

1.8. Generalize Exercise 1.7 as follows. Let F and G be faces of P such that $\dim F = \dim G$, and suppose P has a face F' with $\dim F' < \dim F$. Then there exists a chain $F = F_0, F_1, \ldots, F_n = G$ of faces of P such that $F_i \cap F_{i+1}$ is a facet of F_i and F_{i+1} for $i = 0, \ldots, n-1$. In particular, $\dim F_i = \dim F$ for all i.

1.9. Let P be a polyhedron, $P \neq \operatorname{aff}(P)$. Show that the boundary of P is homeomorphic to the product of an open ball (or an affine space) and a sphere, where each factor may have dimension 0. Characterize these two cases.

1.10. Let P be a polyhedron that has a compact face, let L be the union of the edges of P and L_c the union of the compact edges. Show that L and L_c are connected.

1.11. Let C_1 and C_2 be conical sets in $\mathbb{R}^d$. Prove:

$$(C_1 \cap C_2)^* = C_1^* + C_2^* \qquad \text{and} \qquad (C_1 \cup C_2)^* = (C_1 + C_2)^* = C_1^* \cap C_2^*.$$

1.12. For polytopes $P \subset R^m$ and $Q \subset R^n$ let $\operatorname{Hom}_{\text{aff}}(P, Q)$ denote the set of affine maps from P to Q. Show:

(a) $\operatorname{Hom}_{\text{aff}}(P, Q)$ is a polytope in $\mathbb{R}^{mn+n}$ in a natural way. It is called the *hom-polytope* of P and Q;

(b) $\dim \operatorname{Hom}_{\text{aff}}(P, Q) = (\dim P + 1) \dim Q$;

(c) the facets of $\operatorname{Hom}_{\text{aff}}(P, Q)$ are naturally labeled by the pairs (x, G) where x is a vertex of P and G is a facet of Q;

(d) for the d-simplex Δ_d one has $\operatorname{Hom}_{\text{aff}}(\Delta_d, Q) = Q^{d+1}$.

It is an interesting open problem to compute the f-vector of $\operatorname{Hom}_{\text{aff}}(P_n, P_m)$ where P_j is the regular j-gon; at least in the case $m = n$. The system polymake [139] has a module for the computation of hom-polytopes (called mapping polytopes there).

1.13. Let $C \in \mathbb{R}^d$ be a cone and F a face of C. Show

$$\{y \in (\mathbb{R}^d)^*; y(x) \geq 0 \text{ for all } x \in F\} = C^* - F_C^*.$$

1.14. Let P be a polytope in $\mathbb{R}^d$ of dimension d and suppose that 0 is an interior point of P. Then the *dual* polytope $P^* \subset (\mathbb{R}^d)^*$ is the set

$$P^* = \{y \in (\mathbb{R}^d)^* : y(x) \leq 1 \text{ for all } x \in P\}.$$

(a) Show P^* is indeed a polytope and $P^{**} = P$.

(b) Prove that the normal fan of P consists of all the cones $\mathbb{R}_+ F$ where F is a face of $-P^* = \{-x : x \in P^*\}$.

1.15. The *combinatorial type* of a polyhedral complex Π is the isomorphism class of the partially ordered set of its faces, ordered by inclusion. (Note that the dimensions of the faces are encoded in the combinatorial type if Π is polytopal. Why?)

(a) Show that there is no embedded polytopal complex with the combinatorial type of the Möbius strip as defined in Example 1.36.

(b) Show that the combinatorial type of the (embedded) polytopal complex of Figure 1.5(c) cannot be realized by an embedded complex whose quadrangles are parallelograms.

1.16. Show that the polytopal complex consisting of the boundary of a 3-dimensional cube and one of its space diagonals is not a boundary subcomplex.

1.17. This exercise treats the construction of triangulations of polytopes in a more algorithmic fashion. Compare the proof of Lemma 1.50 and Remark 1.52.

Let $P \subset \mathbb{R}^d$ be a polytope and $x \in \mathbb{R}^d$. Suppose Δ is a triangulation of P. Let $\Delta_<$ be the set of simplices of Δ that lie in facets $P \cap H$ for a support hyperplane H with $x \in H^<$ and define $\Delta_>$ accordingly.

Then define collections $\Sigma_<$ and $\Sigma_>$ of simplices as follows:

(a) If $x \notin \operatorname{aff}(P)$, then $\Sigma_< = \Sigma_> = \Delta \cup \{\operatorname{conv}, \delta, x) : \delta \in \Delta\}$;

(b) otherwise $\Sigma_< = \Delta \cup \{\operatorname{conv}(\delta, x) : \delta \in \Delta_<\}$, $\Sigma_> = \{\operatorname{conv}(\delta, x) : \delta \in \Delta_>\} \cup \{\varnothing\}$.

Show that $\Sigma_<$ and $\Sigma_>$ are triangulations of $\operatorname{conv}(P, x)$. One calls $\Sigma_<$ the *placing* or *lexicographic extension* of P, and $\Sigma_>$ the *pulling* or *reverse lexicographic* extension of Δ. (In general, the latter does not contain Δ.) The terminology will be explained in Section 7.B, p. 267.

1.18. Let X be a finite subset of $\mathbb{R}^d$. The *Voronoi cell* $V(x)$ of $x \in X$ is the set of all $y \in \mathbb{R}^d$ such that $\|y - x\| \le \|y - x'\|$ for all $x' \in X$. Show that the Voronoi cells of $x_1, \dots, x_n$ form the facets of a polyhedral subdivision of $\mathbb{R}^d$, called the *Voronoi diagram* $V(X)$ of X.

1.19. We continue the previous problem, and define a collection $D(X)$ of polytopes as follows: $\operatorname{conv}(Y)$, $Y \subset X$, belongs to $D(X)$ if and only if there exists a face F of the Voronoi diagram such that $Y = \{y \in X : V(y) \supset F\}$.

(a) Show that $D(X)$ is a polytopal subdivision of $P = \operatorname{conv}(X)$, called the *Delaunay subdivision*.

(b) Prove that the Delaunay subdivision is regular. To this end consider the strictly convex function $x \mapsto (x, -\|x\|)$.

(c) Find a condition of type "general position" guaranteeing that the Delaunay subdivision is a triangulation.

1.20. Let P be polytope, X a finite subset of P, and Π a regular subdivision of P with $\operatorname{vert}(\Pi) = X$, obtained as the projection of the bottom of the polytope $Q = \operatorname{conv}\big(P \times \{0\}, (x, w_x) : x \in X\big)$ with $w_x \le 0$ for all x as in Remark 1.58.

Show that for every $\delta > 0$ there exist ε_x, $x \in X$, $0 \le \varepsilon_x < \delta$, such that the bottom of $Q' = \operatorname{conv}\big(P \times \{0\}, (x, w_x - \varepsilon_x) : x \in X\big)$ defines a triangulation Δ with $\operatorname{vert}(\Delta) = X$ that refines Π.

1.21. Derive part (b) of Corollary 1.66 from its part (a) by constructing a suitable support function from a support function of the triangulation of the polytopal complex Π in the proof of Theorem 1.54.

1.22. Convince yourself that the nonregular triangulation on Figure 1.11 can be made regular by perturbing the vertices appropriately, without altering the abstract simplicial complex structure of the triangulation. In particular, nonregularity of a triangulation is not a combinatorial property of the underlying simplicial complex.

More generally, a planar graph G without loops and multiple edges is combinatorially equivalent to the 1-skeleton of a 3-polytope if and only if it is connected and remains so when one or two arbitrary vertices are removed. This is the classical Steinitz theorem; see [381, Lect. 4].

1.23. Let $C \subset \mathbb{R}^d$ be a (rational) cone ($C = \mathbb{R}^d$ is not excluded). Show that every (rational) subdivision Γ of C can be refined into a (rational) regular subdivision. Also prove the analogue for polytopes.

Hint: one can assume that $\dim C = d$. Consider the hyperplanes spanned by all $(d-1)$-dimensional faces of Γ.

1.24. There is a canonical way to associate a conical complex to an arbitrary polytopal complex Π: for every $P \in \Pi$, $P \subset \mathbb{R}^d$ we consider the cone $C(P) = \mathbb{R}_+(P,1) \subset \mathbb{R}^{d+1}$ and then, clearly, these cones define a conical complex $C(\Pi)$. The cones in $C(\Pi)$ are all pointed, and the polytopal complex Π can be considered a *cross-section* of $C(\Pi)$. But not all conical complexes have cross-sections.

Choose six rational noncoplanar points in $\mathbb{R}^3$ as shown in Figure 1.14 where the top and bottom

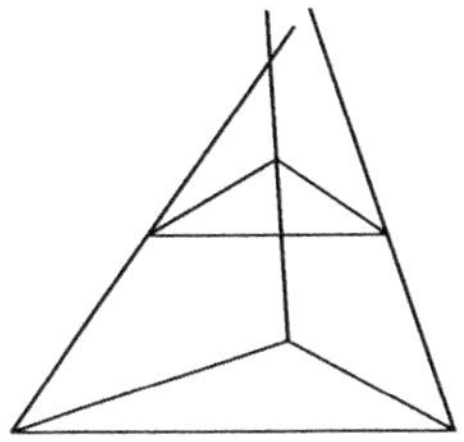

Fig. 1.14. Construction of a conical complex without cross-section

triangles are in parallel planes and at least one of the quadrangles is not flat (indicated by the fact that the three nonhorizontal lines do not meet in a common point). Suppose that $0 \in \mathbb{R}^3$ lies in the interior of the convex hull of these six points so that the cones with common apex 0 that are spanned by the two triangles and the three quadrangles form a fan $\mathscr{F}$ of rational cones in $\mathbb{R}^3$ with $|\mathscr{F}| = \mathbb{R}^3$. Show that $\mathscr{F}$ has no cross-section.

1.25. (a) Show that every positive (rational) cone is isomorphic to the intersection of a (rational) vector subspace with the positive orthant in a suitable space $\mathbb{R}^n$.

(b) Show that every (rational) polytope is isomorphic to the intersection of a (rational) affine subspace and a (rational) simplex in a suitable space $\mathbb{R}^n$.

1.26. Show that the combinatorial type of an arbitrary simplicial polytope is realized by a rational polytope, and that the same is true for the combinatorial types of simple polytopes.

However, in general the combinatorial type of a polytope need not have a rational realization; see [381, Example 6.21] for an 8-dimensional example due to Perles.

Notes

The history of convex and polyhedral geometry is amply documented in the books by Grünbaum [154], Schrijver [312], and Ziegler [381]. In these sources the reader will find references for the classical works of Farkas, Minkowski, Carathéodory, Weyl, Motzkin, and others. A recent exposition of convexity and polyhedra has been given by Barvinok [17]. We have restricted ourselves to the very basics of the theory that are indispensable for the rest of the book.

This applies also to triangulations and polyhedral complexes about which the reader can learn much more from the wonderfully illustrated book (in preparation) by De Loera, Rambau, and Santos [106]. We have not succeeded in finding a single source from which the theory of regular subdivisions has originated. As observed in

Remark 1.57, they have appeared under different names in several works. Their history goes back almost a century when the first examples of nontriangulable nonconvex polytopes were found.

A very important concept not discussed by us is that of shellability for which we refer the reader to [106] and [381].

2

Affine monoids and their Hilbert bases

Affine monoids are the basic structure on which algebras with coefficients in rings will be built later on. Their finiteness properties allow a rich structure theory, both from the combinatorial and the ring theoretic point of view to be pursued in later chapters. An affine monoid defines a cone in a natural way, and therefore the notions of polyhedral convex geometry will be omnipresent in this chapter.

Within the class of affine monoids, the normal ones represent the discrete counterparts to continuous cones since they can be recovered from their cones and the lattice determining the discrete structure. Part of the theory of general affine monoids will be developed by relating them to their normalizations. By Gordan's lemma the normalizations are also affine.

The last two sections deal with the combinatorics of the Hilbert bases of positive normal affine monoids. On the one hand, we will try to bound the degrees of Hilbert basis elements, and on the other we will investigate whether normality can be explained by conditions on the unimodularity of the Hilbert basis.

2.A Affine monoids

In common usage, a monoid is a set M together with an operation $M \times M \to M$ that is associative and has a neutral element. We are mainly interested in a special class of commutative monoids:

Definition 2.1. A monoid is *affine* if it is finitely generated and isomorphic to a submonoid of a free abelian group $\mathbb{Z}^d$ for some $d \geq 0$.

Very often, especially in the commutative algebra literature, affine monoids are called *affine semigroups*—a line of tradition followed in our joint papers. The usage of "monoids" in this book is more compatible with another tradition, that in the K-theoretic literature.

In view of the definition above it is appropriate to use additive notation for the operation in M. The condition on finite generation then just means that there exist $x_1, \dots, x_n \in M$ for which

W. Bruns, J. Gubeladze, *Polytopes, Rings, and K-Thoery*, Springer Monographs in Mathematics,
DOI 10.1007/b105283_2, © Springer Science+Business Media, LLC 2009

$$M = \mathbb{Z}_+x_1 + \cdots + \mathbb{Z}_+x_n = \{a_1x_1 + \cdots + a_nx_n : a_i \in \mathbb{Z}_+\}.$$

We are of course always free to consider an affine monoid as a submonoid of $\mathbb{Z}^d$ for suitable d.

Within the class of commutative monoids, the affine monoids are characterized by being (i) finitely generated, (ii) *cancellative*, and (iii) *torsionfree*.

Cancellativity means that an equation $x + y = x + z$ for $x, y, z \in M$ implies $y = z$. Equivalently, M can be embedded into a group.

More generally, for every commutative monoid M there exist a group $\operatorname{gp}(M)$ and a monoid homomorphism $\iota : M \to \operatorname{gp}(M)$ solving the following universal problem: every monoid homomorphism $\varphi : M \to H$ from M to a group H factors in a unique way as $\varphi = \psi \circ \iota$ with a group homomorphism $\psi : \operatorname{gp}(M) \to H$. Consequently the *group of differences* $\operatorname{gp}(M)$ is unique up to isomorphism. It is constructed as follows. As a set, $\operatorname{gp}(M)$ consists of the equivalence classes $x - y$ of pairs $(x, y) \in M^2$ with $x - y = u - v$ if and only if $x + v + z = u + y + z$ for some $z \in M$. Addition is defined by $(x - y) + (u - v) = (x + u) - (y + v)$. Then $\operatorname{gp}(M)$ is evidently a group, and the map $\iota : M \mapsto \operatorname{gp}(M)$, $\iota(x) = x - 0$, is a monoid homomorphism that satisfies the universality condition. Clearly, when M is cancellative, then ι is injective.

To be torsionfree for a monoid M means that $ax = ay$ for $a \in \mathbb{N}$ and $x, y \in M$ implies $x = y$. If M is cancellative, this condition is equivalent to the torsion freeness of $\operatorname{gp}(M)$, but in general it is not enough to require torsion freeness for $\operatorname{gp}(M)$ (Exercise 2.1).

We agree on the following

Convention 2.2. In the remainder of this book the term *monoid* always means a commutative, cancellative and torsionfree monoid, unless explicitly stated otherwise.

We see that if a monoid M is finitely generated, then it can be embedded into a group which is finitely generated and torsionfree, i. e., isomorphic to a free abelian group $\mathbb{Z}^r$. If M is given as a submonoid of $\mathbb{Z}^d$, then $\operatorname{gp}(M)$ can be identified with the subgroup $\mathbb{Z}M$ of $\mathbb{Z}^d$ generated by M. The monoids that are isomorphic to $\mathbb{Z}_+^r$, $r \in \mathbb{Z}_+$, are called *free monoids*.

Definition 2.3. The *rank* of a monoid M is the vector space dimension of $\mathbb{Q} \otimes \operatorname{gp}(M)$ over $\mathbb{Q}$. In other words, it is the rank of the abelian group $\operatorname{gp}(M)$.

Clearly, if M is affine and $\operatorname{gp}(M) \cong \mathbb{Z}^r$, then $\operatorname{rank} M = r$. However, note that the definition of rank is not restricted to finitely generated monoids.

Remark 2.4. Every submonoid of $\mathbb{Z}$ is finitely generated (Exercise 2.2). It is isomorphic to a submonoid of $\mathbb{Z}_+$, unless it is a subgroup of $\mathbb{Z}$. Submonoids of $\mathbb{Z}_+$ are often called *numerical semigroups*. A vast amount of research has been devoted to them. See Barucci, Dobbs, and Fontana [16].

In contrast, already $\mathbb{Z}^2$ contains submonoids without a finite system of generators, for example $\{(0,0)\} \cup \{(x, y) : x \geq 1\}$.

Subcones C of $\mathbb{R}^d$ are examples of "continuous" monoids. Unless $C = 0$, such a monoid is not finitely generated. Neither is $C \cap \mathbb{Q}^d$ finitely generated if it contains a nonzero vector.

As in ring theory, it is useful to introduce the notion of module.

Definition 2.5. A set N with an (additively written) operation $M \times N \to N$ is called an *M-module* if $(a + b) + x = a + (b + x)$ and $0 + x = x$ for all $a, b \in M$ and $x \in N$.

A subset I of M is an *ideal* if it is a submodule, i. e., if $M + I \subset I$.

A typical example of a module over $M \subset \mathbb{Z}^d$ is $\mathbb{Z}^d$ itself. Then every subset U of $\mathbb{Z}^d$ with $M + U \subset U$ is a submodule. The empty set is considered an M-module (and an ideal).

Remark 2.6. (a) Let $M \subset \mathbb{Z}^d$ be an affine monoid. Then we define the *interior* of M by $\operatorname{int}(M) = M \cap \operatorname{int}(\mathbb{R}_+ M)$. Since $x + y \in \operatorname{int}(\mathbb{R}_+ M)$ for $x \in \operatorname{int}(\mathbb{R}_+ M)$, $y \in \mathbb{R}_+ M$, it follows that $\operatorname{int}(M)$ is an ideal.

One has $0 \in \operatorname{int}(M)$ if and only if M is a group. In that case $\operatorname{int}(M) = M$. In all the other cases $\operatorname{int}(M)$ is not a monoid itself. However, $M_* = \operatorname{int}(M) \cup \{0\}$ is a submonoid of M that will play an important role later on. Note that $M_* = M$ if and only if $\operatorname{rank} M \le 1$ or $M = \operatorname{int}(M)$. Otherwise M_* is not even finitely generated, as the reader may show (Exercise 2.4).

(b) In (a) we have assumed that $M \subset \mathbb{Z}^d$. Nevertheless $\operatorname{int}(M)$ is defined intrinsically in terms of M and does not depend on an embedding of M into $\mathbb{Z}^d$ (or $\mathbb{R}^d$) for some d. The reason is that an isomorphism of monoids $M \subset \mathbb{R}^d$ and $N \subset \mathbb{R}^e$ induces isomorphisms (i) $\operatorname{gp}(M) \cong \operatorname{gp}(N)$, (ii) $\mathbb{R}M \cong \mathbb{R}N$, and (iii) $\mathbb{R}_+ M \cong \mathbb{R}_+ N$. For a coordinate-free definition one views M as a submonoid of the vector space $\mathbb{R} \otimes_{\mathbb{Z}} \operatorname{gp}(M)$.

Alternatively, one can give intrinsic definitions using only the monoid structure; see Exercise 2.3.

(c) Let M be an affine monoid and F a face of the cone $\mathbb{R}_+ M$. Then $F \cap M$, the intersection of M and the monoid F, is a submonoid of M. The submonoids of M that arise in this way are called *extreme submonoids*.

For an arbitrary monoid M and every commutative ring of coefficients R we can form the *monoid algebra* (or *monoid ring*) $R[M]$—the main vehicle in our study of the interactions between discrete geometry, commutative ring theory, and algebraic K-theory. As an R-module $R[M]$ is free with a basis consisting of the symbols X^a, $a \in M$, and the multiplication on $R[M]$ is defined by the R-bilinear extension of $X^a X^b = X^{a+b}$.[1] The elements X^a are called the *monomials* of $R[M]$.

For an M-module N we can analogously define an $R[M]$-module RN that as an R-module is free on the basis X^n, $n \in N$, and on which X^a operates by $X^a X^n = X^{a+n}$. We leave it to the reader to give the proofs of the statements implicitly contained in the descriptions of $R[M]$ and RN.

Proposition 2.7. *Let M be a monoid, N an M-module, and R a ring. Then*

(a) M is finitely generated if and only if $R[M]$ is a finitely generated R-algebra;

[1] Later on, when the use of monoid rings becomes essential, we will switch to a simpler notation identifying the elements $m \in M$ with X^m and writing the monoid operation multiplicatively.

(b) *N is a finitely generated M-module if and only if RN is a finitely generated $R[M]$-module.*

Proof. (a) The implication $\Longrightarrow$ is trivial. For the converse let $f_1, \dots, f_n$ be a system of generators of $R[M]$. There exists a finite subset E of M such that every f_i is an R-linear combination of the elements X^e with $e \in E$. Let $M' = \mathbb{Z}_+ E$. It follows immediately that any R-linear combination of the products $f_1^{a_1} \cdots f_n^{a_n}$ with $a_1, \dots, a_n \in \mathbb{Z}_+^n$ is an R-linear combination of monomials X^b, $b \in M'$. Since $f_1, \dots, f_n$ generate $R[M]$, it follows that $X^a \in R[M']$ for every $a \in M$. This implies $M = M'$.

(b) is proved along the same lines. □

In using Proposition 2.7 for proving assertions about M or N we are free to choose R; for example, we can take R to be a field $\Bbbk$.

Proposition 2.8. *Let M be a finitely generated monoid and N a finitely generated M-module. Then every M-submodule of N is finitely generated.*

Proof. We choose a field $\Bbbk$. Then $\Bbbk[M]$ is a finitely generated $\Bbbk$-algebra and therefore a noetherian ring by Hilbert's basis theorem (see Eisenbud [112, Sect 1.4]). So every submodule of the finitely generated $\Bbbk[M]$-module $\Bbbk N$ is finitely generated. It follows that $\Bbbk U$ is finitely generated for every M-submodule U of X, whence U is finitely generated over M by the previous proposition. □

Gordan's lemma. Very often we will have to round integers:

$$\lfloor x \rfloor = \max\{z \in \mathbb{Z} : z \le x\}, \qquad x \in \mathbb{R},$$

denotes the *floor* of x, and

$$\lceil x \rceil = \min\{z \in \mathbb{Z} : z \ge x\}, \qquad x \in \mathbb{R},$$

denotes the *ceiling* of x. For $y \in \mathbb{R}^n$ we set

$$\lfloor y \rfloor = (\lfloor y_1 \rfloor, \dots, \lfloor y_n \rfloor) \quad \text{and} \quad \lceil y \rceil = (\lceil y_1 \rceil, \dots, \lceil y_n \rceil).$$

The first opportunity to use $\lfloor x \rfloor$ is the proof of *Gordan's lemma*:

Lemma 2.9. *Let C be a rational cone in $\mathbb{R}^d$, and $L \subset \mathbb{Q}^d$ a lattice. Then $C \cap L$ is an affine monoid.*

Proof. Set $C' = C \cap \mathbb{R}L$. Then C' is a rational cone as well. Moreover every element x of $\mathbb{Q}^d \cap \mathbb{R}L$ is a rational linear combination of elements of L and so there exists a positive integer a with $ax \in L$. We choose a finite system of generators $x_1, \dots, x_n$ of C'. As just seen, we can assume that $x_1, \dots, x_n \in L$. Let M' be the affine monoid generated by $x_1, \dots, x_n$.

Every element x of $C' \cap L$ has a representation $x = a_1 x_1 + \cdots + a_n x_n$ with $a_i \in \mathbb{R}_+$. (We can choose $a_1, \dots, a_n \in \mathbb{Q}_+$, but this is irrelevant here.) Therefore

$$x = \big(\lfloor a_1 \rfloor x_1 + \cdots + \lfloor a_n \rfloor x_n\big) + \big(q_1 x_1 + \cdots + q_n x_n\big),$$
$$0 \le q_i = a_i - \lfloor a_i \rfloor < 1, \ i = 1, \dots, n.$$

The first summand on the right hand side is in M', the second is an element of $C' \cap L$ that belongs to a bounded subset B of $\mathbb{R}^n$. It follows that $C' \cap L$ is generated as an M'-module by the finite set $B \cap C' \cap L$. Being a finitely generated module over an affine monoid, the monoid $C \cap L$ is itself finitely generated. □

Another proof of Gordan's lemma will be given in the framework of graded rings; see Lemma 4.12.

The essential point in the hypothesis of Lemma 2.9 is the existence of a lattice L' containing both L and the generators of C: after a change of coordinates we can assume that $L' \subset \mathbb{Z}^d$; then C is rational, and we have arrived at the hypothesis of Lemma 2.9.

The reader may show that the intersection of $\mathbb{Z}^2$ with the cone generated by $(1, 0)$ and $(1, a)$ is a finitely generated monoid if and only if a is rational (Exercise 2.6).

Corollary 2.10. *Let M be a submonoid of $\mathbb{R}^d$, L a lattice in $\mathbb{R}^d$ containing M, and $C = \mathbb{R}_+ M$. Then the following are equivalent:*

(a) M is an affine monoid;
(b) $\widehat{M}_L = C \cap L$ is an affine monoid;
(c) C is a cone.

Moreover, if M is affine, then $\widehat{M}_L$ is a finitely generated M-module.

Proof. It is obvious that both (a) and (b) imply (c), The implication (c) $\Longrightarrow$ (b) is essentially Gordan's lemma. To complete the proof of the equivalence by the implication (c) $\Longrightarrow$ (a) one notes that the affine monoid M' in the proof of the lemma can be chosen as a submonoid of M if the cone under consideration is generated by M. Thus M is a submodule of the finitely generated M'-module $\widehat{M}_L$, and therefore itself finitely generated by 2.8.

Finally, since $\widehat{M}_L$ is finitely generated over M', it is even more so over M. □

The monoids of type $\widehat{M}_L$ will be further investigated in the next subsection.

Corollary 2.11. *Let M and N be affine submonoids of $\mathbb{R}^d$, and let C be a cone generated by elements of $\operatorname{gp}(M)$. Then*

(a) $M \cap N$ is an affine monoid;
(b) $M \cap C$ is an affine monoid;
(c) the extreme submonoids of M are affine.

Proof. (a) The group generated by $M \cup N$ is a finitely generated torsionfree abelian group and therefore isomorphic to $\mathbb{Z}^n$ for some n (possibly, $n > d$). Replacing $\mathbb{R}^d$ by $\mathbb{R}^n$, we may, right from the start, assume that M and N are submonoids of $\mathbb{Z}^d$. The conical set $\mathbb{R}_+ M \cap \mathbb{R}_+ N$ is the intersection of the rational cones $\mathbb{R}_+ M$ and $\mathbb{R}_+ N$. Such an intersection is itself a rational cone by Proposition 1.69. So the claim follows from the previous corollary once we have shown that $\mathbb{R}_+ M \cap \mathbb{R}_+ N = \mathbb{R}_+(M \cap N)$.

The inclusion $\mathbb{R}_+(M \cap N) \subset \mathbb{R}_+M \cap \mathbb{R}_+N$ is trivial. For the converse it is enough to consider rational elements $x \in \mathbb{R}_+M \cap \mathbb{R}_+N$. Then there exist $\alpha, \beta \in \mathbb{Z}$, $\alpha, \beta > 0$, such that $\alpha x \in M$ and $\beta x \in N$. Thus $\alpha\beta x \in M \cap N$, and $x \in \mathbb{R}_+(M \cap N)$.

(b) By Corollary 2.9 $\operatorname{gp}(M) \cap C$ is an affine monoid. Therefore $M \cap C = M \cap \operatorname{gp}(M) \cap C$ is affine by (a).

(c) is an immediate consequence of (b) since every face F of $\mathbb{R}_+M$ is generated by $M \cap F$. □

These results provide us with a wealth of examples of affine monoids. Gordan's lemma has an inhomogeneous generalization. It can be considered as the discrete analogue of Motzkin's theorem 1.27 on the finite generation of polyhedra.

Theorem 2.12. *Let $P \subset \mathbb{R}^d$ be a rational polyhedron, C the recession cone of P, and $L \subset \mathbb{Q}^d$ a lattice. Then $P \cap L$ is a finitely generated module over the affine monoid $C \cap L$.*

Proof. We form the cone $C(P) \subset \mathbb{R}^{d+1}$ over P and extend the lattice L to $L' = L \oplus \mathbb{Z}$. By Gordan's lemma $C(P) \cap L'$ is an affine monoid. Its generators at height 0 generate $C \cap L$, and its generators at height 1 generate $P \cap L$ as a $C \cap L$-module. □

The theorem contains (and is equivalent to) a statement about the set N of solutions of a homogeneous system of linear diophantine inequalities and congruences given as

$$\begin{aligned} &a_{i1}x_1 + \cdots + a_{id}x_d \geq u_i, && i = 1, \dots, m, \quad a_{ij}, u_i \in \mathbb{Z}, \\ &b_{i1}x_1 + \cdots + b_{id}x_d \equiv v_i \quad (w_i), && i = 1, \dots, n, \quad b_{ij}, v_i, w_i \in \mathbb{Z}. \end{aligned} \tag{2.1}$$

(We can split equations into a pair of inequalities.) The inequalities define a rational polyhedron P. The homogeneous congruences $b_{i1}x_1 + \cdots + b_{id}x_d \equiv 0 \ (w_i)$ define a sublattice L of $\mathbb{Z}^d$, and we claim that N is a finite module over L. In fact, the set of solutions of the inhomogeneous system of congruences is an affine lattice of the form $x' + L$ (unless it is empty). Let $P' = P - x'$; then $N = x' + (P' \cap L)$, and to $P' \cap L$ we can apply the theorem: there exists a finite set E generating $P' \cap L$ as an $P \cap L$-module. To find the system of generators of N, we replace each $x \in E$ by $x' + x$. To sum up:

Corollary 2.13. *Let $N \subset \mathbb{Z}^d$ be the set of solutions of the system* (2.1), *and let $M \subset \mathbb{Z}^d$ be the (affine) monoid of solutions of the corresponding homogeneous system. Then N is a finitely generated M-module.*

Irreducible elements, standard map and total degree. The nonzero elements of an integral domain form a commutative cancellative monoid with respect to multiplication. In analogy to number-theoretic nomenclature, let us call an element x of a monoid M a *unit* if x has an inverse in M. Clearly, the units of M form a group, denoted by $\operatorname{U}(M)$. One calls x *irreducible* if in every decomposition $x = y + z$ one of the summands y, z must be a unit.

It is not difficult to analyze affine monoids M in these terms. To this end, and for many other purposes, we introduce the standard map on an affine monoid as follows.

The group $\operatorname{gp}(M)$ is isomorphic to $\mathbb{Z}^r$, $r = \operatorname{rank} M$. We identify $\operatorname{gp}(M)$ and $\mathbb{Z}^r$. Let $C = \mathbb{R}_+ M \subset \mathbb{R}^r$ the cone generated by M. This cone has a representation

$$C = H^+_{\sigma_1} \cap \cdots \cap H^+_{\sigma_s}$$

as an irredundant intersection of halfspaces defined by linear forms on $\mathbb{R}^r$. Each of the hyperplanes H_{σ_i} is generated as a vector space by integral vectors. Therefore we can assume that σ_i is the $\mathbb{Z}^r$-height above H_{σ_i} (with nonnegative values on C; see Remark 1.72). After this standardization we call the σ_i the *support forms* of M and

$$\sigma : M \to \mathbb{Z}^s_+, \qquad \sigma(x) = (\sigma_1(x), \dots, \sigma_s(x))$$

the *standard map* on M. The hyperplanes H_{σ_i} are called the *support hyperplanes* of M.

The standard map has a natural extension to $\mathbb{R}^r$ with values in $\mathbb{R}^s$, also denoted by σ. It restricts to a $\mathbb{Z}$-linear map $\operatorname{gp}(M) \to \mathbb{Z}^s$. Furthermore $\sigma(C) \subset \mathbb{R}^s_+$. Theorem 1.16 implies that $\sigma_1, \dots, \sigma_s$ is a minimal set of generators of the dual cone C^*.

Note that the standard map depends only on the order of $\sigma_1, \dots, \sigma_s$. The σ_i (as $\mathbb{Z}$-linear forms on $\operatorname{gp}(M)$) are defined intrinsically by M; see Remark 2.6(b).

We call $\tau = \sigma_1 + \cdots + \sigma_s$ the *total degree* on M. The term "total degree" is justified: for any ring R of coefficients, σ induces an homomorphism $R[M] \to R[Y_1, \dots, Y_s]$ of R-algebras, namely the R-linear extension of the map $X^m \mapsto Y^{\sigma(m)}$ and $\tau(m)$ is the total degree of the monomial $Y^{\sigma(m)}$.

Proposition 2.14. *Let M be an affine monoid with standard map σ. Then:*

(a) the units of M are precisely the elements x with $\sigma(x) = 0$, or, equivalently, $\tau(x) = 0$;
(b) every element $x \in M$ has a presentation $x = u + y_1 + \cdots + y_m$ in which u is a unit and $y_1, \dots, y_m$ are irreducible;
(c) up to differences by units, there exist only finitely many irreducible elements in M.

Proof. (a) Clearly if $x, -x \in M$, then $\sigma(x) = 0$, since 0 is the only unit in $\mathbb{Z}^s_+$. Conversely, let $\sigma(x) = 0$. Then $\sigma(-x) = 0$ as well, and so $-x \in C$, where $C = \mathbb{R}_+ M$ as above. Thus there exists a positive integer m with $m(-x) \in M$. Then $x' = (m-1)x + m(-x) \in M$, too, and $x + x' = 0$.

(b) Suppose that x is neither a unit nor irreducible. Then there exists a decomposition $x = y + z$ in which neither y nor z is a unit, and since $\tau(x) = \tau(y) + \tau(z)$, assertion (a) implies $\tau(y), \tau(z) < \tau(x)$. So we are done by induction.

(c) By hypothesis, M is finitely generated, say by $x_1, \dots, x_p$. We apply (b) to each of the x_i and obtain a collection H of finitely many irreducible elements. Evidently, every other irreducible element of M has the form $h + u$ with $h \in H$ and a unit u. □

The total degree τ is an example of a *grading* γ on M; we use this notion as a synonym for "monoid homomorphism from M to $\mathbb{Z}$." The choice of this name is motivated by the fact that a grading on M induces a grading on the algebra $R[M]$ in which all elements of R have degree 0 and every monomial X^s has degree $\gamma(s)$ (see Remark 4.5).

Positive monoids. Suppose that 0 is the only unit in M. It follows immediately from Proposition 2.14 that M has only finitely many irreducible elements in this case. Not only do they constitute a system of generators—they must be contained in every system of generators. This observation justifies the definition of the Hilbert basis:

Definition 2.15. A monoid is called *positive* if 0 is its only unit. The unique minimal system of generators of a positive affine monoid M given by its irreducible elements is called the *Hilbert basis* of M and denoted by $\operatorname{Hilb}(M)$.

For positive affine monoids the standard map is injective, and therefore we call it *standard embedding* in this case:

Proposition 2.16. *Let M be an affine monoid with $\operatorname{gp}(M) = \mathbb{Z}^r$ and $C = \mathbb{R}_+ M \subset \mathbb{R}^r$. Then the following are equivalent:*

(a) M is positive;
(b) the standard map σ is injective on M;
(c) $\sigma : \mathbb{R}^r \to \mathbb{R}^s$ is injective;
(d) C is pointed.

Proof. (a) $\Longrightarrow$ (d) Let U be the vector subspace of $\mathbb{R}^r$ consisting of all the elements $x \in C$ with $-x \in C$. It is evidently the intersection of the support hyperplanes of C and therefore a rational vector subspace. It is enough to show that $x = 0$ for $x \in U \cap \mathbb{Q}^r$. For such x there exists $a \in \mathbb{N}$, with $ax, a(-x) \in M$. Hence $x = 0$.

(d) $\Longrightarrow$ (c) If C is pointed, the dual cone C^* has dimension d (see Proposition 1.19). Thus the support forms generate $(\mathbb{R}^r)^*$, and a suitable collection $\sigma_{i_1}, \dots, \sigma_{i_r}$ of r support forms is linearly independent. If $\sigma_{i_j}(x) = 0$ for $j = 1, \dots, r$, then $x = 0$.

The remaining implications (c) $\Longrightarrow$ (b) and (b) $\Longrightarrow$ (a) are trivial. □

The total degree τ on a positive affine monoid is a grading under which only 0 has degree 0, as follows immediately from the injectivity of σ; it is an example of a *positive grading* γ on M, by definition a homomorphism $\gamma : M \to \mathbb{Z}_+$ such that $\gamma(x) = 0$ implies $x = 0$. A positive grading on M induces a positive grading on the algebra $R[M]$ in which all elements of R have degree 0 and every monomial X^s, $s \neq 0$, has positive degree.

We can now characterize positive affine monoids as submonoids of $\mathbb{Z}_+^n$, $n \geq 0$:

Proposition 2.17. *Let M be an affine monoid of rank r and with s support forms. Then the following are equivalent:*

(a) M is positive;
(b) M is isomorphic to a submonoid of $\mathbb{Z}_+^d$ for some d;
(c) M is isomorphic to a submonoid M' of $\mathbb{Z}_+^s$ such that the intersections $H_i \cap \mathbb{R}M'$ of the coordinate hyperplanes $H_1, \dots, H_s$ are exactly the support hyperplanes of M';
(d) M is isomorphic to a submonoid M' of $\mathbb{Z}_+^r$ such that the intersections $H_i \cap \mathbb{R}M'$ of the coordinate hyperplanes $H_1, \dots, H_r$ are among the support hyperplanes of M';
(e) M is isomorphic to a submonoid M' of $\mathbb{Z}_+^r$ with $\operatorname{gp}(M') = \mathbb{Z}^r$;
(f) M has a positive grading.

Proof. Each of (b)–(f) implies (a) for obvious reasons, and each of (c)–(e) implies (b). That (a) $\Longrightarrow$ (f) has already been observed.

For the implication (a) $\Longrightarrow$ (c) we use the injectivity of the standard embedding $\sigma : M \to \mathbb{Z}^s$. Set $M' = \sigma(M)$. Then $M' = \mathrm{gp}(M') \cap \mathbb{R}^s_+$, and since $M' \cong M$ has s support hyperplanes, all the coordinate hyperplanes must be support hyperplanes of M'. For (a) $\Longrightarrow$ (d) we simply choose r linearly independent ones among the support forms, as in the proof of Proposition 2.16(d) $\Longrightarrow$ (c).

For the implication (a) $\Longrightarrow$ (e) we have to find r linear forms $\rho_1, \dots, \rho_r$ with integral coefficients in $(\mathbb{R}_+ M)^*$ that form a basis of $(\mathbb{Z}^r)^*$. The existence of such a basis will be shown in Corollary 2.74. (However, note that the additional conditions in (d) and (e) cannot always be satisfied simultaneously, since $\rho_1, \dots, \rho_r$ cannot always be chosen as support forms of M. Exercise 2.8 asks for an example.) □

Polytopal monoids. We now introduce a special class of affine monoids, whose investigation in combinatorial and algebraic terms will be important for us.

Definition 2.18. Let L be an affine lattice in $\mathbb{R}^d$ and P an L-polytope. The *polytopal affine monoid* $M(P)$ associated with P is the monoid

$$\mathbb{Z}_+\{(x,1) : x \in \mathrm{lat}(P)\}.$$

in $\mathbb{R}^{d+1}$. The set $\{(x,1) : x \in \mathrm{lat}(P)\}$ generating $M(P)$ is denoted by $E(P)$.

Since the set $\mathrm{lat}(P)$ is finite, $M(P)$ is an affine monoid. It is evidently positive and its Hilbert basis is the set $\{(x,1) : x \in \mathrm{lat}(P)\}$.

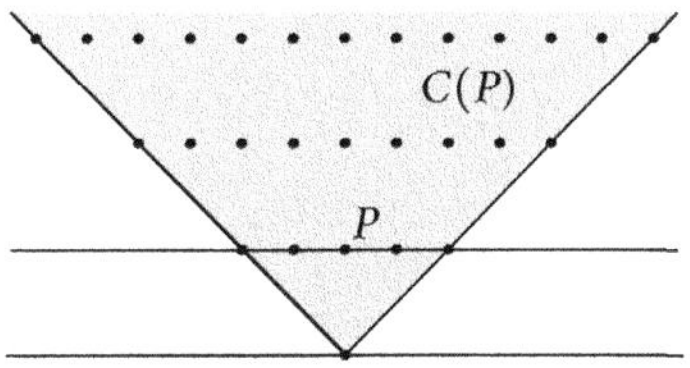

Fig. 2.1. Vertical cross-section of a polytopal monoid

The lattice L is not uniquely determined by P, and therefore the notations $E(P)$ and $M(P)$ are somewhat ambiguous. It will however always be clear which lattice L is to be considered.

Remark 2.19. More generally, we can associate a submonoid of $\mathbb{R}^d \oplus \mathbb{Z}_+ \subset \mathbb{R}^{d+1}$ with every subset X of $\mathbb{R}^d$, replacing the generating set $E(P)$ by $\{(x,1) : x \in X\}$. The structure of this monoid depends only on the affine structure of X. In fact, suppose that α is an affine isomorphism of $\mathbb{R}^d$ onto itself. Then α induces a *linear* isomorphism α' of the vector space $\mathbb{R}^{d+1}$ given as follows:

$$\alpha'(x,h) = (\alpha(x) - \alpha(0), h), \qquad x \in \mathbb{R}^d,\ h \in \mathbb{R}.$$

The restriction of α' to the monoid M over X maps M isomorphically onto the monoid over $\alpha(X)$.

In particular, if $L' = x + L$ is an affine lattice associated with the lattice L, and P' is an L'-polytope, then $P = P' - x$ is an L-polytope such that $M(P') \cong M(P)$. Whenever it should be convenient for the analysis of $M(P)$, we can therefore assume that P has its vertices in a lattice (and not just in an affine lattice).

Polytopal monoids are special instances of *homogeneous* affine monoids; such monoids M are positive and admit a positive grading in which every irreducible element has degree 1. The proof of the following proposition is left to the reader.

Proposition 2.20. *Let M be an affine monoid. Then the following are equivalent:*
(a) *M is homogeneous;*
(b) *there exists a hyperplane H of $\mathbb{R}^d$, not passing through 0, such that M is generated by elements of H;*
(c) *M is positive, and the number of summands in every representation of an element $x \in M$ as a sum of irreducibles is constant.*

Monoids with property (c) are called *half-factorial*; see Geroldinger and Halter-Koch [141] for this notion.

2.B Normal affine monoids

Definition 2.21. Let M be a submonoid of a commutative monoid N. The *saturation* or *integral closure* of M in N is the submonoid

$$\widehat{M}_N = \{x \in N : mx \in M \text{ for some } m \in \mathbb{N}\}$$

of N. One calls M *saturated* or *integrally closed* in N if $M = \widehat{M}_N$.

The *normalization* $\bar{M}$ of a cancellative monoid M is the integral closure of M in $\operatorname{gp}(M)$, and if $\bar{M} = M$, then M is called *normal*.

The terms "integral closure" and "normalization" are borrowed from commutative algebra, to which we will connect them in Section 4.E.

For submonoids of $\mathbb{Q}^d$ the integral closure has a clear geometric interpretation. It shows that the use of the notation $\widehat{M}_L$ in Corollary 2.10 was justified.

Proposition 2.22. *Let $M \subset N$ be submonoids of $\mathbb{Q}^d$ and $C = \mathbb{R}_+ M$. Then $\widehat{M}_N = C \cap N$. If M and N are affine monoids, then so is $\widehat{M}_N$.*

Proof. The inclusion $\widehat{M}_N \subset C \cap N$ is trivial.

For the converse inclusion we choose an element $x \in C \cap N$. Then x has a representation as a $\mathbb{Q}_+$-linear combination of elements of M. Let m be the least common multiple of the denominators of the coefficients. Then $mx \in M$ and, hence, $x \in \widehat{M}_N$.

For the last statement we set $L = \operatorname{gp}(N)$. Then $\widehat{M}_L$ is affine by Corollary 2.10, and the intersection $\widehat{M}_N = N \cap \widehat{M}_L$ is affine by Corollary 2.11. □

Example 2.23. Consider the extension of affine monoids $2\mathbb{Z}_+ \subset \mathbb{Z}_+$. Then $2\mathbb{Z}_+$ is normal and, hence, smaller than its integral closure in $\mathbb{Z}_+$, which is $\mathbb{Z}_+$ itself.

The most important consequence of Proposition 2.22 is the characterization of affine normal monoids. It combines Proposition 2.22 with Gordan's lemma.

Corollary 2.24. *Let $M \subset \mathbb{Z}^r$ be a monoid such that $\operatorname{gp}(M) = \mathbb{Z}^r$. Then $\mathbb{Z}^r \cap \mathbb{R}_+ M$ is the normalization of M.*

Moreover, the following are equivalent:

(a) *M is normal and affine;*
(b) *$\mathbb{R}_+ M$ is finitely generated and $M = \mathbb{Z}^r \cap \mathbb{R}_+ M$;*
(c) *there exist finitely many rational halfspaces $H_i^+ \subset \mathbb{R}^r$ such that $M = \bigcap_i H_i^+ \cap \mathbb{Z}^r$.*

We will sometimes refer to the monoids $H^+ \cap \mathbb{Z}^r$ (with H^+ a rational halfspace) as *discrete halfspaces*. Their structure is given by Corollary 2.27 below.

Before we record another useful corollary, we remind the reader that $N_* = \{0\} \cup \operatorname{int}(N)$ (see Remark 2.6).

Corollary 2.25. *Let M be a (not necessarily affine) integrally closed submonoid of an affine monoid N. If $\operatorname{rank} M = \operatorname{rank} N$, then $\operatorname{gp}(M) = \operatorname{gp}(N)$. In particular, $\operatorname{gp}(N_*) = \operatorname{gp}(N)$.*

Proof. We can assume that $\operatorname{gp}(N) = \mathbb{Z}^r$. Note that $\mathbb{R}M = \mathbb{R}N = \mathbb{R}^r$ since $\operatorname{rank} M = r$. There is nothing to show if $r = 0$. Suppose now that $r > 0$, and choose elements $x_1, \dots, x_r \in M$ generating $\mathbb{R}^r$ as a vector space. Let $M' = \mathbb{Z}_+ x_1 + \cdots + \mathbb{Z}_+ x_r$, and set $M'' = \mathbb{R}_+ M' \cap N$. Then M'' is the integral closure of M' in N and an affine monoid itself by Proposition 2.22. Since $M' \subset M$, we obtain that $M'' \subset M$ by hypothesis on M. We can now replace M by M'' and assume that M is itself affine.

Choose $x \in \operatorname{int}(M)$. Then all support forms of M have positive value on x, and they are indeed linear forms on $\mathbb{R}N = \mathbb{R}M$. For $y \in N$ it therefore follows that $y + kx \in \mathbb{R}_+ M$ for $k \gg 0$. Proposition 2.22 implies that $y + kx$ is integral over M. So $y + kx \in M$ by hypothesis, and $y \in \operatorname{gp}(M)$. □

In the previous subsection we have introduced the standard map $\sigma : M \to \mathbb{Z}^s$ on an affine monoid M. Proposition 2.14(a) shows that $\sigma(x) = 0$ if and only if x is a unit in M. For normal monoids M we can say even more:

Proposition 2.26. *Let M be a normal affine monoid, $\operatorname{U}(M)$ its subgroup of units, and $\sigma : \operatorname{gp}(M) \to \mathbb{Z}^s$ the standard map on M. Then M is isomorphic to $\operatorname{U}(M) \oplus \sigma(M)$.*

Proof. Let $L = \operatorname{gp}(M)$. We claim that $\operatorname{U}(M)$ is the kernel of σ. Clearly $\operatorname{U}(M) \subset \operatorname{Ker} \sigma$. Conversely let $x \in \operatorname{Ker} \sigma$. Then $x \in C \cap L$, where C is the cone generated by M. The normality of M then shows $x \in M$, and so $x \in \operatorname{U}(M)$.

Since $\operatorname{U}(M)$ is a direct summand of L, there exists a projection $\pi : L \to \operatorname{U}(M)$, i. e. a surjective $\mathbb{Z}$-linear map π with $\pi^2 = \pi$. Let $x \in M$. Then $(\pi(x), \sigma(x)) \in \operatorname{U}(M) \oplus \sigma(M)$. Conversely, given $(x_0, y') \in \operatorname{U}(M) \oplus \sigma(M)$ we choose $y \in M$ with $y' = \sigma(y)$. Then $x_0 + y - \pi(y) \in M$, $\pi(x_0 + y - \pi(y)) = x_0$ and $\sigma(x_0 + y - \pi(y)) = y'$. □

Corollary 2.27. *Let H^+ be a rational linear halfspace in $\mathbb{Z}^r$. Then $H^+ \cap \mathbb{Z}^r \cong \mathbb{Z}^{r-1} \oplus \mathbb{Z}_+$.*

Proof. Evidently $M = H^+ \cap \mathbb{Z}^r$ is normal, and $\mathrm{U}(M) = H \cap \mathbb{Z}^r \cong \mathbb{Z}^{r-1}$ since H is rational. Since $\sigma(M) \subset \mathbb{Z}_+$ is normal, we have $\sigma(M) \cong \mathbb{Z}_+$. □

The direct sum of a monoid M with a group G can be considered a trivial extension of M in almost every context treated in this book. Therefore one can usually restrict the discussion of normal affine monoids to the positive ones.

In the proof of Proposition 2.26 the normality of M is only used to the extent that $\mathrm{U}(M)$ is a direct summand of $\mathrm{gp}(M)$. However, if this condition is violated, we cannot expect an isomorphism $M \cong \mathrm{U}(M) \oplus \sigma(M)$. For example, choose $M = \{(x, y) \in \mathbb{Z}^2 : y > 0, \text{ or } y = 0 \text{ and } x \equiv 0 \ (2)\}$.

Polytopal monoids (see Definition 2.18) can be characterized in terms of their normalizations:

Proposition 2.28. *Let M be an affine monoid. Then the following are equivalent:*

(a) *M is polytopal;*

(b) *M is homogeneous and coincides with $\bar{M}$ in degree* 1.

Proof. The implication (a) $\Longrightarrow$ (b) is (almost) the definition of polytopal monoid. In fact, let P be a lattice polytope. The height 1 lattice points of $\mathbb{R}_+M(P)$ are exactly the generators of $M(P)$, and so are contained in $M(P)$.

For the converse let $\mathrm{gp}(M) = \mathbb{Z}^d$. By hypothesis M has a grading γ. We can first extend it to $\mathbb{Z}$-linear form on $\mathbb{Z}^d$ and then to a linear form on $\mathbb{R}^d$. Since M is generated by elements of degree 1, $\mathbb{R}_+M$ is generated by integral vectors of degree 1. Their convex hull is the lattice polytope $P = \{x \in \mathbb{R}_+M : \gamma(x) = 1\}$. Since all the lattice points of P correspond to elements of M by hypothesis, $M \cong M(P)$. □

Normal monoids as pure submonoids. We have seen that every positive affine monoid can be realized as a submonoid of $\mathbb{Z}_+^n$ for suitable n. This statement can be considerably strengthened if M is normal.

Theorem 2.29. *The following are equivalent for a positive affine monoid M:*

(a) *M is normal;*

(b) *there exists $m \in \mathbb{Z}_+$ and a subgroup U of $\mathbb{Z}^m$ of rank r such that M is isomorphic to $\mathbb{Z}_+^m \cap U$;*

(c) *there exist $p, q \in \mathbb{Z}_+$ and a $\mathbb{Z}$-linear map $\lambda : \mathbb{Z}^p \to \mathbb{Z}^q$ such that M is isomorphic to $\mathbb{Z}_+^p \cap \operatorname{Ker} \lambda$.*

If M is normal of rank r and with s support forms, then one can choose $m = r$, $p = s$, and $q = r + s$. Moreover, an isomorphism as in (b) is given by the standard embedding.

Proof. It is obvious that the monoids in (b) and (c) are normal—those in (c) are even integrally closed in $\mathbb{Z}^p$.

Now suppose that M is normal. We identify $\mathrm{gp}(M)$ with $\mathbb{Z}^r$, apply the standard embedding $\sigma : M \to \mathbb{Z}_+^s$, and set $M' = \sigma(M)$, $U = \mathrm{gp}(\sigma(M))$. The monoid isomorphism $\sigma : M \to M'$ extends to an isomorphism $\sigma : \mathrm{gp}(M) \to U$. Proposition 2.22 shows that $x \in M$ if and only if $\sigma(x) \in \mathbb{Z}_+^s$.

For the more difficult implication (b) $\Longrightarrow$ (c) we have to use the *elementary divisor theorem* (for example, see Lang [238, p. 153]): let $L \cong \mathbb{Z}^n$ be a free abelian group, and L' a subgroup of rank m; then there exist a basis $e_1, \dots, e_n$ of L and positive integers $d_1, \dots, d_m$ such that $d_1 e_1, \dots, d_m e_m$ is a basis of L' (and $d_1 \mid d_2 \mid \cdots \mid d_m$).

We can assume that $M = \mathbb{Z}_+^s \cap U$, $U = \operatorname{gp}(M)$, $r = \operatorname{rank} M$, and apply the elementary divisor theorem with $L = \mathbb{Z}^s$ and $L' = U$. Let ρ_i, $i = 1, \dots, s$, be the linear form on $\mathbb{Z}^s$ that assigns each vector its ith coordinate with respect to the basis $e_1, \dots, e_s$. Then an element $x = (x_1, \dots, x_s) \in \mathbb{Z}^s$ belongs to M if and only if (i) $x_1, \dots, x_s \geq 0$, (ii) $\rho_i(x) = 0$ for $i = r+1, \dots, s$, (iii) $\rho_i(x) \equiv 0 \ (d_i)$ for $i = 1, \dots, r$.

In order to achieve our goal we have to convert the congruences into linear equations. We can change the coefficients of $\rho_1, \dots, \rho_r$ (with respect to the given coordinates of $\mathbb{Z}^s$) by adding multiples of d_i, and so we may assume that $\rho_1, \dots, \rho_r$ have nonnegative values on $\mathbb{Z}_+^s$. For elements $x \in \mathbb{Z}_+^s$ the congruence $\rho_i(x) \equiv 0 \ (d_i)$ is then equivalent to the solvability of the equation $\rho_i(x) = y_i d_i$ with $y_i \in \mathbb{Z}_+$, and the solution $y_i = y_i(x)$ is uniquely determined by x.

Define the map $\pi : U \to \mathbb{Z}^{s+r}$ by $x \mapsto \big(x, y_1(x), \dots, y_r(x)\big)$, and set $V = \pi(U)$. Then π is injective, and $\pi(x) \in \mathbb{Z}_+^{s+r}$ if and only if $x \in M$. It remains to show that V is the kernel of a suitable map $\mathbb{Z}^{s+r} \to \mathbb{Z}^s$. This is equivalent (by the elementary divisor theorem) to the torsion freeness of $\mathbb{Z}^{s+r}/V$.

Let $z \in \mathbb{Z}^{s+r}$, $z = (z', z'')$, $z' \in \mathbb{Z}^s$, $z'' \in \mathbb{Z}^r$, and suppose that $mz \in V$ for some positive integer m. Then $x = mz' \in U$. Moreover,

$$\rho_j(z') = m^{-1}\rho_j(x) = m^{-1}y_j(x)d_j = m^{-1}mz_j''d_j = z_j''d_j.$$

It follows that $z' \in U$ and $(z', z'') \in V$. This shows the torsion freeness of $\mathbb{Z}^{s+r}/V$. □

Suppose M is a submonoid of N. Then M is called *pure* in N if $M = N \cap \operatorname{gp}(M)$. Equivalently we can require that $N \setminus M$ is an M-submodule of N. The notion of purity develops its power in the context of monoid algebras, where the purity of M in N means that $R[M]$ is a direct summand of $R[N]$ as an $R[M]$-module. Theorem 2.29 shows that a normal affine monoid can be realized not only as a pure submonoid of $\mathbb{Z}_+^n$ for suitable n (this is (b)), but also as a pure, integrally closed submonoid of a free monoid (this is (c)).

Remark 2.30. Let M be an positive affine normal monoid. The standard embedding $\sigma : M \to \mathbb{Z}^s$ realizes M as a pure submonoid as we have seen in Theorem 2.29(b). The quotient $\mathbb{Z}^s/\sigma(\operatorname{gp}(M))$ is an intrinsic invariant of M, namely the divisor class group $\operatorname{Cl}(M)$ of M (or any monoid algebra $\Bbbk[M]$ in which $\Bbbk$ is a factorial ring). See Corollary 4.56. So $\sigma(\operatorname{gp}(M))$ is a direct summand of $\mathbb{Z}^s$ if and only if $\operatorname{Cl}(M)$ is torsionfree.

Let M be a pure submonoid of the monoid N. Then N decomposes naturally into a disjoint union of M-submodules given by the intersections of N with the cosets of $\operatorname{gp}(M)$ in $\operatorname{gp}(N)$:

$$N = \bigcup \big(y + \operatorname{gp}(M)\big) \cap N,$$

where y runs through a system of representatives of the cosets $z + \operatorname{gp}(M)$, $z \in N$. We call the modules $\big(y + \operatorname{gp}(M)\big) \cap N$ the *coset modules* of M in N. Under suitable finiteness conditions they are finitely generated:

Proposition 2.31. *Let M be a pure submonoid of an affine monoid N. Then M is affine, too, and the coset modules of M in N are finitely generated over M. Moreover, each coset module is isomorphic to an M-submodule of* $\operatorname{gp}(M)$.

Proof. Since M is the intersection of affine monoids, namely N and $\operatorname{gp}(M)$, it is itself affine (Corollary 2.11).

Let us first assume that N is normal. Then M is likewise normal. After the identification of $\operatorname{gp}(N)$ with $\mathbb{Z}^n$, we can understand N as the set of solutions of a homogeneous linear diophantine system of inequalities, and M is cut out from N by a system of such equations and congruences. The coset module is then given as the set of solutions of an inhomogeneous variant of the homogeneous system, and so we can apply Corollary 2.13.

Now let N be arbitrary, and observe that $\bar{M} \subset \bar{N}$ is again pure. Since $\bar{M}$ is a finitely generated M-module, the coset modules of $\bar{M}$ in $\bar{N}$ are finite also over M. But each coset module of M in N is an M-submodule of a coset module of $\bar{M}$ in $\bar{N}$, and the desired finiteness follows.

The last statement is evident: if $w \in (y+\operatorname{gp}(M)) \cap N$, then $-w+(y+\operatorname{gp}(M)) \cap N \subset \operatorname{gp}(M)$. □

The coset modules of the standard embedding play a special role: they represent the divisorial ideals of M; see Theorem 4.62.

Adjoining inverse elements. In the case of a normal affine monoid M, certain extensions of M can be controlled by the support hyperplanes of M. For a monoid M contained in a group G and a subset $N \subset M$ we let $M[-N]$ denote the smallest submonoid of G containing M and all the elements $-x$ with $x \in N$. (If $N = \{x\}$, then we write $M[-x]$ instead of $M[-N]$.) One calls $M[-N]$ the *localization* with respect to N.

Proposition 2.32. *Let M be a normal affine monoid with $\operatorname{gp}(M) = \mathbb{Z}^d$, $x \in M$, and $H_1, \dots, H_s$ its support hyperplanes. Furthermore let F be the face of $\mathbb{R}_+M$ with $x \in \operatorname{int}(F)$. Then*

$$M[-x] = M[-(F \cap M)] = \bigcap \{H_i^+ : x \in H_i\} \cap \mathbb{Z}^d.$$

Furthermore $M[-x]$ splits into a direct sum $L \oplus M'$ where $L \cong \mathbb{Z}^e$, $e = \dim F$.

If M is positive, then M' is positive.

Proof. Let N be the normal affine monoid on the right hand side. Clearly $M[-x] \subset M[-(F \cap M)] \subset N$, since $M \subset N$ and $-(F \cap M) \subset N$. Conversely suppose that $y \in N$. For all the support forms σ_j with $x \notin H_j$ we have $\sigma_j(x) > 0$. Therefore $\sigma_j(y + kx) \geq 0$ for all such j if $k \gg 0$. It follows that $y + kx \in M$ for $k \gg 0$, and so $y \in M[-x]$. Altogether we have $M[-x] = N$.

Let $U = \mathbb{R}F \cap \mathbb{Z}^d$. Then U is a direct summand of $\mathbb{Z}^d$, and, moreover, $U \subset M[-x]$. In fact, $U \subset H_i$ for each support hyperplane H_i of M with $x \in H_i$. By the same argument as in the proof of Proposition 2.26, U splits off M.

For the last assertion one has to show that U is the group of units of $M[-x]$ if M is positive. Let $y \in M[-x]$ be an invertible element, say $y = y' - px$ with $y' \in M$ and $p \in \mathbb{Z}$. Then $-y = y'' - qx$, $y'' \in M$, $q \in \mathbb{Z}$. The equation $y' + y'' - (p+q)x = 0$ shows that $y' = 0$ if $p + q \leq 0$, since y' is a unit in M in this case. If $p + q > 0$, then both y' and y'' must belong to the face F, and we are again done. □

The conductor. Let M be an affine monoid with normalization $\bar{M}$. Then the set of "gaps" $\bar{M} \setminus M$ is, in a sense, small.

Proposition 2.33. *Let M be an affine monoid. Then the ideal*

$$\mathfrak{c}(\bar{M}/M) = \{x \in M : x + \bar{M} \subset M\}$$

is nonempty. Moreover, the set $\bar{M} \setminus M$ is contained in finitely many hyperplanes parallel to the facets of M.

Proof. We have seen in Corollary 2.10 that $\bar{M}$ is a finitely generated M-module. Since $\bar{M} \subset \operatorname{gp}(M)$, there exist elements $z_1, \dots, z_n \in \operatorname{gp}(M)$ with $\bar{M} = (z_1+M) \cup \dots \cup (z_n+M)$. Write $z_i = y_i - x_i$ with $x_i, y_i \in M$, $i = 1, \dots, n$, and set $x = x_1 + \dots + x_n$. Then $x \in \mathfrak{c}(\bar{M}/M)$, and $\mathfrak{c}(\bar{M}/M) \neq \varnothing$.

Pick $x \in \mathfrak{c}(\bar{M}/M)$. If $z \in \bar{M} \setminus M$, then there exists at least one support form σ_i of M with $\sigma_i(z) < \sigma_i(x)$. So z belongs to one of the hyperplanes $\{y : \sigma_i(y) = k\}$, $k \in \mathbb{Z}_+$, $k < \sigma_i(x)$. □

The common name for $\mathfrak{c}(\bar{M}/M)$ is *conductor*. The reader may check that $\mathfrak{c}(\bar{M}/M)$ is the largest ideal of M that is also an ideal of $\bar{M}$ (Exercise 2.9).

Our first corollary shows that monoid elements in the interior of $\mathbb{R}_+M$ are nilpotent modulo any nonempty ideal of M. More precisely:

Corollary 2.34. *Let M be an affine monoid, L a lattice containing* $\operatorname{gp}(M)$, *$x \in M$, and $y \in \widehat{M}_L \cap \operatorname{int}(\mathbb{R}_+M)$. Then $ky \in x + \operatorname{int}(M)$ for some $k \in \mathbb{N}$.*

If, in addition, $y \in \operatorname{gp}(M)$, then $ky \in x + \operatorname{int}(M)$ for $k \gg 0$.

Proof. Suppose first that $y \in \operatorname{gp}(M)$. Then $y \in \bar{M} = \operatorname{gp}(M) \cap \mathbb{R}_+M$. Choose $z \in \mathfrak{c}(\bar{M}/M)$. Then $x + z + \bar{M} \subset x + M$, and after replacing x by $x + z$ we are left with the case in which M is normal. Since $y \in \operatorname{int}(\mathbb{R}_+M)$ we have $\sigma_i(y) > 0$ for all support forms σ_i of M. Thus $\sigma_i(ky) > \sigma_i(x)$ for all i and $k \gg 0$. It follows that $ky - x \in \operatorname{int}(\mathbb{R}_+M) \cap \operatorname{gp}(M) = M$ and therefore $ky \in x + \operatorname{int}(M)$.

In the general case there exists $q \in \mathbb{N}$ with $qy \in M$ for some $q \in \mathbb{N}$. Now we apply to qy what has just been shown. □

One can investigate the set $\bar{M} \setminus M$ in more detail. The next proposition can be derived rather easily from standard facts of commutative algebra. Imitating terminology of this field, we set

$$\operatorname{Rad}(I) = \{x \in M : ax \in I \text{ for some } a \subset \mathbb{N}\}$$

for an ideal I in a monoid M, and call it the *radical* of I.

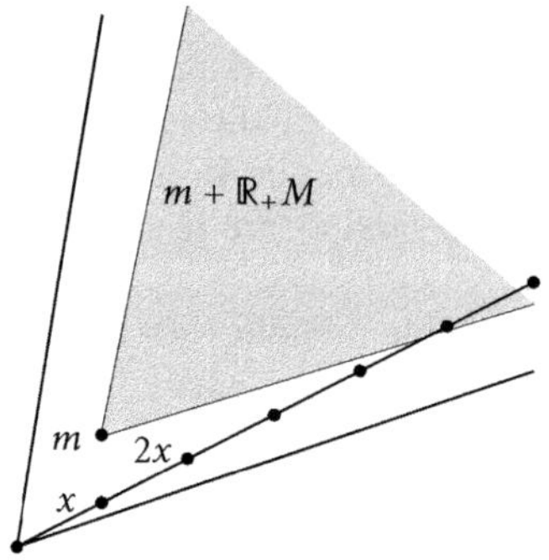

Fig. 2.2. Nilpotence of the interior modulo a nonempty ideal

Proposition 2.35. *Let M be an affine monoid. Then:*

(a) $\operatorname{Rad}(\mathfrak{c}(\bar{M}/M))$ *is the set of all $x \in M$ such that $M[-x]$ is normal.*

(b) $\bar{M} \setminus M$ *is the union of a finite family of sets $x + (F \cap M)$ where $x \in M$ and F is a face such that $F \cap \mathfrak{c}(\bar{M}/M) = \emptyset$. Moreover, if F is maximal among these faces, then at least one set of type $x + (F \cap M)$ must appear.*

We will obtain the proposition as a consequence of a more general result on modules over affine monoids; see Proposition 4.37.

Prime and radical ideals. Let I be an ideal in a monoid M. One calls I a *radical ideal* if $I = \operatorname{Rad}(I)$ and a *prime ideal* if $I \neq M$, and $m + n \in I$ for $m, n \in M$ is only possible if $m \in I$ or $n \in I$. An affine monoid has only finitely many radical ideals and they are completely determined by the geometry of $\mathbb{R}_+ M$:

Proposition 2.36. *Let M be an affine monoid and $I \subset M$ an ideal.*

(a) *I is a radical ideal if and only if I is the intersection of the sets $M \setminus F$ where F is a face of $\mathbb{R}_+ M$ with $F \cap I = \emptyset$.*

(b) *I is a prime ideal if and only if there exists a face F with $I = M \setminus F$.*

Proof. Let F be a face of $\mathbb{R}_+ M$. Then $M \setminus F$ is obviously a prime ideal, and since the intersection of prime ideals is radical, the intersection of sets of type $M \setminus F$ is a radical ideal.

Conversely, suppose that I is a radical ideal, and let $x \in M \setminus I$. There is a unique face F with $x \in \operatorname{int}(F)$, and it is enough to show that $F \cap I = \emptyset$. On the contrary, assume there exists $y \in F \cap I$. Let $N = F \cap M$. Then $kx \in y + N \subset y + M \subset I$ for some $k > 0$ by Corollary 2.34. Since I is a radical ideal, this implies $x \in I$, a contradiction.

For (b) one must show that the set of faces F with $F \cap I = \emptyset$ has a unique maximal element if I is a prime ideal. Assume that $F_1, \dots, F_n$ are the maximal faces in the complement of I. If $n \geq 2$, then a sum $x_1 + \dots + x_n$ with $x_i \in \operatorname{int}(F_i) \cap M$ for all i neither belongs to I if I is prime nor belongs to the union $F_1 \cup \dots \cup F_n$. This is a contradiction. □

By Proposition 2.36(b) the extreme submonoids of M (compare Remark 2.6) are exactly the complements of the prime ideals.

Special instances of prime ideals I are those generated by a single element x; in this case $I = M + x$. Then x is called a *prime element*. We say that prime elements x, y are *nonassociated* if $M + x \neq M + y$, or, in other words, $x - y \notin M$ and $y - x \notin M$.

Proposition 2.37. *Let M be an affine monoid, and N a submonoid generated by prime elements $x_1, \dots, x_n$ that are pairwise nonassociated. Then there exists a single face F of $\mathbb{R}_+M$ such that $x_i \notin F$ for all i, and $M = \mathbb{Z}_+x_1 \oplus \dots \oplus \mathbb{Z}_+x_n \oplus (M \cap F)$.*

Proof. We use induction, starting with the case $n = 1$. So let x be a prime element in M. Since x is not a unit, there exists a facet F not containing x. The ideal $M \setminus F$ is a prime ideal that does not contain any other prime ideal of M, and since $x \notin F$, we must have $M + x = M \setminus F$.

Let y be an element of M. Then there exists $k \in \mathbb{Z}_+$ such that $y - kx \in M$, but $y-(k+1)x \notin M$ (simply because $\sigma_F(x) > 0$). So $y-kx \in F$, for otherwise $y-kx \in M+x$, and $y - (k+1)x \in M$. We conclude that $M = \mathbb{Z}_+x + (M \cap F)$. Since $\mathbb{Z}x \cap F = \{0\}$, we must even have $M = \mathbb{Z}_+x \oplus (M \cap F)$.

Now suppose that $n > 1$. Using x_1 one splits M as $\mathbb{Z}_+x_1 \oplus (M \cap F_1)$. Since $x_2, \dots, x_n \notin M+x$, they all lie in F, and we can conclude by induction since $x_2, \dots, x_n$ are prime elements also in the monoid $M \cap F_1$ and pairwise nonassociated, as the reader may check. □

Unions of normal monoids. A monoid M that is a union of normal submonoids M_i with $\operatorname{gp}(M_i) = \operatorname{gp}(M)$ is evidently itself normal. If M is affine and only finitely many submonoids are brought into play, then a much stronger statement is possible:

Theorem 2.38. *Let M be an affine monoid and $M_1, \dots, M_n$ submonoids of M such that $M = M_1 \cup \dots \cup M_n$ and M_i is normal whenever* $\operatorname{rank} M_i = \operatorname{rank} M$. *Then M is normal, and it is the union of those M_i for which* $\operatorname{rank} M_i = \operatorname{rank} M$.

Proof. We can assume that $\operatorname{gp}(M) = \mathbb{Z}^d$. It is our first goal to reduce the theorem to its special case in which $\mathbb{R}_+M_i = \mathbb{R}_+M$ for all i with $\operatorname{rank} M_i = d$. To this end we consider all hyperplanes H in $\mathbb{R}^d$ that support one of the cones $C_i = \mathbb{R}_+M_i$ with $\operatorname{rank} M_i = d$. This set of hyperplanes induces a dissection Γ of the cone $C = \mathbb{R}_+M$ into rational subcones (see Proposition 1.69). Each cone $D \in \Gamma$ is generated by $D \cap M$, and $M = \bigcup\{D \cap M : D \in \Gamma,\ \dim D = d\}$.

Furthermore, if $D \in \Gamma$ with $\dim D = d$, then $\operatorname{gp}(D \cap M) = \mathbb{Z}^d$ by Corollary 2.25. So the normality of M follows from the normality of the intersections $D \cap M$. Now we can consider all the monoids $D \cap M$, $D \in \Gamma$, $\dim D = d$, and their submonoids $D \cap M_i$ separately. By construction of Γ we may therefore assume that $\mathbb{R}_+M_i = \mathbb{R}_+M$ for all submonoids M_i with $\operatorname{rank} M_i = d$.

Renumbering the M_i if necessary we can further assume that $\operatorname{rank} M_i = d$ for $i = 1, \dots, p$, and $\operatorname{rank} M_i < d$ for $i > p$. Set $G_i = \operatorname{gp}(M_i)$. If our claim does not hold, then we can find $x \in \bar{M} \setminus \bigcup_{i=1}^p M_i = \bar{M} \setminus \bigcup_{i=1}^p G_i$. Set $E = (x + \bigcap_{i=1}^p G_i) \cap C$. Then, on the one hand, $E \subset \bar{M} \setminus \bigcup_{i=1}^p G_i$. On the other hand, E is not contained in the union of finitely many hyperplanes. However, $\bar{M} \setminus M$ (by Proposition 2.33) and each submonoid M_i with rank $M_i < d$ is contained in such a union, and the conclusion is that E contains elements of $M \setminus \bigcup_{i=1}^n M_i$. This is a contradiction. □

Seminormal monoids. A property close to normality and of importance in K-theory (see Section 4.G and in particular Chapter 8) is seminormality:

Definition 2.39. A monoid is *seminormal* if every element $x \in \operatorname{gp}(M)$ with $2x, 3x \in M$ (and therefore $mx \in M$ for $m \in \mathbb{Z}_+$, $m \geq 2$) is itself in M. The *seminormalization* $\operatorname{sn}(M)$ of M is the intersection of all seminormal submonoids of $\operatorname{gp}(M)$ containing M.

It follows immediately from Corollary 2.10 that the seminormalization $\operatorname{sn}(M)$ of an affine monoid M is also affine; in fact, it is contained in the normalization $\bar{M}$, and therefore $\mathbb{R}_+ \operatorname{sn}(M) = \mathbb{R}_+ M$ is a cone.

A normal monoid is obviously seminormal, but the converse does not hold. There even exist seminormal, nonnormal polytopal monoids. We will give an example in Remark 2.56(b). For affine monoids M, the relationship between normality and seminormality is made precise by the next proposition. Recall that $M_* = \operatorname{int}(M) \cup \{0\}$ (Remark 2.6).

Proposition 2.40. *An affine monoid M is seminormal if and only if $(M \cap F)_*$ is a normal monoid for every face F of $\mathbb{R}_+ M$. Thus $M_* = \bar{M}_*$ if M is seminormal.*

Proof. Suppose that M is seminormal. Then $M \cap F$ is obviously seminormal for each face F of $\mathbb{R}_+ M$. Therefore it is enough to show that M_* is normal. By Corollary 2.25, $\operatorname{gp}(M_*) = \operatorname{gp}(M)$. Let $x \in \operatorname{gp}(M_*)$, $x \neq 0$, with $ax \in M_*$ for some $a \in \mathbb{N}$. Then $x \in \operatorname{int}(\mathbb{R}_+ M)$ and, by Corollary 2.34, $kx \in M$ for all $k \gg 0$. Let m be the largest integer for which $mx \notin M$. If $m > 0$, we have $2mx, 3mx \in M$, and so $mx \in M$ by seminormality, an obvious contradiction.

For the converse implication let $x \in \operatorname{gp}(M)$ be such that $2x, 3x \in M$. Let F be the unique face of $\mathbb{R}_+ M$ with $2x \in \operatorname{int}(F)$. Then certainly $3x \in \operatorname{int}(F)$, too, and so $x \in \operatorname{gp}\big(\operatorname{int}(F) \cap M\big)$. By hypothesis $x \in \operatorname{int}(F) \cap M$.

If M is seminormal, then M_* is normal, as just seen, and the normality of M_* implies $M_* = \bar{M}_*$. □

While M_* is almost never finitely generated, it is the filtered union of affine submonoids, and if M is seminormal, these can be chosen to be normal. To be more precise: there exists a family M_i of affine submonoids, indexed by the elements of a set I such that (i) $M = \bigcup_{i \in I} M_i$, and (ii) for all $i, j \in I$ there exists $k \in I$ such that $M_i, M_j \subset M_k$. In such a situation we will simply speak of a *filtered union*. For simplicity we restrict ourselves to positive affine monoids.

Proposition 2.41. *Let M be a positive affine monoid. Then M_* is the filtered union of affine submonoids. If M_* is normal, then these submonoids can be chosen to be normal.*

Proof. We embed M into $\mathbb{Z}^r = \operatorname{gp}(M)$, choose a rational cross-section P of $\mathbb{R}_+M$, and a rational point $z \in \operatorname{int}(P)$. In the affine space $\operatorname{aff}(P)$ we consider the homothety ϑ_λ with center z and factor $\lambda \in (0,1) \cap \mathbb{Q}$. Set $M_\lambda = M \cap \mathbb{R}_+\vartheta_\lambda(P)$. Then M_* is the filtered union of the affine monoids M_λ, and M_λ is normal if M_* is. We leave the detailed proof of this claim to the reader. □

We want to give another characterization of seminormal monoids that will turn out useful in proving the seminormality of their associated algebras and is similar in spirit to the characterization of normal affine monoids in Corollary 2.24.

Proposition 2.42. *Let $M \subset \mathbb{Z}^r$ be a monoid with $\operatorname{gp}(M) = \mathbb{Z}^r$. Then the following are equivalent:*

(a) M is seminormal affine;

(b) there exist finitely many rational halfspaces H_i^+ and subgroups $U_i \subset H_i \cap \mathbb{Z}^r$ such that $\operatorname{rank} U_i = r-1$ and

$$M = \bigcap_i \big(U_i \cup (H_i^> \cap \mathbb{Z}^r)\big).$$

We leave the detailed proof to the reader. Also see Exercises 2.11 and 2.12 for more results on seminormal monoids. However, we want to indicate the construction of the halfspaces and the subgroups U_i: for each *face* F we choose a rational hyperplane H_i intersecting $\mathbb{R}_+M$ exactly in F_i, and set $U_i = \operatorname{gp}(M \cap F) \oplus V_i$ where V_i is a subgroup of $H_i \cap \mathbb{Z}^r$ such that $\operatorname{rank} V_i = r - 1 - \operatorname{rank} U_i$ and $U_i \cap V_i = 0$. Note that the proposition includes the statement that the monoids $U_i \cup (H_i^> \cap \mathbb{Z}^r)$ are affine.

2.C Generating normal affine monoids

Let $C \subset \mathbb{R}^d$ be a pointed rational cone, and L a sublattice of $\mathbb{Q}^d$. Then $C \cap L$ is an integrally closed submonoid of L. In this section we study the Hilbert basis of $C \cap L$. It is no restriction to assume that $\dim C = \operatorname{rank} L$. Otherwise we can replace C by the cone $C \cap \mathbb{R}L$. However, we do not insist on $d = \dim C$.

Simplicial cones and multiplicities. Let $v_1, \dots, v_r$ be linearly independent vectors in the lattice L. By

$$\operatorname{par}(v_1, \dots, v_r) = \{q_1v_1 + \dots + q_rv_r : 0 \le q_i < 1,\ i = 1, \dots, n\}$$

we denote the *semi-open parallelotope* spanned by $v_1, \dots, v_r$ (see Figure 2.3). Let U denote the sublattice generated by $v_1, \dots, v_r$. It is clear that every residue class $x + U$ for an element x in the saturation $\widehat{U}_L = \mathbb{Q}U \cap L = \mathbb{R}U \cap L$ of U in L has a unique representative in $\operatorname{par}(v_1, \dots, v_r)$, namely

$$x' = (a_1 - \lfloor a_1 \rfloor)v_1 + \dots + (a_r - \lfloor a_r \rfloor)v_r,$$

where $x = \sum_{i=1}^r a_iv_i$, $a_i \in \mathbb{Q}$. Furthermore, if $x \in C \cap L$ where C is the cone generated by $v_1, \dots, v_r$, then $a_1, \dots, a_r \ge 0$, and $x \in x' + M$ where $M = \mathbb{Z}_+v_1 + \dots + \mathbb{Z}_+v_r$. We have thus proved:

Proposition 2.43. *With the notation just introduced, the following hold:*
(a) $E = L \cap \operatorname{par}(v_1, \dots, v_r)$ *is a system of generators of the* M*-module* $C \cap L$;
(b) $(x + M) \cap (y + M) = \varnothing$ *for* $x, y \in E$, $x \neq y$;
(c) $\#E = [(\mathbb{Q}U \cap L) : U]$;
(d) $\operatorname{Hilb}(C \cap L) \subset \{v_1, \dots, v_r\} \cup E$.

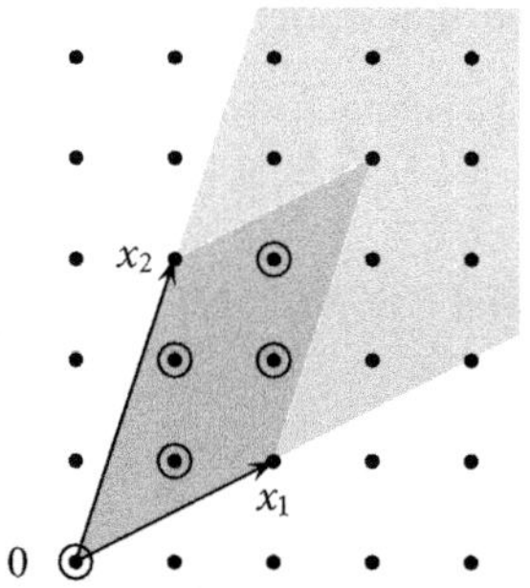

Fig. 2.3. The semi-open parallelotope and the set E

In part (c) we have used $[(\mathbb{Q}U \cap L) : U]$ for the index of the subgroup U in the group $\mathbb{Q}U \cap L$. The proposition can be interpreted as saying that $C \cap L$ is a free module over M with basis E. The rank of this free module, namely $[(\mathbb{Q}U \cap L) : U]$, measures how far U is from $\mathbb{Q}U \cap L$. Before continuing the main theme, namely the generation of normal affine monoids, we want to transfer this measure to lattice simplices, which, unlike the simplex $\operatorname{conv}(0, v_1, \dots, v_r)$ in Proposition 2.43, may not have 0 among their vertices.

Let $L = x + L_0$ be an affine lattice with associated lattice L_0 and U an affine sublattice, say $U = y + U_0$. Then we have $y \in L$ and $U_0 \subset L_0$. Then we set $L/U = L_0/U_0$. Since L_0 and U_0 do not depend on the choice of x and y, this definition is justified. However, it is reasonable beyond the formal justification: L decomposes into a disjoint union of translates of $V + y$ where V runs through the cosets of U_0 in L_0. Consequently we define the index of U in L by

$$[L : U] = [L_0 : U_0].$$

Let L be an affine lattice and Δ an L-simplex (see Definition 1.73). Then the roles of the lattices compared in Proposition 2.43(c) are played by the lattices introduced after Definition 1.73, namely

$$L_\Delta = L \cap \operatorname{aff}(\Delta)$$

and its affine sublattice

$$\mathscr{L}(\Delta) = v_0 + \sum_{v \in \operatorname{vert}(\Delta)} \mathbb{Z}(v - v_0),$$

where v_0 is an arbitrary vertex of Δ.

Definition 2.44. The *multiplicity* of Δ (with respect to L) is the index $\mu_L(\Delta) = [L_\Delta : \mathscr{L}(\Delta)]$.

The multiplicity has a very useful geometric interpretation that will allow us to define it for an arbitrary lattice polytope. In order to simplify the discussion we choose the origin in a vertex v_0 of Δ so that we are back in the situation of Proposition 2.43 and L_Δ and $\mathscr{L}(\Delta)$ are lattices of the same rank r. By the elementary divisor theorem we can find a basis $e_1, \dots, e_r$ of L_Δ and integers $d_1, \dots, d_r \geq 1$ such that $d_1 e_1, \dots, d_r e_r$ is a basis of $\mathscr{L}(\Delta)$. Then 2.43(c) shows that $\mu_L(\Delta) = d_1 \cdots d_r$. On $V = \mathbb{R}L_\Delta$ we introduce the r-dimensional volume function that gives volume 1 to the parallelotope spanned by $e_1, \dots, e_r$. This volume function depends only on L_Δ. We call it the *L-volume* vol_L on V. In fact, every change of basis in L_Δ is by an integral matrix of determinant ± 1. It follows that $\mu_L(\Delta)$ is exactly the L-volume of the parallelotope P spanned by $v_1, \dots, v_r$. By Exercise 3.1 the parallelotope P decomposes into $r!$ simplices that have the same volume as Δ:

Corollary 2.45. *For an L-simplex of dimension r we have* $\mu_L(\Delta) = r! \operatorname{vol}_L(\Delta)$.

Often the multiplicity of Δ is called its *normalized volume*. We can also pass to the monoid over the L-simplex $\Delta \subset \mathbb{R}^d$ and use it to measure the multiplicity of Δ. If $v_0, \dots, v_r$ are the vertices of Δ, then $M(\Delta)$ is generated by the vectors $v_i' = (v_i, 1) \in \mathbb{R}^{d+1}$, $i = 0, \dots, r$. Together with 0 they span again a simplex Δ'. Its multiplicity must be taken with respect to the lattice over L, namely the sublattice L' of $\mathbb{R}^{d+1}$ generated by the vectors $(x, 1)$, $x \in L$. We will write $L' = \mathbb{Z}(L, 1)$. (If $0 \in L$, then $L' = L \oplus \mathbb{Z}$.)

Proposition 2.46. *With the notation introduced, we have* $\mu_L(\Delta) = \mu_{L'}(\Delta')$.

Proof. We can assume that $v_0 = 0$ since all data are invariant with respect to the choice of origin in the affine space containing Δ. Next we are allowed to assume that $\operatorname{aff}(\Delta) = \mathbb{R}^r$, and finally that $L = \mathbb{Z}^r$. Then the L-volume is just the euclidean volume on $\mathbb{R}^r$, and $L' = \mathbb{Z}^{r+1}$. Clearly

$$\begin{aligned} \mu(\Delta) &= r! \operatorname{vol}(\Delta) = |\det(v_1, \dots, v_r)| \\ &= |\det(v_0', \dots, v_r')| = (r+1)! \operatorname{vol}(\Delta') = \mu(\Delta'). \end{aligned} \tag{2.2}$$

Alternatively, one establishes a bijection between $E = \operatorname{par}(v_1, \dots, v_r) \cap \mathbb{Z}^r$ and $E' = \operatorname{par}(v_0', \dots, v_r') \cap \mathbb{Z}^{r+1}$ using $v_0' = (0, 1)$ to lift the elements of E to their proper heights in E'. □

Example 2.47. Let $\Delta \subset \mathbb{R}^n$ be the convex hull of the vectors $x_0 = 0, x_1 = e_1, \dots, x_{n-1} = e_{n-1}$ (here e_i is the ith unit vector) and $x_n = (1, 1, \dots, 1, n)$. Then we have $\mu(\Delta) = n$ (with respect to $L = \mathbb{Z}^n$), as formula (2.2) shows. We consider the simplicial cone generated by $x_i' = (x_i, 1)$. Then $\operatorname{par}(x_1', x_2', \dots, x_n')$ contains exactly n lattice vectors, namely $0 \in \mathbb{R}^{n+1}$ and

$$\begin{aligned} &(1, 1, \dots, 1, a, n-a), && 1 \leq a \leq n/2, \\ &(1, 1, \dots, 1, a, n-a+1), && n/2 < a < n. \end{aligned}$$

The reader should compute this example, including all details. In Example 2.58 it will be used again.

In the situation of Proposition 2.43 the monoid $C \cap L$ depends only on C (and L), and if one wants to determine $\operatorname{Hilb}(C \cap L)$ it is certainly advisable to take $v_1, \dots, v_r$ in such a way that $\#E$ becomes as small as possible. There is a unique such choice: for each extreme ray R of (an arbitrary rational cone) C the monoid $R \cap L$ is normal and of rank 1. Therefore it is generated by a single element e; we call e an *extreme L-generator* of C, or an *extreme integral generator* if $L = \mathbb{Z}^d$. It is obvious that the extreme L-generators are irreducible elements of $C \cap L$ and therefore are elements of the Hilbert basis.

Definition 2.48. Let C be a simplicial rational cone. The simplex $\Delta_L(C)$ with vertices in 0 and the extreme L-generators of C is called the *basic L-simplex* of C. The *L-multiplicity* of C is $\mu_L(C) = \mu_L(\Delta_L(C))$.

Simplices and simplicial cones with the smallest possible multiplicity have a special name:

Definition 2.49. An L-simplex Δ is *unimodular* if $\mu_L(\Delta) = 1$. A simplicial rational cone C is *unimodular* (with respect to L) if $\Delta_L(C)$ is unimodular.

Proposition 2.50. *Let $\Delta = \operatorname{conv}(x_0, \dots, x_r)$ be an L-simplex where $L \subset \mathbb{R}^d$ is an affine lattice, and let $L' = \mathbb{Z}(L, 1)$. Then the following are equivalent:*

(a) Δ is unimodular;
(b) the sublattice generated by $x_1 - x_0, \dots, x_r - x_0$ is a direct summand of $L - x_0$;
(c) the submonoid of L' generated by $x_i' = (x_i, 1)$, $i = 0, \dots, r$, is integrally closed in L';
(d) the cone $C(\Delta)$ is unimodular with respect to L'.

Cones over polytopes. We generalize Proposition 2.43 in two steps, first to cones $C(P)$ where P is an L-polytope, and later on to arbitrary rational cones.

There is a simple strategy of how to make Proposition 2.43 useful for $C(P)$: we triangulate P, consider the simplicial cone over each simplex, and unite the systems of generators obtained in this way. Let us first introduce some special types of triangulations:

Definition 2.51. An L-polytope P is *empty* (with respect to L) if it contains no elements from L other than its vertices. A triangulation Σ of an L-polytope P is *full* (with respect to L) if $\operatorname{vert}(\Sigma) = \operatorname{lat}(P)$, or, equivalently, all its simplices are empty.

A triangulation Σ of an L-polytope is *unimodular* (with respect to L) if all simplices in Σ are unimodular.

It is immediately observed that every unimodular triangulation is full: a unimodular L-simplex contains no point of L different from its vertices, and every point $x \in \operatorname{lat}(P)$ must be contained in one of the simplices of Σ.

The natural degree on $\mathbb{Z}(L, 1) \subset \mathbb{R}^{d+1}$ associated with a lattice $L \subset \mathbb{R}^d$ is given by the last coordinate. If P is an L-polytope, then the natural generators of the polytopal monoid $M(P)$ have degree 1. (Polytopal monoids have been introduced in Definition 2.18.)

Theorem 2.52. *Let P be an L-polytope and $C = C(P)$. Then the integral closure $\widehat{M}(P) = C \cap \mathbb{Z}(L,1)$ of $M(P)$ in $\mathbb{Z}(L,1)$ is generated as an $M(P)$-module by elements of degree $\leq \dim P - 1$.*

Proof. We choose a full triangulation Σ of P, which exists by Theorem 1.51. For each simplex $\Delta \in \Sigma$ we apply Proposition 2.43 to the cone $C(\Delta)$. The union of the corresponding sets E_Δ evidently generates $\widehat{M}(P) = \bigcup_\Delta \widehat{M}(\Delta)$ over $M(P)$.

It remains to show that each element $y \in E_\Delta$ has degree at most $\dim P - 1$. Let $x_1, \dots, x_r$, $r = \dim P + 1$, be the vertices of Δ. Then $y = q_1 x_1' + \cdots + q_r x_r'$, $0 \leq q_i < 1$, $x_i' = (x_i, 1)$, $i = 1, \dots, r$. Clearly

$$\deg y = q_1 + \cdots + q_r < r = \dim P + 1,$$

and if we had chosen a coarse triangulation of Δ we might not have been able to reach the bound $\dim P - 1$. However, the basic simplices of the cones $C(\Delta)$ are empty. If $\deg y = \dim P$, then $y' = (x_1' + \cdots + x_r') - y$ has degree 1 and belongs to the basic simplex of $C(\Delta)$ (more precisely, to its face opposite of 0). This is a contradiction, since $y' \neq x_i'$ for all i. □

Remark 2.53. In the proof of Theorem 2.52 we have constructed a system of generators of $\widehat{M}(P)$ as the union of the sets E_Δ. Note that

$$\sum_\Delta \#E_\Delta = \sum_\Delta \operatorname{vol}_L(\Delta)(\dim P)! = \operatorname{vol}_L(P)(\dim P)!$$

is independent of the triangulation. We call $\mu_L(P) = \operatorname{vol}_L(P)(\dim P)!$ the *multiplicity* of P (with respect to L). Theorem 6.54 will show that the multiplicity of P can indeed be interpreted as the multiplicity of a graded algebra, and this will finally justify the terminology.

Exercise 2.19 contains a variant of Theorem 2.52 taking into account the degrees of the elements in $\operatorname{int}(\widehat{M}(P))$. For $\dim P \leq 2$ the proposition has a strong consequence:

Corollary 2.54. *Let P be an L-polytope of dimension ≤ 2. Then $M(P)$ is integrally closed in $\mathbb{Z}(L,1)$.*

An empty L-simplex of dimension ≤ 2 is unimodular, and every full triangulation of an L-polytope of dimension ≤ 2 is unimodular.

Remark 2.55. In dimension $d \geq 3$ an empty lattice simplex need not be unimodular. For example, let $d = 3$ and consider the simplex

$$\Delta_{pq} = \operatorname{conv}\big((0,0,0), (0,1,0), (0,0,1), (p,q,1)\big)$$

with coprime integers $0 < q < p$. Then Δ_{pq} is empty (with respect to $\mathbb{Z}^3$), but has multiplicity p. It has been proved by White [377] that every empty 3-simplex is isomorphic to Δ_{pq} for some p and q, and that $\Delta_{pq} \cong \Delta_{uv}$ if and only if $p = u$ (an obviously necessary condition) and $v = q$ or $v = p - q$.

A similar classification of empty simplices in higher dimension is unknown. The classification in dimension 3 follows readily if one has shown that there exists a $\mathbb{Z}$-linear form α and an integer m such that Δ is sitting between the hyperplanes given by $\alpha(x) = m$ and $\alpha(x) = m+1$: Δ has *lattice width* 1. This is no longer true in dimension > 3. See Haase and Ziegler [174] and Sebő [313] for a discussion of empty simplices and further references.

Example 2.56. (a) Let Δ_{pq} be the simplex introduced in the preceding remark, $p > 1$. Then the basic simplex of $C(\Delta_{pq})$ is not unimodular, and the monoid $M(\Delta_{pq})$ is generated by the vectors $(x_i, 1)$ where x_i, $i = 1, \dots, 4$, runs through the vertices of Δ_{pq}. By Proposition 2.50 $M(\Delta_{pq})$ is not integrally closed in $\mathbb{Z}^4$. Thus $M(P)$ need not be integrally closed if $\dim P \geq 3$. However, note that Δ_{pq}, being empty, is unimodular with respect to $\mathscr{L}(\Delta_{pq})$. Therefore $M(\Delta_{pq})$ is normal.

(b) Let $v_1, \dots, v_4$ be the vertices of $\Delta = \Delta_{21}$ and $w = (v_1 + \dots + v_4)/2$. Furthermore we set $u_{ij} = (v_i + v_j)$, $1 \leq i \leq j \leq 4$. Then the u_{ij} and w are the lattice points of 2Δ, an integrally closed polytope according to Corollary 2.57 below. Now we form a 4-dimensional polytope P as the convex hull of $F_0 = (\Delta, 0)$ and $F_1 = (2\Delta, 1)$. So Δ and 2Δ are two parallel facets of P.

Set $M = M(P)$. A run of normaliz [74] shows that $\operatorname{Hilb}(\widehat{M})$ consists of the 15 elements corresponding to the lattice points of P, namely $(v_i, 0, 1)$, $(u_{ij}, 1, 1)$, $(w, 1, 1)$, and one more additional element $(w, 0, 2) \notin (P, 1)$. (This element must appear since $M(\Delta)$ is not integrally closed.) We claim that M is seminormal (see Definition 2.39) but not normal. In fact, suppose that $2x, 3x \in M$. Then $x \in \widehat{M}$, and so x can be represented by the 16 elements of $\operatorname{Hilb}(\widehat{M})$. If $(w, 0, 2)$ appears with the coefficient 0, then $x \in M$. If $2x, 3x$ in $\widehat{M}(F_0)$, then $x \in \operatorname{gp}(M(F_0))$, and so $x \in M(F_0) \subset M(P)$ since $M(F_0)$ is normal (albeit not integrally closed). The only possibility remaining is that $(w, 0, 2)$ and at least one of the generators $(u_{ij}, 1, 1)$ or $(w, 1, 1)$ of $M(F_1)$ appear in the presentation of x with positive coefficients. But since $(w, 0, 2) + (u_{ij}, 1, 1) \in M(P)$ as well as $(w, 0, 2) + (w, 1, 1) \in M(P)$ (as the reader may check), we finally conclude that $x \in M(P)$. Therefore the polytopal monoid M is seminormal. It is not normal since $\operatorname{gp}(M) = \mathbb{Z}^4$, and so $\overline{M} = \widehat{M} \neq M$.

(c) There exist 3-dimensional non-(semi)normal polytopes P. Let

$$P = \{x \in \mathbb{R}^3_+ : x_1/2 + x_2/3 + x_3/5 \leq 1\}.$$

Then P is a simplex with vertices $(0,0,0)$, $(2,0,0)$, $(0,3,0)$, and $(0,0,5)$. Obviously $\mathbb{Z}^3$ is the smallest lattice containing the vertices of P. The monoid $M = M(P)$ is not normal: $(1,2,4,2)$ belongs to $C(P)$, but not to M. Already in M_* normality is violated, so M is not even seminormal (see Proposition 2.40).

We refer the reader to [59] for a detailed discussion of the normality of simplices like P.

Nevertheless, Corollary 2.54 can be generalized if we consider multiples of lattice polytopes.

Corollary 2.57. *Let P be an L-polytope and $v \in \operatorname{vert}(P)$. Then $M(cP)$ is integrally closed in $\mathbb{Z}((c-1)v + L, 1)$ and, hence, normal for $c \in \mathbb{N}$, $c \geq \dim P - 1$.*

Proof. After a parallel translation of P we can assume that $v = 0$. Then the surrounding lattice is simply $L \oplus \mathbb{Z}$.

The integral closure of $M(cP)$ in $L \oplus \mathbb{Z}$ is isomorphic to the intersection of the integral closure of $M(P)$ with $L \oplus c\mathbb{Z}$. Under the isomorphism $L \oplus \mathbb{Z} \to L \oplus c\mathbb{Z}$, $(x,h) \mapsto (x,ch)$ the set of generators $E(cP)$ of $M(cP)$ is identified with the set of degree c elements in $\widehat{M}(P)$. Therefore we must show that every element x in $\widehat{M}(P)$ of degree nc, $n \in \mathbb{Z}_+$, is the sum of elements of degree c. There is nothing to show for $n \le 1$. Suppose that $n \ge 2$. Then, by Theorem 2.52, $x = x' + x''$ where $x' \in M(P)$ and $\deg x'' \le c$. Since x' decomposes into a sum of degree 1 elements, we can modify the representation of x to $x = x_1 + x_2$ where x_2 has degree c. The summand x_1 can be subdivided into a sum of degree c elements. □

Example 2.58. Suppose that $P \in \mathbb{R}^n$ is a lattice polytope with $z = 0 \in \operatorname{vert}(P)$. Then $M(P)[-(z,1)] = M' \oplus \mathbb{Z}$ where M' is the submonoid of $\mathbb{Z}^n$ generated by the lattice points in P (considered as vectors). If $M(P)$ is integrally closed in $\mathbb{Z}^{n+1}$, then the monoid M' is integrally closed in $\mathbb{Z}^n$. In other words, the monoid $C \cap \mathbb{Z}^n$, $C = \mathbb{R}_+ P$, is generated by the vectors $x \in C \cap P$.

We apply this argument to $P = c\Delta$ where Δ is the simplex constructed in Example 2.47, $n \ge 3$, $c \le n-2$. The vector $w = (1,\dots,1)$ belongs to C, but it is not contained in M'. First, it is not contained in P, since one has $\sum a_i = n-1$ in a representation $w = a_1x_1 + \dots + a_nx_n$ (take $a = 1$ in 2.47), and, moreover, every lattice vector in C with last coordinate 1 has all its coordinates positive: w is irreducible in $C \cap \mathbb{Z}^n$.

This example shows that Theorem 2.52 or Corollary 2.57 cannot be improved. It is not even possible to improve the bound $\dim P - 1$ if one replaces "integrally closed in $L \oplus \mathbb{Z}$" by "normal." In fact, $\operatorname{gp}(M(c\Delta)) = \mathbb{Z}^{n+1}$ for $c \ge 2$, as the reader may show (take $a = n-1$ in 2.47).

We transfer the attributes "integrally closed" and "normal" from monoids to polytopes:

Definition 2.59. Let P be an L-polytope. We say that P is *L-integrally closed* if $M(P)$ is integrally closed in $\mathbb{Z}(L,1)$ and that it is *normal* if $M(P)$ is normal.

The integral closedness of 2-dimensional lattice polytope P has been derived from the fact that the simplices in a full triangulation of P are unimodular. More generally, by the same argument one has the following:

Proposition 2.60. *Let P be an L-polytope that is covered by its unimodular subsimplices. Then P is integrally closed.*

In the next section and in Chapter 3 we will discuss to what extent the proposition can be reversed: does the integral closedness of P imply the existence of a triangulation into unimodular simplices or at least that P is covered by its unimodular subsimplices? What can be said about cP in this respect?

Pointed rational cones. It is not difficult to generalize the arguments above to arbitrary pointed rational cones. First we have to find a replacement for the polytope P, or rather $(P,1)$, spanning the cone. Now the notion of bottom introduced in Section 1.C is very useful. We set

$$C'_L = \operatorname{conv}(x \in L \cap C,\ x \neq 0\}.$$

Then $C'_L = \operatorname{conv}(\operatorname{Hilb}(C \cap L)) + C$, and so C'_L is a polyhedron by Theorem 1.27. Its bottom $B_L(C)$ is the union of its bounded faces (since C does not contain a line). Moreover, the line segment connecting 0 with a point in C'_L intersects $B_L(C)$ (all the nonbounded faces are faces of C). Let F be a facet of $B_L(C)$ and H_0 the hyperplane through 0 and parallel to F. Then the L-height above H_0 is a linear form; we denote it by β_F and call it a *basic grading*; the *basic F-degree* of C is the constant value $b_F = \beta_F(x)$, $x \in \operatorname{aff}(F)$.

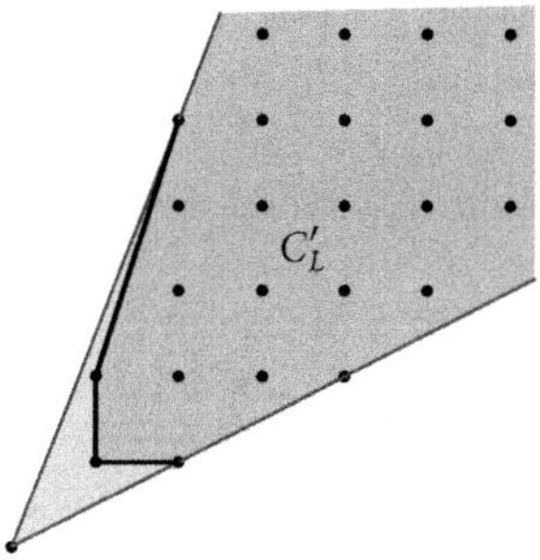

Fig. 2.4. The bottom

Example 2.61. Let $C \subset \mathbb{R}^3$ be generated by $x_1 = (0,1,0)$, $x_2 = (0,0,1)$, and $x_3 = (2,1,1)$. Together with $(0,0,0)$ these vectors span the empty simplex Δ_{21} of multiplicity 2 (see Remark 2.79). Thus the simplicial cone C has a single basic grading, corresponding to the single facet F of the bottom, the triangle spanned by x_1, x_2, x_3. As the reader may check, one has $b_F = 2$.

One needs one more element for $\operatorname{Hilb}(C)$, namely $y = (1,1,1)$: the cones generated by y and two of x_1, x_2, x_3 are unimodular so that C is the union of three unimodular cones.

Proposition 2.62. *Let $C \subset \mathbb{R}^d$ be a pointed rational cone of dimension d, L a lattice in $\mathbb{Q}^d$. Furthermore, let $B = B_L(C) \cap L$ and $M = \mathbb{Z}_+ B$. Then:*

(a) if $x \in C \cap L$, $x \neq 0$, and there exists a facet F of $B_L(C)$ with $\beta_F(x) < 2b_F$, then $x \in \operatorname{Hilb}(C)$; in particular $B \subset \operatorname{Hilb}(C)$;

(b) for each element y in the minimal system of generators G of $C \cap L$ as an M-module there exists a facet F of $B_L(C)$ such that $\beta_F(y) < (d-1)b_F$.

Proof. For (a) it is enough to note that an element $x \in C \cap L$, $x \neq 0$, with $\beta_F(x) < 2b_F$ must be irreducible since b_F is the minimal value of β_F on C'_L.

For (b) we choose a triangulation Σ of the conical complex, formed by the cones over the faces of the bottom $B_L(C)$, such that the rays of Σ constitute the set $\{\mathbb{R}_+x : x \in B\}$ (Theorem 1.54). Let $D \in \Sigma$ be a simplicial cone. Then the basic L-simplex $\Delta_L(D)$ is empty. There is a facet $F \subset B_L(C)$, containing the bottom $B_L(D)$. As in the proof of Theorem 2.52 (where $b_F = 1$) we obtain that $D \cap L$ is generated by elements y with $\beta_F(y) < (d-1)b_F$ as a module over the monoid spanned by $D \cap B$. It only remains to unite all these vectors y to a system of generators of $C \cap L$ over B. □

In dimension 2, the lattice points in the bottom form the Hilbert basis because all the cones over the segments of the unique triangulation Σ of $B_L(C)$ with $\operatorname{vert}(\Sigma) = B_L(C) \cap L$ are unimodular by Corollary 2.54:

Corollary 2.63. *If* $\dim C = 2$, *then* $\operatorname{Hilb}(C \cap L) = B_L(C) \cap L$.

Remark 2.64. There is a remarkable connection between Hilbert bases in dimension 2 and continued fraction expansions, found by van der Corput [360]. Also see Oda [275, 1.6].

In dimension 3, $B_L(C) \cap L$ need not be the Hilbert basis of $C \cap L$, as Example 2.61 shows. Nevertheless one can easily describe the Hilbert basis: $\operatorname{Hilb}(C \cap L)$ consists exactly of $B = B_L(C) \cap L$ and the nonzero elements in the minimal system of generators of $C \cap L$ as a module over $\mathbb{Z}_+B$, as follows immediately from Proposition 2.62. For later application we record

Corollary 2.65. *Let C be a cone of dimension 3, and $D \subset C$ a rational subcone.*
(a) If $B_L(D)$ is contained in $B_L(C)$, then $\operatorname{Hilb}(D \cap L) = D \cap \operatorname{Hilb}(C \cap L)$.
(b) If D is generated by elements $v_1, v_2, v_3 \in B_L(C) \cap L$ that span an empty simplex in a facet F of $B_L(C)$, then

$$\operatorname{Hilb}(D \cap L) = \{v_1, v_2, v_3\} \cup \big(\operatorname{par}(v_1, v_2, v_3) \cap L \setminus \{0\}\big).$$

Proof. (a) An element of $D \cap L$ that is irreducible in $C \cap L$, is certainly irreducible in $D \cap L$. This shows the containment $\supset$, which holds whenever $D \subset C$.

If $\dim D \le 2$, then the converse containment follows from the previous corollary. So let $\dim D = 3$ and consider $x \in \operatorname{Hilb}(D \cap L)$. Then there exists a basic grading β_F of D with $\beta_F(x) < 2b_F$. But β_F is a basic grading of C as well, and so $x \in \operatorname{Hilb}(C)$.

(b) This follows from the argument in the proof of the proposition: $\beta_F(x) < 2b_F$ for each $x \in \operatorname{par}(v_1, v_2, v_3) \cap L$. □

We can give up some precision in measuring the size of the elements of $\operatorname{Hilb}(C)$, using only the extreme L-generators of C. Since the bottom $B_L(C)$ is contained in the convex hull of 0 and the extreme generators, we obtain this corollary:

Corollary 2.66. *Let X be the set of extreme L-generators of the cone C. Then* $\operatorname{Hilb}(C \cap L) \subset (d-1)\operatorname{conv}(0, X)$ *if* $d = \dim C \ge 2$.

Carathéodory rank. So far we have tried to bound the systems of generators or the Hilbert basis in terms of degree. Another type of measure is given by the minimal number of elements in the Hilbert basis of a monoid M that, given $x \in M$, are needed to represent x:

Definition 2.67. Let M be a positive affine monoid. The *representation length* $\rho(x)$ of $x \in M$ is the smallest number k of elements $x_1, \dots, x_k \in \operatorname{Hilb}(M)$ such that x is a $\mathbb{Z}_+$-linear combination of $x_1, \dots, x_k$.

The *Carathéodory rank* of M, $\operatorname{CR}(M)$, is the maximum of the representation lengths $\rho(x)$, $x \in M$.

Since M is finitely generated, $\operatorname{CR}(M)$ is finite, namely $\operatorname{CR}(M) \le \#\operatorname{Hilb}(M)$. Without further hypothesis a better bound for $\operatorname{CR}(M)$ is impossible.

Example 2.68. Let $p_1, \dots, p_n$ be pairwise different prime numbers, and set $q_i = \prod_{j \ne i} p_j$. Then $q_1, \dots, q_n$ are coprime, and every $m \in \mathbb{Z}$, $m \gg 0$ is in the additive monoid generated by $q_1, \dots, q_n$. However, unless m is divisible by one of the prime numbers p_i, all the generators $q_1, \dots, q_n$ appear in every presentation of m.

If M is normal, then one can bound Carathéodory rank by a linear function of rank:

Theorem 2.69 (Sebő). *Let M be a positive normal affine monoid of rank r. Then $\operatorname{CR}(M) = r$ if $r \le 3$, and $r \le \operatorname{CR}(M) \le 2r - 2$ if $r \ge 4$.*

Proof. We identify $\operatorname{gp}(M)$ with $\mathbb{Z}^r$ and set $C = \mathbb{R}_+ M$.

It is easy to see that $\operatorname{CR}(M) \ge r$. In fact, the elements of M that are $\mathbb{Z}$-linear combinations of at most $r-1$ elements of $\operatorname{Hilb}(M)$ are contained in the union of finitely many hyperplanes in $\mathbb{R}^r$, and such a union cannot contain M.

If $r = 1$, then $M \cong \mathbb{Z}_+$, and there is nothing to show. If $r = 2$, then the bottom B of C' (constructed with respect to $L = \mathbb{Z}^2$) decomposes into a sequence of line segments $[x, y]$ such that $(x, y) \cap \mathbb{Z}^2 = \emptyset$. As we have seen in the proof of Proposition 2.62, every lattice point in the cone generated by $[x, y]$ belongs to the monoid generated by x and y. The case $r = 3$ will be discussed in Theorem 2.78.

So let $r \ge 4$ and $x \in M$. As in the proof of 2.62 (and the case $r = 2$ above) we choose a $B_L(C)$-bottom compatible triangulation of C whose rays pass through all lattice points in $B_L(C)$. The line segment $[0, x]$ intersects a facet F of the bottom, and we have seen that we can write $x = x' + x''$ where (i) x' is a linear combination of those (at most r) elements of $B \cap \mathbb{Z}^r$ that generate the smallest cone in Σ containing x, and (ii) $\beta_F(x') \le rb_F - 2$. Therefore x'' can be written as a $\mathbb{Z}_+$-linear combination of at most $r - 2$ elements in $\operatorname{Hilb}(M)$. □

We will continue the discussion of Carathéodory rank in the next section. At this point it is not easy to find monoids M as in the theorem with $\operatorname{CR}(M) > \operatorname{rank} M$.

Remark 2.70. In [58] two variants of CR have been introduced, *asymptotic* and *virtual* Carathéodory rank. The asymptotic Carathéodory rank $\operatorname{CR}^{\mathrm{a}}(M)$ is the smallest number m such that the proportion

$$\frac{\#\{x \in M, \|x\| \le t : \rho(x) > m\}}{\#\{x \in M, \|x\| \le t\}}$$

goes to 0 with $t \to \infty$. Here $\| \ \|$ refers to an arbitrarily fixed norm on the ambient real vector space $M \subset \mathbb{R} \otimes_{\mathbb{Z}} \operatorname{gp}(M)$. The virtual Carathéodory rank $\operatorname{CR}^{\mathrm{v}}(M)$ is the smallest number m such that $\rho(x) \le m$ with only finitely many exceptions.

One can show that $\operatorname{CR}^{\mathrm{a}}(M) \le 2 \operatorname{rank} M - 3$ if M is a normal monoid of rank ≥ 3. For a proof and further results we refer the reader to [58].

2.D Normality and unimodular covering

In the previous sections we used triangulations to construct systems of generators for normal monoids. In this section we will discuss the existence of triangulations into—and, more generally, of covers by unimodular simplices and unimodular cones. While the notions to be introduced below make sense with respect to arbitrary lattice structures, *we will work only with the lattice $\mathbb{Z}^d$ in this section.* This does not restrict the generality in an essential way, but simplifies the formulations somewhat. Consequently, we set $\operatorname{Hilb}(C) = \operatorname{Hilb}(C \cap \mathbb{Z}^d)$ if C is a pointed rational cone.

Definition 2.71. Let $\mathscr{F}$ be a rational fan in $\mathbb{R}^d$ consisting of pointed cones. It is called *unimodular* if all its cones are unimodular.

Theorem 2.72. *Let $\mathscr{F}$ be a rational fan in $\mathbb{R}^d$. Then $\mathscr{F}$ has a regular unimodular triangulation Σ.*

Proof. In the first step $\mathscr{F}$ is triangulated regularly as described in Corollary 1.66 after choosing a rational generating system in each $C \in \mathscr{F}$. By the transitivity of regular subdivisions (the conical analogue of Proposition 1.61) we may then assume that $\mathscr{F}$ consists of rational cones.

It remains to refine the triangulation if there exists a nonunimodular cone $D \in \mathscr{F}$. We use induction on (i) the maximal multiplicity of a d-dimensional cone D in $\mathscr{F}$, and on (ii) the number of such D with maximal multiplicity.

Let $m = \max\{\mu(D) : D \in \mathscr{F}\}$. If $m > 1$, we choose D with $\mu(D) = m$. Let $y_1, \dots, y_d$ be the extreme integral generators of D. Since $M = C \cap \mathbb{Z}^d$ the vectors $y_1, \dots, y_d$ belong to M. By Proposition 2.43 there exists $x \in \operatorname{par}(y_1, \dots, y_d) \cap \mathbb{Z}^d$. Now we apply stellar subdivision by x to $\mathscr{F}$. Figure 2.5 shows a typical situation after two generations of subdivisions in the cross-section of a 3-cone.

In this process every d-dimensional cone $D \in \mathscr{F}$ with $x \in D$ is replaced by the union of subcones $D'' = \mathbb{R}_+ x + D'$, D' a facet of D with $x \notin D'$. Then there exists j such that the y_i with $i \neq j$ generate D'. One has

$$\mu(D'') \le |\det(y_1, \dots, y_{j-1}, x, y_{j+1}, \dots, y_d)| < |\det(y_1, \dots, y_d)| = \mu(D)$$

since the coefficient of y_j in the representation of x is in the interval $[0, 1)$. Since at least one $D \in \mathscr{F}$ with $\mu(D) = m$ is properly subdivided, we are done by induction.

The regularity of the refined triangulation results from the conical analogue of Lemma 1.65 (see also Exercise 1.21). □

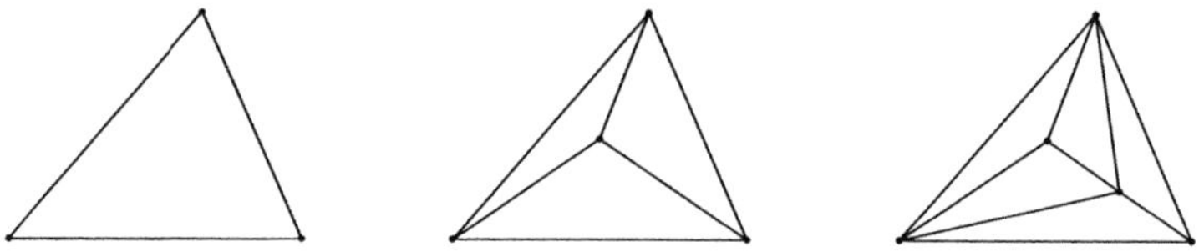

Fig. 2.5. Successive stellar triangulations

In the case in which $\mathscr{F}$ is a subdivision of a (not necessarily pointed) rational cone we can strengthen Theorem 2.72.

Corollary 2.73. *Let C be a rational cone in $\mathbb{R}^d$, and Γ a rational subdivision of C. Then there exists a regular unimodular triangulation of C that refines Γ.*

Proof. By Exercise 1.21 Γ has a rational regular refinement $\mathscr{F}$ to which we can apply Theorem 2.72. The regularity of the resulting triangulation follows from the conical analogue of Proposition 1.61. □

A much weaker consequence that has already been used in the proof of Proposition 2.17 is

Corollary 2.74. *Let C be a rational cone. Then* $\operatorname{int}(C)$ *contains a unimodular subcone.*

For the proof one chooses a simplicial subcone of $\operatorname{int}(C)$ and applies Theorem 2.72 to the fan of its faces. Another byproduct of the theorem is a criterion for the integral closedness of affine monoids:

Theorem 2.75. *Let $M \subset \mathbb{Z}^d$ be a positive affine monoid. Then the following are equivalent:*

(a) M is integrally closed in $\mathbb{Z}^d$;
(b) there exists a (regular) triangulation Σ of $\mathbb{R}_+ M$ such that each $D \in \Sigma$ is a unimodular cone with $\operatorname{Hilb}(D) \subset M$.

Proof. The implication (a) $\implies$ (b) follows immediately from Theorem 2.72 applied to (the fan of faces of) the cone $\mathbb{R}_+M$: the extreme integral generators of the unimodular cones belong to M.

The converse is a (trivial) special case of Theorem 2.38. □

Despite the simplicity of their proofs, 2.72, 2.73, and 2.75 have very important applications, as we will see later on.

Hilbert triangulations and covers. In the previous section we have seen much stronger results for special types of cones. For example, if P is a 2-dimensional lattice polytope, then $C(P)$ has a unimodular triangulation with rays through the Hilbert basis $E(P)$ of $M(P)$. The following definition covers this additional condition.

Definition 2.76. We say that Σ is a *Hilbert triangulation* of a pointed rational cone C if each ray of Σ is of type $\mathbb{R}_+x$ with $x \in \operatorname{Hilb}(C)$. It is a *full* Hilbert triangulation if all rays $\mathbb{R}_+x$ with $x \in \operatorname{Hilb}(C)$ are rays of Σ.

A very strong unimodularity condition for $\operatorname{Hilb}(C)$ is: *Every full Hilbert triangulation of C is unimodular.* All cones of dimension 2 and polytopal cones $C(P)$ with $\dim P = 2$ have this property. Another class is given by the cones over direct products of two unimodular simplices of arbitrary dimension (see Sturmfels [336, Sect. 6]).

Remark 2.77. Relaxing the Hilbert condition, one may ask whether every positive rational cone C contains a finite subset X of $\mathbb{Z}^d$ such that each triangulation using all the rays $\mathbb{R}_+x$, $x \in X$, is unimodular. We leave it to the reader that X has this property if and only if it is *supernormal* in the sense of Hoşten, Maclagan, and Sturmfels [245]: $D \cap X$ contains a Hilbert basis of D for every cone D generated by a subset X' of X. In general, a cone C does not contain a supernormal subset; see [245] for a counterexample.

A substantially weaker and much more important property is

(UHT) *C has a unimodular Hilbert triangulation.*

Theorem 2.78 (Sebő). *Let $C \subset \mathbb{R}^3$ be a positive rational cone. Then C has a unimodular Hilbert triangulation.*

The crucial step in the proof is the following lemma.

Lemma 2.79. *Let v_1, v_2, v_3 be vectors in $\mathbb{Z}^3$, for which $\operatorname{conv}(0, v_1, v_2, v_3)$ is an empty 3-simplex. Furthermore let β be the primitive linear form on $\mathbb{Z}^3$ that has constant value $b > 0$ on v_1, v_2, v_3. Then:*

(a) β is the (unique) basic grading of C;
(b) $\operatorname{Ker} \beta = \mathbb{Z}(v_2 - v_1) + \mathbb{Z}(v_3 - v_1)$;
(c) $b = \mu(C)$;
(d) on $\operatorname{par}(v_1, v_2, v_3) \cap \mathbb{Z}^3$ the linear form β takes exactly the values $0, b+1, \dots, 2b-1$.

Proof. (a) is clear from the choice of β. For (b) we observe that $\mathbb{Z}(v_2-v_1)+\mathbb{Z}(v_3-v_1) \subset \operatorname{Ker}\beta$. On the other hand 0, $v_2 - v_1$, and $v_3 - v_1$ are the vertices of an empty lattice triangle. Such a triangle is unimodular by Corollary 2.54 or, in other words, $v_2 - v_1$ and $v_3 - v_1$ generate a direct summand of $\mathbb{Z}^3$.

Now (c) is clear: since $\operatorname{Ker}\beta \subset U = \mathbb{Z}v_1 + \mathbb{Z}v_2 + \mathbb{Z}v_3$, we have $\mathbb{Z}/\beta(U) \cong \mathbb{Z}^3/U$, and $\mu(C) = \#(\mathbb{Z}^3/U)$. For part (d) one notes that the elements in $\operatorname{par}(v_1, v_2, v_3) \cap \mathbb{Z}^3$ represent the residue class modulo U. On the one hand, they have separate values under β and, simultaneously, only the values given are possible since Δ is empty. □

Proof of Theorem 2.78. We start with a triangulation of the bottom B of C into empty simplices Δ. By Corollary 2.65(a) we have $\operatorname{Hilb}(\mathbb{R}_+\Delta) = \operatorname{Hilb}(C) \cap \mathbb{R}_+\Delta$, and this allows us to replace C by $\mathbb{R}_+\Delta$. In other words, we can assume that C is generated by $v_1, v_2, v_3 \in \mathbb{Z}^3$ for which $\operatorname{conv}(0, v_1, v_2, v_3)$ is an empty simplex.

Assume C is not unimodular. With the notation of the lemma, we choose $w \in \operatorname{par}(v_1, v_2, v_3)$ such that $\beta(w) = b + 1$. Write $w = q_1v_1 + q_2v_2 + q_3v_3$ with rational numbers q_i, $0 \le q_i < 1$. Then $\beta(w) = (q_1 + q_2 + q_3)b$. Furthermore let C_i, $i = 1, 2, 3$ be the cone spanned by w and the v_j with $j \ne i$. Since $\operatorname{conv}(0, v_1, v_2, v_3)$ is empty and $\beta(w)$ is a generator of $\mathbb{Z}^3/U$, all three cones are full-dimensional. We have $\mu(C_i) = q_ib$, and $\sum_{i=1}^{3} \mu(C_i) = b + 1$.

Set $\operatorname{sdiv}(C) = \operatorname{par}(v_1, v_2, v_3) \cap (\mathbb{Z}^3 \setminus \{0\})$. With analogous notation for C_i we claim

$$\operatorname{sdiv}(C) = \{w\} \cup \operatorname{sdiv}(C_1) \cup \operatorname{sdiv}(C_2) \cup \operatorname{sdiv}(C_3).$$

If this equation holds, we are done; the basic simplices of the cones C_i are empty (there is no integer between b and $b+1$), and an application of the same argument to C_i, $i = 1, 2, 3$, shows that all further subdividing vectors can be chosen in $\operatorname{sdiv}(C) \subset \operatorname{Hilb}(C)$ (compare 2.65(b)).

Note that the sets on the right hand side are disjoint. Since $\#\operatorname{sdiv}(C) = b - 1$ and $\#\operatorname{sdiv}(C_i) = q_ib - 1$, the choice of w guarantees that both sides have the same cardinality. Thus it is enough to show that $\operatorname{sdiv}(C)$ is contained in the right hand side R. This holds since $\operatorname{sdiv}(C) \subset \operatorname{Hilb}(C)$ by Corollary 2.65(b), and R, together with v_1, v_2, v_3, generates the monoid $C \cap \mathbb{Z}^3$. □

Remark 2.80. Theorem 2.78 cannot be extended to higher dimension. A polytopal counterexample in dimension 4 can be constructed as follows. We choose an empty 3-simplex Δ_{pq} (see Remark 2.55) with $1 < q < p-1$ (and p, q coprime) and set $P = 2\Delta$. By Corollary 2.57 the polytopal monoid $M(P)$ is integrally closed in $\mathbb{Z}^4$ so that $E(P)$ is a Hilbert basis of $C(P)$. If $C(P)$ admits a unimodular Hilbert triangulation, then P has a triangulation into unimodular simplices. However, such does not exist according to Kantor and Sarkaria [217]. (The first, nonpolytopal counterexample was given by Bouvier and Gonzalez–Sprinberg [40].)

While the example just discussed shows the failure of (UHT) in dimension ≥ 4, it does not exclude that a weaker unimodularity condition is always satisfied, namely:

(UHC) *Let C be a positive rational cone. Then C is the union of the unimodular simplicial cones generated by elements of* $\operatorname{Hilb}(C)$. *That is, a unimodular Hilbert cone.*

As we will see below, (UHC) also fails: there exists an integrally closed polytope P of dimension 5 such that $C(P)$, a cone of dimension 6, violates (UHC); equivalently, P is not covered by its unimodular subsimplices.

While the unimodular subcones generated by elements of $\operatorname{Hilb}(C)$ may fail to cover C, there always exist such unimodular subcones.

Proposition 2.81. *Let C be a pointed rational cone and $0 \subset F_1 \subset \cdots \subset F_d = C$ a complete flag of faces of C. Then there exist $x_1, \dots, x_d \in \operatorname{Hilb}(C)$ with the following properties:*

(a) $\mathbb{R}_+x_1 + \cdots + \mathbb{R}_+x_d$ *is unimodular;*
(b) $x_1, \dots, x_i \in F_i$ *for all* $i = 1, \dots, n$.

Proof. By induction we can assume that $x_1, \dots, x_{d-1} \in F_{d-1}$ have been found such that they satisfy the conditions (a) and (b) for F_{d-1}. Let σ be the support form of C associated with the facet F_{d-1}. Then there exists $y \in C \cap \mathbb{Z}^d$ with $\sigma(y) = 1$. Therefore $\sigma(x) = 1$ for at least one element $x \in \operatorname{Hilb}(C)$. Since $\operatorname{Hilb}(C) \cap F_i = \operatorname{Hilb}(F_i)$, the choice $x_d = x$ satisfies our needs. □

Isomorphism classes of lattice polytopes with given multiplicity. One interesting consequence of Proposition 2.81 (together with Proposition 2.57) is the following finiteness result for polytopes of bounded multiplicity.

Corollary 2.82. *Let d and μ be natural numbers. Then there are only finitely many isomorphism classes of lattice polytopes $P \subset \mathbb{R}^d$ such that $\mu(P) \le \mu$.*

Proof. In view of the definition of morphism of lattice polytopes (see Definition 1.73), we can assume that $\operatorname{aff}(P) = \mathbb{R}^d$, replacing $\mathbb{R}^d$ by $\operatorname{aff}(P)$ and $\mathbb{Z}^d$ by $\operatorname{aff}(P) \cap \mathbb{Z}^d$ if necessary.

It is sufficient to consider only normal d-polytopes P with $\operatorname{gp}(M(P)) = \mathbb{Z}^{d+1}$. In fact, the $(d-1)$th multiple of a lattice d-polytope in $\mathbb{R}^d$ always satisfies these conditions (Corollary 2.57) and, therefore, the multiplication by $d-1$ injects the class of lattice polytopes P in $\mathbb{R}^d$ of multiplicity $\le \mu$ into that of normal lattice polytopes Q in $\mathbb{R}^d$ of multiplicity $\le \mu(d-1)^d$ with $\operatorname{gp}(M(Q)) = \mathbb{Z}^{d+1}$. We leave it to the reader to show that P and P' are isomorphic if and only if $(d-1)P$ and $(d-1)P'$ are isomorphic.

According to Proposition 2.81 every polytope P from our class contains a unimodular d-simplex. This simplex is isomorphic to $\Delta = \operatorname{conv}(0, e_1, \dots, e_d)$, and therefore we can assume that $\Delta \subset P$. But then the condition $\mu(P) \le \mu$ implies that P is contained in the cube

$$\{(x_1, \dots, x_d) : |x_i| \le \mu,\ i = 1, \dots, d\} \subset \mathbb{R}^d. \qquad \square$$

See Bárány and Vershik [15] for an asymptotically sharp bound on the number of isomorphism classes, settling a conjecture of Arnold.

The integral Carathéodory property. If a pointed rational cone $C \subset \mathbb{R}^d$ has (UHC), then it satisfies the *integral Carathéodory property*, a name explained by the obvious analogy to Carathéodory's theorem 1.55:

(ICP) $\operatorname{CR}(C \cap \mathbb{Z}^d) = \dim C$.

Below we will give a counterexample to (ICP) (and therefore to (UHC) as well) and a counterexample to (UHC) that satisfies (ICP). The validity of (UHC) or (ICP) seems to be open in dimensions 4 and 5.

We can formulate all the conditions above for positive affine monoids M and their Hilbert bases: M has (UHT) if $\mathbb{R}_+M$ has a triangulation Σ whose simplices are generated by unimodular subsets of $\operatorname{Hilb}(M)$ (with respect to $\operatorname{gp}(M)$). In a similar manner, (UHC) is transferred to monoids, and (ICP) simply means that $\operatorname{CR}(M) = \operatorname{rank} M$. While the flexibility of the monoid language will be useful in the next subsection, it does not lead to a proper generalization. In fact, the weakest of these properties implies the normality of M:

Proposition 2.83. *Let M be an affine monoid and X a finite subset of M such that every element of M can be represented as a $\mathbb{Z}_+$-linear combination of $r = \operatorname{rank} M$ elements of X. Then M is normal, and every element of M is a $\mathbb{Z}_+$-linear combination of linearly independent elements of X.*

In particular, M is normal if $\operatorname{CR}(M) = r$.

Proof. We consider the submonoids N of M generated by at most r elements of X. If N has rank r, then its Hilbert basis is linear independent, and so N is normal.

By hypothesis, M is the union of all these submonoids N. By Theorem 2.38 it is normal and the union of those N that have rank r. □

Tight cones. We introduce the class of tight cones and monoids, and show that they play a crucial role for (UHC) and (ICP).

Definition 2.84. Let M be a positive normal affine monoid, $x \in \operatorname{Hilb}(M)$, and M' the monoid generated by $\operatorname{Hilb}(M) \setminus \{x\}$. We say that x is *nondestructive* if M' is normal and $\operatorname{gp}(M')$ is a direct summand of $\operatorname{gp}(M)$ (and therefore equal to $\operatorname{gp}(M)$ if $\operatorname{rank}\operatorname{gp}(M) = \operatorname{rank}\operatorname{gp}(M')$). Otherwise x is *destructive*. We say that M is *tight* if every element of $\operatorname{Hilb}(M)$ is destructive. A pointed rational cone $C \subset \mathbb{R}^d$ is *tight* if $C \cap \mathbb{Z}^d$ is tight.

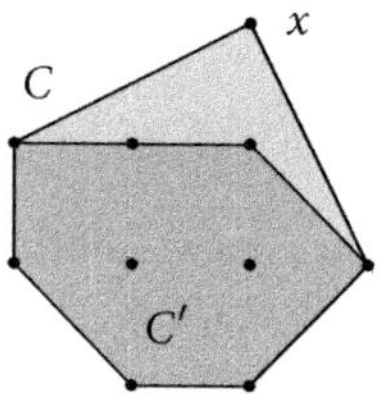

Fig. 2.6. Cross-section of shrinking of a cone

It is clear that only extreme elements of $\operatorname{Hilb}(M)$ can be nondestructive. Suppose that x is an extreme element of $\operatorname{Hilb}(M)$. Then the localization $M[-x]$ splits into a product $\mathbb{Z}x \oplus M_x$ where M_x is again a positive normal affine monoid (see Proposition 2.32).

Lemma 2.85. *Let M be a normal positive affine monoid and $x \in \operatorname{Hilb}(M)$ a nondestructive element. Let M' be the monoid generated by $\operatorname{Hilb}(M) \setminus \{x\}$.*

(a) *If M' and M_x both satisfy* (UHC), *then so does M.*
(b) *One has* $\operatorname{CR}(M) \le \max(\operatorname{CR}(M'), \operatorname{CR}(M_x) + 1)$.

Proof. Suppose M' and M_x both satisfy (UHC). Since $\operatorname{gp}(M')$ is a direct summand of $\operatorname{gp}(M)$ and $\operatorname{Hilb}(M') = \operatorname{Hilb}(M) \setminus \{x\}$ by the hypothesis on x, it is clear that all elements of M' are contained in submonoids of M generated by unimodular subsets X_i of $\operatorname{Hilb}(M)$ (with respect to $\operatorname{gp}(M)$). We have to show that this holds for elements of $M \setminus M'$, too.

Let $z \in M \setminus M'$. By hypothesis on M_x, the residue class of z in M_x has a representation $\bar{z} = a_1\bar{y}_1 + \cdots + a_m\bar{y}_m$ with $a_i \in \mathbb{Z}_+$ and $\bar{y}_i \in \operatorname{Hilb}(M_x)$ for $i = 1, \dots, m$ such that $\bar{y}_1, \dots, \bar{y}_m$ span a direct summand of $\operatorname{gp}(M_x)$. Next observe that $\operatorname{Hilb}(M)$ is mapped onto a system of generators of M_x by the residue class map. Therefore we may assume that the preimages $y_1, \dots, y_m$ belong to $\operatorname{Hilb}(M) \setminus \{x\}$. Furthermore, $z = a_1y_1 + \cdots + a_my_m + bx$ with $b \in \mathbb{Z}$.

It only remains to show that $b \in \mathbb{Z}_+$. There is a representation of z as a $\mathbb{Z}_+$-linear combination of the elements of $\operatorname{Hilb}(M)$ in which the coefficient of x is positive. Thus, if $b < 0$, z has a $\mathbb{Q}_+$-linear representation by the elements of $\operatorname{Hilb}(M) \setminus \{x\}$. This implies $z \in \mathbb{R}_+M'$, and hence $z \in M'$, a contradiction.

This proves (a), and (b) follows similarly. □

For a partial converse of the inequality in (b) see Exercise 2.31. We say that a monoid M as in the lemma *shrinks* to the monoid T if there is a chain $M = M_0 \supset M_1 \supset \cdots \supset M_t = T$ of monoids such that at each step M_{i+1} is generated by $\operatorname{Hilb}(M_i) \setminus \{x\}$ where x is nondestructive. An analogous terminology applies to cones.

Corollary 2.86. *A counterexample to* (UHC) *that is minimal, first with respect to dimension and then with respect to* $\# \operatorname{Hilb}(C)$, *is tight. A similar statement holds for* (ICP).

In fact, suppose that the cone C is a minimal counterexample to (UHC) with respect to dimension, and that C shrinks to D. Then D is also a counterexample to (UHC) according to Lemma 2.85. For (ICP) the argument is the same. It is therefore clear that one should search for counterexamples only among the tight cones. In fact, experiments have shown that such counterexamples are extremely rare, and without the restriction to tight cones such may never have been found.

Remark 2.87. It is not hard to see that there are no tight cones of dimension ≤ 2. However, tight cones exist in all dimensions ≥ 3. The first such example in dimension 3 was found by P. Dueck. The smallest example found by the authors has a Hilbert basis of 19 elements. Its cross-section is a regular hexagon (with respect to the action of the group of 3×3 invertible integer matrices). In dimension 4 there exist plenty of tight cones but none of the examples we have found is of the form C_P with a 3-dimensional lattice polytope P. In dimension ≥ 5 one can easily describe a class of tight cones; see Exercise 2.27.

The shrinking of cones and tests for (UHC) and (ICP) have been implemented as computer programs. See [53], [58], and [64] for a description of the search for counterexamples and the algorithms used for testing (UHC) and (ICP).

Counterexamples to (UHC) and (ICP). Let $C_6 \subset \mathbb{R}^6$, be the cone generated by the vectors $z_1, \dots, z_{10}$:

$$\begin{aligned} z_1 &= (0, 1, 0, 0, 0, 0), & z_6 &= (1, 0, 2, 1, 1, 2), \\ z_2 &= (0, 0, 1, 0, 0, 0), & z_7 &= (1, 2, 0, 2, 1, 1), \\ z_3 &= (0, 0, 0, 1, 0, 0), & z_8 &= (1, 1, 2, 0, 2, 1), \\ z_4 &= (0, 0, 0, 0, 1, 0), & z_9 &= (1, 1, 1, 2, 0, 2), \\ z_5 &= (0, 0, 0, 0, 0, 1), & z_{10} &= (1, 2, 1, 1, 2, 0). \end{aligned}$$

The cone C_6 and the monoid $M_6 = C_6 \cap \mathbb{Z}^6$ have several remarkable properties (see [68] for more detailed information):

(a) $\operatorname{Hilb}(C_6) = \{z_1, \dots, z_{10}\}$.
(b) C_6 has 27 facets, of which 5 are not simplicial.
(c) The automorphism group $\Sigma(M_6)$ of M_6 is the Frobenius group of order 20, the semidirect product of $\mathbb{Z}_5$ with its automorphism group $\mathbb{Z}_5^* \cong \mathbb{Z}^4$. It operates transitively on $\operatorname{Hilb}(M_6)$. In particular, this implies that $z_1, \dots, z_{10}$ are all extreme generators of C_6.
(d) The embedding $C_6 \subset \mathbb{R}^6$ above has been chosen in order to make visible the subgroup U of those automorphisms that map each of the sets $\{z_1, \dots, z_5\}$ and $\{z_6, \dots, z_{10}\}$ to itself; U is isomorphic to the dihedral group of order 10. However, C_6 can even be realized as the cone over a 0-1-polytope in $\mathbb{R}^5$.
(e) The vector of lowest degree disproving (UHC) is $t_1 = z_1 + \cdots + z_{10}$. Evidently t_1 is invariant for $\Sigma(M_6)$, and it can be shown that its multiples are the only such elements.
(f) The Hilbert basis is contained in the hyperplane H given by the equation $-5\zeta_1 + \zeta_2 + \cdots + \zeta_6 = 1$. Thus $z_1, \dots, z_{10}$ are the vertices of a normal 5-dimensional lattice polytope P_5 that is not covered by its unimodular lattice subsimplices (and contains no other lattice points).
(g) If one removes all the unimodular subcones generated by elements of $\operatorname{Hilb}(C_6)$ from C_6, then there remains the interior of a convex cone N. While P_5 has volume $25/120$, the intersection of N and P_5 has only volume $1/(1080 \times 120)$.
(h) The vector
$$z_1 + 3z_2 + 5z_4 + 2z_5 + z_8 + 5z_9 + 3z_{10}$$
cannot be represented by 6 elements of $\operatorname{Hilb}(M)$ (and it is smallest with respect to this property.) Moreover, one has $\operatorname{CR}(C_6) = 7$ (as can be seen from a triangulation containing only two nonunimodular simplices).

In particular, C_6 is even a counterexample to (ICP). If one adds the vectors

$$z_{11} = (0, -1, 2, -1, -1, 2) \quad \text{and} \quad z_{12} = (1, 0, 3, 0, 0, 3),$$

one obtains a tight cone C_6' that satisfies (ICP), but fails (UHC). Remarkably, the Frobenius group of order 20 is the automorphism group of the monoid $M_6' = C_6' \cap \mathbb{Z}^6$, too, but its action on C_6' does not extend the action on M_6! Only a subgroup of order 4 restricts to M_6. Therefore M_6 has 5 conjugate embeddings into C_6', each of which contains $\{z_1, z_5\}$.

Exercises

2.1. Give an example of a monoid with torsion for which $\operatorname{gp}(M)$ is torsionfree.

2.2. Show that every submonoid of $\mathbb{Z}$ is finitely generated.

2.3. The notions of interior and extreme submonoid of a monoid M can be defined without reference to convex geometry. In this exercise we collect the corresponding assertions according to Swan [351, Sect. 5]. Let M be an affine monoid. Prove:

(a) A submonoid E of M is extreme if and only if $x + y \in E$ implies $x, y \in E$ for all $x, y \in M$.

(b) $\operatorname{int}(M) = \{x \in M : \text{for every } y \in M \text{ there exist } n \in \mathbb{N} \text{ and } z \in M \text{ such that } y + z = nx\}$.

For arbitrary monoids one can now use the properties in (a) and (b) to define extreme submonoids and the interior. So let M be an arbitrary monoid. Show:

(c) Every maximal submonoid of $M \setminus \operatorname{int}(M)$ is extreme.

(d) If E is an extreme submonoid and N an arbitrary submonoid. If $E \cap \operatorname{int}(N) \neq \emptyset$, then $N \subset E$.

2.4. Prove the following claims for an affine monoid M:

(a) $M = M_*$ if and only if $\operatorname{rank} M \le 1$.

(b) If $M \neq M_*$, then M_* is not affine.

2.5. Let M be an affine monoid M. Show that the following are equivalent:

(i) M is a group;
(ii) the normalization $\bar{M}$ is a group;
(iii) $\mathbb{R}_+ M$ is a vector space.

2.6. Show that the intersection of $\mathbb{Z}^2$ with the cone generated by $(1,0)$ and $(1,a)$ is a finitely generated monoid if and only if a is rational.

2.7. Let M be a normal affine monoid and $x \in M$. Then $M[-x] \cong \mathbb{Z}^e \oplus M'$ as discussed in Proposition 2.32. Let $\sigma_1, \dots, \sigma_e$ be those support forms of M that vanish on x. Show that the linear map $y \mapsto (\sigma_1(y), \dots, \sigma_t(y))$, $y \in \operatorname{gp}(M)$, restricts to the standard map on M'.

2.8. Show by means of an example that the additional conditions in Proposition 2.17(d) and (e) cannot always be satisfied simultaneously. (Rank 2 is enough.)

2.9. Show that $\mathfrak{c}(\bar{M}/M)$ is the largest ideal of M that is also an ideal of $\bar{M}$.

2.10. Let M be an affine monoid and $x \in \operatorname{int}(M)$. Show:

(a) $mx \in \mathfrak{c}(\bar{R}/R)$ for $m \gg 0$;

(b) $M[-x] = \operatorname{gp}(M)$.

Hint: (a) reduces (b) to the case in which $M = \bar{M}$.

2.11. Let M be a monoid. Show that $\operatorname{sn}(M)$ consists of all $x \in \operatorname{gp}(M)$ such that $mx, nx \in M$ for some coprime $m, n \in \mathbb{Z}_+$.

2.12. Let M be an affine monoid and let N be an overmonoid of M contained in $\operatorname{sn}(M)$. Show that there exists a chain

$$M = M_0 \subset M_1 \subset \cdots \subset M_n = N$$

of monoids such that $M_i = M_{i-1} \cup (M_{i-1} + x)$ for some x with $2x, 3x \in M_{i-1}$, $i = 1, \dots, n$.

Hint: first construct a strictly ascending chain of extensions as required, and note that M-submodules of $\bar{M}$ satisfy the ascending chain condition.

(In ring-theoretic terminology the extension $M_{i-1} \subset M_i$ is called *elementary subintegral* and the extension $M \subset N$ is *subintegral*; see p. 157.)

2.13. Let M be a positive normal affine monoid. We call M *Gorenstein* if there exists $x \in M$ such that $\operatorname{int}(M) = x + M$. (The nomenclature will be justified in Remark 6.34.) Show that M is Gorenstein if and only if there exists $x \in M$ such that $\sigma_F(x) = 1$ for each facet F of $\mathbb{R}_+M$ and the associated support form σ_F, in which case $\operatorname{int}(M) = x + M$.

2.14. Let M be as in Exercise 2.13 and suppose that $\operatorname{gp}(M) = \mathbb{Z}^d$. The *dual* monoid M^* is the set of all linear forms $\gamma \in (\mathbb{Z}^d)^*$ such that $\gamma(y) \geq 0$ for all $y \in M$. (Clearly, $M^* = (\mathbb{R}_+M)^* \cap (\mathbb{Z}^d)^*$.) Show the following are equivalent:

(i) M is polytopal;
(ii) M^* is Gorenstein.

Moreover, prove $M = M^{**}$.

Hint: consider the linear form on $\mathbb{Z}^d$ with respect to which M is polytopal.

2.15. Let $P \subset \mathbb{R}^d$ a lattice polytope containing 0 as an interior lattice polytope. One says P is *reflexive* if the dual polytope is again a lattice polytope (see Exercise 1.14). Show that P is reflexive if and only if 0 has height 1 over each facet of P.

Moreover, prove that the following are equivalent for an arbitrary lattice polytope:

(i) the normalization $\bar{M}(P)$ of $M(P)$ is Gorenstein;
(ii) there exists $x \in \mathbb{Z}^d$ and $k \in \mathbb{N}$ such that $kP - x$ is reflexive.

Batyrev [24] has used reflexive polytopes in the construction of mirror symmetric Calabi–Yau hypersurfaces. See Kreuzer and Skarke [232], [233] for classifications in dimensions 3 and 4. For further results we refer to Haase and Melnikov [172] and Haase and Nill [173].

2.16. (a) Let P be a lattice polygon with exactly one interior lattice point $x = 0$. Show P is reflexive.

(b)Try to find all reflexive lattice polygons. (There are 16 such polygons; see Kreuzer and Skarke [231].)

2.17. Let P be a lattice polygon with no interior lattice point. Show that $\operatorname{int}(M(P))$ is generated by elements of degree 2.

2.18. (a) Let M be a normal positive affine monoid. Suppose $x_1, \dots, x_m \in M$ generate $\operatorname{int}(M)$ as an ideal. Show that for each support form σ_F there exists i with $\sigma_F(x_i) = 1$.

(b) Consider the lattice polytope $P \in \mathbb{R}^3$ with vertices $(1,0,2)$, $(0,1,2)$, $(1,1,0)$, $(-1,-1,-1)$, $(1,1,2)$. Using normaliz or by giving a unimodular triangulation show that P is integrally closed, has exactly one lattice point, but is not reflexive, and $M(P)$ is not Gorenstein.

2.19. Let P be a lattice polytope, $M = \widehat{M}(P)$ be the integral closure of $M(P)$ and $m = \min\{\deg x : x \in \operatorname{int}(M)\}$. Show that M is generated by elements of degree $\leq \dim P + 1 - m$, and formulate a consequence analogous to Corollary 2.57.

Hint: reduce the problem to the simplicial case and apply a similar argument as in the proof of Theorem 2.52.

2.20. With the notation of Exercise 2.19 show that $\operatorname{int}(M)$ is generated by elements of degree $\leq \dim P + 1$ as an $M(P)$-module.

2.21. (a) Show that every positive rational cone is the intersection of finitely many unimodular cones.

(b) Show that every lattice polytope is the intersection of finitely many multiples of unimodular simplices.

2.22. Show that every lattice parallelotope is integrally closed.

2.23. Let $P \subset \mathbb{R}^d$ be a lattice polytope. For a vertex x of P set $C(x) = \mathbb{R}_+(P - x)$. We say that P is *very ample* if for every vertex x of P the vectors $y - x$, $y \in \operatorname{lat}(P)$, generate the monoid $C(x) \cap \mathbb{Z}^d$. (The terminology will be explained in Section 10.B.)

Set $M = M(P) \subset \mathbb{Z}^{d+1}$ and denote by $\widehat{M}$ the integral closure of M in $\mathbb{Z}^{d+1}$. Moreover, for each vertex x of P, let $\tilde{x} = (x,1) \in M(P)$. Prove that the following are equivalent:

(i) P is very ample;
(ii) $k\tilde{x} + \widehat{M} \subset M$ for all $x \in \operatorname{vert}(P)$ and all $k \gg 0$;
(iii) $\operatorname{Rad} \mathfrak{c}(\widehat{M}/M) = M \setminus \{0\}$;
(iv) $M[-\tilde{x}]$ is integrally closed in $\mathbb{Z}^{d+1}$ for all $x \in \operatorname{vert}(P)$;
(v) $\widehat{M} \setminus M$ is a finite set.

It follows that an integrally closed lattice polytope is very ample.

2.24. Let I_i, $i = 1,2,3,4$ be intervals in $\mathbb{Z}$, each containing at least two integers. Define the lattice polytope $P \subset \mathbb{R}^3$ as the convex hull of the set

$$\big((0,0) \times I_1\big) \cup \big((0,1) \times I_2\big) \cup \big((1,1) \times I_3\big) \cup \big((1,0) \times I_4\big).$$

(a) Prove that P is very ample.

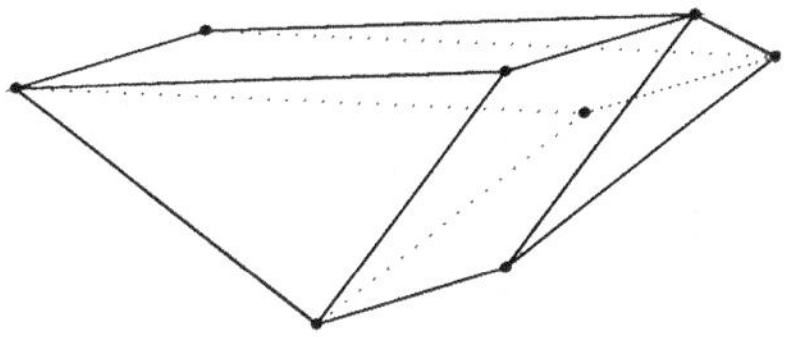

Fig. 2.7. A very ample, not integrally closed polytope

(b) Prove that P is not integrally closed for $I_1 = \{0,1\}$, $I_2 = \{2,3\}$, $I_3 = \{1,2\}$, $I_4 = \{3,4\}$ (see Figure 2.7).

2.25. Let $H \subset \mathbb{R}^d$ be a purely irrational hyperplane through 0 (i.e., containing no lattice point different from 0). Show that $H^>$ is the union of an ascending chain of unimodular subcones.

Hint: in dimension 2 choose a basis $x_1, x_2 \in H^>$ of $\mathbb{Z}^2$. Then exactly one of the cones spanned by $x_1, x_2 - x_1$, and $x_1 - x_2, x_2$ is contained in $H^>$. Choose it as the next member of the chain and iterate. In higher dimension consider pairs of basis elements.

2.26. Let P be a lattice polytope and suppose that $\mathscr{C}_0$ is a collection of unimodular subsimplices of P that covers P (in particular, P is integrally closed). Let $\mathscr{C}$ be the set of all faces of the simplices of $\mathscr{C}_0$, and let us say that $\delta \in \mathscr{C}$ is an *interior face* if $\delta \not\subset \partial P$.

Show that $\min\{\dim \delta : \delta \in \mathscr{C} \text{ is an interior face}\} + 1$ is the minimal degree of an element in $\operatorname{int}(M(P))$ [69, 1.2.5].

2.27. Let W be a cube of dimension d whose lattice points are its vertices and its barycenter. (For example, one can choose the cube with vertices in $\{\pm 1\}^d$ and consider the lattice generated by them and the origin.) Prove:

(a) The polytope W is normal in all dimensions.

(a) The cone $C(W)$ is tight if $d \geq 4$.

Hint: consider the midpoint x of the line segment joining the barycenter and a vertex. It belongs to the convex hull of the other vertices if $d \geq 4$. What follows for $2x$?

2.28. Let M_1 and M_2 be positive affine monoids and set $M = M_1 \oplus M_2$. Prove:
(a) $\mathrm{CR}(M) = \mathrm{CR}(M_1) + \mathrm{CR}(M_2)$.
(b) M satisfies (UHC) if and only if M_1 and M_2 do so.

2.29. Let M_1 and M_2 be homogeneous monoids, and let $M = M_1 \# M_2 = \{(x, y) \in M_1 \oplus M_2 : \deg x = \deg y\}$ be their *Segre product*. Prove:
(a) M is homogeneous;
(b) $\mathrm{CR}(M) \leq \mathrm{CR}(M_1) + \mathrm{CR}(M_2) - 1$;
(c) M satisfies (UHC) if and only if M_1 and M_2 do so.
Hint for (c): use what has been said about products of unimodular simplices. Indirectly it can also be used for (b).

2.30. Let P and Q be lattice polytopes. Prove $M(P \times Q) \cong M(P) \# M(Q)$.
There is also a polytope $\tilde{P}$ such that $M(\tilde{P}) = M(P) \oplus M(Q)$. Identify $\tilde{P}$.

2.31. Let M be a positive affine monoid, x an extreme integral generator, and suppose that $M[-x] \cong \mathbb{Z} \oplus M'$ in such a way that the associated homomorphism maps $\mathrm{Hilb}(M) \setminus \{x\}$ bijectively onto $\mathrm{Hilb}(M')$. Prove:
(a) $\mathrm{CR}(M) \geq \mathrm{CR}(M') + 1$;
(b) if M satisfies (UHC), then M' does so;
(c) a trivial realization of the situation considered in this exercise is $M = \mathbb{Z}_+ \oplus M'$. Show it is the only one if M and M' are homogeneous.

Notes

In 1872 Gordan [145] proved that the monoid of nonnegative solutions of a linear diophantine homogeneous system is finitely generated by its irreducible elements. In the context of invariant theory where it was found, the theorem indeed served as a lemma, namely in Gordan's second, simplified proof of the finite generation of the ring of invariants of binary forms. See M. Noether's obituary of his friend Gordan [274, p. 14]. Hilbert [190, p. 117] reproduced Gordan's theorem and proof and applied it to the same problem (in a much more transparent way).

Like many other theorems, Gordan's theorem has been proved several times. In 1903 Elliott [115] published an algorithm for the computation of the nonnegative solutions of a homogeneous linear diophantine equation. It was generalized to systems of such equations by MacMahon [247, Sect. VIII]. The Elliott–MacMahon algorithm actually yields the multigraded Hilbert series of the monoid of solutions (see Chapter 6). It was analyzed by Stanley in modern terms [323]: the algorithm amounts to computing a unimodular triangulation of the cone C of solutions starting from the unimodular positive orthant containing C.

In the generality of Lemma 2.9, the finite generation of the monoid of lattice points in a rational cone was proved by van der Corput in [360] (and two preceding articles), together with the uniqueness of the minimal system of generators in the positive case.

Normal affine monoids were thoroughly investigated by Hochster [195]. In his terminology pure submonoids are called "full" and integrally closed ones appear as "expanded" submonoids. Purity is a key concept for the investigation of rings of invariants. Results like Theorem 2.29 are modeled after [195]. Theorem 2.38 is a specialization of a theorem proved by the authors [58, Th. 6.1],

In the K-theoretic investigation of monoid rings (Chapter 8), the notion of seminormality, defined by Traverso, suggests itself, and the characterization of seminormal monoids consequently is to be found in [157]. Our treatment is influenced by [160] and Swan's version [351] of [157]. See also Reid and Roberts [299] for a study of seminormality and related properties in monoids.

Theorem 2.52 is essentially due to Ewald and Wessels [119]. It was also proved by Liu, Trotter, and Ziegler [242] and Bruns, Gubeladze, and Trung [69]. Example 2.47 is a slight modification of an example in [119]. An ideal-theoretic variant of Theorem 2.52 was given by Reid, Roberts, and Vitulli [300]. See Bruns, Vasconcelos, and Villarreal [79] for related results.

The notion of Carathéodory rank was suggested by a result of Cook, Fonlupt, and Schrijver [91], later on improved by Sebő [314] to the bound given in Theorem 2.69.

The existence of a unimodular Hilbert triangulation in dimension 3 (Theorem 2.78) was proved independently by Sebő [314], Aguzzoli and Mundici [1], and Bouvier and Gonzalez–Sprinberg [40]. We have reproduced Sebő's proof. In [40] it is furthermore shown that each two such triangulations are connected by a series of "flips."

References for counterexamples to (UHC) and (ICP) and for the strategy of their search have been included in Section 2.D. A weaker version of (UHC) was discussed by Firla and Ziegler [124].

3

Multiples of lattice polytopes

As counterexamples in Section 2.D show, even a normal lattice polytope need not be covered by its unimodular subsimplices. On the other hand, results like Corollary 2.57 show that certain properties improve if one replaces P by a multiple cP. This can be seen as an expression of the fact that the discrete structure of cP approximates the continuous structure of P better and better when $c \to \infty$.

In the last section of this chapter this general principle is confirmed by the unimodular covering of cP for $c \gg 0$, and even in such a way that the range of admissible c depends only on $\dim P$. A companion theorem for rational cones is not only necessary for the proof, but also interesting in its own right.

In comparison, one knows much less about the existence of unimodular triangulations of cP. However, a well-known theorem of Knudsen and Mumford guarantees a unimodular triangulation for at least one $c > 0$ (and all its multiples).

3.A Knudsen–Mumford triangulations

In this section we describe the unimodular triangulation of $\mathbb{R}^d$ constructed by Knudsen and Mumford [221, Chap. III]. It is basic for the results in Sections 3.B and 3.C on the triangulation and covering of multiples of lattice polytopes by unimodular lattice simplices.

Weyl chambers. Let $e_1, \dots, e_d \in \mathbb{R}^d$ be the standard basis of $\mathbb{R}^d$. With each element σ of the permutation group S_d of $\{1, \dots, d\}$ we associate the simplex

$$\Delta_\sigma = \operatorname{conv}\left(0, e_{\sigma(1)}, e_{\sigma(1)} + e_{\sigma(2)}, \dots, e_{\sigma(1)} + e_{\sigma(2)} + \dots + e_{\sigma(d)}\right).$$

Obviously Δ_σ is unimodular (with respect to $\mathbb{Z}^d$) and contained in the unit cube spanned by $e_1, \dots, e_d$. One has

$$\Delta_\sigma = \{(x_1, \dots, x_d) \in \mathbb{R}^d \ : \ 1 \geq x_{\sigma(1)} \geq x_{\sigma(2)} \geq \dots \geq x_{\sigma(d)} \geq 0\}.$$

The order

W. Bruns, J. Gubeladze, *Polytopes, Rings, and K-Thoery*, Springer Monographs in Mathematics,
DOI 10.1007/b105283_3,

$$0 < e_{\sigma(1)} < e_{\sigma(1)} + e_{\sigma(2)} < e_{\sigma(1)} + \cdots + e_{\sigma(d)}$$

is called the *canonical order* of vert(Δ_σ). For the vertices of Δ_1, the simplex corresponding to the identity permutation, we use the notation $z_0 = 0$ and $z_i = e_1 + \cdots + e_i$, $i = 1, \ldots, d$.

We want to show that the simplices Δ_σ tile the entire space $\mathbb{R}^d$ in a natural way. Consider the system of hyperplanes

$$H^{i,j} = \begin{cases} \{(x_1, \ldots, x_d) \in \mathbb{R}^d \,:\, x_{j+1} = x_{i+1}\}, & 0 \le j < i < d, \\ \{(x_1, \ldots, x_d) \in \mathbb{R}^d \,:\, x_{j+1} = 0\}, & 0 \le j < i = d. \end{cases}$$

Note that $H^{i,j}$ is the hyperplane through the origin 0 which is parallel to the complementary faces

$$\operatorname{conv}(z_0, z_1, \ldots, z_j, z_{i+1}, z_{i+2}, \ldots, z_d),\ \operatorname{conv}(z_{j+1}, z_{j+2}, \ldots, z_i) \subset \Delta_1.$$

For instance, $H^{i,i-1} = \operatorname{aff}(z_0, z_1, \ldots, \widehat{z_i}, \ldots, z_d)$, $i = 1, \ldots, d$.

More generally, for every number $k \in \mathbb{Z}$ we put

$$H_k^{i,j} = \begin{cases} \{(x_1, \ldots, x_d) \in \mathbb{R}^d \,:\, x_{j+1} - x_{i+1} = k\}, & 0 \le j < i < d, \\ \{(x_1, \ldots, x_d) \in \mathbb{R}^d \,:\, x_{j+1} = k\}, & 0 \le j < i = d. \end{cases}$$

In particular, $H_0^{i,j} = H^{i,j}$.

Proposition 3.1. *Every point $x = (x_1, \ldots, x_d)$ that is not contained in one of the hyperplanes $H_k^{i,j}$ lies in the interior of exactly one of the integral parallel translates of the simplices Δ_σ, $\sigma \in S_d$. In particular, the hyperplanes $H_k^{i,j}$, $0 \le j < i \le d$, $k \in \mathbb{Z}$, dissect $\mathbb{R}^d$ into simplices that are the integral parallel translates of the simplices Δ_σ, $\sigma \in S_d$.*

Proof. Put $a_i = \lfloor x_i \rfloor$ and $t_i = \{x_i\} = x_i - a_i$. Since $x \notin \bigcup H_k^{i,j}$ for all i, j, k, we have $t_i \neq t_j$ whenever $i \neq j$. Consequently, there is a unique element $\sigma \in S_d$ such that

$$1 > t_{\sigma(1)} > t_{\sigma(2)} > \cdots > t_{\sigma(d)} > 0.$$

Moreover, the connected component of x in $\mathbb{R}^d \setminus \bigcup H_k^{i,j}$ is the set

$$\{(x'_1, \ldots, x'_d) \in \mathbb{R}^d : 1 > x'_{\sigma(1)} - a_{\sigma(1)} > \cdots > x'_{\sigma(d)} - a_{\sigma(d)} > 0\},$$

and this is exactly the interior of the simplex $(a_1, \ldots, a_d) + \Delta_\sigma$. □

The representatives of the infinite collection of unimodular d-simplices thus obtained will be called *Weyl chambers*. Applying integral parallel translations we extend the notion of *canonical order* to the vertex set of any of the Weyl chambers. Figure 3.1 shows the tiling of the plane. We have marked the chambers Δ_1 and $(-2, -1) + \Delta_{(1\,2)}$.

Remark 3.2. Our Weyl chambers are actually the *affine Weyl chambers of type A_d*. In

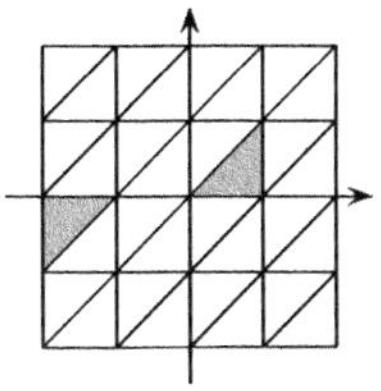

Fig. 3.1. Weyl chambers of the plane

other words, each of the simplices Δ_σ, $\sigma \in S_d$, is bounded by hyperplanes orthogonal to the elements of a root system of type A_d. Since we only need this type of Weyl chamber the attribute "of type A_d" will be omitted. (See [38] or [202] for a general discussion of Weyl groups and Weyl chambers.)

The Weyl chambers do not depend on the order of the basis vectors since the system of hyperplanes $H_k^{i,j}$ is invariant under permutations of the coordinates.

Lemma 3.3. *For all Weyl chambers Δ_1 and Δ_2 the canonical orders of* $\operatorname{vert}(\Delta_1)$ *and* $\operatorname{vert}(\Delta_2)$ *agree on* $\operatorname{vert}(\Delta_1 \cap \Delta_2)$.

Proof. We realize the canonical orders as restrictions of a linear order on $\mathbb{Z}^d$. Then the claim follows immediately.

Consider positive real numbers $\alpha_1, \dots, \alpha_d$ that are linearly independent over $\mathbb{Q}$. Then the linear form $\lambda = \sum \alpha_i X_i$ separates the points of $\mathbb{Z}^d$ and, simultaneously, has positive values at the standard basis vectors e_i. Then the canonical order on the vertices z of every Weyl chamber is induced by the order of the values $\lambda(z) \in \mathbb{R}$. □

To a Weyl chamber $a + \Delta_\sigma$ we associate the mapping

$$\pi : \{1, \dots, d\} \to \mathbb{Z}, \quad i \mapsto a_{\sigma(i)}.$$

Conversely, suppose we are given a mapping $\pi : \{1, 2, \dots, d\} \to \mathbb{Z}$. Then there is a unique permutation $\sigma \in S_d$ satisfying the following conditions:

$$\pi(\sigma(i)) \geq \pi(\sigma(j)) \quad \text{for } i \leq j \tag{3.1}$$

and

$$\pi(\sigma(i)) = \pi(\sigma(j)) \implies \bigl(i < j \iff \sigma(i) < \sigma(j)\bigr). \tag{3.2}$$

Actually, one has the formula

$$\sigma(k) = \#\bigl\{i : \pi(i) > \pi(k) \text{ or } \pi(i) = \pi(k) \text{ and } i \leq k\bigr\}.$$

To π we associate the Weyl chamber $a + \Delta_{\sigma^{-1}}$ where $a = \bigl(\pi(\sigma(1)), \dots, \pi(\sigma(d))\bigr)$. Although different mappings π define different Weyl chambers, the two associations are not mutually inverse. For instance, all chambers Δ_σ, $\sigma \in S_d$, yield the same constant map $\{1, 2, \dots, d\} \to \{0\}$, and all Weyl chambers associated with maps π lie in a certain region in $\mathbb{R}^d$. However, on this region the correspondence is bijective:

Lemma 3.4. *The correspondence above establishes a bijection between the maps* $\pi : \{1, 2, \ldots, d\} \to \mathbb{Z}$ *and the Weyl chambers in the region*

$$Z = \{(x_1, \ldots, x_d) \in \mathbb{R}^d \ : \ x_1 \ge x_2 \ge \cdots \ge x_d\} \subset \mathbb{R}^d$$

Proof. Given a map π, let σ be the permutation derived from it. We want to show that the linear forms $X_{i+1} - X_i$, $i = 1, \ldots, d-1$ (which correspond to the hyperplanes $H^{i,i-1}$), cannot have positive values on the chamber $(\pi(\sigma(1)), \ldots, \pi(\sigma(d))) + \Delta_{\sigma^{-1}}$. But this follows from the conditions (3.1) and (3.2):

$$\big(\pi(\sigma(i+1)) - \pi(\sigma(i))\big) + (x_{i+1} - x_i) = \\ \big(\pi(\sigma(i+1)) - \pi(\sigma(i))\big) + \big(x_{\sigma^{-1}(\sigma(i+1))} - x_{\sigma^{-1}(\sigma(i))}\big) \le 0$$

if $(x_1, \ldots, x_d)$ satisfies the inequalities $1 \ge x_{\sigma^{-1}(1)} \ge \cdots \ge x_{\sigma^{-1}(d)} \ge 0$.

Conversely, given a Weyl chamber $\Delta \subset Z$, $\Delta = (a_1, \ldots, a_d) + \Delta_{\sigma^{-1}}$ for some $a_i \in \mathbb{Z}$ and $\sigma \in S_d$, we have $1 > x_{\sigma^{-1}(1)} > \cdots > x_{\sigma^{-1}(d)} > 0$ for an interior point $(x_1, \ldots, x_d) \in \operatorname{int}(\Delta_{\sigma^{-1}})$. The containment $\Delta \subset Z$ is equivalent to the condition that for each $i \in \{1, \ldots, d-1\}$ either $a_i > a_{i+1}$ or $a_i = a_{i+1}$ and $x_i > x_{i+1}$. But the latter inequality is the same as $x_{\sigma^{-1}(\sigma(i))} > x_{\sigma^{-1}(\sigma(i+1))}$, that is $\sigma(i) < \sigma(i+1)$. Therefore, if π is the map associated to Δ, we indeed recover Δ from π if $\Delta \subset Z$. □

The set Z is bounded by the hyperplanes $H^{i,i-1}$, $i = 1, \ldots, d-1$. In particular, it is tiled by the Weyl chambers contained in it.

Corollary 3.5. *For every natural number* $\mu \ge 1$ *there is a natural bijective correspondence between the mappings* $\pi : \{1, \ldots, d\} \to \{0, 1, \ldots, \mu - 1\}$ *and the Weyl chambers in the multiple* $\mu\Delta_1$*. Moreover, these chambers triangulate the simplex* $\mu\Delta_1$.

The triangulations in this corollary will be called the *canonical triangulations* (with respect to the enumeration $e_1, \ldots, e_d$ of the standard basis vectors e_i).

We introduce a notation for the correspondence between Weyl chambers and maps, and define some additional data. For a map $\pi : \{1, 2, \ldots, d\} \to \mathbb{Z}$ we let

- Δ^π denote the corresponding Weyl chamber;
- $z_0^\pi, z_1^\pi, \ldots, z_d^\pi$ denote the vertices of Δ^π in the canonical order;
- $\sigma^\pi \in S_d$ denote the permutation derived from π;
- $0^\pi = z_0^\pi$ and $e_i^\pi = (e_i)^\pi = z_i^\pi - z_{i-1}^\pi$; $i = 1, \ldots, d$,
- $(x_1^\pi, \ldots, x_d^\pi)$ be the coordinate vector of $(x_1, \ldots, x_d)$ in the new affine coordinate system with origin 0^π and basis $e_1^\pi, \ldots, e_d^\pi$.

The equation $\Delta^\pi = \big(\pi(\sigma^\pi(1)), \ldots, \pi(\sigma^\pi(d))\big) + \Delta_{(\sigma^\pi)^{-1}}$ implies the following rule for the coordinate change:

Lemma 3.6. $x_i = \pi(\sigma^\pi(i)) + x^\pi_{\sigma^\pi(i)}$ *for every* $i = 1, \ldots, d$.

We have used the inverse permutations in defining the correspondence between mappings and Weyl chambers in order to arrive at this nice formula.

Canonical and mixed triangulations. Let $\Delta \subset \mathbb{R}^d$ be a d-simplex together with an order $p_0 < p_1 < \cdots < p_d$ of its vertices. Then the bijection $\Delta_1 \to \Delta$, $z_i \mapsto p_i$, $i = 0, \dots, d$, extends to a unique affine transformation $\mathbb{R}^d \to \mathbb{R}^d$. Using this transformation we can transfer to Δ all data we have associated to Δ_1. The resulting tiling of $\mathbb{R}^d$ is called *canonical* with respect to Δ and the given order on $\operatorname{vert}(\Delta)$.

Let $\delta \subset \Delta$ be a face. The order on $\operatorname{vert}(\delta)$, induced by that of $\operatorname{vert}(\Delta)$, is called *$\Delta$-canonical.* Using this order we can develop similar notions within the affine subspace $\operatorname{aff}(\delta) \subset \mathbb{R}^d$. Furthermore, we can do this in each of the parallel translates $z + \operatorname{aff}(\delta)$, $z \in \mathscr{L}(\Delta)$, where, as usual, $\mathscr{L}(\Delta) \subset \mathbb{R}^d$ is the affine lattice

$$p_0 + \sum_{i=0}^{d} \mathbb{Z}(p_i - p_0) \subset \mathbb{R}^d.$$

The simplices obtained are the *δ-canonical* ones. It follows from Lemma 3.3 that all canonical orders (with respect to Δ) agree in the sense that if δ' and δ'' are two canonical simplices (not necessarily of the same dimension) then $\delta' \cap \delta''$ is again a canonical simplex and the canonical order on $\operatorname{vert}(\delta' \cap \delta'')$ agrees with those on $\operatorname{vert}(\delta')$ and $\operatorname{vert}(\delta'')$.

Let μ and ν be natural numbers. Then we can construct a triangulation of the multiple $\mu\nu\Delta$ as follows. For each member δ of the Δ-canonical triangulation of $\mu\Delta$ we consider the δ-canonical triangulation of $\nu\delta$ (with respect to the Δ-canonical order). Since the canonical orders are compatible, these triangulations can be patched up to a global triangulation $\mathscr{T}$ of $\mu\nu\Delta$. However, we could have obtained $\mathscr{T}$ in a simpler way:

Lemma 3.7. *$\mathscr{T}$ is the Δ-canonical triangulation of $\mu\nu\Delta$.*

Proof. Let δ be a Δ-canonical d-simplex in $\mu\Delta$, and let γ be a δ-canonical d-simplex in $\nu\delta$. Let $\pi : \{1, \dots, d\} \to \{1, \dots, \mu-1\}$ and $\rho : \{1, \dots, d\} \to \{1, \dots, \nu-1\}$ be the corresponding mappings, the latter taken with respect to the coordinates x_i^π (Corollary 3.5). By Lemma 3.6 we then have $x_i = \pi(\sigma^\pi(i)) + x^\pi_{\sigma^\pi(i)}$ and $x_i^\pi = \rho(\sigma^\rho(i)) + x^\rho_{\sigma^\rho(i)}$ for $i = 1, \dots, d$. Thus

$$x_i = \pi(\sigma^\pi(i)) + \rho\big(\sigma^\rho(\sigma^\pi(i))\big) + x^\rho_{\sigma^\rho(\sigma^\pi(i))}, \quad i = 1, \dots, d,$$

and this means that γ is a Δ-canonical simplex in $\mu\nu\Delta$ with corresponding permutation $(\sigma^\rho \circ \sigma^\pi)^{-1}$. □

Let Δ, μ, and ν be as above. Suppose V is an ambient euclidean space strictly containing $\operatorname{aff}(\Delta)$, and q a point in $V \setminus \operatorname{aff}(\Delta)$. Put $\tilde{\Delta} = \operatorname{conv}(\Delta, q)$. We construct a triangulation of the simplex $\mu\nu\tilde{\Delta}$ as follows. By Corollary 3.5 an element from the Δ-canonical triangulation of $\mu\Delta$ is of the type Δ^π for some mapping $\pi : \{1, \dots, d\} \to \{0, \dots, \mu - 1\}$. For each π we consider the canonical triangulation of the multiple $\nu \operatorname{conv}(\Delta^\pi, \mu q)$ with respect to the order $p_0^\pi < p_1^\pi < \cdots < p_d^\pi < \mu q$ on $\operatorname{vert}(\operatorname{conv}(\Delta^\pi, \mu q))$. It follows from Lemma 3.3 that these triangulations can be patched up to a global triangulation, which we denote by $(\nu\mu\tilde{\Delta})^{\mu,\nu}$.

Note that $(\nu\mu\tilde{\Delta})^{\mu,\nu} \neq (\nu\mu\tilde{\Delta})^{\nu,\mu}$ if $\mu \neq \nu$. Nevertheless, by Lemma 3.7, the induced triangulations on $\nu\mu\Delta$ coincide (with the Δ-canonical one).

Next we introduce the concept of *mixed triangulation* for simplicial complexes, generalizing the triangulations $(\nu\mu\tilde{\Delta})^{\mu,\nu}$ constructed above.

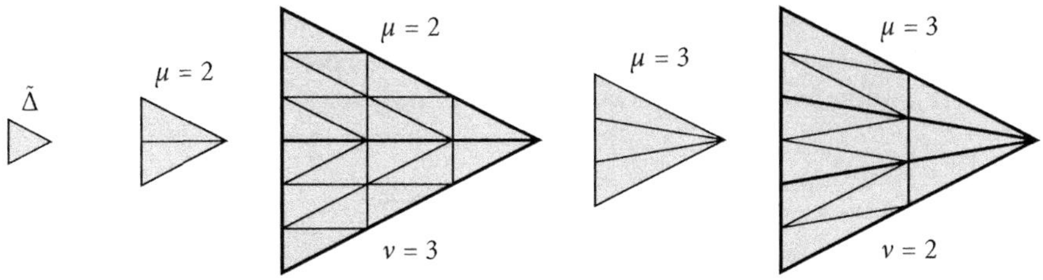

Fig. 3.2. Mixed (μ, ν)-triangulations

Suppose we are given the following data:

- a simplicial complex Π;
- a decomposition $\operatorname{vert}(\Pi) = \{p_1, \ldots, p_r, q_1, \ldots, q_s\}$ such that $\operatorname{openstar}_\Pi(q_i) \cap \operatorname{openstar}_\Pi(q_j) = \emptyset$ for $1 \leq i < j \leq s$;
- an order $p_1 < \cdots < p_r < q_1 < \cdots < q_s$ on $\operatorname{vert}(\Pi)$;
- two systems of natural numbers $\bar{\mu} = (\mu_1, \ldots, \mu_s)$ and $\bar{\nu} = (\nu_1, \ldots, \nu_s)$ such that the products $\mu_i\nu_i$, $i = 1, \ldots, d$, are all equal to, say, μ.

For every simplex $\Delta \in \Pi$ we consider the following triangulation of $\mu\Delta$:

$$\mathscr{T}_\Delta = \begin{cases} (\mu\Delta)^{\mu_i,\nu_i}, & \text{if } q_i \in \Delta \text{ for some } i, \\ \text{the } \Delta\text{-canonical triangulation otherwise.} \end{cases}$$

All the triangulations are formed with respect to the order of vertices we have fixed. By Lemma 3.7 these triangulations can be patched up to a global triangulation of $\mu\Pi$.

Definition 3.8. The global triangulation of $\mu\Pi$ is called the *mixed* $(\bar{\mu}, \bar{\nu})$*-triangulation* with respect to the fixed order of $\operatorname{vert}(\Pi)$ where $\bar{\mu} = (\mu_1, \ldots, \mu_s)$ and $\bar{\nu} = (\nu_1, \ldots, \nu_s)$. This triangulation is denoted by $(\mu\Pi)^{\bar{\mu},\bar{\nu}}$.

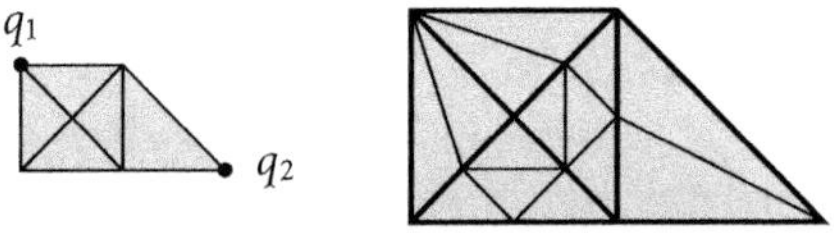

Fig. 3.3. Triangulation $(\mu\Pi)^{\bar{\mu},\bar{\nu}}$ with $\bar{\mu} = (2,2)$, $\bar{\nu} = (1,1)$

3.B Unimodular triangulations of multiples of polytopes

The main result of this section, due to Knudsen and Mumford [221, Chap. III], is that every lattice polytope P has a multiple cP, $c \in \mathbb{N}$, that admits a unimodular triangulation. The only known proof of this result is through its generalization to lattice polytopal complexes (Theorem 3.17).

Let Π be a lattice polytopal complex with lattice structure Λ. Each polytope $Q \in \Pi$ has a multiplicity $\mu(Q) = \mu_{\Lambda_Q}(Q)$ with respect to its associated lattice Λ_Q. The *multiplicity* of Π is

$$\mu(\Pi) = \max\{\mu(Q) : Q \in \Pi\}.$$

For the proof of the main result we have to find $c \in \mathbb{N}$ and a lattice triangulation Σ of cP with $\mu(\Sigma) = 1$.

Comparing multiplicities. Let $L \subset \mathbb{R}^d$ be an affine lattice, $P \subset \mathbb{R}^d$ an L-polytope, and F a facet of P.

We choose $H = \operatorname{aff}(F)$ and set

$$\operatorname{ht}_{L,F}(x) = \operatorname{ht}_{L,H}(x), \qquad x \in \operatorname{aff}(P),$$

where we have chosen the sign in such a way that $P \subset H^+$ (see Remark 1.72).

Proposition 3.9. *Let Δ be an L-simplex, and $v \in \operatorname{vert}(\Delta)$. Let F be the facet of Δ opposite to v. Then*

$$\mu_L(\Delta) = \operatorname{ht}_{L,F}(v)\mu_L(F).$$

If G is an arbitrary face of Δ, then $\mu_L(G) \mid \mu_L(\Delta)$.

Proof. The simplest proof of the formula is by introducing coordinates in a suitable way and computing determinants as in the proof of Proposition 2.46 (which is essentially the case $\operatorname{ht}_{L,F}(v) = 1$). We leave the details to the reader.

The assertion on faces follows by induction and transitivity. □

The simple formula in Proposition 3.9 hides a relationship between the lattices used in the definition of the multiplicities:

$$L_\Delta = L \cap \operatorname{aff}(\Delta), \qquad \mathscr{L}(\Delta) = v_0 + \sum_{v \in \operatorname{vert}(\Delta)} \mathbb{Z}(v - v_0)$$

(v_0 an arbitrary vertex of Δ) and the corresponding lattices defined by F. Since algebraic arguments play a key role in the following, it may be useful to make the relationship explicit.

We choose a vertex v_0 of F as the origin. Then $L_F, L_\Delta, \mathscr{L}(F), \mathscr{L}(\Delta)$ are lattices (and not only affine lattices). Moreover, $\lambda = \operatorname{ht}_{L,F}$ is now a linear form. One has a commutative diagram

$$\begin{array}{ccccccccc}
0 & \longrightarrow & \mathscr{L}(F) & \longrightarrow & \mathscr{L}(\Delta) & \longrightarrow & \mathbb{Z} & \longrightarrow & 0 \\
 & & \downarrow & & \downarrow & & \downarrow{\scriptstyle h} & & \\
0 & \longrightarrow & L_F & \longrightarrow & L_\Delta & \xrightarrow{\lambda} & \mathbb{Z} & \longrightarrow & 0
\end{array}$$

The map $h : \mathbb{Z} \to \mathbb{Z}$ is the multiplication by $h = \operatorname{ht}_{L,F}(v)$. The diagram has exact rows and the vertical maps are injective. (Note that $\mathscr{L}(F) = \mathscr{L}(\Delta) \cap L_F = \mathscr{L}(\Delta) \cap \operatorname{Ker}\lambda$ since Δ is a simplex.) So we obtain an exact sequence of cokernels

$$0 \to L_F/\mathscr{L}(F) \to L_\Delta/\mathscr{L}(\Delta) \to \mathbb{Z}/h\mathbb{Z} \to 0$$

whose numerical essence is the formula $\mu_L(\Delta) = h \cdot \mu_L(F)$.

Lemma 3.10. *Let $L' \subset L$ be lattices in $\mathbb{R}^d$ for which the index $[L : L']$ is finite. Furthermore let Δ, δ be L'-simplices, and suppose that δ is L'-unimodular and that* $\operatorname{aff}(\delta) \subset \operatorname{aff}(\Delta)$. *Then the following hold:*

(a) $\mu_L(\delta) \mid \mu_L(\Delta)$;

(b) $\mu_L(\delta) \mid [L : L']$.

Proof. We can augment δ to an L'-unimodular simplex $\delta' \subset \operatorname{aff}(\Delta)$ such that $\dim \delta' = \dim \Delta$, or equivalently $\operatorname{aff}(\delta') = \operatorname{aff}(\Delta)$. Then $\mathscr{L}(\delta') = L'_{\delta'} = L'_\Delta$, and in view of Proposition 3.9 we may assume that $\delta = \delta'$.

One has $\mu_L(\Delta) = [L_\Delta : \mathscr{L}(\Delta)] = [L_\Delta : L'_\Delta][L'_\Delta : \mathscr{L}(\Delta)]$. The first factor is exactly $\mu_L(\delta')$. This shows (a).

For (b) we note that L_δ/L'_δ is a subgroup of L/L' in a natural way, since $L'_\delta = L' \cap L_\delta$. Furthermore $\mu_L(\delta) = [L_\delta : L'_\delta]$ since $L'_\delta = \mathscr{L}(\delta)$. □

Multiplicity and the distribution of lattice points. Let Δ be an L-simplex with vertices $p_0, \dots, p_d$. In the following we will have to use lattice points in the semi-open parallelotope

$$\operatorname{par}(\Delta, p_0) = p_0 + \operatorname{par}(p_1 - p_0, \dots, p_d - p_0)$$

for the triangulation of suitable multiples $c\Delta$. To this end we need some information about the distribution of these lattice points.

Each point in $\operatorname{aff}(\Delta)$ has a unique representation

$$x = p_0 + \sum_{i=1}^{d} \alpha_i(p_i - p_0), \qquad \alpha_i \in \mathbb{R}.$$

We define the affine form ρ by $\rho(x) = \sum_{i=1}^d \alpha_i$. In other words,

$$\rho(x) = -\operatorname{ht}_{\mathscr{L}(\Delta),F_0}(x) + 1, \tag{3.3}$$

where F_0 is the facet of Δ opposite p_0 and the height is measured with respect to the lattice $\mathscr{L}(\Delta)$.

Note that the heights, measured with respect to L and $\mathscr{L}(\Delta)$ differ only by a positive factor, more precisely

$$\operatorname{ht}_{L,F_0}(x) = \operatorname{ht}_{L,F_0}(p_0)\operatorname{ht}_{\mathscr{L}(\Delta),F_0}(x). \tag{3.4}$$

This follows from the fact that two affine forms that have value 0 on $\operatorname{aff}(F_0)$ differ by a constant factor.

A very useful tool is the *Waterman map* defined by

$$\mathrm{w}(x) = p_0 + \frac{x - p_0}{\lceil \rho(x) \rceil}, \qquad x \neq p_0.$$

Note that $\mathrm{w}(x) \in \Delta$ for all $x \in \mathrm{par}(\Delta, p_0)$. The points $\mathrm{w}(x)$, $x \in \mathrm{lat}(\mathrm{par}(\Delta, p_0))$, $x \neq p_0$, are called p_0-*Waterman points* of Δ.

Let $x \in \mathrm{lat}(\mathrm{par}(p_0, \Delta))$ and suppose that $\mathrm{w}(x)$ does not belong to F_0. In the following we will use that this condition is equivalent to $\rho(x) \notin \mathbb{Z}$.

We construct a triangulation $\mathscr{T}$ of $h\Delta$, $h = \lceil \rho(x) \rceil$, as follows. First we set

$$\tilde{x} = hp_0 + (x - p_0) = (h-1)p_0 + x.$$

Then we choose the Δ-canonical triangulation on $\mathrm{link}_{h\Delta}(\tilde{x})$, and finally we extend it by stellar subdivision with respect to $\tilde{x}$ to a triangulation of $h\Delta$. (The link is formed in the polytopal complex of all faces of $h\Delta$.) Figure 3.4 illustrates $\mathscr{T}$ in dimension 2 (with $h = 2$, $p_0 = 0$, $\tilde{x} = x$).

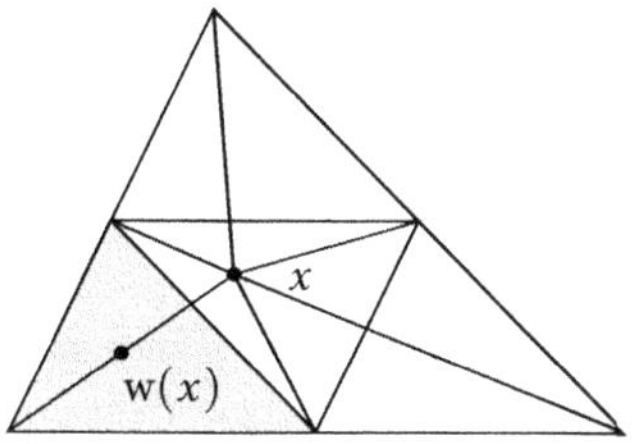

Fig. 3.4. The triangulation $\mathscr{T}$

Lemma 3.11. $\mu(\mathscr{T}) < \mu(\Delta)$.

Proof. Since the construction is invariant under parallel translations we can assume that $p_0 = 0$. Then $\tilde{x} = x$. We let F_i be the facet of Δ opposite p_i.

Consider a d-simplex $\delta \in \mathscr{T}$. By construction x is one of its vertices, and its facet G opposite to x lies in one of the facets hF_i of $h\Delta$. By construction $\mu(G) = \mu(F_i)$, and for the claim it is enough that the L-height of x above G is smaller than the L-height of p_i above F_i; compare Lemma 3.9.

Suppose first that $i = 0$. Then, since $h - \rho(x) < 1$, the height of x above hF_0 is smaller than the height of p_0 above F_0.

Now let $i \geq 1$. Since $x \in \mathrm{par}(\Delta, 0)$, the height of x above $hF_i = F_i$ is smaller than the maximal height of a point of Δ above F_i, and that is attained in p_i. So the height of x above F_i is smaller than that of p_i above F_i. □

For each face F of Δ such that $x \notin \mathrm{aff}(F)$ the triangulation $\mathscr{T}'$ on $\mathrm{conv}(hF, \tilde{x})$ induced by $\mathscr{T}$ is the first step in the construction of the mixed (h, g)-triangulation of $\mathrm{conv}(F, \mathrm{w}(x))$, with $g \in \mathbb{N}$ arbitrary. Since this mixed triangulation is constructed by the canonical extension of $\mathscr{T}'$, we obtain:

Corollary 3.12. *The multiplicity of the mixed triangulation* $(hg\operatorname{conv}(F, \mathrm{w}(x))^{h,g}$ *is* $< \mu(\Delta)$.

Now we compare Δ to its facets.

Lemma 3.13. *Let* $\Delta = \operatorname{conv}(p_0, \dots, p_d)$ *be an L-simplex.*

(a) Let δ *be a face of* Δ. *Then* $\mu(\delta) = \mu(\Delta)$ *if and only if every* p_0*-Waterman point of* Δ *lies in* δ.

(b) Suppose $G = L/\mathscr{L}(\Delta)$ *is a cyclic group and* $\mu(\delta) < \mu(\Delta)$ *for any proper face* $\delta \subset \Delta$. *If* z *is a point in* $\operatorname{lat}(\operatorname{par}(\Delta, p_0))$ *such that the residue class of* $z - p_0$ *generates* G, *then* $\mathrm{w}(z) \in \operatorname{int}(\Delta)$.

Proof. We can assume that $p_0 = 0$ and write $\operatorname{par}(\Delta)$ for $\operatorname{par}(\Delta, 0)$. Moreover, L and $\mathscr{L} = \mathscr{L}(\Delta)$ are groups.

(a) First consider the case $0 \in \delta$. The multiplicity of Δ is the number of lattice points in $\operatorname{par}(\Delta)$, and an analogous statement holds for $\operatorname{par}(\delta)$. Thus both conditions in the lemma are equivalent to $\operatorname{lat}(\operatorname{par}(\Delta)) = \operatorname{lat}(\operatorname{par}(\delta))$.

Now suppose that $0 \notin \delta$, and let $\delta' = \operatorname{conv}(0, \delta)$. In view of the case already settled and since $\mu(\delta) \le \mu(\delta') \le \mu(\Delta)$, we can replace Δ by δ'. In other words, we can assume that δ is the facet F_0 of Δ opposite to 0.

One has $\mathrm{w}(x) \in F_0$ if and only if $\rho(x) \in \mathbb{Z}$. Each point of L differs from a suitable L-point in $\operatorname{par}(\Delta)$ by an element of $\mathscr{L}$. Moreover, $\rho(y) \in \mathbb{Z}$ for all $y \in \mathscr{L}$. So we conclude that $\mathrm{w}(x) \in F_0$ for all L-points $x \neq 0$ in $\operatorname{par}(\Delta)$ if and only if $\rho(y) \in \mathbb{Z}$ for all $y \in L$. By equation (3.3) we obtain the equivalent condition $\operatorname{ht}_{\mathscr{L},F_0}(y) \in \mathbb{Z}$ for all $y \in L$.

In view of equation (3.4) and since there exists $y \in L$ with $\operatorname{ht}_{F_0,L}(y) = 1$, we finally see that our hypothesis is equivalent to $\operatorname{ht}_{L,F_0}(p_0) = 1$, and this in turn is equivalent to $\mu(\Delta) = \mu(F_0)$ by Lemma 3.9.

(b) If $\mathrm{w}(z) \notin \operatorname{int}(\Delta)$, then there exists a facet δ of Δ with $\mathrm{w}(z) \in \delta$.

If $0 \in \delta$, then z and all multiples of z lie in $\mathbb{R}_+\delta \cap L$, and each lattice point in $\operatorname{par}(\Delta)$ is already in $\operatorname{par}(\delta)$. In fact, every $x \in L$ is congruent to a multiple of z modulo $\mathscr{L}$, since z generates G. But then $\mu(\delta) = \mu(\Delta)$.

If $\mathrm{w}(z) \in F_0$, then, by the same argument, all L-points of Δ have integral value under ρ. Part (a) implies $\mu(F_0) = \mu(\Delta)$, contrary to the hypothesis. □

Corollary 3.14. *Let* δ' *and* δ'' *be faces of* Δ. *Then*

$$\mu(\delta') = \mu(\Delta) = \mu(\delta'') \iff \mu(\delta' \cap \delta'') = \mu(\Delta).$$

For the proof of the main theorem we have to analyze the structure of the union of the simplices Δ in a lattice simplicial complex Π with $\mu(\Delta) = \mu(\Pi)$. Let Π be a lattice simplicial complex and put

$$U = \bigcup_{\substack{\Delta \in \Pi \\ \mu(\Delta) = \mu(\Pi)}} \operatorname{int}(\Delta).$$

From Corollary 3.14 we derive the following decomposition result:

Lemma 3.15. *The connected components of U are the open star neighborhoods of the relative interiors of the minimal (with respect to inclusion) simplices $\delta \in \Pi$ with multiplicity $\mu(\delta) = \mu(\Pi)$:*

$$U = \bigcup_{\min\{\delta\in\Pi\,:\,\mu(\delta)=\mu(\Pi)\}} \operatorname{openstar}_{\Pi}(\operatorname{int}(\delta)),$$

the union being disjoint.

Finally, Corollary 3.15 and Lemma 3.13 imply

Corollary 3.16. *Let Π be a lattice simplicial complex such that $\mu(\Pi)$ is a prime number. Suppose $\{\Delta_1, \dots, \Delta_s\} = \min\{\delta \in \Pi : \mu(\delta) = \mu(\Pi)\}$ (defined as in Lemma 3.15). Then for every index $j = 1, \dots, s$ and every vertex $p \in \operatorname{vert}(\Delta_j)$ there exists a p-Waterman point $x \in \operatorname{int}(\Delta_j)$.*

The main theorem. We have completed the preparations for the proof of the central result of this section.

Theorem 3.17 (Knudsen–Mumford). *Let Π be a lattice polytopal complex. Then there exists a natural number c such that $c\Pi$ has a unimodular triangulation.*

Corollary 3.18. *For every lattice polytope P there exists a natural number c such that cP has a unimodular triangulation.*

Remark 3.19. Suppose that $c\Pi$ admits a unimodular triangulation $\mathcal{T}$, and fix an order on $\operatorname{vert}(\mathcal{T})$. By Lemma 3.7 the iterated use of canonical triangulations with respect to this order shows that all multiples $c'c\Pi$, $c' \in \mathbb{N}$, have unimodular triangulations.

Proof of Theorem 3.17. By Theorem 1.51, which guarantees the existence of a lattice triangulation of Π, there is no loss of generality in assuming that Π is a simplicial complex.

The proof is by induction on $\mu(\Pi)$. If $\mu(\Pi) = 1$, then Π is already unimodular. So we can assume that we have proved Theorem 3.17 for simplicial complexes of multiplicity $< k = \mu(\Pi)$.

Since we will consider different lattice structures on the geometric realization $|\Pi|$ we will always specify the lattice structure at issue, $\Lambda = \{\Lambda_\Delta : \Delta \in \Pi\}$ being the original lattice structure on Π.

We fix a total order of the vertices of Π.

Case A: *k is a composite number.* Let $\Delta_1, \dots, \Delta_s \in \Pi$ be the simplices of multiplicity k and $P_j' = \operatorname{par}(\Delta_j, p_j)$, $j = 1, \dots, s$, be the semi-open parallelotope with respect to the minimal vertex of Δ_j ("minimal" refers to the fixed order of the vertices of Π). For each index j we choose a point $z_j \in \operatorname{lat}(\operatorname{par}(\Delta_j, p_j))$ such that the residue class of $z_j - p_j$ has order $r_j < k$ in the quotient group $\Lambda_{\Delta_j}/\mathcal{L}(\Delta_j)$; see p. 68 for the definition of the quotient group. (We can take r_j to be a fixed prime divisor of k independently of j, but that is irrelevant.)

We consider the sequence of lattice structures

$$\Lambda_0, \Lambda_1, \cdots, \Lambda_s, \Lambda$$

on the geometric realization $|\Pi|$, defined recursively as follows:

- Λ_0 is generated by $\text{vert}(\Pi)$; more precisely: $(\Lambda_0)_\Delta = \mathscr{L}(\Delta)$ for each simplex $\Delta \in \Pi$. In particular, (Π, Λ_0) is a unimodular lattice simplicial complex. Therefore (Π, Λ_0) is an embedded lattice polytopal complex. This follows by the same argument that shows that a simplicial complex (without lattice structure) is embedded; see Example 1.43. Let L_0 be the ambient lattice of the embedding.
- For each $j = 1, \ldots, s$ we set $L_j = L_{j-1} + \mathbb{Z}(z_j - p_j)$, and consider the lattice structure Λ_j on Π induced by it: Λ_j is composed of the restrictions $(L_j)_\Delta$ of L_j to the affine spaces $\text{aff}(\Delta)$, $\Delta \in \Pi$. So all lattice simplicial complexes (Π, Λ_0), $(\Pi, \Lambda_1), \ldots, (\Pi, \Lambda_s)$ are embedded. Observe, $[L_j : L_{j-1}] \mid r_j$.

Let Δ be a simplex in Π. By Lemma 3.10(b), the multiplicity $\mu_{L_1}(\Delta)$ divides the index r_1 of L_0 in L_1. Since $r_1 < k$, we can apply the induction hypothesis, and find a factor c_1, for which $c_1\Pi$ has a Λ_1-unimodular triangulation. The same argument can be applied to the extension Λ_2 of Λ_1 etc. We end up with a multiple $(c_1 \cdots c_s)\Pi$ that has a Λ_s-unimodular triangulation.

We claim that every L_s-unimodular simplex δ in $(c_1 \cdots c_s)\Pi$ has multiplicity $< k$ with respect to Λ. If this holds then we are done because we can apply the induction hypothesis again to obtain a multiple $(c' c_1 \cdots c_s)\Pi$ with a Λ-unimodular triangulation.

Suppose first that δ is an L_s-unimodular simplex contained in $\text{aff}(\Delta)$ where Δ is a simplex in Π with $\mu_\Lambda(\Delta) < k$. Applying Lemma 3.10(a) to the lattices $(L_s)_\Delta$ and Λ_Δ, we see that $\mu_\Lambda(\delta) < k$.

Now suppose that $\mu(\Delta) = k$, say $\Delta = \Delta_j$. Then $\mu_\Lambda(\delta)$ divides the index $[\Lambda_\Delta : (L_s)_\Delta]$. Next

$$k = \mu_\Lambda(\Delta) = [\Lambda_\Delta : (L_0)_\Delta] = [\Lambda_\Delta : (L_s)_\Delta][(L_s)_\Delta : (L_0)_\Delta],$$

and $[(L_s)_\Delta : (L_0)_\Delta] > 1$ by the construction of L_s. In fact, $z_j - p_j \notin (L_0)_\Delta$.

Case B: $k = p$ *is a prime number.* Let U be the union of the interiors of the simplices in (Π, Λ) of multiplicity p. By Lemma 3.15 U decomposes into a disjoint union

$$U = \bigsqcup_j \text{openstar}(\text{int}(\Delta_j))$$

and, by Corollary 3.16, each of these Δ_j has an interior Waterman point $q_j = \text{w}(z_j)$ (with respect to a vertex of δ_j). Let (Π', Λ) denote the lattice simplicial complex obtained from Π via stellar triangulation with respect to the new vertices q_j. We order the vertices of Π' in such a way that the order of the vertices of Π remains the same and the points q_j are greater than the vertices of Π.

Let r be a common multiple of the numbers $r_j = \lceil \rho(z_j) \rceil$ and choose natural numbers s_j such that

$$r_1 s_1 = \cdots = r_j s_j = \cdots = r,$$

and let Π'' be the mixed (r_j, s_j)-triangulation of the lattice simplicial complex $(r\Pi', \Lambda)$ with respect to the fixed order of $\text{vert}(\Pi')$. Then (Π'', Λ) is a lattice triangulation of $(r\Pi, \Lambda)$ in which all simplices of multiplicity p have been refined; see

Corollary 3.12. Moreover, no new simplices of multiplicity $\geq p$ have been created, and so the induction hypothesis applies. □

Remark 3.20. (a) One could try to give an effective upper bound for the number c in Theorem 3.17 by tracing its proof. Evidently the bound obtained does not only depend on $\mu(\Pi)$ and $d = \dim \Pi$, but also on the number of simplices in Π.

(b) An easy observation is that c can be chosen in such a way that its prime divisors are bounded by $\lceil d/2 \rceil$. In fact, one only needs to show this in the case where k is prime. Every point $z \in \operatorname{lat}(\operatorname{par}(\Delta, p_0))$ has $\rho(z) < d$. If $\rho(z) > \lceil d/2 \rceil$, then we use the symmetry trick that we have first encountered in the proof of Theorem 2.52. Especially, if $d = 3$ or $d = 4$, then one can take c to be a power of 2.

A lower bound for c is $d - 1$, as is shown by Example 2.58.

Remark 3.21. More precise results are known in dimension 3. Kantor and Sarkaria [217] have shown that 4Π has a unimodular triangulation if Π is a lattice polytopal complex of dimension 3. On the other hand, $2\Delta_{pq}$ does not admit such a triangulation if $q \neq 1$ and $q \neq p - 1$ (see Remark 2.55 for the definition of Δ_{pq}). Moreover, Lagarias and Ziegler [236] have proved that $c\Delta_{pq}$ has a unimodular triangulation for all $c \geq 4$.

But even here some open questions remain: has $c\Pi$ a unimodular triangulation for all $c \geq 4$? What can be said about $3\Delta_{pq}$?

Remark 3.22. It is proved in [221] that the unimodular triangulation of $c\Pi$ constructed above is regular. The proof of the regularity requires a rather technical and involved consideration of different geometric situations.

We only explain where the difficulty lies. In the proof of Theorem 3.17 we have only used stellar and mixed triangulations, and compositions of triangulations already obtained. Therefore, in view of Proposition 1.61 and Lemma 1.65, it is enough to prove the regularity of mixed triangulations.

For the triangulations of the form $(\nu\mu\tilde{\Delta})^{\mu,\nu}$ the regularity condition is the contents of Exercise 3.3. However, the support functions, constructed in the exercise, do not agree on the boundaries when one considers mixed triangulations of general lattice polytopal complexes, and this is makes the proof of regularity very difficult.

It seems there is no known application of the regularity of the triangulations constructed in the proof of Theorem 3.17, neither in the context of *semistable reductions* [221, Chap. II] nor in the context of Koszul properties of polytopal monoid rings (save a few special cases to be considered in Remark 7.39 and, in retrospective, Exercise 3.2). For this reason we have not included the details.

3.C Unimodular covers of multiples of polytopes

Let $P \subset \mathbb{R}^d$ be a lattice polytope. The union of all unimodular d-simplices inside a d-polytope P is denoted by $\operatorname{UC}(P)$. In this section we investigate for which multiples cP one can guarantee that $cP = \operatorname{UC}(cP)$. To this end we let c_d^{pol} denote the infimum of the natural numbers c such that $c'P = \mathrm{UC}(c'P)$ for all lattice d-polytopes P and all natural numbers $c' \geq c$. The following seem to be the only known values of these numbers: $\mathfrak{c}_1^{\mathrm{pol}} = 1$ for trivial reasons, $\mathfrak{c}_2^{\mathrm{pol}} = 1$ by Corollary 2.54, and the equality $\mathfrak{c}_3^{\mathrm{pol}} = 2$ is proved in Kantor and Sarkaria [217]. A priori, it would not be excluded that $\mathfrak{c}_d^{\mathrm{pol}} = \infty$.

The main result of this section is the following polynomial upper bound:

Theorem 3.23. *As $d \to \infty$ one has*

$$\mathfrak{c}_d^{\mathrm{pol}} = O\left(d^6\right).$$

Theorem 3.23 is proved by passage to cones, for which we establish a similar result on covers by unimodular subcones.

We define $\mathfrak{c}_d^{\mathrm{cone}}$ to be the infimum of all natural numbers c such that every rational d-dimensional pointed cone $C \subset \mathbb{R}^d$ admits a unimodular cover $C = \bigcup_{j=1}^k C_j$ for which

$$\mathrm{Hilb}(C_j) \subset c\Delta_C \quad j = 1, \dots, k.$$

Here Δ_C denotes the polytope spanned by 0 and the extreme integral generators of C.

As we already know, the unique full Hilbert triangulation in dimension 2 is unimodular. In particular, $\mathfrak{c}_2^{\mathrm{cone}} = 1$. Since $\mathrm{Hilb}(C) \subset 2\Delta_C$ in dimension 3 and C has a unimodular Hilbert triangulation, we have $\mathfrak{c}_3^{\mathrm{cone}} = 2$. See Corollary 2.66 and Theorem 2.78, which imply $\mathfrak{c}_3^{\mathrm{cone}} \leq 2$. That one has equality is Exercise 3.4.

We can now formulate the main result for unimodular covers of rational cones:

Theorem 3.24. *For all $d \geq 2$ one has*

$$\mathfrak{c}_d^{\mathrm{cone}} \leq \frac{(d+1)d}{2}\left(\lceil \sqrt{d-1}\, \rceil (d-1)\right)^{\mathrm{ld}\, 3}.$$

Here and in the following $\mathrm{ld}\, x$ denotes the base 2 logarithm of x.

Slope independence. As in Section 3.A we consider the system of simplices

$$\Delta_\sigma \subset [0,1]^d, \quad \sigma \in S_d,$$

where S_d is the permutation group of $\{1, \dots, d\}$. We have already seen that the integral parallel translates of the simplices Δ_σ cover the cone

$$\mathbb{R}_+ e_1 + \mathbb{R}_+(e_1 + e_2) + \cdots + \mathbb{R}_+(e_1 + \cdots + e_d) \cong \mathbb{R}_+^d.$$

Suppose we are given a nonzero real linear form

$$\alpha(X_1, \dots, X_d) = a_1 X_1 + \cdots + a_d X_d.$$

The *width* of a polytope $P \subset \mathbb{R}^d$ in direction $(a_1, \dots, a_d)$, denoted by $\mathrm{width}_\alpha(P)$, is defined to be the euclidean distance between the two extreme hyperplanes that are parallel to the hyperplane $a_1 X_1 + \cdots + a_d X_d = 0$ and intersect P. Since $[0,1]^d$ is inscribed in a sphere of radius $\sqrt{d}/2$, we have $\mathrm{width}_\alpha(\Delta_\sigma) \leq \sqrt{d}$ for all linear forms α and all permutations σ. We arrive at the following:

Proposition 3.25. *All integral parallel translates of Δ_σ, $\sigma \in S_d$, that intersect a hyperplane $H \subset \mathbb{R}^d$ are contained in the $\sqrt{d}$-neighborhood of H.*

Let L be an affine lattice. The union of all L-unimodular simplices inside a polytope $P \subset \mathbb{R}^d$ is denoted by $\mathrm{UC}_L(P)$ (we do not require that P be an L-polytope). If $\Delta = \operatorname{conv}(w_0, \dots, w_e) \subset \mathbb{R}^d$ is an e-simplex, then $\mathrm{UC}_\Delta(P)$ denotes $\mathrm{UC}_{\mathscr{L}(\Delta)}(P)$ where, as usual, $\mathscr{L}(\Delta) = w_0 + \sum_{i=0}^{e} \mathbb{Z}(w_i - w_0)$ is the smallest affine lattice containing $\operatorname{vert}(\Delta)$.

Let $\Delta \subset \Delta'$ be two d-simplices in $\mathbb{R}^d$ for which the origin 0 is a common vertex and the two simplicial cones $\mathbb{R}_+\Delta$ and $\mathbb{R}_+\Delta'$ coincide. The following lemma says that the $\mathscr{L}(\Delta)$-unimodularly covered area in a multiple $c\Delta'$, $c \in \mathbb{N}$, approximates $c\Delta'$ with a precision independent of Δ and Δ'. The precision is therefore independent of the slopes of the facets of Δ and Δ' opposite to 0. The lemma will be critical both in the passage to cones (Proposition 3.29) and in the proof of Theorem 3.24.

Lemma 3.26. *For a pair of d-simplices $\Delta \subset \Delta'$, having 0 as a common vertex at which they span the same cone, and a real number $\varepsilon \in (0,1)$, one has*

$$(1-\varepsilon)c\Delta' \subset \mathrm{UC}_\Delta(c\Delta')$$

whenever $c \geq \sqrt{d}/\varepsilon$.

Proof. Let $v_1, \dots, v_d$ be the vertices of Δ different from 0, and let w_i be the vertex of Δ' on the ray $\mathbb{R}_+ v_i$, $i = 1, \dots, d$. By a rearrangement of the indices we can achieve that

$$\frac{\|w_1\|}{\|v_1\|} \geq \frac{\|w_2\|}{\|v_2\|} \geq \dots \geq \frac{\|w_d\|}{\|v_d\|} \geq 1$$

where $\| \ \|$ denotes euclidean norm. Moreover, the assertion of the lemma is invariant under linear transformations of $\mathbb{R}^d$. Therefore we can assume that

$$\Delta = \operatorname{conv}(0, e_1, e_1 + e_2, \dots, e_1 + \dots + e_d).$$

Then $\mathscr{L}(\Delta) = \mathbb{Z}^d$. The ratios above are also invariant under linear transformations. Thus

$$\frac{\|w_1\|}{\|e_1\|} \geq \frac{\|w_2\|}{\|e_1 + e_2\|} \geq \dots \geq \frac{\|w_d\|}{\|e_1 + \dots + e_d\|} \geq 1.$$

Now Lemma 3.28 below shows that the distance h from 0 to the affine hyperplane H through $w_1, \dots, w_d$ is at least 1.

By Proposition 3.25, the subset

$$(c\Delta') \setminus U_{\sqrt{d}}(cH) \subset c\Delta'$$

is covered by integral parallel translates of the simplices Δ_σ, $\sigma \in S_d$ that are contained in $c\Delta'$. ($U_\rho(X)$ is the ρ-neighborhood of X.) In particular,

$$(c\Delta') \setminus U_{\sqrt{d}}(cH) \subset \mathrm{UC}_\Delta(c\Delta'). \tag{3.5}$$

Therefore we have

$$(1-\varepsilon)c\Delta' \subset \left(1 - \frac{\sqrt{d}}{c}\right)c\Delta' \subset \left(1 - \frac{\sqrt{d}}{ch}\right)c\Delta' = \frac{ch - \sqrt{d}}{ch}c\Delta'$$
$$= (c\Delta') \setminus U_{\sqrt{d}}(cH),$$

and the lemma follows from (3.5). □

Remark 3.27. One can derive an analogous result using the trivial tiling of $\mathbb{R}^d_+$ by the integral parallel translates of the unit cube $[0,1]^d$ and the fact that $[0,1]^d$ itself is unimodularly covered. The argument would then get simplified, but the estimate obtained is $c \geq d/\varepsilon$, and thus worse than $c \geq \sqrt{d}/\varepsilon$.

We have formulated the Lemma 3.26 only for full-dimensional simplices, but it is clear that the claim holds for simplices of smaller dimension as well: one simply chooses all data relative to the affine subspace generated by Δ'.

Above we have used the following

Lemma 3.28. *Let $e_1, \dots, e_d$ be the canonical basis of $\mathbb{R}^d$ and set $w_i = \lambda_i(e_1 + \cdots + e_i)$ where $\lambda_1 \geq \cdots \geq \lambda_d > 0$. Then the affine hyperplane H through $w_1, \dots, w_d$ intersects the set $Q = \lambda_d(e_1 + \cdots + e_d) - \mathbb{R}^d_+ = w_d - \mathbb{R}^d_+$ only in the boundary ∂Q. In particular the euclidean distance from 0 to H is $\geq \lambda_d$.*

Proof. The hyperplane H is given by the equation

$$\frac{1}{\lambda_1}X_1 + \left(\frac{1}{\lambda_2} - \frac{1}{\lambda_1}\right)X_2 + \cdots + \left(\frac{1}{\lambda_d} - \frac{1}{\lambda_{d-1}}\right)X_d = 1.$$

The linear form α on the left hand side has nonnegative coefficients and $w_d \in H$. Thus a point whose coordinates are strictly smaller than λ_d cannot be contained in H. □

Passage to cones. In this section we want to relate the bounds for $\mathfrak{c}^{\text{pol}}_d$ and $\mathfrak{c}^{\text{cone}}_d$. This allows us to derive Theorem 3.23 from Theorem 3.24. We will use the *corner cones* of a polytope P: the corner cone C of P at a vertex v is given by

$$C = \mathbb{R}_+(P - v) = \sum_{w \in \text{vert}(P)} \mathbb{R}_+(w - v).$$

Proposition 3.29. *Let d be a natural number. Then $\mathfrak{c}^{\text{pol}}_d$ is finite if and only if $\mathfrak{c}^{\text{cone}}_d$ is finite, and, moreover,*

$$\mathfrak{c}^{\text{cone}}_d \leq \mathfrak{c}^{\text{pol}}_d \leq \sqrt{d}(d+1)\mathfrak{c}^{\text{cone}}_d. \tag{3.6}$$

Proof. Suppose that $\mathfrak{c}^{\text{pol}}_d$ is finite. Then the left inequality is easily obtained by considering the multiples of the polytope Δ_C for a cone C: the cones spanned by those unimodular simplices in a multiple of Δ_C that contain 0 as a vertex constitute a unimodular cover of C.

Now suppose that $\mathfrak{c}^{\text{cone}}_d$ is finite. For the right inequality we first triangulate a polytope P into lattice simplices. Then it is enough to consider a lattice d-simplex $\Delta \subset \mathbb{R}^d$ with vertices $v_0, \dots, v_d$.

Set $c' = \mathrm{c}_d^{\mathrm{cone}}$. For each i there exists a unimodular cover (D_{ij}) of the corner cone C_i of Δ with respect to the vertex v_i such that $c'\Delta - c'v_i$ contains $\Delta_{D_{ij}}$ for all j. Thus the simplices $\Delta_{D_{ij}} + c'v_i$ cover the corner of $c'\Delta$ at $c'v_i$, that is, their union contains a neighborhood of $c'v_i$ in $c'\Delta$.

We replace Δ by $c'\Delta$ and can assume that each corner of Δ has a cover by unimodular simplices. It remains to show that the multiples $c''\Delta$ are unimodularly covered for every number $c'' \geq \sqrt{d}(d+1)$ for which $c''P$ is an integral polytope.

Let

$$\omega = \frac{1}{d+1}(v_0 + \cdots + v_d)$$

be the barycenter of Δ. We define the subsimplex $\Delta_i \subset \Delta$ as follows: Δ_i is the homothetic image of Δ with respect to the center v_i so that ω lies on the facet of Δ_i opposite to v_i. The factor of the homothety that transforms Δ into Δ_i is $d/(d+1)$. In particular, the simplices Δ_i are pairwise congruent. It is also clear that

$$\bigcup_{i=0}^{d} \Delta_i = \Delta. \tag{3.7}$$

The construction of ω and the subsimplices Δ_i commutes with taking multiples of Δ. It is therefore enough to show that $c''\Delta_i \subset \mathrm{UC}(c''\Delta)$ for all i. In order to simplify the use of dilatations we move v_i to 0 by a parallel translation.

In the case in which $v_i = 0$ the simplices $c''\Delta$ and $c''\Delta_i$ are the unions of their intersections with the cones D_{ij}. This observation reduces the inclusion $c''\Delta_i \subset c''\Delta$ to

$$c''(\Delta_i \cap D_{ij}) \subset c''(\Delta \cap D_{ij})$$

for all j. But now we are in the situation of Lemma 3.26, with the unimodular simplex $\Delta_{D_{ij}}$ in the role of the Δ of 3.26 and $\Delta \cap D_{ij}$ in that of Δ'. For $\varepsilon = 1/(d+1)$ we have $c'' \geq \sqrt{d}/\varepsilon$ and so

$$c''(\Delta_i \cap D_{ij}) = c''\frac{d}{d+1}(\Delta \cap D_{ij}) = c''(1-\varepsilon)(\Delta \cap D_{ij}) \subset \mathrm{UC}(c''(\Delta \cap D_{ij})),$$

as desired. (The left inequality in (3.6) has only been stated for completeness; it will not be used later on.) □

The steps in the proof Proposition 3.29 are illustrated by Figure 3.5 where we have marked the barycenter of $c''c'\Delta$.

At this point we can deduce Theorem 3.23 from Theorem 3.24. Using the bound for $\mathrm{c}_d^{\mathrm{cone}}$, given in Theorem 3.24, and that $\lceil\sqrt{d-1}\rceil(d-1) = O(d^{1.5})$ as $d \to \infty$, we get

$$\begin{aligned}
\mathrm{c}_d^{\mathrm{pol}} &\leq \sqrt{d}(d+1)\mathrm{c}_d^{\mathrm{cone}} \\
&\leq \sqrt{d}(d+1)\frac{(d+1)d}{2}\left(\lceil\sqrt{d-1}\,\rceil(d-1)\right)^{\operatorname{ld} 3} \\
&= O\left(d^6\right),
\end{aligned}$$

as desired.

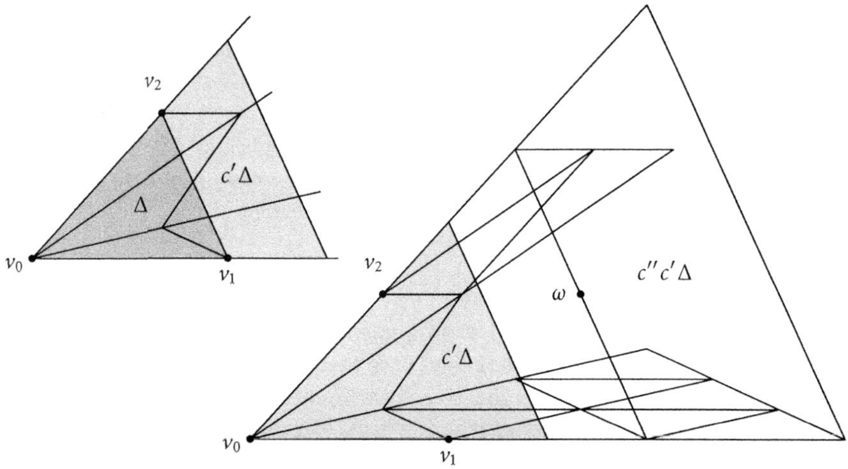

Fig. 3.5. Extension of the corner cover into a multiple of Δ

Bounding covers of simplicial cones. Let C be a simplicial rational d-cone. Recall that the simplex Δ_C is the convex hull of 0 and the extreme integral generators of C. Let $\gamma(C)$ be the minimum of all real numbers c such that C has a cover by unimodular cones D_i, $i \in I$, such that $\operatorname{Hilb}(D_i) \subset c\Delta_C$. Since C can be triangulated into unimodular cones by Theorem 2.75, $\gamma(C)$ is well-defined. Moreover, there exist only finitely many isomorphism classes of simplices Δ_C of given multiplicity and dimension by Corollary 2.82. Thus we can set

$$\gamma_d(\mu) = \max\{\gamma(C) : \dim C = d, \mu(C) \le \mu\}. \tag{3.8}$$

Let C be a simplicial cone with extreme integral generators $v_1, \dots, v_d$. We want to cover C by unimodular subcones. If C is unimodular, there is nothing to do. Otherwise we choose the smallest prime factor p of $\mu(C)$. (Any prime factor would do, but the choice of the smallest makes the proof of Lemma 3.31 easier.) Therefore we can find an element of order p in $\mathbb{Z}^d/(\mathbb{Z}v_1 + \cdots + \mathbb{Z}v_d)$ since $\mu(C)$ is the order of this group. Its representative in $\operatorname{par}(v_1, \dots, v_d)$ has the form

$$x = \sum_{i=1}^{d} \frac{q_i}{p} v_i, \qquad q_i \in \mathbb{Z},\ 0 \le q_i \le p-1. \tag{3.9}$$

For each i with $q_i \ne 0$, multiplying by the inverse of $-q_i$ modulo p and taking again the representative in $\operatorname{par}(v_1, \dots, v_d)$ we obtain a vector

$$x_i' = \sum_{j \ne i} \frac{r_j'}{p} v_j + \frac{p-1}{p} v_i, \qquad r_j' \in \mathbb{Z},\ 0 \le r_j' \le p-1. \tag{3.10}$$

In general x_i' need not have coprime entries, and if it does not, we replace it by the integral generator of $\mathbb{R}_+ x_i' \cap \mathbb{Z}^d$. In this way we arrive at a vector

$$x_i = \sum_{j \neq i} \frac{r_j}{p} v_j + \frac{r_i}{p} v_i, \qquad r_j \in \mathbb{Z},\ 0 \le r_j \le p-1,\ r_i \mid p-1. \tag{3.11}$$

The coefficients r_j/p are bounded above by 1/2 if either $p = 2$ or $p > 2$ and $x_i \neq x_i'$. For each i with $q_i \neq 0$ we set

$$D_i = \sum_{j \neq i} \mathbb{R}_+ v_j + \mathbb{R}_+ x_i.$$

Then

$$\mu(D_i) = \frac{r_i}{p} \mu(C),$$

and we have

$$\mu(D_i) < \mu(C) \qquad \text{and} \qquad \mu(D_i) \le \frac{1}{2}\mu(C) \quad \text{if } p = 2 \text{ or } x_i \neq x_i'. \tag{3.12}$$

The cone C is considered to be of *generation* 0, and the cones D_i, $q_i \neq 0$, produced from it as above, form the first generation. The cones produced from the D_i are of second generation, etc.

Lemma 3.30. *The cone C is covered by the cones D_i, and an iterated application of the procedure above yields a unimodular cover of C.*

Proof. It is enough to show that C is covered by the D_i. Since the multiplicities of the cones are dropping with generation, a unimodular cover is reached after finitely many steps.

Pick $w = \sum_{i=1}^d \alpha_i v_i \in C$. Then $\alpha_1, \dots, \alpha_d \in \mathbb{R}_+$. We choose k such that α_k is minimal among all α_i with $q_i \neq 0$, where q_i is as in (3.9). We claim that $w \in D_k = \sum_{j \neq k} \mathbb{R}_+ v_j + \mathbb{R}_+ x_k'$.

With r_j' as in (3.10) for $i = k$ one has

$$w = \frac{p\alpha_k}{p-1} x_k' + \sum_{j \neq k} \Big(\alpha_j - \frac{r_j' \alpha_k}{p-1}\Big) v_j.$$

The coefficient of x_k' is evidently nonnegative, and the same holds true for the coefficient of v_j if $q_j = 0$ since $r_j' = 0$ in this case as well. But also for $q_j > 0$ we have $\alpha_j - r_j'\alpha_k/(p-1) \ge 0$ since $\alpha_k \le \alpha_j$ by the choice of k and $r_j' \le p-1$. □

We have to bound the size of the vectors that are created by the covering algorithm above.

Lemma 3.31. *Let $C = E_0 \supset E_1 \supset \cdots \supset E_g$ be a sequence of cones such that E_i is of generation i and $\mu(E_g) = 1$. Set*

$$T = \Big\{ j \ge 1 : \mu(E_j) \le \frac{1}{2}\mu(E_{j-1}) \Big\}.$$

(a) Then $\#T \ge g/2$ and, therefore, $g \le 2 \operatorname{ld} \mu(C)$.

(b) For all $i = 0, \dots, g$ one has

$$\operatorname{Hilb}(E_i) \subset \frac{d}{2} \cdot \mu(C)^{\operatorname{ld} 3}.$$

Proof. (a) Suppose that $\mu(E_j) > 1/2\mu(E_{j-1})$. Then, because of (3.12), the smallest prime divisor p of $\mu(E_{j-1})$ is odd and the vector x' chosen as in equation (3.10) that spans E_j over a facet of E_{j-1} has coprime entries. It follows that $\mu(E_j) = (p-1)/p\mu(E_{j-1}) > 1$ and so $j < g$. On the other hand, 2 is the smallest prime divisor of $\mu(E_j)$ (because $p-1$ is even), and therefore $\mu(E_{j+1}) \le \frac{1}{2}\mu(E_j)$. Thus every element of $\{1, \dots, g\} \setminus T$ is followed by an element of T, and so T has at least as many elements as $\{1, \dots, g\} \setminus T$. Using (3.12), we conclude that

$$\mu(C) = \prod_{j=1}^{g} \frac{\mu(E_{j-1})}{\mu(E_j)} \ge \prod_{j \in T} \frac{\mu(E_{j-1})}{\mu(E_j)} \ge 2^{\#T} \ge 2^{g/2}.$$

(b) For $j = 1, \dots, g$ we set

$$a_j = \#\{1, \dots, j\} \setminus T, \qquad b_j = \#\{1, \dots, j\} \cap T, \qquad v_j = \begin{cases} 1, & j \notin T, \\ 2, & j \in T. \end{cases}$$

We define the auxiliary sequence of numbers h_j as follows:

$$h_j = 1,\ j \le 0, \qquad \text{and} \qquad h_j = \frac{d}{2} \cdot 2^{a_j} \left(\frac{3}{2}\right)^{b_j - v_j + 1}, \quad 1 \le j \le g.$$

The sequence (h_j) is nondecreasing, and it is checked by induction that

$$\frac{1}{v_j}(h_{j-d} + \cdots + h_{j-1}) \le h_j, \qquad 1 \le j \le g \tag{3.13}$$

(Exercise 3.5).

Consider the sequence of vectors $y_{-(d-1)} = v_0, \dots, y_0 = v_d, y_1, \dots, y_g$ such that E_j is spanned over a facet of E_{j-1} by y_j, and y_j has been chosen as in (3.11).

We claim that y_j, $j \ge 1$, is contained in $h_j \Delta_C$. To this end we note that

$$y_j = \alpha_1 y_{k_1} + \cdots + \alpha_d y_{k_d}$$

where $\alpha_i \le 1/v_j$ for $i = 1, \dots, d$ and $k_1 < \cdots < k_d \le j-1$. Since the sequence (h_k) is nondecreasing and satisfies the inequality (3.13), one obtains

$$y_j \in \frac{1}{v_j}(h_{k_1} + \cdots + h_{k_d})\Delta_C \subset \frac{1}{v_j}(h_{j-d} + \cdots + h_{j-1})\Delta_C \subset h_j \Delta_C \subset h_g \Delta_C.$$

Moreover,

$$\begin{aligned} h_g &= \frac{d}{2} \cdot 2^{a_g} \left(\frac{3}{2}\right)^{b_g - v_g + 1} \le \frac{d}{2} \cdot 2^{a_g} \left(\frac{3}{2}\right)^{b_g} = \frac{d}{2} \cdot 2^{g - \#T} \left(\frac{3}{2}\right)^{\#T} \\ &\le \frac{d}{2} \cdot 2^{g/2} \left(\frac{3}{2}\right)^{g/2} = \frac{d}{2} \cdot 3^{g/2} \le \frac{d}{2} \cdot 3^{\operatorname{ld} \mu(C)} = \frac{d}{2} \cdot \mu(C)^{\operatorname{ld} 3}. \end{aligned}$$ □

An immediate consequence is a bound on $\gamma_d(\mu)$ that is polynomial and of remarkably low degree.

Corollary 3.32. *Let $d > 1$. Then*

$$\gamma_d(\mu) \le \frac{d\mu^{\operatorname{ld} 3}}{2} < \frac{d\mu^{1.6}}{2}.$$

Remark 3.33. The corollary is due to von Thaden [366], [367]. In [65] weaker bounds for $\mathfrak{c}_d^{\text{pol}}$ and $\mathfrak{c}_d^{\text{cone}}$ were based on triangulations of simplicial cones. The argument, based on covers, uses in a crucial way that the multiplicity of the cones drops by a factor of $1/2$ in at least every second generation. Such a good result for triangulations has not yet been reached. Nevertheless, von Thaden has also improved [65, Th. 4.1] significantly: a simplicial cone C of dimension d and multiplicity μ has a unimodular triangulation by vectors contained in

$$\left(\frac{d^2}{2}\cdot\mu^{\varepsilon_0+\operatorname{ld}(9/4)\operatorname{ld}\mu}\right)\cdot\Delta_C, \qquad \varepsilon_0 = 2+\frac{2}{\ln 3/2}\approx 7.$$

The proof requires considerably more effort than that of Corollary 3.32.

Corner covers. Let C be a rational cone and v one of its extreme generators. We say that a system $\{C_j\}_{j=1}^k$ of subcones $C_j \subset C$ *covers the corner of C at v* if $v \in \operatorname{Hilb}(C_j)$ for all j and the union $\bigcup_{j=1}^k C_j$ contains a neighborhood of v in C.

Lemma 3.34. *Suppose that $\mathfrak{c}_{d-1}^{\text{cone}} < \infty$, and let C be a simplicial rational d-cone with extreme generators $v_1,\dots,v_d$.*

(a) Then there is a system of unimodular subcones $C_1,\dots,C_k \subset C$ covering the corner of C at v_1 such that $\operatorname{Hilb}(C_1),\dots,\operatorname{Hilb}(C_k) \subset (\mathfrak{c}_{d-1}^{\text{cone}}+1)\Delta_C$.

(b) Moreover, each element $w \ne v_1$ of a Hilbert basis of C_j, $j=1,\dots,k$, has a representation $w = \xi_1 v_1+\dots+\xi_d v_d$ with $\xi_1 < 1$.

Proof. For simplicity of notation we set $\mathfrak{c} = \mathfrak{c}_{d-1}^{\text{cone}}$. Let C' be the cone generated by $w_i = v_i - v_1$, $i = 2,\dots,d$, and let V be the vector subspace of $\mathbb{R}^d$ generated by the w_i. We consider the linear map $\pi : \mathbb{R}^d \to V$ given by $\pi(v_1) = 0$, $\pi(v_i) = w_i$ for $i > 0$, and endow V with a lattice structure by setting $L = \pi(\mathbb{Z}^d)$. (One has $L = \mathbb{Z}^d \cap V$ if and only if $\mathbb{Z}^d = \mathbb{Z}v_1 + (\mathbb{Z}^d \cap V)$.) Note that $v_1, z_2,\dots,z_d$ with $z_j \in \mathbb{Z}^d$ form a $\mathbb{Z}$-basis of $\mathbb{Z}^d$ if and only if $\pi(z_2),\dots,\pi(z_d)$ form a $\mathbb{Z}$-basis of L. This holds since $\mathbb{Z}v_1 = \mathbb{Z}^d \cap \mathbb{R}v_1$ and explains the unimodularity of the cones C_j constructed next.

Note that $w_i \in L$ for all i. Therefore $\Delta_{C'} \subset \operatorname{conv}(0, w_2,\dots,w_d)$. The cone C' has a unimodular cover (with respect to L) by cones C_j', $j = 1,\dots,k$, with $\operatorname{Hilb}(C_j') \subset \mathfrak{c}\Delta_{C'}$. We lift the vectors $x \in \operatorname{Hilb}(C_j')$ to elements $\tilde{x} \in C$ as follows. Let $x = \alpha_2 w_2+\dots+\alpha_d w_d$ (with $\alpha_i \in \mathbb{Q}_+$). Then there exists a unique integer $n \ge 0$ such that

$$\begin{aligned}\tilde{x} := nv_1 + x &= nv_1 + \alpha_2(v_2 - v_1)+\dots+\alpha_d(v_d - v_1)\\ &= \alpha_1 v_1 + \alpha_2 v_2 + \dots + \alpha_d v_d\end{aligned}$$

with $0 \le \alpha_1 < 1$. If $x \in c\Delta_{C'} \subset c \cdot \operatorname{conv}(0, w_2, \dots, w_d)$, then $\tilde{x} \in (c+1)\Delta_C$.

We now define C_j as the cone generated by v_1 and the vectors $\tilde{x}$ where $x \in \operatorname{Hilb}(C'_j)$. It only remains to show that the C_j cover a neighborhood of v_1 in C. To this end we intersect C with the affine hyperplane H through $v_1, \dots, v_d$. It is enough that a neighborhood of v_1 in $C \cap H$ is contained in $C_1 \cup \dots \cup C_k$.

For each $j = 1, \dots, k$ the coordinate transformation from the basis $w_2, \dots, w_d$ of V to the basis $x_2, \dots, x_d$ with $\{x_2, \dots, x_d\} = \operatorname{Hilb}(C'_j)$ defines a linear operator on $\mathbb{R}^{d-1}$. Let M_j be its $\| \ \|_\infty$ norm.

Moreover, let N_j be the maximum of the numbers n_i, $i = 2, \dots, d$ defined by the equation $\tilde{x}_i = n_i v_1 + x_i$ as above. Choose ε with

$$0 < \varepsilon \le \frac{1}{(d-1)M_j N_j}, \qquad j = 1, \dots, k,$$

and consider

$$y = v_1 + \beta_2 w_2 + \dots + \beta_d w_d, \qquad 0 \le \beta_i < \varepsilon.$$

Since the C'_j cover C', one has $\beta_2 w_2 + \dots + \beta_d w_d \in C'_j$ for some j, and therefore

$$y = v_1 + \gamma_2 x_2 + \dots + \gamma_d x_d,$$

where $\{x_2, \dots, x_d\} = \operatorname{Hilb}(C'_j)$ and $0 \le \gamma_i \le M_j \varepsilon$ for $i = 2, \dots, d$. Then

$$y = \left(1 - \sum_{i=2}^{d} n_i \gamma_i\right) v_1 + \gamma_2 \tilde{x}_2 + \dots + \gamma_d \tilde{x}_d$$

and

$$\sum_{i=2}^{d} n_i \gamma_i \le (d-1) N_j M_j \varepsilon \le 1,$$

whence $\left(1 - \sum_{i=2}^{d} n_i \gamma_i\right) \ge 0$ and $y \in C_j$, as desired. □

The bound for cones. The next lemma contains the argument that we need to finish the proof of Theorem 3.24. Let us call a simplicial cone C *empty* if Δ_C is an empty simplex.

Lemma 3.35. *Let C be an empty simplicial rational d-cone. Then C has a unimodular cover by unimodular rational subcones D such that*

$$\operatorname{Hilb}(D) \subset (d+1) \cdot \beta_d \cdot \gamma_d(\beta_d), \qquad \beta_d = \left\lceil \sqrt{d-1} \right\rceil (d-1) \tag{3.14}$$

(γ_d has been defined in (3.8)*).*

In Theorem 3.24 we claim the same inequality as in Lemma 3.35, though for arbitrary rational pointed d-cones. Therefore, in order to deduce Theorem 3.24 from Lemma 3.35, it is enough that C can be triangulated into empty simplicial cones C' such that $\Delta_{C'} \subset \Delta_C$. In fact, one first triangulates C into simplicial cones generated by extreme generators of C. After this step one can assume that C is simplicial with extreme generators $v_1, \ldots, v_d$. If Δ_C is not empty, then we use stellar subdivision along a ray through some $v \in \Delta_C \cap \mathbb{Z}^d$, $v \neq 0, v_1, \ldots, v_d$, and for each of the resulting cones C' the simplex $\Delta_{C'}$ has a smaller number of integral vectors than Δ_C. This completes the proof of 3.24.

Before we embark on the proof of Lemma 3.35, we single out a technical step. Let $\{v_1, \ldots, v_d\} \subset \mathbb{R}^d$ be a linearly independent subset. Consider the hyperplane

$$H = \sum_{i=2}^{d} \mathbb{R}(v_1 + (d-1)v_i) \subset \mathbb{R}^d.$$

It cuts a simplex δ off the simplex $\operatorname{conv}(v_1, \ldots, v_d)$ so that $v_1 \in \delta$. Let Φ denote the closure of

$$\mathbb{R}_+\delta \smallsetminus \Big(\big((1 + \mathbb{R}_+)v_1 + \mathbb{R}_+ v_2 + \cdots + \mathbb{R}_+ v_d \big) \cup \Delta \Big) \subset \mathbb{R}^d,$$

where $\Delta = \operatorname{conv}(0, v_1, \ldots, v_d)$. The polytope

$$\Phi' = -\frac{1}{d-1} v_1 + \frac{d}{d-1} \Phi$$

is the homothetic image of the polytope Φ under the dilatation with factor $d/(d-1)$ and center v_1. (See Figure 3.6 for the case $d = 2$.) We will need that

$$\Phi' \subset (d+1)\Delta. \tag{3.15}$$

The easy proof is left to the reader (Exercise 3.6).

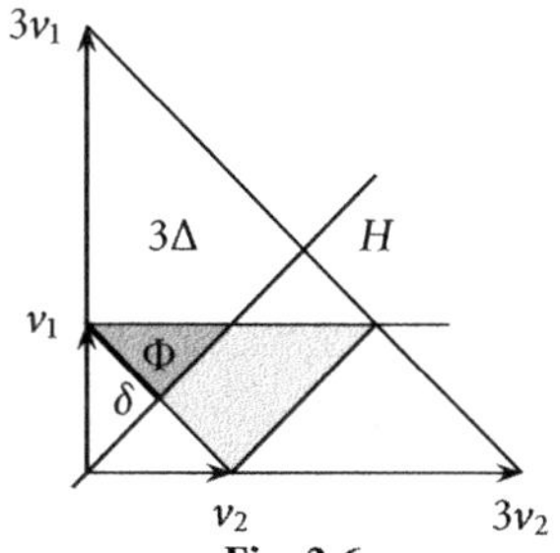

Fig. 3.6.

Proof of Lemma 3.35. The lemma holds for $d = 2$ since empty simplicial 2-cones are unimodular and the right hand side of (3.14) is $3/2$ for $d = 2$. By induction we can now assume that the lemma has been shown for all dimensions $< d$ (the right hand side of (3.14) is increasing with d). We set

$$\beta = \beta_d = \left\lceil \sqrt{d-1} \right\rceil (d-1) \quad \text{and} \quad \kappa = (d+1) \cdot \beta_d \cdot \gamma_d(\beta_d).$$

Let us first outline the course of the somewhat involved arguments that now follow. They are subdivided into four major steps. The first three of them are very similar to their analogues in the proof of Proposition 3.29. In Step 1 we cover the d-cone C by $d+1$ smaller cones, each of which is bounded by the hyperplane that passes through the barycenter of $\operatorname{conv}(v_1,\dots,v_d)$ and is parallel to the facet of $\operatorname{conv}(v_1,\dots,v_d)$ opposite of v_i, $i=1,\dots,d$. We summarize this step in Claim A below.

In Step 2 Lemma 3.34 is applied for the construction of unimodular corner covers. Claim B states that it is enough to cover the constructed subcones of C in direction of the cones forming the corner cover.

In Step 3 we extend the corner cover far enough into C. Lemma 3.26 allows us to do this within a suitable multiple of Δ_C. The most difficult part of the proof is to control the size of all vectors involved.

Lemma 3.26 is applied to simplices $\Gamma = \operatorname{conv}(w_1,\dots,w_e)$ where $w_1,\dots,w_e$ span a unimodular cone of dimension $e \le d$. The cones over the unimodular simplices covering $c\Gamma$ have multiplicity dividing c, and possibly equal to c. Nevertheless we obtain a cover of C by cones with *bounded* multiplicities, so we can apply Corollary 3.32 in Step 4 to obtain a unimodular cover.

A convention: in the course of the proof we will sometimes have to deal with sets of the form $v+C$ where $v \in \mathbb{R}^d$, which will be called a *cone with apex* v.

Step 1. The facet $\operatorname{conv}(v_1,\dots,v_d)$ of Δ_C is denoted by Γ_0. (We use the letter Γ for $(d-1)$-dimensional simplices, and Δ for d-dimensional ones.) For $i=1,\dots,d$ we put

$$H_i = \operatorname{aff}(0,\ v_i+(d-1)v_1,\dots,v_i+(d-1)v_{i-1},\ v_i+(d-1)v_{i+1},\dots,v_i+(d-1)v_d)$$

and

$$\Gamma_i = \operatorname{conv}\big(v_i, \Gamma_0 \cap H_i\big).$$

Observe that $v_1+\dots+v_d \in H_i$. In particular, the hyperplanes H_i, $i=1,\dots,d$, contain the barycenter of Γ_0, i.e., $(1/d)(v_1+\dots+v_d)$. In fact, H_i is the vector subspace of dimension $d-1$ through the barycenter of Γ_0 that is parallel to the facet of Γ_0 opposite to v_i. Clearly, we have the representation $\bigcup_{i=1}^d \Gamma_i = \Gamma_0$, similar to (3.7) above. In particular, each of the Γ_i is homothetic to Γ_0 with factor $(d-1)/d$.

To prove (3.14) it is enough to show the following

Claim A. For each index $i=1,\dots,d$ there exists a system of unimodular cones

$$C_{i1},\dots,C_{ik_i} \subset C$$

such that $\operatorname{Hilb}(C_{ij}) \subset \kappa\Delta_C$, $j=1,\dots,k_i$, and $\Gamma_i \subset \bigcup_{j=1}^{k_i} C_{ij}$.

The step from the original claim to the reduction expressed by Claim A seems rather small—we have only covered the cross-section Γ_0 by the Γ_i, and stated that it is enough to cover each Γ_i by unimodular subcones. The essential point is that these subcones need not be contained in the cone spanned by Γ_i, but just in C. This gives us the freedom to start with a corner cover at v_i and to extend it far enough into C, namely beyond H_i. This is made more precise in the next step.

Step 2. To prove Claim A it is enough to treat the case $i=1$. The induction hypothesis implies $c_{d-1}^{\text{cone}} \le \kappa$ because the right hand side of the inequality (3.14) is an increasing function of d. Thus Lemma 3.34 provides a system of unimodular cones $C_1, \dots, C_k \subset C$ covering the corner of C at v_1 such that

$$\operatorname{Hilb}(C_j) \setminus \{v_1, \dots, v_d\} \subset (\kappa\Delta_C) \setminus \Delta_C, \quad j = 1, \dots, k. \tag{3.16}$$

Here we use the emptiness of Δ_C as it guarantees that $\operatorname{Hilb}(C_j) \cap (\Delta_C \setminus \Gamma_0) = \varnothing$, which will be crucial for the inclusion (3.18) in Step 3.

With a suitable enumeration $\{v_{j1}, \dots, v_{jd}\} = \operatorname{Hilb}(C_j)$, $j = 1, \dots, k$ we have $v_{11} = v_{21} = \dots = v_{k1} = v_1$ and

$$0 \le (v_{jl})_{v_1} < 1, \quad j = 1, \dots, k, \quad l = 2, \dots, d, \tag{3.17}$$

where $(-)_{v_1}$ is the first coordinate of an element of $\mathbb{R}^d$ with respect to the basis $v_1, \dots, v_d$ of $\mathbb{R}^d$ (see Lemma 3.34(b)).

Now we formulate precisely what it means to extend the corner cover beyond the hyperplane H_1. Fix an index $j = 1, \dots, k$ and let $D \subset \mathbb{R}^d$ denote the simplicial d-cone determined by the following conditions:

(i) $C_j \subset D$;
(ii) the facets of D contain those facets of C_j that pass through 0 and v_1;
(iii) the remaining facet of D is in H_1.

Figure 3.7 describes the situation in the cross-section Γ_0 of C.

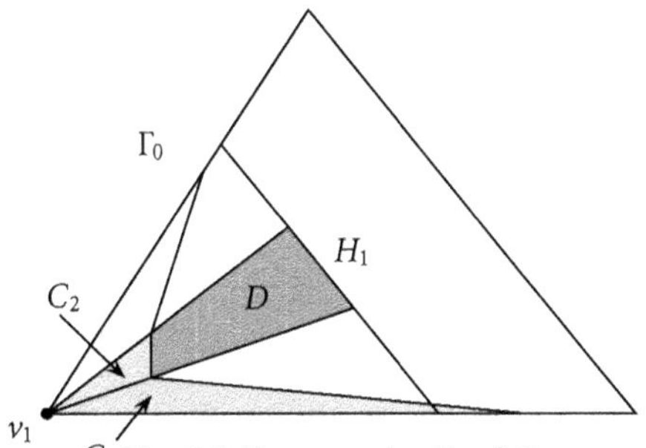

Fig. 3.7. Cross-section Γ_0 of C

By considering all possible values $j = 1, \dots, k$, it becomes clear that to prove Claim A it is enough to prove the following claim.

Claim B. There exists a system of unimodular cones $D_1, \dots, D_T \subset C$ such that

$$\operatorname{Hilb}(D_t) \subset \kappa\Delta_C, \quad t = 1, \dots, T \qquad \text{and} \qquad D \subset \bigcup_{t=1}^{T} D_t.$$

Step 3. For simplicity of notation we put $\Delta = \Delta_{C_j}$, $H = H_1$. (Recall that Δ is of dimension d, spanned by 0 and the extreme integral generators of C_j.) The vertices of Δ, different from 0 and v_1 are denoted by $w_2, \dots, w_d$ in such a way that there exists i_0, $1 \le i_0 \le d$, for which

(i) $w_2, \dots, w_{i_0} \in D \setminus H$ ("bad" vertices, on the same side of H as v_1),
(ii) $w_{i_0+1}, \dots w_d \in \overline{C_j \setminus D}$ ("good" vertices, beyond or on H),

neither $i_0 = 1$ nor $i_0 = d$ being excluded. In the situation of Figure 3.7 the cone C_2 has two bad vertices, whereas C_1 has one good and one bad vertex. (Of course, we see only the intersection points of the cross-section Γ_0 with the rays from 0 through the vertices.)

If all vertices are good, there is nothing to prove since $D \subset C_j$ in this case. So assume that there are bad vertices, i. e., $i_0 \geq 2$. We now show that the bad vertices are caught in a compact set whose size with respect to Δ_C depends only on d, and this fact makes the whole proof work.

Consider the $(d-1)$-dimensional cone

$$E = v_1 + \mathbb{R}_+(w_2 - v_1) + \cdots + \mathbb{R}_+(w_d - v_1).$$

In other words, E is the $(d-1)$-dimensional cone with apex v_1 spanned by the facet $\operatorname{conv}(v_1, w_2, \dots, w_d)$ of Δ opposite to 0. It is crucial in the following that the simplex $\operatorname{conv}(v_1, w_2, \dots, w_d)$ is unimodular (with respect to $\mathbb{Z}^d \cap \operatorname{aff}(v_1, w_2, \dots, w_d)$), as follows from the unimodularity of C_j.

Due to the inequality (3.17) the hyperplane H cuts a $(d-1)$-dimensional (usually nonlattice) simplex off the cone E. We denote this simplex by Γ. Figure 3.8 illustrates the situation by a vertical cross-section of the cone C.

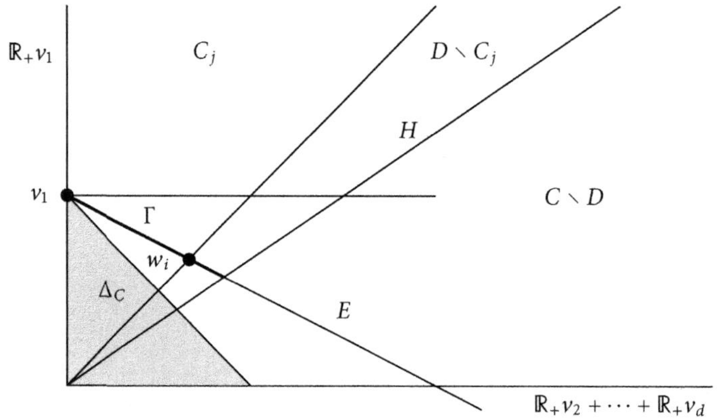

Fig. 3.8.

By (3.16) and (3.17) we have

$$\Gamma \subset \Phi = \overline{\mathbb{R}_+\Gamma_1 \setminus ((v_1 + C) \cup \Delta_C)}.$$

Let ϑ be the dilatation with center v_1 and factor $d/(d-1)$. Then by (3.15) we have the inclusion

$$\vartheta(\Gamma) \subset (d+1)\Delta_C. \tag{3.18}$$

One should note that this inclusion has two aspects: first it shows that Γ is not too big with respect to Δ_C. Second, it guarantees that there is some $\zeta > 0$ *only depending on* d, namely $\zeta = 1/(d-1)$, such that the dilatation with factor $1+\zeta$ and center v_1 keeps Γ inside C. If ζ depended on C, there would be no control on the factor c introduced below.

Let $\Sigma_1 = \operatorname{conv}(v_1, w_2, \dots, w_{i_0})$ and Σ_2 be the smallest face of Γ that contains Σ_1. These are d'-dimensional simplices, $d' = i_0 - 1$. Note that $\Sigma_2 \subset \vartheta(\Sigma_2)$.

We want to apply Lemma 3.26 to the pair

$$\beta v_1 + (\Sigma_1 - v_1) \subset \beta v_1 + (\Sigma_2 - v_1)$$

of simplices with the common vertex βv_1. The lattice of reference for the unimodular covering is

$$L = \mathscr{L}(\beta v_1 + (\Sigma_1 - v_1)) = \beta v_1 + \sum_{j=2}^{i_0} \mathbb{Z}(w_j - v_1).$$

Set

$$\varepsilon = \frac{1}{d} \quad \text{and} \quad c = \frac{d}{d-1}\beta = \left\lceil \sqrt{d-1}\, \right\rceil d.$$

Since $d' \leq d-1$, Lemma 3.26 (after the parallel translation of the common vertex to 0 and then back to βv_1) and (3.18) imply

$$\beta\Sigma_2 \subset \mathrm{UC}_L\big(\beta\vartheta(\Sigma_2)\big) \subset \beta(d+1)\Delta_C. \tag{3.19}$$

Step 4. Consider the i_0-dimensional simplices spanned by 0 and the unimodular $(i_0 - 1)$-simplices appearing in (3.19). Their multiplicities with respect to the i_0-rank lattice $\mathbb{Z}L_{\Sigma_1}$ are all equal to β, since Σ_1, a face of $\operatorname{conv}(v_1, w_2, \dots, w_d)$ is unimodular and, thus, we have unimodular simplices σ on height β. The cones $\mathbb{R}_+\sigma$ have multiplicity dividing β. Therefore, by Corollary 3.32 we conclude that the i_0-cone $\mathbb{R}_+\Sigma_2$ is in the union $\delta_1 \cup \dots \cup \delta_T$ of unimodular (with respect to the lattice $\mathbb{Z}L_{\Sigma_1}$) cones such that

$$\begin{aligned} \operatorname{Hilb}(\delta_1), \dots, \operatorname{Hilb}(\delta_T) &\subset \gamma_d(\beta)\Delta_{\mathbb{R}_+\Sigma_2} \\ &\subset (d+1)\cdot\beta\cdot\gamma_d(\beta)\cdot\Delta_C = \kappa\Delta_C. \end{aligned}$$

In view of the unimodularity of $\operatorname{conv}(v_1, w_2, \dots, w_d)$, the subgroup $\mathbb{Z}L_{\Sigma_1}$ is a direct summand of $\mathbb{Z}^d$. It follows that

$$D_t = \delta_t + \mathbb{R}_+ w_{i_0+1} + \dots + \mathbb{R}_+ w_d, \quad t = 1, \dots, T,$$

is the desired system of unimodular cones. $\square$

Exercises

3.1. Show that the simplices Δ_σ, $\sigma \in S_d$, triangulate the unit d-cube.

3.2. Show that a lattice polytope $P \subset \mathbb{R}^d$, bounded by parallel translates of the hyperplanes $x_i = x_j$, $1 \le i < j \le d$, and $x_i = 0$, $1 \le i \le d$, has a regular unimodular triangulation for which every minimal set of lattice points, not spanning a simplex of the triangulation, has two elements.

Hint: use affine Weyl chambers of type A_d.

3.3. (a) Let $\Delta = \operatorname{conv}(p_0, \dots, p_n) \subset \mathbb{R}^d$ be an n-simplex and $p_0 < \cdots < p_n$. Let $x_0, \dots, x_n$ the corresponding barycentric coordinates in $\operatorname{aff}(\Delta)$. Show that for $\kappa \in \mathbb{N}$ the function

$$f = \sum_{\substack{0 \le j < i \le n \\ 1 \le \rho \le \kappa - 1}} \left| \sum_{j \le s \le i} x_s - \rho \right| : \operatorname{aff}(\Delta) \to \mathbb{R}$$

supports the canonical triangulation of $\kappa\Delta$.

(b) Show that for two natural numbers μ, ν, a point $q \in \mathbb{R}^d \setminus \operatorname{aff}(\Delta)$, and the simplex $\tilde{\Delta} = \operatorname{conv}(p_0, \dots, p_n, q)$ with the order $p_0 < \cdots < p_n < q$ on the vertex set, the corresponding mixed triangulation $(\nu\mu\tilde{\Delta})^{\mu,\nu}$ is regular.

Hint: the functions (like to f in (a)), supporting the canonical triangulations of $\nu \operatorname{conv}(\Delta^\pi, \mu q)$, agree on common faces, where Δ^π be the facet of the canonical triangulation of $\mu\Delta$, corresponding to a mapping $\pi : \{1, \dots, d\} \to \{0, 1, \dots, \mu - 1\}$ (Corollary 3.5). On the other hand, $\nu\mu\tilde{\Delta}$ is triangulated into simplices $\nu \operatorname{conv}(\Delta^\pi, \mu q)$ in a regular way and, hence, Corollary 1.62 applies.

3.4. Prove $\mathfrak{c}_3^{\mathrm{cone}} = 2$.

3.5. Prove that the sequence (h_j) in the proof of Lemma 3.31 is nondecreasing and satisfies (3.31).

3.6. Prove (3.15).

3.7. This exercise shows that it is possible to speed up the triangulation process, used in the proof of Theorem 2.72, little bit.

(a) Let $x_1, \dots, x_d \in \mathbb{Z}^d$ be linearly independent vectors. Denote by $\square(x_1, \dots, x_d)$ the closure in $\mathbb{R}^d$ of the semi-open parallelotope $\operatorname{par}(x_1, \dots, x_n)$. Show that if $x_1, \dots, x_d$ do not form a basis of $\mathbb{Z}^d$ then

$$\frac{d}{d+1} \square(x_1, \dots, x_d) \cap \mathbb{Z}^d \ne \{0\}.$$

(b) Let C be a simplicial cone of dimension d and multiplicity μ. Using claim (a), derive an upper estimate for the number of successive stellar subdivisions, needed to triangulate C into unimodular cones.

(c) Let C be as in (b). Using (a), derive an upper bound for vectors, triangulating C into unimodular simplices. (These bounds are considerably worse than the estimate in Remark 3.33.)

Hint for (a): use induction on d. In the inductive step reduce the general case to the situation when $\{x_1, \dots, x_{i-1}, x_{i+1}, \dots, x_d\}$ is a part of a basis of $\mathbb{Z}^d$ for every i. Then there is a lattice point $x \in \operatorname{int} \square(x_1, \dots, x_d)$ such that for all i one has

$$a, b \in \mathbb{Z}, \quad ax \in bx_i + \sum_{j \ne i} \mathbb{R}x_j \quad \Longrightarrow \quad ax_i \in \sum_{j=1}^{d} \mathbb{Z}x_j.$$

By a rational change of coordinates assume $\square(x_1, \dots, x_d) = \square_d$, the standard unit cube. Then $x = (p_1/q, \dots, p_d/q)$ with $0 < p_i < q$ and $\gcd(p_i, q) = 1$, $i = 1, \dots, d$. One can assume $q \ge d+2$. For the fractional parts we have the equality of sets

$$\left\{\left\{\frac{p_i}{q}\right\},\left\{\frac{2p_i}{q}\right\},\dots,\left\{\frac{(q-1)p_i}{q}\right\}\right\}=\left\{\frac{1}{q},\frac{2}{q},\dots,\frac{q-1}{q}\right\}.$$

The number of elements in $\{1/q,\dots,(q-1)/q\}\cap\left(d(d+1)^{-1},1\right]$ is at most $(q-1)(d+1)^{-1}$. Then for some $1\le c\le q-1$ one has

$$0<\left\{\frac{cp_i}{q}\right\}\le\frac{d}{d+1},\qquad i=1,\dots,d.$$

Therefore, $\left(cx+\mathbb{Z}^d\right)\cap\left(d(d+1)^{-1}\square_d\right)\neq\{0\}$.

Notes

We are not aware of any work preceding Knudsen and Mumford [221, Chap. III] that discusses the existence of unimodular triangulations or covers of multiples of lattice polytopes. The main result of [221, Chap. III] has been reproduced in Section 3.B, except that we have not proved the regularity of the triangulation, as explained in Remark 3.22.

Theorem 3.23 has a predecessor in Bruns, Gubeladze, and Trung [69, Th. 1.3.1], stating that cP is covered by its unimodular subsimplices for every lattice polytope P and all $c\gg 0$, but giving no bound for c. That c can be chosen uniformly for all lattice polytopes of fixed dimension was proved in [65], but with a much weaker bound than obtained in Theorem 3.23. The substantial improvement to a polynomial bound was obtained by von Thaden in his thesis [367]. While all other arguments of [65] have remained unchanged, von Thaden improved the critical lemma on unimodular covers (and triangulations) of simplicial cones.

Part II

Affine monoid algebras

4

Monoid algebras

Already in Chapter 2 we introduced algebras over monoids and profited from the embedding of monoid theory into the much richer theory of rings. In this chapter we study arithmetic properties of monoid algebras versus their monoid analogues.

We start by embedding monoid algebras into the framework of graded rings. Then we investigate homomorphisms of monoid algebras induced by homomorphisms of the underlying monoids. Their kernels are generated by binomials, reflecting the formation of residue classes in monoids modulo congruences.

Aspects of polyhedral geometry will show up in the study of monomial prime ideals in affine monoid algebras: they form a partially ordered set dual to the face lattice of the associated cone, and this gives us an opportunity to relate ideal theory to polyhedral geometry.

Normality of a monoid is equivalent to normality of its monoid algebra (over a normal domain), and, in a relative setting, integral closures correspond with each others. A classical invariant of a normal noetherian domain is its divisor class group. For a normal affine monoid domain it has an elegant description in terms of the support forms of the associated cone. The divisor class group will be an important tool in Chapters 5 and 6.

For the first time in this book, the connection of monoid theory and K-theory will become apparent in the last section when we study seminormal monoid domains and the Picard group.

This chapter is not intended to replace a treatise like Gilmer's well-known book [142], which treats the subject of monoid rings as an end in itself. Nevertheless, the books have a lot of common ground.

4.A Graded rings

In this section we relax our standard convention on monoids: they are only assumed to be commutative, but not necessarily cancellative or torsionfree.

We first introduce graded abelian groups, despite the fact that the grading monoid only plays the role of an index set and its monoid operation is not yet used:

W. Bruns, J. Gubeladze, *Polytopes, Rings, and K-Thoery*, Springer Monographs in Mathematics, DOI 10.1007/b105283_4, © Springer Science+Business Media, LLC 2009

Definition 4.1. Let G be a monoid (written additively). A G-*graded* abelian group A is an abelian group together with a decomposition $A = \bigoplus_{g \in G} A_g$ as a direct sum of subgroups A_g.

The group A_g is called the *degree g graded* or *homogeneous component* of A, and the summand a_g in the decomposition $a = \sum_{g \in G} a_g$, $a_g \in A_g$ for all $g \in G$, is the *degree g homogeneous component* of $a \in A$.

A *graded subgroup* of A is a subgroup U such that $U = \bigoplus_{g \in G} U \cap A_g$.

A homomorphism $\varphi : A \to B$ of G-graded groups is *homogeneous* (or graded) if $\varphi(A_g) \subset B_g$ for all $g \in G$.

Note that we do not require $A_g \neq 0$ for all $g \in G$. Therefore, a homomorphism $f : G \to G'$ induces a G'-grading on any G-graded A group if we set $a_{g'} = \bigoplus_{f(g)=g'} A_g$.

By Exercise 4.1 a subgroup is graded if it contains all the homogeneous components of each of its elements. This criterion can be applied in the very easy proof of the following:

Proposition 4.2. *Let A, B be G-graded groups, and $\varphi : A \to B$ a homogeneous homomorphism. Then* $\operatorname{Ker} \varphi$ *is a graded subgroup of A, and* $\operatorname{Im} \varphi$ *is a graded subgroup of B. Moreover,* $\operatorname{Coker} \varphi$ *is naturally graded with components $B_g/(\operatorname{Im} \varphi)_g$, $g \in G$.*

Definition 4.3. A G-*graded ring* is a ring A whose underlying additive group is G-graded in such a way that $A_g A_h \subset A_{g+h}$ for all $g, h \in G$. A graded A-module is an A-module M together with a decomposition $M = \bigoplus_{g \in G} M_g$ such that $A_g M_h \subset M_{g+h}$ for all $g, h \in G$.

A ring or module homomorphism is *homogeneous* if it is homogeneous as a homomorphism of abelian groups. Subobjects such as ideals, subrings and submodules, are graded if they are graded as subgroups.

Usually our rings are algebras over a ring R of coefficients. A *graded R-algebra* is an R-algebra A such that the structure morphism $R \to A$ maps R to A_0. Note that the homogeneous components of A are R-modules in this situation.

Suppose that A is an abelian group with a fixed decomposition $A = \bigoplus_{i \in I} A_i$, like a graded ring or module. Then we set

$$\operatorname{supp}(a) = \{i \in I : a_i \neq 0\}, \qquad a \in A,\ a = \sum_{i \in I} a_i,$$

and call this set the *support of a*. The *support of a subset* $B \subset A$ is $\operatorname{supp} B = \bigcup_{b \in B} \operatorname{supp}(b)$.

Convention 4.4. The most common gradings are $\mathbb{Z}$-gradings, and therefore we call them simply *gradings*. A *graded* ring or module is $\mathbb{Z}$-graded if not specified otherwise.

While the rings under consideration are often $\mathbb{Z}_+$-graded, it is useful to consider them as $\mathbb{Z}$-graded, since one cannot avoid negative degrees as soon as homogeneous elements are inverted.

Remark 4.5. Note that a grading γ on a monoid M (by definition a monoid homomorphism from M to $\mathbb{Z}$) induces a grading on the monoid algebra $R[M]$; take $\bigoplus_{\gamma(x)=i} Rx$ as the graded component of degree i.

If γ is positive, then $R[M]$ is a *positively graded R-algebra*: all homogeneous elements x have nonnegative degree and $R[M]_0 = R$.

In particular, if M is a positive affine monoid, then M has a positive grading (Proposition 2.17) and $R[M]$ can be considered as a positively graded algebra.

Monomial orders and Newton polytopes. Let G be an abelian group. It is *ordered* with respect to a total order $<$ on G if $gh < g'h$ for all $g, g', h \in G$ with $g < g'$. A suitable order for $G = \mathbb{Z}^r$ is the lexicographic one: $(a_1, \dots, a_r) <_{\text{lex}} (b_1, \dots, b_r)$ if and only if the first nonzero component of $a - b$ is negative. Since the elements of G, especially in the case $G = \mathbb{Z}^r$, often represent monomials, we will call such an order *monomial*.

If A is an abelian group graded by an ordered group G, then we may speak of the *initial term* or *component* of an element of A: the initial term of $a = \sum_{g \in G} a_g$ is a_h with $h = \max(\operatorname{supp} a)$, provided $a \neq 0$. If $M \subset G$ is a monoid and $A = R[M]$, then $a_h = r_h h$ with a uniquely determined element $r_h \in R$. One calls r_h the *initial coefficient* and h the *initial monomial*.

The following lemma is trivial, but very useful:

Lemma 4.6. *Let R be a ring graded by a group G with monomial order $<$. Let M be a graded R-module, r an element of R with initial term r_g, and x an element of M with initial term x_h.*

(a) If $r_g x_h \neq 0$, then $r_g x_h$ is the initial term of rx.

(b) In particular, if R is a domain and M is a torsionfree R-module, then $r_g x_h$ is the initial term of rx.

Remark 4.7. By a theorem of Levi, every torsionfree abelian group can be ordered; see [142, Sect. 3]. We will not need this transfinite fact, since we can usually restrict ourselves to considering finitely generated subgroups of the grading group.

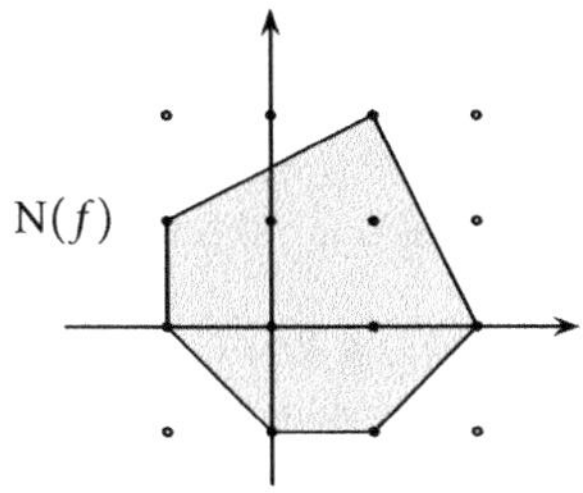

$$f = 1 + X^{-1} + Y^{-1} + XY^{-1} + X^2 + XY^2 + X^{-1}Y$$

Fig. 4.1. A Newton polytope

A more refined approach to the combinatorial structure is provided by Newton polytopes. If a ring R or a module M is $\mathbb{Z}^r$-graded as above, we define the *Newton polytope* of an element $a \in R$ or M by

$$\mathrm{N}(a) = \operatorname{conv}(\operatorname{supp} a).$$

Figure 4.1 shows the Newton polytope of a Laurent polynomial in two variables.

The next lemma connects the Newton polytopes of r and x with the Newton polytope of their product. We use the information on the vertices of Minkowski sums of polytopes given in Exercise 1.5.

Lemma 4.8. *Let G be an abelian group, $G \cong \mathbb{Z}^r$. Let R be a G-graded ring and M a graded R-module, a an element of R and x an element of M.*

(a) *Suppose g is a vertex of $\mathrm{N}(a)$ and h is a vertex of $\mathrm{N}(x)$ such that $g+h$ is a vertex of $\mathrm{N}(a)+\mathrm{N}(x)$. If $a_g x_h \neq 0$, then $g+h$ is a vertex of $\mathrm{N}(ax)$.*

(b) *In particular, if R is a domain and M is a torsionfree R-module, then $\mathrm{N}(ax) = \mathrm{N}(a)+\mathrm{N}(x)$.*

Proof. One has $\operatorname{supp}(ax) \subset \operatorname{supp}(a) + \operatorname{supp}(x)$, and $\operatorname{conv}(\operatorname{supp}(a)+\operatorname{supp}(x)) = \mathrm{N}(a)+\mathrm{N}(x)$. If $g+h = g'+h'$ for $g' \in \operatorname{supp}(a)$ and $h' \in \operatorname{supp}(x)$ and $g+h$ is a vertex of $\mathrm{N}(a)+\mathrm{N}(x)$, then $g' = g$, $h' = h$. Therefore $g+h \in \operatorname{supp}(ax)$, and $g+h$ is a vertex of $\mathrm{N}(ax)$, since it is a vertex of the polytope $\mathrm{N}(a)+\mathrm{N}(x) \supset \mathrm{N}(ax)$.

Every vertex of the Minkowski sum of polytopes is the sum of vertices of the summands. Under the hypothesis all vertices of $\mathrm{N}(a)+\mathrm{N}(x)$ belong to $\mathrm{N}(ax)$, so that $\mathrm{N}(ax) = \mathrm{N}(a)+\mathrm{N}(x)$. □

Prime ideals in graded rings. As usual, $\operatorname{Spec} R$ denotes the set of prime ideals of a ring R. For an R-module M we set

$$\operatorname{Supp} M = \{\mathfrak{p} \in \operatorname{Spec} R : M_{\mathfrak{p}} \neq 0\},$$
$$\operatorname{Ass} M = \{\mathfrak{p} \in \operatorname{Spec} R : \mathfrak{p} = \operatorname{Ann} x \text{ for some } x \in M\}.$$

If $M \neq 0$ is a finitely generated module over a noetherian ring, then $\operatorname{Supp} M \neq \emptyset$, and the minimal elements of $\operatorname{Supp} M$ all belong to $\operatorname{Ass} M$. Moreover, $\operatorname{Supp} M$ consists exactly of the prime ideals that contain the annihilator $\operatorname{Ann} M$. We refer for these basic facts to Matsumura [253, Sect. 4].

For the case $G = \mathbb{Z}$ the next lemma can be found in [72, 1.5.6]. The generalization to arbitrary grading groups is straightforward since only finitely generated subgroups of G enter the proof and all groups isomorphic to $\mathbb{Z}^n$, $n \geq 0$, can be ordered.

Lemma 4.9. *Let G be a torsionfree abelian group and R be a G-graded ring. Let $\mathfrak{p}$ be a prime ideal in R, and $\mathfrak{q}$ the ideal generated by all homogeneous elements in $\mathfrak{p}$.*

(a) *$\mathfrak{q}$ is a prime ideal.*

(b) *Let M be a graded R-module.*
   (i) *If $\mathfrak{p} \in \operatorname{Supp} M$, then $\mathfrak{q} \in \operatorname{Supp} M$.*
   (ii) *If $\mathfrak{p} \in \operatorname{Ass} M$, then $\mathfrak{p} = \mathfrak{q}$. Furthermore $\mathfrak{p}$ is the annihilator of a homogeneous element of M.*

The lemma is the basis for the characterization of graded rings in terms of their graded prime ideals; see [72] for results of this type. Its part (a) can easily be extended to radical ideals and primary ideals (Exercise 4.8).

Groups of units. In cases of particular interest for us one can determine the units of graded rings.

Proposition 4.10. *Let M be a cancellative, torsionfree monoid and A an M-graded ring.*

(a) If M is positive, then $f \in A$ is a unit if and only if f_0 is a unit and the homogeneous components f_x, $x \in M \setminus \{0\}$, are nilpotent.

(b) If A is an integral domain or, more generally, reduced with connected spectrum (in the Zariski topology), then every unit in A is homogeneous.

Proof. (a) That the specified elements are units is obvious: a sum of a unit and a nilpotent element is not contained in any maximal ideal. Conversely, consider a unit f. Evidently f_0 is a unit, and it remains to show that f_x for $x \neq 0$ is contained in all minimal prime ideals. By Lemma 4.9 these are graded, and so we may assume that A is an integral domain. But then we are in the situation of (b), and if M is positive, only homogeneous elements of degree 0 can be units.

(b) Let us first assume that A is an integral domain, and let f be a unit. Since $\operatorname{supp}(f)$ and $\operatorname{supp}(f^{-1})$ are contained in an affine submonoid of M, an argument using Newton polytopes shows that f must be homogeneous.

Now suppose that R is reduced with connected spectrum. Assume that f has at least two nonzero homogeneous components $f_1, \dots, f_n$, and let $U_i \subset \operatorname{Spec} R$ be the set of prime ideals not containing f_i. Since R is reduced, each of the sets U_i is nonempty, and it is an open subset of $\operatorname{Spec} R$ by definition. For a contradiction it is enough to show that the sets U_i are pairwise disjoint. Assume that there exists a prime ideal $\mathfrak{p}$ not containing f_1 and f_2. Then the same holds for the prime ideal $\mathfrak{q}$ generated by the homogeneous elements in $\mathfrak{p}$ (Lemma 4.9). But $R/\mathfrak{q}$ is a again M-graded, and therefore units in $R/\mathfrak{q}$ must be homogeneous by what has been proved already. □

A finiteness theorem. In general it is very difficult to decide whether a subalgebra of a finitely generated algebra is also finitely generated. However, if we define the subalgebra by selecting degrees in a finitely generated submonoid of the grading group, then finite generation always persists:

Theorem 4.11. *Let G be a finitely generated abelian group, M a finitely generated submonoid of G, and $T \subset G$ a finitely generated M-module. Furthermore, let R be a noetherian G-graded ring and N a G-graded finitely generated R-module. Then the following hold:*

(a) R_0 is noetherian ring, and each graded component N_g, $g \in G$, of N is a finitely generated R_0-module.

(b) $A = \bigoplus_{s \in M} R_s$ is a finitely generated R_0-algebra.

(c) $U = \bigoplus_{t \in T} N_t$ is a finitely generated A-module.

Since we allow groups with torsion in the theorem, we must also allow M to have torsion. Let us single out a special case because it gives us an opportunity to re-prove Gordan's lemma:

Lemma 4.12. *Let R be a $\mathbb{Z}$-graded noetherian ring, $R_+ = \bigoplus_{k\ge 0} R_k$, and $R_- = \bigoplus_{k\le 0} R_k$. Then*

(a) R_0 is a noetherian ring, and each graded component R_i is a finitely generated R_0-module;

(b) R, R_+, and R_- are finitely generated R_0-algebras.

Proof. For the noetherian property of R_0 note that $IR \cap R_0 = I$ for all ideals of R_0. Therefore, every ascending chain of ideals in R_0 stabilizes. A similar argument shows that R_i is a noetherian R_0-module for every $i \in \mathbb{Z}$.

For (c) it is enough to consider R_+. Let $\mathfrak{m} = \bigoplus_{i=1}^{\infty} R_i$. We claim that $\mathfrak{m}$ is a finitely generated ideal of R_+. By hypothesis, $\mathfrak{m}R$ has a finite system of generators $x_1, \dots, x_m$, which may certainly be chosen to be homogeneous of positive degrees d_i. Let d be the maximum of $d_1, \dots, d_m$. Then every homogeneous element $y \in \mathfrak{m}$ with $\deg y \ge d$ can be written as a linear combination of $x_1, \dots, x_m$ with coefficients from $R_0 \cup \dots \cup R_{d-1}$ as follows by induction on degree. Thus a set of homogeneous elements spanning $R_1 \oplus \dots \oplus R_d$ over R_0 generates $\mathfrak{m}$ as an ideal of R_+. Such a set can be chosen finite by (a).

Now one observes that an arbitrary homogeneous system of generators $z_1, \dots, z_n$ of the ideal $\mathfrak{m}$ generates R_+ as an R_0-algebra. (Use induction on $k \ge 0$ to show that every element $x \in R_k$ belongs to $R_0[x_1, \dots, x_n]$.) □

In order to derive Gordan's lemma in the form of Corollary 2.11(b) from Lemma 4.12 (or to prove it in the same style) consider an affine monoid M and a rational hyperplane H through the origin. Choosing a linear form defining H, we define a $\mathbb{Z}$-grading on M, and obtain that $M \cap H^+$ is an affine monoid. By induction on the number of support hyperplanes of a rational cone C it follows that $M \cap C$ is affine.

Proof of Theorem 4.11. (a) This follows similarly as Lemma 4.12(a).

(b) First we do the case in which G is torsionfree, $G = \mathbb{Z}^m$, and M is an integrally closed submonoid of $\mathbb{Z}^m$. In this case M is cut out by finitely many halfspaces and the claim follows by an iterated application of Lemma 4.12.

In the general case for G and M we set $G' = G/H$ where H is the torsion subgroup of G, and let $\pi : G \to G'$ denote the natural surjection. Let R' be R with the G'-grading induced by π (its homogeneous components are the direct sums of the components R_g where g is in a fixed fiber of π). Let M' be the integral closure of $\pi(M)$ in G'. Then $A' = \bigoplus_{s' \in M'} R'_{s'}$ is a finitely generated algebra over the noetherian ring R'_0, as we have already shown. But R'_0 is a finitely generated module over R_0 by (a), and so A' is a finitely generated R_0-algebra. In particular, R itself is finitely generated over R_0.

Since A' is a finitely generated R_0-algebra, it is certainly finitely generated over A. It is even a finitely generated module over A: every homogenous element has a power in A. As an A-module, A' is therefore generated by finitely many monomials in a finite homogeneous system of algebra generators. But then a lemma of Artin and Tate (see Eisenbud [112, p. 143]) implies that A is noetherian. As shown above, noetherian G-graded rings are finitely generated R_0-algebras.

(c) By hypothesis, T is the union of finitely many translates $M + t$. Therefore we can assume that $T = M + t$. Passing to the shifted module $N(-t)$ (given by $N(-t)_g =$

N_{g-t}), we can even assume that $M = T$. Now the proof follows the same pattern as that of (b). We leave the details to the reader (see [66, Th. 7.1]). □

4.B Monoid algebras

In this section we relax our standard convention on monoids; they are only assumed to be commutative, but not necessarily cancellative or torsionfree.

Let M be a monoid. Furthermore let R be a commutative ring. In Section 2.A we already introduced the monoid algebra $R[M]$. As an R-module it is free with basis M. Therefore every element $f \in R[M]$ has a unique representation

$$f = \sum_{x \in M} f_x x, \quad f_x \in R,$$

in which $f_x = 0$ for all but finitely many $x \in M$. In order to introduce the multiplication in $R[M]$ we write the monoid operation *multiplicatively*. With this convention,

$$fg = \sum_{x \in M} \sum_{yz=x} f_y g_z yz.$$

Very often the ring R of coefficients will be a field $\Bbbk$. The elements $x \in M$ are called the *monomials* of $R[M]$, whereas the elements of the form rx, $r \in R$, $x \in M$, are called *terms*. A *binomial* is an element $x - y \in R[M]$, $x, y \in M$.

The monoid algebra $R[M]$ has the following universal property, characterizing it uniquely up to isomorphism: for every monoid homomorphism ι from M to the multiplicative monoid of an R-algebra A there exists a unique R-algebra homomorphism $R[M] \to A$ extending ι.

The monoid algebra $R[M]$ is the simplest example of an M-graded R-algebra. Its homogeneous elements are exactly the terms rx with $r \in R$, $x \in M$.

Remark 4.13. Suppose that $\Bbbk$ is a field. Then every M-graded component of $\Bbbk[M]$ has $\Bbbk$-dimension 1, and one may ask oneself whether there exist M-graded $\Bbbk$-algebras A different from $\Bbbk[M]$ such that $\dim A_x = 1$ for all $x \in M$, and if so, how to describe this family of algebras. See Sturmfels [338, Ch. 10] for more information. However, it is easily seen that $A \cong \Bbbk[M]$ if all nonzero homogeneous elements of A are non-zerodivisors of A (Exercise 4.3).

Let M, N be monoids, and $\varphi : M \to N$ a monoid homomorphism. Then φ induces an R-algebra homomorphism $\varphi_R : R[M] \to R[N]$ via

$$\varphi_R\Big(\sum_{x \in M} f_x x\Big) = \sum_{x \in M} f_x \varphi(x).$$

In other words, the R-linear extension of φ is an R-algebra homomorphism.

There is a *natural augmentation* of the R-algebra $R[M]$, i. e., an R-algebra homomorphism $\varepsilon_1 : R[M] \to R$ for which the composition with the *natural embedding* $R \to R[M]$, $r \mapsto r1$, is the identity on R. It is given by

$$\varepsilon_1\left(\sum_{x\in M} f_x x\right) = \sum_{x\in M} f_x.$$

Note that ε_1 is induced by the monoid homomorphism $M \to \{1\}$, $x \mapsto 1$ for all $x \in M$. If M has no nontrivial invertible element, for example if M is a positive affine monoid, then

$$\varepsilon_0\left(\sum_{x\in M} f_x x\right) = f_1$$

is also an R-algebra augmentation of $R[M]$.

We have already shown in Proposition 2.7 that $R[M]$ is finitely generated as an R-algebra if and only if M is a finitely generated monoid. Therefore, if M is affine and R is a noetherian ring, then $R[M]$ is a noetherian ring by the Hilbert basis theorem.

Remark 4.14. The converse is also true: $R[M]$ is noetherian if and only if R is noetherian and M is finitely generated. See [142, Theorem 7.7]. That R is noetherian along with $R[M]$ follows readily from the fact that R is a retract of $R[M]$. One has to work harder to show that M is finitely generated if $R[M]$ is noetherian.

If N is an M-module (see Definition 2.5), then RN, the free R-module with basis N, is an $R[M]$-module in a natural way.

The very easy proof of the next proposition is left to the reader. More or less it uses only that tensor product commutes with direct sum.

Proposition 4.15. *Let R be a ring and M, N be monoids. Then*

$$R[M \oplus N] \cong R[M] \otimes_R R[N] \cong \big(R[M]\big)[N] \cong \big(R[N]\big)[M]$$

as R-algebras.

Monoid domains and reduced monoid algebras. That we call the elements $x \in M$ monomials in $R[M]$ is justified by the following proposition.

Proposition 4.16. *Let M be an affine monoid, $\mathrm{gp}(M) = \mathbb{Z}^r$.*

(a) Then there is an injective R-algebra homomorphism $R[M] \to R[X_1^{\pm1}, \dots, X_r^{\pm1}]$ mapping the elements of M to (Laurent) monomials in the variables $X_1, \dots, X_r$.

(b) If M is positive, then there exists an injective R-algebra homomorphism $R[M] \to R[X_1, \dots, X_r]$ mapping the elements of M to monomials in the variables $X_1, \dots, X_r$.

Proof. Let $e_1, \dots, e_r$ be the unit vectors of $\mathbb{Z}^r$. Then the assignment $e_i \mapsto X_i$ induces an isomorphism of $\mathbb{Z}^r$ and the group of monomials $X_1^{a_1} \cdots X_r^{a_r}$, $(a_1, \dots, a_r) \in \mathbb{Z}^r$. This group isomorphism gives rise to an R-algebra isomorphism $R[\mathrm{gp}(M)] \cong R[X_1^{\pm1}, \dots, X_r^{\pm1}]$, and therefore to an embedding of $R[M]$ into the Laurent polynomial ring.

If M is positive, then there exists an embedding $M \to \mathbb{Z}_+^r$ by Proposition 2.17. It induces an embedding $R[M] \to R[X_1, \dots, X_r]$ that maps monomials to monomials. □

Corollary 4.17. *Let R be an integral domain and M an affine monoid. Then $R[M]$ is an integral domain.*

In fact, the cancellative and torsionfree monoids have a simple characterization in terms of their monoid algebras:

Theorem 4.18. *Let M be a monoid that is not necessarily torsionfree or cancellative, and let R be a ring. Then the following are equivalent:*

(a) R is a domain and M is torsionfree and cancellative.
(b) $R[M]$ is a domain.

Proof. (a) $\Longrightarrow$ (b) Let $f,g \in R[M]$, $f,g \neq 0$. Since $\operatorname{supp}(f)$ and $\operatorname{supp}(g)$ are finite, there exists a finitely generated submonoid N of M such that $f,g \in R[N] \subset R[M]$. Since $R[N]$ is an integral domain by Corollary 4.17, we have $fg \neq 0$.

(b) $\Longrightarrow$ (a) Since $R \subset R[M]$ in a natural way, R is a domain along with $R[M]$. If M is not cancellative, say $xy = xz$ with $x,y,z \in M$, $y \neq z$, then $x(y-z) = 0$ in $R[M]$, but $x, y-z \neq 0$. If M is cancellative, but not torsionfree, say $x^n = y^n$ for $x,y \in M$, $x^k \neq y^k$ for $k < n$, then $0 = x^n - y^n = (x-y)(x^{n-1} + x^{n-2}y + \cdots + y^{n-1})$, and the second factor is nonzero: the monomials appearing in it are pairwise different since M is cancellative.Thus $x = y$. □

For $R[M]$ to be only reduced, one can weaken the hypotheses on R and M. We say that a reduced ring R has *characteristic* 0 if all localizations $R_{\mathfrak{p}}$ with respect to its minimal prime ideals $\mathfrak{p}$ are fields of characteristic 0.

Theorem 4.19. *Let M be a cancellative monoid and R a reduced ring. Suppose that R has characteristic 0 or M is torsionfree. Then $R[M]$ is reduced.*

Proof. A reduced ring R can be embedded into the direct product of its localizations $R_{\mathfrak{p}}$ with respect to its minimal prime ideals $\mathfrak{p}$, and the group ring over this direct product of fields is contained in the direct product of group rings. Therefore it is enough to consider the case in which $R = \Bbbk$ is a field.

Since M is a submonoid of $\operatorname{gp}(M)$ we may assume $M = G$ is a group. Because G is the union of its finitely generated subgroups, it is enough to consider finitely generated groups. Then G is the direct sum of cyclic groups (isomorphic to $\mathbb{Z}$ if G is torsionfree). Hence induction based on Proposition 4.15 reduces the theorem to the case of a cyclic group. (Note that the hypothesis on characteristic 0 is preserved in the intermediate extensions since we start from a field $\Bbbk$.)

One has $\Bbbk[G] = \Bbbk[X^{\pm 1}]$ if $G \cong \mathbb{Z}$ (and only this case is possible if $\operatorname{char} \Bbbk \neq 0$), and $\Bbbk[G] \cong \Bbbk[X]/(X^m - 1)$ if $G \cong \mathbb{Z}/m\mathbb{Z}$. But $X^m - 1$ splits into a product of pairwise prime irreducible polynomials if $\operatorname{char} \Bbbk = 0$. By the Chinese remainder theorem, $\Bbbk[G]$ is a direct product of integral domains, and therefore it is reduced. □

Groups of units. In general it is very difficult to determine the units in monoid algebras and even in group rings (see Roggenkamp [303] for this classical problem), but in the case in which we are mainly interested the task is easy.

Proposition 4.20. *Let R be reduced and M a cancellative, torsionfree monoid.*

(a) If M is positive, then $\mathrm{U}(R[M]) = \mathrm{U}(R)$.

(b) If R is an integral domain or, more generally, $\operatorname{Spec} R$ *is connected, then the units in $R[M]$ are exactly the terms ux where u is a unit in R and x is an invertible element of M, in other words, one has a natural isomorphism* $\mathrm{U}(R[M]) = \mathrm{U}(R) \oplus \mathrm{U}(M)$.

The proposition follows immediately from Proposition 4.10 by specialization, once it has been shown that $R[M]$ is reduced (with connected spectrum) along with R. Reducedness follows from Theorem 4.19, and the very easy argument on connectedness is left to the reader (Exercise 4.4). Part (b) can be generalized; see Exercise 4.5. As a corollary we obtain that the group of units of M is determined by the R-algebra $R[M]$:

Corollary 4.21. *Let R be a ring and M, N be cancellative, torsionfree monoids. If $R[M]$ and $R[N]$ are isomorphic as R-algebras, then* $\mathrm{U}(M) \cong \mathrm{U}(N)$.

Proof. We can first pass to a residue ring of R modulo a maximal ideal, and may therefore assume that R is a field. Then we have an isomorphism $\mathrm{U}(R[M]) \cong \mathrm{U}(R[N])$ that maps $\mathrm{U}(R) \subset \mathrm{U}(R[M])$ to $\mathrm{U}(R) \subset \mathrm{U}(R[N])$. Therefore it induces an isomorphism

$$\mathrm{U}(M) \cong \mathrm{U}(R[M])/\mathrm{U}(R) \cong \mathrm{U}(R[N])/\mathrm{U}(R) \cong \mathrm{U}(N). \qquad \square$$

It would be enough to assume that one of the monoids is cancellative and torsionfree (why?). Corollary 4.21 is the first sign of the isomorphism theorem to be proved in Chapter 5.

Krull dimension. We recall the most important notions of dimension theory. The *Krull dimension* of a ring R is the supremum of the numbers n for which there exists a strictly ascending chain $\mathfrak{p}_0 \subset \mathfrak{p}_1 \subset \cdots \subset \mathfrak{p}_n$. The *height* $\operatorname{ht}\mathfrak{p}$ of a prime ideal $\mathfrak{p}$ is the Krull dimension of the local ring $R_\mathfrak{p}$. The height $\operatorname{ht} I$ of an arbitrary ideal I is the minimum of the heights of the prime ideals $\mathfrak{p} \supset I$. If I is an ideal in a noetherian ring R generated by m elements, then $\operatorname{ht}\mathfrak{p} \le m$ for all minimal prime overideals $\mathfrak{p} \supset I$ by Krull's principal ideal theorem. The *Krull dimension* $\dim N$ of an R-module N is the dimension of $R/\operatorname{Ann} N$. (See Bourbaki [37, Ch. VIII] or Matsumura [253] for the basic notions of dimension theory.)

If the ring R of coefficients is a field $\Bbbk$, then it is very easy to determine the Krull dimension of an affine monoid domain $\Bbbk[M]$:

Proposition 4.22. *Let $\Bbbk$ be a field and M an affine monoid. Then* $\dim \Bbbk[M] = \operatorname{rank} M$.

Proof. We have already shown that $\Bbbk[M]$ is an affine $\Bbbk$-domain. The Krull dimension of such a domain is the transcendence degree of its field of fractions over $\Bbbk$ (see Kunz [235, Ch. 2, §3]). Let $r = \operatorname{rank} M$. Then $\mathrm{QF}(\Bbbk[M]) = \mathrm{QF}(\Bbbk[\mathrm{gp}(M)]) \cong \mathrm{QF}(\Bbbk[\mathbb{Z}^r]) \cong \mathrm{QF}(\Bbbk[X_1, \dots, X_n])$, and so $\dim \Bbbk[M] = r$. $\square$

Theorem 4.23. *Let R be a noetherian ring, and M an affine monoid. Then* $\dim R[M] = \dim R + \operatorname{rank} M$.

Proof. Set $S = R[M]$. Consider a strictly ascending chain of prime ideals $\mathfrak{p}_0 \subset \mathfrak{p}_1 \subset \cdots \subset \mathfrak{p}_n$ in R. Then the ideals $\mathfrak{p}_i S$ form a strictly ascending chain in $R[M]$, too, and they are prime ideals since $S/\mathfrak{p}_i S \cong (R/\mathfrak{p}_i)[M]$.

Suppose that $\mathfrak{m} = \mathfrak{p}_n$ is a maximal ideal, and set $\Bbbk = R/\mathfrak{m}$. Then $\Bbbk[M] \cong S/\mathfrak{m}S$ has Krull dimension $r = \operatorname{rank} M$, and we can extend the chain of prime ideals in S by r more members, which we obtain as preimages in S of the members of a chain $0 \neq \mathfrak{q}_1 \subset \cdots \subset \mathfrak{q}_r$ of prime ideals in $\Bbbk[M]$. Therefore $\dim S \geq \dim R + r$.

It remains to show that $\dim S \leq \dim R + r$ if $\dim R < \infty$. Choose a maximal ideal $\mathfrak{M}$ of S, and set $\mathfrak{p} = \mathfrak{M} \cap R$. Then all elements in $R \setminus \mathfrak{p}$ do not belong to $\mathfrak{M}$, and therefore $S_{\mathfrak{M}}$ is a localization of $R_{\mathfrak{p}}[M]$. We thus have $\dim S_{\mathfrak{M}} \leq \dim R_{\mathfrak{p}} + \dim S_{\mathfrak{M}}/\mathfrak{p}S_{\mathfrak{M}}$. See [253, Theorem 15.1] for this inequality; it follows very easily from Krull's principal ideal theorem.

Therefore it remains to show that $\dim S_{\mathfrak{M}}/\mathfrak{p}S_{\mathfrak{M}} \leq r$. This follows again from Proposition 4.22, since we can view $S_{\mathfrak{M}}/\mathfrak{p}S_{\mathfrak{M}}$ as a localization of $(R_{\mathfrak{p}}/\mathfrak{p}R_{\mathfrak{p}})[M]$. But $R_{\mathfrak{p}}/\mathfrak{p}R_{\mathfrak{p}}$ is a field. □

Remark 4.24. (a) In the proof of Theorem 4.23 we have explicitly constructed chains of prime ideals. A somewhat shorter proof uses the dimension formula for flat local extensions; see Remark 6.8 where we discuss flat extensions in connection with the Cohen–Macaulay property.

(b) Theorem 4.23 cannot be generalized to arbitrary rings of coefficients. In fact, it can happen that $\dim R[X] > \dim R + 1$ (see Seidenberg [315]). For more information on the Krull dimension of monoid rings we refer the reader to Gilmer [142].

4.C Representations of monoid algebras

In this section we relax our standard convention on monoids: they are only assumed to be commutative, but not necessarily cancellative or torsionfree.

So far we have described affine monoids M as submonoids of lattices, and usually this description is very efficient, especially for normal monoids. However, it is often necessary to use a representation of M as a quotient of a free commutative monoid whose basis corresponds to a system N of generators of M. Such a representation of M gives rise to a representation of the monoid algebra $R[M]$ as a residue class ring of a polynomial ring $R[X_x : x \in N]$ over R. The R-algebra homomorphism $\pi : R[X_x : x \in N] \to R[M]$ is induced by the substitution $X_x \mapsto x$. If M is given, say, as a submonoid of a lattice $\mathbb{Z}^n$, then the main task is to calculate the kernel of π. In this section we write the monoid operation multiplicatively, unless indicated otherwise.

Congruences. Quotients of monoids are determined by *congruence relations*. A congruence relation on a monoid M is a set $\mathscr{C}$ of pairs $(x, y) \in M^2$ such that (i) $\mathscr{C}$ is an equivalence relation, and (ii) $(zx, zy) \in \mathscr{C}$ for all $(x, y) \in \mathscr{C}$, $z \in M$. Instead of $(x, y) \in \mathscr{C}$ we will also write $x \sim y$.

The set of congruence classes of $\mathscr{C}$ forms a monoid if we set $\bar{x}\bar{y} = \overline{xy}$. This monoid is denoted by $M/\mathscr{C}$. (It need not be cancellative or torsionfree, even if M is so.)

Let $\varphi : M \to N$ be a monoid homomorphism. Then φ induces a congruence by $x \sim y$ if and only if $\varphi(x) = \varphi(y)$. As in the case of (abelian) groups we have $\operatorname{Im}\varphi \cong M/\mathscr{C}$ in a natural way.

A subset E of $\mathscr{C}$ *generates* $\mathscr{C}$ if $\mathscr{C}$ is the smallest congruence relation containing E. (This definition makes sense since M^2 is a congruence relation, and since the intersection of congruence relations is again a congruence relation.)

The congruence relation generated by E can also be described constructively. In order to capture symmetry let us set $\tilde{E} = E \cup \{(y,x) : (x,y) \in E\}$. The reader may check (Exercise 4.6(a)) that E generates $\mathscr{C}$ if and only if for each $(x,y) \in \mathscr{C}$, $x \neq y$, there exist $n \in \mathbb{N}$, $(u_1,v_1),\dots,(u_n,v_n) \in \tilde{E}$, and $w_1,\dots,w_n \in M$ such that $x = w_1u_1$, $w_1v_1 = w_2v_2,\dots, w_{n-1}v_{n-1} = w_nu_n$, $w_nv_n = y$.

When M is an abelian group, we are on familiar territory:

Proposition 4.25. *Let $\varphi : M \to N$ be a homomorphism of abelian groups and E a subset of G. Then the pairs $(e,1)$, $e \in E$, generate the congruence relation defined by φ if and only if E generates* $\operatorname{Ker}\varphi$.

The proof is left to the reader (Exercise 4.6(b)).

Congruences and binomial ideals. We now relate the congruence of a monoid homomorphism to the kernel of the associated algebra homomorphism φ_R:

Proposition 4.26. *Let $\varphi : M \to N$ be a monoid homomorphism and R a ring. Then the kernel of $\varphi_R : R[M] \to R[N]$ is generated by the binomials $z_1 - z_2$ where $z_1, z_2 \in M$ and $\varphi(z_1) = \varphi(z_2)$.*

Proof. Let $f = \sum_{x\in M} f_x x \in R[M]$. Then we have

$$\varphi_R(f) = \sum_{x\in M} f_x\varphi(x) = \sum_{y\in N}\Bigl(\sum_{\varphi(x)=y} f_x\Bigr)y.$$

Therefore $\varphi_R(f) = 0$ if and only if $\sum_{\varphi(x)=y} f_x = 0$ for all $y \in N$.

It is impossible that $f_x \neq 0$ for a single x. If $f_x \neq 0$ for at least two monomials x_1, x_2, then we use induction, passing from f to $f - f_{x_1}(x_1 - x_2)$. □

The description of $\operatorname{Ker}\varphi_R$ in the proposition depends solely on M, and not on the ring R of coefficients. This observation can be extended further:

Proposition 4.27. *With the notation of Proposition 4.26 let S be a set of binomials $x_1 - x_2 \in \operatorname{Ker}\varphi_R$. Set $\tilde{S} = S \cup -S$. Then the following are equivalent:*

(a) *S generates the ideal* $\operatorname{Ker}\varphi_R$.
(b) *For every binomial $y_1 - y_2 \in \operatorname{Ker}\varphi_R$, $y_1 - y_2 \neq 0$, there exist $n \in \mathbb{N}$, binomials $x_{11} - x_{12},\dots,x_{n1} - x_{n2} \in \tilde{S}$, and monomials $z_1,\dots,z_n$ such that $y_1 = z_1x_{11}$, $z_1x_{12} = z_2x_{21},\dots, z_{n-1}x_{n-1,2} = z_nx_{n1}$, $z_nx_{n2} = y_2$.*
(c) *The pairs (x_1,x_2) with $x_1 - x_2 \in S$ generate the congruence relation on M defined by φ.*

Proof. The implication (b) $\Longrightarrow$ (a) is very easy. First, $-S$ is contained in the ideal generated by S, and, second, we have the telescope sum

$$y_1 - y_2 = \sum_{i=1}^{n} z_i(x_{i1} - x_{i2}).$$

This shows that the ideal generated by S contains all the binomials in $\operatorname{Ker}\varphi_R$, and by Proposition 4.26 this is sufficient.

For the implication (a) $\Longrightarrow$ (b) let us write

$$y_1 - y_2 = \sum_{i=1}^{n} f_i(x_{i1} - x_{i2}) \tag{4.1}$$

with $x_{i1} - x_{i2} \in \tilde{S}$ and $f_i \in R[M]$. Using each binomial several times if necessary, we can assume that every coefficient f_i is in fact a term in $R[M]$, say $f_i = r_i z_i$ with $r_i \neq 0$. Now we define a graph G whose vertices are the monomials $z_i x_{i1}$ and $z_i x_{i2}$, $i = 1, \dots, n$, and whose edges correspond to the binomials $z_i(x_{i1} - x_{i2})$, connecting $z_i x_{i1}$ and $z_i x_{i2}$.

We can split the sum on the right hand side in (4.1) into sums each of which only contains all the binomials corresponding to edges in a connected component of G. If a connected component contains neither y_1 nor y_2, then the corresponding sum is 0 and can be omitted completely. After this simplification only one connected component remains. In fact, if not, then the sum over one component would yield y_1 and that over the other $-y_2$. However, neither y_1 nor y_2 belongs to $\operatorname{Ker}\varphi_R$.

We choose a path from y_1 to y_2 in the graph, and this is exactly what (b) requires us to do. (Since $b, -b \in \tilde{S}$ for $b \in S$, we can use an edge of G in both directions.)

The equivalence of (b) and (c) is evident. In fact, (b) is only a reformulation of (c) in terms of binomials. □

Corollary 4.28. *With the notation of Proposition 4.26, if M is finitely generated, then the congruence defined by φ, equivalently, the ideal* $\operatorname{Ker}\varphi_R$ *is finitely generated.*

Proof. In fact, the proposition shows that we are free in choosing the ring of coefficients in proving the corollary, and so we choose $R = \Bbbk$ to be a field. Then $R[M]$ is a noetherian ring, $\operatorname{Ker}\varphi$ is a finitely generated ideal, and every system of generators of $\operatorname{Ker}\varphi$ contains a finite system of generators. □

Toric ideals. Let $\varphi : M \to N$ be a surjective monoid homomorphism as above. Especially in the case in which $M = \mathbb{Z}_+^n$ (and so $R[M] \cong R[X_1, \dots, X_n]$) one is interested in computing $\ker\varphi_R$: it defines $R[N]$ as a residue class ring of a polynomial ring. In general it is difficult to compute a system of generators of $\operatorname{Ker}\varphi_R$, even if $N \subset \mathbb{Z}^r$, given by a system of generators.

It is important to observe that one can extend φ_R to a homomorphism of group rings $\varphi_R^- : R[\operatorname{gp}(M)] \to R[\operatorname{gp}(N)]$. This extension exists for two (equivalent) reasons: (i) from the ring-theoretic point of view, the monoid elements form a multiplicatively closed set, and so we may pass to the ring of fractions by adjoining their inverses; (ii) from the monoid-theoretic point of view, the monoid homomorphism $\varphi : M \to N$ can be extended to a group homomorphism $\varphi^- : \mathrm{gp}(M) \to \mathrm{gp}(N)$, which then gives rise to the R-algebra homomorphism φ_R^-. (The homomorphism $M \to \mathrm{gp}(M)$ injective only if M is cancellative.)

In the proposition that relates the kernels of φ and φ^- we must use multiplicative notation in M.

Proposition 4.29. *Let $\varphi : M \to N$ be a homomorphism of monoids, and R a ring. Suppose that M is generated by $x_1, \dots, x_n$ and that* $\operatorname{Ker} \varphi_R^-$ *be generated by the elements $w_1, \dots, w_m \in \mathrm{gp}(M)$, $w_i = y_i z_i^{-1}$, with $y_i, z_i \in M$. Let I be the ideal in $R[M]$ generated by the binomials $b_i = y_i - z_i$. Then:*

(a) $IR[\mathrm{gp}(M)]$ is the kernel of the induced homomorphism $\varphi_R^- : R[\mathrm{gp}(M)] \to R[\mathrm{gp}(N)]$.
(b) $\operatorname{Ker} \varphi_R = \{f \in R[M] : (x_1 \cdots x_n)^k f \in I$ *for some* $k \in \mathbb{N}\}$.

Proof. Let us first conclude (b) from (a). By (a) we have $\operatorname{Ker} \varphi_R^- = IR[\mathrm{gp}(M)]$. On the other hand, by general localization arguments, $\operatorname{Ker} \varphi_R^- = (\operatorname{Ker} \varphi_R) R[\mathrm{gp}(M)]$, and $\operatorname{Ker} \varphi_R = \operatorname{Ker} \varphi_R^- \cap R[M]$. Thus $\operatorname{Ker} \varphi_R = IR[\mathrm{gp}(M)] \cap R[M]$, and now (b) follows by the general formula for the contraction of an ideal extended to a localization, since $R[\mathrm{gp}(M)] = R[M][(x_1 \cdots x_n)^{-1}]$.

For (a) we note that $IR[\mathrm{gp}(M)]$ is generated by the binomials $b_i' = w_i - 1$, since we can multiply by z_i^{-1} in $R[\mathrm{gp}(M)]$.

On the other hand, by Proposition 4.25, the pairs $(w_i, 1)$ generate the congruence relation defined by φ^-. Then Proposition 4.27 implies that $b_1', \dots, b_m'$ generate $\operatorname{Ker} \varphi_R^-$.

□

We single out a consequence of part (a) for use in Chapter 6.

Corollary 4.30. *Let R be a ring, and $\varphi : \mathbb{Z}^n \to N$ be a group homomorphism. Then the ideal $I = \operatorname{Ker} \varphi_R$ is generated by binomials $b_1, \dots, b_m$, $m = \operatorname{rank} \operatorname{Ker} \varphi$.*

If $M = \mathbb{Z}_+^n$ and $N \subset \mathbb{Z}^r$ for $n, r \in \mathbb{Z}_+$, the ideal $\operatorname{Ker} \varphi_R^-$ is called a *toric ideal.* The proposition suggests a two-step strategy for the computation of toric ideals. We can easily compute a basis of the kernel U of the group homomorphism $\varphi^- : \mathbb{Z}^n \to \mathbb{Z}^r$, for example, using the Gaussian algorithm over $\mathbb{Q}$ and the elementary divisor algorithm over $\mathbb{Z}$. In this way one finds the binomials $b_1, \dots, b_m$, $m = \operatorname{rank} U$, very quickly. Then it remains the much more difficult task to compute the ideal $I : (X^{x_1} \cdots X^{x_n})^\infty$ where we have used a standard notation from commutative algebra to denote the ideal in 4.29(b). Several special algorithms have been devised for this task; for example, see Bigatti, La Scala, and Robbiano [31] or Hoşten and Sturmfels [201].

Example 4.31. Let N be the submonoid of $\mathbb{Z}^2$ generated by $(3,0), (2,1), (1,2), (0,3)$. Then $\operatorname{rank} N = 2$, and the kernel U of the homomorphism $\mathbb{Z}^4 \to \mathbb{Z}^2$ that maps the unit vectors in $\mathbb{Z}^4$ to the generators of N in the given order has rank 2. Clearly $e_1 - 2e_2 + e_3$, $e_2 - 2e_3 + e_4$ is a basis of U. With the notation of Proposition 4.29 we can choose $I = (X_2^2 - X_1X_3, X_3^2 - X_2X_4)$. However, I is by no means the full kernel of the homomorphism $\Bbbk[X_1, X_2, X_3, X_4] \to \Bbbk[N]$: we have to add $X_1X_4 - X_2X_3$ to I.

We conclude this section by a characterization of those ideals I in polynomial rings $P = \Bbbk[X_1, \dots, X_n]$ for which P/I is an affine monoid algebra $\Bbbk[M]$ in such a way that M is the set of the residue classes of the monomials in P.

Theorem 4.32. *Let $\Bbbk$ be a field and I an ideal in the polynomial ring $P = \Bbbk[X_1, \dots, X_n]$. Let L be the Laurent polynomial ring $\Bbbk[X_1^{\pm 1}, \dots, X_n^{\pm 1}]$.*

Let $\mathcal{M}$ be the set of monomials in P and $\pi : P \to A = P/I$ the natural homomorphism. Then the following are equivalent:

(a) A is an affine monoid algebra with underlying monoid $\pi(\mathcal{M})$.
(b) I is a prime ideal generated by binomials $X^u - -X^v$.
(c) $I = IL \cap P$, and the subgroup U of $\mathbb{Z}^n$ generated by the differences $u - v$, $u, v \in \mathbb{Z}^n_+$, with $X^u - -X^v \in I$ is a direct summand of $\mathbb{Z}^n$.

Proof. (a) $\Longrightarrow$ (b) Since A is a domain, I must be a prime ideal. That I is generated by binomials has been shown above.

(b) $\Longrightarrow$ (c) Consider the congruence relation $\mathcal{C}$ on $\mathcal{M}$ defined by the pairs (u, v) with $X^u - -X^v \in I$. It defines the factor monoid $\mathcal{M}/\mathcal{C}$ (that, a priori, need not be cancellative or torsionfree). Proposition 4.27 shows that I is in fact the kernel of the natural map $P \to \Bbbk[\mathcal{M}/\mathcal{C}]$. We can identify $\mathcal{M}/\mathcal{C}$ with $\pi(\mathcal{M})$. Therefore P/I is the free R-module with basis $\pi(\mathcal{M})$.

The last statement implies that all monomials X^a are nonzero modulo I. Since I is a prime ideal, they are non-zerodivisors modulo I. Therefore $I = J \cap P$ where $J = IL$.

Since I is an ideal, U is a subgroup, as the reader may check. The extension of a prime ideal to a ring of fractions is again prime. Therefore J is a prime ideal in L.

Let $w \in \mathbb{Z}^n$ and $m \in \mathbb{N}$ such that $mw \in U$. We must show $w \in U$. Clearly $X^{mw} - 1 \in J$. Let $p = \operatorname{char} \Bbbk$. Suppose first that $p > 0$, $p \mid m$. Then $X^{mw} - 1 = (X^{mw/p} - 1)^p \in J$, and so $X^{mw/p} - 1 \in J$ since J is a prime ideal. Replacing m by m/p, we argue by induction.

Suppose now that $p = 0$ or $p > 0$, $p \nmid m$. We have

$$X^{mw} - 1 = (X^w - 1)(X^{(m-1)w} + X^{(m-2)w} + \cdots + 1),$$

and the second term is not in the maximal ideal $(X_1 - 1, \dots, X_n - 1) \supset J$. A fortiori, it does not belong to the prime ideal J, and we conclude that $X^w - 1 \in J$. But then $w \in U$.

(c) $\Longrightarrow$ (a) Consider the composition $P \to L \to \Bbbk[\mathbb{Z}^n/U]$. Its kernel is I. Furthermore the image of $\mathcal{M}$ in $\mathbb{Z}^n/U$ is an affine monoid. □

The reader may check that the theorem remains true if P is an arbitrary affine monoid domain and L is replaced by $\Bbbk[\operatorname{gp}(M)]$ (Exercise 4.7).

4.D Monomial prime and radical ideals

The monomial prime or radical ideals in monoid algebras can be described in purely combinatorial terms.

Proposition 4.33. *Let M be a monoid and R an integral domain (a reduced ring). Then a monomial ideal $\mathfrak{p} \subset R[M]$ is a prime (radical) ideal if and only if $I = M \cap \mathfrak{p}$ is a prime (radical) ideal in the monoid M.*

Proof. We give the proof for prime ideals; that for radical ideals is even simpler.

Suppose $\mathfrak{p}$ is a prime ideal. Since it is a monomial ideal, $\mathfrak{p} = \bigoplus_{x \in I} Rx$. Let $y, z \in M$ with $yz \in I$. Then $yz \in \mathfrak{p}$ and so $y \in \mathfrak{p}$ or $z \in \mathfrak{p}$, which in turn implies $y \in I$ or $z \in I$, and so I is a prime ideal.

Conversely, let $I \subset M$ be a prime ideal. Evidently $\mathfrak{p} = \sum_{x \in I} Rx$ is an ideal. Let $f, g \in R[M]$ such that $fg \in \mathfrak{p}$, but $f, g \notin \mathfrak{p}$. Since only finitely many elements of M appear in an equation expressing that $fg \in \mathfrak{p}$, we can pass to a finitely generated submonoid of M, and can therefore assume that M is ordered (by restricting the lexicographic order on $\operatorname{gp}(M) \cong \mathbb{Z}^r$ to M). Stripping off the terms of f and g that belong to monomials in I, we can assume that the smallest monomials of f and g lie not in I. But then the initial monomial of fg is not in I, and it follows that $fg \notin \mathfrak{p}$. □

The prime ideals and radical ideals in affine monoids M have already been determined: they are just the complements of faces and unions of faces of $\mathbb{R}_+ M$, respectively (see Proposition 2.36). For a face F of $\mathbb{R}_+ M$ we set

$$\mathfrak{p}_F = R\{M \setminus F\}.$$

Corollary 4.34. *Let R be an integral domain and M an affine monoid. Then the ideals $\mathfrak{p}_F$ are exactly the monomial prime ideals in $R[M]$, and their intersections are exactly the radical monomial ideals.*

Moreover, for each face F the embedding $R[M \cap F]$ induces the decomposition $R[M] = R[M \cap F] \oplus \mathfrak{p}_F$ as an $R[M \cap F]$-module, and $R[M \cap F]$ is a retract of $R[M]$.

Proof. The first part has been proved already. For the second we note that $R[M] = R[M \cap F] \oplus \mathfrak{p}_F$ as an R-module, and even as an $R[M \cap F]$-module. Since $\mathfrak{p}_F$ is an ideal, the projection $R[M] \to R[M \cap F]$ with kernel $\mathfrak{p}_F$ is a R-algebra morphism that restricts to the identity on $R[M \cap F]$. □

Despite its naturalness the *face projection* $\pi_F : R[M] \to R[M \cap F]$ is *not* induced by a monoid morphism. However, it maps terms of $R[M]$ to terms of $R[M \cap F]$, and is therefore homogeneous with respect to the M-grading.

Corollary 4.35. *With the notation of Corollary 4.34 let I be an ideal in $R[M]$ generated by monomials. Then the minimal prime ideals of I are the ideals $\mathfrak{p}_F$ where F runs through the set of faces that are maximal with respect to $F \cap I = \varnothing$.*

Proof. We only need to combine Lemma 4.9 with Corollary 4.34. □

Finally we determine the heights of the ideals $\mathfrak{p}_F$.

Proposition 4.36. *Let R be an integral domain, and M an affine domain. Then* $\operatorname{ht} \mathfrak{p}_F = \operatorname{rank} M - \dim F$. *If R is noetherian, then* $\dim R[M]/\mathfrak{p}_F = \dim R + \dim F$.

Proof. One has $\operatorname{ht}\mathfrak{p}_F = \dim R[M]_{\mathfrak{p}_F}$. Since $R \cap \mathfrak{p}_F = 0$, we can first invert all nonzero elements of R, and replace R by its field $\Bbbk$ of fractions. This reduces the first equation to a field $\Bbbk$ of coefficients. In an affine algebra A over a field one has $\dim A = \operatorname{ht}\mathfrak{p} + \dim A/\mathfrak{p}$ for all prime ideals ([235, Ch. II, 3.6]), and the first equation follows from the second since $\dim R[M] = \dim R + \operatorname{rank} M$. But the second equation is just this formula, if one uses that $\dim F = \operatorname{rank} M \cap F$ and $R[M]/\mathfrak{p}_F \cong R[M \cap F]$. □

Filtrations and decompositions. Let $N \neq 0$ be a finitely generated module over a noetherian domain. Then N has a filtration

$$0 = N_0 \subset N_1 \subset \cdots \subset N_n = N$$

in which the successive quotients N_i/N_{i-1} are of type $R/\mathfrak{p}_i$ for prime ideals $\mathfrak{p}_i$, $i = 1, \dots, n$ (see [253, Th. 6.4]). Moreover, each associated prime of N appears among the $\mathfrak{p}_i$ (and this fact proves the finiteness of $\operatorname{Ass} N$). In the graded case Lemma 4.9 allows one to find such a filtration in which each N_i is a graded submodule generated by a single homogeneous element modulo N_{i-1}, and the $\mathfrak{p}_i$ are (therefore) graded prime ideals. In fact, one starts the filtration by choosing N_1 being generated by a homogeneous element x whose annihilator is an associated prime of N, and applies the same principle to N/N_1, etc. The ascending chain condition on submodules ensures that the process stops after finitely many steps.

Now let M be an affine monoid, N a finitely generated M-module (see p. 51), and U an M-submodule of N. Then we can choose a field $\Bbbk$ of coefficients and apply the filtration argument to the $\Bbbk[M]$-module $(\Bbbk N)/(\Bbbk U)$.

Proposition 4.37. *Let G be a torsionfree group, $M \subset G$ an affine monoid, $N \subset G$ a finitely generated M-module, and U a submodule of N (possibly $U = \varnothing$). Then $N \setminus U$ decomposes into a disjoint union of subsets of type $x + (M \cap F)$ where $x \in N$ and F is a face of $\mathbb{R}_+M$.*

Moreover, if F is maximal with respect to the condition $F \cap \{y \in M : y + N \subset U\} = \varnothing$, then a subset of type $x + F$ appears in the decomposition.

We leave the detailed verification of the proof to the reader. The essential points to observe are:

(a) $(\Bbbk N)/(\Bbbk U)$ has a filtration by submodules of type $(\Bbbk U_i)/\Bbbk U$ where $U_i \supset U$ is an N-submodule of N;

(b) the filtration induces a decomposition $N \setminus U = \bigcup U_i \setminus U_{i-1}$;

(c) $U_i \setminus U_{i-1}$ is of type $F + x$ if $(\Bbbk U_i)/\Bbbk U_{i-1}$ is of type $\Bbbk[M]/\mathfrak{p}_F$;

(d) the faces F that are maximal with respect to the condition $F \cap \{y \in M : y + N \subset U\} = \varnothing$ correspond to the minimal prime ideals in $\operatorname{Supp}\big((\Bbbk N)/(\Bbbk U)\big)$, and these prime ideals belong to $\operatorname{Ass}\big((\Bbbk N)/(\Bbbk U)\big)$.

When applied to an affine monoid M and its normalization $\bar{M}$, the proposition yields Corollary 2.35(b). Part (a) of 2.35 follows easily from the fact that $N_x = 0$ if and only if $x \in \operatorname{Rad}(\operatorname{Ann} N)$, provided N is a finitely generated module over a ring R. For finitely generated modules N the support $\operatorname{Supp} N$ is the set of prime ideals containing $\operatorname{Ann} N$, and $\operatorname{Supp} N_x = \{\mathfrak{p} \in \operatorname{Supp} N,\ x \notin \mathfrak{p}\}$.

Stanley–Reisner rings. Let $\Bbbk$ be a field, and I a monomial radical ideal in $\Bbbk[X_1, \dots, X_n]$. As we have just seen, the set of monomials in I forms a radical ideal in the monoid $\mathbb{Z}_+^n$, and in its turn a radical ideal in $\mathbb{Z}_+^n$ corresponds to a union of faces of the cone $\mathbb{R}_+^n$. The set of all faces F with $F \cap I = \emptyset$ evidently forms a subfan $\mathscr{F}$ of the face lattice of $\mathbb{R}_+^n$. Each of the cones in $\mathscr{F}$ is simplicial, namely generated by the standard unit vectors e_i it contains. In each face of $\mathscr{F}$ these unit vectors span a simplex, and these simplices together form a simplicial complex Δ embedded in $\mathbb{R}^n$. The simplicial complex Δ is the most efficient combinatorial representation of I.

Conversely, let Δ be an abstract simplicial complex (see Example 1.43) on a subset of $V = \{1, \dots, n\}$. Then Δ determines a monomial radical ideal $I(\Delta)$ in $\Bbbk[X_1, \dots, X_n]$, namely the one generated by all squarefree monomials $X_{i_1} \cdots X_{i_r}$, $1 \le r \le n$, $i_1 < \dots < i_r$, for which $\{i_1, \dots, i_r\}$ is a *nonface* of Δ, i. e., does not belong to Δ. This shows the next proposition:

Proposition 4.38. *The assignment $\Delta \mapsto I(\Delta)$ yields a bijective correspondence between the abstract simplicial complexes whose vertex set is contained in $\{1, \dots, n\}$ and the monomial radical ideals in $\Bbbk[X_1, \dots, X_n]$.*

It is trivial that $I(\Delta)$ is generated by those monomials $X_{i_1} \cdots X_{i_k}$ that are minimal with respect to divisibility among the monomial in $I(\Delta)$. They correspond to the minimal nonfaces of Δ.

One calls $\Bbbk[\Delta] = \Bbbk[X_1, \dots, X_n]/I(\Delta)$ the *face ring* or *Stanley–Reisner ring* of Δ. Stanley–Reisner rings form one of the backbones of combinatorial commutative algebra. We refer to Bruns and Herzog [72], Hibi [188], Miller and Sturmfels [261], and Stanley [332] for their theory.

Stanley–Reisner rings are special instances of algebras defined by monoidal complexes. They will be introduced in Section 7.B.

4.E Normality

We introduced the notions of integral closure and normalization of monoids in Section 2.B. For domains we are using them in the standard way of commutative algebra that we will now outline. A very good basic reference is Bourbaki [37, Ch. V].

Let $R \subset S$ be an extension of commutative rings. Then an element $x \in S$ is *integral* over R if it satisfies an equation

$$x^n + a_{n-1}x^{n-1} + \dots + a_1 x + a_0 = 0, \qquad a_i \in R.$$

Clearly, x is integral over R if and only if the subalgebra $R[x]$ of S is a finitely generated R-module. It is however sufficient for the integrality of x that it is contained in an R-subalgebra of S, that is finitely generated as an R-module. This criterion is fundamental for the theory of integral dependence.

The set of elements $x \in S$ that are integral over R form a subring $\widehat{R}_S$ of S, the *integral closure of R in S*. If $\widehat{R}_S = S$, then S is *integral* over R. If $R = \widehat{R}_S$, then R is *integrally closed in S*. The integral closure $\bar{R}$ of an integral domain R in its field of fractions is called the *normalization* of R, and R is *normal* if $\bar{R} = R$. The classical examples of normal domains are *factorial* domains, i. e., domains in which every nonzero and nonunit element can be written as a product of prime elements. (We call an element *prime* if it generates a prime ideal; a decomposition into prime elements is unique up to order and unit factors.)

We summarize the basic facts about integral closures:

Theorem 4.39.

(a) A reduced ring R is integrally closed in $R[X]$ and $R[X^{\pm 1}]$.
(b) Let $R \subset S \subset T$ be a tower of ring extensions. Then T is integral over R if and only if S is integral over R and T is integral over S.
(c) The integral closure of $R[X]$ in $S[X]$ is $\widehat{R}_S[X]$.
(d) If $T \subset R$ is a multiplicatively closed set, then $T^{-1}\widehat{R}_S$ is the integral closure of $T^{-1}R$ in $T^{-1}S$.
(e) The intersection $\bigcap R_i$ of integrally closed subrings R_i of S is integrally closed in S.
(f) If R is a normal domain, then $R[X]$ is normal.

Fact (a) follows immediately from the definition of integral dependence: a Laurent polynomial over R cannot be a zero of a nonzero polynomial over R (Exercise 4.9). The nontrivial implication of fact (b) is readily derived from the transitivity of module finiteness, whereas (e) is trivial. Although $R[X]$ is a monoid algebra over R with monoid $\mathbb{Z}_+$, we refer for the proof of (c) to [37, Ch. V, §1, Prop. 12]. Together with the (simpler) fact (d), for which we refer to [37, Ch. V, §1, Prop. 16], it is the basis of our discussion below. The last assertion (f) follows from (c). Indeed, let $Q = \mathrm{QF}(R)$. Then $Q[X]$ is integrally closed in its field of fractions $Q(X)$, since $Q[X]$ is factorial, and since R is normal, we conclude from (c) that $R[X]$ is integrally closed in $Q[X]$.

We start with the most basic and most important result:

Theorem 4.40. *Let R be a domain, and let M be a monoid. Then $R[M]$ is normal if and only if R and M are normal.*

Proof. The necessity is very easy. Within the field of fractions of $R[M]$ (which is a domain by Corollary 4.17) we have $R = R[M] \cap \mathrm{QF}(R)$. Therefore, if an element of $\mathrm{QF}(R)$ is integral over R, then it is integral over $R[M]$ and belongs to $R[M]$ if and only if it belongs to R. Moreover, every element $z \in \mathrm{gp}(M)$ with $mz \in M$ for some $m > 0$ represents a monomial $X^z \in R[\mathrm{gp}(M)] \subset \mathrm{QF}(R[M])$ with $(X^z)^m \in R[M]$. If $R[M]$ is normal, then $X^z \in R[M]$; equivalently, $z \in M$.

For the sufficiency we consider first the case in which M is affine. Let $\mathrm{gp}(M) = \mathbb{Z}^r$. We write $\mathbb{R}_+ M \subset \mathbb{R}^r$ as an intersection of rational halfspaces H_i^+ (Theorem 1.15 and Proposition 1.69). Then $M = \bigcap_i \mathbb{Z}^r \cap H_i^+$ by Corollary 2.24(c), and

$$R[M] = \bigcap_i R[\mathbb{Z}^r \cap H_i^+].$$

Since the intersection of normal domains is normal, it is enough to treat the case $M = \mathbb{Z}^r \cap H^+$, where H^+ is a rational halfspace. But then $M \cong \mathbb{Z}^{r-1} \oplus \mathbb{Z}_+$, and $R[M] \cong R[\mathbb{Z}^{r-1} \oplus \mathbb{Z}_+]$. Choosing a basis of $\mathbb{Z}^{r-1}$, we can identify $R[M]$ with the partial Laurent polynomial ring $R[X_1^{\pm 1},\dots,X_{r-1}^{\pm 1},X_r]$, which, as a localization of a polynomial extension of R, is normal (Theorem 4.39(d) and (f)).

Now let M be an arbitrary monoid, and let $f \in \operatorname{QF}(R[M])$ be integral over $R[M]$. Then there exists an affine submonoid N such that $f \in \operatorname{QF}(R[N])$ and f is integral over $R[N]$. Since the normalization of N is contained in M, we are done by the affine case. □

We want to give a relative version that takes into account both ring and monoid extensions. We first prove an auxiliary result.

Lemma 4.41. *Let R be a domain, M an affine submonoid of $\mathbb{Z}^n$ and $\widehat{M}_{\mathbb{Z}^n}$ the integral closure of M in $\mathbb{Z}^n$. Then $R[\widehat{M}_{\mathbb{Z}^n}]$ is the integral closure of $R[M]$ in $R[\mathbb{Z}^n]$.*

Proof. We set $\widehat{M} = \widehat{M}_{\mathbb{Z}^n}$. Since every element of $\widehat{M}$ is integral over $R[M]$ as a monomial, the extension $R[M] \subset R[\widehat{M}]$ is integral. Let U be the smallest direct summand of $\mathbb{Z}^n$ containing M. Then U is the integral closure of $\operatorname{gp}(M)$ in $\mathbb{Z}^n$, and $R[\mathbb{Z}^n]$ is a Laurent polynomial extension of $R[U]$. Therefore $R[U]$ is integrally closed in $R[\mathbb{Z}^n]$. So it remains to show that $R[\widehat{M}]$ is integrally closed in $R[U]$.

Let $\Bbbk = \operatorname{QF}(R)$. If an element of $R[U]$ is integral over $R[\widehat{M}]$, then it is integral over $\Bbbk[\widehat{M}]$, too. Since $R[\widehat{M}] = \Bbbk[\widehat{M}] \cap R[U]$, it is enough to show that $\Bbbk[\widehat{M}]$ is integrally closed in $\Bbbk[U]$. By Corollary 2.25 we have $\operatorname{gp}(\widehat{M}) = U$, and by Theorem 4.40 $\Bbbk[\widehat{M}]$ is integrally closed in its field of fractions. A fortiori it is integrally closed in $\Bbbk[U]$. □

Theorem 4.42. *Let $R \subset S$ be an extension of domains and let $M \subset N$ be an extension of monoids. Then $\widehat{R}_S[\widehat{M}_N]$ is the integral closure of $R[M]$ in $S[N]$.*

Proof. We deal with one extension at a time. Suppose first that $S = R$. Since each element of $\widehat{M} = \widehat{M}_N$ is integral over $R[M]$, it follows immediately that $R[\widehat{M}]$ is integral over $R[M]$. It remains to show that $R[\widehat{M}]$ is integrally closed in $R[N]$.

Consider an element f of $R[N]$ that is integral over $R[M]$. In a monic polynomial over $R[M]$ that has f as a zero only finitely many elements of M can occur. They generate an affine submonoid of M, and together with the monomials appearing in f they generate an affine submonoid of N. Therefore we may at this point assume that M and N are themselves affine. Let M' denote the integral closure of M in $\operatorname{gp}(N)$. Then $R[M']$ is the integral closure of $R[M]$ in $R[\operatorname{gp}(N)]$ by Lemma 4.41.

Altogether we conclude that the integral closure of $R[M]$ in $R[N]$ is contained in $R[N] \cap R[M'] = R[\widehat{M}]$, since $\widehat{M} = M' \cap N$, as is easily checked.

Set $\widehat{R} = \widehat{R}_S$. In the second step we only need to show that $\widehat{R}[M]$ is integrally closed in $S[M]$. Since $\widehat{R}[M] = \widehat{R}[\operatorname{gp}(M)] \cap S[M]$, it is enough that $\widehat{R}[\operatorname{gp}(M)]$ is integrally closed in $S[\operatorname{gp}(M)]$. As above, we can reduce this claim to the case of a finitely generated torsionfree group. Then it amounts to the claim that the integral closure of $R[X_1^{\pm 1},\dots,X_n^{\pm 1}]$ in $S[X_1^{\pm 1},\dots,X_n^{\pm 1}]$ is $\widehat{R}[X_1^{\pm 1},\dots,X_n^{\pm 1}]$. But taking integral closures commutes with polynomial extensions and localizations (Theorem 4.39(c) and (d) above). □

Some of the results above hold for more general rings of coefficients. The reader may explore to what extent Theorem 4.42 can be generalized.

Normality and purity. In Theorem 2.29 we have discussed the embedding of positive affine monoids as pure submonoids of positive orthants $\mathbb{Z}_+^n$. That theorem can be extended to an analogous result on monoid algebras.

Recall that a ring extension $R \to S$ is *pure* if for every R-module M, the map $M \to M \otimes S$, $x \mapsto x \otimes 1$, is injective. Applied to an ideal $I \subset R$, or rather the module R/I, purity yields $I = IS \cap R$. A sufficient condition for purity is that R is a direct summand of S as an R-module.

Theorem 4.43. *Let R be a normal domain, and M a positive affine monoid. Then the following are equivalent:*

(a) M is normal;
(b) there exists n such that $R[M]$ is isomorphic to a pure subalgebra S of $R[X_1, \dots, X_n]$;
(c) there exists n such that $R[M]$ is isomorphic to a subalgebra S of $R[X_1, \dots, X_n]$, for which S is a direct summand of $R[X_1, \dots, X_n]$ as an S-module;
(d) in addition to the properties in (c), S can be chosen to be integrally closed in $R[X_1, \dots, X_n]$.

Proof. Let M be normal. By Theorem 2.29 there is an embedding $\varphi : M \to \mathbb{Z}^n$ such that $\varphi(M)$ is pure in $\mathbb{Z}^n$ and integrally closed. The embedding of monoids extends to an embedding of monoid algebras, and we conclude from Theorem 4.40 that $R[M]$ is integrally closed in $R[\mathbb{Z}_+^n]$.

For the implication (a) $\Longrightarrow$ (d) we can now identify M with its image under φ and have only to show that $R[M]$ is a direct summand of $R[\mathbb{Z}_+^n]$. Set $N = \mathbb{Z}_+^n \setminus M$. Since $M = \operatorname{gp}(M) \cap \mathbb{Z}_+^n$, we conclude that $M + N \subset N$ (writing the monoid operation additively). This amounts to the fact that RN is a module over $R[M]$, and since $R[\mathbb{Z}_+^n] = R[M] \oplus RN$ as an R-module, we conclude that this splitting is in fact one of $R[M]$-modules.

The implications (d) $\Longrightarrow$ (c) $\Longrightarrow$ (b) are trivial, and for (b) $\Longrightarrow$ (a) we only need that normality is passed on to pure subalgebras (and from $R[M]$ to M by Theorem 4.40). See Exercise 4.10. □

Let M be a pure submonoid of N. In Section 2.B we studied the decomposition of N into a disjoint union of coset modules. After the introduction of a ring R of coefficients, one obtains a splitting of $R[N]$ into a direct sum of $R[M]$-submodules, which we call the *coset modules* of $R[M]$ in $R[N]$. Proposition 2.31 then translates as follows:

Proposition 4.44. *Let M be a pure submonoid of an affine monoid N. Then the coset modules of $R[M]$ in $R[N]$ are finitely generated (by monomials) over $R[M]$ and of rank* 1.

Factorial and regular monoid domains. A noetherian local ring R with maximal ideal $\mathfrak{m}$ is *regular* if $\mathfrak{m}$ has a system of generators $x_1, \dots, x_d$, $d = \dim R$. An arbitrary noetherian ring is *regular* if all its localizations with respect to prime ideals are regular local rings. See [72, 2.2] for the theory of regular rings.

By the theorem of Auslander–Buchsbaum–Serre, R is regular if and only if every finitely generated R-module has finite projective dimension over R. By the theorem of Auslander–Buchsbaum–Nagata, regular local rings are factorial.

There exist only trivial examples of factorial or regular affine monoid algebras. We combine both these properties in a single proposition.

Proposition 4.45. *Let R be a domain and M an affine monoid. Then the following are equivalent:*

(a) $R[M]$ is factorial (regular);
(b) R is factorial (regular) and $M \cong \mathbb{Z}^m \oplus \mathbb{Z}_+^n$ for suitable $m, n \in \mathbb{Z}_+$.

The proof is left to the reader (Exercises 4.13, 4.25). The implication (b) $\Longrightarrow$ (a) amounts to the standard fact that (Laurent) polynomial extensions preserve both factoriality and regularity.

Integral closure of monomial ideals. Let R be a commutative ring, and I an ideal of R. An element $x \in R$ is *integral* over I if it satisfies an equation

$$x^n + a_{n-1}x^{n-1} + \cdots + a_1 x + a_0 = 0, \qquad a_j \in I^{n-j},\ j = 0, \ldots, n-1. \tag{4.2}$$

The elements that are integral over I form an ideal $\bar{I}$, the *integral closure* of I in R; see Huneke and Swanson [205].

The integral closure of an affine monoid algebra is determined by the set of lattice points in a convex set, and the same holds true for a monomial ideal in a normal affine monoid algebra:

Theorem 4.46. *Let $M \subset \mathbb{Z}^d$ be a normal affine monoid, $\Bbbk$ a field, and I a monomial ideal in $R = \Bbbk[M]$. Then $\bar{I}$ is the monomial ideal whose monomials correspond to the lattice points in the convex hull of the set of monomials in I.*

Proof. Let x be a lattice point in the convex hull of the lattice points corresponding to elements of I. Then $x = a_1 y_1 + \cdots + a_m y_m$ with $a_j \in \mathbb{Q}_+$, $y_j \in I \cap \mathbb{Z}^d$ for all j and $\sum a_j = 1$. After multiplication with a common denominator n one obtains an equation $x^n = y_1^{b_1} \cdots y_m^{b_m}$ (in multiplicative notation) with $\sum b_j = n$. Thus $x^n \in I^n$, and $x \in \bar{I}$.

For the converse consider an element $f \in R$ whose Newton polytope N is not contained in the convex hull of the monomial basis of I. Then we can find a linear form λ and a vertex z of the Newton polytope such that $\lambda(z) > \lambda(y)$ for all monomials $y \in I$, and $\lambda(z) \geq \lambda(z')$ for all monomials $z' \in N$ (Theorem 1.32). If f would satisfy equation (4.2) for some n, then the Newton polytope of f^n would be contained in the union of the Newton polytopes of the other summands, and this is impossible, as we see by evaluating λ on them. □

Remark 4.47. With Theorem 4.46 we have found an ideal-theoretic classification of the sets of solutions of the inhomogeneous systems of linear diophantine inequalities already discussed in Theorem 2.12. (The reader should justify this statement in detail.)

4.F Divisorial ideals and the class group

Our basic references for this section are [37, Ch. VII] and Fossum [126].

The class group. Let R be an integral domain. A *fractional ideal* in R is a nonzero R-submodule I of $\mathrm{QF}(R)$ for which there exists $a \in R$, $a \neq 0$, such that $aI \subset R$. If R is noetherian, then it is sufficient that I is finitely generated, since we can choose a as a common denominator of the generators. The *inverse ideal* of I is

$$I^{-1} = \{a \in \mathrm{QF}(R) : aI \subset R\}.$$

Clearly I^{-1} is also a fractional ideal, and $I \subset (I^{-1})^{-1}$. For fractional ideals I and J we set

$$I : J = \{a \in \mathrm{QF}(R) : aJ \subset I\},$$

and IJ is the R-submodule of $\mathrm{QF}(R)$ generated by all products xy, $x \in I$, $y \in J$. It is easy to check that $I : J$ and IJ are again fractional. In particular, the fractional ideals form a monoid under multiplication.

A fractional ideal I is called *divisorial* if $I = (I^{-1})^{-1}$. It is *invertible* if $II^{-1} = R$. One checks easily that invertible ideals (to be considered in connection with the Picard group in Section 4.G) are divisorial.

Principal fractional ideals $I = aR$, $a \in \mathrm{QF}(R)$, are evidently invertible, since $I^{-1} = a^{-1}R$.

The next proposition relates the notions just introduced to standard operations and terms of linear algebra. Recall that an R-module is *reflexive* if the natural homomorphism $M \to \mathrm{Hom}_R(\mathrm{Hom}_R(M, R), R)$ is an isomorphism.

Lemma 4.48. *Let R be an integral domain and let I, J be fractional ideals.*

(a) The homomorphism $I : J \to \mathrm{Hom}_R(J, I)$, which maps $a \in I : J$ to the multiplication by a, is an isomorphism.
(b) I is reflexive as an R-module if and only if I is a divisorial ideal.
(c) $(I^{-1})^{-1}$ is the smallest divisorial ideal containing I.
(d) if I is divisorial, then $I : J$ is also divisorial.

Now suppose that R is a *normal noetherian domain*. Though the theory can be developed more generally for Krull domains, we restrict ourselves to the noetherian case. The following theorem characterizes the normal noetherian domains and lays the foundation for the definition of divisors:

Theorem 4.49. *Let R be a noetherian domain. Then the following are equivalent:*

(a) R is normal;
(b) R satisfies the following conditions:
  (i) (Serre's (R_1)) *For each height 1 prime ideal $\mathfrak{p}$ the localization $R_\mathfrak{p}$ is a discrete valuation domain;*
  (ii) (Serre's (S_2))
$$R = \bigcap_{\mathrm{ht}\,\mathfrak{p}=1} R_\mathfrak{p}.$$

Not only normality, but also the conditions (R_1) and (S_2) individually are intrinsic properties of M (over a suitable ring of coefficients). We will come back to this point in Exercises 4.16 and 4.17.

Let us denote the discrete valuation on $\operatorname{QF}(R)$ associated with $R_\mathfrak{p}$ by $v_\mathfrak{p}$. The maximal ideal of $R_\mathfrak{p}$ is generated by a single element t, and if I is a fractional ideal of R, $I_\mathfrak{p} = t^k R_\mathfrak{p}$ for a unique $k \in \mathbb{Z}$. We set $v_\mathfrak{p}(I) = k$. One has $v_\mathfrak{p}(I) = 0$ for all but finitely many $\mathfrak{p}$. (This follows easily from the fact that every element of R is contained in only finitely many height 1 prime ideals.) Therefore we can define the *divisor* of I by

$$\operatorname{div}(I) = \sum_{\mathfrak{p}} v_\mathfrak{p}(I)\operatorname{div}(\mathfrak{p}) \in \operatorname{Div}(R) = \bigoplus_{\mathfrak{p}} \mathbb{Z}\operatorname{div}(\mathfrak{p})$$

where $\operatorname{Div}(R)$ is the free abelian group on the set of height 1 prime ideals and $\operatorname{div}(\mathfrak{p})$ is the basis element representing $\mathfrak{p}$. It follows immediately from the definition that

$$\operatorname{div}(IJ) = \operatorname{div}(I) + \operatorname{div}(J) \quad \text{and} \quad \operatorname{div}(I^{-1}) = -\operatorname{div}(I).$$

Theorem 4.50. *Let R be a normal noetherian domain. Then:*

(a) *A prime ideal of R is divisorial if and only if it has height* 1.
(b) *The divisorial fractional ideals I form a group D with product $(I, J) \mapsto ((IJ)^{-1})^{-1}$.*
(c) *The map $I \mapsto \operatorname{div}(I)$ is an isomorphism of D and* $\operatorname{Div}(R)$.
(d) *Divisorial ideals I and J are isomorphic R-modules if and only if $I : J$ is a principal fractional ideal.*

For the proof of the theorem one observes that $\operatorname{div}\big((I^{-1})^{-1}\big) = \operatorname{div}(I)$, simply because the formation of the inverse ideal commutes with localization, and every ideal of $R_\mathfrak{p}$ is divisorial if $\operatorname{ht}\mathfrak{p} = 1$. Moreover,

$$I = \bigcap_{\operatorname{ht}\mathfrak{p}=1} I_\mathfrak{p} \quad \Longleftrightarrow \quad I \text{ is divisorial.}$$

The *principal divisors*, i. e., divisors of principal ideals, form a subgroup $\operatorname{Princ}(R)$. Therefore we can define the *(divisor) class group* of R by

$$\operatorname{Cl}(R) = \operatorname{Div}(R)/\operatorname{Princ}(R).$$

By Theorem 4.50(d), the divisors $\operatorname{div}(I)$ and $\operatorname{div}(J)$ of divisorial ideals I, J, differ by a principal divisor if and only if I and J are isomorphic as R-modules. Therefore $\operatorname{Cl}(R)$ is the set of isomorphism classes of divisorial fractional ideals of R. We will denote the class of $\operatorname{div}(I)$ by $[I]$. Since $\operatorname{Div}(R)$ is generated by the prime divisors, $\operatorname{Cl}(R)$ is generated by their classes.

The divisor class group measures the deviation of a normal domain from factoriality.

Theorem 4.51. *Let R be a noetherian normal domain. Then the following are equivalent:*

(a) *R is factorial;*

(b) every height 1 *prime ideal of* R *is principal;*
(c) $\operatorname{Cl}(R) = 0$.

The behavior of the divisor class group under polynomial extensions and localization is rather easy to control:

Theorem 4.52. *Let* R *be a normal noetherian domain. Then:*

(a) (Gauss) *The extension* $I \mapsto IR[X]$ *of divisorial ideals induces an isomorphism* $\operatorname{Cl}(R) \cong \operatorname{Cl}(R[X])$.
(b) (Nagata) *Let* T *be a multiplicatively closed subset of* R*. Then the extension* $I \mapsto T^{-1}I$ *induces an exact sequence*

$$0 \to U \to \operatorname{Cl}(R) \to \operatorname{Cl}(T^{-1}R) \to 0$$

in which U *is the subgroup of* $\operatorname{Cl}(R)$ *generated by the classes of the divisorial prime ideals* $\mathfrak{p}$ *such that* $T \cap \mathfrak{p} \neq \varnothing$.

Class groups of monoid algebras. A *fractional monomial ideal* in a monoid domain $R[M]$ is a fractional ideal of type Ix^{-1} where I is a monomial ideal in R and $x \in M$. We first show that the monoid of fractional monomial ideals over an affine monoid domain is closed under the operation $I : J$ and identify the condition under which fractional monomial ideals are isomorphic.

Lemma 4.53. *Let* R *be a domain and* M *an affine monoid. Let* I *and* J *be monomial fractional ideals of* $R[M]$*, and set* $I' = I \cap \operatorname{gp}(M)$, $J' = J \cap \operatorname{gp}(M)$*. Then*

(a) $I : J$ *is a monomial ideal.*
(b) The following are equivalent:
  (i) I *and* J *are isomorphic as* $R[M]$*-modules;*
  (ii) there exists a monomial $x \in \operatorname{gp}(M)$ *such that* $I = xJ$*;*
  (iii) there exists a monomial $x \in \operatorname{gp}(M)$ *such that* $I' = xJ'$.

Proof. We first show that $I : J \subset R[\operatorname{gp}(M)]$. In fact, if $fJ \subset I$, then $fJR[\operatorname{gp}(M)] \subset IR[\operatorname{gp}(M)]$. But $JR[\operatorname{gp}(M)] = R[\operatorname{gp}(M)]$, and similarly for I, so that $f \in R[\operatorname{gp}(M)]$.

Now let $f \in R[M]$ such that $fJ \subset I$, and consider a monomial x representing a vertex of the Newton polytope of f. Let $z \in J'$. Then xz represents a vertex of the Newton polytope of fz. Since $fz \in I$ and I is monomial, it follows that $xz \in I$ as well. Passing from f to $f - f_x x$ and using induction on $\#\operatorname{supp}(f)$, we conclude that all monomials $x \in \operatorname{supp} f$ belong to $I : J$. This proves (a).

For (b) we first observe the trivial implications (iii) $\Longrightarrow$ (ii) $\Longrightarrow$ (i), and for (i) $\Longrightarrow$ (iii) it is enough to find a monomial x with $I = xJ$, since (ii) and (iii) are clearly equivalent.

If I and J are isomorphic as $R[M]$-modules, then, by (a), there exists $f \in R[\operatorname{gp}(M)]$ such that $I = fJ$. Then we also have $J = f^{-1}I$. Using (a) again, we conclude that $f^{-1} \in R[\operatorname{gp}(M)]$, too. The only units in $R[\operatorname{gp}(M)]$ are the terms rx where r is a unit in R (Proposition 4.20). If $rxJ = I$, then $xJ = I$ as well, and we are done. □

For the rest of this subsection R is a normal noetherian domain and M is a normal affine monoid. We denote the set of facets of $\mathbb{R}_+M$ by $\mathscr{F}$. Note that the ideals $\mathfrak{p}_F$, $F \in \mathscr{F}$, introduced in Section 4.D, are exactly the monomial prime ideals of height 1 (see 4.36).

Let us first show that the valuation $v_{\mathfrak{p}_F}$ is the unique extension of the support form σ_F of $\mathbb{R}_+M$ associated with the facet F. More generally let us consider an arbitrary $\mathbb{Z}$-linear τ form on $\mathrm{gp}(M)$. We extend τ to $R[\mathrm{gp}(M)]$ by

$$\tau\Bigl(\sum_{x\in\mathrm{gp}(M)} f_x x\Bigr) = \min\{\tau(x) : f_x \neq 0\}$$

and $\tau(0) = \infty$ ($0 \in R$!). For all f, g one then has

$$\begin{aligned}\tau(fg) &= \tau(f) + \tau(g),\\ \tau(f+g) &\le \min\bigl(\tau(f), \tau(g)\bigr),\\ \tau(f+g) &= \min\bigl(\tau(f), \tau(g)\bigr) \quad \text{if } \tau(f) \neq \tau(g).\end{aligned}$$

This shows that τ is a valuation on $R[M]$. It extends to a valuation on $\mathrm{QF}(R[M])$ by $\tau(f/g) = \tau(f) - \tau(g)$.

Let us return to σ_F. We can now consider it as a valuation on $\mathrm{QF}(R[M])$. In order to identify it with $v_{\mathfrak{p}_F}$ we have only to check that

$$\mathfrak{p}_F = \{f \in R[\mathrm{gp}(M)] : \sigma_F(f) \ge 1\}$$

and that there exists $f \in R[M]$ with $\sigma_F(f) = 1$. Both properties are clearly satisfied. In fact, they imply that the valuation ring of σ_F is really $R[M]_{\mathfrak{p}_F}$, so that $\sigma_F = k v_{\mathfrak{p}_F}$ for some $k \in \mathbb{N}$, and obviously we must have $k = 1$.

Let us now compute the divisor of a monomial $x \in \mathrm{gp}(M)$. (By definition, $\mathrm{div}(a) = \mathrm{div}(Aa)$ for an element $a \in \mathrm{QF}(A)$.) In multiplicative notation, $x = y/z$ with $y, z \in M$. Therefore $\mathrm{div}(x) = \mathrm{div}(y) - \mathrm{div}(z)$. The minimal prime ideals of y and z are all monomial by Corollary 4.35. Therefore

$$\begin{aligned}\mathrm{div}(x) &= \sum_{F\in\mathscr{F}} v_{\mathrm{div}(\mathfrak{p}_F)}(y)\,\mathrm{div}(\mathfrak{p}_F) - \sum_{F\in\mathscr{F}} v_{\mathrm{div}(\mathfrak{p}_F)}(z)\,\mathrm{div}(\mathfrak{p}_F)\\ &= \sum_{F\in\mathscr{F}} \sigma_F(x)\,\mathrm{div}(\mathfrak{p}_F).\end{aligned} \tag{4.3}$$

Similarly we compute the divisor of an arbitrary monomial ideal I:

$$\mathrm{div}(I) = \sum_{F\in\mathscr{F}} \sigma_F(I)\,\mathrm{div}(\mathfrak{p}_F)$$

where

$$\sigma_F(I) = \min\{\sigma_F(x) : x \in I \cap \mathrm{gp}(M)\}. \tag{4.4}$$

On the other hand, let J be the monomial fractional ideal whose monomial basis is determined by the system

$$\sigma_F(x) \geq s_F, \quad x \in \mathrm{gp}(M),$$

of inequalities with $s_F \in \mathbb{Z}$, $F \in \mathscr{F}$. Then one easily checks that J is indeed divisorial: J^{-1} is the monomial ideal with basis determined by the system $\sigma_F(x) \geq -s_F$, and $(J^{-1})^{-1} = J$. (Note that for every F there exists $x \in J$ with $\sigma_F(x) = s_F$.) Equation (4.4) therefore implies the next theorem:

Theorem 4.54. *Let R be a normal noetherian domain and M an affine normal monoid. Then the divisorial monomial ideals of R are exactly the R-submodules of $R[\mathrm{gp}(M)]$ whose monomial basis is determined by a system*

$$\sigma_F(x) \geq s_F, \quad x \in \mathrm{gp}(M),$$

of inequalities with $s_F \in \mathbb{Z}$, $F \in \mathscr{F}$.

In Figure 4.2 we illustrate the formation of monomial divisorial ideals.

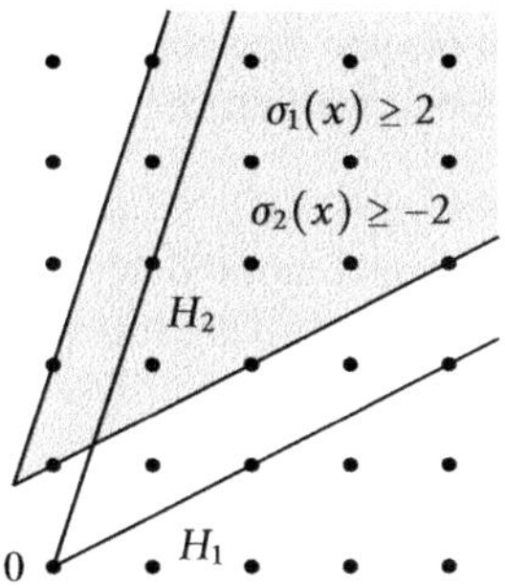

Fig. 4.2. A divisorial monomial ideal

Let $\mathfrak{p}$ be a divisorial prime ideal in a normal domain A, and $v_\mathfrak{p}$ the associated valuation. Then

$$\mathfrak{p}^{(k)} = \{x \in A : v_\mathfrak{p}(x) \geq k\}, \quad k \in \mathbb{Z}_+,$$

is called the kth *symbolic power* of $\mathfrak{p}$. Clearly $\mathrm{div}(\mathfrak{p}^{(k)}) = k\,\mathrm{div}(\mathfrak{p})$ for all $k \in \mathbb{Z}_+$. Then the divisorial ideal I with $\mathrm{div}(I) = k_1 \mathrm{div}(\mathfrak{p}_1) + \cdots + k_m \mathrm{div}(f\,\mathfrak{p}_m)$ is given by $\mathfrak{p}_1^{(k_1)} \cap \cdots \cap \mathfrak{p}_m^{(k_m)}$. (Note that all these divisorial ideals are contained in A.)

Remark 4.55. With Theorem 4.54 we have identified the sets of solutions of the inhomogeneous systems of linear diophantine inequalities whose associated homogeneous system defines a normal monoid irredundantly. Compare this assertion with Remark 4.47.

The theorem shows that the monomial divisorial ideals depend essentially only on M—the ring R just provides the coefficients. Therefore, we call the subsets of $\mathrm{gp}(M)$ that form the R-bases of the monomial divisorial ideals of $R[M]$ *divisorial ideals of* M. Moreover, we are justified to denote the subgroup of $\mathrm{Div}(R[M])$ generated by the monomial divisorial prime ideals simply by $\mathrm{Div}(M)$. Similarly we write $\mathrm{Princ}(M)$ for its subgroup of the principal monomial ideals.

Corollary 4.56 (Chouinard). *The elements of the group*

$$\mathrm{Cl}(M) = \mathrm{Div}(M)/\operatorname{Princ}(M)$$

represent the isomorphism classes of divisorial monomial ideals of $R[M]$.

If $F_1, \dots, F_s$ *are the facets of* $\mathbb{R}_+ M$ *and* $\sigma_1, \dots, \sigma_s$ *are the corresponding support forms, then*

$$\mathrm{Cl}(M) \cong \mathbb{Z}^s/\sigma(\mathrm{gp}(M)),$$

the isomorphism being induced by the map $I \mapsto (\sigma_i(I), \dots \sigma_s(I))$.

Proof. Lemma 4.53 shows that two monomial divisorial ideals I and J are isomorphic if $\mathrm{div}(I) - \mathrm{div}(J)$ is a monomial principal divisor.

For the second statement we have just numbered the basis elements of $\mathrm{Div}(M)$. The identification of $\operatorname{Princ}(M)$ with $\sigma(\mathrm{gp}(M))$ is justified by equation (4.3). □

Remark 4.57. We call $\mathrm{Cl}(M)$ the *class group of* M. This notion is also justified from a purely monoid-theoretic point of view, since the notions of multiplicative ideal theory introduced so far can be developed without coefficients. See Geroldinger and Halter–Koch [141]).

Example 4.58. Let M be the polytopal monoid $M(P)$ where P is the triangle spanned by $(-1,-1), (1,0), (0,1) \in \mathbb{R}^2$. Then $M(P)$ has four generators, the fourth one coming

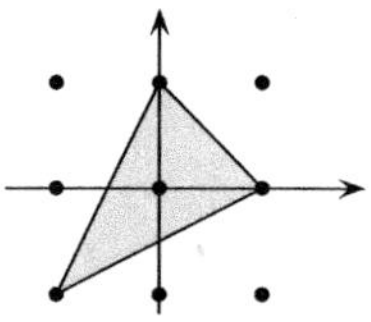

Fig. 4.3. The polytope P

from $(0,0) \in P$. Evaluating the three support forms on the generators of M yields the vectors $u_i = 3e_i$, $i = 1,2,3$ and $u_4 = e_1 + e_2 + e_3$. Clearly $\mathrm{Cl}(M) = \mathbb{Z}^3/\sum \mathbb{Z}u_i \cong (\mathbb{Z}/(3))^2$.

That $\mathrm{Cl}(M)$ is finite in this example is not surprising. In fact, the finiteness of $\mathrm{Cl}(M)$ characterizes the normal monoids that are essentially simplicial: an (arbitrary) affine monoid is called *simplicial* if it generates a simplicial cone.

Corollary 4.59. *Let* M *be a normal affine monoid, and consider the canonical splitting* $M = \mathrm{U}(M) \oplus M'$. *Then:*

(a) $\mathrm{Cl}(M) \cong \mathrm{Cl}(M')$, *and*

$$\operatorname{rank} \mathrm{Cl}(M) = s - \operatorname{rank} M' = s - (\operatorname{rank} M - \operatorname{rank} \mathrm{U}(M)),$$

where s *is the number of facets of* $\mathbb{R}_+ M$.

(b) *In particular,* $\mathrm{Cl}(M)$ *is finite if and only if* M' *is simplicial.*

(c) $\operatorname{Cl}(M) = 0$ *if and only if* $M \cong \mathbb{Z}^m \oplus \mathbb{Z}_+^s$, $m = \operatorname{rank} U(M)$.

Proof. One has $\operatorname{Cl}(N_1 \oplus N_2) \cong \operatorname{Cl}(N_1) \oplus \operatorname{Cl}(N_2)$ for all normal affine monoids (Exercise 4.18). Since $\operatorname{Cl}(G) = 0$ if G is a free group, it follows that $\operatorname{Cl}(M) \cong \operatorname{Cl}(M')$. The rest of (a) and (b) is clear. For (c) we have only to observe that $M' \cong \sigma(\operatorname{gp}(M) \cap \mathbb{Z}_+^s$. If $\operatorname{Cl}(M) = 0$, then $\sigma(\operatorname{gp}(M)) = \mathbb{Z}^s$, and $M' \cong \mathbb{Z}_+^s$. □

Part (c) of the corollary is only a reformulation of Proposition 4.45, at least in combination with the next theorem.

We have now collected all the tools that are necessary for the computation of the class groups of affine monoid algebras.

Theorem 4.60 (Chouinard). *Let R be a normal noetherian domain, and let M be a normal affine monoid. Then*

$$\operatorname{Cl}(R[M]) = \operatorname{Cl}(R) \oplus \operatorname{Cl}(M),$$

where the embedding $\operatorname{Cl}(R) \to \operatorname{Cl}(R[M])$ is induced by the extension of divisorial ideals, and the embedding $\operatorname{Cl}(M) \to \operatorname{Cl}(R[M])$ is induced by the map $I \mapsto [IR]$ on the set of divisorial monomial ideals.

Proof. It is very easy to check that the extension $JR[M] \subset \operatorname{QF}(R)[M]$ of a divisorial ideal J of R is again divisorial. Therefore we obtain an induced map $\operatorname{Cl}(R) \to \operatorname{Cl}(R[M])$. Now let $x \in \operatorname{int}(M)$. Then $R[M, x^{-1}] = R[\operatorname{gp}(M)]$; see Exercise 2.10 or the ring-theoretic argument in the proof of Lemma 5.36. By Nagata's theorem we have an exact sequence

$$0 \to U \to \operatorname{Cl}(R[M]) \to \operatorname{Cl}\big(R[\operatorname{gp}(M)]\big) \to 0.$$

Since $R[\operatorname{gp}(M)]$ is the localization of a polynomial ring over R with respect to a multiplicative set generated by prime elements, The theorems of Gauss and Nagata show that the extension map $J \mapsto JR[\operatorname{gp}(M)]$ induces an isomorphism $\operatorname{Cl}(R) \to \operatorname{Cl}\big(R[\operatorname{gp}(M)]\big)$, which factors through $\operatorname{Cl}(R[M])$. It only remains to identify the group U. By Nagata's theorem it is generated by the classes of the minimal prime ideals of x. These are exactly the monomial prime ideals $\mathfrak{p}_F$, and so U is the subgroup generated by them. Since two monomial divisorial ideals are isomorphic if and only if they represent the same class in $\operatorname{Cl}(M)$, we are done. □

Corollary 4.61. *Every divisorial ideal H in $R[M]$ is isomorphic to one of the ideals IJ where I is a divisorial ideal in R and J is a divisorial monomial ideal.*

Proof. According to the theorem we can find a divisorial ideal I in R and a divisorial monomial ideal J such that H is isomorphic to the smallest divisorial ideal containing IJ. But IJ is itself divisorial, as the reader may check (Exercise 4.14). □

In particular, if R is factorial, for example a field, then every divisorial ideal in $R[M]$ is isomorphic to a monomial divisorial ideal.

The class ring. We have seen in Corollary 4.56 that the standard map determines the class group of a normal monoid M. This is true far beyond the group structure of $\mathrm{Cl}(\Bbbk[M])$. For the sake of a clarity we consider only the case in which the ring of coefficients is a field. (The generalization is immediate.)

Theorem 4.62. *Let $\Bbbk$ be a field, M a normal affine monoid, and $\sigma : \mathrm{gp}(M) \to \mathbb{Z}^s$ the standard map of M. For a class $c \in \mathrm{Cl}(\Bbbk[M]) \cong \mathbb{Z}^s/\sigma(\mathrm{gp}(M))$ let*

$$D_c = \Bbbk\{y \in \mathbb{Z}^s_+ : y \in c\}.$$

Then D_c is a $\Bbbk[M]$-submodule (via σ) of $\Bbbk[\mathbb{Z}^s] \cong \Bbbk[Y_1, \dots, Y_s]$ isomorphic to a divisorial ideal of class $-c$, and

$$\Bbbk[Y_1, \dots, Y_s] = \bigoplus_{c \in \mathrm{Cl}(M)} D_c.$$

In particular, $\Bbbk[Y_1, \dots, Y_s]$ is a $\mathrm{Cl}(M)$-graded ring.

Proof. In view of Corollary 4.59 we may assume $\operatorname{Ker} \sigma = \mathrm{U}(M) = 0$. Then we can consider M to be a pure submonoid of $\mathbb{Z}^s_+$ via σ, and $\mathrm{gp}(M)$ a subgroup of $\mathbb{Z}^s$. As stated in Proposition 4.44, $\Bbbk[Y_1, \dots, Y_s]$ splits into the coset modules of $\Bbbk[M]$ in $\Bbbk[Y_1, \dots, Y_s]$, and by definition these are the modules D_c.

For the rest of the proof it is best one uses exponential notation for the monoid algebras. The monomials in $\mathrm{gp}(M)$ will be denoted by X^x and those in $\mathbb{Z}^s$ by Y^y. The set of all monomials with exponent vector in c is denoted by Y^c.

It only remains to show that D_c is isomorphic as a $\Bbbk[M]$-module to a divisorial ideal of class $-c$. To this end we choose $z \in -c$ and multiply D_c by Y^{-z}. The image I is a $\Bbbk[M]$-submodule of $\Bbbk[\mathrm{gp}(M)]$ spanned by all the monomials X^x such that $\sigma(x) + z \in \mathbb{Z}^s_+$. In other words: by all monomials X^x such that $\sigma(x) \geq z$. Hence I is divisorial with divisor $z_1 \operatorname{div}(\mathfrak{p}_1) + \dots + z_s \operatorname{div}(\mathfrak{p}_s)$, and we are done. □

Thus the polynomial ring $\Bbbk[Y_1, \dots, Y_s]$ bundles all the divisor classes of $\Bbbk[M]$ into a single, well-understood $\Bbbk$-algebra. For analogy with the number-theoretic situation we are justified in calling $\Bbbk[Y_1, \dots, Y_s]$ the *class ring of* $\Bbbk[M]$. See [66] for a study of certain properties of the divisorial ideals of $\Bbbk[M]$ using the class ring.

Remark 4.63. The class group of M limits the pure embeddings of a positive normal affine monoid in the following way: if M is a pure submonoid of $\mathbb{Z}^n_+$, then $\mathrm{Cl}(M)$ is a subquotient of $\mathbb{Z}^n/\mathrm{gp}(M)$. See [66, Corollary 2.3]. This reflects the fact that, roughly speaking, multiples of the support forms must appear in every system of inequalities cutting out M from a group.

Positively graded algebras. Let R be a normal noetherian graded ring, and suppose that the homogeneous nonunits in R generate a proper ideal $\mathfrak{p}$. For example, R is a positively graded algebra over a field $\Bbbk$: $R = \bigoplus_{i=0}^{\infty} R_i$ and $R_0 = \Bbbk$. In this case $\mathfrak{p} = \bigoplus_{i=1}^{\infty} R_i$ is even a maximal ideal. Since monoid algebras $\Bbbk[M]$ with a positive affine monoid have a positive grading, the theorem below applies especially to them. We will need it in Chapter 5.

In the general case $\mathfrak{p}$ is a prime ideal. This follows from Lemma 4.9: by construction all homogeneous elements $\neq 0$ of $R/\mathfrak{p}$ are units, and on the other hand the minimal prime ideals of $R/\mathfrak{p}$ are graded. So we may consider the localization $R_\mathfrak{p}$.

Theorem 4.64. *Under the above hypothesis on R, the extension* $R \to R_\mathfrak{p}$ *induces an isomorphism* $\mathrm{Cl}(R) \to \mathrm{Cl}(R_\mathfrak{p})$.

Proof. Let us first show that every divisorial ideal in R is isomorphic to a graded divisorial ideal. In fact, let T be the multiplicatively closed system of all homogeneous nonzero elements in R. Then it is not hard to see that all homogeneous nonzero elements in $T^{-1}R$ (with its induced grading) are units. It follows readily that $T^{-1}R$ is either a field $\Bbbk$ or of the form $\Bbbk[X, X^{-1}]$, depending on whether there exist homogeneous elements of degree $\neq 0$ (Exercise 4.2). In any case, $T^{-1}R$ is factorial.

Nagata's theorem implies that $\mathrm{Cl}(R)$ is generated by the classes of the graded prime ideals, and so every divisorial ideal is isomorphic to a graded one.

Applying Nagata's theorem to the extension $R \to R_\mathfrak{p}$ yields the surjectivity of the homomorphism $\mathrm{Cl}(R) \to \mathrm{Cl}(R_\mathfrak{p})$. Thus it remains to be shown that a nonprincipal graded ideal I of R extends to a nonprincipal ideal in $R_\mathfrak{p}$. To this end we choose a minimal graded presentation of I. i. e., an exact sequence

$$\mathbb{F} : R^n \xrightarrow{\psi} R^m \xrightarrow{\varphi} I \to 0$$

in which the basis elements of R^m, $m \geq 2$, are mapped to a minimal homogeneous system $x_1, \dots, x_m$ of generators of I and the basis elements of R^n are mapped to a (minimal) homogeneous system of generators of $\operatorname{Ker} \varphi$. The elements of the matrix representing ψ are homogeneous. They must belong to $\mathfrak{p}$ since otherwise one of the x_i was a linear combination of the others. It follows that $(\psi(R_\mathfrak{p}^n) \subset \mathfrak{p}R_\mathfrak{p}^n$. Since $m \geq 2$, $I_\mathfrak{p}$ is not principal—otherwise its homomorphic image $R_\mathfrak{p}^m/\mathfrak{p}R_\mathfrak{p}^m$ would be a cyclic module. □

4.G The Picard group and seminormality

The most elementary description of seminormality for a domain R is through *Hamann's criterion*: R is *seminormal* if every element x of $\mathrm{QF}(R)$ such that $x^2, x^3 \in R$ belongs to R. The description immediately explains the notion of seminormality for monoids. However, it does not reveal the importance of seminormality that rests on its role in K-theory, and for K-theory it is useful to generalize the notion of seminormality as follows (Swan [347]):

Definition 4.65. Let R be a reduced ring. Then R is called *seminormal* if the equation $x^3 = y^2$ for elements $x, y \in R$ is only solvable with $x = z^2$, $y = z^3$ for $z \in R$.

One could omit the requirement of reducedness (Exercise 4.28). The reader may check that z is uniquely determined, and the elementary proof of the next proposition is also left to the reader:

Proposition 4.66. *Let R be seminormal and $S \subset R$ multiplicatively closed. Then $S^{-1}R$ is also seminormal.*

Let R be a reduced ring. As a replacement of the field of fractions of an integral domain we introduce

$$\operatorname{PQF}(R) = \prod_{\mathfrak{p}} \operatorname{QF}(R/\mathfrak{p}),$$

where $\mathfrak{p}$ runs through the minimal prime ideals of R. Since the zero ideal of R is the intersection of the minimal prime ideals, the natural map $R \to \operatorname{PQF}(R)$ that assigns each $a \in R$ the family of its residue classes in $\operatorname{PQF}(R)$ is injective. We consider R to be a subring of $\operatorname{PQF}(R)$.

Proposition 4.67. *Let R be a reduced ring. Then the following are equivalent:*

(a) R is seminormal;
(b) if $u^2, u^3 \in R$ for $u \in \operatorname{PQF}(R)$, then $u \in R$;
(c) if $R \subset S$ is an extension of reduced rings and $u^2, u^3 \in R$ for $u \in S$, then $u \in R$.

Proof. (a) $\Longrightarrow$ (c) Set $x = u^2$ and $y = u^3$. By seminormality we find $z \in R$ with $x = z^2$ and $y = z^3$. The uniqueness of z (in S) implies $u = z \in R$.

(c) $\Longrightarrow$ (b) is trivial, and for (b) $\Longrightarrow$ (a) we let $\pi_{\mathfrak{p}} : R \to R/\mathfrak{p}$ be the natural epimorphism. Suppose that $x^3 = y^2$ for $x, y \in R$. If $\mathfrak{p}$ is a minimal prime of R, then either both $x, y \in \mathfrak{p}$ or $x, y \notin \mathfrak{p}$. Setting $u_{\mathfrak{p}} = 0$ if $x, y \in \mathfrak{p}$, and $u_{\mathfrak{p}} = \pi_{\mathfrak{p}}(y)/\pi_{\mathfrak{p}}(x) \in \operatorname{QF}(R/\mathfrak{p})$ otherwise, we obtain $\pi_{\mathfrak{p}}(x) = u_{\mathfrak{p}}^2$, $\pi_{\mathfrak{p}}(y) = u_{\mathfrak{p}}^3$ for all $\mathfrak{p}$. Set $u = (u_{\mathfrak{p}})$. Then $u^2, u^3 \in R$, and so $u \in R$ by (b). □

Thus $\operatorname{PQF}(R)$ is a replacement for the quotient field of an integral domain, as far as seminormality is concerned. If R has only finitely many minimal primes, for example if R is noetherian, then $\operatorname{PQF}(R)$ is the total ring of fractions of R. This follows easily by prime avoidance. (Prime avoidance will be discussed in Section 6.A.)

The Picard group. Let R be a ring, and P an R-module. We say that P is a *projective* module if there exists an R-module Q such that $P \oplus Q$ is isomorphic to a free R-module. All our projective R-modules will be finitely generated. Therefore we suppress the attribute "finitely generated" when we speak of projective modules. Thus every projective module we will consider is a direct summand of a free R-module of finite rank. Then the complementary module Q is projective (and finitely generated) as well, and it follows immediately that P is finitely presented. For the basic theory of projective modules covering the unproved statements below we refer the reader to Bass [19, Ch. III, §7], Bourbaki [37, Ch. II, §5], and Lam [237].

Projective modules over local rings are free, as follows from Nakayama's lemma. Therefore, given a prime ideal $\mathfrak{p} \subset R$, the localization $P_{\mathfrak{p}}$ is a free $R_{\mathfrak{p}}$-module. In this section we are only interested in projective modules of rank 1, i. e., projective modules for which $\operatorname{rank} P_{\mathfrak{p}} = 1$ for all prime ideals $\mathfrak{p}$. Projective modules of arbitrary rank will be discussed in Chapter 8.

Proposition 4.68. *Let R be a ring and P a projective R-module. Then:*

(a) *if Q is a projective R-module, then $P \otimes Q$ is also projective;*
(b) rank $P = 1$ *if and only if there exists an R-module Q' such that $P \otimes Q' \cong R$. In this case $Q' \cong \operatorname{Hom}_R(P, R)$.*

For part (b) one need not assume that P be finitely generated; see Exercise 4.30. The proposition allows us to introduce the *Picard group* $\operatorname{Pic}(R)$ of R. Its underlying set is the set of isomorphism classes of projective rank 1 modules, and addition is induced by the assignment $[P] + [Q] = [P \otimes Q]$. Clearly $\operatorname{Pic}(R) = 0$ if and only if every projective rank 1 module is isomorphic to R.

The following theorem will allow us to construct nontrivial elements in $\operatorname{Pic}(R)$.

Theorem 4.69. *Let $R \subset S$ be a ring extension and let $I \subset S$ be an R-submodule. Then the following are equivalent:*

(a) *there exists an R-submodule J of S such that $IJ = R$;*
(b) *I is a projective R-module of rank 1 and $IS = S$.*

An R-submodule I of S satisfying the conditions in the theorem is called *invertible* with (uniquely determined) inverse $I^{-1} = J \cong \operatorname{Hom}_R(I, R)$. This notion generalizes that of invertible ideals in the field of fractions of a domain. In the case of a domain all rank 1 projective modules can be realized as invertible ideals:

Proposition 4.70. *Let R be a domain and P a rank 1 projective module. Then:*

(a) *There exists an embedding $P \to \operatorname{QF}(R)$, and every such embedding maps P onto an invertible ideal I of R.*
(b) *Conversely, every invertible ideal of R is a projective module of rank 1.*
(c) *If I and J are invertible ideals, then IJ is invertible, too, and $I \otimes J \cong IJ$ via the multiplication map.*

We want to pursue the question under which conditions the extension $R \to R[M]$ induces an isomorphism of Picard groups. In fact, every extension $R \to S$ induces a natural map $\operatorname{Pic}(R) \to \operatorname{Pic}(S)$ given by the extension $P \mapsto P \otimes S$, and if an S-module Q is of type $P \otimes_R S$ for some R-module P, then we say that Q is *extended from* R. Since we have a retraction $R[M] \to R$, the map $\operatorname{Pic}(R) \to \operatorname{Pic}(R[M])$ is an injection if M is an arbitrary monoid, but the determination of the cokernel is a difficult task.

If R is a normal domain and M is a normal monoid, then the situation is very simple, as we will see now. It is immediate from the definition of the Picard group and Theorem 4.50 that one has a natural embedding $\operatorname{Pic}(A) \to \operatorname{Cl}(A)$ if R is a normal noetherian domain. For $A = R[M]$ this embedding allows us to compute $\operatorname{Pic}(A)$.

Theorem 4.71. *Let R be a noetherian normal domain, and M a normal affine monoid. Then the natural map $\operatorname{Pic}(R) \to \operatorname{Pic}(R[M])$ is an isomorphism.*

Proof. Injectivity has already been stated. So let H be an invertible ideal of $R[M]$. According to Corollary 4.61, $H \cong IJ$ where I is a divisorial ideal in R and J is a divisorial monomial ideal. We can assume that $H = IJ$. After multiplication by (the extension of) I^{-1} we have to show that J is principal.

One checks immediately that J^{-1} is the R-module generated by all monomials y with $xy \in M$ for all $x \in M$. On the one hand, $JJ^{-1} = R[M]$, and on the other hand, it is an R-module generated by the monomials xy, $x \in J$, $y \in J^{-1}$. Thus $xy = 1$ for some $x \in J$, $y \in J^{-1}$, and J is generated by x. □

While we have just seen that in the presence of normality the extension $\operatorname{Pic}(R) \to \operatorname{Pic}(R[M])$ is surjective, let us now show that surjectivity fails if $R[M]$ is not seminormal. The following lemma gives examples of rank 1 nonfree and nonextended projective modules.

Lemma 4.72 (Schanuel). *Let $R \subset T$ be an extension of reduced rings and suppose that $x \in T$, $x \notin R$, but $x^2, x^3 \in R$.*

(a) If $\operatorname{U}(R[x]) = \operatorname{U}(R)$, then the R-submodule $I = (1 + x)R + x^2R \subset T$ is a nonfree projective module of rank 1.

(b) Let M be a monoid, and $t \in M$, $t \neq 1$. Then the $R[M]$-submodule $I = (1+xt)R[M] + (x^2t^2)R[M])$ of $T[M]$ is a projective rank 1 module that is not extended from R.

Proof. We may assume that $T = R[x]$. Then $\operatorname{Spec} R$ and $\operatorname{Spec} T$ are homeomorphic (Exercise 4.27). This fact will play a role in the proof of (b).

(a) We have $IJ = R$ for the R-submodule $J = (1 - x)R + x^2R \subset T$. Thus I is an invertible module, and therefore projective of rank 1.

If $I = cR$ for some element $c \in T$, then $c \in T$ since $I \subset T$. We have $TI = T$ and, hence, $c \in \operatorname{U}(T) = \operatorname{U}(R)$. It follows $I = R$ and $x \in R$—a contradiction.

(b) As in (a), we see that I is invertible with inverse $J = (1 - xt)R[M] + (x^2t^2)R[M])$. Suppose that I is extended from R. In order to derive a contradiction we first choose a localization $R_\mathfrak{p}$ with $x \notin R_\mathfrak{p}$. Replacing R by the localization we may assume that R is local. Then T is local, too, as observed above, and Proposition 4.20(b) shows that $\operatorname{U}(T[M]) = \operatorname{U}(T) \oplus \operatorname{U}(M)$.

If I is extended as a projective module, it is a free module, and therefore of the form $cR[M]$ with $c \in I$. Since $IT[M] = T[M]$, one concludes that c is a unit in T. Thus $c = du$ with $d \in \operatorname{U}(T)$ and $u \in \operatorname{U}(M)$. Now $du(1-xt) = c(1-xt) \in IJ \subset R[M]$, and so $d \in R$ and $dx \in R$ (we have used that $u \neq ut$.)

If $I = cR[M]$, then $J = c^{-1}R[M]$, and $d^{-1} \in R$ follows by similar arguments. But then $x \in R$—again a contradiction. □

By Theorem 4.19, $A[M]$ is a reduced ring for a reduced ring A. Therefore it makes sense to discuss the seminormality of $A[M]$.

Corollary 4.73. *Let A be a reduced ring and M a monoid. If $A[M]$ is not seminormal, then the natural map $\operatorname{Pic}(A) \to \operatorname{Pic}(A[M])$ is not surjective.*

Proof. We will see in Theorem 4.76 that at least one of A and M is not seminormal if $A[M]$ is not seminormal. If A is not seminormal, then Proposition 4.67 and Lemma 4.72(b) yield a nonextended projective module.

Suppose that M is not seminormal, and let $x \in \operatorname{gp}(M)$, $x \notin M$, but $x^2, x^3 \in M$. Then we define I as in part (a) of the lemma. Assume I is extended from A, and choose a maximal ideal $\mathfrak{m}$ of A. Then $IA_{\mathfrak{m}}[M]$ is extended from $A_{\mathfrak{m}}$, and therefore free. However the formation of I commutes with the passage from A to $A_{\mathfrak{m}}$. Therefore it is enough to show that I is always nonfree.

Let $N = x^{\mathbb{Z}_+} \cdot M \subset \operatorname{gp}(M)$ (we use multiplicative notation). Then $\mathrm{U}(M) = \mathrm{U}(N)$ because $N = M \cup (xM)$ and no element in xM can be a unit of N; otherwise we would have $xyz = 1$ or $xyxz = 1$ with $y, z \in M$. In either case x^2 would be a unit of M, so $x = x^3(x^2)^{-1}$ would lie in M. By Proposition 4.20 we have $\mathrm{U}(A[M]) = \mathrm{U}(A[N])$, and Lemma 4.72(a) shows that I is nonfree. □

In Remark 8.44 we will show the converse of Corollary 4.73 for positive monoids and $\mathbb{Q}$-algebras R, and give references for the case in which M is not positive.

Let us now state Swan's version of Traverso's theorem. It is the converse to Corollary 4.73 for the case $M = \mathbb{Z}_+^n$:

Theorem 4.74. *Let R be a reduced ring. Then the following are equivalent:*

(a) R is seminormal;
(b) $\operatorname{Pic}(R) \to \operatorname{Pic}(R[X])$ is an isomorphism;
(c) $\operatorname{Pic}(R) \to \operatorname{Pic}(R[X_1, \dots, X_n])$ is an isomorphism for all n.

We refer to Traverso [357] for the case of a noetherian reduced ring with finite integral closure (in its total ring of fractions) and to Swan [347] for the general case. For an elementary, purely ring-theoretic proof, see Coquand [92].

Dedekind domains. The existence of a nonfree projective rank 1 module in Lemma 4.72 is explained by the lack of (semi)normality, but there also exist classical types of normal domains with nonfree rank 1 projective modules, namely the Dedekind domains that are not principal ideal domains. A regular domain R of Krull dimension ≤ 1 is called a *Dedekind domain*. Thus R is a Dedekind domain if and only if it is a field or if it is a noetherian domain such that $R_{\mathfrak{m}}$ is a discrete valuation domain for every maximal ideal $\mathfrak{m}$ of R. Classical examples of Dedekind domains are the rings of integers in number fields.

Since a finitely generated torsionfree module over a discrete valuation domain is free, every such R-module M is projective if R is Dedekind. Using a theorem of Steinitz, one shows M is of the form *free* $\oplus$ *rank one* (for example, see Milnor [262, §1]). More precisely, M is isomorphic to a direct sum $I_1 \oplus \cdots \oplus I_n$ of invertible fractional ideals $I_1, \dots, I_n \subset R$ and, furthermore, $I_1 \oplus \cdots \oplus I_n \cong R^{n-1} \oplus I_1 \cdots I_n$. In particular, one has $\operatorname{Pic}(R) = \operatorname{Cl}(R)$, and R has nonfree projective modules if and only if it is not a principal ideal domain. A well-known example is $R = \mathbb{Z}[\sqrt{-5}]$.

Seminormality and subintegral extensions. Seminormality has not only been explored from the viewpoint of K-theory, but also as an interesting notion in ring theory, and to some extent it has been developed in analogy to normality. Therefore we want to derive results for seminormality analogous to those for normality.

Let $R \subset S$ be an extension of rings. Then one calls S *subintegral* over R, if (i) S is integral over R, (ii) for each prime $\mathfrak{p}$ of R there exists exactly one prime ideal $\mathfrak{q}$

of S such that $\mathfrak{p} = \mathfrak{q} \cap R$, and, (iii) the extension $R \to S$ induces an isomorphism $R_\mathfrak{p}/\mathfrak{p}R_\mathfrak{p} \cong S_\mathfrak{q}/\mathfrak{q}S_\mathfrak{q}$ if $\mathfrak{p} = \mathfrak{q} \cap R$.

If $R \subset S$ is a ring extension, then there exists a unique maximal R-subalgebra $\operatorname{sn}_S(R)$ which is subintegral over R [357]. It is called the *seminormalization* or *subintegral closure of R in S*. One says that R is *subintegrally closed* or *seminormal in S* if $R = \operatorname{sn}_S(R)$. The absolute *seminormalization* $\operatorname{sn}(R)$ of a reduced ring is its subintegral closure in $\operatorname{PQF}(R)$. By [347, 4.2] this is in accordance with the definition in [347], and according to [347, 4.1] the attribute "absolute" is justified: if $\varphi : R \to S$ is a ring homomorphism to a seminormal ring S, then φ can be extended to a homomorphism $\psi : \operatorname{sn}(R) \to S$; if φ is injective, then so is ψ.

Hamann's criterion [176], [347] gives an elementary description of subintegrality for arbitrary rings: *if $R \subset S$ is a proper subintegral extension, then there exists $x \in S \setminus R$ such that $x^2, x^3 \in R$.* This criterion leads to a constructive description of the subintegral closure $\operatorname{sn}_S(R)$ as follows. Call $R \subset A$ an *elementary subintegral extension* if $A = R[x]$ such that $x^2, x^3 \in R$. (Exercise 4.27 shows that an elementary subintegral extension is indeed subintegral.) Then $\operatorname{sn}_S(R)$ is the filtered union of all subrings of S that arise from R by a finite number of elementary subintegral extensions.

While there seems to be no elementwise description of subintegral dependence in general, such a description is given for $\mathbb{Q}$-algebras by a theorem of Roberts and Singh [302]: an extension $R \subset S$ of $\mathbb{Q}$-algebras is subintegral if and only if for every $b \in S$ there exist $c_1, \dots, c_p \in S$ such that

$$b^n + \sum_{i=1}^{p} \binom{n}{i} c_i b^{n-i} \in R \quad \text{for } n \gg 0.$$

(Compare this to the description of the seminormalization of monoids in Exercise 2.11.)

The basic facts about subintegral closures are summarized as follows:

Theorem 4.75.

(a) A reduced ring R is subintegrally closed in $R[X]$ and $R[X^{\pm 1}]$.
(b) Let $R \subset S \subset T$ be successive ring extensions. Then T is subintegral over R if and only if S is subintegral over R and T is subintegral over S.
(c) The subintegral closure of $R[X]$ in $S[X]$ is $\operatorname{sn}_S(R)[X]$.
(d) If $T \subset R$ is a multiplicatively closed set, then $T^{-1}(\operatorname{sn}_S(R))$ is the subintegral closure of $T^{-1}R$ in $T^{-1}S$.
(e) The intersection $\bigcap R_i$ of subintegrally closed subrings R_i of S is subintegrally closed in S.
(f) If R is seminormal, then $R[X]$ is seminormal.

Fact (a) is obvious since R is even integrally closed in $R[X^{\pm 1}]$. Fact (b) is [347, 2.3], whereas (e) is an easy consequence of Hamann's criterion. For fact (c) we refer to Brewer, Costa and McCrimmon [41, Prop. 1], and (d) is [347, 2.3]. The last assertion (f) follows immediately from the Traverso–Swan theorem.

Returning to monoid rings, we start again with the most basic and most important theorem. Its proof essentially uses only the Traverso–Swan theorem and Proposition 4.66.

Theorem 4.76. *Let R be a reduced ring and M a monoid. Then $R[M]$ is seminormal if and only if R and M are seminormal.*

Proof. The necessity follows immediately from Proposition 4.67 or simply from the definition of seminormality.

For the converse, we consider the affine case first. Let $\mathrm{gp}(M) = \mathbb{Z}^r$. As we have seen in Proposition 2.42, a seminormal affine monoid can be represented in the form

$$M = \bigcap_{i=1}^{s} M_i,$$

where M_i is of the form

$$M_i = U_i \cup (H_i^> \cap \mathbb{Z}^r)$$

with a subgroup $U_i \subset H_i \cap \mathbb{Z}^r$ (of rank $r-1$). It is clearly enough that $R[M_i]$ is seminormal for each i. Therefore we may assume that $s=1$, $M = M_1$, $H = H_1$, $U = U_1$.

We first note that $R[\mathrm{gp}(M)]$ is seminormal since it is a localization of a polynomial extension of R. Suppose that $x^3 = y^2$ for $x, y \in R[M]$. Then we find $f \in R[\mathrm{gp}(M)]$ such that $x = f^2$, $y = f^3$. Write

$$f = f_+ + f_0 + f_-$$

where f_+ collects all terms of f whose monomials lie in $H^>$, f_0 collects all terms with monomials in H, and f_- collects the remaining ones.

Clearly, if $f_- \neq 0$, then $f \notin R[\mathrm{gp}(M) \cap H^+]$. This is impossible, since $R[\mathrm{gp}(M) \cap H^+]$ is even integrally closed in $R[\mathrm{gp}(M)]$.

It follows that $(f_0)^2 = x_0$ and $(f_0)^3 = y_0$. By the seminormality of $R[U]$ we conclude that $f_0 \in R[U] \subset R[M]$. Since $f_+ \in R[M]$ anyway, we are done for M affine.

Now let M be an arbitrary monoid. Then an equation $x^3 = y^2$ for $x, y \in R[M]$ holds already in an affine submonoid N of M. So it is enough to note that $\mathrm{sn}(N) \subset M$ and to use that $\mathrm{sn}(N)$ is affine. □

We draw a consequence for the seminormalization of a monoid ring:

Corollary 4.77. *Let R be a reduced ring and M a monoid. Then $\mathrm{sn}(R)[\mathrm{sn}(M)]$ is the seminormalization of $R[M]$.*

Proof. By the theorem $\mathrm{sn}(R)[\mathrm{sn}(M)]$ is seminormal. On the other hand, it is a subintegral extension of $R[M]$. The claim now follows from [347, 4.2]. □

We give the analogues of Lemma 4.41 and Theorem 4.42 for relative subintegral closures. The proofs are analogous as well, and the details are left to the reader.

Lemma 4.78. *Let R be a reduced ring and M an affine submonoid of $\mathbb{Z}^n$. Then $R[\mathrm{sn}(M)]$ is the subintegral closure of $R[M]$ in $R[\mathbb{Z}^n]$.*

Theorem 4.79. *Let $R \subset S$ be an extension of reduced rings and let $M \subset N$ be an extension of monoids. Set $\mathrm{sn}_N(M) = N \cap \mathrm{sn}(M)$. Then $\mathrm{sn}_S(R)[\mathrm{sn}_N(M)]$ is the subintegral closure of $R[M]$ in $S[N]$.*

Exercises

4.1. Let A be a G-graded abelian group. Show that a subgroup is graded if and only if it contains the homogeneous components of each of its elements.

4.2. Let R be a $\mathbb{Z}^r$-graded ring in which every nonzero homogeneous element is a unit. Show:
(a) $\Bbbk = R_0$ is a field, and $G = \mathrm{U}(R)/\Bbbk^\times$ is isomorphic to the subgroup(!) $\operatorname{supp} R$ of $\mathbb{Z}^r$.
(a) R is isomorphic to $\Bbbk[G]$.
(b) Every finitely generated graded R-module is isomorphic to a direct sum $\bigoplus_i R(-u_i)$, $u_i \in \mathbb{Z}^r$.

4.3. Let $\Bbbk$ be a field and R a $\mathbb{Z}^r$-graded $\Bbbk$-algebra that, as a vector space, is isomorphic to $\Bbbk[M]$ for some submonoid of $\mathbb{Z}^r$. Prove: R and $\Bbbk[M]$ are isomorphic as $\Bbbk$-algebras if (and only if) every nonzero homogeneous element in R is not a zerodivisor.

Hint: pass to $S^{-1}R$ where S is the set of all nonzero homogeneous elements, and use Exercise 4.2.

4.4. Let R be a reduced ring with connected spectrum and M a (cancellative torsionfree) monoid. Show $R[M]$ has connected spectrum.

4.5. Let M be a (cancellative torsionfree) monoid.
(a) Show that Proposition 4.20(b) does not hold for arbitrary reduced rings: let R_1, R_2 be rings, set $R = R_1 \times R_2$, and consider $(1,0)X + (0,1)X^{-1}$ in $R[X^{\pm 1}]$.
(b) Prove $\mathrm{U}(R[M]) = \mathrm{U}(R) \oplus \mathrm{U}(M)^n$ if R is reduced and $\operatorname{Spec} R$ decomposes into a finite number n of connected components.

Hint: by the Chinese remainder theorem, R is the direct product of n rings with connected spectrum.

4.6. (a) Let M be a monoid (not necessarily cancellative or torsionfree), let $E \subset M^2$ and $\tilde{E} = E \cup \{(y,x) : (x,y) \in E\}$. Show that $\mathcal{C} \supset E$ is the congruence relation generated by E if and only if for each $(x,y) \in \mathcal{C}$, $x \neq y$, there exist $n \in \mathbb{N}$, $(u_1,v_1), \dots, (u_n,v_n) \in \tilde{E}$, and $w_1, \dots, w_n \in M$ such that $x = w_1u_1$, $w_1v_1 = w_2v_2, \dots, w_{n-1}v_{n-1} = w_nu_n$, $w_nv_n = y$.
(b) Prove Proposition 4.25.

4.7. Extend Theorem 4.32 to ideals in affine monoid rings $\Bbbk[M]$.

4.8. Extend Lemma 4.9(a) to radical and primary ideals. Deduce that the radical of a graded ideal is graded, and that a graded ideal on a graded noetherian ring is the intersection of graded primary ideals.

4.9. Prove that a reduced ring R is integrally closed in $R[X^{\pm 1}]$.

4.10. Let R be a pure subring of the normal domain S. Show that R is also normal. Hint: $R = S \cap \mathrm{QF}(R)$.

4.11. Let R be an integral domain and M a monoid. Show the following formula for conductor ideals: $\mathfrak{c}(\bar{R}[\bar{M}]/R[M]) = \mathfrak{c}(\bar{R}/R)\mathfrak{c}(\bar{M}/M)$. (By definition, $\mathfrak{c}(S/R) = \{x \in R : xS \subset R\}$.)

4.12. Formulate the very ampleness of a lattice polytope P and the equivalent conditions in Exercise 2.23 in terms of the algebra $\Bbbk[M(P)]$ where $\Bbbk$ is a field.

4.13. Let M be an affine monoid. Prove that $R[M]$ is factorial if and only if R is factorial and $M \cong \mathbb{Z}^m \oplus \mathbb{Z}_+^n$ for suitable m and n.

4.14. Let R be an integral domain, I a divisorial ideal of R, and J a divisorial monomial ideal of $R[M]$. Show IJ is divisorial.

4.15. Let M be an affine monoid, $\Bbbk$ a field, and $R = \Bbbk[M]$. Furthermore let P be the set of height 1 *non-monomial* prime ideals of R. Show:
(a) $R_{\mathfrak{p}}$ is a localization of $L = \Bbbk[\operatorname{gp}(M)]$ and therefore a discrete valuation domain for all $\mathfrak{p} \in P$.
(b) Every height 1 prime ideal $\mathfrak{q}$ of L is of the form $(\mathfrak{q} \cap R)L$. (This is obvious.)
(c) $L = \bigcap_{\mathfrak{p} \in P} R_{\mathfrak{p}}$.

4.16. We keep the notation of Exercise 4.15.
(a) Prove that R satisfies Serre's condition (R_1) if and only if each localization $R_{\mathfrak{p}}$ is a discrete valuation domain for the monomial prime ideals $\mathfrak{p}$.
(b) Let F be a facet of $\mathbb{R}_+ M$, H the associated support hyperplane, and $\mathfrak{p}_F$ the associated prime ideal. Show that $R_{\mathfrak{p}_F}$ is a discrete valuation domain if and only if $M_F = H^+ \cap \operatorname{gp}(M)$ where $M_F = \{x - y : x \in M, y \in M \cap F\}$.
(c) Show that the condition in (b) is equivalent to the following: (i) M contains an element x such that $\sigma_F(x) = 1$, and (ii) $\operatorname{gp}(M \cap F) = \operatorname{gp}(M) \cap H$.

4.17. We still use the notation of Exercises 4.15 and 4.16. Show the following are equivalent:
(i) R satisfies Serre's condition (S_2);
(ii) $R = L \cap \left(\bigcap_F R_{\mathfrak{p}_F}\right)$;
(iii) $M = \bigcap_F M_F$.

4.18. Let M and N be normal affine monoids. Show $\operatorname{Cl}(M \oplus N) = \operatorname{Cl}(M) \oplus \operatorname{Cl}(N)$.

4.19. Let M be a normal affine monoid and let F be a face of $\mathbb{R}_+ M$. Show $\operatorname{Cl}(M \cap F)$ is a subquotient of $\operatorname{Cl}(M)$, i. e., $\operatorname{Cl}(M \cap F) \cong U/V$ where U, V are subgroups of $\operatorname{Cl}(M)$, $V \subset U$. Find an example for which $\operatorname{Cl}(M \cap F)$ is not isomorphic to a subgroup of $\operatorname{Cl}(M)$.

4.20. Let $\lambda_1, \dots, \lambda_n$ be positive integers. We now consider the simplex P with vertices $v_i = \lambda_i e_i$ where $e_1, \dots, e_n$ are the unit vectors in $\mathbb{R}^n$. Let N be the integral closure of $M(P)$ in $\mathbb{Z}^n$.
(a) When is the natural embedding $N \subset \mathbb{Z}^n$ the standard embedding of N? In which way can $N \subset \mathbb{Z}^n$ differ from the standard embedding?
(b) Determine the divisor class group of N.

4.21. Let $\lambda_1, \dots, \lambda_n$ be positive integers and $\Delta \subset \mathbb{R}^n$ be the simplex with vertices 0, $(\lambda_1, 0, \dots, 0)$,..., $(0, \dots, 0, \lambda_n)$. Compute the divisor class group of the normalization of $\Bbbk[\Delta]$. (See [59] for more information on this type of simplices. The theorem describing the divisor class group and its proof can however be simplified.)

4.22. Let $\Bbbk$ be a field and R and S be $\mathbb{Z}$-graded $\Bbbk$-algebras. The *Segre product* of R and S is the subalgebra $\bigoplus_{i=0}^{\infty} R_i \otimes_{\Bbbk} S_i$ of the tensor product $R \otimes_{\Bbbk} S$. (Compare Exercise 2.29 for the Segre product of monoids. How are the notions related?)
Let $P_1, \dots, P_n$ be polynomial rings over the field $\Bbbk$, $P_i = \Bbbk[X_{i1}, \dots, X_{im_i}]$ with $m_i \geq 2$ for all i.
(a) Prove that the Segre product of $P_1, \dots, P_n$ (with their standard grading by total degree) can be identified with $\Bbbk$-subalgebra S of $P = \Bbbk[X_{ij} : i = 1, \dots, n,\ j = 1, \dots, m_i]$ generated by all monomials $X_{1j_1} X_{2j_2} \cdots X_{nj_n}$.
(b) Show that the natural embedding of S into P is the standard embedding of the monoid algebra S.
(c) In the notation of Theorem 4.62 describe the generating sets of monomials for each of the modules D_c and show that $\operatorname{Cl}(S) \cong \mathbb{Z}^{n-1}$.
(d) Let $D(a_1, a_2, \dots, a_{n-1})$, $(a_1, \dots, a_{n-1}) \in \mathbb{Z}^{n-1}$, be the S-submodule of the Laurent polynomial ring in all the indeterminates X^{ij_i} that is generated by the products $y_1 \cdots y_{n-1}$ where y_j is a

monomial of degree a_i in the indeterminates X_{ij_i}. Prove that the assignment $(a_1,\dots,a_{n-1}) \mapsto [D(a_1,a_2,\dots,a_{n-1})]$ is an isomorphism of $\mathbb{Z}^{n-1}$ and $\operatorname{Cl}(S)$.

4.23. Let M be an affine positive normal monoid. Assume $N \subset M$ is an affine submonoid of same rank and integrally closed in M such that every facet of $\mathbb{R}_+M$ contains a facet of $\mathbb{R}_+N$. Then there is an injective group homomorphism from the torsion subgroup of $\operatorname{Cl}(N)$ to that of $\operatorname{Cl}(M)$.

4.24. (a) Let P be a lattice polytope in $\mathbb{R}^d$ having a facet F that is a unimodular simplex. Moreover, suppose that each cone $\mathbb{R}_+(P-v)$, $v \in \operatorname{vert}(F)$, is unimodular. Show that $\operatorname{Cl}(\bar{M}(P))$ is torsionfree.

(b) Give an example of a lattice polygon P ($\dim P = 2$) without the property in (a) such that $\operatorname{Cl}(M(P))$ is nevertheless torsionfree.

4.25. Let R be a noetherian domain and M an affine monoid. Show that $R[M]$ is regular if and only if R is regular and $M \cong \mathbb{Z}^m \oplus \mathbb{Z}_+^n$ for suitable m and n.
Hint: the main point is to show that the regularity of $R[M]$ implies that $M \cong \mathbb{Z}^m \oplus \mathbb{Z}_+^n$. One can assume that R is local, and therefore a domain. Regularity implies normality and so $M = \mathrm{U}(M) \oplus M'$. Replace R by $R[\mathrm{U}(M)]$, and then R by its quotient field. If M is positive, the regularity of $\Bbbk[M]$ for a field $\Bbbk$ implies factoriality by Proposition 4.64. Now use Exercise 4.13.

4.26. Show that every finitely generated abelian group can be realized as the class group of a normal affine monoid.

4.27. Let R be a ring and S an overring such that $S = R + Rx$ for some $x \in S$ with $x^2, x^3 \in R$. Show that the embedding $\iota : R \to S$ induces a homeomorphism $\iota^* : \operatorname{Spec} S \to \operatorname{Spec} R$, and that $R_\mathfrak{p}/\mathfrak{p}R_\mathfrak{p} \cong S_\mathfrak{q}/\mathfrak{q}D_\mathfrak{q}$ if $\mathfrak{p} = \iota^*(\mathfrak{q})$.

4.28. Let R be a ring in which the equation $x^3 = y^2$ is only solvable with $x = z^2$, $y = z^3$ for some $z \in R$. Show R is reduced (Costa [98]).

4.29. Translate the notions related to subintegral ring extensions into the language of monoids and prove directly that every subintegral extension N of M is the filtered union of submonoids N_i that are obtained from M by a finite chain of elementary subintegral extensions.

4.30. Show that for a ring R and an R-module M the following are equivalent:
(i) There exists an R-module N such that $M \otimes_R N \cong R$;
(ii) M is a rank 1 (finitely generated) projective module.

Notes

Many topics of this chapter belong to the classical theory of rings and monoids, and numerous sources have already been mentioned. Gilmer's book [142] contains informative remarks and many references especially for monoid rings. For the theory of commutative monoids and semigroups and their congruences we recommend Grillet's book [153]. Corollary 4.28, a special case of Hilbert's basis theorem, is known as Redei's theorem.

Theorem 4.11 is taken from [66]. The crucial Lemma 4.12 is attributed to Samuel by Rees [298, p. 9]. Swan [351, Th. 4.4] states the monoid version and calls it "Gordan's lemma."

Binomial ideals, generated by polynomials with at most two terms, have been investigated thoroughly by Eisenbud and Sturmfels [114]. The theory of toric ideals has been developed mainly from the viewpoint of initial ideals; we will come back to them in Chapter 7.

The equivalence of normality for an affine monoid and its algebra over a field was stated by Hochster [195, Prop. 1]. The corresponding equivalence for seminormality was proved by Hochster and Roberts [198] and Anderson and Anderson [4].

Divisor class groups of normal monoid domains were computed by Chouinard [84]. Divisor class groups of normal affine monoid domains were also treated in [160].

5

Isomorphisms and automorphisms

Affine monoids and their algebras are closely related to actions of algebraic tori. These form a special class of linear algebraic groups, to be introduced in the first section. We highlight Borel's theorem on the conjugacy of maximal tori since it will be a crucial tool later on.

One of the historical roots of affine monoid algebras is invariant theory: Gordan's lemma [145], proved in 1872 (compare the notes of Chapter 2), can be interpreted as saying that the ring of invariants of a linear action of an algebraic torus is an affine monoid algebra.

In the situation of invariant theory, the torus action vanishes on the object of interest. On the other hand, every affine monoid algebra carries a natural internal torus action from which the monoid structure can be recovered. This is the key idea in the proof of the isomorphism theorem: if $R[M] \cong R[N]$ as R-algebras, then $M \cong N$ as monoids. However, the reduction to the case in which the tori acting on $R[M]$ and $R[N]$ can be viewed as maximal tori in a single linear algebraic group requires some mild hypotheses on the monoids.

The last section is devoted to the group of graded automorphisms of polytopal (and positively graded affine normal) monoid algebras. These are linear algebraic groups and there is a Gaussian algorithm for them—one obtains a complete analogue to the decomposition of invertible matrices over a field into a product of elementary, diagonal and permutation matrices. A critical step is the combinatorial identification of the elementary automorphisms.

5.A Linear algebraic groups

Our standard reference for the theory of linear algebraic groups is Borel [36], to which we refer the reader for proofs of the general facts surveyed below. Since arithmetic issues never come up in our considerations, we remark right away that *in this section the ground field* $\Bbbk$ *is supposed to be algebraically closed.*

An *algebraic group* over $\Bbbk$ is a group endowed with the structure of $\Bbbk$-variety such that the maps $G \times G \to G$, $(g_1, g_2) \mapsto g_1 g_2$, and $G \to G$, $g \mapsto g^{-1}$, are algebraic

W. Bruns, J. Gubeladze, *Polytopes, Rings, and K-Thoery*, Springer Monographs in Mathematics,
DOI 10.1007/b105283_5, © Springer Science+Business Media, LLC 2009

morphisms. A *(homo)morphism of algebraic groups* is a group homomorphism that is also a morphism of varieties. Every (Zariski) closed subgroup of an algebraic group is an algebraic group in a natural way. If G is an affine variety, we denote its coordinate ring by $\mathcal{O}(G)$. (In [36] the coordinate ring is denoted by $\Bbbk[G]$, which we have reserved for the group ring.)

An algebraic group G is the disjoint union of its irreducible components of which exactly one is a subgroup. It is called the *connected component of* G and denoted G^0. Furthermore, G^0 is a normal subgroup of G. Thus the quotient G/G^0 is a finite group. If $G = G^0$ then G is called *connected*. In fact, G is connected (in the topological sense, i. e., there are no proper open-closed subsets of G) if and only if it is irreducible as a variety, but the term "irreducible" is avoided, since it appears with a different meaning in the representation theory of G.

In the discussion that ensues we need the following facts about closedness:

Proposition 5.1. *Let G, G' be algebraic groups.*

(a) If $f : G \to G'$ is a morphism of algebraic groups, then $\operatorname{Ker} f$ *and* $\operatorname{Im} f$ *are closed subgroups of G and G', respectively. Moreover, $f(G^0)$ is a closed connected subgroup of $(G')^0$.*

(b) If $U_1, \dots, U_n$ are closed connected subgroups of G, then the subgroup generated (group theoretically) by their union is closed and connected.

Part (b) follows immediately from [36, 2.2]. Usually it is applied in order to show that a group is connected, but we will need it to prove that certain subgroups are closed.

A classical example of an algebraic group is the *general linear group of order n* for some $n \in \mathbb{N}$, denoted $\mathrm{GL}_n(\Bbbk)$. (More generally, we denote the group of invertible $n \times n$ matrices over a ring R by $\mathrm{GL}_n(R)$). For $R = \Bbbk$ it is naturally viewed as the affine $\Bbbk$-subvariety of the affine space $\Bbbk^{n \times n + 1}$ with coordinates X_{ij}, $i, j = 1, \dots, n$, and Y, cut out by the equation $Y \cdot \det(X_{ij}) = 1$. It follows that the group $\mathrm{GL}(V)$ of $\Bbbk$-automorphisms of a finite-dimensional vector space V carries the structure of a linear algebraic group.

We are mainly interested in the special case of *linear (algebraic) groups*. They are defined as closed subgroups of $\mathrm{GL}_n(\Bbbk)$ (or $\mathrm{GL}(V)$, $\dim V < \infty$).

The subgroup of diagonal matrices in $\mathrm{GL}_n(\Bbbk)$ is an example of a linear group: its defining equations in $\mathrm{GL}_n(\Bbbk)$ are $X_{ij} = 0$ for $i \neq j$. This group is denoted by $\mathbb{T}^n(\Bbbk)$, or simply by $\mathbb{T}^n$ if the ground field is clear from the context. It is naturally identified with $(\Bbbk^\times)^n$. An *algebraic torus* over $\Bbbk$ is an algebraic group that is isomorphic to $\mathbb{T}^n(\Bbbk)$ for some n so that we can write its elements as n-tuples $(\xi_1, \dots, \xi_n)$ with $\xi_i \in \Bbbk^\times$. It is a connected group whose coordinate ring is isomorphic to $\Bbbk[X_1^{\pm 1}, \dots, X_n^{\pm 1}]$.

The *tautological action* of $\mathbb{T}^n$ on $\mathcal{O}(\mathbb{T}^n) = \Bbbk[X_1^{\pm 1}, \dots, X_n^{\pm 1}]$, induced by the group structure on $\mathbb{T}^n$, is given by $(\xi_1, \dots, \xi_n) : X_i \mapsto \xi_i X_i$, $\xi_i \in \Bbbk^\times$, $i = 1, \dots, n$.

We will use the following classical result of Borel [36, 11.3] in order to compare monoid structures on the same ring.

Theorem 5.2. *All maximal (with respect to inclusion) algebraic tori in a linear group G are conjugate.*

Characters and diagonalizable groups. A *character* of an algebraic group is a morphism $\chi : G \to \Bbbk^\times$. The characters form a group $X(G)$ under multiplication, the *character group* of G. Note that the characters are a linearly independent set of functions from G to $\Bbbk$ (Exercise 5.1).

It is not hard to show that the monomials in $\mathcal{O}(\mathbb{T}^n) = \Bbbk[X_1^{\pm 1}, \ldots, X_n^{\pm 1}]$ are exactly the characters of $\mathbb{T}^n$ (Exercise 5.2). Therefore the character group of $\mathbb{T}^n$ is isomorphic to $\mathbb{Z}^n$.

An algebraic group that is isomorphic to a closed subgroup of $\mathbb{T}^n$ is called *diagonalizable*. The following theorem justifies this terminology.

Theorem 5.3. *Let G be an algebraic group. Then the following are equivalent:*

(a) G is diagonalizable;

(b) for every morphism $G \to \mathrm{GL}_n(\Bbbk)$, $n \in \mathbb{N}$, the image of G is conjugate to a subgroup of $\mathbb{T}^n \subset \mathrm{GL}_n(\Bbbk)$;

(c) G is isomorphic to a direct product $\mathbb{T}^m \times A$, where A is a finite abelian group without p-torsion if $p = \operatorname{char} \Bbbk > 0$;

(d) the characters form a $\Bbbk$-basis of $\mathcal{O}(G)$.

The character group of a finite group A as in (c) is isomorphic to A, as follows quickly by decomposition into a product of cyclic groups. Together with the fact that $X(\mathbb{T}^n) \cong \mathbb{Z}^n$ this shows that every finitely generated abelian group B without p-torsion appears as the character group of a diagonalizable group (up to isomorphism). The assignments $G \mapsto X(G)$ and $B \mapsto \mathrm{Hom}(B, \Bbbk^\times)$ induce an anti-equivalence between the category of diagonalizable groups and the category of finitely generated abelian groups without p-torsion. (In Exercise 5.5 the reader is asked to work out the details.)

Corollary 5.4. *A sequence $0 \to G' \to G \to G'' \to 0$ of diagonalizable groups is exact if and only if the induced sequence $1 \to X(G'') \to X(G) \to X(G') \to 1$ is exact.*

The crucial point in the proof of the corollary is the surjectivity of $X(G) \to X(G')$. It follows from the surjectivity of $\mathcal{O}(G) \to \mathcal{O}(G')$ and property 5.3(d) of diagonalizable groups.

Rational representations. A rational representation of an algebraic group G is a morphism of algebraic groups $\rho : G \to \mathrm{GL}(V)$ where V is a finite-dimensional $\Bbbk$-vector space. One says that G *acts rationally* on V and sets

$$gv = (\rho(g))(v), \qquad g \in G,\ v \in V.$$

Since the vector space underlying a $\Bbbk$-algebra is usually of infinite dimension, we extend the notion of rational representation as follows: a group homomorphism $\rho : G \to \mathrm{GL}(V)$ is called a *rational representation* if there exist a partially ordered directed set I and a system V_i, $i \in I$, of finite-dimensional subspaces $V_i \subset V$ such that $V = \bigcup_{i \in I} V_i$ and the action of G on V restricts to a rational representation on V_i for all i.

Suppose that G acts rationally on V, and let χ be a character of G. Then the *weight space* (or space of *semi-invariants*) of the character χ is

$$V_\chi = \{v \in V : gv = \chi(g)v \text{ for all } g \in G\}.$$

In other words, V_χ consists of all those vectors in V that are eigenvectors for each $g \in G$ with corresponding eigenvalue $\chi(g)$. The characters χ such that $V_\chi \neq 0$ are called the *weights of G on V*. (One also says that the character χ occurs in V.)

Lemma 5.5. *The weight spaces of a rational representation are linearly independent. In other words, the linear map*

$$\bigoplus_{\chi \in X(G)} V_\chi \to V, \qquad (v_\chi) \mapsto \sum v_\chi,$$

is injective.

Rational representations of diagonalizable groups split into their weight spaces:

Proposition 5.6. *Let D be a diagonalizable group acting rationally on V. Then*

$$V = \bigoplus_{\chi \in X(D)} V_\chi.$$

In view of Lemma 5.5 it is enough to prove $V = \sum V_\chi$, and this follows from the finite-dimensional case by passage to the limit. By Theorem 5.3 we can then choose an isomorphism $V \cong \Bbbk^n$ such that D is mapped onto a subgroup of the torus $\mathbb{T}^n$. Each weight space is now generated by the basis vectors e_i that belong to it, and each basis vector belongs to a weight space.

Torus actions on monoid rings. In the following we let

$$G_\Bbbk(A)$$

denote the group of $\Bbbk$-algebra automorphisms of a $\Bbbk$-algebra A.

Let M be an affine monoid and $A = \Bbbk[M]$. An element $\tau \in G_\Bbbk(A)$ is called a *toric automorphism* (with respect to M) if $\tau(x) = c_x x$, $c_x \in \Bbbk^\times$ for all $x \in M$. Evidently, the toric automorphisms form a subgroup $\mathbb{T}_\Bbbk(M)$ of $G_\Bbbk(A)$. For τ to be toric it is of course enough that $\tau(x_i) = c_i x_i$ for some system of generators $x_1, \dots, x_n$ of M.

We can describe the toric automorphisms in two ways, first via an isomorphism $\Bbbk[\operatorname{gp}(M)] \cong \Bbbk[Y_1^{\pm1}, \dots, Y_r^{\pm1}]$, $r = \operatorname{rank} M$.

Lemma 5.7. *Let M be an affine monoid of rank r and $e_1, \dots, e_r$ a basis of $\operatorname{gp}(M)$. Consider the $\Bbbk$-algebra isomorphism $\Bbbk[Y_1^{\pm1}, \dots, Y_r^{\pm1}] \cong \Bbbk[\operatorname{gp}(M)]$ induced by the assignment $Y_i \mapsto e_i$. It induces a homomorphism $\mathbb{T}^r \to G_\Bbbk(\Bbbk[M])$, which maps $\mathbb{T}^r$ isomorphically onto $\mathbb{T}_\Bbbk(M)$. Moreover, the torus $\mathbb{T}_\Bbbk(M)$ acts rationally on $\Bbbk[\operatorname{gp}(M)]$.*

Proof. The tautological action of $\mathbb{T}^r$ on $\Bbbk[Y_1^{\pm1}, \dots, Y_r^{\pm1}]$ is transferred via the isomorphism to $\Bbbk[\operatorname{gp}(M)]$. Evidently $\tau(\Bbbk[M]) = \Bbbk[M]$ for all $\tau \in \mathbb{T}^r$, and each monomial in M is an eigenvector for τ since it corresponds to a monomial in the Laurent polynomial ring. This shows that τ acts as a toric automorphism on $\Bbbk[M]$.

Conversely, suppose that σ is a toric automorphism of $\Bbbk[M]$. Then σ extends to an automorphism of $\Bbbk[\mathrm{gp}(M)]$, since $\Bbbk[\mathrm{gp}(M)]$ arises by the inversion of all terms $u\mu$, $u \in \Bbbk^\times$, $\mu \in M$. Moreover, the set of terms is mapped onto itself by σ since σ is toric. But the extension of σ to $\Bbbk[\mathrm{gp}(M)]$ is also toric, and so it is enough to show that every toric automorphism of the Laurent polynomial ring belongs to $\mathbb{T}^r$. This is clear since it is completely determined by its values on the indeterminates. (See Exercise 5.19 for a description of the automorphism group of a Laurent polynomial ring.)

The last statement is evident: we order the subspaces $\Bbbk\mu$, $\mu \in M$, in some way, and write $\Bbbk[M]$ as the union of an ascending chain of finite direct sums of these subspaces. □

The lemma allows us to consider $\mathbb{T}_\Bbbk(M)$ as a (linear) algebraic group. In fact, we can use the isomorphism $\mathbb{T}^r \cong \mathbb{T}_\Bbbk(M)$ to transfer the geometry of $\mathbb{T}^r$ to $\mathbb{T}_\Bbbk(M)$. While the isomorphism depends on the choice of basis for $\mathrm{gp}(M)$, the geometry obtained does not. Every choice of basis is via an element of $\mathrm{GL}_r(\mathbb{Z})$, which can be understood as an algebraic automorphism of $\mathbb{T}^r$.

Second, we characterize the toric automorphisms in terms of the defining ideal of $\Bbbk[M]$ as a residue class ring of a polynomial ring.

Lemma 5.8. *Let M be an affine monoid, $A = \Bbbk[M]$, let $x_1, \dots, x_n$ be a system of generators of M and I the kernel of the epimorphism $\Bbbk[X_1, \dots, X_n] \to A$, $X_i \mapsto x_i$.*

Then an assignment $x_i \mapsto c_i x_i$, $c_i \in \Bbbk^\times$, $i = 1, \dots, n$, can be extended to a (toric) automorphism τ of A if and only if $f(c_1, \dots, c_n) = 0$ for all binomials $f \in I$.

Proof. Suppose first that $\tau \in \mathbb{T}_\Bbbk(M)$. Then $\tau(x_i) = c_i x_i$, $c_i \in \Bbbk^\times$, for all i by the definition of $\mathbb{T}_\Bbbk(M)$. We have to show that $f(c_1, \dots, c_n) = 0$ for all $f \in I$. Pick a binomial $f = X_1^{a_1} \cdots X_n^{a_n} - X_1^{b_1} \cdots X_n^{b_n} \in I$. We have $f(x_1, \dots, x_n) = 0$ and τ is an endomorphism of A. Then also $f(c_1 x_1, \dots, c_n x_n) = 0$, and so

$$c_1^{a_1} \cdots c_n^{a_n} x_1^{a_1} \cdots x_n^{a_n} = c_1^{b_1} \cdots c_n^{b_n} x_1^{b_1} \cdots x_n^{b_n}.$$

This is only possible if $c_1^{a_1} \cdots c_n^{a_n} = c_1^{b_1} \cdots c_n^{b_n}$, whence $f(c_1, \dots, c_n) = 0$.

Let now $c_1, \dots, c_n \in \Bbbk^\times$ be given such that $f(c_1, \dots, c_n) = 0$ for all binomials $f \in I$. Consider the toric automorphism $\sigma \in \mathbb{T}^n$ of $\Bbbk[X_1, \dots, X_n]$ given by the assignment $X_i \mapsto c_i X_i$.

Since $f(c_1 x_1, \dots, c_n x_n) = 0$ for all binomials in I and since I is generated by binomials (see Proposition 4.26), it follows that σ maps I into itself. Hence it induces an endomorphism τ of $A = \Bbbk[X_1, \dots, X_n]/I$. Evidently τ is a toric automorphism of A. □

Corollary 5.9. *With the notation of Lemma 5.8, the map $\mathbb{T}_\Bbbk(M) \to \mathbb{T}^n$, $\tau \mapsto (c_1, \dots, c_n)$, is an algebraic isomorphism of $\mathbb{T}_\Bbbk(M)$ with the closed subgroup of $\mathbb{T}^n$ defined by the ideal $I \cdot \Bbbk[X_1^{\pm 1}, \dots, X_n^{\pm 1}]$.*

It will be very important in the following that we can identify the monoid M by the action of $\mathbb{T}_{\Bbbk}(M)$ on $A = \Bbbk[M]$. Apart from part (f), the following lemma is hardly more than a reformulation of what has already been proved, but it adds a representation-theoretic flavor to it.

Lemma 5.10. *With notation as in Lemma 5.8 we have the following:*

(a) For each $x \in M$ the function $\chi_x : \mathbb{T}_{\Bbbk}(M) \to \Bbbk^{\times}$, $\chi_x(\tau) = \tau(x)/x$, is a character.
(b) If $x \neq y$, then $\chi_x \neq \chi_y$.
(c) Let $z \in A$, $z \neq 0$, and $x \in M$. If $\tau(z) = \chi_x(\tau)z$ for all $\tau \in \mathbb{T}_{\Bbbk}(M)$, then $z = cx$ for some $c \in \Bbbk^{\times}$.
(d) An ideal I of A is stable under the action of $\mathbb{T}_{\Bbbk}(M)$ if and only if it is monomial.
(e) The characters χ_x form a monoid isomorphic to M.
(f) $\mathbb{T}_{\Bbbk}(M)$ is a maximal torus in $\mathrm{G}_{\Bbbk}(A)$.

Proof. (a) We can use the identification of $\mathbb{T}_{\Bbbk}(M)$ with $\mathbb{T}^r$ given by Lemma 5.7. Suppose that $x = e_1^{a_1}\cdots e_r^{a_r}$ with $a_1, \dots, a_r \in \mathbb{Z}$. Then $\chi_x = Y_1^{a_1}\cdots Y_r^{a_r}$ as a function on $\mathbb{T}^r$, and so it is a character.

(b) This is obvious, since x and y correspond to different monomials in $\mathcal{O}(\mathbb{T}^r)$.

(c) If $\tau(z) = \chi_x(\tau)z$ for all $\tau \in \mathbb{T}_{\Bbbk}(M)$, then z belongs to the weight space of the character χ_x on A. It is the 1-dimensional subspace spanned by x.

(d) The nontrivial implication follows immediately from (c), Proposition 5.6 and Lemma 5.7.

(e) The assignment $x \mapsto \chi_x$ is clearly a monoid homomorphism. It is injective by (b).

(f) Suppose g is a $\Bbbk$-automorphism that commutes with all elements $\tau \in T$. Let $x \in M$. Then

$$\tau(g(x)) = g(\tau(x)) = g(\chi_x(\tau)x) = \chi_x(\tau)g(x)$$

for all $\tau \in T$. It follows from (c) that $g(x) = c_x x$ for all $x \in M$, $c_x \in \Bbbk^{\times}$. Therefore g is a toric automorphism. □

An important consequence of Lemma 5.10 is that we can identify the monoid M already from the characters χ_x where x is running through a system of generators of M. This shows that M is completely determined by the action of $\mathbb{T}_{\Bbbk}(M)$ on the vector space spanned by a system of generators of M. The next lemma indicates how Borel's theorem on conjugate tori may be applied to isomorphism problems for monoid algebras.

Lemma 5.11. *Let V be a $\Bbbk$-vector space, and let $T_1, T_2 \subset \mathrm{GL}(V)$ be tori acting rationally on V. If there exists $g \in \mathrm{GL}(V)$ such that $g^{-1}T_1 g = T_2$, then the weights of T_1 and T_2 occurring in V generate isomorphic submonoids of $X(T_1)$ and $X(T_2)$, respectively.*

Proof. The group isomorphism $T_1 \to T_2$, $\tau \mapsto g^{-1}\tau g$, induces a group isomorphism $\Psi : X(T_2) \to X(T_1)$. Since we can similarly consider Ψ^{-1}, it is enough to show that $\Psi(\chi)$ is a character of T_1 occurring in V if χ is a character of T_2 occurring in V. Let χ be a character of T_2 occurring in V.

Let $w \in V_{\chi}$, $w \neq 0$, $\tau \in T_1$, and $\sigma = g^{-1}\tau g$. Then

$$\tau(g(w)) = g(\sigma(w)) = g(\chi(\sigma)w) = \chi(\sigma)g(w) = \big(\Psi(\chi)(\tau)\big)(g(w)).$$

Thus the 1-dimensional space $\Bbbk g(w)$ is invariant under the action of T_1. Since T_1 acts rationally on V, it acts rationally on every subspace of V, and so $\Bbbk g(w)$ must be contained in a weight space of T_1. Its associated character is $\Psi(\chi)$. □

Groups of graded automorphisms. While $G_\Bbbk(A)$ is not a linear algebraic group in general, the property holds for the group

$$\Gamma_\Bbbk(A)$$

of graded automorphisms under suitable hypotheses on A:

Proposition 5.12. *Let $A = \Bbbk \oplus A_1 \oplus A_2 \oplus \cdots$ be a finitely generated, positively graded $\Bbbk$-algebra. Then $\Gamma_\Bbbk(A)$ is a linear group.*

Proof. First of all, the finite generation of A over $\Bbbk$ implies that $\dim_\Bbbk A_i < \infty$ for all $i \in \mathbb{N}$. We choose $d \in \mathbb{N}$ such that $A_1 \oplus \cdots \oplus A_d$ generates A as a $\Bbbk$-algebra. Fix a $\Bbbk$-basis $B = \{z_1, \dots, z_n\}$ of $A_1 \oplus \cdots \oplus A_d$. We have a group embedding $\Gamma_\Bbbk(A) \to \mathrm{GL}_n(\Bbbk)$, defined as follows:

$$\gamma \mapsto (a_{ij}), \qquad \gamma(z_j) = \sum_{i=1}^{n} a_{ij} z_i, \quad j = 1, \dots, n.$$

Let $F_s(X_1, \dots, X_n) \in \Bbbk[X_1, \dots, X_n]$, $s = 1, \dots, t$, be a finite system of polynomials generating the ideal of relations between the elements $z_1, \dots, z_n \in A$.

The entries of the matrix (a_{ij}), corresponding to an element $\gamma \in \Gamma_\Bbbk(A)$, satisfy the condition

$$F_s\left(a_{11}z_1 + \cdots + a_{n1}z_n, \dots, a_{1n}z_1 + \cdots + a_{nn}z_n\right) = 0, \quad s = 1, \dots, t. \tag{5.1}$$

Conversely, if the entries of an invertible $n \times n$ matrix (a_{ij}) satisfy (5.1), then the assignment

$$z_j \mapsto \sum_{i=1}^{n} a_{ij} z_i, \quad j = 1, \dots, n,$$

gives rise to a unique element $\gamma \in \Gamma_\Bbbk(A)$.

The proof is complete once one has converted (5.1) in a system of polynomial equations in the entries of the $n \times n$ matrix formed by $z_1, \dots, z_n$, using a suitable truncation of the algebra A. The details are left to the reader in Exercise 5.11, which generalizes the lemma to a "semigraded" version. □

Together with Lemma 5.10(f) it implies the next corollary:

Corollary 5.13. *Let M be an affine positive monoid. Then a positive grading on M induces a positive grading on $\Bbbk[M]$, and $\mathbb{T}_\Bbbk(M)$ is a maximal torus in the linear group $\Gamma_\Bbbk(A)$.*

5.B Invariants of diagonalizable groups

Let $\Bbbk$ be an algebraically closed field as in the previous section. We consider the group $\mathrm{GL}_n(\Bbbk)$ of invertible $n \times n$ matrices over $\Bbbk$ as the group of homogeneous (with respect to total degree) $\Bbbk$-algebra automorphism of the polynomial ring $R = \Bbbk[X_1, \dots, X_n]$, where each matrix $(a_{ij}) \in \mathrm{GL}_n(\Bbbk)$ acts by a linear substitution:

$$g : R \to R, \qquad X_j \mapsto \sum_{i=1}^{n} a_{ij} X_i. \tag{5.2}$$

Let D be a subgroup of $\mathrm{GL}_n(\Bbbk)$. Then $f \in R$ is an *invariant* of D, if $g(f) = f$ for all $g \in D$. The *ring of invariants* of D is

$$R^D = \{f \in R : g(f) = f \text{ for all } g \in D\}.$$

For fixed f, the equation $g(f) = f$ imposes a system of polynomial equations on the entries of a matrix of g. Therefore we can always assume that D is closed. Then the ring of invariants is nothing but the weight space of the trivial character.

Note that an element $\xi = (\xi_1, \dots, \xi_n) \in \mathbb{T}^n$ transforms the monomial $X_1^{u_1} \cdots X_n^{u_n}$ into the term

$$\xi_1^{u_1} \cdots \xi_n^{u_n} X_1^{u_1} \cdots X_n^{u_n}.$$

Therefore $\xi(f) = f$ for $f \in R$ if and only if each monomial occurring in f is left invariant by ξ.

Let D be a subgroup of $\mathbb{T}^n$. Then it follows that R^D is a monomial subalgebra of R. Much more is true, however:

Theorem 5.14. *Let D be a subgroup of $\mathbb{T}^n$. Then $\Bbbk[X_1, \dots, X_n]^D$ is a normal affine monoid algebra $\Bbbk[M]$ where M is a pure submonoid of $\mathbb{Z}_+^n$.*

Proof. We have already seen that the ring of invariants is of the form $\Bbbk[M]$. We must show that $M = \mathrm{gp}(M) \cap \mathbb{Z}_+^n$. Then M is obviously normal and finitely generated by Gordan's lemma. Let $x, y \in M$ such that $xy^{-1} = z \in \mathbb{Z}_+^s$. If x and y are invariant under D, then z is, too. Thus $z \in M$, and we are done. □

Remark 5.15. Instead of a subgroup of $\mathbb{T}^n$ we can more generally consider a diagonalizable group D that acts rationally on the vector space $\Bbbk X_1 + \cdots + \Bbbk X_n$ since the image of D in $\mathrm{GL}_n(\Bbbk)$ is conjugate to a subgroup of $\mathbb{T}^n$; see Theorem 5.3.

Our next goal is the converse of this theorem, namely to show that every monomial pure subalgebra of $\Bbbk[X_1, \dots, X_n]$ can be realized as a ring of invariants. However, one sees very quickly that this goal can be reached only if $\Bbbk$ has the right characteristic.

Example 5.16. Consider $A = \Bbbk[X^2] \subset R = \Bbbk[X]$. If $\operatorname{char} \Bbbk = 2$, it is impossible to realize A as a ring of invariants of R. The only automorphism α of R leaving X^2 invariant is $\mathbf{1}_A$: set $f = X - \alpha(X)$. Then $\alpha(X^2) = X^2$ implies $f = 0$.

On the other hand, if $\operatorname{char} \Bbbk \neq 2$, then $A = R^{\{1,-1\}}$.

Let G be a subgroup of $\mathbb{Z}^n$. Then we can find a system of diophantine homogeneous linear equations and congruences such that G is its set of solutions: $(x_1, \dots, x_n) \in G$ if and only if

$$\begin{aligned} a_{i1}x_1 + \cdots + a_{in}x_n &= 0, & i &= 1, \dots, p, \quad a_{ij} \in \mathbb{Z}, \\ b_{i1}x_1 + \cdots + b_{in}x_n &\equiv 0 \quad (w_i), & i &= 1, \dots, q, \quad b_{ij}, w_i \in \mathbb{Z}. \end{aligned} \tag{5.3}$$

Moreover, we can assume that $p = n - \operatorname{rank} G$ and that the congruences are chosen in such a way that $w_1 \cdots w_q$ is the order of the torsion group of $\mathbb{Z}^n/G$. Then the monoid $M = G \cap \mathbb{Z}^n_+$ consists of the nonnegative solutions of this system.

Let $\Bbbk$ an algebraically closed field whose characteristic does not divide the order of an element of G—in other words, $\operatorname{char} \Bbbk$ does not divide $w_1 \cdots w_q$. We use the coefficients of the system and the moduli of the congruences in order to define a subgroup D of $\mathbb{T}^n$ by specifying a set of generators. Let ζ_i is a primitive w_ith root of unity. Then the generators of D are given by

$$\begin{aligned} &(\xi^{a_{i1}}, \dots, \xi^{a_{in}}), & i &= 1, \dots, p, \quad \xi \in \Bbbk^\times \\ &(\zeta_j^{b_{j1}}, \dots, \zeta_j^{b_{jn}}), & j &= 1, \dots, q. \end{aligned} \tag{5.4}$$

Theorem 5.17. *Let M be a pure submonoid of $\mathbb{Z}^n_+$, and let $G = \mathbb{Z}^n/\operatorname{gp}(M)$. Suppose that G contains no element of order* char $\Bbbk$. *Then there exists a closed subgroup $D \subset \mathbb{T}^n$ such that $\Bbbk[M] = \Bbbk[X_1, \dots, X_n]^D$.*

More precisely, if M is the set of nonnegative solutions of the system (5.3), *then D can be chosen as the subgroup generated by the elements listed in* (5.4).

Proof. It is enough to check that exactly the monomials in M are invariant under each of the generators of D. The torus element $(\xi^{a_{i1}}, \dots, \xi^{a_{in}})$ sends $X_1^{u_1} \cdots X_n^{u_n}$ to

$$\xi^{a_{i1}u_1 + \cdots + a_{in}u_n} X_1^{u_1} \cdots X_n^{u_n}$$

and the coefficient is equal to 1 for all $\xi \in \Bbbk^\times$ if and only if $a_{i1}u_1 + \cdots + a_{in}u_n = 0$, simply because $\Bbbk$ has infinitely many elements.

To complete the argument one has only to check that $\zeta_j^{b_{i1}u_1 + \cdots + b_{in}u_n} = 1$ if and only if $b_{i1}u_1 + \cdots + b_{in}u_n \equiv 0 \ (w_i)$, and this is clear by the choice of ζ_j. □

Though we have used a specific description of M, the group D only depends on M:

Corollary 5.18. *With the notation of Theorem* 5.17, *the group D is the largest subgroup of $\mathbb{T}^n$ such that $\Bbbk[M] = R^D$. Moreover, $X(D) \cong \mathbb{Z}^n/\operatorname{gp}(M)$.*

Proof. For a monomial $x = X_1^{a_1} \cdots X_n^{a_n}$ let χ_x be the characters of $\mathbb{T}^n$ given by $x \in \mathcal{O}(\mathbb{T}^n)$. That the monomials $x \in M$ are invariants of D can equivalently expressed by the relation

$$D \subset U = \bigcap_{x \in M} \operatorname{Ker} \chi_x \subset \mathbb{T}^n.$$

We have

$$\Bbbk[M] \subset R^U \subset R^D = \Bbbk[M].$$

It follows that $\Bbbk[M] = R^U$, and U is clearly the largest subgroup of $\mathbb{T}^n$ having $\Bbbk[M]$

as its ring of invariants. It remains to show that $D = U$.

The chain $\mathcal{O}(\mathbb{T}^n) \to \mathcal{O}(U) \to \mathcal{O}(D)$ induces a chain of epimorphisms

$$X(\mathbb{T}^n) \cong \mathbb{Z}^n \to X(U) \to X(D).$$

By the construction of D, exactly the characters χ_x, $x \in \mathrm{gp}(M)$, vanish on D, so that $X(D) \cong \mathbb{Z}^n/\mathrm{gp}(M)$ as claimed. On the other hand, all these characters vanish also on U. It follows that the epimorphism $X(U) \to X(D)$ is an isomorphism. Hence $\mathcal{O}(U) \to \mathcal{O}(D)$ is an isomorphism, and so is the embedding $D \subset U$. □

It follows from the corollary in conjunction with Theorem 5.10 that the hypothesis on char $\Bbbk$ in Theorem 5.17 is not only sufficient, but also necessary. Exercise 5.7 treats the case not covered by 5.17.

We know from Theorem 2.29 that every positive affine monoid can be embedded into a monoid $\mathbb{Z}_+^n$ for suitable n as a pure submonoid. The theorem shows that every such embedding yields a realization of $\Bbbk[M]$ as an invariant subring (if the hypothesis on char $\Bbbk$ is satisfied). Applied to the standard embedding we obtain:

Corollary 5.19. *Let M be a positive normal affine monoid, and consider $\Bbbk[M]$ as a subalgebra of $\Bbbk[X_1,\dots,X_s]$ via the standard embedding. Suppose that $\mathrm{Cl}(M)$ contains no element of order* char $\Bbbk$, *and set $D = \mathrm{Hom}(\mathrm{Cl}(M), \Bbbk^\times)$. Then $\Bbbk[M] = \Bbbk[X_1,\dots,X_s]^D$.*

If we embed M in such a way that it is even integrally closed in $\mathbb{Z}_+^n$, and this is likewise possible by Theorem 2.29, then only a torus action is needed for a representation of $\Bbbk[M]$ as a ring of invariants, and the hypothesis on the characteristic of $\Bbbk$ can be dropped.

Corollary 5.20. *Let M be an integrally closed and pure submonoid of $\mathbb{Z}_+^n$. Then there exists a subtorus T of $\mathbb{T}^n$ such that $\Bbbk[M] = \Bbbk[X_1,\dots,X_n]^T$.*

We conclude the section by a statement about weight spaces (or semi-invariants).

Theorem 5.21. *Let $D \subset \mathbb{T}^n$ be a closed subgroup and $R = \Bbbk[X_1,\dots,X_n]$. Then R is the direct sum of the weight spaces R_χ, $\chi \in X(D)$. For every $\chi \in X(D)$ the weight space R_χ is a finite direct sum of finitely generated R^D-modules of rank* 1.

If $X(D) = \mathbb{Z}^n/\mathrm{gp}(M)$ where M is the monoid of D-invariant monomials in R, then R_χ is an R^D-module of rank ≤ 1 for every $\chi \in X(D)$.

If, in addition M contains a monomial $X_1^{a_1}\cdots X_n^{a_n}$ with $a_i > 0$ for all $i = 1,\dots,n$, then rank $R_\chi = 1$ *for all $\chi \in X(D)$.*

Proof. The character group of D is a quotient $\mathbb{Z}^n/U$ where U is a subgroup of $\mathbb{Z}^n$ (see Corollary 5.4). Clearly R_χ is spanned as a $\Bbbk$-vector space by all monomials in R that belong to a single residue class modulo U. Each such residue class is given by the set of solutions of an inhomogeneous version of the system (5.3). Therefore we can apply Corollary 2.13. It implies that R_χ is a finitely generated R^D-module.

Each residue class modulo U splits into residue classes modulo $\mathrm{gp}(M)$, and the subspace spanned by the monomials in a residue class modulo $\mathrm{gp}(M)$ is a rank 1 module over R^D. In fact it appears as a coset module of the pure submonoid M of $\mathbb{Z}_+^n$ (see Proposition 4.44). Clearly, R_χ splits into its coset submodules.

If $U = \operatorname{gp}(M)$, then R_χ has rank 1, provided it is nonzero. If M contains a monomial with strictly positive coefficients, then every residue class modulo $\operatorname{gp}(M)$ contains a monomial in R, and the corresponding weight space is nonzero. □

5.C The isomorphism theorem

In this section we address the following question, known as the *isomorphism problem for monoid rings*: if $R[M]$ and $R[N]$ are isomorphic as R-algebras, are then M and N isomorphic? The analogous question for group rings (not necessarily commutative) is a much studied classical problem.

The isomorphism problem for affine monoids can be solved positively, under mild restrictions:

Theorem 5.22. *Let M and N be affine monoids and R a ring. Suppose $f : R[M] \to R[N]$ is an R-algebra isomorphism. Then $M \cong N$ if one of the following conditions is satisfied:*

(a) M and N are positive and f respects the monomial augmentation ideals;
(b) M is normal or homogeneous;
(c) M is positive and there exists a prime ideal $\mathfrak{n}$ of R such that $\operatorname{char} R/\mathfrak{n} = 0$.

The condition in (a) means the following: let I and J be the ideals in $R[M]$ and $R[N]$, respectively, that are generated by the monomials $\neq 1$; then $f(I) \subset J$. (I and J are the kernels of the augmentation homomorphisms $R[M] \to R$ and $R[N] \to R$ that map all nonunit monomials to 0.) In (a) and (c) it would be enough that M is positive. Then N must also be positive since $\mathrm{U}(M) \cong \mathrm{U}(R[M])/\mathrm{U}(R) \cong \mathrm{U}(R[N])/\mathrm{U}(R) \cong \mathrm{U}(N)$ by Proposition 4.21. The conditions in (b) and (c) can be refined further; see Lemma 5.33.

The first and obvious reduction step in the proof of Theorem 5.22 is the passage to a residue class ring $R/\mathfrak{n}$ with respect to a prime ideal $\mathfrak{n}$ and to an algebraic closure of its field of fractions. Whenever necessary we may therefore assume that $R = \Bbbk$ is an algebraically closed field.

Then the theorem will be reduced to the augmented case. Note that it is very easy to find nonaugmented isomorphisms of monoid algebras: for example, consider $f : \Bbbk[X] \to \Bbbk[X]$, $f(X) = X - 1$. However, it will turn out that nonaugmented isomorphisms can only appear in the presence of polynomial variables, roughly speaking.

After the passage to the augmented case, we will identify $R[M]$ and $R[N]$. Then we are dealing with two monomial structures on the same R-algebra that have the same augmentation ideal $\mathfrak{m}$. Finally Borel's theorem on maximal tori will be applied to a suitable truncation modulo $\mathfrak{m}^d$ for $d \gg 0$.

Remark 5.23. For monoids M of rank 2, the conditions (a), (b), and (c) in Theorem 5.22 are superfluous: if $R[M]$ and $R[N]$ are isomorphic R-algebras, then $M \cong N$; see [162]. We believe that the theorem can be generalized independently of rank M.

Remark 5.24. Theorem 5.22 says that for an affine normal monoid M any *toric structure*[1] on Spec A, $A = \Bbbk[M]$, is conjugate in the automorphism group $G_{\Bbbk}(A)$ of the $\Bbbk$-algebra A to the toric structure given by M. In [28] Berchtold and Hausen prove that if an $(n-1)$-dimensional torus $\mathbb{T}$, $n = \operatorname{rank} M$, acts effectively on Spec A, then $\mathbb{T}$ is conjugate in $G_{\Bbbk}(A)$ to a closed subtorus of $\mathbb{T}^n(\Bbbk) = \operatorname{Hom}(M, \Bbbk^\times)$. The crucial argument is that the action of $\mathbb{T}$ extends to a toric structure on Spec A, so Theorem 5.22 applies.

In the case of tori of dimension smaller than $n-1$ the "toric linearization problem" is highly nontrivial, even when $M = \mathbb{Z}_+^n$. Kaliman, Koras, Makar-Limanov, and Russell [215] have solved the problem for $\mathbb{C}^\times$-actions on $\mathbb{C}^3$.

Remark 5.25. A companion of the isomorphism problem addressed in Theorem 5.22 is the *cancellation problem*: does an isomorphism $R[M] \cong S[M]$ for a monoid M of finite rank and rings R, S imply that $R \cong S$? One should note, though, that the hypotheses of Theorem 5.22 and of the cancellation problem are not really comparable; in Theorem 5.22 we assume much more than a ring isomorphism of $R[M]$ and $R[N]$. (But see Exercise 5.13.)

The cancellation problem has a negative answer already for $M = \mathbb{Z}_+$, $R[X] \cong S[X]$: Hochster [194] has found an affine $\mathbb{R}$-algebra T and a T-algebra S such that $S[X] \cong T[X_1, X_2, X_3]$ as T-algebras, but $S \ncong T[X_1, X_2]$. The construction of T and S is intimately connected with the existence of stably free, but nonfree projective modules and will be indicated in Example 8.3.

Hamann [176] calls a ring R *steadfast* if $R[X]$ is cancellable as an R-algebra: whenever one has an R-algebra isomorphism $(R[X])[Y_1, \dots, Y_n] \cong B[Y_1, \dots, Y_n]$ for an R-algebra B, then $B \cong R[X]$. Results of Hamann [177] and Swan [347] characterize the steadfastness of R by the p-seminormality of R_{red} for all prime numbers p: a reduced ring S is *p-seminormal* for $p \in \mathbb{Z}$ if $x^2 \in S$, $x^3 \in S$ and $px \in S$ imply $x \in S$ for all elements $x \in \operatorname{sn}(S)$; in particular every reduced ring S containing $\mathbb{Q}$ is p-seminormal for all $p \neq 0$, and 0-seminormality is seminormality. Hochster's example shows that it is impossible to obtain cancellation of $R[X_1, X_2]$ without a drastic restriction of R.

A positive result was proved by Miyanishi and Sugie [263]. If $R[X_1, \dots, X_n] \cong \Bbbk[Y_1, \dots, Y_{n+2}]$ for an affine algebra R over an algebraically closed field $\Bbbk$, then $R \cong \Bbbk[Z_1, Z_2]$. However, for $n \geq 3$ the question whether

$$R[X_1, \dots, X_m] \cong \mathbb{C}[Y_1, \dots, Y_{m+n}] \Longrightarrow R \cong \mathbb{C}[Z_1, \dots, Z_n]$$

is one of the famous open problems of affine geometry. See Kraft [228] for a discussion and references. See also Freudenburg and Russell [127] and Finston and Maubach [123].

Two applications of Borel's theorem. Before delving into the technical details of the proof of Theorem 5.22, we explain the relevance of Borel's theorem by proving an instructive special case and a variant for residue class rings of polynomial rings modulo monomial ideals.

[1] Terminology to be made systematic in Section 10.B.

If the isomorphism $R[M] \cong R[N]$ does not only respect the augmentation ideals, but even a grading, then the case is rather simple.

Proposition 5.26. *Let R be a ring and M and N be positive affine monoids, both endowed with a positive grading. If $R[M] \cong R[N]$ as graded R-algebras, then $M \cong N$.*

Proof. As pointed out above, we can replace R by an algebraically closed field and identify the $\Bbbk$-algebras $A = \Bbbk[M]$ and $\Bbbk[N]$ along a graded isomorphism. Then Lemma 5.7 provides us with two (rank M)-dimensional tori in the linear group $\Gamma_{\Bbbk}(A)$ (Proposition 5.12) acting rationally on A. Denote them by T_1 and T_2. By Theorem 5.2 and Corollary 5.13 there exists $g \in \Gamma_{\Bbbk}(\Bbbk[M])$ such that $g^{-1}T_1 g = T_2$. Thus Lemma 5.11 implies that the monoids generated by the weights of T_1 and T_2 occurring in A are isomorphic. But the first of these monoids is isomorphic to M and the second is isomorphic to N, as follows from Lemma 5.10. □

The proof of the next theorem contains a reduction to the augmented case. Such a reduction must be based on the ring-theoretic behavior of monoid elements, which is preserved under ring isomorphism. Here we can use the property of zerodivisor.

Theorem 5.27. *Let $I \subset R[X_1,\dots,X_m]$ and $J \subset R[Y_1,\dots,Y_n]$ be monomial ideals such that $X_i \notin I$ and $Y_j \notin J$ for all i and j. If $R[X_1,\dots,X_m]/I$ and $R[Y_1,\dots,Y_n]/J$ are isomorphic R-algebras, then $m = n$ and there exists a bijective mapping $\vartheta : \{X_1,\dots,X_m\} \to \{Y_1,\dots,Y_m\}$ transforming I into J.*

Proof. As in Proposition 5.26, we can assume that $R = \Bbbk$ is an algebraically closed field. Let $f : \Bbbk[X_1,\dots,X_m]/I \to \Bbbk[Y_1,\dots,Y_n]/J$ be a $\Bbbk$-algebra isomorphism.

The monomial ideal I has a unique minimal set of monomial generators $\mu_1,\dots,\mu_q$. Let $\{i_1,\dots,i_u\}$ be the set of indices $i \in \{1,\dots,n\}$ such that X_i does not divide any μ_j, and let $\{j_1,\dots,j_v\}$ be the complementary set. Then

$$A \cong \Big(\Bbbk[X_{j_1},\dots,X_{j_v}]/(\mu_1,\dots,\mu_q)\Big)[X_{i_1},\dots,X_{i_u}].$$

Clearly, $X_{j_1},\dots,X_{j_v}$ are zerodivisors modulo I, and $X_{i_1},\dots,X_{i_u}$ are not.

Let $\mathfrak{m}$ be the maximal ideal of $A = \Bbbk[X_1,\dots,X_m]/I$ generated by the residue classes x_i of the indeterminates X_i. Then all zerodivisors of A are contained in $\mathfrak{m}$, and the same holds for the maximal ideal $\mathfrak{n}$ of $B = \Bbbk[Y_1,\dots,Y_n]/J$ generated by the residue classes y_i of the Y_i.

Therefore, if $f(x_i) \notin \mathfrak{n}$, then $f(x_i)$ is not a zerodivisor in B, so x_i is not a zerodivisor in A, and $i \in \{i_1,\dots,i_u\}$. Since $X_{i_1},\dots,X_{i_u}$ are polynomial variables in A over its subalgebra $A' = \Bbbk[X_{j_1},\dots,X_{j_v}]/(\mu_1,\dots,\mu_q)$, the substitution $X_i \mapsto X_i - f(x_i)(0)$, $i \in \{1,\dots,n\}$, induces an A'-algebra automorphism g of A. (Here $f(x_i)(0)$ is the constant term of $f(x_i)$ in its monomial representation.)

The $\Bbbk$-algebra isomorphism $fg : A \to B$ maps the maximal ideal $\mathfrak{m}$ onto the maximal ideal $\mathfrak{n}$, both being the kernel of the natural augmentation map in A and B, respectively. Replace f by fg.

Our next goal is to pass to an isomorphism of A and B that is even homogeneous with respect to the standard grading by total degree. To this end we consider the associated graded rings $\operatorname{gr}_{\mathfrak{m}}(A) = \bigoplus_{k=0}^{\infty} \mathfrak{m}^k/\mathfrak{m}^{k+1}$ and $\operatorname{gr}_{\mathfrak{n}}(B)$. Since $f(\mathfrak{m}) = \mathfrak{n}$, we obtain an induced isomorphism $\operatorname{gr}(f) : \operatorname{gr}_{\mathfrak{m}}(A) \to \operatorname{gr}_{\mathfrak{n}}(B)$.

The passage to the associated graded rings is always possible if a ring isomorphism respects the augmentation ideals. But in the present situation both algebras are generated by their homogeneous degree 1 elements. Therefore the natural $\Bbbk$-vector space isomorphism $A \cong \operatorname{gr}_{\mathfrak{m}}(A)$ is even a $\Bbbk$-algebra isomorphism, and similarly for B. Replacing both rings by their associated graded rings and f by $\operatorname{gr}(f)$, we may now assume that our rings are isomorphic under a homogeneous isomorphism.

We identify A and B through f, and let Γ be the group of graded $\Bbbk$-automorphisms of A. This is a linear $\Bbbk$-group by Proposition 5.12. It contains two copies of $\mathbb{T}^n(\Bbbk)$: one consisting of the automorphisms of form $x_i \mapsto a_i x_i$, and the other consisting of the automorphisms of form $y_i \mapsto a_i y_i$, $a_i \in \Bbbk^\times$, $i = 1, \dots, n$. Denote these tori by T_1 and T_2. Both are maximal tori in the group of linear $\Bbbk$-automorphisms of the graded component R_1 (generated by the linearly independent elements $x_1, \dots, x_n$). A fortiori T_1 and T_2 are maximal in the subgroup Γ.

By Borel's theorem we find an automorphism g with $g^{-1}T_1 g = T_2$. It maps each x_i to an element $c_i y_{j(i)}$, $c_i \in \Bbbk^\times$, $i = 1, \dots, n$. After a toric correction and a permutation we can assume $g(x_i) = y_i$. Then $x_{i_1} \cdots x_{i_p} = 0$ (with potential repetition of factors) if and only if $y_{i_1} \cdots y_{i_p} = 0$, and this implies that the monomial ideals I and J correspond to each other under the assignment $X_i \mapsto Y_i$. □

Example 5.28. Let Δ be simplicial complex and $\Bbbk$ a field. The commutative $\Bbbk$-algebras naturally associated with Δ are

(a) its Stanley–Reisner ring $\Bbbk[\Delta] = \Bbbk[X_v : v \in \operatorname{vert}(\Delta)]/I(\Delta)$ (see p. 140), and
(b) $\Bbbk[\Delta]_{\mathrm{sqf}} = \Bbbk[\Delta]/(X_v^2 : v \in \operatorname{vert}(\Delta))$.

The algebra in (b) is the squarefree version of $\Bbbk[\Delta]$. Only squarefree monomials are nonzero in $\Bbbk[\Delta]_{\mathrm{sqf}}$, and the nonzero monomials are in bijective correspondence with the faces of Δ, so that its Hilbert function is given by the f-vector of Δ. In particular, there is no smaller $\Bbbk$-algebra (say, in terms of Hilbert functions) than $\Bbbk[\Delta]_{\mathrm{sqf}}$ from which Δ could be recovered. (Hilbert functions will be discussed in Section 6.D.)

Theorem 5.27 implies that Δ is uniquely determined by the $\Bbbk$-algebra structure of any of these two algebras: if $\Bbbk[\Delta_1] \cong \Bbbk[\Delta_2]$ or $\Bbbk[\Delta_1]_{\mathrm{sqf}} \cong \Bbbk[\Delta_2]_{\mathrm{sqf}}$ as $\Bbbk$-algebras, then $\Delta_1 \cong \Delta_2$ as simplicial complexes. See Exercise 5.12 for an exterior algebra analogue.

Reduction to the augmented case. A key argument for the reduction of Theorem 5.22 to the augmented case is the following lemma.

Lemma 5.29. *Let $\Bbbk$ be a field and $R = \bigoplus_{i=0}^{\infty} R_i$ a positively graded affine normal $\Bbbk$-domain. Then every nonunit x outside the maximal ideal $\mathfrak{m} = \bigoplus_{i=1}^{\infty} R_i$ is a product of prime elements.*

Proof. Let T be the multiplicative system generated by x. The natural map $R \mapsto R_{\mathfrak{m}}$ factors through $T^{-1}R$. By Nagata's theorem 4.52(b) we have a sequence $\operatorname{Cl}(R) \to$

$\mathrm{Cl}(T^{-1}R) \to \mathrm{Cl}(R_{\mathfrak{m}})$ of surjective homomorphisms induced by the extension of divisorial ideals. By Theorem 4.64 the composition is an isomorphism, and Nagata's theorem implies that every divisorial prime ideal containing x must be principal, i. e., generated by a prime element. Up to a unit factor, x is the product of powers of these prime elements. (In a noetherian ring, every nonunit $x \neq 0$ is a product of irreducible elements, and if the minimal prime overideals of (x) are principal, the irreducible factors generate them.) □

Lemma 5.30. *Let M be a positive normal affine monoid, $\Bbbk$ a field, and suppose that $A = \Bbbk[M] = \bigoplus_{i=0}^{\infty} A_i$ is positively graded (independently of any grading of M). Let $\mathfrak{n} = \bigoplus_{i=1}^{\infty} A_i$ and set $M' = M \setminus \mathfrak{n}$. Then $M' = x_1^{\mathbb{Z}_+} \times \cdots \times x_m^{\mathbb{Z}_+}$ is a free monoid, generated by pairwise nonassociated prime elements $x_1, \dots, x_m$. Moreover, $M = M' \oplus M''$ where $M'' = M \cap F$, and F is the unique face of $\mathbb{R}_+ M$ not containing any of $x_1, \dots, x_m$.*

Proof. Note that M' is a submonoid of M since it is the complement of a prime ideal. (Therefore $M' = M \cap G$ for a face G of $\mathbb{R}_+ M$.) In particular M' is finitely generated. It is enough to show that M' is generated by prime elements, since the rest then follows from Proposition 2.37. Note that two different prime elements in M must be coprime since M is positive.

In the previous lemma we have already shown that every nonunit x outside $\mathfrak{n}$ is indeed a product of prime elements, and if x is a monomial then it is evidently the product of monomial prime elements. □

With the next proposition we have reduced Theorem 5.22 to the augmented case if M is positive and normal.

Lemma 5.31. *Let $\Bbbk$ be a field and M and N be two positive normal affine monoids. If $\Bbbk[M]$ and $\Bbbk[N]$ are isomorphic as $\Bbbk$-algebras, then they are isomorphic also as augmented $\Bbbk$-algebras.*

Proof. We identify $A = \Bbbk[M]$ and $\Bbbk[N]$ along the given $\Bbbk$-algebra isomorphism and choose a positive grading γ on N. It defines a positive grading on A. Let $\mathfrak{n}$ be the maximal ideal of A generated by the homogeneous elements of positive degree under γ. Then $\mathfrak{n}$ is of course the augmentation ideal of the monoid algebra $A = \Bbbk[N]$ with respect to its monomial structure defined by N.

It remains to find an automorphism α of A that maps all nonunit elements of M into $\mathfrak{n}$. Since M splits as in Lemma 5.30, we have

$$A = \big(\Bbbk[M'']\big)[\mathbb{Z}_+^m] = \big(\Bbbk[M'']\big)[X_1, \dots, X_m],$$

where we have replaced $x_1, \dots, x_m$ by $X_1, \dots, X_m$ to indicate that they are algebraically independent over $\Bbbk[M'']$. Moreover, the elements of M'' are already contained in $\mathfrak{n}$.

Now we define α as the $\Bbbk[M'']$-automorphism of A given by the substitution $X_i \mapsto X_i - \xi_i$, where ξ_i is the constant coefficient of X_i in its representation as a sum of terms av, $a \in \Bbbk^{\times}$, $v \in N$. □

To finish the reduction in the normal case, we only need to get rid of the units, and this is very easy, since a normal affine monoid splits into its group of units and a positive normal monoid (see Proposition 2.26).

Proposition 5.32. *Let $\Bbbk$ be a field and M and N be two positive normal affine monoids, $M \cong \mathrm{U}(M) \oplus M'$, $N \cong \mathrm{U}(N) \oplus N'$. If $\Bbbk[M]$ and $\Bbbk[N]$ are isomorphic as $\Bbbk$-algebras, then there exists a field L such that $L[M']$ and $L[N']$ are isomorphic also as augmented L-algebras.*

Proof. Let $f : \Bbbk[M] \to \Bbbk[N]$ be an isomorphism. Then f maps the group of units $\mathrm{U}(\Bbbk[M]) = \mathrm{U}(M) \oplus \Bbbk^\times$ (see Proposition 4.20) isomorphically onto $\mathrm{U}(\Bbbk[N]) = \mathrm{U}(N) \oplus \Bbbk^\times$. Therefore f maps $\Bbbk[U(M)]$ isomorphically onto $\Bbbk[U(N)]$. Moreover, taking residue classes modulo $\Bbbk^\times$, we immediately get $\mathrm{U}(M) \cong \mathrm{U}(N)$.

Since $\Bbbk[M] \cong \big(\Bbbk[\mathrm{U}(M)]\big)[M']$ and because of the analogous decomposition for $\Bbbk[N]$, we are allowed to identify the Laurent polynomial rings $\Bbbk[\mathrm{U}(M)]$ and $\Bbbk[U(N)]$ and denote this ring by R. After the identification, $\Bbbk[M] \cong \Bbbk[N]$ as R-algebras. Inverting the nonzero elements of R (or working modulo a maximal ideal), we obtain a field L such that $L[M'] \cong L[N']$. By Lemma 5.31 these two algebras are even isomorphic as augmented L-algebras. □

Now let M and N be positive affine monoids. If $\Bbbk[M]$ and $\Bbbk[N]$ are isomorphic as $\Bbbk$-algebras, then their normalizations $\Bbbk[\bar{M}]$ and $\Bbbk[\bar{N}]$ are isomorphic over $\Bbbk$ as well. (See 4.42 for the normalization of a monoid algebra.) The monoids $\bar{M}$ and $\bar{N}$ are also positive, and so we can apply Lemma 5.31 to show that they are isomorphic as augmented $\Bbbk$-algebras. However, it is not clear that the corrected isomorphism maps $\Bbbk[M]$ to $\Bbbk[N]$.

Lemma 5.33. *Let M and N be positive affine monoids and suppose that $\Bbbk[M]$ and $\Bbbk[N]$ are isomorphic $\Bbbk$-algebras. Furthermore, suppose that for each prime element of $\bar{M}$ there exists $q \in \mathbb{N}$ such that $x^q \in M$ and q is not divisible by the characteristic of $\Bbbk$. Then $\Bbbk[M]$ and $\Bbbk[N]$ are isomorphic as augmented $\Bbbk$-algebras.*

Proof. We choose a positive grading on N and consider $\Bbbk[N]$ as a positively graded $\Bbbk$-algebra. Let f be the given isomorphism from $\Bbbk[M]$ to $\Bbbk[N]$. If its extension to the normalization is augmented, then we are done. Assume it is not. Then we "correct" it as in the proof of Lemma 5.31 by changing the values of those prime elements x of $\bar{M}$ for which $f(x)$ has a nonzero constant term ξ in its representation as a linear combination of monomials in N.

We claim: if $\xi \neq 0$, then $f(x) \in \Bbbk[N]$. Assume the contrary, and let v be an element of least degree in $\bar{N} \setminus N$ such that $v \in \operatorname{supp}(f(x))$, Since $x^q \in M$, we have $f(x)^q \in \Bbbk[N]$. But under the hypothesis on the characteristic of $\Bbbk$, the monomial v belongs also to $\operatorname{supp}(f^q)$, as the reader may check. This is a contradiction.

This shows that we have to correct the values only for such prime elements x for which $f(x) \in \Bbbk[N]$. We can do this successively for the finitely many prime elements, and it remains to show that $(f(x) - \xi)^k f(y) \in \Bbbk[N]$ for $k > 0$, $y \in M$, y not divisible by x, and $x^k y \in M$. This follows immediately if $f(y) \in \Bbbk[N]$.

Now it is an easy exercise to show that $z \in \Bbbk[N]$ if there exists $w \in \Bbbk[\bar{N}]$ with nonzero constant term such that $wz \in \Bbbk[N]$. Apply this to $z = f(y)$ and $w = f(x^q)$. □

The hypothesis of Lemma 5.33 are satisfied if char $\Bbbk = 0$ or $x \in M$ for all prime elements of $\bar{M}$, and especially if all extreme integral generators of $\bar{M}$ belong to M. The latter is the case if M is homogeneous. Together with Proposition 5.32 this covers the reduction to the augmented case under all the hypotheses listed in Theorem 5.22.

The augmented case. The ring-theoretic arguments in the last subsection have put us in a position where we can again apply arguments from the theory of linear algebraic groups that we have prepared in Section 5.A. In the graded case (Proposition 5.26) we had a natural candidate for a linear algebraic group to which we could apply Borel's theorem on maximal tori. If M and N are positive and the isomorphism $A = \Bbbk[M] \cong \Bbbk[N]$ respects the augmentation ideals, then, after the identification of $\Bbbk[M]$ and $\Bbbk[N]$ along the given isomorphism, we can replace $G_{\Bbbk}(A)$ by

$$G_{\Bbbk}(A, \mathfrak{m}) = \{\varphi \in G_{\Bbbk}(A) : \varphi(\mathfrak{m}) = \mathfrak{m}),$$

where $\mathfrak{m}$ is the common augmentation ideal, generated by $M \setminus \{1\}$ as well as by $N \setminus \{1\}$. Clearly, $\mathbb{T}_{\Bbbk}(M), \mathbb{T}_{\Bbbk}(N) \subset G_{\Bbbk}(A, \mathfrak{m})$. They are both maximal tori in $G_{\Bbbk}(A, \mathfrak{m})$ by Lemma 5.10(f). However, $G_{\Bbbk}(A, \mathfrak{m})$ is almost never a linear algebraic group. Therefore it is necessary to pass to a truncation of A.

For every $\varphi \in G_{\Bbbk}(A, \mathfrak{m})$ one has $\varphi(\mathfrak{m}^d) = \mathfrak{m}^d$ for all d. Therefore φ induces a $\Bbbk$-linear map $\bar{\varphi}$ on the vector space $\mathfrak{m}/\mathfrak{m}^d$. The assignment $\varphi \mapsto \bar{\varphi}$ is a group homomorphism from $G_{\Bbbk}(A, \mathfrak{m})$ into $\mathrm{GL}(\mathfrak{m}/\mathfrak{m}^d)$. We denote it by Φ_d.

We want to show that the images of $\mathbb{T}_{\Bbbk}(M)$ and $\mathbb{T}_{\Bbbk}(N)$ are maximal tori in $\Phi_d(G_{\Bbbk}(A, \mathfrak{m}))$ if $d \gg 0$. It is of course enough to consider one of them, since all data so far only depend on $\mathfrak{m}$.

Lemma 5.34. *Let M be a positive affine monoid, $A = \Bbbk[M]$ and $\mathfrak{m}$ the maximal ideal generated by $M \setminus \{1\}$. Then the (isomorphic) image of the torus $T = \mathbb{T}_{\Bbbk}(M) \subset G_{\Bbbk}(A, \mathfrak{m})$ is a maximal torus in $\Phi_d(G_{\Bbbk}(A, \mathfrak{m}))$ for*

$$d \geq 1 + \max\{\gamma(x) : x \in \mathrm{Hilb}(M)\},$$

where γ is a positive grading on M.

Proof. Let $\mathrm{Hilb}(M) = \{x_1, \dots, x_n\}$. By definition we have $d \geq 2$. Already for $d = 2$ the restriction of Φ_d to T is injective since every element τ of T is uniquely defined by the coefficients c_i of the images $\tau(x_i) = c_i x_i$, $i = 1, \dots, n$, and the residue classes of $x_1, \dots, x_n$ form a basis of $\mathfrak{m}/\mathfrak{m}^2$. We may now identify T and $\Phi_d(T)$.

Let $g \in G_{\Bbbk}(A, \mathfrak{m})$ and suppose that $\Phi_d(g)$ commutes with all elements of T. Set $M_{<d} = M \setminus (\{1\} \cup \mathfrak{m}^d)$. Then the residue classes of the elements of $M_{<d}$ form a $\Bbbk$-basis of $\mathfrak{m}/\mathfrak{m}^d$, and, as far as the structure of $\Bbbk$-vector space is concerned, we are allowed to identify the elements of $M_{<d}$ with their images. We have the decomposition

$$\mathfrak{m}/\mathfrak{m}^d = \bigoplus_{x \in M_{<d}} \Bbbk x$$

and the $\Bbbk x$ are subspaces of $\mathfrak{m}/\mathfrak{m}^d$ corresponding to the different characters χ_x, $x \in M_{<d}$; see Lemma 5.10(b). Since $\Phi_d(g)$ commutes with all $\tau \in T$, also $(\Phi_d(g))(x)$ is an eigenvector for the character χ_x. By Lemma 5.10(c) it follows that $(\Phi_d(g))(x) = c_x x$ with $c_x \in \Bbbk^\times$ for all $x \in M_{<d}$.

In particular, we have $(\Phi_d(g))(x_i) = c_i x_i$ with $c_i \in \Bbbk^\times$ for the elements $x_1, \dots, x_n$ of the Hilbert basis. Suppose we can show that

$$f(c_1, \dots, c_n) = 0 \tag{5.5}$$

for all binomials $f \in I$, where I is defined as in Lemma 5.8. Then, by part (b) of that lemma, there exists $\sigma \in T$ such that $\sigma(x_i) = c_i x_i$ for all i, and it follows that $\Phi_d(g) = \sigma \in T$, as desired.

At this point we must lift the data from $\mathfrak{m}/\mathfrak{m}^d$ back to A. By what has been shown,

$$g(x_i) = c_i x_i + y_i, \qquad y_i \in \mathfrak{m}^d, \ i = 1, \dots, n.$$

The choice of d guarantees that $c_i x_i$ is the lowest degree component of $g(x_i)$ with respect to the grading defined by γ. Let $f = X_1^{a_1} \cdots X_n^{a_n} - X_1^{b_1} \cdots X_n^{b_n} \in I$. Then $f(x_1, \dots, x_n) = 0$, and so $f(g(x_1), \dots, g(x_n)) = 0$, too. The lowest degree component of $g(x_1)^{a_1} \cdots g(x_n)^{a_n}$ is $c_1^{a_1} \cdots c_n^{a_n} x_1^{a_1} \cdots x_n^{a_n}$. Considering also the monomial $X_1^{b_1} \cdots X_n^{b_n}$ and comparing lowest degree components, we obtain

$$c_1^{a_1} \cdots c_n^{a_n} x_1^{a_1} \cdots x_n^{a_n} = c_1^{b_1} \cdots c_n^{b_n} x_1^{b_1} \cdots x_n^{b_n}.$$

It follows that $c_1^{a_1} \cdots c_n^{a_n} = c_1^{b_1} \cdots c_n^{b_n}$, and this is equation (5.5). □

We can now conclude the proof of Theorem 5.22. Only part (a) still needs to be proved. So suppose that $\Bbbk[M]$ and $\Bbbk[N]$ with M and N positive are isomorphic as augmented $\Bbbk$-algebras with respect to the augmentation ideals generated by the elements $\neq 1$ in M and N, respectively. We identify the algebras along the isomorphism, and call the algebra A. The augmentation ideals are also identified in the maximal ideal $\mathfrak{m}$. Lemma 5.34 yields that the tori $\mathbb{T}_\Bbbk(M), \mathbb{T}_\Bbbk(N) \subset G_\Bbbk(A, \mathfrak{m})$ are both maximal tori in $\Phi_d(G_\Bbbk(A, \mathfrak{m}))$ for $d \gg 0$. A fortiori, they are maximal tori in every subgroup of $\Phi_d(G_\Bbbk(A, \mathfrak{m}))$ that contains them. Moreover, they are closed subgroups of the linear algebraic group $\mathrm{GL}(\mathfrak{m}/\mathfrak{m}^d)$: with the respect to the basis $M_{<d}$ of $\mathfrak{m}/\mathfrak{m}^d$, each element $\xi \in \mathbb{T}_\Bbbk(M)$ is mapped to a diagonal matrix whose entries are Laurent monomials in the entries of a matrix of ξ with respect to a basis of $\mathrm{gp}(M)$. The image of $\mathbb{T}_\Bbbk(M)$ under a morphism of algebraic groups is closed, and by symmetry the same applies to $\mathbb{T}_\Bbbk(N)$.

Let U be the subgroup of $\mathrm{GL}(\mathfrak{m}/\mathfrak{m}^d)$ generated by $\mathbb{T}_\Bbbk(M)$ and $\mathbb{T}_\Bbbk(N)$. By Proposition 5.1 it is a linear algebraic group, and Borel's theorem applies. It follows that $\mathbb{T}_\Bbbk(M)$ and $\mathbb{T}_\Bbbk(N)$ are conjugate in $\mathrm{GL}(\mathfrak{m}/\mathfrak{m}^d)$. By Lemma 5.11 the monoids generated by the characters of $\mathbb{T}_\Bbbk(M)$ and $\mathbb{T}_\Bbbk(N)$ that occur in V are isomorphic. Since $d \geq 2$, the first of these monoids is isomorphic to M, and the second is isomorphic to N (see Lemma 5.10(e)). In fact, the characters corresponding to the Hilbert bases are present in both cases. Altogether we conclude that M and N are isomorphic.

5.D Automorphisms

Although we do not require $\Bbbk$ to be algebraically closed from now on, the symbols $G_{\Bbbk}(A)$, $\Gamma_{\Bbbk}(A)$, and $\mathbb{T}_{\Bbbk}(M)$ will be retained; they denote the group of automorphisms of a $\Bbbk$-algebra A, that of graded $\Bbbk$-automorphisms when A is positively graded, and the toric automorphisms of an affine monoid ring $\Bbbk[M]$, respectively.

In this section we want to determine the groups of graded automorphisms of polytopal and normal affine monoid algebras. So far we have discussed only the toric automorphisms, but in general the group of graded automorphisms is strictly larger. It is intimately related to the group of automorphisms of projective toric varieties, see Section 10.B. It is remarkable that toric geometry was born in the context of automorphism groups (Demazure [107]). Other applications will be indicated in the notes at the end of this chapter.

Termic automorphisms. Let M be an arbitrary monoid. Then we denote the group of automorphisms of M by $\Sigma(M)$. Clearly every $\sigma \in \Sigma$ induces an automorphism of the monoid algebra $A = \Bbbk[M]$ where $\Bbbk$ is an arbitrary field. Therefore we can consider σ as an element of $G_{\Bbbk}(A)$.

We say that $\gamma \in G_{\Bbbk}(A)$ is *termic* if $\gamma(t)$ is a term for every term $t = ax$, $a \in \Bbbk$, $x \in M$. The termic automorphisms evidently form a subgroup $\operatorname{Tm}_{\Bbbk}(A)$ of $G_{\Bbbk}(A)$. Clearly $\mathbb{T}_{\Bbbk}(M)$ and $\Sigma(M)$ are subgroups of $\operatorname{Tm}_{\Bbbk}(M)$, and they determine its structure as follows.

Proposition 5.35. *For $\gamma \in \operatorname{Tm}_{\Bbbk}(A)$ define $\sigma_\gamma : M \to M$ by $\sigma_\gamma(x) = y$ if $\gamma(x) \in \Bbbk y$. Then:*

(a) $\sigma_\gamma \in \Sigma(M)$, and the assignment $\gamma \mapsto \sigma_\gamma$ defines a surjective homomorphism $\operatorname{Tm}_{\Bbbk}(M) \to \Sigma(M)$ with kernel $\mathbb{T}_{\Bbbk}(M)$;

(b) $\operatorname{Tm}_{\Bbbk}(M)$ is the semidirect product of the normal subgroup $\mathbb{T}_{\Bbbk}(M)$ and the subgroup $\Sigma(M)$.

The proof of the proposition is left to the reader. We will need the following criterion for termic automorphisms. It uses the ideal $I \subset \Bbbk[M]$ generated by all monomials $x \in \operatorname{int}(M)$ (see Remark 2.6.)

Lemma 5.36. *Let M be an affine monoid, $A = \Bbbk[M]$, and $\gamma \in G_{\Bbbk}(A)$. Then the following are equivalent:*

(a) γ is termic;

(b) γ leaves the ideal I invariant;

(c) γ permutes the prime ideals $\mathfrak{p}_F$ where F runs through the facets of $\mathbb{R}_+M$.

Moreover, γ is toric if and only if $\gamma(\mathfrak{p}_F) = \mathfrak{p}_F$ for all facets F.

Proof. First suppose that γ is termic. Then γ permutes the prime ideals $\mathfrak{p}_F$ associated with the facets F of $\mathbb{R}_+M$. In fact, it maps monomial ideals to monomial ideals, prime ideals to prime ideals, and preserves the height of ideals. But then γ leaves the intersection I of the ideals $\mathfrak{p}_F$ invariant.

Conversely, if γ permutes the ideals $\mathfrak{p}_F$, then it leaves their intersection invariant, whence (c) $\Longrightarrow$ (b).

So it remains to show (b) $\Longrightarrow$ (a). Suppose that γ leaves I invariant, and consider a monomial $x \in \operatorname{int}(M)$. By Exercise 2.10, $A_x = A[x^{-1}]$ is the full group ring $\Bbbk[\operatorname{gp}(M)]$.

Next we show that $\Bbbk[\operatorname{gp}(M)] \subset A_{\gamma(x)}$, i. e., every monomial $y \in M$ is a unit in $A_{\gamma(x)}$. If y were a nonunit, it would have to be contained in a height 1 prime ideal. Observe that all minimal prime ideals of A containing y are of the form $\mathfrak{p}_F$ where F is a facet of $\mathbb{R}_+M$ (and $\mathfrak{p}_F$ contains y if and only if $y \notin F$; see Corollary 4.35). Therefore the minimal prime ideals of $A_{\gamma(x)}$ containing y are of the form $(A_{\gamma(x)})\mathfrak{p}_F$. But $(A_{\gamma(x)})\mathfrak{p}_F = A_{\gamma(x)}$ since $\gamma(x) \in \mathfrak{p}_F$ for all facets (and faces) F of $\mathbb{R}_+M$.

The crucial point now is to compare the groups of units of A_x and $A_{\gamma(x)}$. Evidently, these groups are isomorphic via the natural extension of γ to an automorphism of A_x and $A_{\gamma(x)}$, and γ maps the subgroup $\Bbbk^\times$ of $\operatorname{U}(A_x)$ to the same subgroup of $\operatorname{U}(A_{\gamma(x)})$. Therefore, with $r = \operatorname{rank} M$ and in view of Proposition 4.20(b),

$$\mathbb{Z}^r \cong \operatorname{U}(A_x)/\Bbbk^\times \cong \operatorname{U}(A_{\gamma(x)})/\Bbbk^\times.$$

Moreover, the residue classes of the monomials in $\operatorname{U}(A_{\gamma(x)})/\Bbbk^\times$ form a subgroup H of rank r.

Assume that $\gamma(x)$ is not a term. Then none of the powers of $\gamma(x)$ is a term. In other words, none of the multiples of the class of $\gamma(x)$ is in the subgroup H. This shows that $\operatorname{rank}\bigl(\operatorname{U}(A_{\gamma(x)})/\Bbbk^\times\bigr) > r$, a contradiction.

The proof of the last statement is left to the reader (Exercise 5.18). □

Column structures on lattice polytopes. The example $M = \mathbb{Z}_+^n$, $A = \Bbbk[M] = \Bbbk[X_1, \dots, X_n]$ graded by total degree, shows that the group $\Gamma_\Bbbk(A)$ is much larger than $\operatorname{Tm}_\Bbbk(M)$. (In general, $\operatorname{Tm}_\Bbbk(M)$ is not contained in $\Gamma_\Bbbk(A)$, however.) While the toric automorphisms correspond to the diagonal matrices in $\Gamma_\Bbbk(A) = \operatorname{GL}_n(A)$ and the automorphism group $\Sigma(M)$ to the group of permutations of the indeterminates, the analogue of the elementary transformations is still missing. We will construct them for polytopal monoids, using column vectors of the underlying polytope.

Definition 5.37. Let $P \subset \mathbb{R}^n$ be a lattice polytope (with respect to the lattice $\mathbb{Z}^n$). We assume that the lattice points of P generate $\mathbb{Z}^n$ as an affine lattice. Then an element $v \in \mathbb{Z}^n$, $v \neq 0$, is a *column vector* (for P) if there is a facet $F \subset P$ such that $x + v \in P$ for every lattice point $x \in P \setminus F$. The set of column vectors is denoted by $\operatorname{Col}(P)$.

For such P and v we also say that v defines a *column structure* on P. The corresponding facet F is called its *base facet* and is denoted by P_v.

The assumption that $\operatorname{lat}(P)$ generates $\mathbb{Z}^n$ affinely is only for simplicity. If it is not satisfied, then we can replace $\mathbb{Z}^n$ by the lattice affinely generated by $\operatorname{lat}(P)$.

One sees easily that for a column structure on P the set of lattice points in P is contained in the union of rays, visualized as *columns*, parallel to the column vector v and with endpoints in F. This is illustrated by Figure 5.1.

We further illustrate the notion of column vector by Figure 5.2: the polytope P_1 has four column vectors, whereas the polytope P_2 has no column vector.

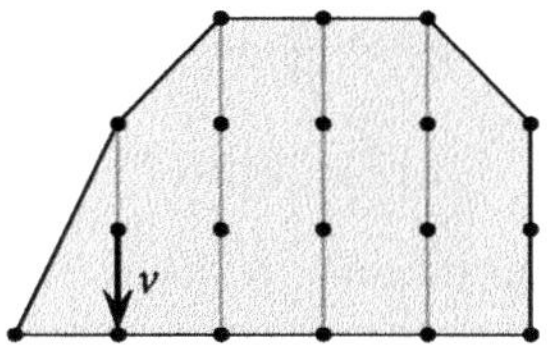

Fig. 5.1. A column structure

Fig. 5.2. Two polytopes and their column structures

Every lattice point x of P has a unique decomposition in the form $x = y - mv$ where $m \in \mathbb{Z}_+$ and $y \in P_v$. The assumption that $\mathbb{Z}^n$ is affinely generated by $\operatorname{lat}(P)$ implies that

$$\operatorname{ht}_{P_v}(x) = m$$

(Exercise 5.20). See Remark 1.72 and Section 3.B for the definition of ht; we can omit the lattice of reference, since it is fixed to be $\mathbb{Z}^n$. It follows easily that v is part of a basis of $\mathbb{Z}^n$. For simplicity we set $\operatorname{ht}_v(x) = \operatorname{ht}_{P_v}(x)$.

Remark 5.38. One can easily control column structures in such formations as homothetic images and direct products of lattice polytopes:

(a) The polytopes P_1 and cP_1 have the same column vectors for every $c \in \mathbb{N}$.
(b) Let P_i be a lattice n_i-polytope, $i = 1, 2$. Then the system of column vectors of $P_1 \times P_2$ is the disjoint union of those of P_1 and P_2 (embedded into $\mathbb{Z}^{n_1+n_2}$).
(c) Actually, (a) is a special case of a more general observation on polytopes: if every facet of P_1 is parallel to a facet of P_2, then $\operatorname{Col}(P_2) \subset \operatorname{Col}(P_1)$.

The proofs of these statements are left to the reader (Exercise 5.21).

Passage to the normalization. We want to analyze the group $\Gamma_{\Bbbk}(\Bbbk[M(P)])$, where $M(P)$ is the polytopal monoid associated with the lattice polytope P. (As in the previous subsection we assume that the lattice points of P generate $\mathbb{Z}^n$ as an affine lattice.) In order to simplify notation we set $\Bbbk[P] = \Bbbk[M(P)]$ and let

$$\Gamma_{\Bbbk}(P) = \Gamma_{\Bbbk}(\Bbbk[P])$$

denote the group of graded $\Bbbk$-automorphisms of $\Bbbk[P]$. The algebras $\Bbbk[P]$ are called *polytopal.*

It is a crucial observation that the group of graded $\Bbbk$-algebra automorphisms does not change if one passes from $\Bbbk[P]$ to its normalization. The normalization is given by $\Bbbk[\bar{P}]$ where $\bar{M}(P)$ is the normalization of $M(P)$. To simplify notation further, we set

$$\Bbbk[\bar{P}] = \Bbbk[\bar{M}(P)],$$

although the symbol $\bar{P}$ has no meaning by itself. There is no need to add a bar to $\Gamma_{\Bbbk}(P)$:

Lemma 5.39. *The natural map* $\Gamma_{\Bbbk}(P) \to \Gamma_{\Bbbk}(\Bbbk[\bar{P}])$, *given by the extension of automorphisms, is a group isomorphism.*

Proof. A $\Bbbk$-algebra automorphism γ of $A = \Bbbk[P]$ extends to a $\Bbbk$-algebra automorphism of the field $\mathrm{QF}(A)$ of fractions, and can then be restricted to a $\Bbbk$-algebra automorphism of the normalization. Moreover, the normalization is contained in the subalgebra B of $\mathrm{QF}(A)$ generated by all fractions of homogeneous elements, and since the extension of a graded automorphism to B is again graded, γ extends to a graded automorphism of the normalization if it is graded.

Therefore it remains to show that any graded $\Bbbk$-algebra automorphism of $\Bbbk[\bar{P}]$ can be restricted to an automorphism of A. In general this is impossible, but the algebras under consideration agree in degree 1, and A is generated by its degree 1 component; see Proposition 2.28. □

The monoids $M(P)$ and $\bar{M}(P)$ are submonoids of $\mathbb{Z}^{n+1}$. We identify a column vector $v \in \mathrm{Col}(P)$ with the vector $(v, 0) \in \mathbb{Z}^{n+1}$, since it represents the difference of lattice points in P, which in $\mathbb{Z}^{n+1}$ have degree 1. Thus v has degree 0 as an element of $\mathbb{Z}^{n+1}$ with respect to the grading given by the last coordinate.

Let F be a facet of P. Then ht_F has a natural extension to $\mathbb{Z}^{n+1}$ by the height with respect to the hyperplane through $C(F)$. In the following it will also be denoted by ht_F. It is of course nothing but the support form of $C(P)$ associated with the facet $C(F)$. Here we prefer the term "height" because of its geometric flavor. If $F = P_v$ for some vector $v \in \mathrm{Col}(P)$ then we will also use ht_v for the extended map $\mathrm{ht}_F : \mathbb{Z}^{n+1} \to \mathbb{Z}$.

Remark 5.40. As elements of $\mathbb{Z}^{n+1}$ the vectors $v \in \mathrm{Col}(P)$ can be characterized as follows:

(i) $\deg v = 0$;
(ii) there exists a facet F of P such that $\mathrm{ht}_F(v) = -1$;
(iii) $\mathrm{ht}_G(v) \geq 0$ for all facets $G \neq F$.

Clearly $F = P_v$.

The proof of the next lemma is now straightforward.

Lemma 5.41. *Let* $v \in \mathrm{Col}(P)$. *Then* $x + v \in M(P)$ *for all* $x \in M(P)$, $x \notin C(P_v)$, *and* $x + v \in \bar{M}(P)$ *for all* $x \in \bar{M}(P)$, $x \notin C(P_v)$.

Elementary automorphisms. At this point we have to switch to *multiplicative notation* since the addition in $\Bbbk[P]$ has to be used. Thus lattice points and column vectors are identified with the corresponding monomials in $\Bbbk[\mathbb{Z}^{n+1}]$. Given $v \in \operatorname{Col}(P)$ and $\lambda \in \Bbbk$, we now define an injective mapping from $M(P)$ to $\operatorname{QF}(\Bbbk[P])$, the quotient field of $\Bbbk[P]$, by the assignment

$$x \mapsto (1+\lambda v)^{\operatorname{ht}_v(x)} x.$$

Since ht_v is a group homomorphism $\mathbb{Z}^{n+1} \to \mathbb{Z}$, our mapping is a homomorphism from $M(P)$ to the multiplicative group of $\operatorname{QF}(\Bbbk[P])$. Now it is immediate from the definition of ht_v and Lemma 5.41 that the (isomorphic) image of $M(P)$ lies actually in $\Bbbk[P]$. Hence this mapping gives rise to a graded $\Bbbk$-algebra endomorphism e_v^λ of $\Bbbk[P]$ preserving the degree of an element. Since e_v^λ is a $\Bbbk$-linear injective map on each (finite-dimensional) graded component, it is an automorphism. In the same way it defines a $\Bbbk$-automorphism of $\Bbbk[\bar{P}]$ by extension from $\Bbbk[P]$ or directly.

One can give an alternative description of e_v^λ. By a suitable integral change of coordinates we may assume that $v = (0,\dots,0,-1)$ and that P_v lies in the subspace $\mathbb{R}^{n-1}$ (thus P is in the upper halfspace). Now consider the standard unimodular n-simplex Δ_n with vertices at the origin and the unit vectors. It is clear that there is a sufficiently large natural number c such that P is contained in a parallel translate of $c\Delta_n$ by a vector from $\mathbb{Z}^{n-1}$. Let Δ denote such a parallel translate. Then we have a graded $\Bbbk$-algebra embedding $\Bbbk[P] \subset \Bbbk[\Delta]$. Moreover, $\Bbbk[\Delta]$ can be identified with the cth Veronese subring of the polynomial ring $\Bbbk[x_0,\dots,x_n]$ in such a way that $v = x_0/x_n$. (The *Veronese subring* $R^{(c)}$ of a graded ring R is the direct sum of those graded components whose degree is divisible by c.) Now the automorphism of $\Bbbk[x_0,\dots,x_n]$ mapping x_n to $x_n + \lambda x_0$ and leaving all the other variables invariant induces an automorphism α of the subalgebra $\Bbbk[\Delta]$, and α in turn can be restricted to an automorphism of $\Bbbk[P]$, which is nothing else but e_v^λ.

It is clear from this description of e_v^λ that it becomes an elementary matrix in the special case when $P = \Delta_n$, after the identification $\Gamma_\Bbbk(P) = \operatorname{GL}_{n+1}(\Bbbk)$. Therefore the automorphisms of type e_v^λ will be called *elementary*.

We now describe the structure of the set of all elementary automorphisms with the same base facet; $G_a(\Bbbk)$ denotes the additive group $(\Bbbk,+)$.

Lemma 5.42. *Let $v_1,\dots,v_q$ be pairwise different column vectors for P with the same base facet $F = P_{v_i}$, $i=1,\dots,q$.*

(a) Then the mapping

$$\varphi : G_a(\Bbbk)^s \to \Gamma_\Bbbk(P), \qquad (\lambda_1,\dots,\lambda_q) \mapsto e_{v_1}^{\lambda_1}\cdots e_{v_q}^{\lambda_q},$$

is an embedding of groups. In particular, $e_{v_i}^{\lambda_i}$ and $e_{v_j}^{\lambda_j}$ commute for any $i,j \in \{1,\dots,q\}$, and the inverse of $e_{v_i}^{\lambda_i}$ is $e_{v_i}^{-\lambda_i}$.

(b) For $x \in \bar{M}(P)$ with $\operatorname{ht}_F(x) = 1$ one has

$$e_{v_1}^{\lambda_1}\cdots e_{v_q}^{\lambda_q}(x) = (1+\lambda_1 v_1 + \cdots + \lambda_q v_q)x.$$

Proof. Set $\psi = \varphi(\lambda_1,\dots,\lambda_q)$. We define a new $\Bbbk$-algebra automorphism ϑ of $\Bbbk[P]$ by first setting

$$\vartheta(x) = (1 + \lambda_1 v_1 + \cdots + \lambda_q v_q)^{\mathrm{ht}_F(x)} x$$

for $x \in M(P)$ and then extending ϑ linearly. Arguments very similar to those above show that ϑ is a graded $\Bbbk$-algebra automorphism of $\Bbbk[P]$ (and $\Bbbk[\bar{P}]$). The lemma is proved once we have verified that $\psi = \vartheta$.

Choose a lattice point $x \in \bar{M}(P)$ such that $\mathrm{ht}_F(x) = 1$. (Such a point exists already in P by Exercise 5.20.) We know that $\mathrm{gp}(M(P)) = \mathbb{Z}^{n+1}$ is generated by x and the lattice points in F. The lattice points in F are left unchanged by both ϑ and φ, and elementary computations show that $\psi(x) = (1 + \lambda_1 v_1 + \cdots + \lambda_q v_q)x = \vartheta(x)$; hence $\psi = \vartheta$. □

The image of the embedding φ given by Lemma 5.42 is denoted by $\mathrm{E}(F)$. Of course, $\mathrm{E}(F)$ may consist only of the identity map of $\Bbbk[P]$, namely if there is no column vector with base facet F. In the case in which P is the unit simplex and $\Bbbk[P]$ is the polynomial ring, $\mathrm{E}(F)$ is the subgroup of all matrices in $\mathrm{GL}_n(\Bbbk)$ that differ from the identity matrix only in the nondiagonal entries of a fixed column. If $\Bbbk$ is algebraically closed, then by Proposition 5.1(a) $\mathrm{E}(F) \cong G_a(\Bbbk)^s$ is a closed subgroup of $\Gamma_\Bbbk(P)$.

Already in the proof of Lemma 5.36 it has become clear that the prime ideals $\mathfrak{p}_F$ play an important role for the automorphism group $\Gamma_\Bbbk(P)$. In the next lemma we describe the action of elementary automorphisms on them. In the following $\mathfrak{p}_F$ will always denote the prime ideal of the normalization $\Bbbk[\bar{P}]$ defined by the facet F.

Lemma 5.43. *Let $v_1, \ldots, v_q$ be column vectors with the common base facet $F = P_{v_i}$, and $\lambda_1, \ldots, \lambda_q \in \Bbbk$. Then*

$$e_{v_1}^{\lambda_1} \cdots e_{v_q}^{\lambda_q}(\mathfrak{p}_F) = (1 + \lambda_1 v_1 + \cdots + \lambda_q v_q)\mathfrak{p}_F$$

and

$$e_{v_1}^{\lambda_1} \cdots e_{v_q}^{\lambda_q}(\mathfrak{p}_G) = \mathfrak{p}_G, \qquad G \neq F.$$

Proof. Note that $(1 + \lambda_1 v_1 + \cdots + \lambda_q v_q)^{\mathrm{ht}_F(x)}(x) \in \mathfrak{p}_F$ for all monomials $x \in \mathfrak{p}_F$. Using the automorphism ϑ from the proof of Lemma 5.42 one concludes that

$$e_{v_1}^{\lambda_1} \cdots e_{v_q}^{\lambda_q}(\mathfrak{p}_F) \subset (1 + \lambda_1 v_1 + \cdots + \lambda_q v_q)\mathfrak{p}_F.$$

The left hand side is a height 1 prime ideal (being an automorphic image of such) and the right hand side is a proper divisorial ideal inside $\Bbbk[\bar{P}]$. Then the inclusion is an equality since the only divisorial ideal of $\Bbbk[\bar{P}]$ strictly containing a height 1 prime ideal is $\Bbbk[\bar{P}]$ itself.

For the second assertion it is enough to treat the case $s = 1$, $v = v_1$, $\lambda = \lambda_1$. One has

$$e_v^\lambda(x) = (1 + \lambda v)^{\mathrm{ht}_F(x)} x,$$

and all the terms in the expansion of the right hand side belong to $\mathfrak{p}_G$ since $\mathrm{ht}_G(v) \geq 0$. As above, the inclusion $e_v^\lambda(\mathfrak{p}_G) \subset \mathfrak{p}_G$ implies equality. □

Lemma 5.44. *Let $F \subset P$ be a facet, $\lambda_0, \ldots, \lambda_q \in \Bbbk^\times$ and $v_0, \ldots, v_q \in \mathbb{Z}^{n+1}$ be pairwise different elements of degree 0. Then $(\lambda_0 v_0 + \cdots + \lambda_q v_q)\mathfrak{p}_F \subset \Bbbk[\bar{P}]$ if and only if v_i is a column vector for P with base facet F for each i with $v_i \neq 0$.*

Proof. Both of the properties that are claimed to be equivalent are themselves equivalent to the fact that $v_i x \in M(P)$ for all i and $x \in M(P)$. Since the v_i are of degree 0, this means $v_i + x \in P$ for all $x \in P$ (in additive notation). □

The Gaussian algorithm for polytopes. The Gaussian algorithm allows us to transform a matrix to a diagonal matrix. Theorem 5.45 claims that such a diagonalization is possible for automorphisms of polytopal algebras, and we will carry it out by a procedure generalizing the Gaussian algorithm.

Clearly, $\mathbb{T}^{n+1} = \mathbb{T}_{\Bbbk}(M(P)) \subset \Gamma_{\Bbbk}(P)$, and since all generators of $M(P)$ are of degree 1, the symmetry group $\Sigma(P) = \Sigma(M(P))$ is contained in $\Gamma_{\Bbbk}(P)$, too. Moreover, the elementary automorphisms are homogeneous. As we will now see, together they generate $\Gamma_{\Bbbk}(P)$.

Theorem 5.45. *Let P be a convex lattice n-polytope and $\Bbbk$ a field. Then there exists an enumeration $F_1, \dots, F_s$ of the facets of $\mathbb{R}_+ M$ such that every element $\gamma \in \Gamma_{\Bbbk}(P)$ has a (not uniquely determined) presentation*

$$\gamma = \alpha_1 \alpha_2 \cdots \alpha_s \tau \sigma,$$

where $\sigma \in \Sigma(P)$, $\tau \in \mathbb{T}^{n+1}$, and $\alpha_i \in \mathrm{E}(F_i)$, $i = 1, \dots, s$.

We isolate the crucial step of the proof in the next lemma. But first we have to specify the enumeration of the facets. To this end we choose a number $g \in \mathbb{N}$ such that for each facet F of P there exists an element $(x, h) \in \mathbb{Z}^{n+1}$, $x \in hP$, with $\operatorname{ht}_F(x) = 1$ and $h \le g$. (Such g exists by Proposition 2.81.) Let N be a finitely generated graded module over $\Bbbk[\bar{P}]$. Then we set

$$v_g(N) = \sum_{i=0}^{g} \dim_{\Bbbk} N_i.$$

In the proof of the next lemma we will use the following properties of v_g:

(a) $v_g(\mathfrak{p}_F \cap \mathfrak{p}_G) < v_g(\mathfrak{p}_F)$ if F and G are different facets of P;

(b) $v_g(\mathfrak{p}_F^{(a)}) < v_g(\mathfrak{p}_F)$ if $a \ge 2$.

(Recall that $\mathfrak{p}_F^{(a)}$ is the ath symbolic power of $\mathfrak{p}_F$.) The first assertion holds for all $g \ge 1$ since $P \setminus (F \cup G)$ contains fewer lattice points than $P \setminus F$. The second holds for our choice of g since in the passage from $\mathfrak{p}_F$ to $\mathfrak{p}_F^{(a)}$ we lose at least one lattice point, namely (x, h) as above. For a graded ideal I of $\Bbbk[\bar{P}]$ and $\gamma \in \Gamma_{\Bbbk}(P)$ one has $v_g(\gamma(I)) = v_g(I)$ since the homogeneous automorphism γ preserves the $\Bbbk$-dimension of the graded components.

Lemma 5.46. *Let $\gamma \in \Gamma_{\Bbbk}(P)$, and choose an enumeration $F_1, \dots, F_s$ of the facets such that $v_g(\mathfrak{p}_{F_1}) \ge \cdots \ge v_g(\mathfrak{p}_{F_s})$. Then there exists a permutation π of $\{1, \dots, s\}$ such that for all i such that*

$$\alpha_s \cdots \alpha_1 \gamma(\mathfrak{p}_{F_i}) = \mathfrak{p}_{F_{\pi(i)}}$$

with suitable $\alpha_i \in \mathrm{E}(F_{\pi(i)})$.

In fact, this lemma implies Theorem 5.45: the resulting automorphism $\delta = \alpha_s\cdots\alpha_1\gamma$ permutes the divisorial prime ideals $\mathfrak{p}_F$. By virtue of Lemma 5.36 we then have $\delta \in \mathrm{Tm}_{\Bbbk}(\Bbbk[P])$. So $\delta = \sigma\tau$ with $\sigma \in \Sigma(P)$ and $\tau \in \mathbb{T}^{n+1}$. Finally one just replaces each α_i by its inverse and each F_i by $F_{\pi(i)}$.

Proof of Lemma 5.46. By Corollary 4.61 the divisorial ideal $\gamma(\mathfrak{p}_F) \subset \Bbbk[\bar{P}]$ is isomorphic to some monomial divisorial ideal Δ, i. e., there is an element $\kappa \in \mathrm{QF}(\Bbbk[\bar{P}])$ such that $\gamma(\mathfrak{p}_F) = \kappa\Delta$. The inclusion $\kappa \in (\gamma(\mathfrak{p}_F) : \Delta)$ shows that κ is a $\Bbbk$-linear combination of some Laurent monomials corresponding to lattice points in $\mathbb{Z}^{n+1}$. We factor out one of the terms of κ, say m, and rewrite the equality as follows:

$$\gamma(\mathfrak{p}_F) = (m^{-1}\kappa)(m\Delta).$$

Then $m^{-1}\kappa$ is of the form $1 + m_1 + \cdots + m_q$ for some Laurent terms $m_1, \ldots, m_q \notin \Bbbk$, while $m\Delta$ is necessarily a divisorial monomial ideal of $\Bbbk[\bar{P}]$ (since 1 belongs to the supporting monomial set of $m^{-1}\kappa$). Now γ is a *graded* automorphism. Hence $(1 + m_1 + \cdots + m_q)(m\Delta) \subset \Bbbk[\bar{P}]$ is a graded ideal. This implies that the terms $m_1, \ldots, m_q$ are of degree 0. Thus there is always a presentation

$$\gamma(\mathfrak{p}_F) = (1 + m_1 + \cdots + m_q)\Delta,$$

where $m_1, \ldots, m_q \notin \Bbbk^\times$ are Laurent terms of degree 0 and $\Delta \subset \Bbbk[\bar{P}]$ is a monomial ideal (we do not exclude the case $q = 0$). A representation of this type is called *admissible.*

For $\gamma \in \Gamma_{\Bbbk}(P)$ consider an admissible representation

$$\gamma(\mathfrak{p}_{F_1}) = (1 + m_1 + \cdots + m_q)\Delta.$$

One has $v_g(\Delta) = v_g(\mathfrak{p}_{F_1})$. Since $\Delta \subset \Bbbk[\bar{P}]$ is a divisorial monomial ideal, there are integers $a_i \geq 0$ such that

$$\Delta = \bigcap_{i=1}^{r} \mathfrak{p}_{F_i}^{(a_i)};$$

see Theorem 4.54. Now apply the function v_g. By its properties (a) and (b) stated above the lemma and because $v_g(\mathfrak{p}_{F_1})$ is maximal among the $v_g(\mathfrak{p}_{F_i})$ it follows that exactly one exponent is 1 and all the others are 0. So $\Delta = \mathfrak{p}_{G_1}$ for some facet G_1. By Lemmas 5.43 and 5.44 there exists $\alpha_1 \in \mathrm{E}(G_1)$ such that

$$\alpha_1\gamma(\mathfrak{p}_{F_1}) = \mathfrak{p}_{G_1}.$$

Now we proceed inductively. Let $1 \leq t < r$. Assume there are facets $G_1, \ldots, G_t$ of P and $\alpha_1 \in \mathrm{E}(G_1), \ldots, \alpha_t \in \mathrm{E}(G_t)$ such that

$$\alpha_t\cdots\alpha_1\gamma(\mathfrak{p}_{F_i}) = \mathfrak{p}_{G_i}, \qquad i = 1, \ldots, t.$$

(Observe that the G_i are automatically different and that $v_g(\mathfrak{p}_{G_i}) = v_g(\mathfrak{p}_{F_i})$.) In view of Lemma 5.43 we need to show there is a facet $G_{t+1} \subset P$, different from $G_1, \ldots, G_t$, and an element $\alpha_{t+1} \in \mathrm{E}(G_{t+1})$ such that

$$\alpha_{t+1}\alpha_t\cdots\alpha_1\gamma(\mathfrak{p}_{F_{t+1}}) = \mathfrak{p}_{G_{t+1}}.$$

For simplicity of notation we put $\gamma' = \alpha_t\cdots\alpha_1\gamma$. Again, consider an admissible representation

$$\gamma'(\mathfrak{p}_{F_{t+1}}) = (1 + m_1 + \cdots + m_q)\Delta.$$

Rewriting this equality in the form

$$\gamma'(\mathfrak{p}_{F_{t+1}}) = \big(m_j^{-1}(1 + m_1 + \cdots + m_q)\big)(m_j\Delta),$$

where $j \in \{0,\ldots,q\}$ and $m_0 = 1$, we get another admissible representation. We will show that by varying j we can obtain a monomial divisorial ideal $m_j\Delta$ such that, in the primary decomposition

$$m_j\Delta = \bigcap_{i=1}^{r} \mathfrak{p}_{F_i}^{(a_i)},$$

there appears a positive power of $\mathfrak{p}_G$ for some facet G different from $G_1,\ldots,G_t$. Then $v_g((m_j\Delta) \le v_g(\mathfrak{p}_G) \le v_g(\mathfrak{p}_{F_{t+1}})$ (due to our enumeration) and the inequality would be strict whenever $\sum_1^r a_i \ge 2$. Thus $m_j\Delta = \mathfrak{p}_G$ and we can proceed as we did for the ideal $\mathfrak{p}_{F_1}$.

Assume to the contrary that in the primary decompositions of all the monomial ideals $m_j\Delta$ there appear only the prime ideals $\mathfrak{p}_{G_1},\ldots,\mathfrak{p}_{G_t}$. We have

$$(1 + m_1 + \cdots + m_q)\Delta \subset \Delta + m_1\Delta + \cdots + m_q\Delta$$

and

$$[(1 + m_1 + \cdots + m_q)\Delta] = [\Delta] = [m_1\Delta] = \cdots = [m_q\Delta]$$

in $\operatorname{Cl}(\Bbbk[\bar{P}])$. Applying $(\gamma')^{-1}$ we arrive at the conclusion that $\mathfrak{p}_{F_{t+1}}$ is contained in a sum of *monomial* divisorial ideals $\Phi_0,\ldots,\Phi_q$, such that the primary decomposition of each of them involves only $\mathfrak{p}_{F_1},\ldots,\mathfrak{p}_{F_t}$. (This follows from the fact that $(\gamma')^{-1}$ maps $\mathfrak{p}_{G_i}$ to the monomial ideal $\mathfrak{p}_{F_i}$ for $i = 1,\ldots,t$; thus intersections of symbolic powers of $\mathfrak{p}_{G_1},\ldots,\mathfrak{p}_{G_t}$ are mapped to intersections of symbolic powers of $\mathfrak{p}_{F_1},\ldots,\mathfrak{p}_{F_t}$, which are automatically monomial.) Furthermore, $\mathfrak{p}_{F_{t+1}}$ has the same divisor class as each of the Φ_i.

Now, if $m \ne 1$ is a degree 0 monomial such that $\mathfrak{p}_{F_{t+1}} = m\Phi$ with $\Phi \subset \Bbbk[\bar{P}]$, then m^{-1} must be a column vector for $\mathfrak{p}_{F_{t+1}}$. Choose a lattice point $x \in P$ for which $\operatorname{ht}_{F_{t+1}}(x)$ is maximal. Since $\operatorname{ht}_{F_{t+1}}(m_i^{-1}y) = \operatorname{ht}_{F_{t+1}}(y) - 1 < \operatorname{ht}_{F_{t+1}}(x)$ for all lattice points $y \in P$, the ideal $\mathfrak{p}_{F_{t+1}}$ is not contained in the sum of the monomial ideals Φ_i, a contradiction. □

The structure of the graded automorphism group. In order to give more information on $\Gamma_\Bbbk(P)$ we have to introduce a subgroup of $\Sigma(P)$ defined as follows: assume v is an *invertible column vector*, i.e., $-v$ is also a column vector. Then for every point $x \in \operatorname{lat}(P)$ there is a unique $y \in \operatorname{lat}(P)$ such that $\operatorname{ht}_v(x) = \operatorname{ht}_{-v}(y)$ and $x - y$ is parallel to v. The mapping $x \mapsto y$ gives rise to a monoid automorphism of $M(P)$: it inverts columns that are parallel to v. It is easy to see that these automorphisms generate a normal subgroup of $\Sigma(P)$, which we denote by $\Sigma(P)_{\text{inv}}$.

Theorem 5.47. *Let P be a lattice n-polytope and let $\Bbbk$ be an algebraically closed field.*

(a) The connected component $\Gamma_{\Bbbk}(P)^0 \subset \Gamma_{\Bbbk}(P)$ is generated by the subgroups $\mathrm{E}(F_i)$ and $\mathbb{T}^{n+1}$.

(b) $\Gamma_{\Bbbk}(P)^0$ consists precisely of those graded automorphisms of $\Bbbk[P]$ which induce the identity map on the divisor class group of the normalization of $\Bbbk[P]$.

(c) $\dim \Gamma_{\Bbbk}(P) = \#\operatorname{Col}(P) + n + 1$.

(d) $\Gamma_{\Bbbk}(P)^0 \cap \Sigma(P) = \Sigma(P)_{\mathrm{inv}}$ and $\Gamma_{\Bbbk}(P)/\Gamma_{\Bbbk}(P)^0 \cong \Sigma(P)/\Sigma(P)_{\mathrm{inv}}$.

Proof. (a) Since $\mathbb{T}^{n+1}$ and the $\mathrm{E}(F_i)$ are connected groups they generate a connected subgroup H of $\Gamma_{\Bbbk}(P)$ (Proposition 5.1(b)). This subgroup acts trivially on $\operatorname{Cl}(\Bbbk[\bar{P}])$ by Lemma 5.43 and the fact that the classes of the $\mathfrak{p}_{F_i}$ generate the divisor class group. Furthermore, H has finite index in $\Gamma_{\Bbbk}(P)$ bounded by $\#\Sigma(P)$. Therefore $H = \Gamma_{\Bbbk}(P)^0$.

(b) Suppose $\gamma \in \Gamma_{\Bbbk}(P)$ acts trivially on $\operatorname{Cl}(\Bbbk[\bar{P}])$. We want to show that $\gamma \in H$. Let E denote the connected subgroup of $\Gamma_{\Bbbk}(P)$, generated by the elementary automorphisms. Any automorphism that preserves the divisorial ideals $\mathfrak{p}_{F_i}$ is automatically toric (Lemma 5.36). Therefore we have only to show that there is an element $\varepsilon \in E$ such that

$$\varepsilon\gamma(\mathfrak{p}_{F_i}) = \mathfrak{p}_{F_i}, \qquad i = 1, \dots, s.$$

From Lemma 5.46 we know that there are $\varepsilon_1 \in E$ and a permutation π of $\{1, \dots, s\}$ such that

$$\varepsilon_1\gamma(\mathfrak{p}_j) = \mathfrak{p}_{\pi(j)}, \qquad j = 1, \dots, s. \tag{5.6}$$

Since ε_1 and γ both act trivially on $\operatorname{Cl}(\Bbbk[\bar{P}])$, we get

$$\mathfrak{p}_{\pi(j)} = m_j\mathfrak{p}_j, \qquad j = 1, \dots, s,$$

for some monomials m_j of degree 0.

By Lemma 5.44 we conclude that if $m_j \neq 1$ (in additive notation, $m_j \neq 0$), then both m_j and m_j^{-1} are column vectors with the base facets F_j and $F_{\pi(j)}$, respectively. Observe that the automorphism

$$\varepsilon_{\pi(j)} = e^1_{m_j} e^{-1}_{m_j^{-1}} e^1_{m_j} \in E$$

interchanges the ideals $\mathfrak{p}_j$ and $\mathfrak{p}_{\pi(j)}$, provided $m_j \neq 1$. Now we can complete the proof by successively "correcting" the equations (5.6).

(c) In the enumeration of the facets used for Theorem 5.45 we have the surjective algebraic map

$$\mathrm{E}(F_1) \times \cdots \times \mathrm{E}(F_r) \times \mathbb{T}^{n+1} \times \Sigma(P) \to \Gamma_{\Bbbk}(P)$$

induced by composition. The left hand side has dimension $\#\operatorname{Col}(P) + n + 1$ (Lemma 5.42). Hence $\dim \Gamma_{\Bbbk}(P) \le \#\operatorname{Col}(P) + n + 1$.

To derive the opposite inequality we can additionally assume that P contains an interior lattice point. Indeed, Remark 5.38(a) and Theorem 5.45 show that the natural group homomorphism $\Gamma_{\Bbbk}(P) = \Gamma_{\Bbbk}(\Bbbk[\bar{P}]) \to \Gamma_{\Bbbk}(cP)$, induced by restriction, is surjective for every $c \in \mathbb{N}$. (The surjectivity of $\mathbb{T}_{\Bbbk}(M(P)) \to \mathbb{T}_{\Bbbk}(M(cP))$ follows from the fact that $\Bbbk$ is closed under taking roots). So we can work with cP, which contains an interior point provided c is large.

Let $x \in P$ be an interior lattice point and let $v_1, \dots, v_q$ be different column vectors. Then the supporting monomial set of $e_{v_i}^{\lambda_i}(x)$, $\lambda \in \Bbbk^\times$, is not contained in the union of those of $e_{v_j}^{\lambda_j}(x)$, $j \neq i$ (just look at the projections of x through v_i into the corresponding base facets). This shows that we have $\#\operatorname{Col}(P)$ linearly independent tangent vectors of $\Gamma_\Bbbk(P)$ at $1 \in \Gamma_\Bbbk(P)$. Since the tangent vectors corresponding to the elements of $\mathbb{T}^{n+1}$ clearly belong to a complementary subspace and $\Gamma_\Bbbk(P)$ is a smooth variety.[2]

(d) Assume v and v^{-1} both are column vectors. Then the automorphism

$$\varepsilon = e_v^1 e_{v^{-1}}^{-1} e_v^1 \in \Gamma_\Bbbk(P)^0$$

maps monomials to terms; more precisely, ε inverts up to scalars the columns parallel to v: every $x \in \bar{M}(P)$ is sent either to the appropriate $y \in \bar{M}(P)$ or to $-y \in \Bbbk[\bar{P}]$ (Exercise 5.25). Then it is clear that there is an element $\tau \in \mathbb{T}^{n+1}$ such that $\tau\varepsilon$ is a generator of $\Sigma(P)_{\mathrm{inv}}$. Hence $\Sigma(P)_{\mathrm{inv}} \subset \Gamma_\Bbbk(P)^0$.

Conversely, if $\sigma \in \Sigma(P) \cap \Gamma_\Bbbk(P)^0$ then σ induces the identity map on $\operatorname{Cl}(\Bbbk[\bar{P}])$. Hence $\sigma(\mathfrak{p}_{F_i}) = m_j\mathfrak{p}_{F_i}$ for some monomials m_j, and the very same arguments we have used in the proof of (b) show that $\sigma \in \Sigma(P)_{\mathrm{inv}}$. Thus $\Gamma_\Bbbk(P)/\Gamma_\Bbbk(P)^0 = \Sigma(P)/\Sigma(P)_{\mathrm{inv}}$. □

Observe that we do not claim the existence of a normal form as in Theorem 5.45 for the elements of $\Gamma_\Bbbk(P)^0$ if $\Sigma(P)_{\mathrm{inv}}$ is excluded from the generating set. In fact, in general such a normal form does not exist (Exercise 5.24).

Remark 5.48. The graded retractions, i. e., idempotent endomorphisms, of the rings $\Bbbk[P]$ have been considered in [63]. At least the retractions ρ with $\dim \rho(\Bbbk[P]) = \dim \Bbbk[P]-1$ can be classified. For a discussion of graded homomorphisms of polytopal rings see also [62].

Normal affine monoids. While we have introduced the notion of column vector only for polytopes P, the proofs of Theorems 5.45 and 5.47 uses the data of the normalization $\Bbbk[\bar{P}]$ of the monoid ring $\Bbbk[P]$. It is not difficult to see that both theorems can be generalized from the polytopal to the normal case. (In view of Lemma 5.39 this change is indeed a generalization.)

Let M be an arbitrary affine monoid with a positive grading denoted by deg (and extended to $\operatorname{gp}(M)$). Then $v \in \operatorname{gp}(M)$, $v \neq 0$, is a *column vector* of M (with respect to deg) if $\deg v = 0$ and there exists a facet F of $\mathbb{R}_+M$ such that $x + v \in M$ for all $x \in M$, $x \notin M \setminus F$. It follows immediately from Lemma 5.41 that this general notion of column vector is consistent with the polytopal case. Moreover, if v is a column vector for M, then it is also a column vector for $\bar{M}$ (but if M is not polytopal, the converse does not hold in general).

Using the same formula as in the polytopal case one now defines the elementary automorphisms $e_v^\lambda \in \Gamma_\Bbbk(M) = \Gamma_\Bbbk(\Bbbk[M])$. Lemma 5.42 remains true without changes, and Lemma 5.44 remains true if one considers column vectors of $\bar{M}$.

After these generalizations one obtains the following:

[2] A rigorous argument invokes the concepts of semisimple and unipotent elements in linear groups.

Theorem 5.49. *Let M be a normal affine monoid with a positive grading, and $\Bbbk$ a field. Then there exists an enumeration $F_1, \dots, F_s$ of the facets of $\mathbb{R}_+M$ such that every element $\gamma \in \Gamma_\Bbbk(M)$ has a (not uniquely determined) presentation*

$$\gamma = \alpha_1\alpha_2\cdots\alpha_s\tau\sigma,$$

where $\sigma \in \Sigma(M)$, $\tau \in \mathbb{T}_\Bbbk(M)$, and $\alpha_i \in \mathrm{E}(F_i)$. Moreover, Theorem 5.47 holds accordingly.

A determinantal application. As an application to rings and varieties outside the class of monoid algebras we determine the groups of graded automorphisms of the determinantal rings, a result which goes back to Frobenius [129, p. 99] and has been reproved many times since then. See, for instance, [373] for a group-scheme theoretical approach for general commutative rings of coefficients, covering also the classes of generic symmetric and alternating matrices. (The generic symmetric case can be done by the same method as the one below.)

In plain terms, Corollary 5.50 answers the following question: let $\Bbbk$ be an infinite field, $\varphi : \Bbbk^{mn} \to \Bbbk^{mn}$ a $\Bbbk$-automorphism of the vector space $\Bbbk^{mn}$ of $m \times n$ matrices over $\Bbbk$, and r an integer, $1 \le r < \min(m,n)$; when is $\operatorname{rank}\varphi(A) \le r$ for all $A \in \Bbbk^{mn}$ with $\operatorname{rank} A \le r$? This holds obviously for transformations $\varphi(A) = SAT^{-1}$ with $S \in \mathrm{GL}_m(\Bbbk)$ and $T \in \mathrm{GL}_n(\Bbbk)$, and for the transposition if $m = n$. Indeed, these are the only such transformations. For simplicity we have stated the corollary only for algebraically closed fields:

Corollary 5.50. *Let $\Bbbk$ be an algebraically closed field, X an $m \times n$ matrix of indeterminates, and let $R = \Bbbk[X]/I_{r+1}(X)$ be the residue class ring of the polynomial ring $\Bbbk[X]$ in the entries of X modulo the ideal generated by the $(r+1)$-minors of X, $1 \le r < \min(m,n)$.*

(a) The connected component $G^0 \subset \Gamma_\Bbbk(R)$ is the image of $\mathrm{GL}_m(\Bbbk) \times \mathrm{GL}_n(\Bbbk)$ in $\mathrm{GL}_{mn}(\Bbbk)$ under the map $(S,T) \mapsto (A \mapsto SAT^{-1})$, and is isomorphic to $\mathrm{GL}_m(\Bbbk) \times \mathrm{GL}_n(\Bbbk)/\Bbbk^\times$ where $\Bbbk^\times$ is embedded diagonally.
(b) If $m \neq n$, the group G is connected, and if $m = n$, then G^0 has index 2 in $\Gamma_\Bbbk(R)$, and $\Gamma_\Bbbk(R) = G^0 \cup \tau G^0$ where τ is the transposition.

Proof. The singular locus of $\operatorname{Spec} R$ is given by $V(\mathfrak{p})$ where $\mathfrak{p} = I_r(X)/I_{r+1}(X)$; $\mathfrak{p}$ is a prime ideal in R (see Bruns and Vetter [80, (2.6), (6.3)]). It follows that every automorphism of R must map $\mathfrak{p}$ onto itself. Thus a linear substitution on $\Bbbk[X]$ for which $I_{r+1}(X)$ is stable also leaves $I_r(X)$ invariant and therefore induces an automorphism of $\Bbbk[X]/I_r(X)$. This argument reduces the corollary to the case $r = 1$.

For $r = 1$ one has the isomorphism

$$R \to \Bbbk[Y_iZ_j : i = 1,\dots,m,\ j = 1,\dots,n] \subset \Bbbk[Y_1,\dots,Y_m,Z_1,\dots,Z_n]$$

induced by the assignment $X_{ij} \mapsto Y_iZ_j$. Thus R is just the Segre product of $\Bbbk[Y_1,\dots, Y_m]$ and $\Bbbk[Z_1,\dots,Z_n]$, or, equivalently, $R \cong \Bbbk[P]$ where P is the direct product of the unit simplices Δ_{m-1} and Δ_{n-1}. (Compare Exercises 2.29 and 4.22.) Part (a) follows now from an analysis of the column structures of P (see Remark 5.38(b)) and the torus actions.

For (b) one observes that $\mathrm{Cl}(R) \cong \mathbb{Z}$; ideals representing the divisor classes 1 and -1 are given by $(Y_1Z_1, \ldots, Y_1Z_n)$ and $(Y_1Z_1, \ldots, Y_mZ_1)$ [80, 8.4]. If $m \neq n$ these ideals have different numbers of generators; therefore every automorphism of R acts trivially on the divisor class group. In the case $m = n$, the transposition induces the map $s \mapsto -s$ on $\mathrm{Cl}(R)$. Now the rest follows again from the theorem above. (Instead of the divisorial arguments one could also discuss the symmetry group of $\Delta_{m-1} \times \Delta_{n-1}$.) □

Exercises

In the following, $\Bbbk$ denotes an algebraically closed field whenever an exercise uses notions that we have introduced only for such fields.

5.1. Show that the characters of an algebraic group G over $\Bbbk$ form a linearly independent set of functions from G to $\Bbbk$.

5.2. Show that the monomials in $\Bbbk[X_1^{\pm 1}, \ldots, X_n^{\pm 1}]$ are the characters of $\mathbb{T}^n(\Bbbk)$.

5.3. Let $G \subset \Bbbk^\times$ be a finite group. Show that the character group $X(G)$ is isomorphic to G.

5.4. Prove Corollary 5.4. Hint: if G is a closed subgroup of G', then the natural morphism $\mathcal{O}(G') \to \mathcal{O}(G)$ is surjective.

5.5. Show that the assignments $G \mapsto X(G)$ and $B \mapsto \mathrm{Hom}(B, \Bbbk^\times)$ induce an equivalence between the category of diagonalizable groups and the category of finitely generated abelian groups without p-torsion, $p = \operatorname{char} \Bbbk$.

5.6. Let D be a diagonalizable group. Show that the natural pairing $D \times X(D) \to \Bbbk^\times$, induces an isomorphism $D \cong \mathrm{Hom}(X(D), \Bbbk^\times)$.

5.7. Let M be a normal monoid of monomials in $R = \Bbbk[X_1, \ldots, X_n]$ where $\Bbbk$ is an algebraically closed field of characteristic $p > 0$. Set $M' = \{x \in R : x^{p^e} \in M \text{ for some } e \in \mathbb{N}\}$. Show that $\Bbbk[M']$ is the smallest subalgebra of R that contains $\Bbbk[M]$ and is of type R^D for some subgroup $D \subset \mathbb{T}^n(\Bbbk)$.

5.8. (a) Let $R = \Bbbk[X_1, \ldots, X_n]$. Represent the Veronese subalgebras $\Bbbk[R_k]$ as rings of invariants of R if $\operatorname{char} \Bbbk = 0$. ($R_k$ denotes the vector space spanned by the monomials of total degree k.)
(b) Represent $\Bbbk[R_k]$ as a ring of invariants of a torus action in arbitrary characteristic.

5.9. Let the torus $\mathbb{T}^1(\Bbbk)$ act on $\Bbbk[Y_1, \ldots, Y_m, Z_1, \ldots, Z_n]$ by $Y_i \mapsto \xi Y_i$, $i = 1, \ldots, m$, and $Z_j \mapsto \xi^{-1} Z_j$, $j = 1, \ldots, n$. Compute the ring S of invariants and all weight spaces. Find minimal systems of generators of the weight spaces as S-modules.

5.10. Give an example of an action of $\mathbb{T}^1(\Bbbk)$ on $\Bbbk[X_1, X_2]$ whose weight spaces do not all have rank 1. (This is very easy.)

5.11. Prove Lemma 5.12 in the following generalized form. Let A be a $\Bbbk$-algebra, and Suppose that A is generated over $\Bbbk$ by a finite-dimensional $\Bbbk$-vector subspace V. Then the group $\Gamma_\Bbbk(A, V)$ of $\Bbbk$-algebra automorphisms $\gamma : A \to A$ with $\gamma(V) = V$ is a linear group.

5.12. Let Δ be an abstract simplicial complex on the vertex set $V = \{v_1, \ldots, v_n\}$. For a field $\Bbbk$ consider the exterior algebra $E = \Bbbk\langle e_1, \ldots, e_n\rangle$, and let $J(\Delta)$ be the ideal generated by all monomials $e_{i_1} \wedge \cdots \wedge e_{i_s}$ such that $\{v_{i_1}, \ldots, v_{i_s}\} \notin \Delta$. One calls $E/J(\Delta)$ the *exterior face ring of* Δ.

Formulate and prove the analogue of the statements in Example 5.28 for the exterior face ring.

5.13. Let R be a ring, M and N affine monoids, and suppose that $\varphi : R[M] \to R[N]$ is a ring isomorphism. Show that the restriction of φ to R is an automorphism of R in the following cases:

(a) R is a field;

(b) R is reduced and an integral extension of its subring $\mathbb{Z}1_R$;

(c) M is positive, R is reduced and integral over the subring generated by its units.

5.14. Give an example of an artinian ring R with a nontrivial embedding $R \to R[X]$.

5.15. Find the missing details in the proof of Lemma 5.33.

5.16. (a) Let $f \in \Bbbk[X_1, \ldots, X_n]$ be an irreducible polynomial, not associated to any of the X_i. Show that f has infinitely many pairwise nonassociated images under the action of $\mathbb{T}^n(\Bbbk)$.
(b) Let A be a finitely generated $\Bbbk$-subalgebra of $\operatorname{QF}(\Bbbk[X_1, \ldots, X_n])$ that is closed under the action of $\mathbb{T}^n(\Bbbk)$. Show $A \subset \Bbbk[X_1^{\pm 1}, \ldots, X_n^{\pm 1}]$.

5.17. Let $\mathscr{V}$ be an affine variety over an algebraically closed field $\Bbbk$, containing the torus $\mathbb{T}^n(\Bbbk)$ as an Zariski open subset. Suppose the group structure of $\mathbb{T}^n(\Bbbk)$ extends to an algebraic group action $\mathbb{T}^n(\Bbbk) \times \mathscr{V} \to \mathscr{V}$. Show that $\mathscr{O}(\mathscr{V}) \cong \Bbbk[M]$ for an affine monoid M of rank n.
Hint: use the fact that $\mathscr{O}(\mathbb{T}^n(\Bbbk)) = \Bbbk[X_1^{\pm 1}, \ldots, X_n^{\pm 1}]$ is a localization of the affine algebra $\mathscr{O}(\mathscr{V})$, and the latter is stable under the extension of the toric automorphisms of $\mathbb{T}^n(\Bbbk)$.
This gives an alternative description of affine monomial algebras and, more importantly, indicates what the right nonaffine analogues of the varieties associated with affine monoid algebras should be. These so-called toric varieties will be discussed in Chapter 10.

5.18. Let M be an affine monoid, $A = \Bbbk[M]$, and $\gamma \in G_{\Bbbk}(A)$. Show γ is toric if and only if $\gamma(\mathfrak{p}_F) = \mathfrak{p}_F$ for all facets F of $\mathbb{R}_+ M$.

5.19. Let M be an affine monoid and $\Bbbk$ an arbitrary field.

(a) Prove Proposition 5.35.

(b) Let $A = \Bbbk[\mathbb{Z}^n]$. Show that every automorphism of A is termic and that $G_{\Bbbk}(A)$ is a semidirect product of $\mathbb{T}^n$ and $\operatorname{GL}_n(\mathbb{Z})$.

5.20. Let $P \subset \mathbb{R}^n$ be a lattice polytope for which $\operatorname{lat}(P)$ generates $\mathbb{Z}^n$ affinely. Furthermore, let v be a column vector of P. If $x = y - mv$ for $x \in \operatorname{lat}(P)$ and $y \in P_v$, show $m = \operatorname{ht}_v(x)$.

5.21. Prove the statements in Remark 5.38.

5.22. Let P be a lattice polytope, $v \in \operatorname{Col}(P)$, and let G be a facet with $\operatorname{ht}_G(v) > 0$. Prove: if there exists a column vector w with base facet G and $\operatorname{ht}_F(w) > 0$, then $w = -v$.

5.23. Using the arguments in the proof of Lemma 5.46, show that a column vector v with base facet F is invertible if $v_g(\mathfrak{p}_F) \geq v_g(\mathfrak{p}_G)$ for all facets G.

5.24. Let γ be the automorphism of $\Bbbk[X_1, X_2]$ exchanging X_1 and X_2. Show that γ cannot be represented by a product $\varepsilon_1 \varepsilon_2 \tau$ where $\varepsilon_1, \varepsilon_2$ are elementary and τ is toric.

5.25. Show that the automorphism

$$\varepsilon = e_v^1 e_{v^{-1}}^{-1} e_v^1 \in \Gamma_{\Bbbk}(P)^0,$$

appearing in the proof of Theorem 5.47, inverts up to scalars the columns parallel to v: every $x \in \bar{M}(P)$ is sent either to the appropriate $y \in \bar{M}(P)$ or to $-y \in \Bbbk[\bar{P}]$. Proceed as follows:
(a) Reduce the claim to a single column, i. e., to the polytopal algebra $\Bbbk\big[[0,m]\big]$ of a lattice line segment.
(b) Show that the algebra in (a) is isomorphic to $\Bbbk[X^m, X^{m-1}Y, \dots, XY^{m-1}, Y^m]$.
(c) Now reduce the claim to $\Bbbk[X, Y]$.

5.26. Let $A = \Bbbk[X, Y, Z]$ and $n \in \mathbb{N}$, $n \geq 3$. Find a grading of A for which $\dim \Gamma_{\Bbbk}(A) = n$.

5.27. Let $\Bbbk$ be a field and $\Gamma_{\Bbbk}(\mathbb{R}^n_+)$ be the group of graded automorphisms of the *infinite polytopal* $\Bbbk$*-algebra* $\Bbbk[\mathbb{R}^n_+]$ whose underlying "polytope" is the infinite polyhedron $\mathbb{R}^n_+$, defined in a natural way (work out the details!). Let 0 denote the origin of $\mathbb{R}^n_+$ and e_i denote the ith standard basis element.
(a) Prove: for every element $\xi = (\xi_0, \xi_1, \dots, \xi_n) \in \mathbb{T}^{n+1}(\Bbbk)$ the assignments

$$(0,1) \mapsto \xi_0(0,1),$$
$$(e_i,1) \mapsto \xi_i(e_i,1), \quad i = 1, \dots, n$$

extend in a unique way to an element of $\Gamma_k(\mathbb{R}^n_+)$. (Vectors denote their associated monomials!)
(b) The action of $\mathbb{T}^{n+1}(\Bbbk)$ on $\Bbbk[\mathbb{R}^n_+]$ introduced in (a) defines a group embedding $\mathbb{T}^n(\Bbbk) \to \Gamma_{\Bbbk}(\mathbb{R}^n_+)$. Show that the image is a maximal torus.
(c) Identifying $\Bbbk^\times$ with the image in $\Gamma_{\Bbbk}(\mathbb{R}^n_+)$ of the first coordinate subgroup of $\mathbb{T}^n(\Bbbk)$, show that $\Gamma_{\Bbbk}(\mathbb{R}^n_+)/\Bbbk^\times$ is isomorphic to the group of augmented $\Bbbk$-algebra automorphisms of the polynomial algebra $\Bbbk[X_1, \dots, X_n]$ (i. e., those leaving the ideal $(X_1, \dots, X_n)$ invariant).
(d) Introduce the notion of a column vector for $\mathbb{R}^n_+$ in exact analogy to Definition 5.37 with the following extra condition: if a column vector v is associated with a facet $F \subset \mathbb{R}^n_+$, then for every lattice point $x \in \mathbb{R}^n_+ \setminus F$ there is a natural number k for which $c + vk \in F$. (This condition is automatic for polytopes!) Then a pair of elements $v \in \operatorname{Col}(\mathbb{R}^n_+)$ and $\lambda \in \Bbbk$ defines an element $e_v^\lambda \in \Gamma_k(\mathbb{R}^n_+)$, called an elementary automorphism. Moreover, the exact analogue of Lemma 5.42 is true. (Hint: use the same argument as in the polytopal case.)
(e) The subgroup of $\Gamma_{\Bbbk}(\mathbb{R}^n_+)$, generated by the elementary automorphisms, intersects $\Bbbk^\times$ trivially (as introduced in (c)). As a consequence, this subgroup is isomorphic to a subgroup of the group of augmented $\Bbbk$-automorphisms of $\Bbbk[X_1, \dots, X_n]$, and we get the notion of an elementary automorphism of $\Bbbk[X_1, \dots, X_n]$.
(f) Any automorphism of $\Bbbk[X_1, \dots, X_n]$ of the form

$$X_j \mapsto X_j, \quad j \neq i,$$
$$X_i \mapsto X_i + g_{i-1}, \quad g_{i-1} \in \Bbbk[X_1, \dots, X_{i-1}], \quad g_{i-1}(0, \dots, 0) = 0,$$

is a composition of elementary automorphisms.
(g) The subgroup of the group of augmented $\Bbbk$-automorphisms of $\Bbbk[X_1, \dots, X_n]$, generated by the elementary automorphisms, is exactly the group of *tame* automorphisms of $\Bbbk[X, \dots, X_n]$. Here "tame" means to be a composition of linear changes of the variables and automorphisms of the form

$$X_1 \mapsto X_1, \; X_2 \mapsto X_2 + f_1, \dots, X_n \mapsto X_n + f_{n-1}$$

where $f_i \in \Bbbk[X_1, \dots, X_i]$. The latter are called *triangular* automorphisms. (Hint: show that any triangular automorphism is a composite of automorphisms of the form in (f). Then show that every elementary automorphism is tame.)

However, the analogue of Theorem 5.45 fails for $\Gamma_{\Bbbk}(\mathbb{R}_+^n)$. In other words, there are (augmented) automorphisms of $\Bbbk[X_1,\dots,X_n]$ that are not tame, contrary to what was known as the *tame generation conjecture*. This conjecture was finally disproved by Shestakov and Umirbaev [320] using Nagata's candidate for a wild automorphism.

Notes

In 1982 Demushkin claimed [108] to have proved the isomorphism theorem for normal affine monoids. He attributes the crucial idea of using Borel's theorem on maximal tori to Danilov. However, there were serious gaps in Demushkin's approach, and it needed several substantial changes. This was done in [163] with an extension of the result to not necessarily normal monoids.

Campillo, Grabowski, and Müller [81] obtained results, related to the isomorphisms and automorphisms of affine monoid rings. Two special cases of the isomorphism theorem proved relevant in topological applications: Theorem 5.27 was used by Farber, Hausman, and Dirk [120] to address Kevin Walker's conjecture in topological robotics; Sabalka [309] used the combinatorial invariance of the exterior face ring (Exercise 5.12) in the isomorphism problem for tree braid groups. González Pérez and Gonzalez–Sprinberg [143] use Theorem 5.22 to attach analytically invariant monoids to the germs of not necessarily normal affine toric varieties at their closed orbits, as well as to quasi-ordinary hypersurface singularities.

The column vectors for a lattice polytope P are equivalent to Demazure's roots [107] of the normal fan $\mathcal{N}(P)$, provided the projective variety $\operatorname{Proj}(\Bbbk[P])$ is smooth. The relationship between Theorem 5.47 and Demazure's result in [107] is that the former leads to a description of the automorphism group an arbitrary projective toric variety while the latter considers the automorphism group of general smooth complete toric varieties. All this will be explained in Chapter 10. Theorems 5.45 and 5.47 were extended to lattice polytopal complexes in [60], where the automorphism groups of the corresponding arrangements of projective toric varieties are also studied.

The combinatorial description of the dimension of $\Gamma_{\Bbbk}(P)$, given in Theorem 5.47(c), was used by Castryck and Voight [82] to compute the dimension of the locus $\mathcal{M}_g^{\mathrm{nd}}$ of nondegenerate curves inside the moduli space $\mathcal{M}_g$ of curves of genus g. Here nondegeneracy of a curve is understood as a smoothness condition along the edges of the Newton polygon of the associated Laurent polynomial $f \in \Bbbk[X^{\pm 1}, Y^{\pm 1}]$.

The graded automorphism groups of polytopal monoid rings are polytopal extensions of $\mathrm{GL}_n(\Bbbk)$ where elementary automorphisms correspond to elementary matrices. So we have polytopal extensions of the building blocks of the classical Bass–Whitehead group K_1 (the latter is formally introduced in Chapter 9). The analogy can be pushed further in both K-theoretical directions: (i) the idempotent endomorphisms of polytopal monoid rings [63] (this corresponds to K_0), and (ii) higher syzygies between elementary automorphisms [61], [67] (this corresponds to higher K-groups). The resulting higher polyhedral K-groups are modeled after Quillen's + construction [295].

6

Homological properties and Hilbert functions

Objects of combinatorial geometry define affine monoids in a natural way, as we have seen in previous chapters, and affine monoids give rise to monoid algebras. The Hilbert functions of such algebras therefore count solutions to linear diophantine systems of equations and inequalities. This connection, developed by Stanley about 30 years ago, is still one of the most beautiful applications of commutative algebra to combinatorics.

The rationality of Hilbert series is already a powerful result, but some of its refined properties reflect homological conditions for the rings and modules under consideration. The most remarkable phenomenon is the reciprocity between a Cohen–Macaulay ring and its canonical module, in algebraic as well as in combinatorial terms.

In the first four sections of this chapter we develop some notions of homological algebra over graded rings and the theory of Hilbert functions. While we will apply them mainly to affine monoid algebras, they are of importance for a much larger class of graded rings. Therefore monoid algebras and polytopes will play a less prominent role for a while.

In the last section we discuss divisorial ideals over normal affine monoid rings from the combinatorial and homological viewpoint.

In this chapter we assume that the reader is familiar with elementary homological algebra, namely the functors Ext and Tor and their basic properties.

6.A Cohen–Macaulay rings

A central notion of homological commutative algebra is that of Cohen–Macaulay ring and module. It plays an important role in various parts of commutative algebra, and especially in its combinatorially influenced areas.

We refer the reader to Bruns and Herzog [72] for the systematic study of regular sequences and the Cohen–Macaulay property, and restrict ourselves to recalling the most important notions and facts.

W. Bruns, J. Gubeladze, *Polytopes, Rings, and K-Thoery*, Springer Monographs in Mathematics,
DOI 10.1007/b105283_6, © Springer Science+Business Media, LLC 2009

Regular sequences and systems of parameters. Let R be a ring and M an R-module. We say that $x_1, \dots, x_n \in R$ form a *(regular) M-sequence* if

(i) $M/(x_1, \dots, x_n)M \neq 0$;
(ii) x_i is a non-zerodivisor on $M/(x_1, \dots, x_{i-1})M$ for $i = 1, \dots, n$.

In dealing with regular sequences it is best to directly relate them to homological invariants. We quote from [72, 1.2.5 and 1.2.10]:

Theorem 6.1 (Rees). *Let R be a noetherian ring, I and ideal and M a finitely generated R-module. Then all the maximal M-sequences have the same finite length, given by*

$$\operatorname{grade}(I, M) = \min\{k : \operatorname{Ext}_R^k(R/I, M) \neq 0\}.$$

More generally, if N is a finitely generated R-module with $\operatorname{Supp} N = \{\mathfrak{p} : I \subset \mathfrak{p}\}$, *then*

$$\operatorname{grade}(I, M) = \min\{k : \operatorname{Ext}_R^k(N, M) \neq 0\}.$$

One basis for the proof of the proposition is *prime avoidance* [72, 1.2.2]: if an ideal J is not contained in any of finitely many prime ideals $\mathfrak{p}_1, \dots, \mathfrak{p}_n$, then there exists $x \in I$, $x \notin \mathfrak{p}_1 \cup \dots \cup \mathfrak{p}_n$. In the presence of a grading one is of course interested in choosing x as a homogeneous element. This is not always possible, but can be done in an important special case:

Lemma 6.2. *Let R be a $\mathbb{Z}$-graded ring and I an ideal. Let $\mathfrak{p}_1, \dots, \mathfrak{p}_n$ be prime ideals such that $I \not\subset \mathfrak{p}_i$ for $i = 1, \dots, n$.*

(a) If I is generated by elements of positive degree, then there exists a homogeneous element $x \in I$, $x \notin \mathfrak{p}_1 \cup \dots \cup \mathfrak{p}_n$.
(b) If R contains an infinite field and I is generated by elements of constant degree g, then x can be chosen of degree g.

Part (a) is [72, 1.5.10], and part (b) follows from the simple fact that a vector space over an infinite field cannot be the union of finitely many proper subspaces.

If R is a local noetherian ring of Krull dimension d and with maximal ideal $\mathfrak{m}$, then elements $x_1, \dots, x_d \in \mathfrak{m}$ are called a *system of parameters* if $\mathfrak{m} = \operatorname{Rad}(x_1, \dots, x_d)$, or equivalently, $\dim R/(x_1, \dots, x_d) = 0$. A *system of parameters* of a finitely generated R-module M is a sequence $x_1, \dots, x_e \in \mathfrak{m}$, $e = \dim M$, such that $\dim M/(x_1, \dots, x_e)M = 0$, or equivalently, the residue classes of $x_1, \dots, x_e$ form a system of parameters in $R/\operatorname{Ann} M$.

The analogue in graded rings is a *homogeneous system of parameters*. Let $R = \bigoplus_{i=0}^{\infty} R_i$ be a finitely generated positively graded algebra over a field $\Bbbk = R_0$, and set $\mathfrak{m} = \bigoplus_{i>0}$. Then homogeneous elements form a homogeneous system of parameters if $\mathfrak{m} = \operatorname{Rad}(x_1, \dots, x_d)$. The basic properties and the existence of such systems are given by [72, 1.5.17]:

Theorem 6.3. *Let $\Bbbk$ be a field and R a finitely generated positively $\mathbb{Z}$-graded $\Bbbk$-algebra of Krull dimension d.*

(a) The following are equivalent for homogeneous elements $x_1, \dots, x_d$:

(i) $x_1, \dots, x_d$ *is a homogeneous system of parameters;*
(ii) R *is an integral extension of* $\Bbbk[x_1, \dots, x_d]$;
(iii) R *is a finitely generated* $\Bbbk[x_1, \dots, x_d]$*-module.*

(b) *There exist homogeneous elements* $x_1, \dots, x_d$ *satisfying one, and therefore all, of the conditions in (a). Moreover, such elements are algebraically independent over* $\Bbbk$.

(c) *If* R *is generated by elements of degree 1 and* $\Bbbk$ *is infinite, then such* $x_1, \dots, x_d$ *can be chosen to be of degree 1.*

If $x_1, \dots, x_d$ is a homogeneous system of parameters, then $\Bbbk[x_1, \dots, x_d]$ is called a *graded Noether normalization* of R. In other words, a graded Noether normalization is a graded subalgebra S, isomorphic to a polynomial ring over $\Bbbk$ such that R is a finitely generated S-module. The existence of (graded) Noether normalizations makes it possible to treat finitely generated algebras over $\Bbbk$ as modules over polynomial rings, a fact we will use several times in the following.

More generally, suppose M is a finitely generated graded R-module of Krull dimension d. Then $x_1, \dots, x_d$ form a *homogeneous system of parameters for* M if $\operatorname{Supp} M/(x_1, \dots, x_d)M = \{\mathfrak{m}\}$. Equivalently one can require that the residue classes form a homogeneous system of parameters in $R/\operatorname{Ann} M$. This shows that $x_1, \dots, x_d$ are again algebraically independent and that M is a finitely generated module over $\Bbbk[x_1, \dots, x_d]$.

As a consequence of prime avoidance one obtains [72, 1.5.11 and 1.5.12]:

Proposition 6.4. *Let* R *be a Noetherian* $\mathbb{Z}$*-graded ring, and let* I *be an ideal in* R. *Set* $h = \operatorname{ht} I$ *and* $g = \operatorname{grade}(I, M)$ *where* M *is a finitely generated* R*-module.*

(a) *If* I *is generated by homogeneous elements of positive degree, then there exist sequences* $x_1, \dots, x_h$ *and* $y_1, \dots, y_g$ *of homogeneous elements of* I *such that* $\operatorname{ht}(x_1, \dots, x_i) = i$ *for* $i = 1, \dots, h$, *and* $y_1, \dots, y_g$ *is an* M*-sequence.*

(b) *If* R *contains an infinite field and* I *is generated by elements of constant degree* g, *then the* x_i *and* y_j *can be chosen of degree* g.

Depth and Cohen–Macaulay modules. Suppose that R is noetherian local with maximal ideal $\mathfrak{m}$, and M is finitely generated. Then $\operatorname{depth} M$ denotes the length of a maximal M-sequence in $\mathfrak{m}$. This is a well-defined invariant since all maximal M-sequences have the same length, as has already been stated in Theorem 6.1.

Proposition 6.5. *Let* R *be a noetherian local ring with maximal ideal* $\mathfrak{m}$, *and* M *a nonzero* R*-module.*

(a) *Then* $\operatorname{depth} M = \min\{k : \operatorname{Ext}_R^k(R/\mathfrak{m}, M) \neq 0\}$.

(b) *If* $x_1, \dots, x_u$ *is an* M*-sequence, then*

$$\operatorname{depth} M/(x_1, \dots, x_u)M = \operatorname{depth} M - u,$$
$$\dim M/(x_1, \dots, x_u)M = \dim M - u.$$

One calls M *Cohen–Macaulay* if $\operatorname{depth} M = \dim M$. (The zero module is Cohen–Macaulay by convention.) If N is an arbitrary finitely generated R-module, then depth $N \le \dim R/\mathfrak{p}$ for all $\mathfrak{p} \in \operatorname{Ass} N$ [72, 1.2.13]. Therefore, $\dim R/\mathfrak{p} = \dim M$ for all prime ideals $\mathfrak{p} \in \operatorname{Ass} M$ if M is Cohen–Macaulay. By prime avoidance one concludes that a module M is Cohen–Macaulay if and only if some (equivalently, every) system of parameters of M is M-regular.

For a general noetherian ring R, a finitely generated R-module M is Cohen–Macaulay if all its localizations $M_{\mathfrak{m}}$ with respect to maximal ideals $\mathfrak{m}$ of R are Cohen–Macaulay. This definition is justified since the Cohen–Macaulay property localizes: if M is Cohen–Macaulay, then $M_{\mathfrak{p}}$ is Cohen–Macaulay for every $\mathfrak{p} \in \operatorname{Spec} R$ [72, 2.1.3].

A noetherian ring R is *Cohen–Macaulay* if it is a Cohen–Macaulay module over itself. One can describe Cohen–Macaulay rings by the equality of grade and height:

Proposition 6.6. *A noetherian ring R is a Cohen–Macaulay ring if and only if* $\operatorname{ht} I = \operatorname{grade}(I, R)$ *for all proper ideals I of R.*

In the graded case the Cohen–Macaulay property can be characterized without resorting to localizations.

Proposition 6.7. *Let R be a field or a discrete valuation domain, A a positively $\mathbb{Z}$-graded finitely generated R-algebra with $A_0 = R$, and M a nonzero finitely generated graded A-module that is torsionfree as an R-module. Let $\mathfrak{p}$ be the maximal ideal of R, with generator t if R is a discrete valuation domain.*

Moreover, let $\mathfrak{m}$ be the maximal ideal of A generated by $\mathfrak{p}$ and the homogeneous elements of positive degree. Finally, suppose that $x_1, \dots, x_d$ are elements of A forming a homogeneous system of parameters of $M/\mathfrak{p}M$ and set $S = R[x_1, \dots, x_d]$. Then the following are equivalent:

(a) M is a Cohen–Macaulay module;
(b) $M_{\mathfrak{m}}$ is a Cohen–Macaulay $A_{\mathfrak{m}}$-module;
(c) $x_1, \dots, x_d$ or $t, x_1, \dots, x_d$, resp., is an $M_{\mathfrak{m}}$-sequence;
(d) $x_1, \dots, x_d$ or $t, x_1, \dots, x_d$, resp., is an M-sequence;
(e) M is a free S-module.

Proof. Note that $M/\mathfrak{m}M \neq 0$. This observation takes care of condition (i) in the definition of M-sequence for (c) and (d).

(a) $\Longrightarrow$ (b) is trivial. For (b) $\Longrightarrow$ (a) we have to show that $A_{\mathfrak{q}}$ is Cohen–Macaulay for all prime ideals $\mathfrak{q}$. Note that all homogeneous nonunits of A are contained in $\mathfrak{m}$. Therefore (b) together with the localization of the Cohen–Macaulay property implies that $M_{\mathfrak{q}}$ is Cohen–Macaulay if $\mathfrak{q}$ is graded. Otherwise let $\mathfrak{q}^*$ be the prime ideal generated by all homogeneous elements in $\mathfrak{q}$ (Lemma 4.9). Then one has

$$\dim M_{\mathfrak{q}} = \dim M_{\mathfrak{q}^*} + 1 \quad \text{and} \quad \operatorname{depth} M_{\mathfrak{q}} = \operatorname{depth} M_{\mathfrak{q}^*} + 1$$

see [72, 1.5.8, 1.5.9]. These equations imply that $M_{\mathfrak{q}}$ is Cohen–Macaulay, too.

(b) $\Longleftrightarrow$ (c) For this equivalence we observe that the sequence under consideration is a system of parameters for $M_{\mathfrak{m}}$. (Note that t is a non-zerodivisor of M by hypothesis.) It is an $M_{\mathfrak{m}}$-sequence if and only if $M_{\mathfrak{m}}$ is Cohen–Macaulay.

(c) $\Longleftrightarrow$ (d) A homogeneous element of A is a non-zerodivisor of a graded A-

module N if and only if it is a non-zerodivisor of $N_{\mathfrak{m}}$. This follows easily from the fact that all associated prime ideals of N are contained in $\mathfrak{m}$ (Lemma 4.9).

(d) $\iff$ (e)Note that $\dim S = d + \dim R$ since M is torsionfree over R. Thus S can be considered a polynomial ring over R. So the equivalence of (d) and (e) is a statement about graded S-modules. It is left to the reader (Exercise 6.2). □

We have included discrete valuation domains as ground rings in Proposition 6.7 since this is necessary for reduction to characteristic p in Proposition 6.11.

Remark 6.8. (a) For the investigation of the Cohen–Macaulay property of monoid rings, it is sufficient to consider fields of coefficients. In fact, suppose R is a noetherian ring, let $\mathfrak{q}$ be a prime ideal in $R[M]$, and set $\mathfrak{p} = R \cap \mathfrak{q}$. Since $R[M]$ is a free R-module, $R[M]_{\mathfrak{q}}$ is a flat local extension of $R_{\mathfrak{p}}$, and one has

$$\dim R[M]_{\mathfrak{q}} = \dim R_{\mathfrak{p}} + \dim R[M]_{\mathfrak{q}}/\mathfrak{p}R[M]_{\mathfrak{q}},$$
$$\operatorname{depth} R[M]_{\mathfrak{q}} = \operatorname{depth} R_{\mathfrak{p}} + \operatorname{depth} R[M]_{\mathfrak{q}}/\mathfrak{p}R[M]_{\mathfrak{q}};$$

see Matsumura [253, 15.1] and [72, 1.2.16]. Furthermore, $R[M]_{\mathfrak{q}}/\mathfrak{p}R[M]_{\mathfrak{q}}$ is a localization of the monoid algebra $(R_{\mathfrak{p}}/\mathfrak{p}R_{\mathfrak{p}})[M]$ over the field $R_{\mathfrak{p}}/\mathfrak{p}R_{\mathfrak{p}}$, and it follows that $R[M]$ is Cohen–Macaulay if and only if R and all the monoid algebras $(R_{\mathfrak{p}}/\mathfrak{p}R_{\mathfrak{p}})[M]$, $\mathfrak{p} \in \operatorname{Spec} R$, are Cohen–Macaulay.

(b) In order to make use of Lemma 6.2(b) it is often desirable to extend the base field, and conversely it is sometime useful if one can restrict it, say for a monoid algebra. Neither restriction or extension of the base field destroys the Cohen–Macaulay property [72, 2.1.10,].

(c) Similar statements apply to all properties and invariants that behave well under flat local extensions.

Normal affine monoid rings. First we consider a very simple case:

Proposition 6.9. *Let M be a normal simplicial affine monoid and $\Bbbk$ a field. Then $\Bbbk[M]$ is Cohen–Macaulay.*

Proof. Since M is positive, it has a positive grading that induces a positive $\mathbb{Z}$-grading on $\Bbbk[M]$. Let $x_1, \dots, x_d$, $d = \operatorname{rank} M$, be the extreme integral generators of M. Then M is a finitely generated module over the submonoid N generated by $x_1, \dots, x_d$, and it decomposes into the disjoint union of the sets $y + N$, $y \in \operatorname{par}(x_1, \dots, x_d) \cap M$ (see Proposition 2.43).

Translating this statement into ring-theoretic terms we obtain that $\Bbbk[M]$ is a free $\Bbbk[N]$-module with basis $M \cap \operatorname{par}(x_1, \dots, x_d)$. □

The crucial point in the proof is the existence of a *monomial* homogeneous system of parameters for $\Bbbk[M]$ in the simplicial case. But the converse is also true (Exercise 6.1), and we have to work much harder to generalize Proposition 6.9 to all normal affine monoids. (In the simplicial case, Serre's condition (S_2) is already sufficient for the Cohen–Macaulay property; see Exercise 6.3.)

Theorem 6.10 (Hochster). *Let M be a normal affine monoid. Then $\Bbbk[M]$ is Cohen–Macaulay for every field $\Bbbk$.*

The reduction of Theorem 6.10 to positive monoids is easy. In fact, $M = \mathrm{U}(M) \oplus M'$ where M' is positive and normal. So $\Bbbk[M]$ is a Laurent polynomial extension of $\Bbbk[M']$, and the Cohen–Macaulay property of $\Bbbk[M]$ follows from that of $\Bbbk[M']$: it is destroyed neither by a polynomial extension nor by the inversion of elements. (Conversely, the Cohen–Macaulay property of $\Bbbk[M']$ follows from that of $\Bbbk[M]$).

For a positive affine monoid M the algebra $\mathbb{Z}[M]$ certainly satisfies the hypothesis of the next proposition, the *reduction to characteristic p*.

Proposition 6.11. *Let R be a positively graded, finitely generated and torsionfree $\mathbb{Z}$-algebra. If $R \otimes \mathbb{Z}/(p)$ is Cohen–Macaulay for at least one prime number p, then $R \otimes \Bbbk$ is Cohen–Macaulay for every field $\Bbbk$ of characteristic* 0.

Proof. In view of Remark 6.8(b) it is enough to prove that $R \otimes \mathbb{Q}$ is Cohen–Macaulay.

By hypothesis, $R/pR \cong R \otimes \mathbb{Z}/(p)$ is Cohen–Macaulay. Choose a homogeneous system of parameters $x_1, \dots, x_d$ for R/pR. Since R is torsionfree over $\mathbb{Z}$, the sequence $p, x_1, \dots, x_d$ is R-regular. Thus it is $R_{(p)}$-regular where we have localized with respect to the prime ideal (p) in $\mathbb{Z}$. Proposition 6.7 implies that $R_{(p)}$ is Cohen–Macaulay. But then the further localization $R_{(0)} = R \otimes \mathbb{Q}$ is also Cohen–Macaulay. □

We want to derive Hochster's theorem from a more general result. Despite the fact that we will prove it only in characteristic p, the theorem is stated without this restriction.

Theorem 6.12 (Hochster-Roberts-Kempf). *Let $\Bbbk$ be a field and let S be a finitely generated graded $\Bbbk$-subalgebra of a polynomial ring $R = \Bbbk[X_1, \dots, X_n]$. If there exists an S-submodule T of R such that $R = S \oplus T$, then S is Cohen–Macaulay.*

We know that a positive normal affine monoid ring is a direct summand of a polynomial ring (Theorem 4.43), and so Hochster's theorem follows.

Theorem 6.12 is itself proved by reduction to characteristic p. In general, the hypothesis $R = S \oplus T$ would not survive the reduction. However, the following lemma does so, and it pinpoints the equational content of Theorem 6.12.

Lemma 6.13. *Let $\Bbbk$ be a field, and let $f_1, \dots, f_s$ be algebraically independent homogeneous elements of positive degree in $R = \Bbbk[X_1, \dots, X_n]$. Suppose that S is a module-finite graded $\Bbbk[f_1, \dots, f_s]$-algebra such that there exists a homogeneous homomorphism $\psi : S \to R$ of $\Bbbk[f_1, \dots, f_s]$-algebras. If $g_{r+1}f_{r+1} = g_1 f_1 + \dots + g_r f_r$ with $g_1, \dots, g_{r+1} \in S$ for some r, $0 \le r \le s-1$, then $\psi(g_{r+1}) \in (f_1, \dots, f_r)R$.*

For the derivation of the Hochster–Roberts–Kempf theorem we choose $f_1, \dots, f_s$ as a homogeneous system of parameters for S and $\psi : S \to R$ as the given embedding. We want to show that $f_1, \dots, f_s$ is S-regular. If it should fail, then we have an equation $g_{r+1}f_{r+1} = g_1 f_1 + \dots + g_r f_r$ with $g_{r+1} \notin (f_1, \dots, f_r)S$ for some r. On the other hand, Lemma 6.13 shows that $g_{r+1} \in (f_1, \dots, f_r)R$, and this is a contradiction: since $R = S \oplus T$, one has $IR \cap S = I$ for every ideal I of S.

The following proof of Lemma 6.13 is due to Knop [223]. It is a condensation of a tight closure argument of Hochster and Huneke.

Proof of Lemma 6.13. It needs two steps:

(i) the case of positive characteristic;
(ii) the reduction from characteristic 0 to positive characteristic.

Under the general hypothesis of the lemma, step (ii) is technically demanding (see [72, p. 294]). Therefore we restrict ourselves to step (i), which covers our needs.

So let $\Bbbk$ be a field of characteristic $p > 0$. We can replace $\Bbbk$ by its algebraic closure L. The hypothesis certainly survives the base field extension to L, and so does the conclusion since $IL[X_1,\dots,X_n] \cap \Bbbk[X_1,\dots,X_n] = I$ for every ideal I of $\Bbbk[X_1,\dots,X_n]$. We may therefore assume that $\Bbbk = \Bbbk^p$. (In characteristic p the subset $\{x^p : x \in \Bbbk\}$ of $\Bbbk$ is a subfield and denoted by $\Bbbk^p$.)

There exist a finitely generated free $\Bbbk[f_1,\dots,f_s]$-submodule $F \subset S$ and a nonzero element $c \in \Bbbk[f_1,\dots,f_s]$ such that $cS \subset F$. (This holds because $\Bbbk[f_1,\dots,f_s]$ is a domain and S is finitely generated: choose $\operatorname{rank} S$ linearly independent elements in S and consider the submodule F generated by them.)

Let $q = p^e$ be a power of the characteristic p, take the qth power of the equation $g_{r+1}f_{r+1} = g_1f_1 + \cdots + g_rf_r$ and multiply by c to obtain

$$f_{r+1}^q(cg_{r+1}^q) = \sum_{i=1}^r f_i^q(cg_i^q).$$

The elements cg_i^q, $i = 1,\dots,r+1$, are in the free $\Bbbk[f_1,\dots,f_s]$-module F, and in $\Bbbk[f_1,\dots,f_s]$ the elements $f_1,\dots,f_s$ are indeterminates. Thus they form an F-sequence, and there exist $h_{iq} \in F$ such that $cg_{r+1}^q = \sum_{i=1}^r f_i^q h_{iq}$. Applying the $\Bbbk[f_1,\dots,f_s]$-homomorphism $\psi : S \to \Bbbk[X_1,\dots,X_n]$ one has

$$c\psi(g_{r+1})^q = \sum_{i=1}^r f_i^q \psi(h_{iq}).$$

Let M be the set of monomials $\mu = X_1^{\mu_1}\cdots X_n^{\mu_n}$ with $\mu_i < q$ for $i = 1,\dots,n$. Taking qth powers in the algebraically closed field $\Bbbk$ is bijective. Therefore every element $h \in \Bbbk[X_1,\dots,X_n]$ has a *unique* representation $h = \sum_{\mu\in M}(h_\mu)^q\mu$; in particular,

$$\psi(h_{iq}) = \sum_{\mu\in M}(h_{iq\mu})^q\mu.$$

Thus

$$\sum_{i=1}^r f_i^q\psi(h_{iq}) = \sum_{\mu\in M}\Big(\sum_{i=1}^r h_{iq\mu}f_i\Big)^q\mu = \sum_{\mu\in M}(h_{q\mu})^q\mu$$

with $h_{q\mu} \in (f_1,\dots,f_r)R$.

The crucial point is that c does not depend on q. We choose q so large that $c = \sum_{\mu\in M} c'_\mu\mu$ with $c'_\mu \in \Bbbk$. Let $c'_\mu = (c_\mu)^q$. Then

$$\sum_{\mu \in M} (c_\mu \psi(g_{r+1}))^q \mu = \sum_{\mu \in M} (h_{q\mu})^q \mu.$$

Since $c \neq 0$ there exists μ with $c_\mu \neq 0$, and so

$$\psi(g_{r+1}) = \frac{1}{c_\mu} h_{q\mu} \in (f_1, \dots, f_r)R. \qquad \square$$

Remark 6.14. The Cohen–Macaulay property of direct summands of regular rings in characteristic $p > 0$ was proved by Hochster and Roberts [197] in connection with the Cohen–Macaulay property for rings of invariants of linearly reductive groups. The characteristic 0 case of Theorem 6.12 was proved by Kempf [220].

The Hochster–Roberts–Kempf theorem can be strengthened. In characteristic 0 direct summands S of polynomial rings (or the affine varieties defined by such S) have rational singularities by a theorem of Boutot [39]. In characteristic p Hochster and Huneke showed that S is F-regular; see [72, 10.1.3]. These properties include that R is Cohen–Macaulay.

Hochster and Huneke [196] even proved that direct summands of regular rings containing a field are Cohen–Macaulay rings; also see [72, 10.4.1].

6.B Graded homological algebra

While the Cohen–Macaulay property discussed in the previous section is defined ring-theoretically without reference to the grading, certain objects associated with a graded ring depend on the grading. Therefore we have to discuss some aspects of graded homological algebra. For $\mathbb{Z}$-graded rings it has been developed in [72].

Graded Ext and Tor functors. Over every $\mathbb{Z}^r$-graded ring one can do graded homological algebra. The objects of the underlying category are the $\mathbb{Z}^r$-graded modules, and the morphisms are the homogeneous R-linear maps.

In addition, one must allow homomorphisms that shift the grading (though these may fail to be morphisms in the category). But first we have to shift gradings in the modules themselves: $M(u)$ is the graded module with homogeneous components

$$M(u)_v = M_{u+v}. \tag{6.1}$$

For example, in $R(-1)$ the element $1 \in R$ has degree 1. Despite the fact that M and $M(u)$ cannot be distinguished as R-modules, they are in general nonisomorphic graded modules. However, if there exists a homogeneous unit of degree u, then M and $M(u)$ are isomorphic graded modules (with respect to which isomorphism ?).

Now one sets

$$^*\mathrm{Hom}_R(M, N) = \bigoplus_{u \in \mathbb{Z}^r} \{f : M \to N(u) \text{ homogeneous}\}.$$

Every graded R-module M has a *graded free resolution*: this is an exact complex

$$\mathbb{F} : \cdots \to F_{m+1} \xrightarrow{\varphi_m} F_m \to \cdots \to F_1 \xrightarrow{\varphi_1} F_0 \xrightarrow{\pi} M \to 0$$

in which F_i is a graded module that is free as an R-module with a basis of homogeneous elements, and moreover, each homomorphism is homogeneous. A graded free resolution is constructed as follows: one chooses a homogeneous system of generators of M, say $(g_i)_{i \in I}$ and sets

$$F_0 = \bigoplus_{i \in I} R(-\deg g_i).$$

Then the basis element $e_i = 1 \in R(-\deg g_i)$ (in degree $\deg g_i$) is mapped to g_i. The homomorphism π resulting from this assignment has a graded kernel (Proposition 4.2) to which we can apply the same procedure. If we write out the matrices of the linear maps φ_i with respect to the homogeneous bases chosen, then their entries are homogeneous elements of R.

A graded free module is clearly a projective object in the category of graded modules, which, as just seen, has enough projectives. Therefore we can consider the right derived functors of $^*\mathrm{Hom}(-, N)$, called $^*\mathrm{Ext}^i_R(-, N)$.

One has a natural embedding $^*\mathrm{Hom}_R(M, N) \to \mathrm{Hom}_R(M, N)$. In general it is not surjective, but if M is finitely generated, then it is an isomorphism. If R is noetherian and M is finitely generated, then M has a resolution by finitely generated graded free modules. Therefore, in the category of all R-modules,

$$^*\mathrm{Ext}^i_R(M, N) = \mathrm{Ext}^i_R(M, N)$$

for all i and all N; in other words, $\mathrm{Ext}^i_R(M, N)$ has a natural grading in this case.

Similarly one defines graded Tor functors $\mathrm{Tor}^R_i(M, -)$ as the left derived functors of $M \otimes -$ or $- \otimes N$. Graded Tor need not be decorated by a star, since the tensor product of arbitrary graded modules is graded in a natural way. Therefore the graded Tor modules are always the Tor modules in the category of all R-modules. See [72, 1.5.19].

***Local rings.** Let $\Bbbk$ be a field and R be a finitely generated $\mathbb{Z}^r$-graded $\Bbbk$-algebra. We call R *positively* $\mathbb{Z}^r$*-graded* if $R_0 = \Bbbk$ and the degrees occurring in $\Bbbk$ form a positive submonoid of $\mathbb{Z}^r$. The prime example of a positively $\mathbb{Z}^r$-graded ring is a polynomial ring $R = \Bbbk[X_1, \dots, X_n]$ in which the indeterminates are homogeneous elements of nonzero degrees that generate a positive monoid $M \subset \mathbb{Z}^r$.

Suppose R is positively $\mathbb{Z}^r$-graded. Then there is a unique maximal member among its proper graded ideals, namely the ideal generated by all homogeneous elements of degree $\neq 0$. Let $\mathfrak{p}$ be a prime ideal in R, and T be the set of all homogeneous elements $x \notin \mathfrak{p}$. The ring $T^{-1}R$, denoted

$$R_{(\mathfrak{p})}$$

in the following, cannot be called positively graded in any reasonable way (unless $T = \{1\}$), but again we find a single maximal member among the proper graded ideals, namely $T^{-1}\mathfrak{p}^*$ where $\mathfrak{p}^*$ is the prime ideal generated by all monomials in $\mathfrak{p}$.

Definition 6.15. Let R be a $\mathbb{Z}^r$-graded ring. Then we say that R is **local* with **maximal ideal* $\mathfrak{m}$ if $\mathfrak{m}$ is the single maximal member in the set of all proper graded ideals of R, or, in other words, the homogeneous nonunits of R generate the proper ideal $\mathfrak{m}$.

Suppose R is a *local ring. Then R_0 itself is a local ring with maximal ideal $\mathfrak{m}_0 = R_0 \cap fm$, and $R/\mathfrak{m}$ is a Laurent polynomial ring over $R_0/\mathfrak{m}_0$. Moreover, every graded module over $R/\mathfrak{m}$ is free (Exercise 4.2).

For a finitely generated graded module M over a noetherian *local ring we set

$$^*\dim M = \dim M_{\mathfrak{m}} \quad \text{and} \quad ^*\operatorname{depth} M = \operatorname{depth} M_{\mathfrak{m}}.$$

If $\mathfrak{m}$ is a true maximal ideal, then $^*\dim = \dim M$.

In the following we will frequently use that the homological invariants of a finitely generated graded R-module can be measured at the localization $M_{\mathfrak{m}}$. For example, M is Cohen–Macaulay if and only if $M_{\mathfrak{m}}$ is Cohen–Macaulay. In the positively $\mathbb{Z}$-graded case this was proved in Proposition 6.7, but the argument there covers the *local $\mathbb{Z}$-graded case as well. The $\mathbb{Z}^r$-graded case can be treated by induction on r. We discuss the principle in Exercise 6.7.

One can again define homogeneous systems of parameters, but these do not always exist.

A noetherian *local ring is called *regular* if the localization $R_{\mathfrak{m}}$ is regular. As just indicated, regularity then spreads to R itself (see [72, 2.2.24] for the $\mathbb{Z}$-graded case). But we will not need this fact.

Minimal graded free resolutions. Suppose R is *local with *maximal ideal $\mathfrak{m}$. Then every finitely generated graded module N over R has a *minimal graded free resolution*

$$\mathbb{F} : \cdots \to F_{m+1} \xrightarrow{\varphi_m} F_m \to \cdots \to F_1 \xrightarrow{\varphi_1} F_0 \xrightarrow{\pi} N \to 0$$

characterized by the property that $\varphi_i \otimes R/\mathfrak{m} = 0$ for all $i \geq 1$. It is constructed as outlined above, but at each step we choose a minimal homogeneous system of generators, first of N, then of $\operatorname{Ker} \pi$, etc. If we write out the matrices representing the maps with respect to the bases of the free modules constructed in this process, then all entries are homogeneous and in $\mathfrak{m}$. Therefore, as required, $\varphi_i \otimes R/\mathfrak{m} = 0$ for all i. In fact, if there were an entry $\notin \mathfrak{m}$, it would be a unit, in contradiction to the minimality of the system of generators at each step. (We have already used this argument in the proof of Theorem 4.64.)

While minimal graded free resolutions are unique up to isomorphism, it is useful to assume that $\mathfrak{m}$ is a true maximal ideal if one wants to define graded Betti numbers. In this case all homogeneous units have degree 0. It is customary to write minimal graded free resolution in the form

$$\cdots \to \bigoplus_{u \in \mathbb{Z}^r} R(-u)^{\beta_{iu}} \to \cdots \to \bigoplus_{u \in \mathbb{Z}^r} R(-u)^{\beta_{0u}} \to N \to 0$$

where the direct summands of F_i with shift $-u \in \mathbb{Z}^r$ have been collected in the free module $R(-u)^{\beta_{iu}}$. The *graded Betti numbers* $\beta_{iu} = \beta_{iu}(N)$ are well-defined since minimal graded free resolutions are unique up to isomorphism and, moreover, $R(-u) \ncong R(-v)$ if $u \neq v$ if all homogeneous units have degree 0. The *total ith Betti number* of M is $\beta_i(M) = \sum_u \beta_{iu}(M)$.

Graded Betti numbers can be described homologically. By construction of the minimal graded free resolution one has $\mathbb{F} \otimes_R \Bbbk = 0$ $\Bbbk = R/\mathfrak{m}$. Therefore

$$\beta_{iu}(N) = \dim_{\Bbbk} \operatorname{Tor}_i^R(N, \Bbbk)_u \tag{6.2}$$

for all i and u.

It is immediately clear that the minimal graded free resolution of a graded module M over a *local ring R with *maximal ideal $\mathfrak{m}$ is converted into a minimal free resolution of $M_\mathfrak{m}$ over the local ring $R_\mathfrak{m}$ when tensored with $R_\mathfrak{m}$. This observation allows one to exploit linear algebra over the truly local ring $R_\mathfrak{m}$ for the investigation of graded modules over R.

Remark 6.16. (a) Every graded projective module M over a *local ring R is free with a homogeneous basis. In fact, the minimal graded free resolution $\mathbb{F}$ of M localizes to the minimal free resolution of $M_\mathfrak{m}$ over $R_\mathfrak{m}$. But $M_\mathfrak{m}$ is a free module, and its minimal free resolution has length 0. The same holds for $\mathbb{F}$, and so M is free.

(b) Minimal graded free resolutions remain such under flat *local extensions of *local $\mathbb{Z}^r$-graded rings where an extension $f : R \to S$ of such rings is **local* if f maps the *maximal ideal of R into that of S.

Via the passage to $R_\mathfrak{m}$ we obtain *Hilbert's syzygy theorem* for regular *local rings.

Theorem 6.17. *Let R be a regular *local ring with *maximal ideal $\mathfrak{m}$. Then $\mathfrak{m}$ is generated by a homogeneous regular sequence.*

Moreover, every finitely generated graded R-module has a minimal free resolution

$$0 \to F_p \to \cdots \to F_1 \to F_0 \to N \to 0, \qquad p = {}^*\dim R - {}^*\operatorname{depth} N.$$

Proof. Let $x_1, \dots, x_n$ be a minimal homogeneous system of generators of $\mathfrak{m}$. It is also a minimal system of generators of the maximal ideal $\mathfrak{m}R_\mathfrak{m}$ of the regular local ring $R_\mathfrak{m}$. But then $x_1, \dots, x_n$ is an $R_\mathfrak{m}$-sequence, and therefore an R-sequence, as follows by the argument in the proof of Proposition 6.7.

For the second part we just localize the minimal graded free resolution $\mathbb{F}$. Then we obtain a finite free resolution by Hilbert's syzygy theorem for regular local rings [72, 2.2.14]. The length of the resolution is given by the Auslander–Buchsbaum formula [72, 1.3.3]. □

In addition to the hypotheses of Theorem 6.17 assume that $\mathfrak{m}$ is a maximal ideal of R. Then the minimal free resolution of N takes the form

$$\mathbb{F} : 0 \to \bigoplus_{u \in \mathbb{Z}^r} R(-u)^{\beta_{pu}} \to \cdots \to \bigoplus_{u \in \mathbb{Z}^r} R(-u)^{\beta_{0u}} \to N \to 0,$$
$$p = {}^*\dim R - {}^*\operatorname{depth} N. \tag{6.3}$$

Now we strengthen the hypothesis further and assume that N is Cohen–Macaulay. First, regardless of the Cohen–Macaulay property, ${}^*\mathrm{Ext}^k_R(N,R) = 0$ for all $k < \operatorname{grade}(\operatorname{Ann} N, R)$ by Theorem 6.1. Second, in the regular *local ring R one has $\dim R/\operatorname{Ann} M = \dim R - \operatorname{grade}(\operatorname{Ann} M, R) = {}^*\operatorname{depth} M$. Thus, ${}^*\mathrm{Hom}_R(\mathbb{F}, R)$ is acyclic, and (read as a chain complex) resolves ${}^*\mathrm{Ext}^p_R(N,R)$, $p = \operatorname{proj\,dim} N$.

Theorem 6.18. *Let R be a regular *local ring of *dimension n, and $\mathbb{F}$ be the minimal graded free resolution of a $\mathbb{Z}^r$-graded module N of *dimension d. Then N is Cohen–Macaulay if and only if $\mathbb{F}$ has length $n - d$.*

Now suppose that N is Cohen–Macaulay, and choose $g \in \mathbb{Z}$.

(a) *Then* $\mathrm{Hom}_R(\mathbb{F}, R(-g))$ *is the minimal graded free resolution of the Cohen–Macaulay module* $N' = {}^*\mathrm{Ext}^{n-d}_R(N, R(-g))$. *Moreover,*

$$ {}^*\mathrm{Ext}^{n-d}_R(N', R(-g)) \cong N, $$

and $\operatorname{Ann} N' = \operatorname{Ann} N$.

(b) *If the *maximal ideal $\mathfrak{m}$ is maximal, then*

$$ \beta_{iu}(N', R(-g)) = \beta_{n-d-i,g-u}(N). $$

Proof. The assertion on the Cohen–Macaulay property is an immediate consequence of the Auslander–Buchsbaum formula and the fact that M is Cohen–Macaulay if and only if its localization with respect to the *maximal ideal is Cohen–Macaulay.

The duality for the modules is obtained if we apply ${}^*\mathrm{Hom}_R(-, R(-g))$ twice. It implies the equality of annihilators since one has $\operatorname{Ann} \mathrm{Ext}^k_R(N,Q) \subset \operatorname{Ann} Q$ for all modules N and Q and all $k \in \mathbb{Z}_+$. The equality of annihilators shows that N and N' have the same *dimension, and it follows from the first assertion that N' is also Cohen–Macaulay.

For the duality of Betti numbers one compares $\mathbb{F}$ and ${}^*\mathrm{Hom}_R(\mathbb{F}, R(-g))$ directly. □

Remark 6.19. Set $E = {}^*\mathrm{Ext}^{n-d}_R(-, R(-g))$. It is useful to observe that the identity $E(E(N)) \cong N$ in Theorem 6.18 represents a natural equivalence of functors on the category of graded Cohen–Macaulay R-modules of *dimension d. In other words, if $f : N_1 \to N_2$ is a homogeneous homomorphism of such modules, then $E(E(f)) = f$.

The Koszul complex. The most basic construction of (graded) free resolutions relies on the Koszul complex associated with a sequence of elements $x_1, \dots, x_n$ in a ring R. Let F be the free R-module with basis $e_1, \dots, e_n$. Then one considers the linear map $\partial : F \to R$, $\partial(e_i) = x_i$. It induces homomorphisms $\bigwedge^k F \to \bigwedge^{k-1} F$ given by

$$ \partial(e_{i_1} \wedge \cdots \wedge e_{i_k}) = \sum_{j=1}^{k} (-1)^{i+1} a_{i_j} e_{i_1} \wedge \cdots \wedge \widehat{e_{i_j}} \wedge \cdots \wedge e_{i_k} $$

where $\widehat{\ }$ indicates an element to be omitted. Then $\partial \circ \partial = 0$, so that

$$\mathbb{K}(x) : 0 \to \bigwedge^n F \xrightarrow{\partial} \bigwedge^{n-1} F \xrightarrow{\partial} \dots \xrightarrow{\partial} \bigwedge^2 F \xrightarrow{\partial} F \xrightarrow{\partial} R \to 0$$

is a complex, the *Koszul complex* associated with the sequence $x_1, \dots, x_n$. By construction, $H_0(\mathbb{K}(x)) = R/(x_1, \dots, x_n)$.

If R is graded and $x_1, \dots, x_n$ are homogeneous, then we can make the Koszul complex graded by setting

$$\deg e_{i_1} \wedge \dots \wedge e_{i_k} = \sum_{j=1}^{k} \deg x_{i_j}.$$

A remarkable property of the Koszul complex is its duality. One has an isomorphism $\mathbb{K}(x) \cong \operatorname{Hom}_R(\mathbb{K}(x), R)$ in which, up to sign, the basis element $e_{i_1} \wedge \dots \wedge e_{i_k}$ of $\bigwedge^k F$ given above is mapped to the basis element $(e_{j_1} \wedge \dots \wedge e_{j_{n-k}})^*$ of $\operatorname{Hom}_R(\bigwedge^{n-k} F, R)$ where * denotes the dual basis and $\{1, \dots, n\} = \{i_1, \dots, i_k, j_1, \dots, j_{n-k}\}$.

In order to obtain duality also in the graded case, we have to insert a shift. Set $g = \sum_{i=1}^n \deg x_i$. Then the last free module in $\mathbb{K}(x)$ is generated by an element of degree g so that its dual is generated by an element of degree $-g$. In order to get a homogeneous isomorphism to R, we have to shift this degree to 0, and this is enough for the homogeneous isomorphism

$$\mathbb{K}(x) \cong \operatorname{Hom}_R(\mathbb{K}(x), R)(-g). \tag{6.4}$$

Theorem 6.20. *Suppose the ideal I is generated by a homogeneous regular sequence $x_1, \dots, x_n$ on the graded ring R. Let $g = \sum_{i=1}^n \deg x_i$. Then $\mathbb{K}(x)$ is a graded free resolution of R/I, and one has isomorphisms*

$$\operatorname{Tor}_i^R(N, R/I) \cong {}^*\operatorname{Ext}_R^{n-i}(R/I, N)(-g), \qquad i = 0, \dots, n,$$

for every graded R-module N.

For the crucial acyclicity of $\mathbb{K}(x)$ see [72, 1.6.14]. The isomorphisms result immediately from the isomorphism

$$N \otimes \mathbb{K}(x) \cong \mathbb{K}(x) \otimes N \cong ({}^*\operatorname{Hom}(\mathbb{K}(x), R)(-g)) \otimes N \cong {}^*\operatorname{Hom}(\mathbb{K}(x), N)(-g),$$

implied by equation (6.4).

Remark 6.21. (a) Let M be a positive affine monoid and write $R = \Bbbk[M]$ as a residue class ring of a polynomial ring S by a toric ideal (see Theorem 4.32). The graded Betti numbers $\beta_{uk} = \dim_{\Bbbk} \operatorname{Tor}_u^S(R, \Bbbk)$ of R over S can be computed from the homology of $R \otimes \mathbb{K}(X)$ where X stands for the sequence of indeterminates. This approach leads to the computation of the graded Betti numbers of S via so-called *squarefree divisor complexes* of M. See Stanley [332, 7.9] or Bruns and Herzog [71].

(b) If M is a normal affine monoid, then $R = \Bbbk[M]$ is Cohen–Macaulay for all fields $\Bbbk$ (Theorem 6.10), and in particular the length of a minimal free resolution of R as a residue class ring of a polynomial ring does not depend on $\Bbbk$. However, this is no longer true for arbitrary graded (or total) Betti numbers. The connection with squarefree divisor complexes quickly yields counterexamples [71, 2.1], and even some classical monoid rings are counterexamples; see Hashimoto [180].

Change of rings. Especially for the construction of the (graded) canonical module in Section 6.C one has to compare Ext modules over different rings. A basic result of this type is *Rees' lemma.*

Lemma 6.22. *Let R be a $\mathbb{Z}^r$-graded ring, and Q and N graded R-modules. If $x \in R$ is a homogeneous element of degree g which is a non-zerodivisor of R and Q and annihilates N, then ${}^*\mathrm{Ext}_R^{i+1}(N,Q) \cong {}^*\mathrm{Ext}_{R/(x)}^i(N,Q/xQ)(g)$.*

The non-graded version is proved in [72, 3.1.16], and the graded version is proved in the same way: using the axiomatic characterization of derived functors, one shows that ${}^*\mathrm{Ext}_R^{i+1}(-,Q)$ is the ith derived functor of ${}^*\mathrm{Hom}_{R/(x)}(-,Q/xQ)$; see Rotman [307, Th. 7.22 and Ex. 7.27]. We indicate the technique in the proof of the following

Proposition 6.23. *Let S and R be $\mathbb{Z}^r$-graded ring, and $\varphi : S \to R$ a homogeneous homomorphism. Moreover, let Q be a graded S-module and $c \in \mathbb{Z}_+$. Suppose that ${}^*\mathrm{Ext}_S^j(R,Q) = 0$ for all $j \in \mathbb{Z}_+$, $j \neq c$. If $\mathrm{Ann}_S R$ contains a homogeneous sequence $x_1,\dots,x_c$ that is regular on Q and S, then*

$$ {}^*\mathrm{Ext}_R^k(N, {}^*\mathrm{Ext}_S^c(R,Q)) \cong {}^*\mathrm{Ext}_S^{k+c}(N,Q). $$

Proof. Consider the functors $T_k = {}^*\mathrm{Ext}_S^{k+c}(-,Q)$, $k \in \mathbb{Z}$, on the category of graded R-modules. ($T^k(N)$ is indeed an R-module in a natural way if N is a graded R-module.)

It follows from Theorem 6.1 that $T_k(N) = 0$ for all $k < 0$ and all graded R-modules N. Next, the hypothesis on ${}^*\mathrm{Ext}_S^j(R,Q) = 0$ implies that $T_k(F) = 0$ for $k > 0$ if F is free. Since the T_k evidently form a connected sequence of functors, it only remains to show that

$$ {}^*\mathrm{Hom}_R(-, {}^*\mathrm{Ext}_S^c(R,Q)) \cong {}^*\mathrm{Ext}_S^c(-,Q). $$

For $c = 0$ this amounts to the adjoint isomorphism [307, 2.11]

$$ {}^*\mathrm{Hom}_R(N, {}^*\mathrm{Hom}_S(R,Q)) \cong {}^*\mathrm{Hom}_S(N \otimes_R R, Q) \cong {}^*\mathrm{Hom}_S(N,Q) $$

for R-modules N. The case $c > 0$ is reduced to the case $c = 0$ by an iterated application of Rees' lemma along the exact sequence of length c in $\mathrm{Ann}_S R$. □

Remark 6.24. The assumption of the existence of the sequence $x_1,\dots,x_c$ in Proposition 6.23 is superfluous. However, without it one has to use a more advanced argument, namely the change of rings spectral sequence

$$ {}^*\mathrm{Ext}_R^p(N, {}^*\mathrm{Ext}_S^q(R,Q)) \underset{p}{\Longrightarrow} {}^*\mathrm{Ext}_S^n(N,Q) $$

stated in [307, 11.66] for the nongraded case. The hypothesis on ${}^*\mathrm{Ext}_S^j(R,Q)$ implies that the nonzero $E_2^{p,q}$ terms are concentrated in a single column and the spectral sequence collapses. Then the change of rings formula follows.

6.C The canonical module

The canonical module of a Cohen–Macaulay ring R rightly carries its name. As we will see, its structure reflects interesting properties of R, and when it can be computed in the graded case, important numerical data of R can be determined easily.

***Canonical modules.** After the introduction of graded Ext we can define the notion of graded canonical module:

Definition 6.25. Let R be a $\mathbb{Z}^r$-graded *local Cohen–Macaulay ring with *maximal ideal $\mathfrak{m}$, and set $d = {}^*\dim R$. Then a finitely generated graded R-module ω_R is **canonical* if

(i) $\operatorname{Ann} \omega_R = 0$,
(ii) ω_R is a Cohen–Macaulay module,
(iii) $^*\operatorname{Ext}_R^d(R/\mathfrak{m}, \omega_R) = R/\mathfrak{m}$.

Note that (ii) (in conjunction with (i)) is equivalent to $^*\operatorname{Ext}_R^j(R/\mathfrak{m}, \omega_R) = 0$ for $j = 0, \dots, d-1$.

Proposition 6.26. *Let R be a $\mathbb{Z}^r$-graded regular *local ring, and let $x_1, \dots, x_n$ be a regular sequence generating the *maximal ideal $\mathfrak{m}$. Then the *canonical module ω_R is given by $x_1 \cdots x_n R \cong R(-g)$, $g = \sum_{i=1}^n \deg x_i$, and it is unique up to isomorphism.*

Proof. The Koszul complex $\mathbb{K}(x)$ resolves $R/\mathfrak{m}$ by Theorem 6.20, and therefore $^*\operatorname{Ext}_R^n(R/\mathfrak{m}, R) \cong (R/\mathfrak{m})(g)$. Passing to $R(-g)$, we correct the shift. The uniqueness follows if we realize that ω_R, a Cohen–Macaulay module over the regular *local ring R, must be free by Hilbert's syzygy theorem. Such a module is uniquely determined by the value of $^*\operatorname{Ext}_R^n(R/\mathfrak{m}, -)$ on it (why?). □

Example 6.27. (a) A simple example for Proposition 6.26 is the Laurent polynomial ring $R = \Bbbk[X_1, \dots, X_m, Y_1^{\pm 1}, \dots, Y_n^{\pm 1}]$ with its $\mathbb{Z}^{m+n}$-grading: $\omega_R = X_1 \cdots X_n R = R(-\sum \deg X_i)$.

(b) Let R be as in (a) and consider a $\mathbb{Z}^r$-grading on R under which it is a *local ring and all indeterminates are homogeneous. Furthermore let $\mathfrak{p}$ be a graded prime ideal. We claim that $R_{(\mathfrak{p})}(-\sum \deg X_i)$ is still the *canonical module.

In fact, the shifts by $-\deg X_i$ with $X_i \in \mathfrak{p}$ can inductively be eliminated by Rees' lemma, and those outside $\mathfrak{p}$ represent homogeneous units. Therefore we can assume $m = 0$. But in this case there exist units in all degrees, and all shifts can be omitted.

Now we prove the existence and uniqueness in the positively $\mathbb{Z}^r$-graded case. Let us first indicate the reduction to the positively $\mathbb{Z}$-graded case. The advantage of the latter is the existence of homogeneous regular sequences by Proposition 6.4, and they enable us to apply the change of rings formulas in 6.22 and 6.23.

Let M be the grading monoid of a ring R, and consider a positive grading $\gamma : M \to \mathbb{Z}$. Let $w \in \mathbb{Z}$. By taking the direct sum over the homogeneous components N_z, $z \in \mathbb{Z}^r$, $\gamma(z) = w$, we obtain a $\mathbb{Z}$-grading on N, and for $N = R$, on R itself. In the case, in which R is noetherian and N, N' are finitely generated $\mathbb{Z}^r$-graded R-modules, the change of grading commutes with the formation of $^*\mathrm{Hom}$ and $^*\mathrm{Ext}$. In other words, the $\mathbb{Z}$-grading on $\mathrm{Hom}_R(N, N')$ is obtained from the $\mathbb{Z}^r$-grading in exactly the way described above. If we want to show that $^*\mathrm{Ext}_R(N, N') = R/\mathfrak{m}$, then it is enough to prove this for a coarsening of the $\mathbb{Z}^r$-grading by γ, provided we can vary γ over a basis of $\mathrm{Hom}_{\mathbb{Z}}(\mathrm{gp}(M), \mathbb{Z})$. For positive affine monoids the interior of the dual cone $C^*(M)$ always contains such a basis (Corollary2.74), and this reduces the verification of the condition on $^*\mathrm{Ext}$ in Definition 6.25 to the positively $\mathbb{Z}$-graded case if R is positively $\mathbb{Z}^r$-graded. The condition on the annihilator is independent of the grading anyway.

Theorem 6.28. *Let $\Bbbk$ be a field and R a positively $\mathbb{Z}^r$-graded Cohen–Macaulay $\Bbbk$-algebra of Krull dimension d with *maximal ideal $\mathfrak{m}$. Let $S = \Bbbk[X_1, \dots, X_n] \to R$ is a homogeneous homomorphism that makes R a finitely generated S-module. Then $\omega_R = {}^*\mathrm{Ext}_S^{n-d}(R, \omega_S)$ is a *canonical module, and it is unique up to isomorphism. Moreover,*

$$^*\mathrm{Ext}_R^j(R/\mathfrak{m}, \omega_R) = \begin{cases} R/\mathfrak{m}, & j = d, \\ 0, & j \neq d. \end{cases} \tag{6.5}$$

Proof. The canonical module of S has been determined in Example 6.27. Set $\omega_R = {}^*\mathrm{Ext}_S^{n-d}(R, \omega_S)$. This a $\mathbb{Z}^r$-graded R-module. As pointed out above, it is enough to consider the positively $\mathbb{Z}$-graded case in order to prove that ω_R is a *canonical module. For simplicity of notation we identify $\Bbbk$ and $R/\mathfrak{m}$. Let $I = \mathrm{Ann}_S R$.

By graded prime avoidance, I contains a homogeneous regular S-sequence of length $n-d$. Therefore we can use Proposition 6.23 in order to compute $^*\mathrm{Ext}_S^k(\Bbbk, \omega_R)$:

$$^*\mathrm{Ext}_S^k(\Bbbk, \omega_R) = {}^*\mathrm{Ext}_R^k(\Bbbk, {}^*\mathrm{Ext}_S^{n-d}(R, \omega_S)) = {}^*\mathrm{Ext}_S^{n-d+k}(\Bbbk, \omega_S).$$

It vanishes for $k \neq d$ and is equal to $\Bbbk$ for $k = d$.

Theorem 6.18 implies first that ω_R is a Cohen–Macaulay module of dimension d. But in connection with Remark 6.19 it shows also that $\mathrm{Ann}_R\, \omega_R = 0$. In fact, if multiplication by (homogeneous) $a \in R$ vanishes on ω_R, then it must also vanish on R itself, as follows by a second application of $^*\mathrm{Ext}_S^{n-d}(-, \omega_S)$.

For the uniqueness of ω_R one chooses a representation of R as a residue class ring of a polynomial ring S. Uniqueness follows if $^*\mathrm{Ext}_S^{n-d}(\omega', \omega_S) = S$ for every *canonical R-module ω' because then $\omega' = {}^*\mathrm{Ext}_S^{n-d}(S, \omega_S)$. We claim that the total Betti number of ω' over S is 1 and that $\beta_{n-d,g}(\omega') = 1$. This claim reduces to the positively $\mathbb{Z}$-graded case in the same way as (6.5). Assume it has been shown. Then $\beta_{0,0}({}^*\mathrm{Ext}_S^{n-d}(\omega', \omega_S)) = 1$, and all the other graded Betti numbers β_{0j} vanish. Together with the fact that $^*\mathrm{Ext}_S^{n-d}(\omega', \omega_S)$ has the same annihilator as ω', namely I, it follows that $^*\mathrm{Ext}_S^{n-d}(\omega', \omega_S) = R$.

By graded prime avoidance (Proposition 6.4) we can find homogeneous elements $y_1, \dots, y_d \in S$ that form a regular sequence for R and S. Let us denote residue classes modulo $(y_1, \dots, y_d)$ by $\bar{\ }$, and set $h = \sum \deg y_i$. Using Rees' lemma 6.22 twice (and successively for the whole sequence $y_1, \dots, y_d$) we obtain

$$^*\mathrm{Ext}^d_S(\Bbbk,\omega') = {}^*\mathrm{Hom}_{\bar S}(\Bbbk,\bar\omega')(h) = {}^*\mathrm{Hom}_{\bar R}(\Bbbk,\bar\omega')(h) = {}^*\mathrm{Ext}^d_R(\Bbbk,\omega_S) = \Bbbk.$$

The duality of the Koszul complex (Theorem 6.20) for the sequence of indeterminates then yields

$$\mathrm{Tor}^S_{n-d}(\omega',\Bbbk) = {}^*\mathrm{Ext}^d_R(\Bbbk,\omega_S)(-g) = \Bbbk(-g).$$

Applying the duality for Betti numbers in Theorem 6.18 we obtain the desired graded Betti numbers for $^*\mathrm{Ext}^d_S(\Bbbk,\omega')$. □

Remark 6.29. (a) For the computation of the canonical modules it is useful to note that it is (isomorphic to) a divisorial ideal over a normal noetherian domain R. In fact, by change of rings we see that $\mathrm{Hom}_R(\mathrm{Hom}_R(R,\omega_R),\omega_R) = R$. This implies $\operatorname{rank}\omega_R = 1$. It remains to show that ω_R is a reflexive module. But this holds true for all Cohen–Macaulay modules N of full dimension over normal noetherian domains: if $\mathfrak p$ is a height 1 prime, then $\operatorname{depth} N_{\mathfrak p} = 1$ since $R_{\mathfrak p}$ is a regular local ring. For all other primes $\mathfrak p$ one has $\operatorname{depth} N_{\mathfrak p} = \dim R_{\mathfrak p} \ge 2$, and altogether this implies reflexivity [72, 1.4.1].

(b) As in [72] one can define the *canonical module by equation (6.5). In order to show that it implies $\operatorname{Ann}\omega_R = 0$ one first shows that ω_R is a canonical module in the ordinary sense, i. e., $(\omega_R)_{\mathfrak p}$ is a canonical module of the local ring $R_{\mathfrak p}$ for all $\mathfrak p \in \operatorname{Spec} R$ (see Exercise 6.7). Then the condition on the annihilator follows from the corresponding local theorem [72, 3.3.4].

(c) Using the change of rings argument in Remark 6.24, one can generalize Theorem 6.28 to residue class rings of regular *local rings.

(c) The reader may have noted that Theorem 6.28 and its proof are valid (and even simpler) in the nongraded local case as well if one assumes that the Cohen–Macaulay ring under consideration is the residue class ring of a regular local ring.

Suppose R is a positively $\mathbb{Z}^r$-graded ring written as a residue class ring of a polynomial ring (or just as a finitely generated module over S). Since $\omega_S = S(-g)$ for suitable g and $\omega_R = {}^*\mathrm{Ext}^c_S(R,\omega_S)$, $c = \dim S - \dim R$, Theorem 6.18 immediately gives the graded Betti numbers of ω_R:

$$\beta_{iu}(\omega_R) = \beta_{c-i,g-u}(R). \tag{6.6}$$

The **canonical module of a normal affine monoid domain.* As for the Cohen–Macaulay property, the simplicial case is guiding us.

Proposition 6.30. *Let M be a normal simplicial affine monoid and $\Bbbk$ a field. Then the ideal generated by* $\operatorname{int}(M)$ *is the *canonical module of $\Bbbk[M]$.*

Proof. We choose $x_1,\dots,x_d$ as the extreme integral generators of M, and let N be the monoid generated by them. Then $R = \Bbbk[N]$ is a positively graded polynomial ring, and $\omega_R = x_1\cdots x_d R$. In other words, ω_R is the R-submodule of $S = \Bbbk[M]$ spanned by the monomials in $x_1\cdots x_d N$ (multiplicative notation). Set $x = x_1\cdots x_d$.

Let $z_1,\dots,z_m$ be the monomial basis of the R-module S given by the monomials corresponding to the lattice points in $\operatorname{par}(x_1,\dots,x_d)$. Then $z_i' = x/z_i$ is a monomial in the interior of $\mathbb{R}_+N$, and, as is easily checked, the z_i' form a basis of I as an R-module. Now we identify $\operatorname{Hom}_R(S, \omega_R) = \omega_{\Bbbk[M]}$ (Proposition 6.8) with the R-module I as follows: $z_i'(z_j) = x$ if $i = j$, and 0 otherwise.

One must of course check that $\operatorname{Hom}_S(R, \omega_S)$ and I are isomorphic as R-modules. This is left to the reader. □

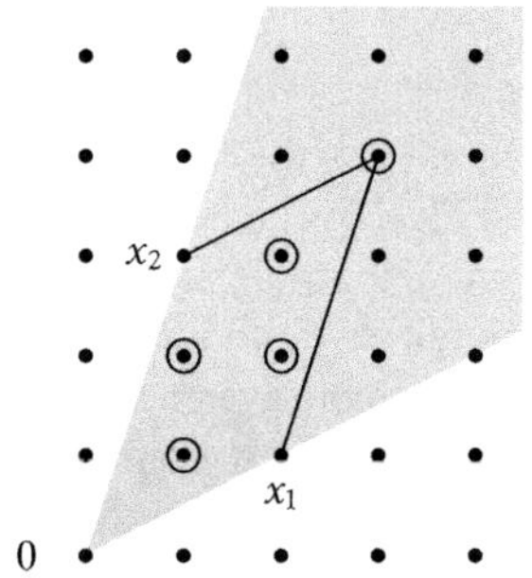

Fig. 6.1. The basis of the interior ideal

Now we turn to the general case. By Remark 6.29 the canonical module of a normal noetherian Cohen–Macaulay domain is (isomorphic to) a divisorial ideal. Let M be a positive normal affine monoid and set $R = \Bbbk[M]$. Then R is a normal noetherian domain, is Cohen–Macaulay by Hochster's theorem, and has a uniquely determined *canonical module by Theorem 6.28. It follows from Theorem 4.54 that the * canonical module of R is given by a uniquely determined fractional monomial divisorial ideal

$$\omega_R = \Bbbk\{x \in \operatorname{gp}(M) : \sigma_F(x) \ge \nu_F,\ F \in \mathscr{F}\},$$

where $\mathscr{F}$ is the set of facets of $\mathbb{R}_+M$, σ_F is the support form associated with the facet F, and the ν_F are uniquely determined integers. Setting

$$V_F = \Bbbk\{x \in \operatorname{gp}(M) : \sigma_F(x) \ge \nu_F\},$$

we have $\omega_R = \bigcap V_F$. In order to compute ω_R we have to find the numbers ν_F.

Let $\mathfrak{p}_F$ be the monomial divisorial prime ideal of R associated with F and set $R_F = R_{(\mathfrak{p}_F)}$. We know the localization R_F very well: in order to obtain it, we have to invert all monomials in the facet F. Therefore $R_F = \Bbbk[\operatorname{gp}(M) \cap H_F^+]$ (compare the proof of Theorem 4.40).

Evidently $(V_F)_{(\mathfrak{p}_F)} = V_F$. On the other hand, $(V_G)_{(\mathfrak{p}_F)} = \Bbbk[\operatorname{gp}(M)]$ for $G \ne F$. Suppose we have shown that the graded canonical module localizes (in this case). Then

$$\omega_{R_F} = (\omega_R)_{(\mathfrak{p}_F)} = V_F, \qquad F \in \mathscr{F}.$$

But we can also compute ω_{R_F} directly: $R_F \cong \Bbbk[Z_1^{\pm 1}, \dots, Z_{r-1}^{\pm 1}, W]$, $r = \operatorname{rank} M$. Here W is a monomial with value 1 under σ_F and $Z_1, \dots, Z_{r-1}$ form a basis of the group $\mathrm{gp}(M) \cap F$ of monomial units. By Proposition 6.26, $\omega_{R_F} = WR_F$. Comparing the two expressions for ω_{R_F}, we conclude that $v_F = 1$.

It remains to prove that the formation of the *canonical module commutes with the inversion of the monomials in the facet F. To this end we consider an epimorphism $S = \Bbbk[X_1, \dots, X_m, Y_1, \dots, Y_n] \to \Bbbk[M]$ that sends the indeterminates $X_1, \dots, X_m$ to the Hilbert basis elements in M that are contained in F, and the indeterminates $Y_1, \dots, Y_n$ to the remaining ones. The grading on S is defined as the pullback of the grading on R. The canonical module of S is then given by $\omega_S = X_1 \cdots X_m Y_1 \cdots Y_n S$ (Example 6.27(b)) and $\omega_R = {}^*\mathrm{Ext}^c_R(R, \omega_S)$ with $c = n - r$.

Let $\mathfrak{q}$ be the preimage of $\mathfrak{p}_F$ in S and set $S_F = S_{(\mathfrak{q})}$. Example 6.27(b) shows that we get the *canonical module of S_F by localizing that of S, and we are done by change of rings, as soon as a regular sequence of length c in $I = \mathrm{Ann}_{S_F} R_F$ has been found. (If we allowed ourselves to use Remark 6.24, this would be superfluous.)

Set $L = \Bbbk[X_1^{\pm 1}, \dots, X_m^{\pm 1}] \subset S_F$. We can assume that Y_1 is mapped to W. Then I contains binomials

$$Y_j - -Y_1^{h_j}\mu_j, \qquad \mu_j \in L, j = 2, \dots, n. \tag{6.7}$$

Moreover, $I \cap L$ is also spanned by a regular sequence as follows from Corollary 4.30. In fact, this intersection is the kernel of a homomorphism of Laurent polynomial rings $L \to \Bbbk[Z_1^{\pm 1}, \dots, Z_{r-1}^{\pm 1}]$ that arises by a homomorphism of the underlying groups. Actually, I is generated by a regular sequence, formed by the binomials in (6.7) and the regular sequence generating $I \cap L$. This claim and the remaining details are left to the reader.

To sum up, in the positive case we have proved

Theorem 6.31 (Danilov, Stanley). *Let M be a normal affine monoid and $\Bbbk$ a field. Then the ideal generated by* $\mathrm{int}(M)$ *is the *canonical module of* $\Bbbk[M]$.

For the general case one splits$M$ in the form $\mathrm{U}(M) \oplus M'$ and writes $\Bbbk[M] = (\Bbbk[\mathrm{U}(M)])[M']$. Then it follows readily from the definition of the * canonical module that $\omega_{\Bbbk[M]} = \omega_{\Bbbk[M']} \otimes_{\Bbbk} \Bbbk[\mathrm{U}(M)]$.

Remark 6.32. Let M be a seminormal affine monoid, and $\bar{M}$ its normalization. Since $\mathrm{int}(M) = \mathrm{int}\,\bar{M})$ by Proposition 2.40, $\mathrm{int}(M)\Bbbk[\bar{M}]$, the canonical module of $\Bbbk[\bar{M}]$, is an ideal of $\Bbbk[M]$. This observation provides the entry to an investigation of seminormal affine monoid algebras from the homological side; see Bruns, Li, and Römer [77]. However $\mathrm{int}(M)\Bbbk[\bar{M}]$ is *not* the canonical module of $\Bbbk[M]$, unless $M = \bar{M}$ (Exercise 6.11).

Gorenstein rings. A Cohen–Macaulay ring is called *Gorenstein* if R is its own canonical module. In the graded case this amounts to the fact that the *canonical module is graded isomorphic to a shifted copy $R(h)$ of R (compare Exercise 6.7).

In view of Theorem 6.31 it is very easy to determine the Gorenstein rings among the normal affine monoid algebras:

Theorem 6.33. *Let M be a normal affine monoid and $\Bbbk$ a field. Then the following are equivalent:*

(a) $\Bbbk[M]$ is Gorenstein;
(b) there exists $x \in \operatorname{int}(M)$ such that $\operatorname{int}(M) = x + M$;
(c) there exists $x \in M$ such that $\sigma_i(x) = 1$ for all support forms σ_i of $\mathbb{R}_+M$.

Proof. By the definition of Gorenstein rings and by Theorem 6.31, R is Gorenstein if and only if $\Bbbk(\operatorname{int}(M))$ is generated by a single element x. Clearly x must be a monomial. This yields the equivalence of (a) and (b).

Suppose that (c) holds. Then $y - x \in M$ for every $y \in \operatorname{int}(M)$ since $\sigma_i(y) > 0$ for all i. Thus $\operatorname{int}(M) = x + M$. Conversely, if (b) holds, we see that $\sigma_i(x) = 1$ since $\operatorname{int}(M)$ contains an element y with $\sigma_i(y) = 1$, and then $y - x \in M$ implies $\sigma_i(x) \le 1$. But since $x \in \operatorname{int}(M)$, we must have $\sigma_i(x) = 1$. □

Remark 6.34. Theorem 6.33 justifies our terminology in Exercise 2.13: M is Gorenstein if and only if $\Bbbk[M]$ is (normal and) Gorenstein.

In particular, the Gorenstein property of $\Bbbk[M]$ is independent of $\Bbbk$ if M is normal. For arbitrary affine monoids the Cohen–Macaulay and Gorenstein properties may very well depend on $\Bbbk$; see Bruns, Römer, and Li[77] and Hoa [193].

6.D Hilbert functions

Hilbert functions are the most important numerical invariants of graded modules, and they form the bridge from commutative algebra to its combinatorial applications.

In this section we will treat only algebras R over a field $\Bbbk$ that are graded by positive affine monoids with $R_0 = \Bbbk$. They are *local with *maximal ideal $\mathfrak{m}$ generated by all homogeneous elements of degree $\neq 0$. Note that $^*\dim M = \dim M_{\mathfrak{m}} = \dim M$ for such a module. Therefore we will henceforth simplify notation by omitting *.

Hilbert functions and their generating functions. Let R be a $\mathbb{Z}^r$-graded $\Bbbk$-algebra and N a graded R-module such that $\dim_\Bbbk N_u < \infty$ for all $u \in \mathbb{Z}^r$. Then we can consider its *Hilbert function*

$$H(N,-) : \mathbb{Z}^r \to \mathbb{Z}, \qquad H(N,u) = \dim_\Bbbk N_u.$$

For enumerative functions like Hilbert functions it is extremely useful to introduce their (ordinary) generating function. In our case it is the *Hilbert series*

$$H_N(\boldsymbol{t}) = \sum_{u \in \mathbb{Z}^r} H(N,u)\boldsymbol{t}^u.$$

Here $\boldsymbol{t}$ stands for a family $t_1, \dots, t_r$ of indeterminates and, as usual, $\boldsymbol{t}^u$ denotes the monomial $t_1^{u_1} \cdots t_r^{u_r}$.

The set $\mathbb{C}[[\mathbb{Z}^r]]$ of all series

$$\sum_{u \in \mathbb{Z}^r} c_u \boldsymbol{t}^u, \qquad c_u \in \mathbb{C},$$

carries an additive structure and a multiplicative structure that, however, is only partially defined: the product

$$\sum_w \Big(\sum_{u+v=w} a_u b_v t^{u+v} \Big)$$

of the series $\sum a_u t^u$ and $\sum b_v t^v$ can only be formed if for each $w \in \mathbb{Z}^r$ there exists only finitely many decompositions $w = u + v$ such that $a_u, b_v \neq 0$. Apart from the fact that products do not always exist, all ring axioms are satisfied. In particular, $\mathbb{C}[[\mathbb{Z}^r]]$ is a module over the Laurent polynomial ring $\mathbb{C}[\mathbb{Z}^r]$ which is naturally contained in $\mathbb{C}[[\mathbb{Z}^r]]$. However, note that it is not torsionfree: for example,

$$(1 - t^w) \sum_{u \in \mathbb{Z}^r} t^u = 0, \qquad w \in \mathbb{Z}^r.$$

For $f = \sum c^u t^u \in \mathbb{C}[[\mathbb{Z}^r]]$ we set $\operatorname{supp} f = \{u \in \mathbb{Z}^r : c_u \neq 0\}$. Let $M \subset \mathbb{Z}^r$ be a positive affine monoid and

$$\mathbb{C}[[M]] = \{f \in \mathbb{C}[[\mathbb{Z}^r]] : \operatorname{supp} f \subset M\}.$$

Proposition 6.35. *$\mathbb{C}[[M]]$ is a noetherian local integral domain with maximal ideal $\mathfrak{m} = \{\sum_{u \in M} c_u t^u : c_0 = 0\}$. It is the completion of $\mathbb{C}[M]$ with respect to the maximal ideal $\mathfrak{m} \cap \mathbb{C}[M]$.*

Proof. Set $R = \mathbb{C}[[M]]$. Since M is positive, every element of M has only finitely many decompositions as a sum of two elements of M. Therefore the product fg is well-defined for all $f, g \in R$, and it evidently belongs to R. So R is indeed a ring. Choosing a grading on M, we can split every element f of $\mathbb{C}[[M]]$ into the (infinite) sum $\sum_{n=0}^{\infty} f_n$ of its homogeneous components $f_n = \sum_{\deg x = n} f_x$. Since the product of the initial (with respect to degree) components of f and g is the initial component of fg. This implies that $\mathbb{C}[[M]]$ is an integral domain.

The degree induces a filtration $(I_k)_{k \in \mathbb{Z}_+}$ on R in which I_k consists of all elements of initial degree $\geq k$. This filtration is cofinal with the $\mathfrak{m}$-adic one, and we may use it instead. Evidently $\mathbb{C}[M]$ is dense in R: $f = \lim \sum_{n=0}^{m} f_n$. Likewise evidently, R is complete. The completion of a noetherian ring with respect to an ideal-adic topology is again noetherian. (See [253, §8] for the theory of completion.)

The inverse of $f = z + g$, $z \in \mathbb{C}$, $z \neq 0$, $g \in \mathfrak{m}$, is given by the geometric series

$$z^{-1} \sum_{k=0}^{\infty} (zg)^k,$$

which converges in R. □

In order to allow shifts outside M also, we enlarge $\mathbb{C}[[M]]$ by adjoining all Laurent monomials:

$$\mathscr{L}[[M]] = \{f \in \mathbb{C}[[\mathbb{Z}^r]] : t^u f \in \mathbb{C}[[M]] \text{ for some } u \in \mathbb{Z}^r\}.$$

Evidently $\mathscr{L}[[M]]$ is an overring of $\mathbb{C}[[M]]$ and an integral domain itself.

Proposition 6.36. *Let $M \subset \mathbb{Z}^r$ be a positive affine monoid, R a finitely generated M-graded $\Bbbk$-algebra, and N a finitely generated graded R-module. Then $H_N(\mathbf{t}) \in \mathscr{L}[[M]]$.*

Proof. Choose a system $x_1, \dots, x_m$ of generators of N. Then $\operatorname{Hilb}_{Rx_i}(\mathbf{t}) \in \mathscr{L}[[M]]$ for all i, and $\operatorname{supp} H_N(\mathbf{t}) \subset \bigcup \operatorname{supp} H_{Rx_i}(\mathbf{t}) = \operatorname{supp} \sum H_{Rx_i}(\mathbf{t})$. □

Hilbert series and rational functions. The real power of the Hilbert series lies in the fact that it is a rational function, as we will see below. However, "is" has to be made precise. Let $M \subset \mathbb{Z}^r$ be a positive affine monoid. When we write

$$f = \frac{g}{h} \qquad f \in \mathscr{L}[[M]],\ g, h \in \mathbb{C}[\mathbb{Z}^r],$$

then it is of course to be understood that h is a unit in $\mathscr{L}[[M]]$ and $hf = g$. The notation hides that g/h as an element of $\mathbb{C}[[\mathbb{Z}^r]]$ depends on M! For example,

$$\frac{1}{1-t} = \sum_{k=0}^{\infty} t^k \in \mathscr{L}[[\mathbb{Z}_+]], \quad \text{but} \quad \frac{1}{1-t} = \frac{-t^{-1}}{1-t^{-1}} = \sum_{k=1}^{\infty} -t^{-k} \in \mathscr{L}[[\mathbb{Z}_-]].$$

Fortunately, addition and multiplication (if possible) do not depend on M.

Theorem 6.37 (Hilbert–Serre). *Let M be a positive affine monoid, R a finitely generated M-graded $\Bbbk$-algebra with homogeneous system $x_1, \dots, x_d$ of generators, and N a finitely generated graded R-module. Then there exists $Q \in \mathbb{C}[\mathbb{Z}^r]$ such that*

$$H_N(\mathbf{t}) = \frac{Q}{(1-\mathbf{t}^{\deg x_1})\cdots(1-\mathbf{t}^{\deg x_d})}.$$

Proof. Note that all factors in the denominator are units in $\mathscr{L}[[M]]$ so that the right hand side does indeed denote an element of $\mathscr{L}[[M]]$.

The theorem is proved by induction on d. It is trivially true for $d = 0$ in which case N is a graded finite-dimensional $\Bbbk$-vector space. In the general case consider the multiplication by x_d on N and let U be its kernel. Then we have an exact sequence

$$0 \to U(-\deg x_d) \to N(-\deg x_d) \to N \to N/x_dN \to 0$$

that decomposes into its homogenous components, each of which is an exact sequence of finite-dimensional $\Bbbk$-vector spaces. Since $\Bbbk$-dimension is additive along exact sequences, the same holds true for Hilbert functions and Hilbert series. Thus

$$H_N(\mathbf{t}) - H_{N(-\deg x_d)}(\mathbf{t}) = H_{N/x_dN}(\mathbf{t}) - H_{U(-\deg x_d)}(\mathbf{t}).$$

Both N/x_dN and U are finitely generated over $\Bbbk[x_1, \dots, x_{d-1}]$ since they are annihilated by x_d (and R is noetherian). Thus we can apply induction to them and, moreover,

$$H_N(\mathbf{t}) - H_{N(-\deg x_d)}(\mathbf{t}) = (1-\mathbf{t}^{\deg x_d})H_N(\mathbf{t}),$$

and in $\mathscr{L}[M]$ we are allowed to divide by $1-\mathbf{t}^{\deg x_d}$. □

For an alternative proof, following Hilbert [189], one uses Hilbert's syzygy theorem. We replace $\Bbbk[x_1,\dots,x_d]$ by the polynomial ring $S = \Bbbk[X_1,\dots,X_d]$ with $\deg X_i = \deg x_i$. Over S the module N has a finite free resolution $\mathbb{F}$ given as in (6.3), and then

$$H_N(t) = \sum_{i=1}^{p}\sum_{u\in\mathbb{Z}^r}(-1)^i t^u \beta_{iu}(N) H_S(t) = \frac{\sum_{i=1}^{p}\sum_{u\in\mathbb{Z}^r}(-1)^i\beta_{iu}(N)t^u}{(1-t^{\deg X_1})\cdots(1-t^{\deg X_d})} \tag{6.8}$$

since $H_S(t) = 1/(1-t^{\deg X_1})\cdots(1-t^{\deg X_d})$. This proof is less elementary, but the formula just derived is extremely useful since it links the Hilbert series to the homological properties of N that are encoded in the graded free resolution.

The $\mathbb{Z}$-graded case. This is the classical case occurring in projective algebraic geometry and also in many combinatorial applications. Its main advantage is that one can control the denominator of the Hilbert series by a homogeneous system of parameters and that the Hilbert function is a quasipolynomial for sufficiently large degrees. A function $q:\mathbb{Z}\to\mathbb{C}$ is called a *quasipolynomial* of period π if

$$q(n) = a_d(n)n^d + \cdots + a_1(n)n + a_0(n), \qquad n\in\mathbb{Z},$$

with functions $a_i:\mathbb{Z}\to\mathbb{C}$ that satisfy $a_i(n) = a_i(m)$ whenever $m\equiv n\ (\pi)$, and moreover, π is the smallest positive integer for which these conditions hold. (So a polynomial is a quasipolynomial of period 1.)

Hilbert functions of $\mathbb{Z}$-graded modules have been treated extensively in [72, Ch. 4]; also see [52]. Therefore we restrict ourselves to stating the main results and a brief discussion.

Theorem 6.38. *Let R be a $\mathbb{Z}$-graded $\Bbbk$-algebra generated by homogeneous elements of positive degrees. Furthermore let N be a nonzero, finitely generated graded R-module with a homogeneous system of parameters $x_1,\dots,x_d$, $d=\dim N$, and set $e_i = \deg x_i$.*

Then there exists a Laurent polynomial $Q\in\mathbb{Z}[t,t^{-1}]$ such that

$$H_N(t) = \frac{Q(t)}{(1-t^{e_1})\cdots(1-t^{e_d})}, \qquad Q(1)>0.$$

Moreover there is a (unique) quasipolynomial P_N of period dividing $\operatorname{lcm}(e_1,\dots,e_d)$, *such that*

$$\begin{aligned} H(N,i) &= P_N(i), & i &> \deg H_N,\\ H(N,i) &\neq P_N(i), & i &= \deg H_N. \end{aligned}$$

That $H_N(t)$ is a rational function with denominator $(1-t^{e_1})\cdots(1-t^{e_d})$ follows immediately from Theorem 6.37 if we use that N is finitely generated over $S = \Bbbk[x_1,\dots,x_d]$.

We only sketch the proof of the remaining assertions. For the inequality $Q(1)>0$ the notion of rank of a module is useful: a module M over a ring A has *rank* r if $M\otimes T$ is free of rank r over the total ring of fractions T of A. Therefore every finitely generated module over an integral domain has a well-defined rank. See [72, 1.4] for a discussion of this notion.

As an S-module of dimension $d = \dim S$, N has positive rank, and $\operatorname{rank} N = Q(1)$. This follows from equation (6.8) since $\operatorname{rank} N$ is the alternating sum of the total Betti numbers [72, 1.4.6].

The existence of the quasipolynomial P_N can be derived from the form of the denominator of the rational function; see Stanley [330]. It can also be derived from the case of algebras generated in degree 1; the steps will be outlined in connection with Theorem 6.48. The quasipolynomial P_N is called the *Hilbert quasipolynomial* of N.

Specializing further, we consider algebras that are generated in degree 1. By passing to an infinite extension of $\Bbbk$, we may assume that there exist homogeneous systems of parameters in degree 1 (Lemma 6.3). So we obtain as a direct consequence of Theorem 6.38 the following:

Theorem 6.39. *Let R be a $\mathbb{Z}$-graded $\Bbbk$-algebra generated by homogeneous elements of degree 1. Then Theorem 6.38 can be strengthened as follows:*
(a) the denominator of $H_N(t)$ can be chosen to be $(1-t)^d$;
(b) P_N is a polynomial.
Moreover, let $e(N) = Q(1) > 0$. If $d \geq 1$, then

$$P_N = \frac{e(N)}{(d-1)!} X^{d-1} + \text{terms of lower degree.}$$

The number $e(N)$ is called the *multiplicity* of N.

Hilbert series of Cohen–Macaulay rings and modules. It is customary to write the numerator polynomial of $H_N(t)$ in Theorem 6.38 in the form

$$h_v t^v + \cdots + h_u t^u, \qquad h_v, h_u \neq 0,\ v \leq u.$$

Evidently $v = \min\{i : N_i \neq 0\}$.

The numbers h_i can be easily interpreted if N is a Cohen–Macaulay module. Choose a homogeneous system of parameters $x_1, \ldots, x_d$ and set $e_i = \deg x_i$. Then N is free over $S = \Bbbk[x_1, \ldots, x_d]$. Let $y_1, \ldots, y_m$ be a basis of N consisting of homogeneous elements, and let $\tilde{h}_i$ be the number of degree i elements among $y_1, \ldots, y_m$. Then, as an S-module,

$$N = \bigoplus_i S(-i)^{\tilde{h}_i}, \qquad H_N(t) = \frac{\sum_i \tilde{h}_i t^i}{\prod_{i=1}^d (1-t^{e_i})}.$$

It follows immediately that $h_i = \tilde{h}_i$.

Theorem 6.40. *Under the hypotheses of Theorem 6.38 suppose, in addition, that N is a Cohen–Macaulay module, and let h_i be the coefficient of t^i in the numerator polynomial of $H_N(t)$ as in Theorem 6.38 with respect to a homogeneous system of parameters $x_1, \ldots, x_d$.*

Then h_i is the number of degree i elements in a homogeneous basis of the module N over $\Bbbk[x_1,\dots,x_d]$. In particular $h_i \geq 0$ for all i.

Moreover, if R is generated in degree 1 over $\Bbbk$ and $R = N$, then $h_0 = 1$ and $h_i > 0$ for all i, $1 \leq i \leq d + \deg H_R(t)$.

Observe for the last part that one can prove the nonnegativity of the h_i in a different way: $h_v t^v + \dots + h_u t^u$ is the Hilbert series of $N/(x_1,\dots,x_d)N$. For R itself, $R/(x_1,\dots,x_d)R$ is a $\Bbbk$-algebra generated in degree 1. It can have no gaps in the sequence $h_0 = 1,\dots,h_u$.

For an algebra R generated in degree 1, we will call $(1, h_1,\dots,h_u)$ the *h-vector* of R. (In the general positively graded case, $(1, h_1,\dots,h_u)$ depends on the denominator for which there may be no canonical choice.) Note that the h-vector fulfills much stronger conditions than just positivity; as the h-vector of a $\Bbbk$-algebra generated in degree 1, it satisfies *Macaulay's inequalities*; see [72, 4.2.10].

The Hilbert function of the canonical module. The duality of a Cohen–Macaulay ring and its canonical module has a striking interpretation in terms of Hilbert series:

Theorem 6.41 (Stanley). *Let M be a positive affine monoid and R a finitely generated Cohen–Macaulay M-graded $\Bbbk$-algebra. Then*

$$H_{\omega_R}(t) = (-1)^d H_R(t^{-1}), \qquad d = \dim R.$$

Before we prove the theorem, an explanation of the term $H_R(t^{-1})$ is advisable. For a rational function $f \in \mathscr{L}[[M]]$ we consider an expression $f = g/h$ as a quotient of Laurent polynomials with a unit $h \in \mathscr{L}[[M]]$. In g and h we substitute t^{-u} for every occurrence of t^u, $u \in \mathbb{Z}^r$, obtaining the Laurent polynomials $g(t^{-1})$ and $h(t^{-1})$. If $h(t^{-1})$ is a unit in $\mathscr{L}[[M]]$, then we are justified to consider $f(t^{-1}) = g(t^{-1})/h(t^{-1})$ as an element of $\mathscr{L}[[M]]$. (Give an example for which $h(t^{-1})$ is not a unit!)

Despite of the fact that it is tempting (and perhaps more suggestive) to replace t^u by t^{-u} in the series expansion of f, this is *not* what we do! In fact, under this operation we would pass from $\mathscr{L}[[M]]$ to $\mathscr{L}[[-M]]$, obtaining the expansion of $f(t^{-1})$ in $\mathscr{L}[[-M]]$.

Proof of Theorem 6.41. Writing $1 - t^{-u}$, $u \in M$, in the form $-t^{-u}(1 - t^u)$, one sees that $H_R(t^{-1}) \in \mathscr{L}[[M]]$.

Represent R as a residue class ring of an M-graded polynomial ring $P = \Bbbk[X_1,\dots,X_n]$, and let β_{iu} denote the graded Betti numbers of R as a P-module. Then the graded Betti numbers of $\omega_R = {}^*\operatorname{Ext}_P^s(R, \omega_P))$, $s = \dim P - \dim R$, are given by

$$\beta_{iu}(\omega_R) = \beta_{s-i,g-u}(R)$$

where $g = \sum \deg X_i$.

For the Hilbert series of R, equation (6.8) yields

$$H_R(t) = \frac{\sum_{i=1}^{p} \sum_{w \in \mathbb{Z}^r} (-1)^i \beta_{iu} t^w}{(1 - t^{\deg X_1})\cdots(1 - t^{\deg X_n})},$$

and for the Hilbert series of ω_R it yields

$$H_{\omega_R}(\boldsymbol{t}) = \frac{\sum_{i=1}^{p}\sum_{w\in\mathbb{Z}^r}(-1)^i\beta_{s-i,g-u}\boldsymbol{t}^w}{(1-\boldsymbol{t}^{\deg X_1})\cdots(1-\boldsymbol{t}^{\deg X_n})}.$$

In order to obtain the desired result, one factors out $(-1)^n\boldsymbol{t}^g$ from the numerator and denominator of $H_{\omega_R}(\boldsymbol{t}^{-1})$. □

From the theorem one can draw various consequences, especially in the $\mathbb{Z}$-graded case.

Corollary 6.42. *Let R be a positively $\mathbb{Z}$-graded Cohen–Macaulay $\Bbbk$-algebra. Then* $\deg H_R(t) = -\min\{i : (\omega_R)_i \neq 0\}$.

In fact, by Theorem 6.41, the numerator of $H_{\omega_R}(t)$ is given by $\sum_{j=0}^{u} h_j t^{-e-j}$ if $\sum_{j=0}^{u} h_i t^i$ is the numerator of $H_R(t)$ and $e = \sum \deg e_i$.

For Gorenstein rings we obtain:

Corollary 6.43 (Stanley). *Let R be a positively $\mathbb{Z}$-graded, finitely generated $\Bbbk$-algebra of Krull dimension d, and let $H_R(t) = (h_0 + \cdots + h_u t^u)/\prod_{i=1}^{d}(1-t^{e_i})$ with $h_u \neq 0$ be the Hilbert series of R. Suppose that R is a Gorenstein ring. Then*

(a) $\omega_R \cong R(a)$, $a = \deg H_R(t)$;
(b) $h_i = h_{u-i}$ for $i = 0,\dots,u$ (we say that the h-vector is palindromic*);*
(c) $H_R(t^{-1}) = (-1)^d t^{-a} H_R(t)$.

Conversely, if R is a Cohen–Macaulay integral domain such that

$$H_R(t^{-1}) = (-1)^d t^{-h} H_R(t)$$

for some $h \in \mathbb{Z}$, then R is Gorenstein, and $h = \deg H_R(t)$.

The assertions (a), (b), and (c) are very easily proved. For the characterization of the Gorenstein rings among the graded integral domains one chooses a nonzero element y from ω_b, $b = \min\{i : \omega_i \neq 0\}$. Since ω_R is torsionfree, the map $R \to \omega$, $x \mapsto xy$ is injective. Under the condition on the Hilbert function, it must be an isomorphism.

The Hilbert quasipolynomial of the canonical module. One can rewrite the reciprocity between $H_R(t)$ and $H_{\omega_R}(t)$ as a functional equation relating their Hilbert quasipolynomials. To this end we study the substitution

$$I : \mathbb{C}[[\mathbb{Z}]] \to \mathbb{C}[[\mathbb{Z}]], \qquad \sum_k c_k t^k \mapsto \sum_k c_k t^{-k}.$$

Both the rings $\mathscr{L}[[\mathbb{Z}_+]] = \mathbb{C}[[t]][t^{-1}]$ and $\mathscr{L}[[\mathbb{Z}_-]] = \mathbb{C}[[t^{-1}]][t]$ are fields; in both cases we have adjoined the inverse of the generator of the maximal ideal of a discrete valuation domain. The field $\mathbb{C}(t)$ of rational functions in one variable is contained in both of them, but the resulting two embeddings $\mathbb{C}(t) \subset \mathbb{C}[[\mathbb{Z}]]$ must be clearly distinguished. They are the Laurent expansions

$$\mathcal{L}_0(f) \in \mathcal{L}[[\mathbb{Z}_+]] \qquad \text{and} \qquad \mathcal{L}_\infty(f) \in \mathcal{L}[[\mathbb{Z}_-]]$$

at 0 and ∞. But no analysis is needed to define them: $\mathcal{L}_0$ is the extension of the natural embedding $\mathbb{C}[t] \to \mathbb{C}[[t]]$ to the fields of fractions, and $\mathcal{L}_\infty$ can be described analogously.

Lemma 6.44. *The restriction of I to $\mathbb{C}[[t]][t^{-1}]$ is a field isomorphism to $\mathbb{C}[[t^{-1}]][t]$, and*

$$I(\mathcal{L}_0(f)) = \mathcal{L}_\infty(f(t^{-1})), \qquad f \in \mathbb{C}(t).$$

Proof. Evidently, the restriction of I to $\mathbb{C}[[t]]$ maps this integral domain isomorphically onto $\mathbb{C}[[t^{-1}]]$, and we only extend the map to the fields of fractions. Let $f \in \mathbb{C}(t)$, $f = g/h$, $f, g \in \mathbb{C}[t]$. Then

$$I(\mathcal{L}_0(f)) = I\left(\frac{g}{h}{}_0\right) = \frac{I(g)}{I(h)}{}_\infty = \frac{g(t^{-1})}{h(t^{-1})}{}_\infty = \mathcal{L}_\infty(f(t^{-1})).$$

We have marked the fractions so that it becomes clear where we are taking them. □

In the $\mathbb{Z}$-graded case of Theorem 6.41 we have $H_{\omega_R}(t) = (-1)^d H_R(t^{-1})$. In order to relate the Hilbert quasipolynomials of R and ω_R, we have to understand the relationship between the Laurent expansions of f and $f(t^{-1})$. It is given by the lemma:

Lemma 6.45. *Let $P : \mathbb{Z} \to \mathbb{C}$ be a quasipolynomial. Then there exists a (unique) rational function $f \in \mathbb{C}(t)$ such that*

$$\mathcal{L}_0(f) = \sum_{k=0}^{\infty} P(k) t^k.$$

Moreover, $\mathcal{L}_0(f(t^{-1})) = \sum_{k=1}^{\infty} -P(-k)t^k$.

Proof. Let $H = \sum_{k=0}^{\infty} P(k)t^k$ and $J = \sum_{k=1}^{\infty} -P(k)t^{-k}$. Both H and J are elements of $\mathbb{C}[[\mathbb{Z}]]$, and

$$H - J = \sum_{k=-\infty}^{\infty} P(k)t^k.$$

Let e be the period of P and d its degree, and set $Q = (1 - t^e)^{d+1}$. Then

$$Q(H - J) = QH - QJ = 0.$$

This equation is proved by induction on d. In fact, $(1 - t^e) \sum P(k)t^k = \sum \tilde{P}(k)t^k$ with the quasi-polynomial $\tilde{P}$ of period e given by $\tilde{P}(j) = P(j) - P(j - e)$, $j \in \mathbb{Z}$. Therefore $\deg \tilde{P} \le \deg P - 1$.

Analyzing QH and QJ, we find that the coefficients of QH vanish for $k < 0$ and those of QJ vanish for $k > (d + 1)e$. Therefore the equation $QH = QJ$ can only hold if both QJ and QH are the same Laurent polynomial K. Set $f = K/Q \in \mathbb{C}(t)$. Then

$$Q\mathcal{L}_0(f) = \mathcal{L}_0(Qf) = \mathcal{L}_0(K) = K,$$

and we conclude that $\mathcal{L}_0(f) = H$. Similarly we see that $\mathcal{L}_\infty(f) = J$.

Now we can apply Lemma 6.44. It yields $\mathcal{L}_0(f(t^{-1})) = I(\mathcal{L}_\infty(f))$. □

Remark 6.46. The proof shows that $(1-t^e)^{d+1}$ is a denominator for f. This polynomial encodes the linear difference equation satisfied by the values of P. One could therefore start with a sequence $(c_k)_{k\in\mathbb{Z}}$ of coefficients satisfying such an equation, and obtain a rational function with corresponding denominator. (The reader should fill in the details.)

Proposition 6.47. *Let R be a positively graded Cohen–Macaulay $\Bbbk$-algebra of Krull dimension d with Hilbert quasipolynomial P. Set $a = \deg H_R(t)$.*
(a) Then $k \mapsto (-1)^{d-1}P(-k)$ is the Hilbert quasipolynomial of ω_R, and

$$H(\omega_R, k) = (-1)^{d-1}P(-k) \qquad \text{for all } k \geq 1.$$

(b) If R is Gorenstein, then $P(-k) = (-1)^{d-1}P(k+a)$ for all $k \in \mathbb{Z}$.

Proof. We write the Hilbert series of R in the form

$$H_R(t) = \sum_{k=0}^{\infty} P(k)t^k + \sum_{k=0}^{a}\big(H(R,k) - P(k)\big)t^k \tag{6.9}$$

where the second sum is empty if $a < 0$. Thus

$$H_{\omega_R}(t) = (-1)^{d-1}\sum_{k=1}^{\infty} P(-k)t^k + (-1)^d \sum_{k=0}^{a}\big(H(R,k) - P(k)\big)t^{-k}.$$

This yields (a), and (b) now follows from Theorem 6.43. □

The coefficients of the Hilbert quasipolynomial. Suppose that R is a positively $\mathbb{Z}$-graded $\Bbbk$-algebra and N a finitely generated graded R-module. That the Hilbert function of N is given by a quasipolynomial in sufficiently high degrees can be reduced to the case of an algebra generated in degree 1 as follows. Let S be the subalgebra of R generated by all elements of degree $e = \operatorname{lcm}(e_1, \ldots, e_d)$. Then N is finitely generated over S (Exercise 6.12), and decomposes into the direct sum

$$N = N^0 \oplus \cdots \oplus N^{e-1}, \qquad N^j = \bigoplus_{i \equiv j\ (e)} N_i.$$

Dividing degrees in S by e and normalizing those in N^j analogously, we can consider S as a $\Bbbk$-algebra generated in degree 1, and N^j as a graded S-module. The Hilbert polynomial of N^j over S then gives the jth component of the Hilbert quasipolynomial of R.

Suppose that there exists a degree 1 homogeneous non-zerodivisor of N in R. Then multiplication by x yields a chain $N^0 \to N^1 \to \cdots \to N^{d-1} \to N^0$ of *injective* S-linear maps that allow us to compare the Hilbert polynomials of the N^j: they must all have the same degree and leading coefficient. Therefore the leading coefficient of the Hilbert quasipolynomial of N is constant. More generally, we could have used a non-zerodivisor of degree coprime to e, and, in a sense, the homogeneous elements of degree coprime to e constitute the "glue" between the "slices" N^j resulting from the R-module structure on their direct sum N.

Let $P = \sum a_i(n)n^i$ be a quasipolynomial. Then the *grade* of P is the smallest integer $g \geq -1$ such that the coefficient a_i is constant for all $i > j$.

Theorem 6.48. *Suppose that $R = \Bbbk[x_1, \dots, x_n]$ is a positively $\mathbb{Z}$-graded $\Bbbk$-algebra and N a finitely generated graded R-module. Let π be the period of the Hilbert quasipolynomial $P = P_N$. Let I be the ideal of R generated by all homogeneous elements x of R such that* $\gcd(\deg x, \pi) = 1$. *Then*

$$\operatorname{grade} P < \dim N/IN.$$

Proof. We use induction on $\dim N$. If $\dim N = 0$, then $P = 0$ and there is nothing to show.

Suppose that $\dim N > 0$. If $\dim N = \dim N/IN$, then the assertion holds since $\operatorname{grade} P \leq \deg P = \dim N$. If $\dim N/IN < \dim N$, there exists a homogeneous element $x \in I$ of degree coprime to π such that $x \notin \mathfrak{p}$ for all prime ideals $\mathfrak{p} \in \operatorname{Supp} N$ with $\dim R/\mathfrak{p} = \dim N$. In fact, I is not contained in their union. So the existence of x follows from the prime avoidance lemma [72, 1.5.10], except that the lemma makes no assertion on $\deg x$. The reader may check that this extra condition can indeed be satisfied or look up Bruns and Ichim [75].

Then we examine the exact sequence

$$0 \to U(-\deg x) \to N(-\deg x) \to N \to N/xN \to 0,$$

where U is the kernel of the multiplication by x. Evidently the induction hypothesis applies to $N' = N/xN$ and also to U since $\operatorname{Supp} U \subset \operatorname{Supp}(N/xN)$. Moreover, $\dim(N'/IN') = \dim N/IN$ and $\dim(U/IU) \leq \dim(N/IN)$.

It follows that $P_N(k) - P_N(k - \deg x) = P_{N'}(k) - P_U(k - \deg x)$ is a quasipolynomial of grade $< \dim N/IN$. The next lemma completes the proof. □

Lemma 6.49. *Let $Q(n) = \sum a_k(n)n^k$ be a quasipolynomial. If $Q(n) - Q(n-g)$ is of grade $< \gamma$ for some g coprime to the period π of Q, then* $\operatorname{grade} Q < \gamma$.

Proof. Let $u = \deg Q$ and let us first compare the leading coefficients. We can assume $\gamma \leq u$. Then one has $a_u(n) - a_u(n-g) = C$ for some constant C and all n, and so $a_u(n) - a_u(n - \pi g) = \pi C$. Since π is the period, we conclude that $C = 0$, and $a_u(n) = a_u(n-g)$. But g is coprime to π, and it follows that a_u is constant.

The descending induction being started, one argues as follows for the lower coefficients. Suppose that $k \geq \gamma$. Then $a_k(n) - a_k(n-g)$ is a polynomial in the coefficients a_j for $j > k$ and g. Since the higher coefficients are constant by induction, it follows that $a_k(n) - a_k(n-g)$ is constant, too, and the rest of the argument is as above. □

6.E Applications to enumerative combinatorics

Having developed the theory of Hilbert functions in the general framework of graded rings, we now specialize them to the objects of our prime interest.

Hilbert series of normal affine monoids. In the first theorem we relate the Hilbert series of a normal affine monoid algebra and that of its interior ideal. Since the Hilbert series do not depend on $\Bbbk$, we may simply write $H(M,-)$ for $H(\Bbbk[M],-)$, etc.

Theorem 6.50. *Let M be a normal affine monoid of rank d, and let $x_1,\dots,x_e$ be the extreme integral generators of $\mathbb{R}_+M$ in $\operatorname{gp}(M)=\mathbb{Z}^d$. Then*

$$H_M(\boldsymbol{t}) = \sum_{x\in M} \boldsymbol{t}^x = \frac{Q(\boldsymbol{t})}{(1-\boldsymbol{t}^{x_1})\cdots(1-\boldsymbol{t}^{x_e})},$$

with a polynomial $Q \in \mathbb{Z}[M]$. Moreover, Q and the denominator have no common factor in $\mathbb{C}[\operatorname{gp}(M)]$.

The Hilbert series of the ideal $\operatorname{int}(M)$ is given by

$$H_{\operatorname{int}(M)}(\boldsymbol{t}) = (-1)^{e-d}\frac{\boldsymbol{t}^{x_1+\dots+x_e}Q(\boldsymbol{t}^{-1})}{(1-\boldsymbol{t}^{x_1})\cdots(1-\boldsymbol{t}^{x_e})}.$$

Proof. Note that M is a finite module over its submonoid generated by the extreme integral generators (or any other submonoid generating $\mathbb{R}_+M$). Therefore we may chose the denominator as specified.

That $Q \in \mathbb{Z}[M]$ (and not just in $\mathbb{Z}[\operatorname{gp}(M)]$) is easy to see: all modules that arise in the proof of Theorem 6.37 are subquotients of $\Bbbk[M]$ and have their support (as graded modules) in M.

An extreme integral generator generates a direct summand of $\operatorname{gp}(M)$. Therefore we may assume that $\boldsymbol{t}^{x_i}$ is a variable of the Laurent polynomial ring $\mathbb{C}[\operatorname{gp}(M)]$. Then $1-\boldsymbol{t}^{x_i}$ generates a prime ideal. Therefore, if Q and the denominator had a nontrivial common factor, one of the factors $1-\boldsymbol{t}^{x_i}$ would cancel. It would then follow that M is contained in a union of finitely many shifted copies of the submonoid generated by the remaining extreme integral generators, and this is evidently false.

To obtain the Hilbert series of the interior ideal, one simply connects Theorems 6.31 and 6.41. □

Counting lattice points in polytopes. A classical problem in the geometry of numbers is counting lattice points in convex bodies, and especially the behavior of this number under refinement of the lattice. For rational convex polytopes $P \subset \mathbb{R}^n$, the corresponding enumerative function

$$E(P,k) = \#\left(P \cap \frac{1}{k}\mathbb{Z}^n\right)$$

can be interpreted as the Hilbert function of an affine monoid algebra. Note that

$$E(P,k) = \#\operatorname{lat}(kP),$$

and set $M = C(P)\cap\mathbb{Z}^{n+1}$. (Recall $C(P)$ is the cone generated by $P\times\{1\}$ in $\mathbb{R}^{n+1}$.) We grade M by setting

$$\deg(x_1,\dots,x_{n+1}) = x_{n+1}.$$

With respect to this grading,

$$E(P,k) = H(M,k),$$

and all the results on Hilbert functions can be applied.

Evidently the Hilbert function of $\operatorname{int}(M)$ counts the lattice points in the interior of M. In analogy with the definition of E we set

$$E^+(P,k) = \#\left(\operatorname{int}(P) \cap \frac{1}{k}\mathbb{Z}^n\right), \qquad k \geq 1.$$

Therefore the theory of the canonical module allows us to relate $E(P,k)$ and $E^+(P,k)$. The corresponding generating functions are given by

$$E_P(t) = \sum_{k=0}^{\infty} E(P,k)t^k \quad \text{and} \quad E_P^+(t) = \sum_{k=1}^{\infty} E^+(P,k)t^k.$$

Theorem 6.51 (Ehrhart, Stanley). *Let P be a rational polytope of dimension d and let v be the smallest number k for which kP has an interior lattice point. Then the following hold:*

(a) *$E_P(t)$ is the expansion of a rational function at 0:*

$$E_P(t) = \frac{1 + h_1 t + \cdots + h_u t^u}{(1 - t^{e_1})\cdots(1 - t^{e_{d+1}})}, \qquad h_u \neq 0,$$

with suitable positive integers $e_1, \ldots, e_{d+1}$.

(b) *As a rational function, $E_P(t)$ has degree $-v$.*

(c) *There exists a quasipolynomial of degree d and period dividing $\operatorname{lcm}(e_1, \ldots, e_{d+1})$ such that $E(P,k) = q_P(k)$ for all $k > -v$, in particular, $E(P,k) = q_P(k)$ for all $k \geq 0$.*

(d) *$E_P^+(t) = (-1)^{d+1}E_P(t^{-1})$ and $E^+(P,k) = (-1)^d q_P(-k)$ for all $k \geq 1$.*

(e) *The coefficients h_i are nonnegative.*

All statements in the theorem are specializations of results in the last section, and we leave it to the reader to collect the arguments.

The function $E(P,-)$ is the *Ehrhart function* of P, $E_P(t)$ is its *Ehrhart series* and q_P the *Ehrhart quasipolynomial*. Part (d) is *Ehrhart's reciprocity theorem*.

If P is even a lattice polytope, then Theorem 6.51 can be strengthened: the exponents e_i in the denominator can all be chosen to be 1, and q is a polynomial. The next theorem generalizes this statement.

Theorem 6.52 (Ehrhart, McMullen, Stanley). *Let P be a d-dimensional rational convex polytope and g be a natural number. Suppose that the affine hull of every g-dimensional face of P contains a lattice point. Then* grade $q_P < g$.

Proof. Let F be a g-dimensional face of P. Since the affine hull of F contains a point with integer coordinates, kF contains such a point for all $k \gg 0$. We chose k big enough so that kF contains a lattice point x_F for every g-dimensional face F, and, moreover, $\gcd(k, \pi) = 1$ where π is the period of q_P.

Set $M = C(P) \cap \mathbb{Z}^{n+1}$, $R = \Bbbk[M]$, and let $J \subset R$ be the ideal generated by the monomials corresponding to the lattice points (x_F, k). If $\dim R/J \leq g$, then we are done by Theorem 6.48 because the ideal I in Theorem 6.48 contains J.

Since J is a monomial ideal, the minimal prime overideals $\mathfrak{p}$ of J are monomial. By Corollary 4.34, for each such $\mathfrak{p}$ there is a face $G_\mathfrak{p}$ of P such that $\mathfrak{p}$ is generated by all monomials outside the face $\mathbb{R}_+(G_\mathfrak{p} \times \{1\})$ of $C(P)$. One has $\dim R/\mathfrak{p} = \dim G_\mathfrak{p} + 1$. By the choice of J, the dimension of $G_\mathfrak{p}$ is bounded by $g - 1$. Altogether we conclude that $\dim R/J \leq g$. □

If P is a lattice polytope, then we can normalize the denominator of $E_P(t)$ to be $(1-t)^{d+1}$, and the Ehrhart function can be considered to be uniquely determined by the h-vector $(h_0 = 1, h_1, \dots, h_u)$. A complete characterization of the potential h-vectors of lattice polytope seems to be unknown. However, see [72, 6.3.15] for inequalities it has to satisfy (in addition to positivity).

The reciprocity law in Theorem 6.51(d) can be strengthened if the monoid $M = C(P) \cap \mathbb{Z}^{n+1}$ is Gorenstein. We will state an analogue in Theorem 6.53. If, in addition, P is integrally closed and M is (therefore) generated in degree 1, then one is very much justified in asking for further properties of the h-vector. We will come back to this question below.

Nonnegative solutions of homogeneous linear diophantine equations. We fix a number n and consider $n \times n$ matrices $A = (x_{ij})$ with entries in $\mathbb{Z}_+$ whose row and column sums all have the same value k. The problem of determining the number $\mathscr{M}(n,k)$ of such *magic squares* was a motivating example for Stanley [323], [332]. Anand, Dumir, and Gupta [3] had previously made some conjectures about $\mathscr{M}(n,k)$ as a function of the *magic sum* k, which were then proved by Stanley.

8	1	6
3	5	7
4	9	2

16	3	2	13
5	10	11	8
9	6	7	12
4	15	14	1

Fig. 6.2. Two famous magic squares

Let us consider all $n \times n$ matrices $A = (x_{ij})$ with entries in $\mathbb{Z}_+$ that satisfy the system of equations

$$\sum_{j=1}^{n} x_{ij} = \sum_{k=1}^{n} x_{kl}, \qquad i, l = 1, \dots, n. \tag{6.10}$$

They evidently form a monoid M_n obtained as the intersection of $\mathbb{Z}^{n\times n}$ with the rational cone $\mathbb{R}_+^{n\times n}$ and the rational subspace of $\mathbb{R}^{n\times n}$ of all solutions to the system (6.10). Thus M_n is the intersection of $\mathbb{Z}^{n\times n}$ with a rational cone, and therefore it is an integrally closed affine submonoid of $\mathbb{Z}^{n\times n}$. The assignment $A \mapsto \sum_{j=1}^{n} x_{ij}$ is evidently a positive grading on M_n, and the elements of degree k are exactly the magic squares of magic sum k. So $\mathscr{M}(n,-)$ is "just" the Hilbert function of an integrally closed affine monoid.

Let us more generally consider graded monoids M that arise as the set of nonnegative solutions of a homogeneous system

$$a_{i1}x_1 + \cdots + a_{in}x_n = 0, \qquad a_{ij} \in \mathbb{Z},\ i = 1,\dots,m. \tag{6.11}$$

of diophantine linear equations.

Instead of formulating the analogue of Theorem 6.51 we just indicate how to translate the problem of counting solutions of (6.11) to counting lattice points in polytopes. Simply define P as the intersection of the cone of nonnegative real solutions of (6.11) with the hyperplane defined by the equation $\deg x = 1$; then $H(M,k) = E(P,k)$.

Let us return to the magic squares. For them the polytope P in degree 1 is in fact integral: it is the *Birkhoff* polytope of doubly stochastic matrices whose vertices are the permutation matrices (clearly of magic sum 1). But more is true: the permutation matrices generate the monoid M_n (Exercise 6.16), and therefore $\mathscr{M}(n,-)$ is the Hilbert function of a monoid (or algebra) generated in degree 1. It follows that $\mathscr{M}(n,-)$ is given by a polynomial, and the entries of the h-vector are strictly positive.

A trivial magic square is the matrix e all of whose entries are 1. The support forms of M_n are just the coordinate linear forms, and so e is an element satisfying the condition in Theorem 6.33(c): M_n is Gorenstein and e generates its canonical module. This fact allows one to compute $\deg H_{M_n}(t)$, and therefore the length of the h-vector (see Exercise 6.16).

We conclude our discussion of linear diophantine equations with a discussion of the Gorenstein case. In order to identify the set of strictly positive solutions with the integral points in the relative interior of the space of all nonnegative solutions, we must assume that the system is *nondegenerate*, i. e., its set of solutions is not contained in any of the coordinate hyperplanes. The assumptions in the following theorem imply that the system is nondegenerate.

Theorem 6.53. *Let M be the monoid of its nonnegative integral solutions of the system* (6.11), *and let q be the Hilbert quasipolynomial of M.*

(a) Suppose $\mathsf{e} = (1,\dots,1) \in M$ *and set* $v = \deg(\mathsf{e})$. *Then*
  (i) M is Gorenstein;
  (ii) the h-vector $(h_0,\dots,h_u)$ is palindromic;
  (iii) one has $q(-k) = (-1)^{r-1}q(k-v)$ for all $k \in \mathbb{Z}$.

(b) If the coordinate hyperplanes (intersected with $\operatorname{aff}(M)$*) are the support hyperplanes of $\mathbb{R}_+M$, then each of (i), (ii), and (iii) implies that* e *is a solution to the system* (6.11).

Proof. By the definition of M, a subset of the set of coordinate hyperplanes (after intersection with $\operatorname{aff}(M)$) constitutes the set of support hyperplanes of $\mathbb{R}_+M$. If e is a solution, then it generates int(M). By Theorem 6.33, M is Gorenstein, and (ii) and (iii) follow from Corollary 6.43 and Proposition 6.47.

Part (b) follows from Exercise 6.15 and Corollary 6.43. □

The reader to whom the elements of M_n do not seem magic enough can introduce further equations, for example, by requiring that the sum over the diagonals of the matrices also equal the magic sum. As long as e is a solution, the monoid of solutions remains Gorenstein. Of course, it will almost always fail to be generated in degree 1.

Volumes and multiplicities. Already in Section 2.C we have used the term "multiplicity" for a magnitude closely related to the volume of a polytope. We can now justify this terminology.

Theorem 6.54. *Let $P \subset \mathbb{R}^d$ be a d-dimensional lattice polytope, E a subset of* $\operatorname{lat}(P)$ *containing* $\operatorname{vert}(P)$, *M the monoid generated by $E \times 1 \subset \mathbb{Z}^{d+1}$, and $\hat{M} = C(P) \cap \mathbb{Z}^{d+1}$. Furthermore let $\Bbbk$ be a field, $R = \Bbbk[M]$, and $\hat{R} = \Bbbk[\hat{M}]$. Then, with $m = [\mathbb{Z}^{d+1} : \operatorname{gp}(M)]$,*

$$m \cdot e(R) = e(\hat{R}) = \mu(P) = d!\ \operatorname{vol}(P).$$

Proof. First of all we should note that R is generated by its degree 1 elements and $\hat{R}$ is a finitely generated R-module of rank m. Since multiplicity can be computed as rank over a Noether normalization of R, the first equation follows.

For the second equation let a_d be the leading coefficient of the Ehrhart polynomial of P; then $e(\hat{R}) = d!\, a_d$. On the other hand, elementary arguments of measure theory show that

$$\operatorname{vol}(P) = \lim_{k\to\infty} \frac{E(P,k)}{k^d} = a_d. \qquad \square$$

The term multiplicity is not only used for algebras, but also for ideals. Let R be a local Noetherian ring with maximal ideal $\mathfrak{m}$, $d = \dim R$, let I be an $\mathfrak{m}$-primary ideal, and denote the length of a module M by $\lambda(M)$. Then

$$e(I,R) = d! \lim_{k\to\infty} \frac{\lambda(R/I^k)}{k^d} \tag{6.12}$$

is the *multiplicity of R with respect to I*. One can also define $e(I,R)$ as the multiplicity of the associated graded ring $\operatorname{gr}_I(R) = \bigoplus_{k\ge 0} I^k/I^{k+1}$, and the only reason we do not is that we have considered Hilbert functions only for algebras whose zeroth component is a field. However, this approach guarantees the existence of the limit; see [72, 4.8].

Equation (6.12) also defines the multiplicity of an ideal I that is primary to the graded maximal ideal $\mathfrak{m}$ of a positively graded algebra R over a field $\Bbbk$. It does not matter whether one passes first to the localizations $R_\mathfrak{m}$ and $I_\mathfrak{m}$. Moreover, length can be replaced by vector space dimension over $\Bbbk$.

Suppose $M \subset \mathbb{Z}^d$ is a positive affine monoid of rank d, and I is a monomial ideal in $R = \Bbbk[M]$ primary to the maximal ideal generated by all monomials $\neq 1$. Let $\mathscr{P}(I)$ be the convex hull of the lattice points corresponding to the monomials in I. It is indeed a polyhedron (prove this). Since I is primary to $\mathfrak{m}$, it contains a power of each of the extreme generators of M, and therefore $\mathbb{R}_+M \setminus \mathscr{P}(I)$ is the union of finitely many polytopes that are obtained as the pyramids with apex 0 whose bases are the compact facets of $\mathscr{P}(I)$. If $I = \mathfrak{m}$, then the union of the compact facets is just the bottom of M; see Figure 2.4.

Theorem 6.55. *Suppose $M \subset \mathbb{Z}^d$ is a positive affine monoid of rank d, and I is a monomial ideal in $R = \Bbbk[M]$ primary to the maximal ideal generated by all monomials $\neq 1$. Furthermore let $\hat{M}$ be the integral closure of M in $\mathbb{Z}^d$, and $\hat{R} = \Bbbk[\hat{M}]$. Then, with $m = [\mathbb{Z}^d : \operatorname{gp}(M)]$,*

$$m \cdot e(I,R) = e(I\hat{R},\hat{R}) = d!\operatorname{vol}(\mathscr{P}(I)).$$

Proof. Since m is the rank of $\hat{R}$ as an R-module, the first equation is a purely ring-theoretic result; see [72, 4.7.9]. From now on we may assume that $M = \hat{M}$.

Since $\mathbb{R}_+M \setminus \mathscr{P}(I)$ is the union of finitely many polytopes, one has

$$\operatorname{vol}(\mathscr{P}(I)) = \lim_{k\to\infty} \frac{\#\bigl(\mathbb{R}_+M \setminus k\mathscr{P}(I)) \cap \mathbb{Z}^d\bigr)}{k^d}.$$

On the other hand, the number of lattice points appearing in the numerator under the limit is the $\Bbbk$-vector space dimension of $R/\overline{I^k}$, the residue class ring with respect to the integral closure of I^k (Theorem 4.46). While $\overline{I^k} \neq I^k$ in general, it is close enough to I^k for our purpose.

First one can replace I by an arbitrary power I^q since $e(I^q,R) = q^d e(I,R)$, and since the corresponding volumes are related by the same factor. Second, we can replace I^q by its integral closure $\overline{I^q}$ without changing the multiplicity (Huneke and Swanson [205, 11.2.1]). Third, I has a power I^q such that

$$\overline{\bigl(I^q\bigr)}^{\,r} = \overline{\bigl(I^q\bigr)^r} \quad \text{for all } r. \tag{6.13}$$

In order to finish the proof we have only to justify this assertion.

The normalization of the Rees algebra $\mathscr{R} = \mathscr{R}(I,R) = \bigoplus_{r\geq 0} I^r$ is $\overline{\mathscr{R}} = \bigoplus_{r\geq 0} \overline{I^r}$. In our case one can derive this fact from Theorem 4.46, or one can look it up [205, 5.2.1] for the general case. The crucial point is that $\overline{\mathscr{R}}$ is a finitely generated module over $\mathscr{R}$, which in our case is an affine monoid algebra itself: the monoid is generated by all elements $(x,0)$ $x \in M$, and all $(y,1)$ where y runs through a monomial system of generators of I. Now the finite generation follows from Corollary 2.10 and Theorem 4.42.

Pick q so large that $\overline{\mathscr{R}}$ is generated over $\mathscr{R}$ by monomials of degree $\leq q$. Then equation (6.13) follows. (Compare the proof of Corollary 2.57.) □

Hilbert functions and triangulations. Let $C \subset \mathbb{R}^d$ be a rational cone of dimension d, and Γ a rational subdivision of C. Then C decomposes into the disjoint union of the interiors of the cones $\gamma \in \Gamma$, and therefore the Hilbert series of $M = C \cap \mathbb{Z}^d$ is just the sum of the Hilbert series of $\operatorname{int}(\gamma) \cap \mathbb{Z}^d$ (viewed as a module over $\gamma \cap \mathbb{Z}^d$). We consider the lattice $\mathbb{Z}^d$ as being fixed and simplify notation by writing H_C for $H_{C\cap\mathbb{Z}^d}$, etc. With this convention,

$$H_M(t) = \sum_{\gamma\in\Gamma} H_{\text{int}(\gamma)}(t). \tag{6.14}$$

Similarly we can decompose the Hilbert series of $\text{int}(M)$, but this time we must sum only over all γ for which $\text{int}(\gamma) \subset \text{int}(C)$. Let us denote this subset by $\text{int}(\Gamma)$. Then

$$H_{\text{int}(C)}(t) = \sum_{\gamma\in\text{int}(\Gamma)} H_{\text{int}(\gamma)}(t). \tag{6.15}$$

Since $\text{int}(\gamma) \cap \mathbb{Z}^d$ defines the canonical module of $\gamma \cap \mathbb{Z}^d$,

$$H_{\text{int}(\gamma)}(t) = (-1)^{\dim\Gamma} H_\gamma(t^{-1}).$$

Applying the same argument to C, we obtain inclusion-exclusion type formulas for $H_C(t)$ and $H_{\text{int}(C)}(t)$:

Theorem 6.56 (Stanley). *With the notation introduced,*

$$H_C(t) = \sum_{\gamma\in\text{int}(\Gamma)} (-1)^{\dim C-\dim\gamma} H_\gamma(t),$$
$$H_{\text{int}(C)}(t) = \sum_{\gamma\in\Gamma} (-1)^{\dim C-\dim\gamma} H_\gamma(t).$$

The theorem takes a particularly elegant form if all γ are unimodular. In this case $\gamma \cap \mathbb{Z}^d$ is the free abelian monoid generated by its extreme integral generators $e_1,\dots,e_d$, $d = \dim\gamma$, and therefore

$$H_\gamma(t) = \frac{1}{(1-t^{e_1})\cdots(1-t^{e_d})},$$
$$H_{\text{int}(\gamma)}(t) = \frac{t^{e_1+\cdots+e_d}}{(1-t^{e_1})\cdots(1-t^{e_d})} = (-1)^d H_\gamma(t^{-1}).$$

So far we have used commutative algebra to derive purely combinatorial results on lattice points in cones, but at this point we can turn the tables. From the decomposition formulas (6.14) and (6.15) it follows immediately that $H_C(t)$ and $H_{\text{int}(C)}(t)$ are rational functions. However, for the reciprocity law $H_{\text{int}(C)}(t) = (-1)^{\dim C} H_C(t)$, one has to prove one of the formulas in Theorem 6.56 directly. The main tool is the Euler–Poincaré characteristic, and the details can be found in Stanley [323] and [330].

Once the equation $H_{\text{int}(C)}(t) = (-1)^{\dim C} H_C(t)$ has been established, it follows that $\text{int}(C) \cap \mathbb{Z}^d$ defines the canonical module of $C \cap \mathbb{Z}^d$ since it has the right $\mathbb{Z}^d$-graded Hilbert function, and a monomial ideal is uniquely determined by its $\mathbb{Z}^d$-graded Hilbert function. (This argument was used in [325]).

***h*-vectors of Gorenstein polytopes.** Let R be a positively $\mathbb{Z}$-graded Gorenstein $\Bbbk$-algebra generated in degree 1. Then

$$H_R(t) = \frac{1 + h_1 t + \cdots + h_u t^u}{(1-t)^d}, \qquad d = \dim R,\ h_u \neq 0,$$

with a palindromic h-vector $h = (h_0 = 1, h_1, \ldots, h_u)$: $h_i = h_{u-i}$, $i = 0, \ldots, u$. It is an open question whether h is *unimodal* whenever R is a Gorenstein *integral domain*. (For arbitrary Gorenstein rings there exist counterexamples; see Stanley [325, p. 70].) Unimodality means that

$$h_0 \le \cdots \le h_{\lfloor u/2 \rfloor}.$$

An even stronger question is whether the *g-vector*

$$g = (h_0, h_1 - h_0, h_2 - h_1, \ldots, h_{\lfloor u/2 \rfloor} - h_{\lfloor u/2 \rfloor - 1})$$

is a *Macaulay sequence*, i. e., the Hilbert function of a 0-dimensional graded $\Bbbk$-algebra generated in degree 1. The terminology reminds us of Macaulay's theorem characterizing such Hilbert functions in terms of binomial expansions (see [72, 4.2]).

There is a class of Gorenstein rings for which the questions just posed have a positive answer. Let R be the Stanley–Reisner ring of the boundary of a simplicial polytope; then the g-vector of R is a Macaulay sequence. This is the content of McMullen's g-conjecture, made a theorem by Stanley [328]. (The converse has been shown by Billera and Lee [32].) The g-theorem and Stanley's proof will be discussed on Chapter 10; see Theorem 10.41.

Let us call a lattice polytope P *Gorenstein* if it is normal and $\Bbbk[P]$ is Gorenstein. Then the Hilbert function of $\Bbbk[P]$ is just the Ehrhart function of P. Even for the h-vector of this Hilbert function unimodality is not known in general, but one has

Theorem 6.57. *Let P be a Gorenstein polytope with a regular unimodular triangulation. Then the g-vector of (the Ehrhart series of) P is a Macaulay sequence.*

This theorem was proved by Bruns and Römer [78]. We restrict ourselves to an outline of the proof. Let x be the generator of $\operatorname{int}(M)$. We decompose x into a sum of elements of $\operatorname{Hilb}(M)$. Note that $\operatorname{Hilb}(M)$ is just given by $\operatorname{lat}(P) \times \{1\}$. Therefore $x = y_1 + \cdots + y_t$ where $\deg y_i = 1$ for all i. Let σ be the standard embedding of M. Then $\sigma(x) = (1, \ldots, 1)$ and, therefore, the $\sigma(y_i)$ are pairwise disjoint 0-1-vectors.

The first essential step is the construction of a regular unimodular triangulation of $\mathbb{R}_+ M$ whose core is the unimodular cone generated by $y_1, \ldots, y_t$: the *core* of a triangulation is the intersection of its facets. By hypothesis, P has a unimodular triangulation. It induces a unimodular triangulation Θ of C. We restrict Θ to those faces of C that do not contain any of the irreducible elements y_i, and let Σ denote the restriction.

Lemma 6.58. *Let*

$$\Lambda = \Sigma \cup \bigcup_{j=1}^{t} \{\mathbb{R}_+(G \cup \{y_{i_1}, \ldots, y_{i_j}\} : G \in \Sigma,\ 1 \le i_1 < \cdots < i_j \le t\}.$$

Then Δ is a regular unimodular triangulation of C with core $\mathbb{R}_+ y_1 + \cdots + \mathbb{R}_+ y_t$.

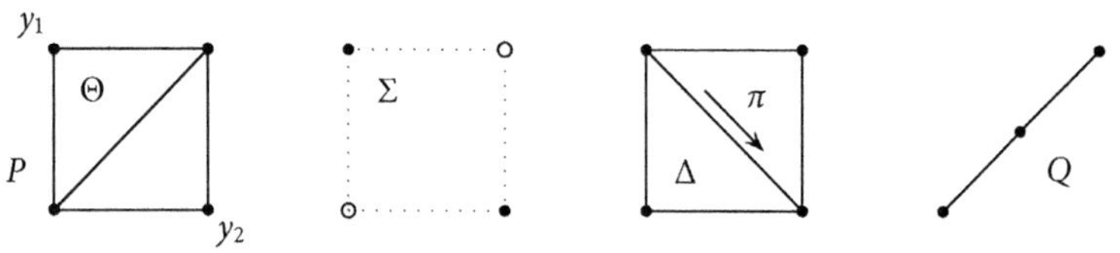

Fig. 6.3. Change of the triangulation and projection

In order to obtain just a triangulation by this construction it is enough that the sets of facets not containing y_i, $i = 1, \dots, t$, be pairwise disjoint. The unimodularity of Δ follows from the unimodularity of Σ and the fact that $\sigma(y_1 + \dots + y_t) = (1, \dots, 1)$. Finally the regularity of Σ implies the regularity of Δ. Figure 6.3 illustrates the change of triangulation for the unit lattice square, including the projection π along the subspace U generated by $y_2 - y_1, \dots, y_t - y_{t-1}$ onto $\mathbb{R}^n/U$ (with associated lattice $\mathbb{Z}^n/(U \cap \mathbb{Z}^n)$):

Lemma 6.59. *The images of the cones in Δ form a regular unimodular triangulation of $\pi(C)$ which is nothing but the cone over the lattice polytope $Q = \pi(P)$. Moreover, the core of $\pi(\Delta)$ is the image of the core of Δ, namely the ray from 0 through $\pi(y_1)$.*

In order to obtain a triangulation by projection it is sufficient to project along a subspace spanned by elements in the core. The unimodularity follows from the unimodularity of Δ and the fact that the vectors $y_2 - y_1, \dots, y_t - y_{t-1}$ generate a direct summand of $\mathbb{Z}^n$. (Compare Exercise 1.6.)

Restricting the triangulation $\pi(\Delta)$ to the polytope Q we obtain a regular unimodular triangulation Γ of Q.

Lemma 6.60. *The h-vector of the Ehrhart series of P coincides with the corresponding h-vector of Q. Moreover, the triangulation Γ is regular and unimodular.*

Thus we have reached a Gorenstein polytope with a single interior lattice point $\pi(y_1)$ that has the same h-vector as P, and moreover, has a regular unimodular triangulation Γ whose core is the single interior lattice point. Restricting the unimodular triangulation to the boundary of Q, we get a simplicial complex whose Stanley–Reisner ring has the same h-vector as P.

Now the regularity of the triangulation becomes important. After a translation we may assume that $0 = \pi(y_1)$ is the unique lattice point of Q. Then the simplices of Γ span a simplicial projective fan, and Proposition 1.67 yields a simplicial polytope Q' whose boundary is combinatorially equivalent to $\Gamma|\partial Q$. Finally we are in a position to apply the g-theorem and to conclude the proof.

Remark 6.61. (a) Exploiting the ideas of the proof of Theorem 6.57 more carefully, one obtains the following statements:

(i) Let P be a Gorenstein polytope. Then there exists an integrally closed Gorenstein polytope Q such that $\operatorname{int}(Q)$ contains a unique lattice point and such that the h-vectors of the Ehrhart series of P and Q coincide.
(ii) If P has a (not necessarily regular) unimodular triangulation, then there exists a simplicial sphere S whose Stanley–Reisner ring has the same h-vector as P.

Of course, for (ii) we simply drop the regularity of the triangulation and stop at Lemma 6.60. For (i) we use that the conclusion can already be reached if we start from a unimodular triangulation of C, which always exists (Theorem 2.75).

(b) If we start directly from a Gorenstein monoid M that it not necessarily defined by a lattice polytope, then we arrive at the following theorem [78, Theorem 3]: let $\operatorname{int}(M) = x + M$ and $x = y_1 + \cdots + y_t$. Then the binomials $X^{y_2} - X^{y_1}, \dots, X^{y_t} - X^{y_{t-1}}$ form a regular sequence in $\Bbbk[M]$, and $\Bbbk[M]/(X^{y_2} - X^{y_1}, \dots, X^{y_t} - X^{y_{t-1}})$ is again a Gorenstein normal affine monoid domain.

(c) It is essential in Theorem 6.57 that P be integrally closed. Otherwise there exist counterexamples; see Mustaţă and Payne [273].

6.F Divisorial linear algebra

In this section we assume that the reader is familiar with Section 4.F, and omit detailed cross-references to the results it contains.

Conic divisor classes. A certain class of divisorial ideals in normal affine monoid algebras plays a special role, and its members turn out to be always Cohen–Macaulay. Let M be a normal affine monoid embedded into $\mathbb{Z}^d = \operatorname{gp}(M)$, and let $F_1, \dots, F_s$ be the facets of $\mathbb{R}_+ M$ with support forms $\sigma_1, \dots, \sigma_s$. We have seen in Theorem 4.54 and Corollary 4.61 that every divisorial ideal is isomorphic to a monomial divisorial ideal, and that the monomial divisorial ideals are given in terms of their monomial bases as follows. We choose $u \in \mathbb{Z}^s$ and set

$$\mathscr{P}(u) = \{x \in \mathbb{R}^d : \sigma(x) \geq u\}.$$

Here $\sigma(x) = (\sigma_1(x), \dots, \sigma_s(x))$, and the inequality must be read componentwise. Then we obtain a divisorial ideal $\mathscr{D}(u)$ whose monomial basis corresponds to the lattice points in the polyhedron $\mathscr{P}(u)$:

$$\mathscr{D}(u) = \Bbbk \cdot \{x \in \mathbb{Z}^d : \sigma(x) \geq u\}, \qquad u \in \mathbb{Z}^s.$$

Let us first identify the torsion divisor classes:

Proposition 6.62. *The class of $\mathscr{D}(u)$, $u \in \mathbb{Z}^s$, is a torsion element in* $\operatorname{Cl}(R)$ *if and only if there exists $y \in \mathbb{Q}^d$ such that $\mathscr{P}(u) = y + \mathbb{R}_+ M$.*

Proof. The formation of $\mathscr{D}(u)$ is compatible with taking integral multiples of divisors: $k \operatorname{div}(\mathscr{D}(u)) = \operatorname{div}(\mathscr{D}(ku))$. The class of $\mathscr{D}(u)$ is torsion if and only if there exists k such that $\mathscr{D}(ku) = x + \mathscr{D}(0)$ for some $x \in \mathbb{Z}^d$, and the latter condition is equivalent to $\mathscr{P}(u) = x/k + \mathscr{P}(0)$. □

Thus, if $\mathscr{D}(u)$ represents a torsion class, the monomial basis of $\mathscr{D}(u)$ is cut out from $\mathbb{Z}^d$ by a translate of the cone $\mathbb{R}_+ M$. This property applies to more divisorial ideals if we relax the condition that $u \in \mathbb{Z}^d$: a monomial divisorial ideal is *conic* if its monomial basis is given by the lattice points in $x + \mathbb{R}_+ M$ for some $x \in \mathbb{R}^d$. In our previous notation, the conic divisorial ideals are defined as

$$\mathcal{D}(\sigma(x)), \qquad x \in \mathbb{R}^d.$$

The divisor class of a conic ideals is also called *conic*.

Remark 6.63. (a) Among the conic divisorial ideals are the ideals $\mathfrak{q}_G$ defined by a face G and the facets F of $\mathbb{R}_+ M$ in the following way:

$$\mathfrak{q}_G = \bigcap_{F \supset G} \mathfrak{p}_F.$$

This includes the divisorial prime ideals $\mathfrak{p}_F$ and their intersection $\omega = \mathfrak{p}_1 \cap \cdots \cap \mathfrak{p}_s$. Therefore, whenever $\operatorname{Cl}(R)$ is not a torsion group, then there exist conic divisor classes that are not torsion elements in $\operatorname{Cl}(R)$.

In order to prove that $\mathfrak{q}_G$ is conic we have to find $y \in \mathbb{R}^d$ such that $0 < \sigma_F(y) \le 1$ if $F \supset G$ and $-1 < \sigma_F(y) \le 0$ otherwise. Choose z' in the interior of $\mathbb{R}_+ M$ and z'' in the interior of $-G$. Then $\sigma_F(z' + az'')$ has the desired sign for $a \gg 0$, and we can take $y = a'(z' + az'')$ for $a' > 0$ sufficiently small.

(b) We will see below that every conic divisorial ideal $\mathfrak{q}$ is Cohen–Macaulay, and for every Cohen–Macaulay ideal $\mathfrak{q}$ the ideal $\omega : \mathfrak{q} = \operatorname{Hom}_R(\mathfrak{q}, \omega)$ is also Cohen–Macaulay (see Exercise 6.10 or [72, 3.3.10]). Nevertheless, it is interesting to observe that $\omega : \mathfrak{q}$ is conic if $\mathfrak{q}$ is so (Exercise 6.18). From (b) we conclude that the divisorial ideals

$$\mathfrak{r}_G = \bigcap_{F \not\supset G} \mathfrak{p}_F$$

are also conic, since $\operatorname{div}(\mathfrak{r}_G) + \operatorname{div}(\mathfrak{q}_G) = \sum_F \operatorname{div}(\mathfrak{p}_F) = \operatorname{div}(\omega)$.

It follows immediately from the definition that two conic ideals $\mathcal{D}((\sigma(x))$ and $\mathcal{D}(\sigma(y))$ coincide if and only if $\lceil \sigma(x) \rceil = \lceil \sigma(y) \rceil$ (recall that $\lceil x \rceil$ is the ceiling of x). This condition partitions $\mathbb{R}^d$ into a set $\mathcal{S}$ of equivalence classes that we want to describe geometrically.

First we decompose $\mathbb{R}^s$ into the union of the semi-open cubes

$$Q_z = \{y \in \mathbb{R}^s : \lceil y \rceil = z\}, \qquad z \in \mathbb{Z}^s,$$

and then we take their preimages in $\mathbb{R}^d$ with respect to σ. In other words, if we consider $\mathbb{R}^d$ as a subspace of $\mathbb{R}^s$ via σ, we intersect $\mathbb{R}^d$ with the semi-open cubes Q_z. The intersection is a semi-open polytope, called a *compartment* in the following.

Evidently the decomposition $\mathcal{S}$ of $\mathbb{R}^d$ into compartments is invariant under the action of $\mathbb{Z}^d$ by translation:

$$X \in \mathcal{S} \iff X + y \in \mathcal{S}, \qquad y \in \mathbb{Z}^d.$$

Therefore $\mathcal{S}$ induces a decomposition $\mathcal{T} = \mathcal{S}/\mathbb{Z}^d$ of the n-dimensional torus $\mathbb{R}^d/\mathbb{Z}^d$. Instead we can also decompose a fundamental domain of the action of $\mathbb{Z}^d$,

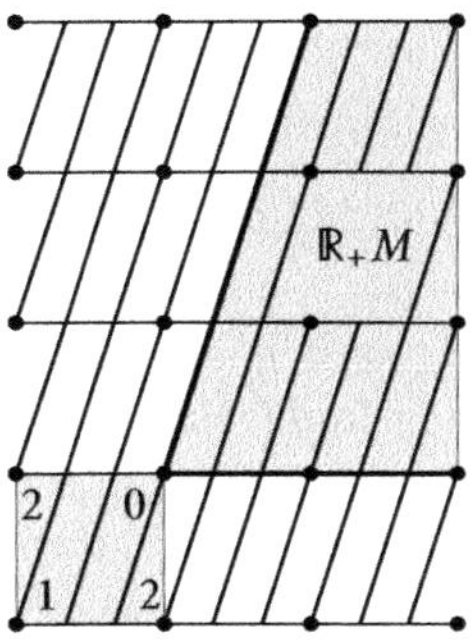

Fig. 6.4. The decomposition $\mathscr{T}$

for example the semi-open cube $(-1,0]^d$. Figure 6.4 illustrates the construction for a very simple example. In this case $\mathscr{T}$ has three elements, marked 0, 1, 2 in the figure.

It is clear that $\lceil\sigma(x)\rceil$ takes only finitely many values on $(-1,0]^d$, and therefore $\mathbb{R}^d/\mathbb{Z}^d$ decomposes into finitely many subsets.

Theorem 6.64. *Let M be a positive normal affine monoid. Then there exist only finitely many conic divisor classes for the monoid algebra $\Bbbk[M]$, and they are in bijective correspondence with the decomposition $\mathscr{T}$ of $\mathbb{R}^d/\mathbb{Z}^d$.*

Proof. Two divisorial ideals $\mathscr{D}(u)$ and $\mathscr{D}(v)$ are isomorphic if and only if $\lceil u\rceil-\lceil v\rceil \in \sigma(\mathbb{Z}^d)$. For $u=\sigma(y)$ and $v=\sigma(z)$ this holds if and only if the compartments of $\mathscr{S}$ containing y and z, respectively, differ by translation by an element of $\mathbb{Z}^d$. □

In terms of the decomposition of $\mathbb{R}^s$ into the semi-open cubes Q_z we can rephrase the definition of conic divisorial ideal as follows: $\mathscr{D}(u)$ is conic if and only if $Q_{\lceil u\rceil}\cap \sigma(\mathbb{R}^d)\neq\varnothing$ (and it is torsion if $\lceil u\rceil\in\sigma(\mathbb{Z}^d)$). An equivalent condition is given by the next proposition. Its easy proof is left to the reader.

Proposition 6.65. *The divisorial ideal $\mathscr{D}(u)$ is conic if and only if the coset $\sigma(\mathbb{R}^d)-\lceil u\rceil$ in $\mathbb{R}^s/\sigma(\mathbb{R}^d)$ contains a point in the semi-open cube $Q=(-1,0]^s$.*

Let T be the torsion subgroup of $\operatorname{Cl}(R)\cong\mathbb{Z}^s/\sigma(\mathbb{Z}^d)$. Then $\operatorname{Cl}(R)/T$ is a free abelian group of rank $m=s-d$. It can be identified with a rank m lattice L in the m-dimensional vector space $\mathbb{R}^s/\sigma(\mathbb{R}^d)$. Proposition 6.65 shows that the conic classes are given by the lattice points in the image of the semi-open cube $Q=(-1,0]^s$ in $\mathbb{R}^s/\sigma(\mathbb{R}^d)$, and therefore form the set of lattice points in a lattice polytope in this vector space. It follows from Remark 6.63(c) that the this polytope is centrally symmetric with respect to the point $[\omega]/2$. For a special case this idea will be shown in Figure 6.5.

Frobenius endomorphisms of monoid algebras. Monoid algebras ave a special class of endomorphisms. Let M be a not necessarily cancellative or torsionfree monoid and $c\in\mathbb{N}$. Then

$$c_* : M \to M, \qquad c_*(x) = x^c,$$

is evidently a monoid endomorphism and it induces an endomorphism

$$c_* : R[M] \to R[M], \qquad \sum_{x \in M} r_x x \mapsto \sum_{x \in M} r_x x^c,$$

of the monoid algebra $R[M]$. (As usual, R is a commutative ring.) The endomorphism c_* (of M or $R[M]$) is called a *monoidal Frobenius endomorphism* of $R[M]$. Therefore we can view $R[M]$ as an algebra over itself via c_*. In order to denote the change in algebra structure, we denote $R[M]$ by $R[M^{1/c}]$ if we view it as an $R[M]$-algebra via c_*. Thus, the product of $a \in R[M]$ and $b \in R[M^{1/c}]$ is given by $c_*(a) \cdot b$ where $\cdot$ denotes the ordinary multiplication in $R[M]$. If M is affine, c_* is injective, and $R[M]$ is a subalgebra of $R[M^{1/c}]$. In this case we can simply define $M^{1/c}$ to be the monoid $\{x/c : x \in M\} \subset \mathbb{Q} \otimes \operatorname{gp}(M)$. Note that the monoidal Frobenius endomorphism c_* induces an embedding $c_* : \operatorname{gp}(M) \to \operatorname{gp}(M^{1/c})$.

Suppose now that M is affine. Then the extension $\operatorname{gp}(M) \subset \operatorname{gp}(M^{1/c})$ is of finite index c^r, $r = \operatorname{rank} M$. We denote the quotient $\operatorname{gp}(M^{1/c})/\operatorname{gp}(M)$ by G_c.

The M-module $M^{1/c}$ decomposes into the disjoint union

$$M^{1/c} = \bigcup_{N \in G_c} M^{1/c} \cap N$$

of the intersections of $M^{1/c}$ with the residue classes of $\operatorname{gp}(M^{1/c})$ modulo $\operatorname{gp}(M)$. Each of the intersections is nonempty. The decomposition of $M^{1/c}$ induces a corresponding decomposition of $R[M]$:

Proposition 6.66. *For each $N \in G_c$ let D_N denote the free R-module with basis $M^{1/c} \cap N$. Then*

$$R[M^{1/c}] = \bigoplus_{N \in G_c} D_N.$$

For a normal affine monoid M we can identify the summands D_N and count how often each isomorphism class occurs among them. In part (c) of the next theorem, $\operatorname{vol}(\tau)$ denotes the volume of a compartment of class τ. For simplicity we restrict ourselves to a field of coefficients.

Theorem 6.67. *Let M be a normal affine monoid of rank d, and consider the decomposition of $\Bbbk[M^{1/c}]$ in the $\Bbbk[M]$-modules D_N.*

(a) Then each of the modules D_N is isomorphic to a conic divisorial ideal.

(b) For $\tau \in \mathscr{T}$ let $v_\tau(c)$ be the multiplicity with which the conic divisor class corresponding to τ occurs among the direct summands D_N of $\Bbbk[M^{1/c}]$. Then

$$v_\tau(c) = \#\left(\tau \cap \frac{1}{c}\mathbb{Z}^d\right).$$

(c) In particular, there exists a quasipolynomial $q_\tau : \mathbb{Z} \to \mathbb{Z}$ with rational coefficients such that $v_\tau(c) = q_\tau(c)$ for all $c \geq 1$. One has

$$q_\tau(c) = \operatorname{vol}(\tau)c^d + a_\tau c^{d-1} + \tilde{q}_\tau(c), \qquad c \in \mathbb{Z},$$

where $a_\tau \in \mathbb{Q}$ is constant and $\tilde{q}_\tau$ is a quasipolynomial of degree $\leq d-2$.

(d) In particular, every conic class occurs among the summands D_N if $c \gg 0$.

Proof. (a) As before, $\mathrm{gp}(M) = \mathbb{Z}^d$. Then $\mathrm{gp}(M^{1/c}) = (1/c)\mathbb{Z}^d$, and $M^{1/c} = C \cap (1/c)\mathbb{Z}^d$, $C = \mathbb{R}_+ M$. The monomial basis of the module D_N is now given by $C \cap N$. Choose an element $x \in N$. Then the parallel translation by $-x$ maps $C \cap N$ bijectively onto $(C - x) \cap \mathbb{Z}^d$, and thus induces an isomorphism of D_N with the conic divisorial ideal $\mathcal{D}(\sigma(x))$.

(b) The points of $(-1, 0] \cap (1/c)\mathbb{Z}^d$ represent each residue class of G_c exactly once. Therefore the sets $((C - x) \cap \mathbb{Z}^d) + x$ cover $C \cap (1/c)\mathbb{Z}^d$. After introducing coefficients from $\Bbbk$, we obtain the decomposition of $\Bbbk[M^{1/c}]$ into $\Bbbk[M]$-modules along the residue classes of $(1/c)\mathbb{Z}^d$ modulo $\mathbb{Z}^d$, and it is clear that each class occurs as often as there are points in a compartment representing τ.

(c) Each compartment is the disjoint union of open rational polytopes γ, called *cells*. By Theorem 6.51 the number of the points of $\gamma \cap (1/c)\mathbb{Z}^d$, $c \geq 1$, is given by a quasipolynomial of degree $\dim \gamma$. For the full-dimensional cell γ_d the leading coefficient is constant and equals the volume of γ, which coincides with the volume of τ. The lower-dimensional cells only contribute to the lower terms in the quasipolynomial.

It remains to show that the second largest coefficient is also constant. But this follows from Theorem 6.52: the facets of γ_d are translates of the support hyperplanes of C by integral vectors and therefore contain lattice points, and the same argument shows that the $(d-1)$-dimensional cells have constant leading coefficient in their Ehrhart quasipolynomial.

(d) is now trivial. □

Theorem 6.52 contains sufficient conditions under which further coefficients of q_v are constant. In particular, if the origin is the only 0-dimensional cell in $\mathcal{T}$, then all coefficients are constant.

Corollary 6.68. *Every conic divisorial ideal of the algebra $\Bbbk[M]$ is Cohen–Macaulay.*

Proof. By Hochster's theorem, $\Bbbk[M^{1/c}]$ is a Cohen–Macaulay ring since, after all, it is just $\Bbbk[M]$ as a $\Bbbk$-algebra. But it is also a finitely generated $\Bbbk[M]$-module, and as such it is likewise Cohen–Macaulay. Therefore each of its direct summands is Cohen–Macaulay. □

Hilbert–Kunz function and multiplicity. Suppose $\Bbbk$ has characteristic $p > 0$ and R is a $\Bbbk$-algebra. Then the *Frobenius endomorphism* $F : R \to R$ is defined by $F(x) = x^p$. If $R = \Bbbk[M]$, then F differs from p_* in its treatment of the coefficients, and the R-module structures of R over itself via F and p_* cannot be identified. Nevertheless they are very close to each other if $\Bbbk$ is a perfect field or even an algebraically closed field. In this case $F : \Bbbk \to \Bbbk$ is an automorphism, and the R-submodule structures of R via F and p_* coincide.

Therefore we can apply the analysis of $\Bbbk[M^{1/c}]$ in order to study the behavior of the Frobenius endomorphism and its iterates. An important invariant of a finitely generated graded module N over positively graded $\Bbbk$-algebras R with irrelevant maximal ideal $\mathfrak{m} = \bigoplus_{k=1}^{\infty} R_k$ is its *Hilbert–Kunz function* defined by

$$\mathrm{HK}_R(N,e) = \dim_{\Bbbk} N/F^e(\mathfrak{m})N.$$

In analogy to the usual multiplicity the *Hilbert–Kunz multiplicity* is given by

$$e_{\mathrm{HK},R}(N) = \lim_{e\to\infty} \frac{\dim_{\Bbbk} N/F^e(\mathfrak{m})N}{p^{de}}, \qquad d = \dim N.$$

See Huneke [203] for a general introduction to Hilbert–Kunz multiplicity.

For a finitely generated graded module N over a positively graded $\Bbbk$-algebra R we let

$$\mu_R(N), \quad \text{or simply} \quad \mu(R)$$

denote its minimal number of generators. The same notation is used for modules over local rings, and $\mu(N) = \mu(N_{\mathfrak{m}})$ for the maximal ideal $\mathfrak{m}$ generated by the elements of positive degree in R. (One could more generally consider *local graded rings.) By Nakayama's lemma,

$$\mu(N) = \dim_{R/\mathfrak{m}} N/\mathfrak{m}N.$$

Therefore $\mathrm{HK}_R(R,e) = \mu_{F^e(R)}(R)$. In the situation of an affine monoid algebra $R = \Bbbk[M] = N$ this amounts to $\mathrm{HK}_R(R,e) = \mu_R(R^{1/p^e})$. As an immediate consequence of Theorem 6.67 we obtain the following:

Theorem 6.69. *Let $\Bbbk$ be a perfect field of characteristic $p > 0$ and M a positive normal affine monoid of rank d. Set $R = \Bbbk[M]$ and let $\mathfrak{m}$ be the maximal ideal of R generated by the monomials different from 1. Moreover, let D_τ denote a conic divisorial ideal of type $\tau \in \mathscr{T}$. Then*

$$\mathrm{HK}_R(R,e) = \sum_{\tau\in\mathscr{T}} \nu_\tau(p^e)\mu_R(D_\tau), \qquad e \in \mathbb{Z},\ e \geq 1,$$

is the value of the quasipolynomial $q_{\mathrm{HK}} = \sum_\tau \mu_R(D_\tau)\nu_\tau$ at p^e. It has constant leading coefficient

$$e_{\mathrm{HK},R}(R) = \sum_\tau \mathrm{vol}(\tau)\mu_R(D_\tau) \in \mathbb{Q},$$

and also the coefficient of its degree $d-1$ term is constant and rational.

The theorem follows immediately from Theorem 6.67 since the minimal number of generators of a direct sum simply sum up. See Exercise 6.20 for a generalization.

That the coefficient of the degree $d-1$ term is constant, as we have derived from combinatorial arguments, is in fact true in much more generality; see Huneke, McDermott, and Monsky [204].

Ehrhart reciprocity extended. Remark 6.63 allows us to extend Ehrhart's reciprocity theorem. We have derived it from the equation $H_\omega(t) = (-1)^d H_R(t^{-1})$ relating the Hilbert series of a graded ring and its canonical module. More generally,

$$H_{M'}(t) = (-1)^d H_M(t^{-1}), \qquad M' = \mathrm{Hom}_R(M,\omega),$$

if M is a graded Cohen–Macaulay R-module of dimension equal to $\dim R$ (see Exercise 6.10). Applying this equation to the conic ideals in Remark 6.63 we obtain an extension of Ehrhart reciprocity.

Let G be a face of the rational polytope P. By D_G we denote the union of all facets containing G, and by U_G the union of the remaining facets. Then we set

$$E(P,G,k) = \#\big((P \setminus D_G) \cap \mathbb{Z}^d\big), \qquad E^+(P,G,k) = \#\big((P \setminus U_G) \cap \mathbb{Z}^d\big),$$

and define the generating functions $E_{P,G}(t)$ and $E^+_{P,G}(t)$ accordingly.

Since $E(P,P,-) = E(P,-)$ and $E^+(P,P,-) = E^+(P,-)$ the following theorem is an extension of Ehrhart reciprocity:

Theorem 6.70 (Ehrhart, Stanley).

$$E^+_{P,G}(t) = (-1)^d E_{P,G}(t^{-1}).$$

Proof. Let H denote the face of the cone $\mathbb{R}_+(P \times \{1\})$ generated by $G \times \{1\}$. Then, with the notation of Remark 6.63, $E(P,G,-)$ is the Hilbert function of the ideal $\mathfrak{q}_H$ and $E^+(P,G,-)$ is the Hilbert function of $\mathfrak{r}_H$. Moreover, $\mathfrak{r}_H = \operatorname{Hom}_R(\mathfrak{q}_H, \omega)$. □

For the theorem it is only important that $\mathfrak{q}_H$ is Cohen–Macaulay and that $\mathfrak{r}_H = \operatorname{Hom}_R(\mathfrak{q}, \omega)$. Similarly one can apply reciprocity to the nonnegative solutions of an *inhomogeneous* diophantine linear system of equations, provided they constitute the monomial basis for a Cohen–Macaulay ideal over the monoid algebra defined by the solutions to the associated homogeneous system; see [332].

Conic versus Cohen–Macaulay classes. In certain cases all Cohen–Macaulay divisor classes are conic (see [66]). In general however, there exist more Cohen–Macaulay classes than conic ones. An example is given by the Segre product R of three copies of the polynomial ring $\Bbbk[X_1, X_2]$. Its underlying monoid is the monoid of the unit cube in $\mathbb{R}^3$, and the divisor class group is isomorphic to $\mathbb{Z}^2$. See Exercise 4.22 for the isomorphism. Figure 6.5 shows the Cohen–Macaulay and conic classes. The conic classes are those in the shaded polytope. See [51] for the details. Baeţica [14] contains more such computations. They show that one cannot expect a regular structure for the set of Cohen–Macaulay classes: if $\operatorname{Cl}(R) = \mathbb{Z}$, the Cohen–Macaulay classes need not form the lattice points in a segment.

Finiteness of the Cohen–Macaulay classes. Despite of the fact that Cohen–Macaulay divisorial ideals are not conic in general, their number is always finite for normal affine monoid algebras.

Theorem 6.71. *Let M be a positive normal affine monoid, $\Bbbk$ a field, $R = \Bbbk[M]$, and $m \in \mathbb{Z}_+$. Then there exist only finitely many $c \in \operatorname{Cl}(R)$ such that a divisorial ideal D_c of class c has $\mu(D_c) \le m$.*

As a consequence of Theorem 6.71, the number of Cohen–Macaulay classes is also finite:

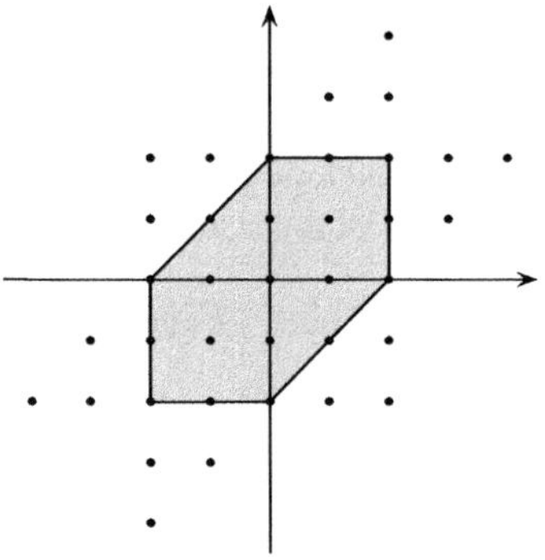

Fig. 6.5. Conic and Cohen–Macaulay classes

Corollary 6.72. *There exist only finitely many $c \in \mathrm{Cl}(R)$ for which a divisorial ideal of class c is a Cohen–Macaulay module.*

We first derive the corollary from the theorem. Let $\mathfrak{m}$ be the irrelevant maximal ideal of R. If D_c is a Cohen–Macaulay module, then $(D_c)_\mathfrak{m}$ is a Cohen–Macaulay module (and conversely). Furthermore,

$$e((D_c)_\mathfrak{m}) \geq \mu((D_c)_\mathfrak{m}) = \mu(D_c).$$

By Serre's numerical Cohen–Macaulay criterion (for example, see [72, 4.7.11]), the rank 1 $R_\mathfrak{m}$-module $(D_c)_\mathfrak{m}$ is Cohen–Macaulay if and only if its multiplicity $e((D_c)_\mathfrak{m})$ coincides with $e(R_\mathfrak{m})$.

Proof of Theorem 6.71. we use our standard conventions: $\mathrm{gp}(M) = \mathbb{Z}^d$ and $\mathbb{R}_+M$ has s facets with support forms $\sigma_1, \ldots, \sigma_s$. Let D be a monomial divisorial ideal of R. There exists $u \in \mathbb{Z}^s$ such that the lattice points in the set

$$\mathscr{P}(u) = \{x \in \mathbb{R}^d : \sigma(x) \geq u\}$$

give a $\Bbbk$-basis of D. The polyhedron $\mathscr{P}(u)$ is uniquely determined by its extreme points since each of its facets is parallel to one of the facets of $\mathbb{R}_+M$ and passes through such an extreme point. (Otherwise $\mathscr{P}(u)$ would contain a full line, and this is impossible if M is positive.)

Moreover, D is of torsion class if and only if $\mathscr{P}(u) = x + \mathbb{R}_+M$ for some $x \in \mathbb{Q}^d$; in other words: if and only if $\mathscr{P}(u)$ has a single compact face, namely $\{x\}$.

Suppose that D is not of torsion class. We form the line complex $\overline{\mathscr{L}}$ consisting of all 1-dimensional faces of the polyhedron $\mathscr{P}(u)$. Then $\overline{\mathscr{L}}$ is connected, and each extreme point is an endpoint of a 1-dimensional face. Since there is more than one extreme point, all extreme points are endpoints of compact 1-dimensional faces, and the line complex $\mathscr{L}(D)$ formed by the *compact* 1-dimensional faces is also connected. Since each facet passes through an extreme point, D is uniquely determined by $\mathscr{L}(D)$ (as a subset of $\mathbb{R}^d$).

Let $\mathscr{C}$ be an infinite family of divisor classes and choose a divisorial ideal D_c of class c for each $c \in \mathscr{C}$. Assume that the minimal number of generators $\mu(D_c)$, $c \in \mathscr{C}$, is bounded above by a constant B. By Lemma 6.73 below, the Euclidean length of all the line segments $\ell \in \mathscr{L}(D_c)$, $c \in \mathscr{C}$, is then bounded by a constant B'.

It is now crucial to observe that the endpoints of all the line segments under consideration lie in an overlattice $L = (1/m)\mathbb{Z}^d$ of $\mathbb{Z}^d$. In fact, each such point is the unique solution of a certain system of linear equations composed of equations $\sigma_i(x) = u_i$, and therefore can be solved over $(1/m)\mathbb{Z}$ where $m \in \mathbb{Z}$ is a suitable common denominator.

Let us consider two line segments ℓ and ℓ' in $\mathbb{R}^d$ as equivalent if there exists $z \in \mathbb{Z}^d$ such that $\ell' = \ell + z$. Since the lengths of all the line segments under consideration are bounded and their endpoints lie in $(1/m)\mathbb{Z}^d$, there are only finitely many equivalence classes of line segments $\ell \in \mathscr{L}(D_c)$, $c \in \mathscr{C}$.

Similarly we consider two line complexes $\mathscr{L}(D)$ and $\mathscr{L}(D')$ as equivalent if $\mathscr{L}(D') = \mathscr{L}(D) + z$, $z \in \mathbb{Z}^d$. Let the monomial basis of D' be given by the lattice points in $\mathscr{P}(u')$, $u' \in \mathbb{Z}^s$. Then the equation $\mathscr{L}(D') = \mathscr{L}(D) + z$ holds if and only if $\mathscr{P}(u') = \mathscr{P}(u) + z$, or, in other words, the divisor classes of D and D' coincide.

Since there are only finitely many equivalence classes of line segments and the number of lines that can appear in a complex $\mathscr{L}(D)$ is globally bounded (for example, by 2^s), one can only construct finitely many connected line complexes that appear as $\mathscr{L}(D)$, up to equivalence of line complexes. This contradicts the infinity of the family $\mathscr{C}$. □

Lemma 6.73. *Let M be a positive normal monoid, $\Bbbk$ a field, and $\mathscr{D}(u)$ a monomial divisorial ideal whose class is not torsion. Then there exists a constant $B > 0$, only depending on M, such that $\mu(\mathscr{D}(u)) \geq B\lambda$ where λ is the maximal Euclidean length of a compact 1-dimensional face of the polyhedron $\mathscr{P}(u)$.*

Proof. We assume that $\mathbb{Z}^d = \operatorname{gp}(M)$ so that the cone $\mathbb{R}_+M$ and the polyhedron $\mathscr{P}(u)$ are subsets of $\mathbb{R}^d$. Let ℓ be a 1-dimensional compact face of $\mathscr{P}(u)$.

There exists $\varepsilon > 0$ such that $U_\varepsilon(x) \cap \mathscr{P}(u)$ contains a lattice point for each $x \in \mathscr{P}(u)$. (In fact, $\mathbb{R}_+M$ contains a unit cube, and $x + \mathbb{R}_+M \subset \mathscr{P}(u)$ for $x \in \mathscr{P}(u)$.) Let $x \in \ell$. We can assume that

$$\sigma_i(x) \begin{cases} = u_i, & i = 1, \dots, m, \\ > u_i, & i > m. \end{cases}$$

Let $\rho = \sigma_1 + \dots + \sigma_m$. There exists $B' > 0$ such that $\rho(y) < B'$ for all $y \in \mathbb{R}^n$ with $|y| < \varepsilon$.

Furthermore we have $\rho(z) > 0$ for all $z \in \mathbb{R}_+M$, $z \neq 0$. Otherwise the facets $F_1, \dots, F_m$ would meet in a halfline contained in $\mathbb{R}_+M$, and this is impossible if ℓ is compact. In particular, there are only finitely many lattice points $z \in M$ such that $\rho(z) < B'$, and so there exists $\delta > 0$ such that $\rho(z) < B'$ for $z \in M$ is only possible with $|z| < \delta$.

Now suppose that $\mathscr{D}(u)$ is generated by $y_1, \dots, y_q$. For $x \in \ell$ we choose a lattice point $p \in U_\varepsilon(x) \cap \mathscr{P}(u)$. By assumption there exists $z \in \mathbb{R}_+M$ such that $p = y_i + z$. Then

$$\rho(z) = \rho(p) - \rho(y_i) \le \rho(p) - \rho(x) = \rho(p - x) < B.$$

Thus $|z| < \delta$, and therefore $|x - y_i| < \delta + \varepsilon$.

It follows that the Euclidean length of ℓ is bounded by $2q(\delta + \varepsilon)$. Of course δ depends on ρ, but there exist only finitely many choices for ρ if one varies ℓ. □

Let $b_{\max}$ be the maximal dimension of a compact face of $\mathscr{P}(u)$. The argument of the proof shows that the minimal number of generators of $\mathscr{D}(u)$ grows of order $c^{b_{\max}}$ when $c \to \infty$. The compact faces of $\mathscr{D}(u)$ do not only determine the growth of the number of generators of $\mathscr{P}(cu)$ as $c \to \infty$; they also determine the asymptotic value of depth $\mathscr{D}(cu)$:

$$\lim_{c\to\infty} \operatorname{depth} \mathscr{D}(cu) = \dim R - b_{\max}.$$

See [66] for this and related results.

Exercises

6.1. Let M be a positive affine monoid, $\Bbbk$ a field and S a subalgebra of $\Bbbk[M]$ generated by monomials. Show that $\Bbbk[M]$ is a finitely generated S-module if and only if S contains a monomial from each extreme ray of $\mathbb{R}_+M$.

6.2. Prove the equivalence (d) $\Longleftrightarrow$ (e) of Proposition 6.7.

Hint: for the implication (d) $\Longrightarrow$ (e) choose a minimal homogeneous system of generators of M and show that it is linearly independent, using induction on d.

6.3. Let M be a simplicial affine monoid and $\Bbbk$ a field, choose a monoid element x_i from each extreme ray of $\mathbb{R}_+M$ and let N be the submonoid generated by the x_i. Show:

(a) $\Bbbk[M]$ is Cohen–Macaulay if and only if there exist $y_1, \dots, y_n \in M$ such that M is the disjoint union of the subsets $y_j + N$.

(b) (Goto–Suzuki–Watanabe [147]) $\Bbbk[M]$ is Cohen–Macaulay if (and only if) it satisfies Serre's condition (S_2).

Hint for (b): a sequence of monomials is regular if and only if each two of its members form a regular sequence.

6.4. Let P be the simplex with vertices $(0,0,0)$, $(2,0,0)$, $(0,3,0)$, and $(0,0,5)$. It was shown in Example 2.56(c) that $M(P)$ is not normal. Nevertheless $R = \Bbbk[M(P)]$ is Cohen–Macaulay and even Gorenstein. Prove this. (You may want to use a computer algebra system like CoCoA [86], Macaulay2 [150], or Singular [152].) Find the generator of the canonical module of R.

6.5. Consider the same type of simplex as in Exercise 6.4, but replace 2, 3, 5 by 3, 11, 31. Show that $\Bbbk[M(P)]$ is not normal (for example by normaliz), but satisfies Serre's condition (R_1) (see Exercise 4.16). Is $\Bbbk[M(P)]$ Cohen–Macaulay? (Also see [59] where (R_1) appears in the disguise of "almost 1-normality.")

6.6. Show that ${}^*\operatorname{Ext}^i_R(M(u), N(v)) \cong {}^*\operatorname{Ext}^i_R(M,N)(v-u)$.

6.7. Let M be a $\mathbb{Z}^r$-graded module over a $\mathbb{Z}^r$-graded ring. We define a $\mathbb{Z}^{d-1}$-grading grad′ on R and M by forgetting the last component of $\deg x \in \mathbb{Z}^r$ for $x \in R$ or $x \in M$, and a $\mathbb{Z}$-grading grad″ by selecting the last component.

(a) Show: if N' is a graded submodule of M with respect to grad′ and N'' is the submodule of N' generated by all its elements that are homogeneous with respect to grad″, then N'' is a $\mathbb{Z}^r$-graded submodule.

(b) In the situation of Remark 6.29(b) show that $\omega = \omega_R$ is indeed a canonical module of $R_\mathfrak{m}$ for the *maximal ideal $\mathfrak{m}$.

(c) Let R be a $\mathbb{Z}^r$-graded ring, and suppose C is a module such that $C_\mathfrak{p}$ is a canonical module of $R_\mathfrak{p}$ for all $\mathbb{Z}^r$-graded prime ideals of R. Show that $C_\mathfrak{q}$ is a canonical module for all prime ideals that are graded with respect to grad′ and that C is indeed a canonical module for R.

Hint: the induction step included in (c) can be done as in [72, 3.6.9].

The strategy outlined in this problem can be used to derive several statements about $\mathbb{Z}^r$-graded rings from their $\mathbb{Z}$-graded specializations.

6.8. Let $\Bbbk$ be a field, $\varphi : S \to R$ be a homomorphism of positively $\mathbb{Z}^r$-graded Cohen–Macaulay algebras such that R is a finitely generated S-module. Then $\omega_R = {}^*\operatorname{Ext}_S^{t-d}(R, \omega_S)$, $t = \dim S$, $d = \dim R$.

6.9. Let R be a *local Cohen–Macaulay ring with canonical module ω_R and $x_1, \dots, x_n$ a homogeneous R-regular sequence. Show

$$\omega_{R/(x_1,\dots,x_n)} = (\omega_R/(x_1,\dots,x_n)\omega R)(\textstyle\sum \deg x_i).$$

6.10. Let R be a positively $\mathbb{Z}^r$-graded $\Bbbk$-algebra of dimension d, ω its *canonical module, and N a graded Cohen–Macaulay R-module of dimension e.

Show:

(a) $N' = {}^*\operatorname{Ext}_R^{d-e}(N, \omega)$ is also Cohen–Macaulay;

(b) $H_{N'}(\mathbf{t}) = (-1)^e H_N(\mathbf{t}^{-1})$;

(c) $N'' = N$.

Hint: write R as a residue class ring of a polynomial ring S as in the proof of Theorem 6.28, and convince yourself that it is sufficient to do the positively $\mathbb{Z}$-graded case. Then all data can be lifted to S, and Theorem 6.18 applies. (See [72, 4.4.5] for the $\mathbb{Z}$-graded case and a proof by reduction to Krull dimension 0.)

6.11. Let $R = \Bbbk[M]$ be a normal affine monoid algebra. Show the following are equivalent:

(i) R is normal;

(ii) R is Cohen–Macaulay and $\omega_R = \operatorname{int}(M)R$.

6.12. With the notation developed above Theorem 6.48, show that R is a finitely generated S-module.

6.13. Under the hypotheses of Theorem 6.48 let ρ be a divisor of the period π, and J the ideal generated by all homogeneous elements whose degree is coprime to π. What information on the Hilbert quasipolynomial P_M does $\dim M/JM$ carry?

6.14. Suppose that the indeterminates X_i of $R = \Bbbk[X_1, \dots, X_n]$ have pairwise coprime positive degrees e_i in $\mathbb{Z}$. Show that only the zeroth coefficient of P_R is nonconstant and that it has period $e_1 \cdots e_n$.

The Hilbert function of R is an algebraic disguise of the enumerative function appearing in the *Frobenius coin problem*; see Beck and Robbins [26].

What can be said if three of the e_i each generate $\mathbb{Z}$ as a group?

6.15. Let q be a quasipolynomial, and $F(t) = \sum_{k=0}^{\infty} q(k)t^k = (h_0 + \cdots + h_u t^u)/(\prod_{i=1}^{d}(1 - t^{e_i})$ be the associated generating function. Show the following are equivalent:

(i) the h-vector $(h_0, \dots, h_u)$ is palindromic;
(ii) $q(-k) = (-1)^{r-1} q(k+a)$ for all $k \in \mathbb{Z}$;
(iii) $H_R(t^{-1}) = (-1)^d t^{-a} H_R(t)$.

6.16. (a) Prove that the monoid of magic squares is generated by the permutation matrices.
(b) Show that coordinate hyperplanes are the support hyperplanes of the monoid of magic squares (after intersection with its affine hull).
(c) Compute the dimension of the monoid of magic squares and the degree of its Hilbert series.

6.17. Prove that the volume of an n-dimensional integral polytope P in $\mathbb{R}^n$ is

$$\operatorname{vol}(P) = \frac{1}{2}(E(P,1) + E^+(P,1) - 2) \qquad \text{for } n = 2, \text{ and}$$

$$\operatorname{vol}(P) = \frac{1}{6}(E(P,2) - 3E(P,1) - E^+(P,1) + 3) \qquad \text{for } n = 3.$$

For $n = 2$ this is *Pick's formula*.
Hint: the coefficients of a polynomial can be determined by interpolation.

6.18. Let D be a conic divisorial ideal of $\Bbbk[M]$ and ω the canonical module. Then $\omega : D$ is also conic.

6.19. (a) Let $M \subset \mathbb{Z}^2$ be the monoid with Hilbert basis $(1,0)$, $(1,1)$, $(1,2)$, $(1,3)$. (This is the monoid of Figure 6.4.) Compute the Hilbert–Kunz function and multiplicity.
(b) Give a nontrivial example for which all coefficients of the Hilbert–Kunz function are constant.
(c) Prove the following toy case of a theorem of Watanabe and Yoshida [372]: if $e_{\mathrm{HK}}(\Bbbk[M]) = 1$, then $M \cong \mathbb{Z}_+^d$, $d = \operatorname{rank} M$.

6.20. Suppose S is a graded subalgebra of R such that R is a finite S-module and the normalization of S. Show $e_{\mathrm{HK},S}(S) = \sum_{\tau \in \mathcal{T}} \operatorname{vol}(\tau)\mu_S(D_\tau) \in \mathbb{Q}$.

6.21. Let P be the unit square in $\mathbb{R}^2$, $\mathcal{F}$ a subset of its facets, and $\mathfrak{q}_{\mathcal{F}}$ the intersection of the prime ideals $\mathfrak{p}_F$, $F \in \mathcal{F}$. When is $\mathfrak{q}_{\mathcal{F}}$ Cohen–Macaulay and when does the analogue of Theorem 6.70 hold?

Notes

For the development of the notions of Cohen–Macaulay ring, Gorenstein ring, and canonical module we refer the reader to notes of Chapters 1–3 of [72] and its extensive bibliography. For the sake of a compact treatment we have chosen to avoid the use of injective modules and resolutions so that Hilbert's syzygy theorem becomes the cornerstone for the existence and uniqueness of the canonical module.

An important topic for (combinatorial) commutative algebra is local cohomology. It is covered in detail in [72] and Brodmann and Sharp [47]. The toric case is also discussed by Miller and Sturmfels [261]. After Stanley's work [329], local cohomology of toric rings has been investigated in several articles, for example Goto and Watanabe [148], Trung and Hoa [358], Ishida [209] and Schäfer and Schenzel [310]. It should be noted that seminormal affine monoid domains have a characterization in terms of local cohomology given by Bruns, Li, and Römer [77]. The connection between seminormality and local cohomology had already been established by Hochster and Roberts [198].

Proofs for the Cohen–Macaulay property of normal affine monid algebras via local cohomology as well as Hochster's original proof [195] require more intricate combinatorial arguments then the arithmetic approach via the Hochster–Roberts–Kempf theorem. On the other hand, the computation of local cohomology yields the canonical module as a byproduct via local duality (for example, see [72, 6.3.5]). Our divisorial computation is close to Danilov's argument using differentials [102].

The connection between enumerative combinatorics and commutative algebra, discovered by Stanley in his proof of the Anand–Dumir–Gupta conjectures [323], sparked renewed interest in the theory of Hilbert functions from the side of commutative algebra. Stanley's article [325] contains the basic results, including the reciprocity between the Hilbert function of a Cohen–Macaulay ring and its canonical module. Theorem 6.48 is taken from Bruns and Ichim [75]. For the sources of Lemma 6.44, going back to Polya and Popoviciu, we refer the reader to Stanley [332]

Independently of Stanley's work, and preceding it, Ehrhart found the main properties of functions counting lattice points in polytopes and solutions of linear diophantine systems. A comprehensive account of his work is the book [111]. We refer the reader to Beck and Robbins [26] for this elegant topic. For combinatorial proofs of Theorem 6.52 see Ehrhart [111], McMullen [255], and Stanley [327]. The commutative algebra approach to Theorem 6.70 was given by Miller and Reiner [260]. It extends a theorem of Stanley [324] on "reciprocal domains." Theorem 6.57, due to Bruns and Römer [78], has a more special predecessor in Athanasiadis [9].

Section 6.F follows [66] and [51]. The Cohen–Macaulay property of conic divisor classes (without this name) was proved by Stanley; see [326], [329], [332], where they represent sets of solutions of inhomogeneous linear diophantine systems. Their Cohen–Macaulay property was also proved by Dong [110] via a topological approach to squarefree divisor complexes (compare Remark 6.21(b)). The rationality of the Hilbert–Kunz multiplicity of toric rings was proved by Watanabe [371]; see also Eto [116].

7

Gröbner bases, triangulations, and Koszul algebras

In this chapter we investigate the algebraic data of toric ideals that correspond to regular subdivisions of the cones generated by affine monoids: monomials orders and initial ideals. Though the correspondence is not strong enough for unconditional implications in either direction, it yields powerful results in cases where the unimodularity of triangulations or the squarefreeness of the initial ideals can be established. For example, the polytopal algebras defined by lattice polygons turn out to be Koszul algebras under the obvious necessary conditions.

In studying initial ideals of toric ideals one is naturally led to the introduction of the class of algebras associated with monoidal complexes. They generalize monoid algebras in the same way as fans generalize cones.

The chapter's last section is a continuation of our study of multiples of lattice polytopes. As it will turn out, there are no algebraic obstructions to the existence of "good" unimodular triangulations of high multiples.

7.A Gröbner bases and initial ideals

We recall the definitions and some important properties of Gröbner bases, monomial orders, and initial ideals. For further information on the theory of Gröbner bases we refer the reader to Eisenbud [112], Kreuzer and Robbiano [230], and Sturmfels [338]. A compact treatment, on which this section is based, can be found in Bruns and Conca [54], [55].

Gröbner bases and initial ideals. Throughout this section let $\Bbbk$ be a field and R be the polynomial ring $\Bbbk[X_1, \dots, X_n]$. The set of monomials of R will be denoted by $\mathcal{M}$ in this section. We have introduced monomial orders, initial terms and initial monomials already on p. 125. In this chapter we impose the further condition

$$1 < y$$

for all $y \in \mathcal{M}$, $y \neq 1$, and in the following it is tacitly understood that it is satisfied. Since 1 is the smallest monomial, the order $<$ refines divisibility in $\mathcal{M}$. As an important

W. Bruns, J. Gubeladze, *Polytopes, Rings, and K-Thoery*, Springer Monographs in Mathematics,
DOI 10.1007/b105283_7, © Springer Science+Business Media, LLC 2009

consequence, every nonempty subset of $\mathcal{M}$ has a minimal element; equivalently, there are no infinite descending chains in $\mathcal{M}$. This follows from the fact that every (monomial) ideal in R is finitely generated (Hilbert's basis theorem). Therefore a subset N of $\mathcal{M}$ has only finitely many elements that are minimal with respect to divisibility in $\mathcal{M}$. One of them is the minimal element of N in the monomial order.

We list the most important monomial orders. Let deg be the ordinary total degree on R. For monomials $y_1 = X_1^{a_1} \cdots X_n^{a_n}$ and $y_2 = X_1^{b_1} \cdots X_n^{b_n}$ one defines

(a) the *lexicographic order* (lex) by $y_1 <_{\text{lex}} y_2$ if for some k one has $a_k < b_k$ and $a_i = b_i$ for $i < k$;
(b) the *degree lexicographic order* (deglex) by $y_1 <_{\text{deglex}} y_2$ if $\deg(y_1) < \deg(y_2)$ or $\deg(y_1) = \deg(y_2)$ and $y_1 <_{lex} y_2$;
(c) the *(degree) reverse lexicographic order* (revlex) by $y_1 <_{\text{revlex}} y_2$ if $\deg(y_1) < \deg(y_2)$ or $\deg(y_1) = \deg(y_2)$ and for some k one has $a_k > b_k$ and $a_i = b_i$ for $i > k$.

These three monomial orders satisfy $X_1 > X_2 > \cdots > X_n$.

For an arbitrary $\Bbbk$-vector subspace V of R the *initial subspace* $\operatorname{in}(V)$ is the $\Bbbk$-subspace generated by the initial monomials of the elements of V. If I is an ideal, the $\operatorname{in}(I)$ is an ideal itself (why?), called the *initial ideal* of I.

A *Gröbner basis* of an ideal I is a subset G of I such that the initial monomials $\operatorname{in}(g)$, $g \in G$, generate $\operatorname{in}(I)$. Since $\operatorname{in}(I)$ is finitely generated, I has a finite Gröbner basis. The *Buchberger algorithm* computes a Gröbner basis from an arbitrary system of generators of I. It follows from the *reduction algorithm* that a Gröbner basis generates I.

Proposition 7.1. *Let $V_1 \subset V_2$ be $\Bbbk$-vector subspaces of R.*

(a) For each monomial $y \in \operatorname{in}(V_2) \setminus \operatorname{in}(V_1)$ choose a polynomial $f_y \in V_2$ with $\operatorname{in}(f_y) = y$. Then the residue classes of the f_y form a $\Bbbk$-basis of V_2/V_1.
(b) In particular, for $V_2 = R$ the set $\mathcal{M} \setminus \operatorname{in}(V_1)$ is a $\Bbbk$-basis of R/V_1.
(c) If $\operatorname{in}(V_1) = \operatorname{in}(V_2)$, then $V_1 = V_2$.
(d) Suppose R is endowed with a $\mathbb{Z}^r$-grading for which all monomials are homogeneous and V_1 is a graded subspace of R. Then R/V_1 and $R/\operatorname{in}(V_1)$ have the same (multigraded) Hilbert functions.

A *weight vector* for R is an element $w \in \mathbb{N}^n$. It assigns the indeterminate X_i the *weight* w_i. Of course, the choice of a weight vector is nothing but giving R a positive $\mathbb{Z}$-grading under which the monomials are homogeneous, and therefore the whole terminology of graded rings with the prefix (or index) w can be applied. In particular,

$$\deg_w X^a = \sum_{i=1}^{n} a_i w_i = \langle a, w \rangle,$$

where $\langle -, - \rangle$ denotes the standard scalar product.

A weight vector w determines a *monomial preorder* if one sets

$$X^a \le_w X^b \iff \langle a, w \rangle \le \langle b, w \rangle.$$

The only axiom of a monomial order not satisfied is antisymmetry: for $n > 1$ there always exist distinct monomials X^a and X^b such that simultaneously $X^a \le_w X^b$ and $X^b \le_w X^a$.

The *w-initial component* $\operatorname{in}_w(f)$ of a polynomial f is simply its w-homogeneous component of highest degree, and the *w-initial subspace* of $V_1 \subset R$ is the subspace generated by the polynomials $\operatorname{in}_w(f)$, $f \in V_1$. Then Proposition 7.1(c) and (d) hold analogously. The best way to see this is to refine the w-preorder by an auxiliary monomial order in the same way as deglex refines the preorder with respect to $(1,\dots,1)$.

The Gröbner deformation. From the structural point of view it is of utmost importance that one can compare the rings R/I and $R/\operatorname{in}(I)$ and transfer good properties of $R/\operatorname{in}(I)$ to R/I. In order to construct the connection for w-initial ideals we introduce a further variable T and define the *w-homogenization* of $f = \sum_a c_a X^a$, $c_a \in \Bbbk$, by

$$\hom_w(f) = \sum_a c_a X^a T^{m-\langle a,w\rangle} \in R[T], \qquad m = \deg_w(f).$$

The *w-homogenization of an ideal I* is the ideal in $R[T]$ generated by all polynomials $\hom_w(f)$, $f \in I$.

Proposition 7.2. *With the notation introduced, set $S = R[T]/\hom_w(I)$, and let t be the residue class of T in S. Then $t-u$ is a non-zerodivisor in S for every $u \in \Bbbk$. Moreover, for $u \neq 0$*

$$R/I \cong S/(t-u) \quad \textit{and} \quad R/\operatorname{in}_w(I) \cong S/(t).$$

In a more geometric language, the homomorphism $\Bbbk[T] \to S$ induces a morphism $\operatorname{Spec} S \to \operatorname{Spec} \Bbbk[T]$ whose *generic fiber* over the closed point $u \in \Bbbk \setminus \{0\}$ is R/I, and whose special fiber over $u = 0$ is $R/\operatorname{in}_w(I)$. Since t is a non-zerodivisor, $\operatorname{Spec} S$ is flat over $\operatorname{Spec} \Bbbk[T]$, and $\operatorname{Spec} R/I$ deforms to $\operatorname{Spec} R/\operatorname{in}(I)$ when $t \to 0$ (over $\mathbb{C}$).

While Proposition 7.2 seems to be restricted to weight preorders, its implications hold for monomial orders as well since a monomial order can always be replaced by a weight preorder if only finitely many pairs of monomials must be compared:

Proposition 7.3. *Let $<$ be a monomial order on R.*

(a) Let $\{(y_1,z_1),\dots,(y_k,z_k)\}$ be a finite set of pairs of monomials such that $y_i > z_i$ for all i. Then there exists a weight $w \in \mathbb{N}^n$ such that $\deg_w(y_i) > \deg_w(z_i)$ for all i.

(b) Let I be an ideal of R. Then there exists a weight $w \in \mathbb{N}^n_+$ such that $\operatorname{in}_<(I) = \operatorname{in}_w(I)$.

Proof. We include the proof since it uses polyhedral geometry. Note that it is enough to prove (a) in the case in which all y_i are the same monomial y. In fact, we can multiply both y_1 and z_1 by $y_2 \cdots y_k$, etc. without changing the hypothesis or the conclusion.

For simplicity we identify the monomials with their exponent vectors. Let P be the polytope spanned by $y, z_1, \dots, z_k$, and let $Q = y + \mathbb{R}^n_+$. We claim that $P \cap Q = \{y\}$. If not, then, after multiplication by a sufficiently large positive integer (or taking the corresponding powers of the monomials), we can assume that $P \cap Q$ contains an integral point $x \neq y$. On the one hand, $x > y > z_i$ in the monomial order, and on the other hand, x is a convex combination of $z_0 = y, z_1, \dots, z_k$ with rational coefficients:

$x = \sum(a_i/b_i)z_i$ with $a_i, b_i \in \mathbb{Z}_+$ and $\sum a_i/b_i = 1$. This leads to a binomial relation $x^b = y^{a'_0} z_1^{a'_1} \cdots z_k^{a'_k}$, contradicting the fact that x is larger than every factor on the right hand side and the number of factors coincide.

By Theorem 1.32 there exists a linear form λ such that $\lambda(x) \geq \lambda(y) > \lambda(z_i)$ for all $x \in Q$ and $i = 1, \dots, k$. The first inequality implies that λ has positive coefficients with respect to the canonical basis of $\mathbb{R}^n$, and they form the desired weight vector.

Part (b) is an easy consequence of the fact that I has a finite Gröbner basis, and is left to the reader. □

Theorem 7.4. *Let I be an ideal in $R = \Bbbk[X_1, \dots, X_n]$ and suppose that* $\operatorname{in}(I)$ *is formed with respect to a monomial order or a weight preorder. If $R/\operatorname{in}(I)$ has one of the properties of being reduced, a normal domain, Cohen–Macaulay, or Gorenstein, then R/I has it, too.*

The essential point is that all these properties ascend from $S/(t)$ to S if t is a homogeneous non-zerodivisor in a positively graded noetherian $\Bbbk$-algebra S, and that they descend from S to the dehomogenization $S/(1-t)$ if t has degree 1 [72, 1.5.18].

Theorem 7.4 opens an avenue for the application of Gröbner bases and initial ideals to structural questions beyond efficient computation. The initial ideal with respect to a monomial order is a monomial ideal and can therefore be investigated with combinatorial methods. See [54] for a paradigmatic case.

Remark 7.5. (a) It is clear that the theory of initial ideals can be extended to ideals in finitely generated monomial subalgebras of $\Bbbk[X_1, \dots, X_n]$, and therefore to arbitrary positive affine monoid algebras.

(b) One can even go one step farther: for a subalgebra A of $\Bbbk[X_1, \dots, X_n]$ the initial subspace $\operatorname{in}(A)$ is a subalgebra $\operatorname{in}(A)$, and thus one can study A via $\operatorname{in}(A)$ and ideals I in A via their initial ideals $\operatorname{in}(I) \subset \operatorname{in}(A)$. However, this approach is limited by the fact that $\operatorname{in}(A)$ need not be finitely generated over $\Bbbk$, even if A is so. On the other hand, if $\operatorname{in}(A)$ is finitely generated, then Theorem 7.4 remains valid.

Not all properties of a ring R satisfy the conditions given after Theorem 7.4, and those cannot be expected to pass from $\operatorname{in}(A)$ to A (or from $R/\operatorname{in}(I)$ to R/I). A notable example is the nonexistence of nonfree projective modules; see Example 9.24.

For the study of initial subalgebras and *toric deformation*, i. e., the passage from A to $\operatorname{in}(A)$, we refer the reader to Conca, Herzog, and Valla [88], Robbiano and Sweedler [301], Sturmfels [338, Chapter 11], and [54].

Koszul algebras and graded Betti numbers. Let R be a positively graded $\Bbbk$-algebra that is generated by its degree 1 elements, and let $\mathfrak{m}$ be the maximal ideal generated by its homogeneous elements of positive degree. One says that R is a *Koszul algebra* if $\Bbbk \cong R/\mathfrak{m}$ has a linear free resolution

$$\cdots \to R(-(n+1))^{\beta_{n+1}} \to R(-(n))^{\beta_n} \to \cdots R(-1)^{\beta_1} \to R \to \Bbbk \to 0$$

as an R-module. If $R = \Bbbk[X_1, \dots, X_n]$, then $\Bbbk$ is resolved by the Koszul complex [72, 1.6], and so R is a Koszul algebra. (However, it follows from the Auslander–Buchsbaum–Serre theorem [72, 2.2.7] and [72, 2.2.25] that $\Bbbk$ has an infinite resolution in all other cases.)

Clearly $\beta_1 = \dim_{\Bbbk} R_1$ is the minimal number of generators of $\mathfrak{m}$. We can rephrase the linearity of the resolution by saying that the graded Betti numbers $\beta_{ij}(\Bbbk)$ of $\Bbbk$ as an R-module vanish, unless $i = j$.

Koszul algebras were introduced by Priddy [293]. They have been intensively investigated. We refer the reader to Polishchuk and Positselski [289], Backelin and Fröberg[13], and Fröberg [131] for more information on this interesting and important class of algebras.

Set $m = \beta_1$, let $x_1, \dots, x_m$ be a system of generators of $\mathfrak{m}$, and consider the epimorphism $\Bbbk[X_1, \dots, X_m] \to R$, $X_i \mapsto x_i$, with kernel I. If R is Koszul, then I is generated by homogeneous polynomials of degree 2, but the converse does not hold (see Remark 7.38(a)). However, Gröbner bases of degree 2 are strong enough to force the Koszul property:

Theorem 7.6. *Let $S = \Bbbk[X_1, \dots, X_m]$ with its standard grading, and I be a graded ideal in R that, with respect to some monomial order, has a Gröbner basis of degree 2 polynomials. Then S/I is a Koszul algebra.*

The theorem is proved in two steps. First, one uses a theorem of Fröberg [130] by which a residue class ring of R by an ideal generated by monomials of degree 2 is Koszul. Second, one compares S/I and $S/\operatorname{in}(I)$.

Proposition 7.7. *Let $S = \Bbbk[X_1, \dots, X_m]$ be endowed with a $\mathbb{Z}^r$-grading, and J, J_1, J_2 be graded ideals of $S = \Bbbk[X_1, \dots, X_m]$ with $J \subset J_1$ and $J \subset J_2$. Furthermore let w be a weight vector. Then*

$$\dim_{\Bbbk} \operatorname{Tor}_i^{R/J}(R/J_1, R/J_2)_u \le \dim_{\Bbbk} \operatorname{Tor}_i^{R/\operatorname{in}_w(J)}(R/\operatorname{in}_w(J_1), R/\operatorname{in}_w(J_2))_u$$

for all $i \ge 0$ and all $u \in \mathbb{Z}^r$.

As always, the graded structure on the Tor modules is inherited from the graded structure of their arguments. We refer the reader to [54, 3.13] for a proof in the $\mathbb{Z}$-graded case that is easily generalized. Theorem 7.6 follows immediately if we note that

$$\beta_{ij}(\Bbbk) = \dim_{\Bbbk} \operatorname{Tor}_i^{S/I}(S/\mathfrak{n}, S/\mathfrak{n})_j, \qquad \mathfrak{n} = (X_1, \dots, X_m)$$

(see equation (6.2)). A different proof of Theorem 7.6 is given in Bruns, Herzog, and Vetter [73]. Proposition 7.7 appears in a more special from (but for a larger class of rings) in Anick [8, 4.1].

One should note that Proposition 7.7 has many more applications. In particular, one can take $J = 0$, $J_1 = I$, and $J_2 = \mathfrak{n}$. Then the proposition compares the graded Betti numbers of S/I and $S/\operatorname{in}(I)$:

Corollary 7.8. *Let $S = \Bbbk[X_1, \dots, X_m]$ be endowed with an M-grading where M is a positive affine monoid and w a weight vector. Moreover, let I be a graded ideal. Then*

$$\beta_{iu}(S/I) \le \beta_{iu}(S/\operatorname{in}_w(I))$$

for all $i \in \mathbb{Z}$ and all $u \in M$.

7.B Initial ideals of toric ideals

In this section we establish the connection between monomial orders for affine monoid algebras $\Bbbk[M]$, or rather their defining ideals, and regular subdivisions of the cone $\mathbb{R}_+M$. First we introduce a structure that naturally accompanies such subdivisions.

Monoidal complexes and their algebras. Roughly speaking, a monoidal complex is a conical complex (see Definition 1.35) enriched with monoids in a compatible way. More precisely:

Definition 7.9. A *monoidal complex* $\mathbb{M}$ consists of a conical complex $\Gamma = \{c, \pi_c : C_c \to c\}$ and a system of monoids (M_c), $c \in \Gamma$, satisfying the conditions:

(a) $M_c \subset C_c$ for all $c \in \Gamma$;
(b) $\mathbb{R}_+M_c = C_c$ for all $c \in \Gamma$ (*cone generation*);
(c) $\operatorname{gp}(M_c)$ is a lattice in $\mathbb{R}C_c$;
(d) for all $c, d \in \Gamma$ the map $\pi_d^{-1} \circ \pi_c$ restricts to a monoid isomorphism between $M_c \cap \pi_c^{-1}(c \cap d)$ and $M_d \cap \pi_d^{-1}(c \cap d)$ (*compatibility*).

Less formally: in each cone C_c we have a monoid M_c generating it, and the intersection of M_c with a face $F = C_f$ of C_c is M_f.

For a ring R and a monoidal complex $\mathbb{M}$ one defines the algebra $R[\mathbb{M}]$ as the inverse limit of the R-algebras $R[M_c]$:

$$R[\mathbb{M}] = \varprojlim\big(f_{cd} : R[M_c] \to R[M_{c\cap d}] \,\big|\, f_{cd} \text{ the face projection}\big). \tag{7.1}$$

Recall that the inverse limit is the subring of the direct product $\prod_{c\in\Gamma} R[M_c]$ consisting of those elements (a_c) with $f_{cd}(a_c) = f_{dc}(a_d)$ for all c, d.

There is another way to construct $R[\mathbb{M}]$, leading directly to a monomial structure. The monoids M_c form a direct system of sets with respect to the embeddings

$$\pi_d^{-1} \circ \pi_c : M_c \to M_d, \qquad c \subset d \text{ a face.}$$

The direct limit

$$|\mathbb{M}| = \varinjlim\big(\pi_d^{-1} \circ \pi_c : M_c \to M_d \mid c \subset d \text{ a face}\big)$$

of the sets M_c is in general not a monoid, but it carries a partial monoid structure: the sum $x + y$ is defined whenever there exists $c \in \Gamma$ with $x, y \in M_c$, and it is independent of the choice of c. We consider each M_c as a subset of $|\mathbb{M}|$ in the natural way, and M_c is then a (full) submonoid of the partial monoid $|\mathbb{M}|$.

The partial monoid structure allows us to define an algebra structure on the free R-module

$$R[|\mathbb{M}|] = \bigoplus_{x\in|\mathbb{M}|} RX^x.$$

We set

$$X^xX^y = \begin{cases} X^{x+y}, & x, y \in M_c, \\ 0, & x + y \text{ is not defined.} \end{cases}$$

The R-bilinear extension of this product makes $R[|\mathbb{M}|]$ an R-algebra. The elements of $|\mathbb{M}|$ are called *monomials*.

Example 7.10. We turn the Möbius strip of Figure 1.5 into a monoidal complex $\mathbb{M}$ by considering each quadrangle as a unit square and choosing the monoid over it as the corresponding polytopal monoid. Together with the compatibility conditions this fixes the structure of $\mathbb{M}$ completely. With $U = X^u$, etc. we have

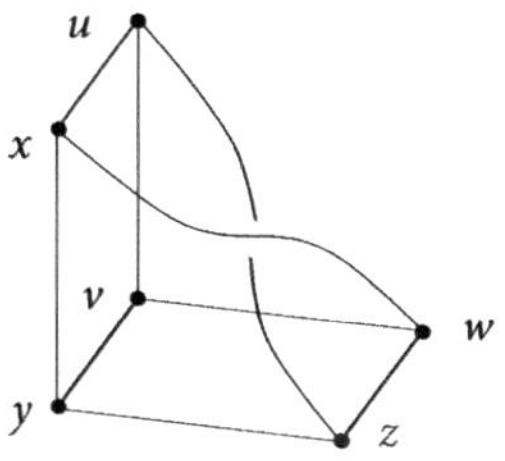

Fig. 7.1. Möbius strip as a monoidal complex

$$UVW = UVZ = UWY = VWX = UYZ = VXZ = WXY = XYZ = 0,$$

and the generators satisfy three binomial equations

$$XZ = UW, \quad YW = VZ, \quad XV = UY$$

resulting from the unit squares. As we will see in the next proposition, these relations do indeed define $R[|\mathbb{M}|] = R[\mathbb{M}]$.

Let $c \in \Gamma$. Then the ideal $\mathfrak{p}_c$ of $R[|\mathbb{M}|]$ generated by all monomials $x \notin M_c$ is evidently a prime ideal, and the composition $R[M_c] \to R[|\mathbb{M}|] \to R[|\mathbb{M}|]/\mathfrak{p}_c$ is an isomorphism. Therefore we can identify $R[M_c]$ and $R[|\mathbb{M}|]/\mathfrak{p}_c$. Moreover, if $c \subset d$, then the natural epimorphism $R[|\mathbb{M}|]/\mathfrak{p}_d \to R[|\mathbb{M}|]/\mathfrak{p}_c$ is nothing but the face projection f_{dc}. This observation allows us to compute the inverse limit $R[\mathbb{M}]$.

Proposition 7.11. *With the notation introduced,* $R[\mathbb{M}] = R[|\mathbb{M}|]$.

Proof. As observed, the inverse system given by the face projections f_{cd} can be identified with the inverse system of natural epimorphisms defined by the system of residue class rings $R[|\mathbb{M}|]/\mathfrak{p}_c$, $c \in \Gamma$.

Each of the ideals $\mathfrak{p}_c$ has a monomial basis, and this implies that addition and intersection of such ideals are distributive over each other. Thus

$$\varprojlim R[|\mathbb{M}|]/\mathfrak{p}_c \cong R[|\mathbb{M}|]/\bigcap_{c \in \Gamma} \mathfrak{p}_c \cong R[|\mathbb{M}|]$$

by Exercise 7.2 and since $\bigcap_{c\in\Gamma} \mathfrak{p}_c = 0$. □

In view of Proposition 7.11 we no longer need to distinguish between $R[\mathbb{M}]$ and $R[|\mathbb{M}|]$, keeping the simpler notation $R[\mathbb{M}]$.

Proposition 7.12. *With the notation introduced, the following hold:*

(a) $R[\mathbb{M}]$ *is reduced.*

(b) Let c *be a facet of* Γ*, and let* $\mathfrak{p}_c$ *be the* $\Bbbk$*-vector space in* $R[\mathbb{M}]$ *generated by all monomials* $x \in |\mathbb{M}|$, $x \notin M_c$. *Then* $\mathfrak{p}_c$ *is a minimal prime ideal of* $R[\mathbb{M}]$*, and all minimal prime ideals of* $R[\mathbb{M}]$ *arise in this way.*

(c) The monomial prime ideals of $\Bbbk[\mathbb{M}]$ *are exactly the* $\Bbbk$*-vector spaces generated by the monomials not in* M_c *where* c *is a face of* Γ.

(d) Let $(x_e)_{e\in E}$ *be a family of elements of* $|\mathbb{M}| \cup \{0\}$ *generating* $R[\mathbb{M}]$ *as an* R*-algebra. (Equivalently,* $\{x_e : e \in E\} \cap M_c$ *generates* M_c *for each* $c \in \Gamma$*.) Then the kernel* I *of the surjection*

$$\varphi : R[X_e : e \in E] \to R[\mathbb{M}], \qquad \varphi(X_e) = x_e,$$

is generated by

(i) all monomials $\prod_{e\in H} X_e$ *where* H *is a subset of* E *for which* $\{x_h : h \in H\}$ *is not contained in any monoid* M_c, $c \in \Gamma$*; and*

(ii) all binomials $\prod_{h\in H} X_h^{a_h} - \prod_{k\in K} X_k^{b_k}$ *where* $H, K \subset E$*, all* x_h, x_k *are contained in a monoid* M_c *for some* $c \in \Gamma$*, and* $\sum_{h\in H} a_h x_h = \sum_{k\in K} b_k x_k$.

Proof. We have already observed that the zero ideal of $R[\mathbb{M}]$ is the intersection of the prime ideals $\mathfrak{p}_c$. Therefore $R[\mathbb{M}]$ is reduced, and its minimal prime ideals are exactly the minimal ones among the $\mathfrak{p}_c$. This proves (a) and (b). Since every prime ideal contains a minimal prime ideal, (b) reduces (c) to the case of an affine monoid for which we have already found the monomial prime ideals (Corollary 4.34).

It is also clear that I contains the ideal J generated by all the monomials and binomials listed in (d). For the converse, let f be a polynomial such that $\varphi(f) = 0$. Then we can assume that all monomials of f map to elements of $|\mathbb{M}|$ since all other monomials belong to J. Now let $c \in \Gamma$, and define f_c to be the polynomial that arises as the sum of those terms of f whose monomials are mapped to elements of $M_c \subset |\mathbb{M}|$. Then $\varphi(f_c) = 0$ as well, and it follows that f_c belongs to the ideal in $R[X_e : e \in E]$ generated by all those binomials in (ii) for which $x_h, x_k \in M_c$; see Proposition 4.26. Therefore we may replace f by $f - f_c$, and finish the proof by induction on the number of terms of f. □

Example 7.13. (a) Let $\mathscr{F}$ be a rational fan in $\mathbb{R}^d$, and set $M_C = C \cap \mathbb{Z}^d$ for each cone $C \in \mathscr{F}$. Then the monoids M_C evidently form a monoidal complex $\mathbb{M}_{\mathscr{F}}$, and the algebra $R[\mathbb{M}_{\mathscr{F}}]$ is the toric face ring introduced by Stanley [331].

(b) Let Δ be an abstract simplicial complex on the vertex set $\{1, \dots, n\}$ (see Example 1.43). Then Δ has a geometric realization by the simplices $\operatorname{conv}(e_{i_1}, \dots, e_{i_m})$ such that $\{i_1, \dots, i_m\}$ belongs to Δ (here $e_1, \dots, e_n$ is the canonical basis of $\mathbb{R}^n$). The cones over the faces of the geometric realization form a fan $\mathscr{F}$, and its toric face ring S given by (a) is no more than the Stanley–Reisner ring of Δ. In fact, according to Proposition

7.12, the kernel of the natural epimorphism $R[X_1, \dots, X_n] \to S$ is generated by those monomials $X_{j_1} \cdots X_{j_r}$ such that $\{j_1, \dots, j_r\} \notin \Delta$.

Algebras associated with monoidal complexes therefore generalize Stanley–Reisner rings by allowing arbitrary conical complexes as their combinatorial backbone and, consequently, monoid algebras as their ring-theoretic flesh.

(c) The polyhedral algebras of [60] are another special case of the algebras associated with monoidal complexes. For them the cones are generated by lattice polytopes P of a lattice polytopal complex with corresponding monoids $M(P)$.

Remark 7.14. From the construction principle of the algebras $R[\mathbb{M}]$ it is clear that their properties depend (at least) on the combinatorial properties of Γ and the algebraic properties of the rings $R[M_c]$. For results of this type (beyond Stanley–Reisner rings) we refer the reader to Brun, Bruns, and Römer [48], Brun and Römer [50], Bruns, Koch, and Römer [76], Ichim and Römer [206], [207], Yuzvinsky [379] and the authors' article [60].

Initial ideals of toric ideals. Let M be an affine monoid and $(x_e)_{e \in E}$ a finite system of generators of M. For every field (or commutative ring) $\Bbbk$ one has an associated surjective $\Bbbk$-algebra homomorphism

$$\varphi : \Bbbk[X_e : e \in E] \to \Bbbk[M], \qquad X_e \mapsto x_e.$$

The kernel of φ will be denoted by I_E. It is the toric ideal defining $\Bbbk[M]$ with respect to the system of generators $(x_e)_{e \in E}$.

For every monomial order or weight preorder on the polynomial ring we obtain an initial ideal $\operatorname{in}(I_E)$. Let us first observe that $\operatorname{in}(I_E)$ is generated by monomials and binomials.

Lemma 7.15. *Let $I \subset \Bbbk[X_1, \dots, X_n]$ be an ideal generated by monomials and binomials and w a weight vector. Then $\operatorname{in}_w(I)$ is generated by the monomials in I and the initial components of the binomials in I (either a monomial or the binomial itself).*

Proof. Let $g \in \operatorname{in}_w(I)$, $g = \operatorname{in}_w(f)$, $f \in I$. We must show that g can be written as a linear combination of initial components of monomials and binomials in I. We refine the weight preorder to a monomial order $<$ and use induction on $x = \operatorname{in}_<(g) = \operatorname{in}_<(f)$. We may assume that f has initial coefficient 1.

If x is a multiple of a monomial in I, we can immediately pass to $f - x$. Otherwise, as in the proof of Proposition 4.27 one finds a binomial $x - y \in I$ such that both x and y belong to the monomial support of f. (This is the crucial point.) Then $x > y$ by the definition of x. Set $f' = f - (x - y)$ and $g' = \operatorname{in}_w(f')$, unless $f' = 0$ and we are already done.

The reader may check that $\operatorname{in}_w(f) = \operatorname{in}_w(x - y)$ if $\deg_w f' < \deg_w f$, and $\operatorname{in}_w(f) = \operatorname{in}_w(x - y) + g'$ else. Clearly $\operatorname{in}_<(g') < \operatorname{in}_<(g)$ so that induction applies to g', and we are again done. □

Ideals generated by monomials and binomials have occurred in connection with monoidal complexes, and there is a strong relationship, as we will see now.

Let us first consider an arbitrary rational subdivision Γ of $C = \mathbb{R}_+ M$. Then Γ together with the system of monoids $F \cap M$, $F \in \Gamma$, forms a monoidal complex $\mathbb{M}_\Gamma$ for which $|\mathbb{M}_\Gamma| = M$. In this construction we have only broken the monoid structure of M, but not lost any elements.

Now suppose that $(x_e)_{e \in E}$ is a system of generators of M as above, and that the extreme rays of the cones in Γ go through elements x_e. Then we can consider the monoid $M_{F,E} = \mathbb{Z}_+(F \cap \{x_e : e \in E\})$ for each $F \in \Gamma$. With these monoids we obtain again a monoidal complex $\mathbb{M}_{\Gamma,E}$. Since $F \cap M$ need not be generated by the elements $x_e \in F$, it may very well happen that $|\mathbb{M}_{\Gamma,E}| \subsetneq M$.

According to Proposition 7.3, an initial ideal is always determined by a weight vector $(w_e)_{e \in E}$ for the system of indeterminates $(X_e)_{e \in E}$ of the polynomial ring mapping onto the monoid algebra $R = \Bbbk[M]$ via the assignment $X_e \mapsto x_e$. The weight vector w determines a conical subdivision Γ of the cone $C = \mathbb{R}_+ M$ as follows. Let $D \subset \mathbb{R}^{d+1}$ be the cone generated by the vectors (x_e, w_e), $e \in E$. The projection onto the first d coordinates maps D onto C. The *bottom B of D with respect to C* consists of all points $(x, h_x) \in D$ such that the line segment $[(x, 0), (x, h_x)]$ intersects D only in (x, h_x). In other words, $h_x = \min\{h' : (x, h') \in D\}$. Clearly $h_x > 0$ for all $x \in C$, $x \neq 0$. We set

$$\operatorname{ht}(x) = h_x.$$

(For simplicity of notation we suppress the dependence of ht on w.) The bottom is a subcomplex of the boundary of D (or D itself if the weight vector represents a grading on M). Its projection onto C defines a regular conical subdivision Γ_w of C with support function ht (see Section 1.F):

Lemma 7.16.

(a) *For all $y_1, \dots, y_m \in C$ and $\alpha_1, \dots, \alpha_m \in \mathbb{R}_+$ one has*

$$\operatorname{ht}\Big(\sum_{i=1}^{m} \alpha_i y_i\Big) \le \sum_{i=1}^{m} \alpha_i \operatorname{ht}(y_i). \tag{7.2}$$

In particular, ht *is a convex function on C.*

(b) *Its domains of linearity are the facets F of Γ_w: suppose $\alpha_i > 0$, $i = 1, \dots, m$, in (7.2); then equality holds if and only if there is a facet of Γ_w containing $y_1, \dots, y_m$.*

Proof. (a) Consider a linear combination $\sum_{i=1}^m \alpha_i y_i$ of elements of C with $\alpha_i > 0$. Then $\sum_{i=1}^m \alpha_i (y_i, \operatorname{ht}(y_i)) \in D$, and so $\operatorname{ht}(\sum_{i=1}^m \alpha_i y_i) \le \sum_{i=1}^m \alpha_i \operatorname{ht}(y_i)$.

(b) Let G be the face of Γ_w such that $\sum_{i=1}^m \alpha_i y_i \in \operatorname{int}(G)$. If $\operatorname{ht}(\sum_{i=1}^m \alpha_i y_i) = \sum_{i=1}^m \alpha_i \operatorname{ht}(y_i)$, then $\sum_{i=1}^m \alpha_i (y_i, \operatorname{ht}(y_i))$ lies in the face G' of the bottom of D that is projected onto G. But then all $(y_i, \operatorname{ht}(y_i))$ belong to G', too, and so all y_i are in G.

The converse: if all y_i lie in the same facet F of Γ_w, then the points $(y_i, \operatorname{ht}(y_i))$ lie in the corresponding facet F' of the bottom of D. So $\sum_{i=1}^m \alpha_i (y_i, \operatorname{ht}(y_i)) \in F'$, too, and this implies $\operatorname{ht}(\sum_{i=1}^m \alpha_i y_i) = \sum_{i=1}^m \alpha_i \operatorname{ht}(y_i)$. □

Since the weights w_e are positive, D is a pointed cone, even if C is not, and therefore all faces of Γ_w are pointed, too.

For each face F of Γ_w we let $M_{F,w}$ be the monoid generated by all $x_e \in F$ for which $\operatorname{ht}(x_e) = w_e$. The faces F and the monoids $M_{F,w}$ form a monoidal complex $\mathbb{M}_w$, the *monoidal complex defined by* w. Note that each extreme ray of a face F of Γ_w is the image of an extreme ray of D. The latter contains a point (x_e, w_e), and therefore $w_e = \operatorname{ht}(e)$. This implies $F = \mathbb{R}_+ M_{F,w}$. The remaining conditions for a monoidal complex are fulfilled as well.

It is important to note that the monoidal complex $\mathbb{M}_w$ is not only dependent on Γ_w or on the pair (Γ_w, E). The simple example in Figure 7.2 illustrates this fact. The cone C is generated by the vectors $x_i = (i, 1)$, $i = 1, \dots, 4$. The figure shows cross-sections of C and D for three choices of weights.

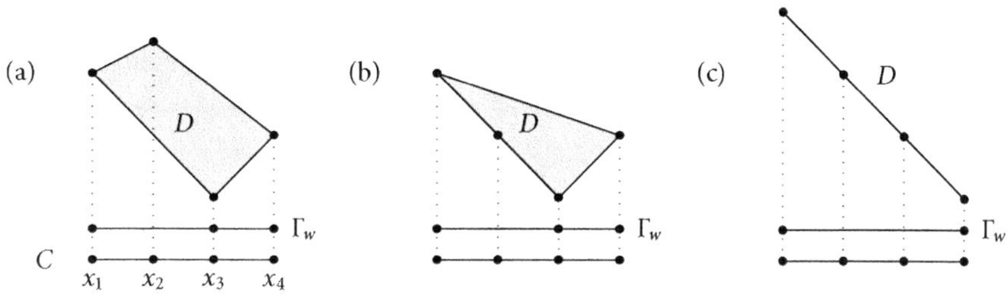

Fig. 7.2. Weights on a monoid

The ideal I_E is generated by $X_1X_4 - X_2X_3$, $X_1X_3 - X_2^2$, $X_2X_4 - X_3^2$. It is equal to the initial ideal in case (c), whereas $\operatorname{in}_w(I_E)$ is generated by X_1X_4, X_2^2, X_2X_4 in case (a), and $X_1X_3 - X_2^2$, X_1X_4, X_2X_4 in case (b). In all cases it is clear that the listed polynomials generate an ideal contained in the initial ideal, and the converse can be proved by Hilbert function arguments (see Exercise 7.9).

The algebra $\Bbbk[\mathbb{M}_w]$ is again a residue class ring of $\Bbbk[X_e : e \in E]$ under the assignment

$$X_e \mapsto \begin{cases} x_e, & x_e \in |\mathbb{M}_w|, \\ 0, & \text{else.} \end{cases}$$

The kernel of this epimorphism is denoted by J_w. In the example of Figure 7.2 systems of generators of J_w are given by X_1X_4, X_2 in case (a), and $X_1X_3 - X_2^2$, X_1X_4, X_2X_4 in case (b) whereas $J_w = I_E$ in case (c).

The following lemma is crucial for the relation between the initial ideal $\operatorname{in}_w(I_E)$ and the ideal J_w:

Lemma 7.17. *Let K be a subset of E such that the elements x_k, $k \in K$, belong to the monoid $M_{F,w}$ for a facet F of Γ_w. Moreover, let H be another subset of E, and suppose that $\sum_{h \in H} \alpha_h x_h = \sum_{k \in K} \beta_k x_k$ with positive coefficients α_h, β_k. Then*

$$\sum_{h \in H} \alpha_h w_h \geq \sum_{k \in K} \beta_k w_k,$$

and equality holds if and only if all x_h, $h \in H$, belong to $M_{F,w}$, too.

Proof. By Lemma 7.16(a), (b), and since $\operatorname{ht}(x_k) = w_k$ for all k, we have

$$\sum_{h\in H} \alpha_h w_h \geq \sum_{h\in H} \alpha_h \operatorname{ht}(x_h) \geq \operatorname{ht}\Big(\sum_{h\in H} \alpha_h x_h\Big) = \operatorname{ht}\Big(\sum_{k\in K} \beta_k x_k\Big)$$
$$= \sum_{k\in K} \beta_k \operatorname{ht}(x_k) = \sum_{k\in K} \beta_k w_k.$$

Now suppose that equality holds. Then the argument in the proof of Lemma 7.16(b) shows that all x_h belong to the face G of Γ_w in whose interior the point $\sum_{h\in H} \alpha_h x_h = \sum_{k\in K} \beta_k x_k$ lies. This implies $G \subset F$. Moreover, $w_h = \operatorname{ht}(x_h)$ for all h, and so $x_h \in M_{F,w}$.

On the other hand, if $x_h \in M_{F,w}$ for all h, then equality holds since $w_h = \operatorname{ht}(x_h)$ for all h and the x_h belong to a domain of linearity of ht. □

One cannot expect that $\operatorname{in}_w(I_E) = J_w$ since J_w is always a radical ideal, and $\operatorname{in}_w(I_E)$ need not be radical as seen above. However, this is the only obstruction:

Theorem 7.18 (Sturmfels). *Let M be an affine monoid, $(x_e)_{e\in E}$ a finite system of generators of M, and $(w_e)_{e\in E}$ a weight vector. Moreover, let $\mathbb{M}_w$ be the monoidal complex defined by w. Then the ideal J_w is the radical of the initial ideal $\operatorname{in}_w(I_E)$.*

Proof. Let us first show $J_w \subset \operatorname{Rad} \operatorname{in}_w(I_E)$. Recall that J_w is generated by monomials and binomials (Proposition 7.12). The monomial generators of J_w have the form $\prod_{h\in H} X_h$ such that there is no face of Γ_w for which $M_{F,w}$ contains all x_h simultaneously. On the other hand there exists such a face F containing $\sum x_h$ (we use additive notation in the monoid M). In each extreme ray of F lies an element x_e, $e \in E$. Therefore one has an equation

$$\sum_{h\in H} x_h = \sum_{k\in K} \beta_k x_k, \qquad \beta_k \in \mathbb{Q}, \beta_k > 0 \tag{7.3}$$

with $x_k \in M_{F,w}$ for all $k \in K$. Moreover, in view of Lemma 7.16(b) we have

$$\sum_{h\in H} w_h > \sum_{k\in K} \beta_k w_k. \tag{7.4}$$

Clearing the denominators in equation (7.3) and passing to multiplicative notation yields a binomial relation

$$\Big(\prod X_h\Big)^a - \prod X_k^{b_k} \in I_E, \qquad a\beta_k = b_k.$$

Inequality (7.4) shows that the weight of the first monomial is strictly larger than the weight of the second. Therefore $\prod_{h\in H} X_h \in \operatorname{Rad} \operatorname{in}_w(I_E)$.

The situation with the binomials $b = \prod X_h^{a_h} - \prod X_k^{b_k}$ for which all elements x_h and x_k lie in the same monoid $M_{F,w}$ is much simpler: in this case both monomials have the same weight by Lemma 7.16(b), and so b itself belongs to $\operatorname{in}_w(I_E)$.

For the converse inclusion, by Lemma 7.15 it is enough to consider the initial components of the binomials

$$b = \prod_{h\in H} X_h^{\alpha_h} - \prod_{k\in K} X_k^{\beta_k} \in I_E.$$

Passage to additive notation leads to an equation $\sum \alpha_h x_h = \sum \beta_k x_k$. The point $\sum \alpha_h x_h$ lies in a facet F of Γ_w. Therefore we find a subset L of E such that $x_e \in M_{F,w}$ for $e \in L$ and rational coefficients $\gamma_l > 0$ with

$$\sum_{h\in H} \alpha_h x_h = \sum_{k\in K} \beta_k x_k = \sum_{l\in L} \gamma_l x_l.$$

Clearing denominators and replacing the monomials of b by suitable powers we obtain two binomials in I_E:

$$\prod_{h\in H} X_h^{\alpha_h} - \prod_{l\in L} X_l^{\gamma_l} \quad \text{and} \quad \prod_{k\in K} X_k^{\beta_k} - \prod_{l\in L} X_l^{\gamma_l}.$$

Suppose first that $\prod X_h^{\alpha_h}$ is the initial monomial of b, having strictly larger weight than the second. Then it has also strictly larger weight than $\prod X_l^{\gamma_l}$ by Lemma 7.16(b). Therefore it is impossible to find a facet G of Γ_w (not even $G \neq F$) such that $x_h \in M_{G,w}$ for all $h \in H$. It follows that $\prod X_h \in J_w$ since J_w is a radical ideal.

Suppose now that both monomials of b have the same weight. If it is strictly larger than the weight of $\prod X_l^{\gamma_l}$, then both monomials of b belong to J_w individually, as just seen.

In the only case remaining all three monomials have the same weight. But in this case all x_h, $h \in H$, and x_k, $k \in K$, lie in the same monoid $M_{F,w}$, and we conclude that the binomial b belongs to J_w. □

Corollary 7.19. *The following are equivalent:*

(a) $\operatorname{in}_w(I_E)$ *is a radical ideal;*

(b) $M_{F,w} = M \cap F$ *for all facets* F *of* Γ_w.

Proof. Consider the M-grading on $S = \Bbbk[X_e : e \in E]$ in which each indeterminate X_e has degree x_e. According to Proposition 7.1(d), $R = S/I_E$ and $S/\operatorname{in}_w(I_E)$ have the same Hilbert function, given by $\dim_{\Bbbk} R_x = 1$ for all $x \in M$.

The theorem implies that $\operatorname{in}_w(I_E)$ is contained in J_w with equality if and only if $\operatorname{in}_w(I_E)$ is a radical ideal. Because of the inclusion, $\operatorname{in}_w(I_E) = J_w$ if and only if $S/\operatorname{in}_w(I_E)$ and S/J_w have the same Hilbert function. But this is equivalent to $|\mathbb{M}_w| = M$, which in its turn is just condition (b). □

We say that a monoidal complex is *free* if all its monoids are free commutative monoids. Evidently this implies that the associated conical complex is simplicial, but not conversely. The free monoidal complexes are exactly those derived from abstract simplicial complexes (compare Example 7.13(b)).

A monoidal complex $\mathbb{M}_w$ arising from an affine monoid M is *unimodular* with respect to $\operatorname{gp}(M)$ if it is free and $\operatorname{gp}(M_{F,w})$ is a direct summand of $\operatorname{gp}(M)$ for all faces F of Γ_w. Equivalently, we could require that the associated conical complex of $\mathbb{M}_w$ is a unimodular triangulation of $\mathbb{R}_+M$ and that $M_{F,w}$ is generated by the extreme integral generators of $\mathbb{R}_+F$.

Corollary 7.20.

(a) $\operatorname{Rad} \operatorname{in}_w(I_E)$ *is a monomial ideal if and only if* $\mathbb{M}_w$ *is a free monoidal complex.*

(b) $\operatorname{in}_w(I_E)$ *is a monomial radical ideal if and only if* $\mathbb{M}_w$ *is unimodular with respect to* $\operatorname{gp}(M)$.

(c) M *is normal if and only if it admits a system of generators* $(x_e)_{e\in E}$ *and a weight vector* $(w_e)_{e\in E}$ *for which* $\operatorname{in}_w(I_E)$ *is a monomial radical ideal.*

Proof. The equivalence in (a) is a consequence of Theorem 7.18 and the fact that the defining ideal of an algebra $\Bbbk[\mathbb{M}]$ is monomial if and only if $\mathbb{M}$ is free (Proposition 7.12(d)).

For (b), suppose first that $\operatorname{in}_w(I_E)$ is a monomial radical ideal. Then $\mathbb{M}_w$ is free by (a), and $M_{F,w} = M \cap F$ for all facets F of Γ_w by Corollary 7.19. In other words, the monoids $M_{F,w} = M \cap F$ are integrally closed in M (Proposition 2.22). Since unimodularity passes to faces, it is enough to consider the facets F of Γ_w. In this case $\operatorname{rank} M_{F,w} = \operatorname{rank} M$, and $\operatorname{gp}(M_{F,w}) = \operatorname{gp}(M)$ follows from Corollary 2.25.

The converse is simpler since unimodularity evidently implies integral closedness. Then one applies the converse implications of (a) and 7.19.

The implication $\Longleftarrow$ in (c) results from (b) and Theorem 2.38. For the converse we use Theorem 2.75 in order to construct a regular unimodular triangulation Γ of $C = \mathbb{R}_+M$ by elements of M, which we choose as a system of generators. The weights are then given by the values of a rational support function, multiplied by a suitable common denominator. □

Remark 7.21. One can use Corollary 7.20(c) for a proof of Hochster's theorem 6.10 by which normal affine monoid algebras are Cohen–Macaulay. We can assume that M is positive. In view of Theorem 7.4 it is enough to show that the Stanley–Reisner ring $\Bbbk[\Delta]$ ring of a regular triangulation Δ of a cross-section of $C = \mathbb{R}_+M$ is Cohen–Macaulay. By a theorem of Munkres [72, 5.4.6], the Cohen–Macaulay property of $\Bbbk[\Delta]$ only depends on the topological type of $|\Delta|$. A cross-section of a pointed cone is homeomorphic to a simplex whose Stanley–Reisner ring is certainly Cohen–Macaulay. (One can avoid the theorem of Munkres by using the shellability of Δ. See [72, Ch. 5] for this concept.)

Remark 7.22. Suppose I is an ideal in a polynomial ring R such that R/I is Gorenstein. Then one may ask whether there exists a monomial order on R such that $R/\operatorname{in}(I)$ is also Gorenstein. The answer to this question is unknown even for some classical ideals. However, if I is a toric ideal defined by a Gorenstein lattice polytope with a unimodular regular triangulation, then such a monomial order exists. In fact, a monomial order corresponding to the triangulation in Theorem 6.57 yields such an initial ideal; see Bruns and Römer [78]. For special cases such initial ideals have been explicitly constructed by Conca, Hoşten, and Thomas [89].

In the investigation of a normal monoid M one is usually not interested in an arbitrary system of generators of M, but in $\operatorname{Hilb}(M)$. As we have seen in Section 2.D, one can not always find a (regular) unimodular triangulation by elements of $\operatorname{Hilb}(M)$, and this limits the value of Corollary 7.20 considerably. Nevertheless, it is very powerful when the unimodularity of certain triangulations is given automatically, and such a case will be studied in the next section.

The multiplicities of the minimal primes. We have found the radical of $\operatorname{in}_w(I_E)$, and can describe its minimal prime ideals. As a crude measure of the equality of $\operatorname{in}_w(I_E)$ and its radical, one can use the multiplicities of the minimal prime ideals of $\Bbbk[X_e : e \in E]/\operatorname{in}_w(I_E)$: the *multiplicity* of a minimal prime ideal $\mathfrak{p}$ in a noetherian ring R is the length of $R_\mathfrak{p}$ as a module over itself. It is in fact a finite number since $R_\mathfrak{p}$ is an artinian ring. If R is reduced then the localizations with respect to the minimal prime ideals are fields, or equivalently, the multiplicities of the minimal prime ideals are all equal to 1. The converse is not true: R may have nonminimal associated prime ideals.

There is another way to find the multiplicities of the minimal prime ideals $\mathfrak{p}$: R has a filtration

$$\mathcal{F} : 0 = I_0 \subset I_1 \subset \cdots \subset I_m = R$$

by ideals such that each quotient I_{i+1}/I_i is isomorphic to a residue class ring $R/\mathfrak{q}$ with respect to a prime ideal $\mathfrak{q}$. We have already discussed the existence of such a filtration and its consequences for the proof of Proposition 4.37 and follow similar ideas again. Localization of the filtration with respect to $\mathfrak{p}$ yields a filtration of $R_\mathfrak{p}$ in which all quotients $R/\mathfrak{p}$ are replaced by the residue class field $\Bbbk(\mathfrak{p}) = R_\mathfrak{p}/\mathfrak{p}R_\mathfrak{p}$ and all quotients $R/\mathfrak{q}$ with $\mathfrak{q} \neq \mathfrak{p}$ by 0 (note that $\mathfrak{p}$ is a minimal prime ideal). The residue class field $\Bbbk(\mathfrak{p})$ has length 1 as an $R_\mathfrak{p}$-module, and therefore the multiplicity of $\mathfrak{p}$ in R is just the multiplicity with which $R/\mathfrak{p}$ appears as a quotient in the filtration $\mathcal{F}$.

We return to the setting of Theorem 7.18, and apply these considerations to $R = \Bbbk[X_e : e \in E]/\operatorname{in}_w(I_E)$. As we have observed, it is an M-graded ring, and as a graded $\Bbbk$-vector space it can be identified with $\Bbbk[M]$. For R the filtration $\mathcal{F}$ can be chosen in such a way that all ideals I are graded prime ideals. Therefore the successive quotients have the form $(R/\mathfrak{q})(-x)$ with a monomial prime ideal $\mathfrak{q}$ and an element $x \in M$. In the vector space structure, the filtration becomes a direct sum decomposition, and it yields a decomposition of the set M into pairwise disjoint subsets representing the monomial bases of the quotients. Each of them has the form $x + M_{F,w}$ where F is a face of Γ_w (compare Proposition 7.12(c)). All the subsets $x + M_{G,w}$ for which G is not a facet of Γ_w are contained in finitely many hyperplanes. These combinatorial conditions determine the number of subsets $x + M_{F,w}$ for the facets F:

Theorem 7.23. *Let M be an affine monoid, and $M_1, \dots, M_n$ submonoids of M with $\operatorname{rank} M_i = \operatorname{rank} M$ such that $\mathbb{R}_+M$ is the union of the cones $\mathbb{R}_+M_i$, $i = 1, \dots, n$, but not the union of any $n-1$ of them. Moreover, let x_{ij}, $i = 1, \dots, n$, $j = 1, \dots, k_i$, be elements of $\operatorname{gp}(M)$. Then the following are equivalent:*

(a) there exist finitely many hyperplanes $H_1, \dots, H_m$ in $\operatorname{aff}(M)$ such that

$$M \subset \Bigl(\bigcup_{i,j} x_{ij} + M_i\Bigr) \cup H_1 \cup \cdots \cup H_m,$$

and for each $i = 1, \dots, n$ the sets $x_{ij} + M_i$, $j = 1, \dots, k_i$, are pairwise disjoint;

(b) the elements x_{ij} form a system of representatives of $\operatorname{gp}(M)/\operatorname{gp}(M_i)$ for each $i = 1, \dots, n$.

Proof. (a) $\Longrightarrow$ (b) Let us first show that the x_{ij} belong to pairwise different residue classes modulo $\mathrm{gp}(M_i)$. Suppose that two elements x_{ij}, say x_{i1} and x_{i2}, lie in the same residue class. The sets M_i and $(x_{i2}-x_{i1})+M_i$ are again disjoint, and $x_{i2}-x_{i1} \in \mathrm{gp}(M_i)$. Note that there exists $y \in M$ such that $y + \bar{M}_i \subset M_i$: the conductor $\mathfrak{c}(\bar{M}_i/M_i)$ is nonempty (Proposition 2.33). It follows that $\bar{M}_i$ and $x + \bar{M}_i$ are disjoint for some $x \in \mathrm{gp}(M_i)$. But this is impossible.

It remains to show that for each i, say $i = n$, all residue classes of $\mathrm{gp}(M)$ modulo M_i are represented. Observe that $\mathbb{R}_+M$ is not contained in the union U of the cones $\mathbb{R}_+M_1, \dots, \mathbb{R}_+M_{n-1}$ and those among the hyperplanes H_u that pass through 0, and we can find a point z that belongs to $\mathrm{int}(\mathbb{R}_+M)$ outside U. The ray $\mathbb{R}_+z$ intersects each of the remaining hyperplanes in at most one point, and for sufficiently large a the point az has a neighborhood V satisfying the following conditions:

(1) $V \cap H_u = \varnothing$ for $u = 1, \dots, m$;
(2) $V \cap (x_{ij} + \mathbb{R}_+M_i) = \varnothing$ for all $i = 1, \dots, n-1$ and $j = 1, \dots, k_i$;
(3) $V \cap M$ contains a point of each residue class of $\mathrm{gp}(M)$ modulo $\mathrm{gp}(M_n)$.

For (2) we use that there exists a support form σ_k of $\mathbb{R}_+M_i$ with $\sigma_k(z) < 0$. Therefore $\sigma_k(az) < \sigma_k(x_{ij})$ for $a \gg 0$, and so $az \notin x_{ij} + \mathbb{R}_+M_i$. Enlarging a further, we can find an arbitrarily large open ball V around az satisfying (1) and (2). If the radius is big enough, an open ball meets all residue classes modulo a subgroup of full rank, independently of its center.

It follows immediately from (1), (2), and (3) that all residue classes modulo $\mathrm{gp}(M_n)$ must be represented by the x_{nj}.

(b) $\Longrightarrow$ (a) It is enough to note that $M \cap \mathbb{R}_+M_i$ is covered by the sets $x_{ij} + M_i$ up to finitely many hyperplanes, and for that it suffices to consider each residue class of M modulo $\mathrm{gp}(M_i)$ individually. The claim then reduces to the fact that M_i and $x + M_i$ differ only in finitely many hyperplanes if $x \in \mathrm{gp}(M)$. We leave the rest to the reader. Again the nonemptyness of the conductor is helpful. □

Corollary 7.24. *In the situation of Theorem 7.18 let F be a facet of Γ_w. Then the multiplicity of the minimal prime ideal $\mathfrak{p}_F$ of $\mathrm{in}_w(I_E)$ corresponding to F is the index* $[\mathrm{gp}(M) : \mathrm{gp}(M_{F,w})]$.

In an important case the multiplicity can be interpreted as the multiplicity (or normalized volume) of a simplex.

Corollary 7.25. *In the situation of Theorem 7.18 let F be a facet of Γ_w for which $M_{F,w}$ is free, generated by $x_1, \dots, x_d$. Then the multiplicity of the prime ideal $\mathfrak{p}_F$ is given by* $\mu_{\mathrm{gp}(M)}(\mathrm{conv}(0, x_1, \dots, x_d))$.

In particular, if M is homogeneous and all elements $x_1, \dots, x_d$ have degree 1, then the multiplicity of $\mathfrak{p}_F$ equals the multiplicity of the simplex $\mathrm{conv}(x_1, \dots, x_d)$ *with respect to* $\mathrm{gp}(M)$.

Lex and revlex monomial orders. If $<$ is a monomial order, then the initial ideal

$\text{in}_<(I_E)$ induces a free monoidal complex, and in particular the subdivision Γ_w is a triangulation (the weight vector w has been chosen according to Proposition 7.3). We conclude this section by describing the triangulations obtained from the lex and the revlex orders in the case in which the system of generators (x_e) is contained in a hyperplane. Then the triangulations of the cone $\mathbb{R}_+ M$ through the vectors (x_e) correspond bijectively to the triangulations of the lattice polytope $\operatorname{conv}(x_e : e \in E)$. As it turns out, both types of triangulations have already been discussed in Remark 1.52(a), Lemma 1.50, and Exercise 1.17.

Let the indeterminates be numbered $X_1, \dots, X_n$ such that $X_1 > \cdots > X_n$. The corresponding elements in M are $x_1, \dots, x_n$.

We start with lex. The first element to be inserted is x_n, and we can suppose that $P_{i+1} = \operatorname{conv}(x_n, \dots, x_{i+1})$ has already been triangulated with triangulation Λ_{i+1}. Consider x_i. If $x_i \in P_{i+1}$, then we simply proceed to x_{i-1}. Otherwise we extend Λ_{i+1} by all simplices $\operatorname{conv}(\delta, x_i)$ for which δ is a simplex of Λ_{i+1} that is visible from x_i. The passage from Λ_{i+1} to Λ_i is the lexicographic extension by x_i introduced in Exercise 1.17.

For revlex, we start with x_1. Again, we assume that $P_{i-1} = \operatorname{conv}(x_1, \dots, x_{i-1})$ has already been triangulated with triangulation Δ_{i-1}. Then we restrict Δ_{i-1} to the link of x_i in the boundary of $P_i = \operatorname{conv}(x_1, \dots, x_i)$, obtaining Δ'_{i-1}, and insert the simplices $\operatorname{conv}(\delta, x_i)$, $\delta \in \Delta'_{i-1}$.

Figure 7.3 shows the triangulations of a hexagon derived from a lex and a revlex order. The generators $x_1, \dots, x_6$ correspond to the vertices. In the passage to the tri-

Fig. 7.3. A lex and a revlex triangulation

angulations a great deal of information is lost. For example, in dimension 2 the revlex triangulation shows only which vertex is the smallest.

7.C Toric ideals and triangulations

To simplify notation, we will always use the Hilbert basis of a positive affine monoid M as a system of generators, and consider the representation $S = \Bbbk[X_x : x \in \operatorname{Hilb}(M)] \to \Bbbk[M]$. Toric ideals, their Gröbner bases, graded free resolutions of $\Bbbk[M]$, etc. are always taken with respect to this representation, unless explicitly indicated otherwise.

Bounding degrees in toric ideals. The algebra $\Bbbk[M]$ for a normal affine monoid is Cohen–Macaulay, and we know its canonical module. This allows us to prove bounds for its graded Betti numbers (introduced on p. 208).

Theorem 7.26. *Let M be a normal affine monoid, $\Bbbk$ a field, and $R = \Bbbk[M]$. Let*

$$\mathbb{F} : 0 \to \bigoplus_{u \in \mathbb{Z}^r} S(-u)^{\beta_{pu}} \to \cdots \to \bigoplus_{u \in \mathbb{Z}^r} S(-u)^{\beta_{0u}} \to R \to 0,$$
$$p = \#\operatorname{Hilb}(M) - \operatorname{rank} M,$$

be the minimal M-graded free resolution of R over the polynomial ring S introduced above, with $\beta_{iu} = \beta_{iu}(R)$. If $\beta_{jv} \neq 0$, then there exist nonzero elements $y_1, \dots, y_{p-j} \in M$ and $z \in \operatorname{int}(M)$ such that

$$v + y_1 + \cdots + y_{p-j} + z = g, \qquad g = \sum_{x \in \operatorname{Hilb}(M)} x \tag{7.5}$$

(with additive notation in M).

Proof. We use induction on $p - j$. By equation (6.6) (p. 215) one has $\beta_{pv}(R) = \beta_{0,g-v}(\omega_R)$. Therefore, if $\beta_{pu}(R) \neq 0$, then $g - v$ (as an element of M) belongs to the minimal monomial system of generators of ω_R, and so lies in $\operatorname{int}(M)$ (Theorem 6.31). Suppose that $j < p$. We claim that there exists $y \in M$, $y \neq 0$, such that $\beta_{j+1,v+y}(R) \neq 0$.

This is most easily seen by writing the map

$$\bigoplus_{u \in \mathbb{Z}^r} S(-u)^{\beta_{j+1,u}} \to \bigoplus_{u \in \mathbb{Z}^r} S(-u)^{\beta_{ju}}$$

as a matrix. We let the image be generated by its rows. The column corresponding to a basis element of shift v must have a nonzero entry, and we choose y as its degree. If all entries were zero, then in the complex $\operatorname{Hom}_S(\mathbb{F}, S(-g))$ an element of a basis of a free module would be mapped to zero, and this is impossible since $\operatorname{Hom}_S(\mathbb{F}, S(-g))$ is a minimal graded free resolution of ω_R (compare Theorem 6.18). □

In the following we will restrict ourselves to homogeneous affine monoids M, generated by their degree 1 elements. Whenever such a monoid is normal, it is of type $M(P)$ for the lattice polytope P spanned by its degree 1 elements, and P is normal. For an arbitrary lattice polytope$P$ we let $\Bbbk[P]$ again denote the monoid algebra $\Bbbk[M(P)]$.

Theorem 7.26 bounds the shifts in a minimal graded free resolution rather efficiently. In the homogeneous case one obtains this consequence:

Corollary 7.27 (Sturmfels). *Let P be a normal lattice polytope, $M = M(P)$ be the corresponding polytopal monoid, and m be the minimal degree of an element in $\operatorname{int}(M)$. If $\beta_{jk}(\Bbbk[M]) \neq 0$, then $k \leq j + \dim P + 1 - m$.*

In particular, the toric ideal I defining $\Bbbk[M]$ is generated by elements of degree $\leq \dim P + 2 - m \leq \dim P + 1$.

Proof. Equation (7.5) yields

$$k = \deg v = n - \sum_{l=1}^{p-j} \deg y_l - \deg z \leq n - (n - (\dim P + 1) - j) - m. \qquad \square$$

Remark 7.28. It was conjectured by Sturmfels [339, 2.8] that Corollary 7.27 can be strengthened to a bound for the initial ideal of I with respect to a suitable term order. The general case seems to be unknown, but for the class of Δ-*normal* polytopes it was proved by O'Shea and Thomas [280]. Moreover, I has an initial ideal that defines a Cohen–Macaulay ring in this case. One calls P Δ-normal if P has a regular triangulation for which each of the simplices is again normal.

For further results on Gröbner bases in the (not necessarily normal) simplicial case see Hellus, Hoa, and Stückrad [182].

Consider the lattice triangle spanned by $(-1,-1)$, $(1,0)$, and $(0,1)$ with its four lattice points. Then the corresponding toric ideal is generated by $X_4^3 - X_1X_2X_3$. This simple example shows that Corollary 7.27 cannot be strengthened without a further hypothesis.

Theorem 7.29 (Ogata). *Let P be a normal lattice polytope that contains at least* $\dim P + 2$ *lattice points. Then the following are equivalent:*

(a) *∂P contains at least* $\dim P + 2$ *lattice points;*
(b) *the toric ideal I defining $\Bbbk[P]$ is generated by elements of degree* $\le \dim P$.

We restrict ourselves to showing that (a) is necessary for (b). The converse implication needs methods of projective algebraic geometry not treated in this book; see Ogata [276]. Suppose that P has $\ge \dim P + 2$ lattice points, but only $\dim P + 1$ in its boundary. Then these lattice points are the vertices $x_0, \dots, x_d$. There is a further lattice point in the interior, say y. Writing y as a convex linear combination of $x_0, \dots, x_d$ and clearing denominators yields a binomial

$$X_0^{a_0} \cdots X_d^{a_d} - Y^b \in I.$$

It does not lie in the subideal of I generated by elements of degree $\le d$, since there is no binomial in I that has a monomial in only d of the variables $X_0, \dots, X_d$.

The degree of a triangulation. Let Δ be an abstract simplicial complex with vertex set V. If Δ is not (the set of faces of) a full simplex, we set

$$\deg \Delta = \max\{\dim \sigma : \sigma \text{ is a minimal nonface of } \Delta\} + 1,$$

and call it the *degree* of Δ. The degree of Δ is nothing but the maximal degree (in the ordinary sense) of the elements in the minimal monomial generating set of the Stanley–Reisner ideal of Δ.

Let M be an affine monoid. In proving degree bounds for the toric ideal J defining $\Bbbk[M]$ one can try the following strategy: find a regular unimodular triangulation Δ of $\mathbb{R}_+M$, and an estimate for the degree of Δ. The same estimate holds for the elements of a Gröbner basis of J whose initial ideal is given by $I(\Delta)$, and therefore for a system of generators of J.

The unimodularity of the triangulation implies the normality of M, but even in the presence of normality the main obstruction to this strategy is the potential lack of unimodularity. A weaker condition is that Δ is full (Definition 2.51): all $x \in \text{Hilb}(M)$ span rays of Δ. Clearly, a unimodular triangulation must be full.

In the following we restrict ourselves to lattice polytopes P (and the monoids $M(P)$). They have regular full lattice triangulations by Corollary 1.66, and we now show there even exist such lexicographic triangulations. Let $x_1, \dots, x_n$ be the lattice points of P, and consider the lexicographic triangulation with respect to $x_1 > \dots > x_n$. In general it is not full, namely if $x_i \in \text{conv}(x_{i+1}, \dots, x_n)$, and x_i is discarded from the triangulation. However, rearranging the order if necessary we can assume that x_i is a vertex of $\text{conv}(x_i, x_{i+1}, \dots, x_n)$ for all i (start with a vertex x_1 of the polytope). Such an order is called *exterior*. The lexicographic triangulation with respect to an exterior order is full.

Remark 7.30. For the construction of the full lexicographic triangulations in Theorems 7.33 and 7.35 it is irrelevant that the set of lattice points of P is used. One can replace it by an arbitrary finite set spanning P.

For an arbitrary simplicial complex one has $\deg \Delta \le \dim \Delta + 2$. For triangulations of polytopes there is a better bound. Moreover, each minimal nonface can be considered as a simplex in the geometric sense.

Lemma 7.31. *Let Δ be a triangulation of a polytope P of dimension d.*

(a) *If $\{x_0, \dots, x_n\} \subset \text{vert}(\Delta)$ is a minimal nonface, then it is an affinely independent set. In particular,* $\deg \Delta \le d + 1$.
(b) *If $\{x_0, \dots, x_d\}$ is a minimal nonface, then the simplices of Δ that are contained in* $P' = \text{conv}(x_0, \dots, x_d)$ *triangulate P'.*
(c) *Furthermore, $\{x_0, \dots, x_d\}$ is a minimal nonface if and only if each subset spans a simplex of Δ and the d-simplex* $\text{conv}(x_0, \dots, x_d)$ *contains a point $y \in \text{vert}(\Delta)$ in its interior.*

Proof. (a) Assume $\{x_0, \dots, x_n\}$ is not affinely independent. Since it is a minimal nonface of Δ (as an abstract simplicial complex), all its proper subsets are affinely independent and span simplices of Δ. Let δ_i be the convex hull of the points x_j, $j \ne i$. All the δ_i span the same affine space A. As $\text{int}(\delta_0) \cap \text{int}(\delta_i) = \varnothing$ for $i \ne 0$, x_0 and x_i must lie on opposite sides of the hyperplane H_i of A through the points x_j, $j \ne 0, i$. Since these hyperplanes are the support hyperplanes of δ_0 in A, the intersection of the halfspaces $H_i^< \subset A$ is empty (why?). On the other hand, it must contain x_0, a contradiction.

(b) As just shown, $\{x_0, \dots, x_d\}$ is an affinely independent set. Therefore P' is a simplex of dimension d. The boundary of P' is triangulated by its facets, which all belong to Δ. Choose $x \in \text{int}(P')$ and $\delta \in \Delta$ such that $x \in \delta$. If $\delta \not\subset P'$, then even $\text{int}(\delta) \not\subset P'$, and we choose $y \in \text{int}(\delta) \setminus P'$. The line segment $[x, y]$ meets a facet F of P'. The intersection point z lies in $\text{int}(\delta)$ since $x \ne y$ and $y \in \text{int}(\delta)$. But this is a contradiction: $F \cap \delta$ is a proper face of δ since both simplices belong to Δ.

(c) It is only to be shown that $\text{conv}(x_0, \dots, x_d)$ contains a point $y \in \text{vert}(\Delta)$ in its interior if $\{x_0, \dots, x_d\}$ is a minimal nonface. But this follows immediately from (b). □

Remark 7.32. If a lattice polytope P has a regular unimodular triangulation, then the corresponding toric ideal I has an initial ideal generated in degree $\leq \dim P + 1$, as follows from Corollary 7.20(b) and Lemma 7.31(a). In particular, I itself satisfies the same degree bound.

In the same way that Theorem 7.29 improves Corollary 7.27, one can improve Lemma 7.31 if there is an extra lattice point in the boundary.

Theorem 7.33. *Let P be a lattice polytope of dimension n containing at least $n+2$ lattice points. Then P has a regular full lattice triangulation Δ of degree $\leq n$ if and only if ∂P contains at least $n+2$ lattice points.*

The condition is certainly necessary: P has at least $n+1$ vertices, and if they are the only lattice points in ∂P, then P is a simplex with at least one lattice point in its interior. All the facets of P must belong to Δ, but P itself cannot be in Δ. Therefore P is a minimal nonface, and $\deg \Delta \geq n+1$.

The sufficiency of the condition is easy to see if P has no interior lattice point. In this case it follows immediately from Lemma 7.31(c); every full lattice triangulation has degree $\leq n$. If P has interior lattice points, then the situation is rather complicated, and we refer the reader to Bruns, Gubeladze, and Trung [69, 3.3.1]. However, the case of dimension 2 will be treated in detail below.

The next lemma gives a precise inductive condition under which the lexicographic triangulation relative to an exterior order has degree $\leq \dim P$. Let Q be a union of faces of a polytope P, $\dim P = d$. We say that a triangulation Δ of P is *d-restricted* on Q if each $(d-1)$-simplex $\tau \in \Delta$ with $\partial\tau \subset Q$ is contained in Q.

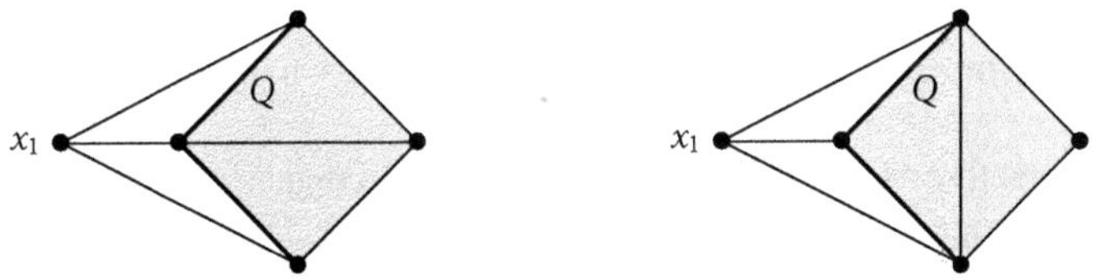

Fig. 7.4. Lexicographic extensions of degree 2 and 3 respectively

Lemma 7.34. *Let Δ' be a triangulation of $P' = \operatorname{conv}(x_2, \dots, x_n)$, and Δ its lexicographic extension to $P = \operatorname{conv}(x_1, \dots, x_n)$. Set $d = \dim P$ and let Q be the set of points of P' that are visible from x_1. Then $\deg \Delta \leq d$ if and only if*

(a) $\deg \Delta' \leq d$, and
(b) Δ' is d-restricted on Q.

Proof. Suppose that the conditions (a) and (b) are satisfied. Assume that δ is a minimal nonface of Δ of dimension d. Then x_1 must be a vertex of δ because of (a). The remaining vertices $y_1, \dots, y_d$ of δ belong to Q and $\tau = \operatorname{conv}(y_1, \dots, y_d) \in \Delta'$. All the other facets of δ contain x_1, and are therefore contained in Q. Hence all the faces of δ that have dimension $\dim \delta - 2$ and do not contain x_1 are also contained in Q. They form the boundary of τ. Because of (b), $\tau \subset Q$. So δ contains no vertex of the triangulation Δ in its interior, a contradiction to Lemma 7.31(c).

Now suppose $\deg \Delta \leq d$. Then condition (a) is satisfied because every minimal nonface of Δ' is a minimal nonface of Δ. Let $\tau \in \Delta$ be a $(d-1)$-simplex with $\partial\tau \subset Q$, and let H be the hyperplane (in $\operatorname{aff}(P)$) through τ, oriented such that $x_1 \in H^<$. If $\tau \not\subset Q$, then there exists a vertex x_j of P' with $x_j \in H^<$. Since the full boundary of τ is visible from x_1, the point x_j must lie in the interior of the simplex δ spanned by τ and x_1. Then δ is a minimal nonface of Δ, a contradiction. □

Lattice polygons. In dimension 2, condition (b) of Lemma 7.34 just says that every edge of Δ that connects two vertices in Q is contained in Q. In the following a *path* in the boundary ∂P of a polygon P is a connected union of edges of P different from ∂P. Two paths are *disjoint* if they share at most one point.

Theorem 7.35. *Let P be a lattice polygon with $\#\operatorname{lat}(\partial P) \geq 4$. Let $\partial P = C_1 \cup \dots \cup C_r$ be a decomposition of ∂P into $r \geq 3$ disjoint paths. Then there exists an exterior order $>$ on $\operatorname{lat}(P)$ such that the corresponding lexicographic triangulation Δ is unimodular and satisfies the following conditions:*

(i) $\deg \Delta = 2$;

(ii) every edge of Δ with vertices on C_i lies on C_i, $i = 1, \dots, r$.

In particular, there exists an exterior order $>$ on $\operatorname{lat}(P)$ such that the corresponding triangulation Δ is unimodular and of degree 2.

Proof. We have noticed already that the lexicographic triangulation with respect to an exterior order is full, and a full lattice triangulation of a lattice polygon is automatically unimodular (Corollary 2.54). Therefore it is not necessary to mention unimodularity any further.

Suppose first that $\#\operatorname{lat}(\partial P) = 4$. Then P is either a triangle $\langle x_1, x_2, x_3\rangle$ with a lattice point x_4 on the edge $[x_1, x_3]$ or a quadrangle $\langle x_1, x_2, x_3, x_4\rangle$. Since the number of the paths $C_1, \dots, C_r$ is at least three, we may assume that x_2 and x_4 do not belong to the same path.

If $\#\operatorname{lat}(P) = 4$, i. e., P has no interior point, we obtain the lexicographic triangulation of P which corresponds to the exterior order $x_1 > x_2 > x_3 > x_4$ by connecting x_2 and x_4.

If $\#\operatorname{lat}(P) > 4$, then P has an interior lattice point. Without restriction we may assume that the triangle $\langle x_1, x_2, x_4\rangle$ contains an interior point of P. Let P' be the convex hull of the lattice points in P except x_1, and Q the part of the boundary of P' that is visible from x_1. Then $\#\operatorname{lat}(Q) \geq 3$, hence $\#\operatorname{lat}(\partial P') \geq 4$. Moreover, $Q \cup [x_2, x_3] \cup [x_3, x_4]$ is a decomposition of $\partial P'$ into three disjoint paths. By induction on the number $\#\operatorname{lat}(P)$ we may assume that there is an exterior order $>$ on $\operatorname{lat}(P')$ such that the corresponding lexicographic triangulation Γ of P' satisfies the conditions:

(1) $\deg \Gamma = 2$;

(2) every edge of Γ with vertices on Q lies on Q.

By Lemma 7.34, the resulting exterior order $>$ on $\operatorname{lat}(P)$ with x_1 as the maximal element induces a lexicographic triangulation Δ of P with $\deg \Delta = 2$. Neither $[x_2, x_4]$ nor $[x_1, x_3]$ are faces of Δ so that condition (ii) of the theorem is trivially satisfied. In fact, x_3 is not visible from x_1, and $[x_2, x_4]$ has both its vertices in Q. This finishes the case $\# \operatorname{lat}(\partial P) = 4$.

Assume now that $\# \operatorname{lat}(\partial P) > 4$. Choose x_1 to be the common vertex x of C_1 and C_r, and define P' and Q as above. We have $\# \operatorname{lat}(P') \geq \# \operatorname{lat}(P) - 1 \geq 4$. If $\# \operatorname{lat}(C_i) = 2$, i. e., C_i has no lattice points in its interior, for all $i = 1, \ldots, r$, then $r = \# \operatorname{lat}(\partial P)$ and $\partial P'$ has a decomposition into $r - 1 \geq 3$ disjoint paths $D_1 \cup \cdots \cup D_{r-1}$ with

$$D_i = \begin{cases} Q, & i = 1, \\ C_i, & i = 2, \ldots, r-1. \end{cases}$$

If there exists a path C_i with $\# \operatorname{lat}(C_i) > 2$, we may assume that it is C_1. Set

$$D_i = \begin{cases} C_1 \cap \partial P', & i = 1, \\ C_i, & i = 2, \ldots, r-1, \\ C_r \cap \partial P', & i = r, \\ Q, & i = r+1. \end{cases}$$

If $\# \operatorname{lat}(C_r) = 2$, then $\partial P'$ has the decomposition $D_1 \cup \cdots \cup D_{r-1} \cup D_{r+1}$, and if $\# \operatorname{lat}(C_r) > 2$, it has the decomposition $D_1 \cup \cdots \cup D_{r+1}$ into disjoint paths. In any case, by induction on the number $\# \operatorname{lat}(P)$ we may assume that there is an exterior order $>$ on $\operatorname{lat}(P')$ such that the corresponding lexicographic triangulation Γ of P' satisfies the following conditions:

(1) $\deg \Gamma = 2$;
(2) every edge of Γ with vertices on a path D_i lies on D_i.

Note that Q is one of the paths D_i. By Lemma 7.34, the resulting exterior order $>$ on $\operatorname{lat}(P)$ with x_1 as its maximal element induces a lexicographic triangulation Δ of P with $\deg \Delta = 2$. Due to the definition of D_i and the property (2), every edge of Δ with vertices on a path C_i, $i = 2, \ldots, r-1$, must lie on C_i. For $i = 1, r$, such an edge must have x_1 as a vertex. The other vertex must be the only lattice point of C_i visible from x_1. Hence this edge lies on C_i. □

Remark 7.36. For certain classes of lattice polygons one can explicitly describe monomial orders that yield unimodular lexicographic triangulations of degree 2. For example let P be a rectangular lattice triangle with the vertices $(0,0)$, $(a_1, 0)$, and $(0, a_2)$. By symmetry we may assume $a_1 \geq a_2$. Then we define the order $\prec$ on $\mathbb{Z}^2$ by setting $(x_1, x_2) \prec (y_1, y_2)$ if $x_1 > y_1$ or $x_1 = y_1$, $x_2 < y_2$. It can be shown that the associated lexicographic triangulation is unimodular and of degree 2. For $a_1 = 4$ and $a_2 = 3$ it is given by Figure 7.5.

Corollary 7.37. *Let P be a lattice polygon with at least four lattice points. Let $R = \Bbbk[P]$ and I the corresponding toric ideal. Then the following are equivalent:*

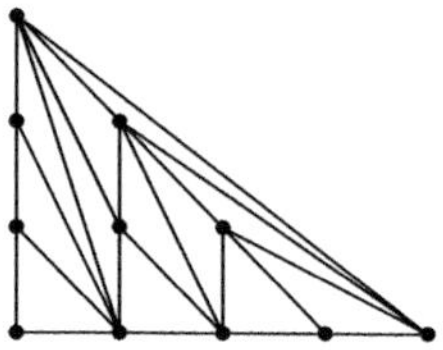

Fig. 7.5. A lexicographic triangulation of a rectangular triangle

(a) *∂P contains at least four lattice points;*
(b) *P has a unimodular (lexicographic) triangulation of degree 2;*
(c) *I has a quadratic Gröbner basis;*
(d) *R is a Koszul algebra;*
(e) *I is generated by elements of degree 2.*

Proof. The crucial implication (a) $\Longrightarrow$ (b) is Theorem 7.35. (b) $\Longrightarrow$ (c) follows from Corollary 7.20 since the degree of triangulation is the maximal degree of a monomial generator of the Stanley–Reisner ideal. The implication (c) $\Longrightarrow$ (d) is covered by Theorem 7.6, whereas (d) $\Longrightarrow$ (e) is an elementary property of Koszul algebras. Finally, (e) $\Longrightarrow$ (a) is the implication of Theorem 7.29 that we have proved. □

Remark 7.38. It is very reasonable to ask whether the implications (c) $\Longrightarrow$ (d) $\Longrightarrow$ (e) of Corollary 7.37 can be reversed for graded $\Bbbk$-algebras in general, but right away in the class of affine monoid algebras this is not possible.

(a) An example of an affine monoid algebra R that is defined by binomials of degree 2, but not Koszul, has been given by Roos and Sturmfels [305]. It does not even have a rational *Poincaré series*

$$P_R(t) = \sum_{i=0}^{\infty}\Big(\sum_j \beta_{ij}(\Bbbk)\Big)t^i$$

whose coefficients are the total Betti numbers $\sum_j \beta_{ij}(\Bbbk)$ of $\Bbbk$ over R: if R were Koszul, then $H_R(t)P_R(-t) = H_{\Bbbk}(t) = 1$, and the rationality of the Hilbert series would imply the rationality of the Poincaré series. (The relation between the Hilbert series of R and $\Bbbk$ and the Poincaré series of R is proved in the same way as equation (6.8).)

The problem whether the Poincaré series of a local (or graded) ring is rational was posed by Serre [318, p. 114], and the first counterexample was given by Anick [7]. See Laudal and Sletjøe [240], Gasharov, Peeva, and Welker [138], and Peeva and Sturmfels [288] for further results on the rationality of Poincaré series of toric rings.

Ohsugi and Hibi [278] have given an example of a normal affine monoid ring that is defined by degree 2 binomials, but not Koszul.

(b) An example of an affine monoid algebra that is Koszul, but has no Gröbner basis of degree 2 (with respect to its monomial system of generators) is the algebra generated by all lattice points in the triangle $\operatorname{conv}\big((0,0),(3,0),(0,3)\big)$, except the point $(1,1)$. One can use CaTS [212] to show that it has no Gröbner basis of degree 2. The example had been suggested as a candidate by Peeva and Sturmfels, and its Koszul property has been proved by Caviglia [83].

Another example with the same properties has been found by Ohsugi and Hibi [278]. The underlying monoid is generated by 0-1-vectors, but is not normal. This excludes the existence of a quadratic monomial initial ideal; see Exercise 7.11.

(c) We do not know of a normal affine monoid algebra which is Koszul but has no quadratic monomial initial ideal. Likewise it seems to be an open question whether there exists such algebras with a quadratic initial monomial ideal, but without a squarefree such ideal.

(d) It does not seem to be known whether there exist affine monoid algebras $\Bbbk[M]$ for which the Koszul property depends on the characteristic of $\Bbbk$. However, Roos [304] has constructed a torsionfree $\mathbb{Z}$-algebra R such that $R \otimes \mathbb{Q}$ is Koszul, but $R \otimes \mathbb{Z}/(p)$ is not Koszul for any prime number p.

Moreover, there exist non–Koszul affine monoid algebras whose Poincaré series depends on $\Bbbk$. The monoid generated by the 10 vertex-facet incidence vectors of the minimal triangulation of the real projective plane yields an example.

Remark 7.39. The construction of degree 2 unimodular regular triangulations is a difficult task. Except for polygons, it has been successful only in special cases. In [69] it has been constructed for all lattice polytopes $P \subset \mathbb{R}^d$ that are bounded by parallel translates of the hyperplanes with the equations $x_i = x_j$, $1 \le i < j \le d$, and $x_i = 0$, $i = 1, \dots, d$. This includes the multiples of the unit simplex spanned by 0, e_1, $e_1 + e_2, \dots, e_1 + \dots + e_d$.

The proof uses the Knudsen–Mumford triangulations of Chapter 3. For them regularity and degree 2 follows immediately from the fact that they are defined by hyperplane dissections (Section 1.F and Exercise 7.12).

In the special case of the multiples $c\Delta$ of the unit simplex it follows that the defining ideals of the Veronese subrings of a polynomial ring admit binomial quadratic Gröbner bases with a squarefree initial ideal. Also see Remark 7.42.

Remark 7.40. There are numerous classes of monoid algebras for which monomial initial ideals have been computed. For example, see Hoşten [200], Maclagan and Thomas [246], and Sturmfels [338].

It is a very interesting open question whether the monoids generated by the base vectors of matroids define Koszul algebras. White [378] has shown that they are normal, but the Koszul property (and even quadratic generation of the defining ideals) has only been proved for special classes of matroids. See Blum [35] and Herzog and Hibi [184].

Further techniques for Koszul algebras. So far we have used a single argument in order to derive the Koszul property for a $\Bbbk$-algebra R, namely the construction of a quadratic initial ideal. However, as mentioned in Remark 7.38, there exist such algebras without a quadratic monomial initial ideal (at least, as long as we do not allow coordinate changes).

A property that is weaker than the existence of a quadratic monomial initial ideal is the existence of a *Koszul filtration*, formally introduced by Conca, Trung, and Valla [90] after it had been discussed informally by several authors. This technique is used to prove the Koszul property for the examples in Remark 7.38.

Another approach uses *divisor complexes* for the computation of the graded Betti numbers of $\Bbbk$ over R in a similar way as squarefree divisor complexes can be used for the graded Betti numbers of R over a polynomial ring (see Remark 6.21). We refer the reader to Herzog, Reiner, and Welker [186] and Peeva, Reiner, and Sturmfels [287].

Payne [282] uses a combinatorial characterization of diagonal Frobenius splitting for toric varieties to derive new classes of normal lattice polytopes $P \subset \mathbb{R}^d$ such that $\Bbbk[P]$ is Koszul. Namely, $\Bbbk[P]$ is Koszul if the facet normals of P are spanned by a subset of a root system $\Phi \subset \mathbb{R}^d$ whose irreducible factors are of type A, B, C, D, or G. Type A gives rise to the class of polytopes mentioned in Remark 7.39. It is not known whether the lattice polytopes arising from the other types of root systems also have regular unimodular triangulations (of degree 2).

Peeva [286] gives an overview of the various techniques used to prove the Koszul property for monoid algebras. The article also discusses stronger versions of this notion and contains numerous references.

7.D Multiples of lattice polytopes

In this section we will show that high multiples of lattice polytopes yield Koszul algebras. This result will turn out to be a special case of a theorem on Veronese subrings of Cohen–Macaulay rings that we treat first.

Initial ideals for high Veronese subrings. We consider a positively graded $\Bbbk$-algebra $R = \bigoplus_{k=0}^{\infty} R_k$ that as a module over its subalgebra $\Bbbk[R_1]$ is finitely generated by homogeneous elements $y_1, \dots, y_n$. For $c \geq \max_i \deg y_i$ it is easy to see that the Veronese subring $R^{(c)}$ is generated by the homogeneous elements of degree c as a $\Bbbk$-algebra, and it is justified to ask whether $R^{(c)}$ is a Koszul algebra after the degree has been normalized by division by c. This is indeed the case for $c \gg 0$, and we will discuss a general bound for c in Remark 7.44.

The typical example in our context is the algebra $R = \Bbbk[\bar{M}(P)]$ where P is a lattice polytope. Then R is a Cohen–Macaulay ring by Theorem 6.10. For the questions discussed in this section we may extend $\Bbbk$ if necessary, and therefore assume that R has a Noether normalization S generated by algebraically independent elements of degree 1 (Theorem 6.3; see also Remark 6.8). The Cohen–Macaulay ring R is a free S-module, and this fact simplifies the situation considerably.

Theorem 7.41. *Let R be a positively graded Cohen–Macaulay $\Bbbk$-algebra, finitely generated over its subalgebra $\Bbbk[R_1]$. Moreover, let S be a Noether normalization by elements of degree 1, and let $y_1, \dots, y_n, y_{n+1} = 1$ be a homogeneous basis of the (free) S-module R. Suppose that $c \geq \max_i \deg y_i$.*

Then $R^{(c)}$ has a presentation as a residue class ring of a polynomial ring A over $\Bbbk$ by an ideal with a quadratic Gröbner basis. In particular, $R^{(c)}$ is a Koszul algebra.

Proof. Let $x_1, \dots, x_d$ be a system of algebraically independent elements of degree 1 generating S. Then we have a presentation

$$\varphi : Q = \Bbbk[X_1, \dots, X_d, Y_1, \dots, Y_n] \to R, \qquad X_i \mapsto x_i,\ Y_j \mapsto y_j.$$

Of course, we set $\deg X_i = 1$ for all i and $\deg Y_j = \deg y_j$ for all j. The restriction of φ to $\Bbbk[X_1, \dots, X_d]$ maps this ring bijectively onto $\Bbbk[x_1, \dots, x_d]$. For simplicity of notation we will therefore identify X_i and x_i, $i = 1, \dots, d$. Moreover, we set $I = \operatorname{Ker} \varphi$.

On Q we introduce a monomial order $<$ that satisfies the following conditions:

(i) $z_1 < z_2$ if $\#_Y z_1 < \#_Y z_2$, where $\#_Y$ denotes the number of factors Y_j;
(ii) The induced monomial order on $\Bbbk[X_1, \dots, X_d]$ is the revlex order with respect to $X_1 > \dots > X_d$;
(iii) $Y_1 > \dots > Y_n$.

Next we want to replace Q by a small monomial subalgebra that is mapped surjectively onto $R^{(c)}$: let $Q^{[c]}$ be the $\Bbbk$-subalgebra generated by all monomials $X_{i_1} \cdots X_{i_c}$ and all monomials $X_{i_1} \cdots X_{i_k} Y_j$ with $k + \deg Y_j = c$. Evidently the restriction $\varphi^{[c]}$ of φ to $Q^{[c]}$ is still surjective. On $Q^{[c]}$ we normalize degrees in the same way as on $R^{(c)}$.

We claim that $I^{[c]} = I \cap Q^{[c]} = \operatorname{Ker} \varphi^{[c]}$ has a Gröbner basis of degree 2 elements with respect to the restriction of the monomial order to $Q^{[c]}$:

The initial ideal of $I^{[c]}$ is generated by all monomials $(uY_j)(vY_k)$ where $j, k = 1, \dots, n$ (including the case $j = k$) and u and v are monomials in $\Bbbk[X_1, \dots, X_d]$ such that $\deg u + \deg Y_j = \deg v + \deg Y_k = c$.

First, all these monomials belong to $\operatorname{in}(I^{[c]})$: since $1, y_1, \dots, y_n$ generate R as an S-module, there is an equation

$$(uy_j)(vy_k) = w \cdot 1 + \sum_{l=1}^{n} z_l y_l, \qquad w, z_l \in \Bbbk[X_1, \dots, X_d].$$

The corresponding element $(uY_j)(vY_k) - w \cdot 1 - \sum_{l=1}^{n} z_l Y_l \in I^{[c]}$ has $(uY_j)(vY_k)$ as its initial monomial (because of rule (i)).

Second, a polynomial f of $I^{[c]}$ that has only terms with at most one Y-factor must be zero since it would express a linear dependence of $1, y_1, \dots, y_n$ over $\Bbbk[X_1, \dots, X_d]$. Therefore it contains a term with at least two Y-factors, and $\operatorname{in}(f)$ satisfies this property, too. But then $\operatorname{in}(f)$ is divisible by one of the monomials $(uY_j)(vY_k)$.

Now the polynomial ring A, of which $R^{(c)}$ will be a residue class ring, has to be constructed: for each of the irreducible monomials z generating $Q^{[c]}$ we introduce a variable Z_z, and let A be the polynomial ring in all these variables Z_z. Let $\psi : A \to Q^{[c]}$ be given by the assignment $Z_z \mapsto z$. The presentation of $R^{(c)}$ is given by the composition

$$A \xrightarrow{\psi} Q^{[c]} \xrightarrow{\varphi^{[c]}} R^{(c)}.$$

We must analyze its kernel J.

Consider a monomial $Z_{z_1} \cdots Z_{z_k}$. We list all the factors X_i and Y_j of its image $\psi(Z_{z_1} \cdots Z_{z_k}) = z_1 \cdots z_k$ as follows:

$$X_{i_1}, \dots, X_{i_s}, \quad i_1 \le \dots \le i_s, \qquad Y_{j_1}, \dots, Y_{j_t}, \quad j_1 \le \dots \le j_t.$$

Then we arrange them in the following sequence. The first factor is Y_{j_1}. It is followed by $X_{i_1}\cdots X_{i_u}$ with $u = c - \deg Y_{j_1}$. Then we proceed with Y_{j_2} followed by $X_{i_{u+1}}\cdots X_{i_v}$ with $v = u + c - \deg Y_{j_2}$ etc. If we have run out of factors Y_j, then only factors X_i are used. Finally the rearranged sequence is cut into monomials $w_1, \dots, w_k$ of degree c. We call $Z_{w_1}\cdots Z_{w_k}$ a *standard monomial*, and set $\tau(Z_{z_1}\cdots Z_{z_k}) = Z_{w_1}\cdots Z_{w_k}$ in order to refer the standard monomial derived from $Z_{z_1}\cdots Z_{z_k}$.

It follows immediately from Proposition 4.26 that $H = \operatorname{Ker}\psi$ is generated by all binomials $a - \tau(a)$ where a is a monomial in A. Moreover, the (residue classes of) the standard monomials form a $\Bbbk$-vector space basis of $A/H \cong Q^{[c]}$. If $Z_{z_i}Z_{z_j}$ is standard for all $i, j = 1, \dots, k$, then $Z_{z_1}\cdots Z_{z_k}$ is itself standard. In other words, every nonstandard monomial is divisible by a nonstandard monomial of degree 2.

Since we aim at a Gröbner basis of J of degree 2, it is useful to choose a monomial order on A in which a is the initial monomial of $a - \tau(a)$, provided $a \neq \tau(a)$: we order the monomials of A by setting $a \prec b$ if (1) $\psi(a) < \psi(b)$, or (2) a precedes b in the reverse-lexicographic monomial order on A. Note that (1) orders the variables Z_z. Therefore (2) makes sense. We leave it to the reader to check that indeed $a \succeq \tau(a)$ for all monomials a of A. The rules (ii) and (iii) above that have not been used yet come now into play.

The initial ideal $\operatorname{in}(H)$ contains all nonstandard monomials. It is in fact generated by them since the standard monomials form a basis of $A/H \cong Q^{[c]}$, and the monomials outside $\operatorname{in}(H)$ do the same. Hence

$\operatorname{in}(H)$ *is generated by the nonstandard monomials of degree* 2.

Now the final claim proves the theorem:

The initial ideal of J is generated by all nonstandard monomials of degree 2 *and all monomials Z_uZ_v where $\#_Y u = \#_Y v = 1$.*

Choose $f \in J$. If the initial monomial $a = \operatorname{in}(f)$ is nonstandard, then it is divisible by a nonstandard monomial of degree 2. So we can assume that a is standard. Since $a' \prec a$ for all other monomials a' of f, it follows that $\psi(a) \neq \psi(a')$. Therefore $\psi(a)$ is the initial monomial of $\psi(f)$, and $\psi(f) \neq 0$ (compare the definition of $\prec$). If $\#_Y(\psi(a)) \geq 2$, then a is divisible by one of the monomials Z_uZ_v. The remaining case $\#_Y(\psi(a)) \leq 1$ leads to a contradiction: $I^{[c]}$ would contain a nonzero polynomial whose initial monomial has at most one Y-factor, a case already ruled out above. □

Remark 7.42. (a) Even if $n = 0$ and $R = \Bbbk[X_1, \dots, X_d]$ Theorem 7.41 is an interesting result: all Veronese subalgebras of R are defined by ideals with a Gröbner basis of degree 2 binomials.

In particular, the Veronese subrings of $\Bbbk[X_1, \dots, X_d]$ are Koszul algebras. Actually this holds true for the Veronese subalgebras of an arbitrary Koszul algebra [13].

(b) However simple the proof of Theorem 7.41 is for $R = \Bbbk[X_1, \dots, X_d]$, so is it disappointing for the construction of triangulations: $Z_z^2 \in \operatorname{in}(J)$ unless $z = X_i^c$ for some i. In other words, the induced triangulation on the cth multiple $c\Delta$ of the unit simplex Δ is the trivial one by $c\Delta$ itself. In Remark 7.39 we have seen that one can do much better: the Knudsen–Mumford triangulation of $c\Delta$ is regular, unimodular and of degree 2.

In the following we need the notion of *Castelnuovo–Mumford regularity*. Let M be a finitely generated graded module over the polynomial ring $\Bbbk[X_1, \dots, X_n]$ with its standard grading ($\deg X_i = 1$ for all i). Then

$$\operatorname{reg}(M) = \max\{j - i : \beta_{ij}(M) \neq 0\}.$$

Here the β_{ij} are the graded Betti numbers of M (see p. 208. For a detailed treatment of Castelnuovo–Mumford regularity we refer the reader to [72, 4.3] and Eisenbud [112, 20.5].

Under Cohen–Macaulay conditions $\operatorname{reg}(M)$ can be expressed in terms of several other invariants:

Proposition 7.43. *Let M be a graded Cohen–Macaulay module of dimension d over the polynomial ring $A = \Bbbk[X_1, \dots, X_n]$ with its standard grading. Set $k = \max\{j : \beta_{n-d,j} \neq 0\} - n + d$, $l = \min\{j : {}^*\operatorname{Ext}_A^{n-d}(M, \omega_A)_j \neq 0\}$, let m be the degree of the numerator polynomial of the Hilbert series of M, and m' the maximal degree of an element of a homogeneous basis of M over a Noether normalization of degree 1. Then*

$$\operatorname{reg}(M) = k = d - l = m = m'.$$

Proof. For the first equation it is enough to note that the maximal shifts of the free modules grow at least by one with each step. This follows by the same argument that we have used in the proof of Theorem 7.26. The second equation follows from the duality between the minimal graded free resolutions of M and ${}^*\operatorname{Ext}_A^{n-d}(M, \omega_A)$ stated in Theorem 6.18. For the third equation one uses Corollary 6.42, at least in the case that $M = R$ is a Cohen–Macaulay ring; in the general case the analogous consequence of Exercise 6.10 must be used (or see [72, 4.4.5]). The last equation finally results from the fact that the coefficients of the numerator polynomial count the number of basis elements of M over a Noether normalization in the corresponding degree (Theorem 6.40). □

Remark 7.44. (a) In the case that $R = \Bbbk[R_1]$, Theorem 7.41 can be considerably strengthened and extended to non-Cohen–Macaulay algebras: $R^{(c)}$ is defined by an ideal with a degree 2 Gröbner basis for all $c \geq (\operatorname{reg}(R) + 1)/2$; see Eisenbud, Reeves and Totaro [113].

(b) The extension to non-Cohen–Macaulay algebras is also possible under the remaining assumptions of Theorem 7.41. Let I_j be the annihilator in $S = \Bbbk[R_1]$ of y_j modulo the S-submodule of R generated by $y_{j+1}, \dots, y_{n+1}$, $j = 1, \dots, n+1$, and set $m = \max_j \deg y_j + \operatorname{reg}(S/I_j)$. Then the conclusion of the theorem holds for $c \geq m + 1$. See [69, 1.4.1].

Multiples of polytopes. As an immediate consequence of Theorem 7.41 we obtain the following:

Corollary 7.45. *Let $P \subset \mathbb{R}^d$ be a lattice polytope and set $m = \min\{k : \operatorname{int}(kP) \cap \mathbb{Z}^d \neq \varnothing\}$. Then, with respect to a suitable presentation as a residue class ring of a polynomial ring, the ideal defining $\Bbbk[cP]$ admits a quadratic Gröbner basis, and $\Bbbk[cP]$ is a (normal) Koszul algebra for $c \geq \dim P + 1 - m$.*

Proof. We choose $M = C(P) \cap \mathbb{Z}^{d+1}$. Then $R = \Bbbk[M]$ satisfies the hypotheses of Theorem 7.41. Clearly $R^{(c)} = \Bbbk[cP]$. Moreover $\operatorname{int}(M)$ is the monomial basis of ω_R by Theorem 6.31, and R is generated over a Noether normalization of degree 1 by elements of degree $\le \dim R - m = \dim P + 1 - m$ according to Proposition 7.43. Thus Theorem 7.41 can be applied. □

If P is not a simplex, then there is no monomial Noether normalization, and therefore Corollary 7.45 does not imply that $\Bbbk[cP]$ is defined by a binomial ideal with a Gröbner basis of degree ≤ 2. However, for simplices it does so:

Corollary 7.46. *Let $P \subset \mathbb{R}^d$ be a lattice simplex. Then, with the notation of Corollary 7.45, $R^{(c)}$ has a presentation as a residue class ring of a polynomial ring by a toric ideal with a Gröbner basis of degree 2.*

Remark 7.47. (a) Corollary 7.45 shows that there is no algebraic obstruction to a positive answer to the following question: *Does cP always have a regular unimodular triangulation of degree 2 for $c \ge \dim P$?* The answer seems to be unknown even in dimension 3. In dimension 2, Theorem 7.35 answers the question positively.

For arbitrary dimension, Theorem 3.17 guarantees the existence of a (regular) unimodular triangulation for some c (see p. 103 for the discussion of regularity), and the construction does not allow us to bound the degree of the triangulation.

(b) If P is an integrally closed polytope, then R is generated in degree 1 as a $\Bbbk$-algebra. Therefore the theorem of Eisenbud–Reeves–Totaro (compare Remark 7.44) applies and yields the conclusion of Corollary 7.45 already for $c \ge (\operatorname{reg}(R)+1)/2 = (\dim P + 1 - m)/2$. (However, even in the simplicial case coordinate transformations may be necessary.)

Remark 7.48. (a) Let R be a finitely generated graded $\Bbbk$-algebra. Following Green and Lazarsfeld [151] we say that R has N_0 if it is generated in degree 1, and R has N_p, $p \ge 1$, if the graded Betti numbers of R (as a residue class ring of a polynomial ring) satisfy the following conditions:

(i) $\beta_{1i} = 0$ for $i \ne 2$; in other words, the defining ideal is generated in degree 2;
(ii) $\beta_{ij} = 0$ for $i = 2, \dots, p$ and $j \ne i+1$; in other words, the matrices in the maps φ_i for $i = 2, \dots, p$ have linear entries.

In this terminology, Corollary 7.45 says that $\Bbbk[cP]$ satisfies N_1 for $c \ge \dim P + 1 - m$.

Hering, Schenck, and Smith have shown that $\mathbb{C}[cP]$ satisfies N_p for $p \ge 1$ and $c \ge \dim P - m + p$, provided $c \ge p$ [183, 1.3]. See also Bruns, Conca, and Römer [56] for results on the N_p property of Veronese subrings.

Ogata [277] has given a bound that is sharper for $m = 1$: if $\dim P \ge 3$ and $p \ge 1$, then $\Bbbk[cP]$ satisfies N_p. For $p = 1$ one obtains that $\Bbbk[(\dim P - 1)P]$ (generated in normalized degree 1 by Corollary 2.57) is defined by quadratic binomials. In dimension 2, Corollary 7.45 says precisely when R is defined by quadratic binomials.

(c) Corollary 7.37 has been improved by Schenck [311]: let P be a lattice polygon; then $\mathbb{C}[P]$ satisfies N_p if ∂P contains at least $p+3$ lattice points.

Exercises

7.1. Let R_1 and R_2 be $\mathbb{Z}$-graded algebras over the field $\Bbbk$ that are generated by their degree 1 elements. Suppose $\varphi_i : P_i \to R_i$, $i = 1, 2$, is a presentation of R_i as a residue class ring of polynomial ring $P_i = \Bbbk[X_{i1}, \dots, X_{i,n_i}]$ over $\Bbbk$ by a homogeneous homomorphism.

Show if $\operatorname{Ker}\varphi_1$ and $\operatorname{Ker}\varphi_2$ are generated by homogeneous polynomials of degree 2, then the same holds for the kernel of the presentation of the Segre product $R_1\#R_2$ as a residue class ring of the polynomial ring $\Bbbk[Y_{jk} : j = 1, \dots, n_1, k = 1, \dots, n_2]$ by the homomorphism φ, $\varphi(Y_{ij}) = \varphi_1(X_{1i})\varphi_2(X_{2j})$.

The Segre product of Koszul algebras is again Koszul; see [13].

7.2. Let R be a ring and $I_1, \dots, I_n$ a set of ideals, partially ordered by inclusion.

(a) Show that the inverse limit of the system of natural epimorphisms $R/I_i \to R/I_j$, $I_i \subset I_j$, is isomorphic to $R/(I_1 \cap \dots \cap I_n)$ if $I_i \cap (I_j + I_k) = I_i \cap I_j + I_i \cap I_k$ and $I_i + I_j \cap I_k = I_i \cap I_j + I_i \cap I_k$ for all i, j, k.

(b) Give an example showing that the distributivity in (a) is a crucial condition.

7.3. Characterize the monomial radical ideals of $R[\mathbb{M}]$ similarly to Proposition 7.12(c).

7.4. Let $\mathbb{M}$, $\mathbb{M}'$ be a monoidal complexes in which every face of the underlying conical complexes Γ and Γ' is an intersection of facets. Suppose $\Bbbk[\mathbb{M}]$ and $\Bbbk[\mathbb{M}']$ are isomorphic rings. Show that Γ and Γ' are combinatorially equivalent.

Hint: consider minimal prime ideals and their sums.

7.5. Let $\mathbb{M}$ be a monoidal complex with associated conical complex Γ. We say that a subcomplex Δ of Γ is *convex* if the following holds: whenever a subset X of $|\Delta|$ is contained in a face F of Γ, then it is contained in a face of Δ. Prove the following are equivalent:

(i) Δ is a convex subcomplex;

(b

(ii) the natural embedding $\Bbbk[\mathbb{M}|\Delta] \to \Bbbk[\mathbb{M}]$ induces an isomorphism $\Bbbk[\mathbb{M}|\Delta] \cong \Bbbk[\mathbb{M}]/I$ where I is the ideal generated by all monomials outside $|\Delta|$.

7.6. Let $\mathbb{M}$ be a monoidal complex with associated conical complex Γ such that Γ is a subdivision of a conical complex Σ. For each face c of Σ we let Γ_c denote the subcomplex of all faces of Γ that belong to c, and define $\mathbb{M}_c$ accordingly. Show that Γ_c is a convex subcomplex of Γ for each c, and relate the defining ideal of $R[\mathbb{M}]$ (as in Proposition 7.12) to the defining ideals of the algebras $R[\mathbb{M}_c]$, c a facet of Σ.

7.7. Let $R = \Bbbk[X_1, \dots, X_n]$ and $f \in R$, $f \neq 0$. Prove the following are equivalent for a monomial x of f:

(i) there exists a monomial order $<$ on R with $x = \operatorname{in}_<(f)$;

(ii) x is a vertex of the Newton polytope $\mathrm{N}(f)$ and no other monomial of f is a multiple of x.

7.8. Consider a monomial order and a minimal prime ideal $\mathfrak{p}_F$ of $\operatorname{in}(I_E)$ as in Corollary 7.25. We define a homomorphism $\varphi : \Bbbk[X_e : e \in E] \to \Bbbk[X_e : x_e \in F]$ by the substitution $\varphi(X_e) = X_e$ for $x_e \in F$ and $\varphi(X_e) = 1$ for $x_e \notin F$. Show that $\Bbbk[X_e : x_e \in F]/\varphi(\operatorname{in}(I_E))$ has finite $\Bbbk$-dimension equal to the multiplicity of $\mathfrak{p}_F$.

Hint: in the passage from $\Bbbk[X_e : e \in E]/\operatorname{in}(I_E)$ to its localization we stop after the inversion of the X_e with $x_e \notin F$. Then we arrive at the Laurent polynomial extension $\big(\Bbbk[X_e : x_e \in F]/\varphi(\operatorname{in}(I_E))\big)[X_e^{\pm 1} : x_e \notin F]$. As a module over $\Bbbk[X_e^{\pm 1} : x_e \notin F]$ it is free and its rank equals the multiplicity of $\mathfrak{p}_F$.

7.9. Prove all claims in the text about the example in Figure 7.2. Let $A = \Bbbk[X_e : e = 1, \dots, 4]$. By in_0 we denote the initial ideal with respect to a suitable monomial order that the reader has to find, and in_a and in_b denote the initial ideals with respect to the weight vectors in Figure 7.2(a) and (b).

(a) Set $R = \Bbbk[M]$ and show $H_R(t) = (1+2t)/(1-t)^2$, for example by using the decomposition into graded $\Bbbk$-vector spaces along the unique unimodular triangulation of the lattice segment $[1, 4]$.

(b) Now set $J = (X_1X_4 - X_2X_3,\ X_1X_3 - X_2^2,\ X_2X_4 - X_3^2)$ and $I_0 = (X_1X_4, X_1X_3, X_2X_4)$. Then $I_0 \subset \operatorname{in}_0(J)$. Moreover, A/I_0 has the same decomposition into graded $\Bbbk$-vector spaces as $\Bbbk[M]$. Show $H_{A/I_0}(t) = H_R(t)$, and conclude that $I_0 = \operatorname{in}_0(J)$, $\Bbbk[M] = A/J$, and $J = I_E$.

(c) Next note that $I_0 \subset \operatorname{in}_0(X_1X_3 - X_2^2, X_1X_4, X_2X_4) \subset \operatorname{in}_b(J)$. What can you say without any further computation?

(d) Finally set $I_1 = (X_1X_4, X_2^2, X_2X_4)$. Then $I_1 \subset \operatorname{in}_a(J)$. But A/I_1 has again the right Hilbert function as a decomposition into graded $\Bbbk$-vector spaces shows.

7.10. Prove the claims in Remark 7.36 and give an example showing that the condition $a_1 \geq a_2$ is essential.

7.11. Let M be an affine monoid with system of generators $(x_e)_{e \in E}$. Let $<$ be a monomial order on $\Bbbk[X_e : e \in E]$ whose associated triangulation Δ is full with respect to (x_e). Prove that M is normal if $\Bbbk[M]$ admits a quadratic initial ideal with respect to $<$.

In particular, the existence of a quadratic initial ideal implies normality if each of the x_e is a vertex of the polytope spanned by them.

7.12. Let P be a polytope and suppose that the dissection of P by finitely many hyperplanes is a triangulation Δ. Show $\deg \Delta = 2$.

7.13. Let $P \subset \mathbb{R}^d$ be a lattice polytope and consider the monoid $M = C(P) \cap \mathbb{Z}^{d+1}$.

(a) Show that $\operatorname{int}(M)$ is generated over $M(P)$ by elements of degree $\leq \dim P + 1$, and give an example for which this bound is sharp.

One can use either combinatorial arguments or the Hilbert series. The second choice reveals that, after the introduction of coefficients, ω_R is generated in degree $\leq \dim P + 1$ over a degree 1 Noether normalization of $\Bbbk[M]$.

(b) Prove that $\operatorname{int} M(cP)$ is generated by degree 1 elements for $c \geq \dim P + 1$ [69, 1.3.3]. (In the terminology of commutative algebra, $\Bbbk[M(cP)]$ is *level*, i. e., the elements in the minimal system of generators of the canonical module have the same degree. See Exercise 2.17 for another such case.)

Notes

Numerous references have been inserted into the text of this chapter. Therefore we confine ourselves to some additional comments and sources in these notes.

The correspondence between regular triangulations of "vector configurations" and the radicals of their initial ideals was established by Sturmfels [336]. His book [338] has sparked a quite considerable body of research on this and related topics. See also Hoşten [200] and Maclagan and Thomas [246] for references; the latter contains a rather elementary introduction to the subject. In our presentation we have avoided assumptions on "genericity" so that regular triangulations had to be generalized to regular subdivisions, and Stanley–Reisner rings to algebras associated with monoidal complexes. In this generality the connection between subdivisions and initial ideals was also treated by Brun and Römer [49].

Theorem 7.35 and its proof are taken from Bruns, Gubeladze, and Trung [69]. It has a predecessor in the work of Koelman [225], [226], who proved that the toric ideal is generated in degree 2 under the conditions of 7.35. For a special class of toric ideals the same bound was proved by Ewald and Schmeinck [118].

Like for several other classes of ideals, the problem of set theoretic generation of toric ideals up to radical has been discussed intensively. Bruns, Gubeladze, and Trung [70] contains several references for results of this type.

Affine monoid algebras generated by pairwise incomparable squarefree monomials are presented by simplicial complexes in a natural way: one considers the monomial generators as the facets of the simplicial complex (and vice versa). A related class is given by Rees algebras of squarefree monomial ideals. Numerous papers have been devoted to this class of monoid algebras, linking their algebraic properties to the combinatorics of the complex. Villarreal's book [364] gives a very good introduction to this circle of ideas. A noteworthy more recent paper is Herzog and Hibi [185].

Part III

K-theory

8

Projective modules over monoid rings

Quillen and Suslin have shown that projective modules over polynomial extensions $R[X_1, \ldots, X_n] \cong R[\mathbb{Z}_+^n]$ of principal ideal domains R are extended from R, thus solving a famous problem of Serre. The main objective of this chapter is the generalization of the Quillen–Suslin theorem from monoid algebras corresponding to free monoids to those defined by arbitrary seminormal monoids.

In the first sections we develop the necessary background material and outline the strategy of the proof, whose main steps form the central part of the chapter.

The chapter concludes with converse results showing that seminormality is the weakest possible condition for the main theorem, and with generalizations concerning the homotopy invariance of K_0 for seminormal monoid extensions of regular rings, the Picard group of $R[M]$, and the passage from $R[M]$ to a residue class ring modulo a monomial ideal. The time for K-theory to enter the scene has come.

8.A Projective modules

Recall from Section 4.G that a module P over a (commutative) ring R is *projective* if it is a direct summand of a free R-module. We keep our convention that projective modules are assumed to be finitely generated, or equivalently, they are direct summands of free modules of finite rank. This class of projective R-modules is denoted by $\mathbb{P}(R)$.

The restriction to finitely generated modules is justified since, by a theorem of Bass [18], nonfinitely generated projective modules over a noetherian domain are free.

For a ring homomorphism $f : R \to R'$ and an R'-module M' we say that M' is *extended from R* if there exists an R-module M such that $M' \cong R' \otimes_R M$ as R'-modules. Occasionally we use the notation $f_*(M) = R' \otimes_R M$.

Recall that a module P is *finitely presented* if there exists an exact sequence $R^m \to R^n \to P \to 0$ of R-modules with $m, n \in \mathbb{N}$. The following basic property of finitely presented modules, an easy exercise on the right exactness of the functor $R \otimes -$, will frequently be used later on.

W. Bruns, J. Gubeladze, *Polytopes, Rings, and K-Thoery*, Springer Monographs in Mathematics,
DOI 10.1007/b105283_8, © Springer Science+Business Media, LLC 2009

Lemma 8.1. *Let I be a filtered partially ordered set and let $\{f_{ij} : R_i \to R_j \ : \ i, j \in I,\ i < j\}$ be a family of ring homomorphisms such that $f_{jk} \circ f_{ij} = f_{ik}$ whenever $i < j < k$. Let R denote the direct limit $\varinjlim R_i$. Then for every finitely presented R-module P there exist an index $i \in I$ and a finitely presented R_i-module P_i such that $P \cong R \otimes_{R_i} P_i$.*

If P is projective, then one can choose $i \in I$ such that P_i is projective.

Of course, over a noetherian ring all our modules are finitely presented. But later on it will be crucial that we work with nonnoetherian rings as well. Our projective modules are direct summands of free modules of finite rank and, therefore, are finitely presented whether the ground ring is noetherian or not.

This is not the place to develop the general theory of projective modules. Rather we confine ourselves to stating some important basic properties. For proofs see Bass [19, Ch. III, §7], Bourbaki [37, Ch. II, §5], or Lam [237, Ch. I].

Lemma 8.2. *Let R be a ring and P be an R-module.*

(a) P is projective if and only if P is a finitely presented R-module and $P_{\mathfrak{m}}$ is free over $R_{\mathfrak{m}}$ for every maximal ideal $\mathfrak{m} \subset R$.

(b) If P is projective, then the map $\operatorname{rank}_P : \operatorname{Spec} R \to \mathbb{Z}$, *given by* $\mathfrak{p} \mapsto \operatorname{rank}_{R_{\mathfrak{p}}}(P_{\mathfrak{p}})$, *is continuous with respect to the Zariski topology on* $\operatorname{Spec} R$ *and the discrete topology on* $\mathbb{Z}$.

Assume $f : R \to R'$ is a ring homomorphism such that there exists a ring retraction $g : R' \to R$ with $gf = \mathbf{1}_R$. Let an R'-module M' be extended from R. Then the module M such that $M' = R' \otimes_R M$ is unique up isomorphism: $M \cong R \otimes_{R'} M'$. Obviously, M inherits both projectivity and finite presentability from M' (and conversely).

Example 8.3. We have already constructed nonfree projective modules of rank 1 in Lemma 4.72. A very interesting, completely different example of a nonfree projective module is given by the following construction due to Kaplansky [237, Ch. 1, §4]. Consider the $\mathbb{R}$-algebra $R = \mathbb{R}[x_1, x_2, x_3]$ generated by three generators x_1, x_2, x_3, subject to the relation $x_1^2 + x_2^2 + x_3^2 = 1$; we consider R as the ring of the real-valued polynomial functions on the standard unit sphere $S^2 \subset \mathbb{R}^3$. The R-module P is defined by the split exact sequence

$$0 \longrightarrow P \longrightarrow R^3 \xrightarrow{\ f\ } R \longrightarrow 0\,, \qquad f(\zeta, \eta, \vartheta) = x_1\zeta + x_2\eta + x_3\vartheta.$$

It is a projective module because $P \oplus R \cong R^3$. In other words, P is even a *stably free* R-module: there exist $m, n \in \mathbb{N}$ such that $P \oplus R^m \cong R^n$.

We claim that P is not free. Assume to the contrary that P is free. Then there exists an R-automorphism of $\alpha : R^3 \to R^3$ fitting into the commutative diagram with exact rows

$$\begin{array}{ccccccccc}
0 & \longrightarrow & P & \longrightarrow & R^3 & \xrightarrow{\ f\ } & R & \longrightarrow & 0 \\
 & & \downarrow & & \downarrow{\scriptstyle \alpha} & & \| & & \\
0 & \longrightarrow & R^2 & \xrightarrow[\iota]{} & R^3 & \xrightarrow[\pi]{} & R & \longrightarrow & 0
\end{array}$$

where $\iota(\zeta,\eta) = (\zeta,\eta,0)$ and $\pi(\zeta,\eta,\vartheta) = \vartheta$. This is equivalent to the existence of an invertible 3×3 matrix over R having the form

$$\begin{pmatrix} x_1 & x_2 & x_3 \\ y_1 & y_2 & y_3 \\ z_1 & z_2 & z_3 \end{pmatrix}.$$

In particular, for no point $z = (z_1, z_2, z_3) \in S^2$ is the vector

$$F(z) = \big(y_1(z), y_2(z), y_3(z)\big) \in \mathbb{R}^3$$

parallel to z. We obtain a continuous vector field on S^2 if we assign to each point $z \in S^2$ the orthogonal projection of $F(z)$ onto the tangent plane of S^2 at z. As just seen, this vector field vanishes nowhere, in contradiction to the theorem that every such vector field must have a zero (for example, see Munkres [267, pp. 367–368]). This shows that P is not free. In geometric terms, the tangent vector bundle (over the reals) to S^2, represented by the module P, is nontrivial because it does not have a nonvanishing global section, let alone a nonvanishing algebraic global section.[1]

Actually, R has strong ring theoretical properties: it is regular and a unique factorization domain (see [237, Ch. 1, §4] for the details). Therefore, P does not split into a direct sum of two nonzero modules, since each of the two summands would be a rank 1 projective module, and over a unique factorization domain such modules are free.

As Hochster [194] has shown, one can derive a counterexample to the cancellation problem for polynomial extensions from R and P (compare Remark 5.25). Since $P \oplus R \cong R^3$, the symmetric algebras of $P \oplus R$ and R^3 are isomorphic R-algebras. Thus

$$\operatorname{Sym}(P \oplus R) \cong \operatorname{Sym}(P)[X] \cong \operatorname{Sym}(R^3) \cong R[X_1, X_2, X_3].$$

However, $\operatorname{Sym}(P) \not\cong R[X_1, X_2]$.

For a detailed study of the relationship between projective modules and vector bundles on spheres we refer the reader to Swan [350].

8.B The main theorem and the plan of the proof

The central result of this chapter is

Theorem 8.4. *Let R be a principal ideal domain (PID) and M be a seminormal monoid. Then all finitely generated projective $R[M]$-modules are free.*

Note that we do not require that the monoid M be finitely generated or without nontrivial units. (However, a seminormal monoid is cancellative and torsionfree by definition.) That the condition of seminormality is inevitable has already been shown in Corollary 4.73. In Theorem 8.27 we will obtain an even stronger converse result.

Theorem 8.4 generalizes the classical theorem of Quillen [297] and Suslin [341] that solved a problem posed by Serre [316]:

[1] The corresponding algebraic geometry terminology will be introduced in Chapter 10.

Theorem 8.5. *Let R be a PID and m a nonnegative integer. Then all finitely generated projective modules over the polynomial ring $R[\mathbb{Z}_+^m]$ are free.*

Various extensions and applications of Theorem 8.4 will be discussed in Sections 8.I and 10.G. The proof of the theorem is given in Sections 8.D–8.G, while Section 8.C contains the classical results on projective modules on which the Quillen–Suslin theorem is based, to be extended and adapted to monoid rings in general.

Before delving into the technical details we sketch the course of the proof of Theorem 8.4.

The key case. When $\operatorname{rank} M = 0$ there is nothing to prove—all (i. e., finitely generated) torsionfree modules over a PID are free. The case of $\operatorname{rank} M = 1$ is easily reduced to the Quillen–Suslin theorem; see the discussion following Lemma 8.17. Therefore, without loss of generality, we can assume that the following conditions hold:

(a) $\operatorname{rank} M \geq 2$;
(b) Theorem 8.4 has been shown for monoids N such that $\operatorname{rank} N < \operatorname{rank} M$.

Later on we will see that the key case of Theorem 8.4 is the one in which the following additional conditions hold:

(c) R is a local PID (in other words, a discrete valuation domain);
(d) M is an affine, positive, and normal monoid.

The reduction to this case (see Corollary 8.19) is based on classical results on projective modules to be recapitulated in Section 8.C. We mention the inductive hypothesis (b) explicitly since some of the reduction steps rely on it.

In the rest of this section we assume that the conditions (a)–(d) all hold.

Pyramidal descent. As proved in Proposition 1.21, every pointed cone has a cross-section: if α is a linear form in the (absolute) interior of the dual cone C^*, then the intersection of the hyperplane H with equation $\alpha(x) = 1$ is a polytope.

For a positive affine monoid M we choose α as a rational linear form. Then we obtain a rational cross-section of $\mathbb{R}_+M$. For a subset N of $\mathbb{R}M$ we set

$$\Phi(N) = \operatorname{conv}\left\{\frac{x}{\alpha(x)} : x \in N, \alpha(x) \neq 0\right\}.$$

This notation will be used frequently in this chapter and the next. Clearly $\Phi(M)$ is the cross-section of $\mathbb{R}_+M$ defined by α.

In the sequel it will be crucial that we can shrink the monoid M in a way that is controlled by a (rational) subpolytope of $\Phi(M)$. Therefore we introduce the following notation that can be applied to every submonoid M of $\mathbb{R}^d$ and every subset $W \subset \mathbb{R}^d$: we set

$$M|W = M \cap \mathbb{R}_+W.$$

If M is an affine monoid and W is a rational polytope, then $M|W$ is again affine. This follows from Lemma 2.9 and Corollary 2.11.

If a rational polytope P decomposes into two rational polytopes Δ and Γ such that Δ is a pyramid of dimension $\dim P$ with apex v, and Γ meets Δ in the facet opposite to v, then we say that P is a *pyramidal extension* of Γ and that $\Delta \cup \Gamma$ is a *pyramidal decomposition* of P. We do not exclude the case in which $P = \Delta$ and Γ is the facet of Δ opposite of v. When $\dim \Delta = \dim \Gamma$ we call P a *nondegenerate pyramidal extension* of Γ, and $\Delta \cup \Gamma$ a *nondegenerate pyramidal decomposition* of P.

The K-theoretical heart of the proof of Theorem 8.4 is the following theorem.

Theorem 8.6 (Pyramidal descent). *Let $v \in \Phi(M)$ be a vertex and $\Phi(M) = \Delta \cup \Gamma$ a nondegenerate pyramidal decomposition of $\Phi(M)$. Then every projective $R[M]$-module is extended from $R[M|\Gamma]$.*

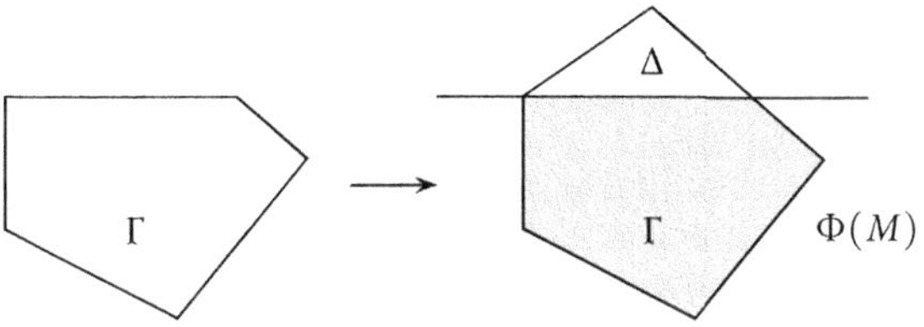

Fig. 8.1. Pyramidal descent

We remark that in the proof of Theorem 8.6 we will make heavy use of projective modules over *nonnoetherian* rings of type $R[M_*]$ (see Remark 2.6 for the definition of M_*), and here we leave the realm of noetherian rings—M_* is never a finitely generated monoid unless $\operatorname{rank} M = 1$ (which is a trivial case in the context of projective modules).

Shrinking the cross-section. The rest of the proof of Theorem 8.4 is a purely geometric argument.

Theorem 8.7. *Let P be a (rational) polytope and $z \in \operatorname{int}(P)$ a rational point. Then there exists a sequence $(P_i)_{i \in \mathbb{N}}$ of (rational) polytopes with the following properties:*

(a) For all $i \in \mathbb{N}$ one has
- *(i) $P_i \subset P$;*
- *(ii) $P_{i+1} \subset P_i$ or $P_{i+1} \supset P_i$;*
- *(iii) if $P_{i+1} \subset P_i$, then P_i is a nondegenerate pyramidal extension of P_{i+1} (in particular, $\dim P_i = \dim P$ for all i).*

(b) For every $\varepsilon > 0$ there exists an $i \in \mathbb{N}$ such that $P_i \subset U_\varepsilon(z) \cap P$.

This theorem will be proved in Section 8.G. Let us show here that Theorem 8.6 and Theorem 8.7 reduce Theorem 8.4 to the Quillen–Suslin theorem. We can assume $\operatorname{rank} M = d$. The cone $\mathbb{R}_+ M$ has a unimodular triangulation by Theorem 2.75, and we choose a rational point z in the interior of an intersection $\Phi(M) \cap D$ where D is a d-dimensional unimodular cone of the triangulation. Next we choose a sequence of rational polytopes $P_i \subset P = \Phi(M)$ as in Theorem 8.7. Then there exists $\varepsilon > 0$ such that $D \cap P$ contains $U_\varepsilon \cap P$. Figure 8.2 illustrates this construction. It shows members in the sequence (P_i) that are homothetic images of P with respect to the center z, and the inclusion of such images in the sequence will be a crucial argument in its construction.

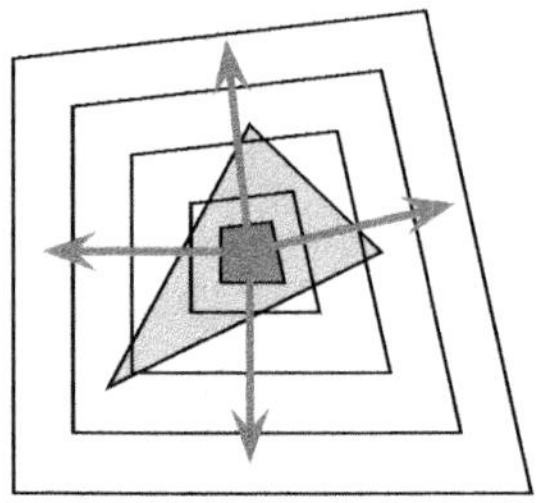

Fig. 8.2. Extending modules from shrunk polytopes

Let Q be a projective $R[M]$-module. The extension $P_1 \subset P$ is pyramidal, and therefore there exists a projective $R[M|P_1]$-module Q_1 such that Q is extended from Q_1. Now we can define a projective $R[M|P_{i+1}]$-module Q_{i+1}, $i \geq 1$, recursively by (i) $Q_{i+1} = Q_i \otimes R[M|P_{i+1}]$ if $P_{i+1} \supset P_i$, or (ii) as a projective $R[M|P_{i+1}]$-module that extends to Q_i if $P_{i+1} \subset P_i$. Clearly, this construction is only possible because of Theorem 8.6. Note that Q is extended from each of the Q_i.

For ε as above there exists $j \in \mathbb{N}$ such that $P_j \subset U_\varepsilon(z) \cap P$, and we can consider the extension $Q' = Q_j \otimes R[D \cap \mathbb{Z}^d]$. Since D is unimodular, $D \cap \mathbb{Z}^d \cong \mathbb{Z}^d_+$, and so Q' is a free module over $R[D \cap \mathbb{Z}^d]$ by the Quillen–Suslin theorem. But Q is extended from Q', and so Q is free.

8.C Projective modules over polynomial rings

For the proof of Theorem 8.4 we must recall several theorems that have been developed in the course of the solution of Serre's problem on projective modules over polynomial rings.

Milnor patching. A diagram of commutative rings

$$\begin{array}{ccc} A & \longrightarrow & A_1 \\ \downarrow & & \downarrow f_1 \\ A_2 & \xrightarrow[f_2]{} & A' \end{array} \tag{8.1}$$

is said to have the *Milnor patching property* if the following condition is satisfied: whenever we are given projective modules P_1 over A_1 and P_2 over A_2 and an isomorphism $A' \otimes_{A_1} P_1 \cong A' \otimes_{A_2} P_2$ of A'-modules, the pullback P of the diagram

$$\begin{array}{ccc} & & P_1 \\ & & \downarrow \\ P_2 & \longrightarrow & A' \otimes P_1 \cong A' \otimes P_2 \end{array}$$

(i. e., the set of pairs $(x, y) \in P_1 \times P_2$ such that x and y are sent to the same element in $A' \otimes P_2$) is a finitely generated projective A-module and the natural maps $A_i \otimes_A P \to P_i$ are isomorphisms.

We will need the following special cases in which (8.1) has the Milnor patching property:

(A) The diagram (8.1) is *cartesian*, i. e., $A = \{(x, y) \in A_1 \times A_2 : f_1(x) = f_2(y)\}$, and $A_1 \to A'$ is surjective.

(B) *Karoubi squares.* The diagram has the form

$$\begin{array}{ccc} A & \xrightarrow{f} & B \\ \downarrow & & \downarrow \\ S^{-1}A & \longrightarrow & f(S)^{-1}B \end{array}$$

where $S \subset A$ is regular on A and B, i. e., S consists of non-zerodivisors of A and $f(S)$ consists of non-zerodivisors of B, and $A/sA \to B/f(s)B$ is an isomorphism for every $s \in S$. (Actually, it suffices if we only require that $A/sA \to B/f(s)B$ is a surjective homomorphism.)

(C) *Localization squares.* The diagram has the form

$$\begin{array}{ccc} A & \longrightarrow & A_s \\ \downarrow & & \downarrow \\ A_t & \longrightarrow & A_{st} \end{array}$$

where $As + At = A$.

Case (A) is proved in Milnor [262, §2], (B) is considered in Swan [348, App. A], and case (C) is just standard sheaf patching on Spec A (see, for instance, [237, Ch. 1, Cor. 3.12]).

As an immediate consequence of the definition of Milnor patching we obtain:

Lemma 8.8. *Suppose the diagram* (8.1) *has the Milnor patching property and let* $P \in \mathbb{P}(A_1)$. *If* $A' \otimes P$ *is extended from* A_2, *then* P *is extended from* A.

The following lemma is an application of localization squares.

Lemma 8.9. *Let* P *be a projective* $R[X, X^{-1}]$*-module. Let* $f \in R[X^{-1}]$ *be monic in* X^{-1}. *If* P_f *is extended from* R, *then* P *is extended from* $R[X]$.

Proof. Write $f = X^{-n}g$ where $g \in R[X]$ and $g(0) = 1$. Then

$$\begin{array}{ccc} R[X] & \longrightarrow & R[X, X^{-1}] \\ \downarrow & & \downarrow \\ R[X]_g & \longrightarrow & R[X, X^{-1}]_g = R[X, X^{-1}]_f \end{array}$$

is a localization square. Since P_f is extended from R, it is extended from $R[X]_g$ and we are done by Milnor patching. □

Quillen's local-global principle. Quillen's derivation [297] of Theorem 8.5 is based on the following local-global principle, which we will use without proof in the following.

Theorem 8.10. *Let R be a commutative ring and P be a finitely presented $R[X]$-module. Then P is extended from R if and only if $P_\mathfrak{m}$ is extended from $R_\mathfrak{m}$ for every maximal ideal $\mathfrak{m} \subset R$.*

We need the extension of this theorem to general graded rings. In the following theorem $P_\mathfrak{m}$ denotes the localization of P with respect to the multiplicative subset $R_0 \setminus \mathfrak{m} \subset R$.

Theorem 8.11. *Let $R = R_0 \oplus R_1 \oplus \cdots$ be a graded ring and P a finitely presented R-module. Then P is extended from R_0 if and only if $P_\mathfrak{m}$ is extended from $(R_0)_\mathfrak{m}$ for every maximal ideal $\mathfrak{m} \subset R_0$.*

It is possible to simply adapt Quillen's original argument to the more general situation of graded rings. However, following Swan [351] we explain how Theorem 8.11 results from the special case of polynomial rings, using a variant of the so-called Swan–Weibel homotopy trick, Lemma 8.39 below.

Proof of Theorem 8.11. Consider the following ring homomorphisms:

(i) the inclusion maps $i : R_0 \to R$ and $j : R \to R[X]$; moreover, for maximal ideals $\mathfrak{m} \subset R_0$ and $\mathfrak{n} \subset R$ we will also use i and j for the inclusion maps $(R_0)_\mathfrak{m} \to R_\mathfrak{m}$ and $R_\mathfrak{n} \to R_\mathfrak{n}[X]$;
(ii) the map $w : R \to R[X]$ defined by $r \to rX^n$ for $r \in R_n$;
(iii) the augmentation $\varepsilon : R \to R_0$ sending R_n to 0 for $n > 0$;
(iv) $e_k : R[X] \to R$ for $k = 0, 1 \in R$ by sending X to k.

Then $e_0 j = e_1 j = e_1 w = 1_R$ and $e_0 w = i\varepsilon$ (in this proof we suppress the symbol $\circ$ for composition). Now let $W = w_*(P)$. If $\mathfrak{m} \subset R_0$ is a maximal ideal, then $P_\mathfrak{m}$ is extended from $(R_0)_\mathfrak{m}$ and so has the form $P_\mathfrak{m} = i_*(V)$ for some finitely presented $(R_0)_\mathfrak{m}$-module V. Therefore $W_\mathfrak{m} \cong w_*(P_\mathfrak{m}) \cong w_* i_*(V) \cong j_* i_*(V)$. If $\mathfrak{n} \subset R$ is a maximal ideal, choose a maximal ideal $\mathfrak{m} \supset \mathfrak{n} \cap R_0$ of R_0. Then $W_\mathfrak{n}$ is a localization of $W_\mathfrak{m}$ and so is extended via j. By the original Quillen local-global principle, W is also extended from R via j: $W \cong j_*(Q)$ for some Q, and this is the crucial point. So

$$P \cong e_{1*}(W) \cong e_{1*} j_*(Q) = e_{0*} j_*(Q) \cong e_{0*}(W) \cong e_{0*} w_*(P) \cong i_* \varepsilon_*(P),$$

showing that P is extended from R_0. □

Inverting monic polynomials. For a ring R and a variable X we let $R(X)$ denote the localization of the polynomial ring $R[X]$ with respect of the multiplicative set of all monic polynomials, i. e., those with invertible leading coefficient.

Although based on substantially different approaches, both Quillen's and Suslin's solutions to Serre's problem use the following criterion for a projective module over a polynomial ring to be extended from the ring of coefficients.

Theorem 8.12. *Let R be a ring and P be a projective $R[X]$-module. Then P is extended from R if and only if $R(X) \otimes_{R[X]} P$ is extended from R.*

Theorem 8.12, also known as the *affine Murthy–Horrocks theorem*, will be proved at the end of this section. Next we explain how it implies the Quillen–Suslin theorem (Theorem 8.5). Actually, we do so for its extension to Laurent polynomial rings, due to Swan [349]:

Theorem 8.13. *Let R be a PID and m, n nonnegative integers. Then all finitely generated projective modules over the Laurent polynomial ring $R[\mathbb{Z}_+^m \oplus \mathbb{Z}^n]$ are free.*

We need the following two lemmas. The first one is proved in [237, Ch. 4, Cor. 1.3].

Lemma 8.14.

(a) $\dim R(X) = \dim R$.
(b) If R is a PID then so is $R(X)$.

Lemma 8.15. *Let P be a projective $R[X, X^{-1}]$-module. If $R(X) \otimes P$ and $R(X^{-1}) \otimes P$ are extended from R, then P is extended from R.*

Proof. There exists a monic polynomial f in X^{-1} such that P_f is extended from R (since $R(X^{-1})$ is a filtered union of extensions $R[X, X^{-1}]_f$). By Lemma 8.9 P is extended from $R[X]$: $P \cong R[X, X^{-1}] \otimes Q$ for some projective $R[X]$-module Q. Then $R(X) \otimes Q = R(X) \otimes P$ is extended from R. By Theorem 8.12 Q is extended from R and, hence, so is P. □

Proof of Theorem 8.13. We have to show that projective modules over the Laurent polynomial ring $R[\mathbb{Z}_+^m \oplus \mathbb{Z}^n]$ are free if R is a PID. We use induction on $m + n$. The case $m + n = 0$ is obvious.

Let P be a projective $R[\mathbb{Z}_+^m \oplus \mathbb{Z}^n]$-module. First assume that $m > 0$. By the induction hypothesis and Lemma 8.14 the module $R(X)[\mathbb{Z}_+^{m-1} \oplus \mathbb{Z}^n] \otimes P$ is free over $R(X)[\mathbb{Z}_+^{n-1} \oplus \mathbb{Z}^n]$, where we make the identification $R[\mathbb{Z}_+^{m-1} \oplus \mathbb{Z}^n][X] = R[\mathbb{Z}_+^m \oplus \mathbb{Z}^n]$. Since $R[\mathbb{Z}_+^{m-1} \oplus \mathbb{Z}^n](X)$ is a further localization of $R(X)[\mathbb{Z}_+^{m-1} \oplus \mathbb{Z}^n]$, we are done by Theorem 8.12.

Now assume that $n > 0$. By Lemma 8.14 and the induction hypothesis the modules $R(X)[\mathbb{Z}_+^m \oplus \mathbb{Z}^{n-1}] \otimes P$ and $R(X^{-1})[\mathbb{Z}_+^m \oplus \mathbb{Z}^{n-1}] \otimes P$ are free over $R(X)[\mathbb{Z}_+^m \oplus \mathbb{Z}^{n-1}]$ and $R(X^{-1})[\mathbb{Z}_+^m \oplus \mathbb{Z}^{n-1}]$, respectively. Here we make the identification $R[\mathbb{Z}_+^m \oplus \mathbb{Z}^{n-1}][X, X^{-1}] = R[\mathbb{Z}_+^m \oplus \mathbb{Z}^n]$. Since $R[\mathbb{Z}_+^m \oplus \mathbb{Z}^{n-1}](X)$ and $R[\mathbb{Z}_+^m \oplus \mathbb{Z}^{n-1}](X^{-1})$ are further localizations of $R(X)[\mathbb{Z}_+^m \oplus \mathbb{Z}^{n-1}]$ and $R(X^{-1})[\mathbb{Z}_+^m \oplus \mathbb{Z}^{n-1}]$, respectively, we are done be Lemma 8.15. □

Roberts' theorem. The next theorem, Lam's axiomatization of a theorem of Roberts, will play a crucial role for pyramidal descent. Let us first introduce two important subgroups of $\mathrm{GL}_n(R)$ for a commutative ring R. The group of $n \times n$ matrices of determinant 1 over R is called the *special linear group of order n* and denoted by $\mathrm{SL}_n(R)$. The the *group of elementary $n \times n$ matrices* $\mathrm{E}_n(R)$ is the subgroup generated by all elementary $n \times n$ matrices, i. e., those matrices that have the entry 1 on the diagonal and exactly one off-diagonal nonzero element.

Note that a surjective ring homomorphism $R \to S$ induces a surjective group homomorphism $\mathrm{E}_n(R) \to \mathrm{E}_n(S)$. Therefore, if $\mathrm{E}_n(S) = \mathrm{SL}_n(S)$, the natural homomorphism $\mathrm{SL}_n(R) \to \mathrm{SL}_n(S)$ is surjective, too. By the Gauß algorithm, one has $\mathrm{E}_n(S) = \mathrm{SL}_n(S)$ if S is a local ring (Exercise 8.3).

In the theorem and its proof we use the following notation: $\mathrm{J}(R)$ is the *Jacobson radical* of a ring R, i. e., the intersection of all its maximal ideals,

Theorem 8.16. *Let $(L, \mathfrak{m})$ be a local ring, A an L-algebra, P a finitely generated A-module, and $S \subset A$ a multiplicative subset consisting of non-zerodivisors of A and P. Suppose the following conditions hold for the natural number n:*

(a) A/fA is a finitely generated L-module for all $f \in S$;

(b) every matrix $\alpha \in \mathrm{GL}_n(\overline{S^{-1}A})$ can be decomposed as a product $\beta\gamma$ with β in the image of $\mathrm{GL}_n(S^{-1}A)$ and γ in the image of $\mathrm{GL}_n(\bar{A})$, where $\bar{A} = A/\mathfrak{m}A$ and $\overline{S^{-1}A} = S^{-1}A/\mathfrak{m}S^{-1}A$;

(c) there is an L-subalgebra $B \subset S^{-1}A$ with $S^{-1}A = A + B$ and $\mathfrak{m}B \subset \mathrm{J}(B)$;

(d) $S^{-1}P \cong (S^{-1}A)^n$ and $\bar{P} \cong \bar{A}^n$.

Then $P \cong A^n$.

This is proved in [237, Ch. 4, §4]. We now explain how Theorem 8.12 follows from Theorems 8.10 and 8.16. The reader may check that Theorem 8.10 and Theorem 8.16 are the only results on projective modules that we have not proved, apart from Milnor patching.

Proof of Theorem 8.12. By Theorem 8.10 we can assume that R is local. Let $\mathfrak{m} \subset R$ be the maximal ideal. We apply Theorem 8.16 to $L = R$ and $A = R[X]$. Let S be the set of monic polynomials. Conditions (a) and (d) are clearly satisfied because $S^{-1}A = R(X)$ and $\bar{A} = \Bbbk[X]$ where $\Bbbk = R/\mathfrak{m}$.

Furthermore $\overline{S^{-1}A} = \Bbbk(X)$, so $\mathrm{SL}_n(\overline{S^{-1}A}) = \mathrm{E}_n(\overline{S^{-1}A})$ and $\mathrm{SL}_n(S^{-1}A) \to \mathrm{SL}_n(\overline{S^{-1}A})$ is surjective. To prove (b) we observe that $\mathrm{U}(\Bbbk(X))$ is generated by the images of S and $\mathrm{U}\,\Bbbk^\times$. Condition (b) follows, since every element α of $\mathrm{GL}_n(\overline{S^{-1}A})$ can now be decomposed into a product $\alpha'\beta\gamma$ where α' is in the image of $\mathrm{SL}_n(S^{-1}A)$, β is the image of an invertible diagonal matrix over $S^{-1}A$, and γ is such a matrix over $\Bbbk(X)$.

For (c) we choose B to be the set of all f/g where g is monic and $\deg(f) \le \deg(g)$. Then $S^{-1}A = A + B$ follows easily if we divide f by g when $\deg(f) > \deg(g)$. Finally, if $m \in \mathfrak{m}$, then $1 + mf/g = (g + mf)/g$ is a unit of B and this implies $\mathfrak{m}B \subset \mathrm{J}(B)$. □

8.D Reduction to the interior

The first reduction in the proof of Theorem 8.4 is given by the following:

Lemma 8.17. *It is sufficient to prove Theorem 8.4 for the case in which R is a local PID and M is an affine positive seminormal monoid.*

Proof. Let $N = (M \setminus \mathrm{U}(M)) \cup \{1\}$ (writing the monoid operation multiplicatively). It is clear that N is also a seminormal monoid. We have the cartesian square of R-algebras

$$\begin{array}{ccc} R[N] & \longrightarrow & R[M] \\ \downarrow & & \downarrow \\ R & \longrightarrow & R[\mathrm{U}(M)] \end{array}$$

in which the vertical maps send the elements of $N \setminus \{1\}$ to $0 \in R$. This square is of Milnor type (A). Since $\mathrm{U}(M)$ is the filtered union of finitely generated torsionfree and, hence, free abelian groups, all projective $R[\mathrm{U}(M)]$-modules are free by Lemma 8.1 and Theorem 8.5. Now it suffices to have all projective $R[N]$-modules free.

The monoid N is the filtered union of positive affine monoids M_i. The seminormalizations of the M_i are also affine (see p. 66), so that we can replace each M_i by its seminormalization. By Lemma 8.1 it is enough that the projective $R[M_i]$-modules are free for all i.

For the lemma we still have to show that R can be replaced by its localizations if M is a positive affine seminormal monoid. By Proposition 2.17 the ring $R[M]$ admits a grading $R[M]$ in which the elements of M are homogeneous. But then Theorem 8.10 implies that the projective $R[M]$-modules are free if the projective $R_{\mathfrak{m}}[N]$-modules are free for all maximal ideals $\mathfrak{m}$ of R. □

Lemma 8.17 finishes the proof of Theorem 8.4 for rank 1 monoids: the monoids M_i are all seminormal of rank 1, and therefore isomorphic to $\mathbb{Z}_+$, as is easily seen. So the Quillen–Suslin theorem implies that all projective $R[M_i]$-modules are free.

The next lemma shows two facts simultaneously. Namely, it further reduces the general case to that of normal monoids and, moreover, it reduces the problem to the interior of M.

Lemma 8.18. *Let M be an affine positive seminormal monoid of rank d and R a PID. Suppose that the projective $R[N]$-modules are free for every affine positive normal monoid N of rank $< d$. Then every projective $R[M]$-module is extended from $R[M_*]$.*

Proof. Let $F_1, \dots, F_n$ be the faces of $\mathbb{R}_+M$, labeled in such a way that $F_i \supset F_j$ implies $i \le j$. In particular, $F_1 = \mathbb{R}_+M$ and $F_n = \{0\}$. Set $M_i = M \cap F_i$ and $U_i = \operatorname{int}(M_i)$. Then the M_i are the extreme submonoids of M, and the U_i form a partition of M (see Theorem 1.10(d)). Let $W_i = U_1 \cup \dots \cup U_i$ and $W_0 = \varnothing$.

We claim that

$$MW_i \subset W_i. \tag{8.2}$$

In fact, let $x \in M$ and $y \in W_i$. Then $y \in U_j$ for some $j \leq i$. Since the U_k form a partition of M, we have $xy \in U_q \subset M_q \subset F_q$ for some q. It follows that $y \in M_q$. So $U_j \cap M_q \neq \emptyset$. Since $\operatorname{int}(F_j) \cap F_q \neq \emptyset$, we get $M_j \subset M_q$ by Theorem 1.10(d). Therefore, $q \leq j$ and $xy \in U_q \subset W_i$.

Now let J_i be the R-submodule of $R[M]$ generated by W_i. It is an ideal of $R[M]$ by (8.2). Set $A_i = R[M]/J_i$. The kernel of the natural homomorphism $A_{i-1} \to A_i$ is the free R-module with basis U_i. So for $i < n$ we have the Milnor squares

$$\begin{array}{ccc} R[(M_i)_*] & \longrightarrow & A_{i-1} \\ \downarrow & & \downarrow \\ R & \longrightarrow & A_i. \end{array} \tag{8.3}$$

Since $(M_i)_*$ is the filtered union of affine positive normal monoids by Proposition 2.41, all finitely generated projective $R[(M_i)_*]$-modules, $i > 1$, are free by our inductive hypothesis. Since $A_{n-1} = R$, descending induction on i, in conjunction with Lemma 8.8, shows that all projective A_i-modules are extended from R for $1 \leq i < n$, and are therefore free. The lemma follows by a final application of Milnor patching because $A_0 = R[M]$. □

If N is just seminormal, then N_* is the filtered union of normal affine monoids (Propositions 2.40, 2.41). Using Lemma 8.18, we finally reach the following reduction of Theorem 8.4.

Corollary 8.19. *It is enough to prove Theorem 8.4 under the following additional assumptions:*

(a) M is affine, positive, and normal;
(b) rank $M \geq 2$;
(c) R is local;
(d) (Induction hypothesis on rank) projective $R'[N]$-modules are free for every PID R' and all positive normal monoids N with rank $N <$ rank M;
(e) all projective $R[M]$-modules are extended from $R[M_*]$.

8.E Graded "Weierstrass preparation"

In the next section we will need a graded variant of the formal Weierstraß preparation theorem.

Theorem 8.20. *Let $A = A_0 \oplus A_1 \oplus \cdots$ be a graded ring and let $v \in A_d$. Suppose $M = M_0 \oplus M_1 \oplus \cdots$ is a graded A-module satisfying the condition*

$$v : M_i \to M_{i+d}, \quad x \mapsto vx, \quad \textit{is an isomorphism for } i \geq 0. \tag{8.4}$$

Choose $f \in A$, $f = \sum_{i \geq 0} f_i$, with $f \equiv f_0 + f_1 + \cdots + f_{nd-1} + v^n \mod \operatorname{nil}(A)$. Then every $z \in M$ can be written as $z = fq + r$ with $q \in M$ and $r \in M_0 + \cdots + M_{nd-1}$. Moreover, q and r are unique.

Proof. Let $f = f_0 + \cdots + f_m$. Then $f_{nd+1}, \dots, f_m$ and $f_{nd} - v^n$ are nilpotent and therefore generate a homogeneous nilpotent ideal J. Suppose $J^h = 0$. We use induction on h. If $h = 0$, the usual division algorithm clearly applies. Let

$$N = \{x \in M : v^k x \in J^{h-1}M \text{ for some } k \geq 0\} = \bigcup_{k\geq 0} (J^{h-1}M : v^k).$$

Then N satisfies (8.4) and so does M/N. Moreover, M/N is a module over A/J^{h-1} and N is a module over A/J since $x \in N$ implies $v^k Jx = 0$ and v is regular on M by (8.4). By induction, the conclusion of the theorem applies to the residue classes of f in A/J^{h-1} and A/J and the modules M/N and to N respectively. Writing $\bar{z} = f\bar{q} + \bar{r}$ in M/N and lifting it to M, we get $z = fq + r + w$ where $w \in N$. An application of the induction hypothesis to N yields $w = fq' + r'$ so that $z = f(q+q') + (r+r')$. For the uniqueness it is enough to consider $z = 0$. Then we must have $\bar{q} = \bar{r} = 0$ by induction applied to M/N. Hence q and r lie in N, and so $q = r = 0$ by induction applied to N. □

Corollary 8.21. *Let $A = A_0 \oplus A_1 \oplus \cdots$ be a graded ring and let v be an element of A_d, satisfying the condition*

$$v : A_i \to A_{i+d}, \quad x \mapsto vx, \quad \text{is an isomorphism for } i \geq 0. \tag{8.5}$$

Let $f \in A$ with $f \equiv f_0 + f_1 + \cdots + f_{nd-1} + v^n \mod (\operatorname{nil}(A))$. Then $f = (1+\eta)(v^n - r)$ where $\eta \in \operatorname{nil}(A)$ and $r \in A_0 + \cdots + A_{nd-1}$.

Proof. Apply the theorem with $M = A$ and $z = v^n$ to get $v^n = fq + r$.

But the theorem also applies to $M = A/\operatorname{nil}(A)$. In fact, the surjectivity in (8.4) for this choice of M follows from that in (8.5) for A, while the injectivity follows from the fact that v is regular on A. Therefore x is nilpotent if vx is. In $M = A/\operatorname{nil}(A)$ we have

$$\bar{v}^n = \bar{f} - (\bar{f}_0 + \bar{f}_1 + \cdots + \bar{f}_{nd-1}) = \bar{q}\bar{f} + \bar{r}$$

by the theorem. Thus the uniqueness of $\bar{q}$ implies $\bar{q} = 1$, and so $q = 1 + \mu$ with μ nilpotent. Therefore, q is invertible and $q^{-1} = 1 + \eta$ with $\eta \in \operatorname{nil}(A)$. □

8.F Pyramidal descent

Now we have completed the preparations for the proof of Theorem 8.6. During the entire section M is supposed to be an affine, positive, normal monoid of rank $d \geq 2$. We may assume that $\operatorname{gp}(M) = \mathbb{Z}^d$ and fix a rational hyperplane $H \subset \mathbb{R}^d$ that defines a cross-section $\Phi(M)$ of the cone $\mathbb{R}_+ M \subset \mathbb{R}^d$.

Local descent. The first step is a local version of pyramidal descent:

Theorem 8.22. *Let $(R, \mathfrak{m})$ be a local PID, and $M \subset \mathbb{Z}^d$ a positive normal monoid with $\operatorname{gp}(M) = \mathbb{Z}^d$. Assume that projective $R'[H]$-modules are free for every positive normal monoid H with $\operatorname{rank} H < d$ and every PID R'. Suppose $\Phi(M) = \Delta \cup \Gamma$ is a pyramidal* *decomposition of $\Phi(M)$ with pyramid Δ, and $N = M|\Gamma$. Let $\mathfrak{M} = (\mathfrak{m}, N^+) \subset R[N]$ be the maximal ideal generated by $\mathfrak{m}$ and the noninvertible elements of N.*

Then for every projective $R[M]$-module Q the module $Q_{\mathfrak{M}} = (R[N] \setminus \mathfrak{M})^{-1} Q$ is free over $R[M]_{\mathfrak{M}} = (R[N] \setminus \mathfrak{M})^{-1} R[M]$.

The proof mimics the derivation of Theorem 8.12 from Theorem 8.16. However, the situation is more complicated now.

Let us first define and examine all the data that depend only on the monoid structures. The polytopes Δ and Γ meet in a common face of dimension $d-2$. It spans a rational vector subspace H' of dimension $d-1$, and we choose the unique primitive integral form δ vanishing on H' and having $\delta(x) \geq 0$ for all $x \in \Delta$. This linear form will serve as a degree function. Note that it separates N from $M|\Delta$, having values ≤ 0 on N and values ≥ 0 on $M|\Delta$.

Let v be the vertex of $\Phi(M)$ that serves as the apex of the pyramid Δ and t be the extreme integral generator of $\mathbb{R}_+M$ in the extreme ray $\mathbb{R}_+v$ of $\mathbb{R}_+M$.

Lemma 8.23. *Let $x \in M$ with $\delta(x) \geq a\delta(t)$, $a \in \mathbb{Z}_+$. Then $x - at \in M$ (the monoid operation written additively).*

Proof. Since $\delta(x) > 0$, we have $x \in M|\Delta$. Since Δ is a pyramid with apex v, all support linear forms of Δ vanish on t, except δ, which defines the facet opposite to v. Therefore we only need to check that $\delta(x - at) \geq 0$, and this is obvious. □

Since $\mathbb{R}_+N + \mathbb{R}_+t = \mathbb{R}_+M$, the monoid M is integral over $N' = N + \mathbb{Z}_+t$. In particular it is finitely generated as an N'-module, say by $m_1 = 0$ and $m_2, \dots, m_n \in M|\Delta$. Then, as an N-module, it is generated by all sums $kt + m_i$, $k \in \mathbb{Z}_+$, $i = 1, \dots, n$. Note that, given $b \in \mathbb{Z}_+$, only finitely many of these monoid elements have value $\leq b$ under δ.

From now on we use multiplicative notation for monoids. The ideal of $R[M]$ generated by N^+ (the set of noninvertible elements of N) is the free R-module with basis $X = \{yz \ : \ y \in N^+,\ z \in M\}$. Therefore, the R-algebra $C = R[M]/(N^+)$ is free with basis $Y = M \setminus X$; the multiplication on C is the R-bilinear extension of

$$y \cdot z = \begin{cases} yz, & yz \in Y, \\ 0, & \text{otherwise}. \end{cases}$$

The linear form δ defines a grading on C since C is the quotient of R-modules graded by δ. With $Y_k = \{x \in Y \ : \ \delta(x) = k\}$ we have $C_k = RY_k$. Moreover, $C_k = 0$ for $l < 0$ because $N^+ \subset X$. So C is positively graded, $C = C_0 \oplus C_1 \oplus \cdots$, and $C_0 = R$.

Lemma 8.24. *The R-linear map $C_i \to C_{i+\delta(t)}$, $c \mapsto tc$, is an isomorphism for all $i \geq 0$.*

Proof. It is enough to show that the multiplication by t maps Y_i bijectively onto $Y_{i+\delta(t)}$. We first show that the image lies in $Y_{i+\delta(t)}$. Let $y \in Y_i$ and assume that $ty \notin Y_{i+\delta(t)}$. Then $ty = xz$ with $x \in N^+$, $z \in M$. Now $\delta(y) > 0$ and $\delta(x) \leq 0$ so $\delta(z) > \delta(t)$. By Lemma 8.23 $z = st$ for some $s \in M$. Therefore $y = xs$, contradicting the fact that $y \in Y_i$.

The map is clearly injective, and it is easily seen that it is surjective, too: if $y \in Y_{i+\delta(t)}$, then $y = tx$ for some $x \in M$ by Lemma 8.23 and x must lie in Y_i, since $y \in X$ if $x \in X$. □

We prove Theorem 8.22 by applying Roberts' theorem 8.16 to the rings $L = R[N]_{\mathfrak{M}}$ and $A = R[M]_{\mathfrak{M}}$. We have to define a multiplicative subset $S \subset R[M]_{\mathfrak{M}}$ and an $R[N]_{\mathfrak{M}}$-subalgebra $B \subset S^{-1}R[M]_{\mathfrak{M}}$ so that the conditions of Theorem 8.16 are satisfied.

We choose $S \subset A$ as follows. First we set $\deg(m) = \delta(m)$ for the elements of M. Next we write $\deg(f) \leq c$ for an element $f \in A$ if $f = r_1w_1 + \cdots + r_pw_p$ where $r_i \in L$ and $w_i \in M$ such that $\delta(w_i) < c$ for all i. Then we call an element $g \in A$ *monic* if $g = t^a + f$ with $\deg(f) < a\deg(t)$ and set $\deg(g) = a\deg(t)$. Now we choose S as the set of monic elements. Evidently S is a multiplicatively closed set. (The reader may worry whether the degree is well-defined—we will justify the definition below.)

One can imitate the usual division algorithm since A is generated as an L-module by the elements of M:

Lemma 8.25. *Let $f \in A$ and $g \in S$. Then we can write $f = gq + r$ in A with* $\deg(r) < \deg(g)$.

Finally, we put

$$B = \{\alpha + (f/g) \ : \ \alpha \in L,\ g \in S,\ f \in A,\ \deg(f) < \deg(g)\}.$$

This is easily seen to be a subring of $S^{-1}A$.

Now we check the conditions of Theorem 8.16. Let $P = Q_{\mathfrak{M}}$ (notation as in Theorem 8.22). Since A is a domain, it is clear that S is regular on A and P.

(a) *If $f \in S$, then A/fA is finite over L.*

By Lemma 8.25 A/fA is generated as an L-module by the monomials m_it^k where m_i runs through the generating set of M over N, $k \in \mathbb{Z}_+$, and $\deg(m_it^k) < \deg(f)$. As already remarked, there are only finitely many such monomials.

(b) *Every matrix $\alpha \in \mathrm{GL}_n(\overline{S^{-1}A})$ can be decomposed as a product $\beta\gamma$ with β in the image of $\mathrm{GL}_n(S^{-1}A)$ and γ in the image of $\mathrm{GL}_n(\bar{A})$, where $\bar{A} = A/\mathfrak{M}A$ and $\overline{S^{-1}A} = S^{-1}A/\mathfrak{M}S^{-1}A$.*

Since $L/\mathfrak{M} = R/\mathfrak{m} = \Bbbk$ is a field, we see that

$$\bar{A} = A/\mathfrak{M}A = R[M]/(\mathfrak{m}, N^+) = \Bbbk[M]/(N^+).$$

If $x \in M$, $x \neq 1$, then $x^m = t^b y$ with $b \geq 0$ and $y \in N$. So either x belongs to

$$t^{\mathbb{Z}_+} = \{t^r : r \in \mathbb{Z}^+\}$$

or x is nilpotent modulo the ideal generated by N^+. Since $t \in M$ generates an extreme submonoid, $M \setminus t^{\mathbb{Z}_+}$ is an ideal of M (see p. 65), and $R[M]/(M \setminus t^{\mathbb{Z}_+}) \cong R[t]$. Therefore,

$$R[M]/(N^+)_{\text{red}} = R[t]$$

and so $\bar{A}_{\text{red}} = \Bbbk[t]$.

Let us examine the image of S in $\Bbbk[t]$. Let $g = t^a + \sum s_i m_i$ with $m_i \in M$, $\deg(m_i) < a\deg(t)$ and $s_i \in L$. Since $(M \setminus t^{\mathbb{Z}_+})$ goes to 0, only terms $s_i m_i$ with $m_i = t^j$ for some $j \in \mathbb{Z}_+$ can survive in $\Bbbk[t]$. Since $\deg(m_i) < a\deg(t)$, it follows that $j < a$. Furthermore the coefficient s_i goes to its residue class in $\Bbbk$. Therefore the residue class of g is indeed a monic polynomial in $\Bbbk[t]$. This observation justifies the notion of degree introduced above. Moreover, the image of S is exactly the set of monic polynomials, as every monic polynomial in $\Bbbk[t]$ can clearly be lifted to an element of S.

Consequently $(\overline{S^{-1}A})_{\text{red}} = \Bbbk(t)$. It follows that $\overline{S^{-1}A}$ is a local ring (having exactly one prime ideal), so that $\mathrm{SL}_n(\overline{S^{-1}A}) = \mathrm{E}_n(\overline{S^{-1}A})$, and $\mathrm{SL}_n(S^{-1}A) \to \mathrm{SL}_n(\overline{S^{-1}A})$ is surjective. Therefore, we only need to prove that $\mathrm{U}(S^{-1}A) \oplus \mathrm{U}(\bar{A}) \to \mathrm{U}(\overline{S^{-1}A})$ is surjective. (Write the matrix α to be decomposed as a product $\upsilon\xi$ with $\xi \in \mathrm{SL}_n(\overline{S^{-1}A})$ and υ a diagonal matrix.)

Now we use the graded structure of $R[M]/(N^+)$ that we have introduced above. It follows that $\bar{A} = C/\mathfrak{M}C = \bar{A}_0 \oplus \bar{A}_1 \oplus \cdots$ where $\bar{A}_i = C_i/\mathfrak{M}C_i$ and Lemma 8.24 implies that the multiplication by t is an isomorphism $\bar{A}_i \to \bar{A}_{i+e}$, $e = \deg(t)$, for all $i \geq 0$. A unit of $\overline{S^{-1}A}$ has the form f/s where s is in the image of $S \subset \mathrm{U}(S^{-1}A)$ and $f \in \bar{A}$ divides some element of the image of S. Since $\bar{A}_{\text{red}} = \Bbbk[t]$ we see that, up to a factor from $\mathrm{U}(\Bbbk)$, f maps to a monic polynomial in $\Bbbk[t]$. By Corollary 8.21, $f = (1+\eta)g$ where η is nilpotent and $g = t^m + a_{me-1} + \cdots + a_0$ with $a_i \in \bar{A}_i$. Since $1+\eta \in \mathrm{U}(\bar{A})$, it will suffice to show that g lies in the image of S. We can lift g to h in A, choosing $h = t^m + b_{me-1} + \cdots + b_0$ with b_i a linear combination over L of elements of Y_i. Since the elements of Y_i have smaller degree than t^m for $i < me$, we have indeed lifted g to a monic element.

(c) $S^{-1}A = A + B$ *and* $\mathfrak{M}B \subset \mathrm{J}(B)$.

Every element of $S^{-1}A$ has the form f/g with g monic. Writing $f = gq + r$ as in Lemma 8.25 we see that $f/g = q + r/g \in A + B$ so $S^{-1}A = A + B$.

If $m_i \in \mathfrak{M}$ and $\alpha_i + f_i/g \in B$, then $u = 1 + \sum m_i(\alpha_i + f_i/g) = \varepsilon + f/g$ where $\varepsilon = 1 + \sum m_i\alpha_i$ is a unit in L and $f = \sum m_i f_i$. Therefore, $u = \varepsilon(g + \varepsilon^{-1}f)/g$ which is a unit of $S^{-1}A$ since $g + \varepsilon^{-1}f$ is monic. This shows that $\mathfrak{M}B \subset \mathrm{J}(B)$.

(d) $S^{-1}P \cong (S^{-1}A)^n$ *and* $\bar{P} \cong (\bar{A})^n$.

Since $\bar{A}_{\text{red}} = \Bbbk[t]$, the projective module $\bar{P}/\mathrm{nil}(\bar{A})\bar{P}$ is free over $\bar{A}_{\text{red}}$ and hence $\bar{P}$ is free over $\bar{A}$, say $\bar{P} \cong (\bar{A})^n$. Suppose P_t is free over A_t. Since $t \in S$, $S^{-1}P \cong (S^{-1}A)^{n'}$ for some n'. Both rings $S^{-1}A$ and $\bar{A}$ map to the same ring $\overline{S^{-1}A}$ (which is a nilpotent extension[2] of $\Bbbk(t)$). This proves $n = n'$.

So it only remains to show that P_t is free over A_t. In its turn, this would follow if Q_t were free over $R[M]_t$. By Proposition 2.32 we have $R[M]_t \cong R[M'][\mathbb{Z}] \cong R[M'][X, X^{-1}]$ where M' is an affine positive normal monoid of rank $d-1$, and

[2] A ring S is a *nilpotent extension* of a ring R if $R = S/I$ with an ideal $I \subset \mathrm{nil}(S)$.

$R[M'](X)$ is a localization of $R(X)[M']$. Therefore, Lemma 8.14 and Theorem 8.12, together with the induction assumption on rank in Theorem 8.22, give us the desired freeness of Q_t.

Global descent. The derivation of Theorem 8.6 from Theorem 8.22, to be given now, is *not* a local-to-global passage in the sense of patching local data. Rather we will apply Theorem 8.22 to an infinite family of rings distinct from the original $R[M]$ to derive the analogue of Theorem 8.6 for M_*. The final step to the monoid M is taken with the help of Lemma 8.18. We keep the notation introduced in Theorem 8.22.

Proof of Theorem 8.6. By Corollary 8.19 we can additionally assume that (i) $R = (R, \mathfrak{m})$ is a local PID and (ii) projective $R[M]$-modules are extended from $R[M_*]$.

We want to prove that for every projective $R[M]$-module P there exists a projective $R[M|\Gamma]$-module P_Γ such that

$$P \cong R[M] \otimes_{R[M|\Gamma]} P_\Gamma. \tag{8.6}$$

By Lemma 8.18 we find $P_* \in \mathbb{P}(R[M_*])$ such that

$$P \cong R[M] \otimes_{R[M_*]} P_*.$$

Fix such P_* and consider the maximal ideal $\mathfrak{M}_* = \big(\mathfrak{m}, \operatorname{int}(M|\Gamma)\big) \subset R[M|\Gamma_*]$. Note that $\operatorname{int}(M|\Gamma)$ is the ideal of noninvertible elements in $M|\Gamma_* = (M|\Gamma)_*$ (for simplicity of notation).

First we prove that

$$(P_*)_{\mathfrak{M}_*} = \big(R[M|\Gamma_*] \setminus \mathfrak{M}_*\big)^{-1} P_* \text{ is free over}$$
$$R[M_*]_{\mathfrak{M}_*} = \big(R[M|\Gamma_*] \setminus \mathfrak{M}_*\big)^{-1} R[M_*]. \tag{8.7}$$

There are rational equidimensional polytopes $\Gamma_i \subset \operatorname{int}(\Gamma)$ and rational points $v_i \in \operatorname{int}(\Delta)$, $i \in \mathbb{N}$ such that the following conditions are satisfied:

(i) $v_i \to v$ as $i \to \infty$;
(ii) $\Gamma_1 \subset \Gamma_2 \subset \cdots$;
(iii) $\operatorname{int}(\Gamma) = \bigcup_{i \in \mathbb{N}} \Gamma_i$;
(iv) for every $i \in \mathbb{N}$ exactly one facet of Γ_i is visible from v_i.

This approximation of Γ is illustrated by Figure 8.3. For example, it can be realized as follows: we choose a rational point in the interior of Γ, and consider the homotheties ϑ_i with center z and factor $i/(i+1)$; then we set $v_i = \vartheta_i(v)$ and $\Gamma_i = \vartheta_i(\Gamma)$.

Consider the affine positive normal submonoids $M_i = M|\operatorname{conv}(\Gamma_i, v_i)$. By Lemma 8.1 there exists i such that $P_* \cong R[M_*] \otimes_{R[M_i]} P_i$. Then

$$(P_*)_{\mathfrak{M}_*} = R[M_*]_{\mathfrak{M}_*} \otimes_{R[M_i]_{\mathfrak{M}_i}} (P_i)_{\mathfrak{M}_i}$$

where $\mathfrak{M}_i = \big(\mathfrak{m}, (M_i|\Gamma_i)'\big) \subset R[M_i]$. Therefore, it is enough to show that $(P_i)_{\mathfrak{M}_i}$ is free over $R[M_i]_{\mathfrak{M}_i}$. But now we are exactly in the situation of Theorem 8.22, the roles of M and Q being played by M_i and P_i. This proves (8.7).

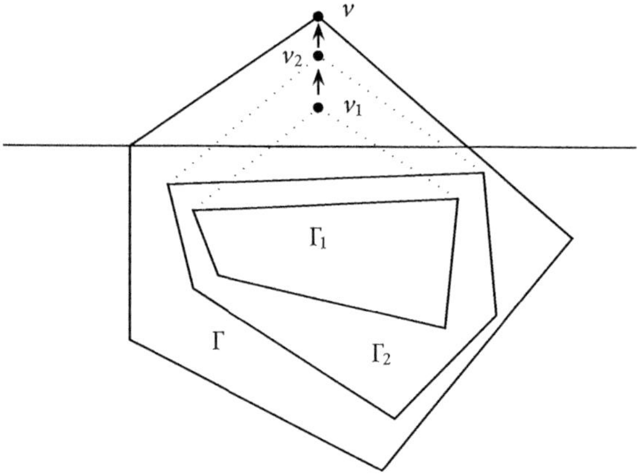

Fig. 8.3. The approximation of Γ

Now we prove that P_* is extended from $R[M|\Gamma_*]$. Then it follows that P is extended from $R[M|\Gamma]$, and we are done. By Lemma 8.8 and equation (8.7) it suffices to show that the commutative diagram

$$\begin{array}{ccc} R[M|\Gamma_*] & \longrightarrow & R[M_*] \\ \downarrow & & \downarrow \\ R[M|\Gamma_*]_{\mathfrak{M}_*} & \longrightarrow & R[M_*]_{\mathfrak{M}_*} \end{array}$$

is a Karoubi square. Therefore, it is enough to show that for all elements $s \in R[M|\Gamma_*] \setminus \mathfrak{M}_*$ the map

$$R[M|\Gamma_*]/s \to R[M_*]/s$$

is an isomorphism. This can be checked locally on $R[M|\Gamma_*]$, i. e., it suffices to show that

$$\alpha_{\mathfrak{N}} : (R[M|\Gamma_*]/s)_{\mathfrak{N}} \to (R[M_*]/s)_{\mathfrak{N}} \tag{8.8}$$

is an isomorphism for every maximal ideal $\mathfrak{N} \in \max(R[M|\Gamma_*])$.

Suppose first that $\mathfrak{N} \cap M|\Gamma_* = \varnothing$. Then $M|\Gamma_* \subset R[M|\Gamma_*] \setminus \mathfrak{N}$ and $\alpha_{\mathfrak{N}}$ is a further localization of $(M|\Gamma_*)^{-1}R[M|\Gamma_*] \to (M|\Gamma_*)^{-1}R[M_*]$. But by Corollary 2.25 we have

$$\operatorname{gp}(M|\Gamma) = \operatorname{gp}(M|\Gamma_*) \subset (M|\Gamma_*)^{-1}M_* \subset \operatorname{gp}(M_*) = \operatorname{gp}(M) = \operatorname{gp}(M|\Gamma)$$

and $\alpha_{\mathfrak{N}}$ turns out to be a localization of an isomorphism.

Now suppose that $\mathfrak{N} \cap M|\Gamma_* \neq \varnothing$. We will see that $\mathfrak{N} = \mathfrak{M}_*$. This implies that the source and target of $\alpha_{\mathfrak{N}}$ are both the zero module.

In fact, $\mathfrak{N} \cap M|\Gamma_*$ is a prime ideal of the monoid $M|\Gamma_*$. But there is no other prime ideal in $M|\Gamma_*$ except $\operatorname{int}(M|\Gamma)$[3]. This follows as in the proof of Proposition 2.36: every

[3] It is here where we need the fact that the cone $\mathbb{R}_+M|\Gamma_* \subset \mathbb{R}^d$ is open. This explains the necessity of the passage to interior submonoids.

$x \in \operatorname{int}(M|\Gamma)$ is nilpotent over the ideal (in the monoid $M|\Gamma_*$) generated by an arbitrary $y \in \operatorname{int}(M|\Gamma)$. Moreover, since R is local, the inclusion $\operatorname{int}(M|\Gamma) \subset \mathfrak{N}$ implies that $\mathfrak{m} \subset \mathfrak{N}$ and, thus, $\mathfrak{N} = \mathfrak{M}_*$. □

8.G How to shrink a polytope

For Theorem 8.4 it remains to show the purely geometric Theorem 8.7. We first prove that a pyramidal decomposition of a face can be extended to the whole polytope, an argument obviously necessary for an inductive attack.

Lemma 8.26. *Let $P \subset \mathbb{R}^d$ be a rational polytope and $F \subset P$ be a proper face (not necessarily a facet). Suppose $G \subset F$ is a pyramidal extension. Then there exists a pyramidal extension $P' \subset P$ such that $P' \cap F = G$.*

Proof. By induction we can assume that F is a facet of P. Let $v \in F$ be the vertex not in G, and Δ be the pyramid in the decomposition $F = G \cup \Delta$. Its apex is v (Figure 8.4). We claim that there exists a rational affine function $\alpha : \operatorname{aff}(P) \to \mathbb{R}$ satisfying the following conditions:

(i) $\alpha(x) < 0$ for all $x \in G \setminus \Delta$ and $\alpha(x) > 0$ for $x \in \Delta \setminus G$;
(ii) $\alpha(w) < 0$ for every vertex w of P outside Δ.

Then the extension of polytopes

$$Q = \{z \in P : \alpha(z) \leq 0\} \subset P$$

is pyramidal. In fact, P decomposes into Q and $\Delta' = \{z \in P : \alpha(z) \geq 0\}$, and

Fig. 8.4. Extending a pyramidal decomposition

the hyperplane H_α separating Q and Δ' intersects the interior of P. Moreover, Δ' is a pyramid. Namely, the ray R from v through any point $x \in \Delta'$, $x \neq v$, leaves P through a facet not containing v, and $\alpha(y) \leq 0$ at the last point y of R in P. Since $\alpha(v) > 0$, the line segment $[v, y]$ meets H_α. Thus R leaves Δ' through H_α.

In order to construct α, we choose β as an affine form vanishing on the facet F and having negative values on the rest of P. Furthermore we choose γ as an extension of an affine form on $\operatorname{aff}(F)$ that defines the given pyramidal decomposition of F (and has positive value in v). We then set $\alpha = \beta + \varepsilon\gamma$ for a rational $\varepsilon > 0$ small enough such that $\alpha(w) < 0$ for all vertices w of P outside Δ. □

Let P be a rational polytope. We say that a (finite or infinite) sequence (P_i) of rational subpolytopes is *admissible* if it satisfies the following conditions for all i:

(i) $P_i \subset P$;
(ii) $P_{i+1} \subset P_i$ or $P_{i+1} \supset P_i$;
(iii) if $P_{i+1} \subset P_i$, then P_i is a pyramidal extension of P_{i+1}.

As we saw in Section 8.B, the proof of Theorem 8.4 will be complete once we have shown Theorem 8.7: given a rational point $z \in \operatorname{int}(P)$, there exists an admissible sequence (P_i) such that $P_i \subset U_\varepsilon(z) \cap P$ for every $\varepsilon > 0$ and $i \gg 0$. If this condition is satisfied, we say that (P_i) *contracts* to z. Observe that an admissible sequence (P_i) that contracts to z is necessarily infinite. (However, we do not require that z be contained in the polytopes P_i.)

Proof of Theorem 8.7. We use induction on $\dim P$. It is enough to prove that there is a finite admissible sequence $P = P_1, \dots, P_n$ such that

$$z \in \operatorname{int}(P_n) \text{ and } P_n \subset \operatorname{int}(P). \tag{8.9}$$

Then we can find a rational number $\lambda \in (0,1)$ such that $P_n \subset \vartheta_\lambda(P)$ where ϑ_λ is the homothety with center z and factor λ. Now we set $P_{n+i} = \vartheta_\lambda(P_i)$, $i = 1, \dots, n$. Then $P_{2n} \subset \vartheta_{\lambda^2}(P)$. By iteration we obtain an admissible sequence of polytopes contracting to the point z, and are done.

Now we want to show the existence of an admissible sequence satisfying (8.9). We can assume that $\dim P \geq 1$, for otherwise P is a point and, thus, coincides with its own interior. The case $\dim P = 1$ is obvious because we can construct P_2 by cutting off a small (rational) subsegment from P_1, and P_3 by cutting off a small rational subsegment at the other endpoint, etc.

Let $d = \dim P \geq 2$. Assume that for every d-polytope P and each facet Q of P we can find an admissible sequence $P_1, \dots, P_n$ such that

$$z \in \operatorname{int}(P_n) \text{ and } P_n \cap Q = \varnothing. \tag{8.10}$$

Then we apply the same argument to P_n and cut off $Q' \cap P_n$ where Q' is another facet of P. Iterating the process we remove all facets of the original polytope P.

By the previous step all we need to show is that for every facet $Q \subset P$ there exists an admissible sequence $P = P_1, \dots .P_n$ satisfying (8.10).

Let $\Delta \subset Q$ be a rational simplex and $z' \in \operatorname{int}(\Delta)$ a rational point. By the inductive hypothesis we can find an admissible sequence (Q_i) in Q that contracts to z'. In particular, $Q_m \subset \Delta$ for some m. We can extend (Q_i) to an admissible sequence in P: whenever we pass from Q_i to Q_{i+1} by removing a pyramid, Lemma

8.26, applied to P_i and the extension $Q_{i+1} \subset Q_i$, yields P_{i+1}. If $Q_{i+1} \supset Q_i$, we choose $P_{i+1} = \operatorname{conv}(P_i \cup Q_{i+1})$.

Now replace P_m by $\operatorname{conv}(P_m \cup \Delta)$. The intersection of the new P_m with the facet Q of the original polytope P is the simplex Δ. Choose a vertex v of Δ. Then Δ is a pyramidal extension of its facet Δ' opposite to v. Applying Lemma 8.26 once more, we can cut off a pyramid from P_m in such a way that all points of Δ, except those in Δ', have been removed. Again we have reached a polytope whose intersection with Q is a simplex, but now of smaller dimension. Iterating the procedure we can cut off Δ completely. In particular, there is an admissible sequence $P_1, \dots, P_{n-1}$ such that $P_{n-1} \cap Q = \varnothing$. Finally, we choose $P_n = \operatorname{conv}(P_{n-1}, R)$ where R is a rational polytope such that $z \in \operatorname{int}(R)$ and $R \subset \operatorname{int}(P_1)$. □

8.H Converse results

In this section we relax our standard convention on monoids: they are only assumed to be commutative, but not necessarily cancellative or torsionfree.

Now we prove the following converse to Theorem 8.4. Its part (b) is a special case of Corollary 4.73, and only mentioned here because we want to discuss its validity under a relaxation of the conditions on M.

Theorem 8.27.

(a) *Let M be a cancellative, but not necessarily torsionfree, monoid. If* $\operatorname{Pic}(R[M]) = 0$ *for every PID R of characteristic 0, then M is torsionfree and seminormal.*
(b) *If M is cancellative and torsionfree and* $\operatorname{Pic}(\Bbbk[M]) = 0$ *for some field $\Bbbk$, then M is seminormal.*

The condition that M be cancellative cannot be dropped, and we cannot restrict ourselves in Theorem 8.27(a) only to fields. This is explained by the following examples.

Example 8.28. Consider the commutative noncancellative monoid M generated by three elements x, y, z subject to the relations $x^2 = xy = y^2$. Then all projective $R[M]$-modules are free for every PID R; in particular $\operatorname{Pic}(R) = 0$. In fact, $R[M]_{\mathrm{red}} \cong R[X,Y]/(X-Y)$ is a polynomial ring over R.

Example 8.29. Let $M = (\mathbb{Z}_+ \oplus \mathbb{Z}_2) \setminus \{(0,1)\}$. Then all projective $\Bbbk[M]$-modules are free for every field $\Bbbk$. The reader may check that $\Bbbk[M] \cong \Bbbk[X,Y]/(X^2 - Y^2)$. If $\operatorname{char} \Bbbk \neq 2$, then $\Bbbk[M] = \Bbbk[X,Y]/(X+Y)(X-Y) \cong \Bbbk[X,Y]/(XY)$, and the freeness of projective $\Bbbk[M]$-modules follows from Theorem 8.48 below. But if $\operatorname{char} \Bbbk = 2$, then $\Bbbk[M]_{\mathrm{red}} \cong \Bbbk[X]$.

The proof of Theorem 8.27(a) requires some preparation.

Proposition 8.30. *Let G be an abelian group. Then the group ring $R[G]$ is seminormal for every PID R of characteristic 0 if and only if G is torsionfree.*

Proof of Theorem 8.27(a). Theorem 4.19 implies that $R[M]$ is reduced since char $R = 0$.

By Lemma 8.14, $R(X)$ is a PID if R is, so $\operatorname{Pic}(R(X)[M]) = 0$ for every PID R of characteristic 0. Therefore, the composite map

$$\operatorname{Pic}(R[M][X]) = \operatorname{Pic}(R[X][M]) \to \operatorname{Pic}(R(X)[M]) \to \operatorname{Pic}(R[M](X))$$

is zero. Since $\operatorname{Pic}(A[X]) \to \operatorname{Pic}(A(X))$ is always injective by Theorem 8.12, we get $\operatorname{Pic}(R[M][X]) = \operatorname{Pic}(R[M]) = 0$. By Theorem 4.74 $R[M]$ is seminormal. Since $R[\operatorname{gp}(M)]$ is a localization of $R[M]$ by a multiplicatively closed set of non-zerodivisors, $R[\operatorname{gp}(M)]$ is seminormal. So Proposition 8.30 implies that M is torsionfree. □

The proof of Proposition 8.30, which we postpone until the end of this section, is based on a series of auxiliary results.

Lemma 8.31. *Let R be a commutative ring, let G be an abelian group, and suppose that $R[G]$ is seminormal. Let T be the torsion subgroup of G. Then $R[T]$ is also seminormal.*

Proof. The subring $R[T]$ of $R[G]$ is itself reduced. Suppose $x^3 = y^2$ in $R[T]$. Then there is $z \in R[G]$ such that $z^2 = x$ and $z^3 = y$. We can find a subgroup $H \subset G$, containing T, such that $z \in R[H]$ and H/T is finitely generated. Since H/T is torsionfree, it is free, and we can write $H = T \oplus F$. Since H retracts to T, the group ring $R[H]$ retracts onto $R[T]$, sending z to $w \in R[T]$ with $w^2 = x$ and $w^3 = y$. □

Lemma 8.32. *Let T be an abelian torsion group and assume that $R[T]$ is seminormal for a PID R with an infinite number of distinct residue characteristics. Let H be a finite subgroup of T. Then $R[H]$ is also seminormal.*

Proof. Suppose that $x^3 = y^2$ in $R[H]$. Then there is $z \in R[T]$ such that $z^2 = x$ and $z^3 = y$. We can find a finite subgroup $H' \subset T$ containing H such that $z \in R[H']$. Choose a maximal ideal $\mathfrak{m} \subset R$ such that the characteristic of $\Bbbk = R/\mathfrak{m}$ is prime to the order of H'. Consider the diagram

$$\begin{array}{ccc} R[H] & \longrightarrow & R[H'] \\ \downarrow & & \downarrow \\ \Bbbk[H] & \longrightarrow & \Bbbk[H'] \end{array}$$

and let $\bar{x}, \bar{y}, \bar{z}$ denote the images of x, y, z in $\Bbbk[H']$. Since $\gcd(\operatorname{char}\Bbbk, |H'|) = 1$, $\Bbbk[H]$ and $\Bbbk[H']$ are finite products of finite extensions of $\Bbbk$ by Maschke's theorem, and hence reduced and seminormal. Since $\overline{z^2}, \overline{z^3} \in \Bbbk[H]$, it follows that $\bar{z} \in \Bbbk[H]$. Therefore, $z \in R[H] + \mathfrak{m}R[H']$. Now $R[H'] = R[H] \oplus F$ as an R-module where F is the free R-module on $H' \setminus H$. This shows that $z \in R[H] \oplus \mathfrak{m}F$. Since there are an infinite number of choices for $\mathfrak{m}$, we have $\bigcap \mathfrak{m}F = 0$. Therefore $z \in \bigcap (R[H] \oplus \mathfrak{m}F) = R[H] \oplus 0 = R[H]$. □

Lemma 8.33 (Rush [308]**).** *Let R be a domain of characteristic 0 and let H be a finite group of order n. If $R[H]$ is seminormal and $n \notin \mathrm{U}(R)$ then R/nR is reduced.*

Proof. Let $N \in R[H]$ be the sum of elements of H and let $S = R[H]/(N)$. Since $\Bbbk[H] = \Bbbk \times \Bbbk[H]/(N)$ where $\Bbbk = \mathrm{QF}(R)$, we see easily that S is reduced and the diagram

$$\begin{array}{ccc} R[H] & \longrightarrow & S \\ {\scriptstyle f}\downarrow & & \downarrow \\ R & \longrightarrow & R/nR, \end{array}$$

$f(H) = \{1\}$, is cartesian. So we are done by the following lemma. □

Lemma 8.34. *Let*

$$\begin{array}{ccc} A & \longrightarrow & B \\ \downarrow & & \downarrow \\ C & \longrightarrow & D \end{array}$$

be a cartesian diagram with B and C reduced and $B \to D$ surjective. If A is seminormal, then D is also reduced.

Proof. We think of A as the set of pairs $(b, c) \in B \times C$ such that b and c have the same image in D. Clearly, A is reduced. If D were not reduced, we could find $t \in B$ mapping to $\bar{t}$ with $\bar{t} \neq 0$, but $\bar{t}^2 = 0$. Let $x = (t^2, 0)$, $y = (t^3, 0)$ in A. Then $x^3 = y^2$, but there is no $z \in A$ with $z^2 = x$ and $z^3 = y$. If there were such a $z = (b, c)$, then $b = t$ and $c = 0$ since B and C are reduced, and it would follow that $\bar{t} = 0$. □

Proof of Proposition 8.30. If the torsion subgroup $T \subset G$ is not trivial we choose a finite nontrivial subgroup H of T. Let A be the ring of integers of $\mathbb{Q}(\sqrt{n})$ where $n = |H|$, and let $\mathfrak{p}_1, \dots, \mathfrak{p}_m$ be the prime ideals of A containing n. Consider $B = S^{-1}A$ where $S = A \setminus (\mathfrak{p}_1 \cup \dots \cup \mathfrak{p}_m)$. Then B is a semilocal Dedekind domain, and therefore a PID [37, Ch. VII, §2, Prop. 1]. Since the class group of A is finite, we can find prime ideals $\mathfrak{q}_1, \dots, \mathfrak{q}_p$, all different from $\mathfrak{p}_1, \dots, \mathfrak{p}_m$, that generate the class group of A. Now we choose an element $s \in (\mathfrak{q}_1 \cap \dots \cap \mathfrak{q}_p) \setminus (\mathfrak{p}_1 \cup \dots \cup \mathfrak{p}_m)$. It follows that $R = A[s^{-1}]$ is a PID with infinitely many residue characteristics. By Lemma 8.31 $R[T]$ is seminormal. But, since R/nR is not reduced, $R[H]$ is not seminormal by Lemma 8.33—a contradiction with Lemma 8.32. □

8.I Generalizations

For Theorem 8.4 the base ring R is required to be a PID. If one weakens this condition to regularity, then one can still prove the homotopy invariance of the Grothendieck functor K_0 under the passage to $R[M]$ for seminormal M, and homotopy invariance does indeed characterize seminormal monoids as we will see.

Another type of generalization is considered in the last part of the section, namely residue class rings modulo monomial ideals.

The Grothendieck group K_0. First we give an overview of the basic facts on the Grothendieck functor K_0. For the general theory the reader is referred to the classical books [19] and Swan [345].

We start with some observations on an arbitrary functor $\mathsf{F} : \mathsf{Rings} \to \mathsf{AbGroups}$ from the category of rings to that of abelian groups. (Recall that we consider only commutative rings.)

(i) Let $f : R \to R'$ be a ring homomorphism such that there exists a ring retraction $g : R' \to R$ with $gf = \mathbf{1}_R$. Then $\mathsf{F}(R)$ is a direct summand of $\mathsf{F}(R')$, and we identify $\mathsf{F}(R)$ with its isomorphic image in $\mathsf{F}(R')$.

For us, the most relevant examples of such ring homomorphisms are the inclusion maps $R \to R'$ where the graded ring R' is a monoid algebra $R[M]$ of an affine positively graded monoid M. However, R is always a retract of $R[M]$, whether the monoid M is affine and positive or not: the surjective R-homomorphism $\varepsilon_1 : R[M] \to R$, $M \mapsto 1$, splits the inclusion map $R \to R[M]$ (see Section 4.B).

(ii) To a commutative diagram of rings

$$\begin{array}{ccc} A & \xrightarrow{f_1} & B \\ {\scriptstyle f_2}\downarrow & & \downarrow{\scriptstyle g_1} \\ C & \xrightarrow[g_2]{} & D \end{array}$$

we usually associate the sequence of group homomorphisms

$$\mathsf{F}(A) \xrightarrow{\gamma} \mathsf{F}(B) \oplus \mathsf{F}(C) \xrightarrow{\beta} \mathsf{F}(D),$$

where

$$\gamma(x) = \big(\mathsf{F}(f_1)(x), \mathsf{F}(f_2)(y)\big), \qquad x \in \mathsf{F}(A),$$
$$\beta(y,z) = \mathsf{F}(g_1)(y) - \mathsf{F}(g_2)(z), \qquad (y,z) \in \mathsf{F}(B) \oplus \mathsf{F}(C).$$

The commutativity of the diagram implies that $\beta\gamma = 0$.

(iii) If a ring homomorphism $f : A \to B$ induces an isomorphism $\mathsf{F}(f) : \mathsf{F}(A) \to \mathsf{F}(B)$ we simply write $\mathsf{F}(A) = \mathsf{F}(B)$, assuming the ring homomorphism f is clear from the context.

The functor $K_0 : \mathsf{Rings} \to \mathsf{AbGroups}$ assigns to a ring R the group

$$K_0(R) = F_X/G,$$

where F_X is the free abelian group on the set X of isomorphism classes (P) of projective R-modules and G is the subgroup generated by the elements of type $(P \oplus Q) - (P) - (Q)$. For a ring homomorphism $f : R \to R'$ the functorial homomorphism $K_0(f) : K_0(R) \to K_0(R')$ is given by $[P] \to [R' \otimes_R P]$.

Two projective R-modules P and Q define the same element $[P] = [Q]$ in $K_0(R)$ if and only if P and Q are *stably isomorphic*: $P \oplus R^n \cong Q \oplus R^n$ for some n. In particular, $K_0(R) = \mathbb{Z}$ means that every projective R-module P is stably free, i. e., $P \oplus R^m \cong R^n$ for some $m, n \in \mathbb{N}$. If M is seminormal and R is a PID, then $K_0(R[M]) = \mathbb{Z}$ by Theorem 8.4. A generalization of this equality for higher-dimensional regular rings is given by Theorem 8.37 below.

The tensor product of two projective modules induces a commutative ring structure on $K_0(R)$, and we have a natural embedding $\operatorname{Pic}(R) \to \operatorname{U}(K_0(R))$: the homomorphism is injective since the map $\bigwedge^{\text{rank}}$ introduced below is a section. In general, $\operatorname{Pic}(R) \neq \operatorname{U}(K_0(R))$; see Example 9.25.

Consider the ring $H_0(R)$ of continuous functions $\operatorname{Spec} R \to \mathbb{Z}$ where $\operatorname{Spec} R$ carries the Zariski topology and $\mathbb{Z}$ is a discrete set. It follows easily from Lemma 8.2(b) that the assignment $[P] \mapsto \operatorname{rank}_P$ induces a well-defined ring homomorphism $K_0(R) \to H_0(R)$.

Next we define a group homomorphism $\bigwedge^{\text{rank}} : K_0(R) \to \operatorname{Pic}(R)$. Consider a module $P \in \mathbb{P}(R)$. Because $\operatorname{Spec} R$ is quasicompact (every open cover contains a finite subcover) the image of $\operatorname{rank}_P \in H_0(R)$ is a finite subset of $\mathbb{Z}_+$. (This follows also from the fact that P is finitely generated.) Then $(\operatorname{rank}_P)^{-1}(\mathbb{N}) \subset \operatorname{Spec} R$ is the disjoint union of certain subsets that are simultaneously open and closed. This decomposition of $\operatorname{Spec} R$ induces a decomposition of the form $R = e_1 R \times \cdots \times e_n R$ where $e_1, \dots, e_n$ are mutually orthogonal idempotents, and $e_i P \in \mathbb{P}(e_i R)$ is of constant rank for every $i = 1, \dots, n$. (Here we view $e_i R$ as a ring whose unit element is e_i.) We have $\bigwedge^{\operatorname{rank}(e_i P)}(e_i P) \in \operatorname{Pic}(e_i R)$, $i = 1, \dots, n$, where the exterior powers are taken over $e_i R$. Then the direct sum of the rank 1 modules over the rings $e_i R$ is a rank 1 module over R which we denote by $\bigwedge^{\text{rank}}(P)$. It is easily checked that the assignment $[P] \to \bigwedge^{\text{rank}}(P)$ gives rise to a well-defined group homomorphism, denoted by $\bigwedge^{\text{rank}} : K_0(R) \to \operatorname{Pic}(R)$.

The diagram below with exact rows and columns serves as the definition of further groups associated to $K_0(R)$, namely $\tilde{K}_0(R)$ and $SK_0(R)$:

$$
\begin{array}{ccccccccc}
 & & 0 & & & & & & \\
 & & \downarrow & & & & & & \\
 & & SK_0(R) & & & & & & \\
 & & \downarrow & & & & & & \\
0 & \longrightarrow & \tilde{K}_0(R) & \longrightarrow & K_0(R) & \xrightarrow{\text{rank}} & H_0(R) & \longrightarrow & 0 \\
 & & \downarrow & & \downarrow{\scriptstyle \bigwedge^{\text{rank}}} & & & & \\
 & & \operatorname{Pic}(R) & = & \operatorname{Pic}(R) & & & & \\
 & & \downarrow & & \downarrow & & & & \\
 & & 0 & & 0 & & & &
\end{array}
$$

(For the surjectivity of the map rank see Exercise 8.6.) It is not hard to see that the restriction of $\bigwedge^{\mathrm{rank}}$ to $\tilde{K}_0(R)$ is still surjective: $\bigwedge^{\mathrm{rank}}([P]-[Q]) = \bigwedge^{\mathrm{rank}}([P])$ where $Q \in \mathbb{P}(R)$ is the direct sum of the modules $(e_i R)^{\mathrm{rank}\, e_i P}$.

When $\operatorname{Spec} R$ is connected, then $H_0(R) = \mathbb{Z}$. In this case we have $SK_0(R) = 0$ if and only if every projective R-module stably has the form "free ⊕ rank 1."

For every Milnor square of rings

$$\begin{array}{ccc} A & \longrightarrow & B \\ \downarrow & & \downarrow{\scriptstyle f} \\ C & \longrightarrow & D \end{array}$$

there is the exact K_0-*Mayer–Vietoris sequence*

$$K_0(A) \to K_0(B) \oplus K_0(C) \to K_0(D); \tag{8.11}$$

see [262, §3]. It is easily observed that we have a similar exact sequence for H_0. Therefore, the cartesian square above yields the following commutative diagram with exact rows and columns —the right half of the diagram in [19, Ch. IX, Cor. 5.12]:

$$\begin{array}{ccccc} 0 & & 0 & & 0 \\ \downarrow & & \downarrow & & \downarrow \\ SK_0(A) & \longrightarrow & SK_0(B) \oplus SK_0(C) & \longrightarrow & SK_0(D) \\ \downarrow & & \downarrow & & \downarrow \\ \tilde{K}_0(A) & \longrightarrow & \tilde{K}_0(B) \oplus \tilde{K}_0(C) & \longrightarrow & \tilde{K}_0(D) \\ \downarrow & & \downarrow & & \downarrow \\ \mathrm{Pic}(A) & \longrightarrow & \mathrm{Pic}(B) \oplus \mathrm{Pic}(C) & \longrightarrow & \mathrm{Pic}(D) \\ \downarrow & & \downarrow & & \downarrow \\ 0 & & 0 & & 0. \end{array} \tag{8.12}$$

Finally, we have the following

Lemma 8.35. *If $I \subset A$ is an ideal contained in the Jacobson radical $\mathrm{J}(A)$ then $K_0(A) \to K_0(A/I)$ is a monomorphism and $\mathrm{Pic}(A) \to \mathrm{Pic}(A/I)$ is an isomorphism.*

In particular, the homomorphisms $\tilde{K}_0(A) \to \tilde{K}_0(A/I)$ and $SK_0(A) \to SK_0(A/I)$ are monomorphisms.

The proof is based on the following consequence of the Nakayama lemma: modules $P, Q \in \mathbb{P}(A)$ are isomorphic if and only if their reductions mod I are isomorphic over A/I; see [19, Ch. IX, Prop. 1.3].

Homotopy invariance. One of the first results in algebraic K-theory is Grothendieck's theorem [19, Ch. XII, §3] (originally published by Serre [317]) on the *K_0-homotopy invariance* of a regular ring of arbitrary Krull dimension:

Theorem 8.36. *For every regular ring R the inclusion map into the polynomial and Laurent polynomial rings $R \to R[X] \to R[X,X^{-1}]$ induces natural isomorphisms $K_0(R) = K_0(R[X]) = K_0(R[X,X^{-1}])$.*[4]

By iterated use of Theorem 8.36 we get the equality $K_0(R) = K_0(R[\mathbb{Z}^m \oplus \mathbb{Z}_+^n])$ for every regular ring R and all $m, n \in \mathbb{Z}_+$. In view of Theorem 8.4 it is natural to ask to what extent the same equality holds for general monoid rings. The complete answer to this question is given by the following:

Theorem 8.37. *Let R be a regular ring and M be a monoid. Then*

(a) the following conditions are equivalent:
  (i) $K_0(R) = K_0(R[M])$,
  (ii) $\operatorname{Pic}(R) = \operatorname{Pic}(R[M])$,
  (iii) M is seminormal;

(b) one has $SK_0(R) = SK_0(R[M])$.

In the proof of Theorem 8.37 we will use the following stable version of Theorem 8.11 for projective modules:

Theorem 8.38. *Let $R = R_0 \oplus R_1 \oplus \cdots$ be a graded ring and P be a projective R-module. Then P is stably extended from R_0, i. e., $[P] \in \operatorname{Im}(K_0(R_0) \to K_0(R))$, if and only if $P_\mathfrak{m}$ is stably free for every maximal ideal $\mathfrak{m} \subset R_0$.*

In the special case when $R = R_0[X]$ this is proved in Vorst [368], along with the same equality for all higher K-groups. Then the general case of graded rings is derived in the same way as Theorem 8.11 was derived from Theorem 8.10; see Section 8.C. The *Swan–Weibel homotopy trick* to be used here is the following:

Lemma 8.39. *Let $\mathsf{F} : \mathsf{Rings} \to \mathsf{AbGroups}$ be a functor and $A = A_0 \oplus A_1 \oplus \cdots$ be a graded ring. If $\mathsf{F}(A[X]) = \mathsf{F}(A)$ then $\mathsf{F}(A) = \mathsf{F}(A_0)$.*

The proof can be found in [237, p. 183]. It is based on the same ideas as Theorem 8.11, and in Exercise 8.1 the reader is asked to find it.

One more result we will use in the proof of Theorem 8.37 is the following theorem for which we refer to [19, Ch. IV, Cor. 3.5].

Theorem 8.40 (Bass' cancellation). *Let A be a noetherian ring of finite Krull dimension d. Then two modules $P, Q \in \mathbb{P}(A)$ of rank $> d$ (i. e., $\operatorname{rank}_P(\mathfrak{p}) > d$ for all $\mathfrak{p} \in \operatorname{Spec} A$) are stably isomorphic if and only if they are isomorphic.*

[4] Grothendieck's theorem is true for regular schemes, which are not necessarily affine, where the group K_0 is defined in terms of locally free sheaves, see Section 10.A.

Proof of Theorem 8.37(a). (i) $\Longrightarrow$ (ii) The inclusion map $R \to R[M]$ is split by the augmentation $\varepsilon_1 : R[M] \to R$, $M \to 1$. By functoriality, the map $\operatorname{Pic}(R) \to \operatorname{Pic}(R[M])$ is injective.

On the other hand, let P be an invertible $R[M]$-module. The invertible $R[M]$-modules $(P \otimes_{R[M]} R) \otimes_R R[M]$ and P are stably isomorphic since $K_0(R) \to K_0(R[M])$ is an isomorphism. But we have already observed that stably isomorphic invertible modules are necessarily isomorphic.

(ii) $\Longrightarrow$ (iii) Since regular rings are seminormal, this is an immediate consequence of Corollary 4.73 and Theorem 4.76.

(iii) $\Longrightarrow$ (i) This is similar to the proof of Theorem 8.4. We describe the steps that correspond to each other in more detail.

(1) Since the homomorphism $K_0(R) \to K_0(R[M])$ is split injective, it is enough to show its surjectivity.

(2) The functor K_0 commutes with filtered direct limits, and so the general case reduces to the situation when M is finitely generated.

(3) Using the inductive hypothesis on $\operatorname{rank} M$ and the same cartesian squares as in Section 8.D together with Theorem 8.36, we can assume that M is an affine positive normal monoid.

(4) For any module $P \in \mathbb{P}(R[M])$ we have the equivalence

$$[P] \in \operatorname{Im}\bigl(K_0(R) \to K_0(R[M])\bigr)$$
$$\iff \qquad [P \oplus R[M]^{\dim R + \operatorname{rank} M + 1}] \in \operatorname{Im}\bigl(K_0(R) \to K_0(R[M])\bigr).$$

Therefore, it is enough to prove the following stronger claim: *all projective $R[M]$-modules of rank* $> \dim R + \operatorname{rank} M$ *are extended from* R.

(5) By Theorem 8.11 we can further assume that R is a regular local ring. Then all projective R-modules are free. We want to prove that all projective $R[M]$-modules are of rank $> \dim R + \operatorname{rank} M$ are free. This will be done by induction on $\operatorname{rank} M$, the case $\operatorname{rank} M = 1$ following from Theorems 8.36 and 8.40.

(6) Using the inductive hypothesis on $\operatorname{rank} M$ and the same cartesian squares as in Section 8.D, we are left to show that all projective $R[M_*]$-modules of rank $> \dim R + \operatorname{rank} M$ are free. Observe that the monoid ring $R[M]$ is a domain since R is local and regular. In particular, the projective modules, involved in the Milnor patching associated to any of the cartesian squares mentioned above, have the same constant rank.

(7) Next we go on with *exactly* the same argument as in the proof of Theorem 8.6—the pyramidal descent. The only difference that we encounter in this process is the verification of the last condition in Roberts' theorem for the corresponding objects, namely that $S^{-1}P$ is free over $S^{-1}A$; see p. 302. We have to show that all projective $R[M'][\mathbb{Z}]$-modules of rank $> \dim R + \operatorname{rank} M$ are free where M' is an affine positive normal monoid with $\operatorname{rank} M' = \operatorname{rank} M - 1$. By Lemma 8.15 it is enough to show that all projective modules over $R(X)[M']$ and $R(X^{-1})[M']$, having rank $> \dim R + \operatorname{rank} M$,

are extended correspondingly from $R(X)$ and $R(X^{-1})$. Since $\dim R = \dim R(X) = \dim R(X^{-1})$ [237, Ch. 4, Cor. 1.3], the result follows from Theorem 8.11 and induction on the monoid rank.

(8) Using global descent as in Section 8.F, we conclude that all projective $R[M_*]$-modules are extended from an R-algebra isomorphic to $R[\mathbb{Z}_+^{\operatorname{rank} M}]$. But then Theorems 8.36 and 8.40 give the desired freeness. □

Proof of Theorem 8.37*(b).* Theorem 8.36 implies $SK_0(R) = SK_0(R[\mathbb{Z}^n])$ for $n \in \mathbb{N}$. Moreover, the functor SK_0, like K_0, commutes with filtered direct limits. Therefore the Mayer–Vietoris SK_0-sequence (the upper row of the diagram (8.12)), associated to the cartesian square of rings

$$\begin{array}{ccc} R[N] & \longrightarrow & R[M] \\ \downarrow & & \downarrow \\ R & \longrightarrow & R[\mathrm{U}(M)], \end{array}$$

with $N = (M \setminus \mathrm{U}(M)) \cup \{1\}$ and $N \setminus \{1\} \to 0$, reduces the general case to the situation in which M is an affine positive monoid.

Applying the Mayer–Vietoris SK_0-sequences to cartesian squares of the form (8.3), we see that it is enough to show that $SK_0(R) = SK_0(R[M_*])$ where M is an affine positive monoid. (Observe that here we consider squares of the form (8.3) for arbitrary, not necessarily normal monoids.)

Proposition 2.33 guarantees the existence of $x \in M$ with $x\bar{M} \subset M$ (in multiplicative notation). Corollary 2.34 implies

$$R = \big(R[M_*]/Rx\bar{M}\big)_{\mathrm{red}} = \big(R[\bar{M}_*]/Rx\bar{M}\big)_{\mathrm{red}}. \tag{8.13}$$

With $N = x\bar{M} \cup \{1\}$ we have the cartesian squares

$$\begin{array}{ccc} R[N] & \hookrightarrow & R[M_*] \\ \downarrow & & \downarrow \\ R & \hookrightarrow & R[M_*]/Rx\bar{M} \end{array} \qquad \begin{array}{ccc} R[N] & \hookrightarrow & R[\bar{M}_*] \\ \downarrow & & \downarrow \\ R & \hookrightarrow & R[\bar{M}_*]/Rx\bar{M} \end{array} \tag{8.14}$$

where the vertical arrows represent the reduction modulo the ideal $Rx\bar{M} \subset R[M]$ and the horizontal arrows represent the inclusion maps. For simplicity of notation we set $I = Rx\bar{M}$ in the following.

By (8.13), the bottom homomorphisms in the diagrams (8.14) are *split* nilpotent extensions. Lemma 8.35 yields

$$SK_0(R) = SK_0(R[M_*]/I) = SK_0(R[\bar{M}_*]/I).$$

Suppose we have shown that the embedding $R \to R[N]$ induces an isomorphism of SK_0-groups. Then we apply the Mayer–Vietoris sequence associated to the left diagram in (8.14) to get an exact sequence

$$SK_0(R) \to SK_0(R) \oplus SK_0(R[M_*]) \to SK_0(R).$$

In this sequence the map on the left sends $SK_0(R)$ diagonally to $SK_0(R) \oplus SK_0(R) \subset SK_0(R) \oplus SK_0(R[M_*])$, whereas the homomorphism on the right, when restricted to $SK_0(R) \oplus SK_0(R)$ is the codiagonal map $(u, u') \mapsto u - u'$. It follows easily that $SK_0(R[M_*]) = SK_0(R)$, as desired.

In order to show the crucial equation $SK_0(R) = SK_0(R[N])$ we have to borrow the following Mayer–Vietoris sequence; it is the SK-version of the K_1-K_0-sequence from Section 9.A: it extends the sequence for SK_0 to the left. In particular, the right square in (8.14) gives rise to the following exact sequence

$$\begin{aligned} SK_1(R) \oplus SK_1(R[\bar{M}_*]) \to SK_1(R[\bar{M}_*]/I) \to SK_0(R[N]) \\ \to SK_0(R) \oplus SK_0(R[\bar{M}_*]) \to SK_0(R[\bar{M}_*]/I). \end{aligned} \tag{8.15}$$

Theorem 8.4 implies that $SK_0(R) = SK_0(R[\bar{M}_*])$. We also have $SK_1(R) = SK_1(R[\bar{M}_*]/I)$ because $R \to R[\bar{M}_*]/I$ is a nilpotent extension (see Section 9.A). More precisely, the restriction of the map $SK_1(R) \oplus SK_1(R[\bar{M}_*]) \to SK_1(R[\bar{M}_*]/I)$ is an isomorphism $SK_1(R) \to SK_1(R[\bar{M}_*]/I)$. Therefore, (8.15) yields the exact sequence

$$0 \to SK_0(R[N]) \to SK_0(R) \oplus SK_0(R) \to SK_0(R) \to 0.$$

By the same arguments on diagonal and codiagonal maps as above we conclude $SK_0(R) = SK_0(R[N])$, and are done. □

Remark 8.41. Using the same technique as above, but applying cancellation results for Laurent polynomial rings over Noetherian coefficient rings [349] that are stronger than Theorem 8.40, Swan [351, Cor. 1.4] deduces the following improvement of Theorem 8.37: for a regular ring R of finite Krull dimension d and an arbitrary monoid M, all projective $R[M]$-modules of rank d are extended from R. Combining this technique with the work of Bhatwadekar and Rao [30], Swan furthermore shows that for every localization R of a regular affine algebra over a field and every *positive* monoid M all projective $R[M]$-modules are actually extended from R. See Theorem 1.2 and the discussion following Conjecture Q_n in Section 3 of [351].

Picard groups. For positive affine monoids M one can describe the Picard group $\operatorname{Pic}(R[M])$. In combination with Theorem 8.37 the following result then gives a complete characterization of the additive structure of $K_0(R)$ when R is a regular ring and $\mathbb{Q} \subset R$. (The restriction to reduced rings of coefficients is irrelevant since Pic is stable under the passage to a residue class ring modulo a nilpotent ideal.)

Theorem 8.42. *Let R be a seminormal ring such that $\mathbb{Q} \subset R$ and let M be an affine positive monoid. Then*

$$\operatorname{Pic}(R[M]) \cong \operatorname{Pic}(R) \oplus \Bigl(\bigoplus_{\operatorname{sn}(M) \setminus M} R \Bigr).$$

We need two results on units.

Lemma 8.43. *Let A be commutative ring containing the field $\mathbb{Q}$ and $\mathfrak{n} \subset A$ be a nilpotent ideal. Then the multiplicative group $1+\mathfrak{n} = \{1+x : x \in \mathfrak{n}\}$ is isomorphic to the additive group $\mathfrak{n}$.*

Proof. A pair of mutually inverse homomorphisms is given by

$$\exp : \mathfrak{n} \to 1+\mathfrak{n}, \quad x \mapsto \sum_{n\geq 0} \frac{x^n}{n!},$$

$$\log : 1+\mathfrak{n} \to \mathfrak{n}, \quad 1+x \mapsto -\sum_{n>0} \frac{(-x)^n}{n}. \qquad \square$$

Let $A = A_0 \oplus A_1 \oplus \cdots$ be a graded ring. Then, by Proposition 4.10,

$$\mathrm{U}(A) = \{u + a_1 + \cdots + a_n : u \in \mathrm{U}(A_0),\ n \in \mathbb{N},\ a_i \in A_i \text{ is nilpotent},\ i = 1,\dots,n\}. \qquad (8.16)$$

Proof of Theorem 8.42. Note that $\mathrm{sn}(M)$ is an affine monoid (see the discussion immediately after Definition 2.39). Therefore, $\mathrm{sn}(M)$ is also an affine positive monoid and by Proposition 2.17(f) we can fix a grading $R[\mathrm{sn}(M)] = R \oplus R_1 \oplus \cdots$ making the monomials homogeneous elements. Then we have induced gradings on $R[M']$ for any submonoid $M' \subset \mathrm{sn}(M)$ as well as on monomial quotients of $R[M']$.

First we consider the special case when $M = \mathrm{sn}(M)$. By Theorem 4.76 $R[\mathrm{sn}(M)]$ is seminormal. Therefore, by Theorem 4.74(b), we have $\mathrm{Pic}(R[M]) = \mathrm{Pic}(R[M][X])$. Because of the grading on $R[M]$, Lemma 8.39 implies $\mathrm{Pic}(R) = \mathrm{Pic}(R[M])$.

Now suppose that $M \neq \mathrm{sn}(M)$. Then there exists $x \in \mathrm{sn}(M) \setminus M$ such that $x^2, x^3 \in M$. Therefore, $I = R[M]x^2 + R[M]x^3$ is an ideal in both rings $R[M]$ and $R[M']$, where we have set $M' = Mx^{\mathbb{Z}_+}$ (we are using multiplicative notation). In particular, we have the Milnor square

$$\begin{array}{ccc} R[M] & \longrightarrow & R[M'] \\ \downarrow & & \downarrow \\ R[M]/I & \longrightarrow & R[M']/I. \end{array} \qquad (8.17)$$

The U-Pic Mayer–Vietoris sequence, associated to (8.17) (see [19, Ch. IX, Cor. 5.12])[5], contains the exact sequence

$$\{1\} \to \mathrm{U}(R[M]) \to \mathrm{U}(R[M]/I) \oplus \mathrm{U}(R[M']) \xrightarrow{f} \mathrm{U}(R[M']/I) \to \mathrm{Pic}(R[M])$$
$$\to \mathrm{Pic}(R[M]/I) \oplus \mathrm{Pic}(R[M']) \xrightarrow{g} \mathrm{Pic}(R[M']/I). \qquad (8.18)$$

By Proposition 4.20 one has $\mathrm{U}(R[M']) = \mathrm{U}(R)$. Therefore, by Lemmas 8.43 and equation (8.16), we get

[5] This sequence is simply obtained by combination of the lowest horizontal rows of the corresponding diagrams (8.12) and (9.2).

$$\operatorname{Coker} f \cong \frac{\mathrm{U}(R) \oplus (1 + \operatorname{nil}(R[M']))}{\mathrm{U}(R) \oplus (1 + \operatorname{nil}(R[M]))} \cong \frac{\operatorname{nil}(R[M']/I)}{\operatorname{nil}(R[M]/I)} \cong R(M' \setminus M). \tag{8.19}$$

The isomorphism on the right is an isomorphism of R-modules, and seen as follows. Both algebras $R[M']/I$ and $R[M]/I$ are $\operatorname{gp}(M)$-graded, and therefore their nilradicals are graded ideals (Exercise 4.8). But a graded ideal in one of these rings is a free module over R on the basis formed by the (residue classes of) the monomials it contains. So the quotient is free on the difference of the bases, and that is formed by the set $M' \setminus M$. It cannot be larger, and is not smaller since all elements in $M' \setminus M$ are nilpotent mod I.

We have also to compute $\operatorname{Ker} g$. The last observation implies $(R[M]/I)_{\mathrm{red}} = (R[M']/I)_{\mathrm{red}}$. In view of this equation and Lemma 8.35, the map $\operatorname{Pic}(R[M]/I) \to \operatorname{Pic}(R[M']/I)$ is an isomorphism. Since g is the codiagonal of the two homomorphisms

$$\operatorname{Pic}(R[M]/I) \to \operatorname{Pic}(R[M']/I) \quad \text{and} \quad \operatorname{Pic}(R[M']) \to \operatorname{Pic}(R[M']/I),$$

we arrive at the isomorphism

$$\operatorname{Ker} g \cong \operatorname{Pic}(R[M']). \tag{8.20}$$

Therefore, (8.18), (8.19), and (8.20) yield a short exact sequence of the form

$$0 \to R(M' \setminus M) \to \operatorname{Pic}(R[M]) \to \operatorname{Pic}(R[M']) \to 0.$$

Since $\mathbb{Q} \subset R$ (used for the second time!), the additive group of the free R-module $R(M' \setminus M)$ is injective. In particular, the short exact sequence above splits, and we have $\operatorname{Pic}(R[M]) \cong \operatorname{Pic}(R[M']) \oplus R(M' \setminus M)$.

Note that $\operatorname{sn}(M)$ is reached by a finite number of extensions as above (Exercise 2.12). Thus, by accumulation along the chain of extensions, we obtain

$$\operatorname{Pic}(R[M]) = \operatorname{Pic}\big(R[\operatorname{sn}(M)]\big) \oplus \Big(\bigoplus_{\operatorname{sn}(M) \setminus M} R \Big).$$

But we have already shown that $\operatorname{Pic}(R[\operatorname{sn}(M)]) = \operatorname{Pic}(R)$. □

Remark 8.44. (a) One can drop the condition that M be finitely generated. In fact, if M is a seminormal monoid without nontrivial units then it is a filtered union of its affine positive seminormal submonoids (see the beginning of the proof of Lemma 8.17.) So one can use the fact that the functor $\operatorname{Pic} : \mathsf{Rings} \to \mathsf{AbGroups}$ commutes with filtered direct limits (Exercise 8.2).

(b) Using a different technique, Singh and Roberts [302] obtained a very general result, containing Theorem 8.42 as a special case. Our proof is "monoid ring friendly", so to speak.

(c) The theorem of Singh and Roberts holds for $\mathbb{Q}$-algebras, and not even Theorem 8.42 can be generalized to positive characteristics; see Dayton [104, 3.9] for a counterexample. However, Singh [321] has found a modification that is valid in all characteristics.

(d) Theorem 8.42 hints at the existence of a natural R-module structure on $\mathrm{Pic}(R[M])/\mathrm{Pic}(R)$ under more general conditions. It does indeed exist; see [104], [302]. The source of such R-module structures for K-theoretical functors is a natural action of the ring of big Witt vectors over R, containing the *ghost copy* of R. For details see Section 10.F where big Witt vectors will play an important role.

(e) The situation is much more complicated if M is not positive, since the group of units grows in the passage from R to $R[M]$. The Picard group of $R[X^{\pm 1}]$, where R is an arbitrary commutative ring, has been computed by Weibel [375]. For a seminormal monoid M and an integral domain of coefficients the computation of $\mathrm{Pic}(R[M])$ can be reduced to Weibel's result; see Anderson [6].

Monoid rings over Dedekind domains. The local-global principle (Theorem 8.11) and the arguments in the proof of Lemma 8.17 readily imply the following extension of Theorem 8.4: for a Dedekind domain R and a seminormal monoid M every projective $R[M]$-module is extended from R. Since every projective R-module is of the form *free* $\oplus$ *rank* 1, one has $SK_0(R) = 0$.

By Theorem 8.37(b) we know that $SK_0(R[M]) = 0$ when M is a monoid and R is a Dedekind domain. In other words, projective $R[M]$-modules stably have the form *free* $\oplus$ *rank* 1 for such R and M. Motivated by this observation, Swan proved [351, Th. 1.5] the following stronger nonstable result, which answers a question of Murthy positively:

Theorem 8.45. *For a Dedekind domain R and a monoid M every projective $R[M]$-module is of the form* free $\oplus$ rank 1.

Swan's proof is based on Theorem 8.4 and on a theorem about projective modules of general interest. Since the argument is beyond the scope of interaction of discrete geometry and K-theory, we confine ourselves to stating the theorem. (See p. 157 for the definition of subintegral extensions.)

Theorem 8.46 (Swan). *Let $A \subset B$ be a subintegral extension of rings. Let P be a projective A-module. Then:*

(a) If $B \otimes P \cong Q_1 \oplus \cdots \oplus Q_n$, then $P \cong P_1 \oplus \cdots \oplus P_n$ with $Q_i \cong B \otimes P_i$.
(b) If $B \otimes P$ has the form free $\oplus$ rank 1, *then the same holds true of P.*

Theorem 8.45 follows from Theorem 8.46. In fact, if M is a monoid with seminormalization M', then for any ring R the monoid ring extension $R[M] \subset R[M']$ is subintegral while, as already observed, projective $R[M']$-modules are extended from R, provided R is a Dedekind domain.

Monomial Quotients. The following theorem of Vorst [369] extends the Quillen–Suslin theorem to monomial quotients of polynomial algebras.

Theorem 8.47. *Let R be a ring such that the projective modules over the polynomial ring $R[X_1, \dots, X_n]$ are free. Let $I \subset R[X_1, \dots, X_n]$ be an ideal generated by monomials. Then the projective $R[X_1, \dots, X_n]/I$-modules are also free.*

The R-algebras mentioned in Theorem 8.47 have sometimes been called *discrete Hodge algebras* (De Concini, Eisenbud, and Procesi [105]). Now we show that this theorem can be generalized to arbitrary monoid rings:

Theorem 8.48. *Let R be a ring and M be an affine monoid such that the projective $R[M]$-modules are extended from R. Then the projective $R[M]/RI$-modules are extended from R for every proper ideal $I \subset M$.*

Remark 8.49. (a) Theorem 8.47 extends to arbitrary monoids of *finite rank* [160, §3.2]. But here we prefer to avoid the involved technical subtleties. In [351] Swan presents the following version which strengthens both the hypothesis and the conclusion: if the projective $R[M]$-modules are extended from R for *all* seminormal monoids M, then the same is true for the projective $R[M]/RI$-modules. These results are based on the idea used already in the proof of Lemma 8.18.

(b) Although Theorem 8.48 and its proof do not refer to seminormality, it follows from Corollary 4.73 that M has to be seminormal.

(c) For another generalization of Theorem 8.47 see Yuzvinsky [380].

We need the following:

Lemma 8.50. *Let R be a ring and*

$$\begin{array}{ccc} A & \longrightarrow & A_1 \\ \downarrow & & \downarrow f \\ A_2 & \longrightarrow & A' \end{array}$$

be a cartesian diagram of R-algebras. Suppose f is split surjective and $R \to A'$ is a split injective. Let P be a projective A-module such that $A_1 \otimes_A P$ and $A_2 \otimes_A P$ are extended from R. Then P is extended from R.

Proof. Let $A_i \otimes_A P \cong P_i \cong A_i \otimes_R Q_i$ for projective R-modules Q_i, $i = 1, 2$. Since the composite $R \to A_i \to A' \to R$ is $\mathbf{1}_R$ and $A' \otimes P_1 \cong A' \otimes P_2$ we have $A' \otimes_R Q_1 \cong A' \otimes_R Q_2$. Therefore we can assume $Q_1 = Q_2 = Q$. Let $Q' = A' \otimes_R Q$. Then P is obtained by Milnor patching of P_1 and P_2 along some automorphism α of Q' , while $A \otimes_R Q$ is obtained by patching P_1 and P_2 along $\mathbf{1}_{Q'}$. Since the splitting map $A' \to A_1$ is an R-algebra homomorphism, $P_1 \cong A_1 \otimes_{A'} Q'$ and we can lift α to an automorphism $1 \otimes \alpha$ of $A_1 \otimes_{A'} Q'$. This shows that the two patching diagrams are isomorphic and so $P \cong A \otimes_R Q$. □

Proof of Theorem 8.48. By the Nakayama lemma, projective modules are isomorphic if they are modulo an ideal contained in the Jacobson radical. Since the kernel of the natural homomorphism $R[N]/RI \to R[N]/R\sqrt{I}$ is nilpotent, there is no loss of generality in assuming that I is a radical ideal. Then $I = \mathfrak{p}_1 \cap \dots \cap \mathfrak{p}_n$ is the intersection of prime ideals $\mathfrak{p}_i$ of M; see Proposition 2.36. We will use induction on n.

In the case $n = 1$ we have $I = M \setminus F$ where F is a face of $\mathbb{R}_+ M$ (Proposition 2.36). Therefore the embedding $N = M \cap F \to M$ induces an isomorphism $R[N] \cong$

$R[M]/RI$. On the other hand, $R[N]$ is an R-retract of $R[M]$ (Corollary 4.34). So if P is a projective $R[N]$-module then we have

$$P \cong R[N] \otimes_{R[M]} R[M] \otimes_{R[N]} P \cong R[N] \otimes R[M] \otimes_R P_0 \cong R[N] \otimes_R P_0$$

where $P_0 = R \otimes_{R[M]} P$ with respect to the augmentation $R[M] \to R$ determined by $M \setminus \mathrm{U}(M) \mapsto 0$, $U(M) \mapsto 1$.

Now assume $n > 1$. Consider the ideals $\mathfrak{p} = \mathfrak{p}_1$ and $J = \mathfrak{p}_2 \cap \cdots \cap \mathfrak{p}_n$. Then $I = \mathfrak{p} \cap J$, and $L = \mathfrak{p} \cup J$ is a proper ideal of M since $\mathfrak{p}, J \subset M \setminus \mathrm{U}(M)$. The diagram

$$\begin{array}{ccc} R[M]/RI & \longrightarrow & R[M]/RJ \\ \big\downarrow & & \big\downarrow f \\ R[M]/R\mathfrak{p} & \longrightarrow & R[M]/RL \end{array}$$

is cartesian. Moreover, the homomorphism f is a split surjective R-algebra homomorphism. In fact, let $K = M \setminus \mathfrak{p}$ and $S = K \cap L = K \cap J$. Since $M = K \cup L$, $R[K]/RS \cong R[M]/RL$. Now the inclusion $K \to M$ takes S into J and induces a map $R[K]/RS \to R[M]/RJ$ splitting f. So Lemma 8.50 applies. □

Exercises

8.1. Prove Lemma 8.39.

8.2. Let I be a filtered partially ordered set and let $\{f_{ij} : R_i \to R_j \ : \ i, j \in I,\ i < j\}$ be a family of ring homomorphisms such that $f_{jk} \circ f_{ij} = f_{ik}$ whenever $i < j < k$. Then $\operatorname{Pic}(\varinjlim_I R_i) = \varinjlim_I \operatorname{Pic}(R_i)$.

8.3. Show that $\mathrm{SL}_n(R) = \mathrm{E}_n(R)$ for any semilocal ring R and any $n \geq 2$.

8.4. The *dual module* of a module N over a ring A is defined by $N^* = \operatorname{Hom}_A(N, A)$.
(a) Let R be a PID containing $\mathbb{Q}$ and M a positive affine monoid. Show that M is seminormal if (and only if) $P \cong P^*$ for all projective $R[M]$-modules P.
(b) Show that for any ring R and any module $P \in \mathbb{P}(R)$ we have $P \cong P^{**}$.

8.5. Let $R = R_0 \oplus R_1 \oplus \cdots$ be a graded (commutative) ring. Prove that every (finitely generated) graded projective R-module is extended from R_0.

8.6. Prove for a ring R:
(a) The homomorphism rank : $K_0(R) \to H_0(R)$ has a right inverse $[R^-] : H_0(R) \to K_0(R)$, defined as follows. Every element $\sigma \in H_0(R)$ is a difference $\sigma_1 - \sigma_2$ for some $\sigma_1, \sigma_2 \in H_0(R)$ with *nonnegative* values. Put

$$[R^{\sigma}] = [R^{\sigma_1}] - [R^{\sigma_2}], \quad f \in H_0(R),$$

where the module $R^{\sigma_1} \in \mathbb{P}(R)$ is determined by the condition that if σ_1 has constant value r on $\operatorname{Spec}(eR) \subset \operatorname{Spec}(R)$, $e \in R$ an idempotent, then $e\,(R^{\sigma_1}) = (eR)^r$, and similarly for R^{σ_2}.
(b) Every element of $\tilde{K}_0(R)$ is of the form $[P]-[Q]-[R^{\operatorname{rank}_P}]+[R^{\operatorname{rank}_Q}]$ for some $P, Q \in K_0(R)$. Moreover, $[P] - [Q] - [R^{\operatorname{rank}_P}] + [R^{\operatorname{rank}_Q}] = 0 \in K_0(R)$ if and only if there exists a module $L \in \mathbb{P}(R)$ for which $P \oplus L$ and $Q \oplus L$ are both free R-modules.

(c) If modules $P, Q, L \in \mathbb{P}(R)$ are such that $P \oplus L$ and $Q \oplus L$ are both of the form free $\oplus$ rank 1, then

$$[P] - [Q] - [R^{\operatorname{rank}_P}] + [R^{\operatorname{rank}_Q}] \in SK_0(R) \Longrightarrow$$
$$[P] - [Q] - [R^{\operatorname{rank}_P}] + [R^{\operatorname{rank}_Q}] = 0.$$

8.7. This problem establishes properties of the pair (SK_0, K_0), which will be used in the next exercise.

Let $\mathsf{F}, \mathsf{G} : \mathsf{Rings} \to \mathsf{AbGroups}$ be covariant functors. We call the pair (F, G) *admissible* if the following conditions are satisfied:

(i) F commutes with the direct limits of filtered diagrams;

(ii) for a Milnor square of rings

$$\begin{array}{ccc} A & \longrightarrow & B \\ \downarrow & & \downarrow \\ C & \longrightarrow & D \end{array}$$

the associated sequence $\mathsf{F}(A) \to \mathsf{F}(B) \oplus \mathsf{F}(C) \to \mathsf{F}(D)$ is exact;

(iii) for a subintegral ring extension $A \subset B$ the homomorphism $\mathsf{F}(A) \to \mathsf{F}(B)$ is injective;

(iv) there exists a natural transformation of functors $\mathsf{F} \to \mathsf{G}$ such that for any ring R the homomorphism $\mathsf{F}(R) \to \mathsf{G}(R)$ is injective;

(v) for a ring extension $A \subset B$, making B a free A module of rank $n < \infty$, there is a homomorphism $T_{BA} : \mathsf{G}(B) \to \mathsf{G}(A)$ such that the composite map $\mathsf{G}(A) \to \mathsf{G}(B) \xrightarrow{T_{BA}} \mathsf{G}(A)$ is multiplication by n.

(Functors F satisfying the conditions (i) and (ii) are correspondingly called *continuous* and *semiexact*.)

Show that the pair (SK_0, K_0) is an admissible pair of functors.

Hint: use Exercise 8.6 and Theorem 8.46.

8.8. Let (F, G) be an admissible pair of functors in the sense of Exercise 8.7. Prove the following implication for any natural number n and all affine simplicial monoids $M \subset \mathbb{Z}^n$:

$$\mathsf{F}(R) = \mathsf{F}(R[X_1, \dots, X_n]) \quad \Longrightarrow \quad \mathsf{F}(R) = \mathsf{F}(R[M]).$$

(This gives an almost categorial proof of Theorem 8.37(b).)

Outline of the proof:

(1) Without loss of generality we can assume that M is seminormal by condition (iii) in Exercise 8.7.

(2) Using Corollary 2.74 and Proposition 2.40, show that there exists a free basis $B = \{x_1, \dots, x_n\}$, $n = \operatorname{rank} M$, of $\operatorname{gp}(M)$ such that $B \subset \operatorname{int}(M)$.

(3) Let $c \geq 2$ be a natural number and consider the affine positive submonoid

$$M' = M + \sum_{i=1}^{n} \mathbb{Z}_+ \frac{x_i}{c} \subset \mathbb{Q} \otimes \operatorname{gp}(M).$$

Show that $R[M']$ is a free $R[M]$ module of rank c^n, with a basis (using multiplicative notation):

$$\{x_1^{a_1/c} \dots x_n^{a_n/c} \mid 0 \leq a_1, \dots, a_n \leq c - 1\}.$$

(4) Using conditions (iv) and (v), show that the group $\operatorname{Ker}\big(\mathsf{F}(R[M]) \to \mathsf{F}(R[M'])\big)$ is of c^n-torsion.

(5) Using again condition (iii) in Exercise 8.7, show $\operatorname{Ker}\big(\mathsf{F}(R[M]) \to \mathsf{F}(R[\operatorname{sn}(M')])\big)$ is also of c^n-torsion.

(6) Let $\mathbb{Z}[1/c]M_*$ be the *c-divisible hull* of M_*, i.e., $\mathbb{Z}[1/c]M_* = \{m/c^a \mid m \in M_*,\ a \in \mathbb{Z}_+\}$ (in additive notation). Show that $\operatorname{sn}(M') = M + \mathbb{Z}[1/c]M_*$. In particular we have the following Milnor square of R-algebras:

$$\begin{array}{ccc} R[\mathbb{Z}[1/c]M_*] & \longrightarrow & R[\operatorname{sn}(M')] \\ \downarrow & & \downarrow \\ R & \longrightarrow & R[\operatorname{sn}(M')]/R\big(\mathbb{Z}[1/c]\operatorname{int}(M)\big) \end{array}$$

(7) Show that $\mathbb{Z}[1/c]M_*$ is a filtered union of rank n free monoids.

(8) Using the continuity and semiexactness of F (conditions (i) and (ii) in Exercise 8.7), show the implication

$$\mathsf{F}(R) = R[\operatorname{sn}(M')]/R\big(\mathbb{Z}[1/c]\operatorname{int}(M)\big) \quad\Longrightarrow\quad \mathsf{F}(R[M])/\mathsf{F}(R) \text{ is of } c^n\text{-torsion.}$$

(9) Apply an inductive argument, similar to the one in the proof of Lemma 8.18, involving the monoids $\mathbb{Z}[1/c](M \cap F)_*$ where F runs through the faces of $\mathbb{R}_+M$, and arrive at the conclusion that $\mathsf{F}(R[M])/\mathsf{F}(R)$ is of c^n-torsion.

(10) Complete the proof by consideration of another natural number $c' \geq 2$ coprime to c.

8.9. Assume the following is true: all projective R_f-modules are free for any regular local ring $(R, \mathfrak{m})$ and all elements $f \in \mathfrak{m} \smallsetminus \mathfrak{m}^2$. Prove: then every projective $R[\mathbb{Z}_+^n]$-module is extended from R for all regular rings R and natural numbers n.

Hint: reduce the general case to $R[\mathbb{Z}_+] = R[X]$ where $(R, \mathfrak{m})$ is local. Use the equality $R(X) = B_T$ where $T = X^{-1}$ and B denotes the regular local ring $R[T]_{(\mathfrak{m},T)}$.

The background of this exercise will be outlined in the notes below.

8.10. Show that for a ring R and a monic polynomial $f \in R[X]$ the group homomorphism $K_0(R[X]) \to K_0(R[X]_f)$ is injective. (This is the K_0 part of the Murthy–Pedrini theorem [270, Th. 1.3].)

8.11. Let $M \subset \mathbb{Q}^n$ be an integrally closed positive submonoid such that the convex subset $\mathbb{R}_+M \subset \mathbb{R}^n$ is a cone. Let $\mathsf{F} : \mathsf{Rings} \to \mathsf{AbGroups}$ be a continuous semiexact functor. For every ring R such that $\mathsf{F}(R) = \mathsf{F}(R[X_1, \dots, X_n])$, $n = \operatorname{rank} M$, show the equivalence

$$\mathsf{F}(R[M])/\mathsf{F}(R) \text{ is a finitely generated group} \iff \mathsf{F}(R) = \mathsf{F}(R[M]).$$

Outline of the proof:

(1) Fix a rational hyperplane $H \subset \mathbb{R}^n$ such that $\mathbb{R}_+M$ is spanned over 0 by the intersection $\Phi(M) = \mathbb{R}_+M \cap H$. Using induction on dimension and Milnor squares similar to those in the proof of Lemma 8.18, show that the inclusion map $R[\mathbb{R}_+ \operatorname{int}(\Phi(M)) \cap \mathbb{Q}^n] \to R[M]$ induces a surjective homomorphism $\mathsf{F}\big(R[\mathbb{R}_+ \operatorname{int}(\Phi(M)) \cap \mathbb{Q}^n]\big) \to \mathsf{F}(R[M])$. Informally, this can be phrased as "all elements of $\mathsf{F}(R[M])$ are extended from the interior of the rational polytope $\Phi(M)$."

(2) Fix a rational point $z \in \operatorname{int}\Phi(M)$. For a rational number $0 < \lambda < 1$ let $\Phi(M)_\lambda$ denote the homothetic image of $\Phi(M)$ with factor λ and center z. Show that if λ is sufficiently close to 1, then the inclusion map $R[\mathbb{R}_+\Phi(M)_\lambda \cap \mathbb{Q}^n] \to R[M]$ induces a surjective homomorphism

$\mathsf{F}\big(R[\mathbb{R}_+\Phi(M)_\lambda \cap \mathbb{Q}^n]\big) \to \mathsf{F}(R[M])$. On the other hand the monoids M and $\mathbb{R}_+\Phi(M)_\lambda \cap \mathbb{Q}^n$ are isomorphic.

(3) Iterate the process and use the idea of a "sandwiched simplex" from the proof of Theorem 8.4.

8.12. Let $M \subset \mathbb{Q}^3$ be an integrally closed positive submonoid such that $\mathbb{R}_+M$ is a cone with four extreme rays. Let $H \subset \mathbb{R}^3$ be a rational hyperplane such that $\mathbb{R}_+M$ is spanned over 0 by the intersection $\Phi(M) = \mathbb{R}_+M \cap H$. Denote the midpoints of the edges of the quadrangle $\Phi(M) = \mathbb{R}_+M \cap H$ by A, B, C, D and set $P = \operatorname{conv}(A, B, C, D)$.

Show the following implication for all functors $\mathsf{F} : \mathsf{Rings} \to \mathsf{AbGroups}$ and any ring R:

$$\mathsf{F}(R) = \mathsf{F}(R[X_1, X_2, X_3, X_4]) \Longrightarrow$$
$$\operatorname{Im}\big(\mathsf{F}(R[\mathbb{R}_+P \cap \mathbb{Q}^3]) \to \mathsf{F}(R[M])\big) = \mathsf{F}(R).$$

Hint: the embedding of quadrangles $P \to \Phi(M)$ factors through a rational 3-simplex.

8.13. Let R be a ring and $\mathbb{M}$ a monoidal complex with underlying conical complex Γ, as introduced in Definition 7.9. Show that the results of this chapter extend to $R[\mathbb{M}]$ as follows:

(a) If R is a PID and the monoids M_c are seminormal for all $c \in \Gamma$, then all projective $R[\mathbb{M}]$-modules are free.

(b) If R is regular, then $SK_0(R) = SK_0(R[\mathbb{M}])$.

(c) The seminormalizations $\operatorname{sn}(M_c)$, $c \in \Gamma$, form a monoidal complex, which we denote by $\operatorname{sn}(\mathbb{M})$. On the other hand, the normalizations $\bar{M}_c$, $c \in \Gamma$, may fail to form a monoidal complex.

(d) If R is seminormal and all monoids M_c, $c \in \Gamma$, are seminormal, then $\operatorname{Pic}(R) = \operatorname{Pic}(R[\mathbb{M}])$.

(e) If R is seminormal, then

$$\operatorname{Pic}(R[\mathbb{M}])/\operatorname{Pic}(R) \cong \operatorname{Pic}(R) \oplus \Big(\bigoplus_{|\operatorname{sn}(\mathbb{M})| \setminus |\mathbb{M}|} R\Big).$$

Hint: in (a), (b), and (d) use the corresponding statement for monoid rings in combination with Milnor squares of the form

$$\begin{array}{ccc} R[(M_c)_*] & \longrightarrow & R[\mathbb{M}] \\ \downarrow & & \downarrow \\ R & \longrightarrow & R[\mathbb{M}]/R\operatorname{int}(M_c) \end{array}$$

where $c \in \Gamma$ is a maximal element with respect to inclusion. Successive applications of these squares reduce the claim in question to the restriction of $\mathbb{M}$ to the codimension 1 skeleton of Γ. Then induction on $\max(\dim C_c : c \in \Gamma)$ applies.

Notes

The question whether all projective modules over $\Bbbk[x_1, \dots, x_n]$ are free was raised by Serre in his seminal work [316], which initiated the systematic use of sheaves in algebraic geometry. (Sheaves will make their appearance in Chapter 10.) The question quickly acquired the status of "Serre's problem", also known as "Serre's conjecture."[6] The twenty years of unceasing efforts towards a final solution culminated in the works

[6] Serre himself often objected to "conjecture."

of Quillen and Suslin, both published in 1976 and putting the problem to rest. These efforts played an important role in the development of the classical part of K-theory that focuses on projective modules and Grothendieck groups for general rings.

Serre's problem has had a lasting impact on a wider scope of research areas. For instance, the observation in [316] of a relationship between complete intersection and projective modules sparked an extensive investigation. The theory is summarized in the books of Ischebeck and Rao [208], Mandal [248], and Murthy's survey [269].

For affine normal monomial subrings of $\mathbb{k}[X_1, \ldots, X_n]$ Theorem 8.4 was conjectured by Anderson [5] shortly after the Quillen–Suslin theorem had been proved. Anderson established the case $n = 2$, extending Seshadri's proof of the 2-dimensional case of Serre's problem [319]. One cannot do better than to consult Lam's book [237] for the vast panorama of chronologically ordered results that preceded and followed the final solution by Quillen and Suslin.

The special case of of Theorem 8.4 for simplicial normal monoids was proved in 1982 independently by Chouinard [85] and in [156]. Both works relied on Quillen's technique. The approach in [156] uses Theorem 8.11 and Theorem 8.22 (for simplicial normal monoids M and "degenerate" pyramidal extensions). Theorem 8.4 was finally proved in [157].

A very natural extension of Serre's problem is the Bass–Quillen conjecture that projective $R[X]$-modules are extended from R if the latter is a regular ring. Horrocks [199] and Murthy [268] had long shown the validity of this claim for 2-dimensional local regular rings. Then it follows from Quillen's local-global principle and the affine Horrock's theorem that all projective $R[X_1, \ldots, X_n]$-modules are extended from R for any regular ring R of dimension ≤ 2. This result is contained in both papers [297] and [341].

An outstanding "post Quillen–Suslin" result is Lindel's proof [241] of the geometric case of the Bass–Quillen conjecture [20], [297], i. e., when R is a localization of a finitely generated regular algebra over a field. By Popescu's approximation theorem [292] for regular rings containing a field, Lindel's result immediately extends to this class of coefficient rings. (See [353] for a systematic exposition of Popescu's result.)

Using Lindel's technique, Bhatwadekar and Rao [30] answered in the positive the geometric case of Quillen's question [297] whether projective R_f-modules are free when $(R, \mathfrak{m})$ is a regular local ring and $f \in \mathfrak{m} \setminus \mathfrak{m}^2$. As explained in Exercise 8.9, a general positive answer to the question would imply the Bass–Quillen conjecture in full generality. Although [30] does not imply new cases of the conjecture, not covered by Lindel's work, it has lead to further results on projective modules over seminormal monoid rings; see Remark 8.41.

With the advent of effective methods in commutative algebra and algebraic geometry, many theorems underwent an algorithmic analysis. Fitchas–Galligo [125] and Lopez–Sturmfels [244] gave algorithms that find free bases of projective modules over a polynomial ring. The module can be given, for instance, in the form of an idempotent matrix with polynomial entries. Laubenbacher and Woodburn [239] applied a similar algorithmic analysis to Theorem 8.4.

9

Bass–Whitehead groups of monoid rings

The last chapter has shown that projective modules and the Grothendieck groups K_0 are trivial for seminormal monoid rings. The K-theoretic study of monoid rings in full swing, we now turn to the analogous problem for K_1. Surprisingly, the invariance of K_1 under the passage from R to $R[M]$ breaks down completely already for simplicial monoids that are not free.

The chapter concludes with a discussion of positive results on the action of elementary matrices on unimodular rows and the so-called nilpotence of higher K-groups.

9.A The functors K_1 and K_2

The discussion of K_1 requires the use of K_2. Therefore we introduce both these functors. Although they are defined for rings in general as well as for not necessarily affine schemes, we consider only the case of commutative rings.

The functor K_1. For a ring R the *stable general linear group* $\mathrm{GL}(R)$ is the direct limit of the diagram

$$\mathrm{GL}_1(R) \to \cdots \to \mathrm{GL}_n(R) \to \mathrm{GL}_{n+1}(R) \to \cdots, \qquad * \mapsto \begin{pmatrix} * & 0 \\ 0 & 1 \end{pmatrix}.$$

The elements of $\mathrm{GL}(R)$ can be thought of as invertible infinite square matrices with only finitely many entries of the main diagonal $\neq 1$ and only finitely many nondiagonal entries $\neq 0$. The *stable special linear group* $\mathrm{SL}(R)$ is defined similarly, using only matrices of determinant 1. The determinant maps $\det : \mathrm{GL}_n(R) \to \mathrm{U}(R)$ are compatible with the diagram above, and we get a map $\mathrm{GL}(R) \to \mathrm{U}(R)$, denoted again by det. It is a surjective homomorphism, split by the embedding $\mathrm{U}(R) = \mathrm{GL}_1(R) \to \mathrm{GL}(R)$.

The embeddings $\mathrm{GL}_n(R) \to \mathrm{GL}_{n+1}(R)$ respect the subgroups of elementary matrices, and by passing to the limit we get the *stable subgroup of elementary matrices*

W. Bruns, J. Gubeladze, *Polytopes, Rings, and K-Thoery*, Springer Monographs in Mathematics,
DOI 10.1007/b105283_9, © Springer Science+Business Media, LLC 2009

$E(R) \subset SL(R)$. By definition, $E_n(R)$ is generated by the standard elementary matrices e_{ij}^a where $a \in R$ is the entry on the position (i, j), $i \neq j$, the main diagonal entries are equal to 1, and all other entries are zero. The easily checked relation $\left[e_{ij}^a, e_{jl}^b\right] = e_{il}^{ab}$ for $i \neq l$ and the *Whitehead lemma* [19, Ch. V, Prop. 1.7] imply

$$E(R) = [E(R), E(R)] = [GL(R), GL(R)] \tag{9.1}$$

where $[g, h]$ denotes the commutator $ghg^{-1}h^{-1}$ of g and h and $[G, G]$ is the commutator subgroup of a group G. In particular, $E(R)$ is a normal subgroup of $GL(R)$ and $GL(R)/E(R)$ is the abelianization of $GL(R)$. Since the determinant of an elementary matrix is 1, we have the induced homomorphism $\det : GL(R)/E(R) \to U(R)$.

The *Bass functor* $K_1 : \mathsf{Rings} \to \mathsf{AbGroups}$ is defined by $R \mapsto GL(R)/E(R)$ (and in the obvious way for ring homomorphisms). The group $K_1(R)$ is also called the *Bass–Whitehead group* of R.

The determinant homomorphism $\det : GL(R) \to U(R)$ gives rise to the functorial splitting

$$K_1(R) = U(R) \oplus SK_1(R)$$

where $SK_1(R) = SL(R)/E(R)$. This defines another functor $SK_1 : \mathsf{Rings} \to \mathsf{AbGroups}$.

Consider a Milnor square of rings

$$\begin{array}{ccc} A & \longrightarrow & B \\ \downarrow & & \downarrow{\scriptstyle f} \\ C & \longrightarrow & D. \end{array}$$

The stable groups and homomorphisms introduced above assemble in a natural way into the following commutative diagram with exact rows and columns, called the (K_1-) *Mayer–Vietoris sequence* [19, Ch. IX, Cor. 5.12]:

$$\begin{array}{ccccccc} & & 0 & & 0 & & 0 \\ & & \downarrow & & \downarrow & & \downarrow \\ & & SK_1(A) & \longrightarrow & SK_1(B) \oplus SK_1(C) & \longrightarrow & SK_1(D) \\ & & \downarrow & & \downarrow & & \downarrow \\ & & K_1(A) & \longrightarrow & K_1(B) \oplus K_1(C) & \longrightarrow & K_1(D) \\ & & \downarrow{\scriptstyle \det} & & \downarrow{\scriptstyle \det} & & \downarrow{\scriptstyle \det} \\ 0 & \longrightarrow & U(A) & \longrightarrow & U(B) \oplus U(C) & \longrightarrow & U(D) \\ & & \downarrow & & \downarrow & & \downarrow \\ & & 0 & & 0 & & 0. \end{array} \tag{9.2}$$

Moreover, there is a *connecting homomorphism* $\partial : K_1(D) \to \tilde{K}_0(A)$ that connects the diagrams (8.12) and (9.2), yielding the *K_1-K_0-Mayer–Vietoris sequence* (of size 3×6) whose middle row is

$$K_1(A) \to K_1(B) \oplus K_1(C) \to K_1(D) \to \tilde{K}_0(A) \to \tilde{K}_0(B) \oplus \tilde{K}_0(C) \to \tilde{K}_0(D).$$

One defines ∂ as follows: for an element $z \in K_1(D)$ pick a representative $\alpha \in \mathrm{GL}_n(D)$, $n \in \mathbb{N}$, and apply Milnor patching to the following diagram of free modules

$$\begin{array}{ccc} & & B^n \\ & & \downarrow \\ C^n & \longrightarrow & D^n \cong_\alpha D^n \end{array}$$

to obtain a projective A-module P; then set $\partial(z) = [P] - [A^n] \in \tilde{K}_0(A)$. That ∂ is well-defined and possesses the properties stated above is easy to verify. For the details see [19, Ch. IX, §5].

For a Milnor square as above we also have the exact sequence

$$K_1(A) \to K_1(B) \oplus K_1(C) \to K_1(D) \to K_0(A) \to K_0(B) \oplus K_0(C) \to K_0(D)$$

whose connecting map $K_1(D) \to K_0(A)$ is the composite of ∂ and the inclusion map $\tilde{K}_0(A) \to K_0(A)$.

Here is an SK_1-analogue of Lemma 8.35.

Lemma 9.1. *For any ideal $I \subset R$, contained in the Jacobson radical* $\mathrm{J}(R)$, *we have*

$$SK_1(R) = SK_1(R/I).$$

Proof. A matrix $\alpha \in \mathrm{SL}(R)$ belongs to $\mathrm{E}(R)$ whenever $\alpha \equiv \mathbf{1} \mod I$ as follows from the standard Gauss algorithm. This shows that the preimage of $\mathrm{E}(R/I)$ in $\mathrm{SL}(R)$ is $\mathrm{E}(R)$. In other words, the homomorphism $SK_1(R) \to SK_1(R/I)$ is a monomorphism. But it is also an epimorphism because, first, any element $\beta \in \mathrm{SL}(R/I)$ lifts to some element $\gamma \in \mathrm{GL}(R)$ —a consequence of the fact that the preimage of $\mathrm{U}(R/I)$ in R coincides with $\mathrm{U}(R)$, and, second, dividing the first row of γ by $\det(\gamma)$ we get a preimage of β in $\mathrm{SL}(R)$. □

The functor K_2. For a ring R and $a, b \in R$ one has the following *Steinberg relations*:

$$e_{ij}^a e_{ij}^b = e_{ij}^{a+b}, \quad i \neq j,$$

$$[e_{ij}^a, e_{kl}^b] = \begin{cases} e_{il}^{ab}, & j = k,\ i \neq l, \\ 1, & j \neq k \text{ and } i \neq l. \end{cases}$$

The *Steinberg group* $\mathrm{St}(R)$ is defined as the group generated by symbols x_{ij}^a, $i, j \in \mathbb{N}$, $i \neq j$, $a \in R$, subject to the relations

$$x_{ij}^a x_{ij}^b = x_{ij}^{a+b}, \quad i \neq j,$$

$$[x_{ij}^a, x_{kl}^b] = \begin{cases} x_{il}^{ab}, & j = k, i \neq l, \\ 1, & j \neq k \text{ and } i \neq l. \end{cases}$$

A surjective group homomorphism $f : H \to G$ is called a *universal central extension of* G if it satisfies the following conditions:

(i) $\mathrm{Ker}(f)$ is contained in the center of H;
(ii) for every surjective homomorphism $f' : H' \to G$ with $\mathrm{Ker}(f')$ contained in the center of H', there exists a unique homomorphism $h : H \to H'$ such that $f'h = f$.

Theorem 9.2 (Milnor [262, §5]**).** *Let R be a ring. The surjective group homomorphism $\psi : \mathrm{St}(R) \to \mathrm{E}(R)$, $\psi\left(x_{ij}^a\right) = e_{ij}^a$, is a universal central extension of $\mathrm{E}(R)$. Moreover, $\mathrm{Ker}(\psi)$ coincides with the center of $\mathrm{St}(R)$.*

The *Milnor group* $K_2(R)$ is defined to be $\mathrm{Ker}(\psi)$, with notation as in Theorem 9.2. Clearly, we get a functor $K_2 : \mathsf{Rings} \to \mathsf{AbGroups}$.

Theorem 9.3 (Milnor [262, §6]**).** *Let*

$$\begin{array}{ccc} A & \longrightarrow & B \\ \downarrow & & \downarrow \\ C & \longrightarrow & D \end{array}$$

be a cartesian square of rings in which all homomorphisms are surjective. Then there is a natural connecting homomorphism $K_2(D) \to K_1(A)$ that gives rise to an exact sequence

$$\begin{aligned} K_2(A) \to K_2(B) \oplus K_2(C) \to K_2(D) \to \\ \to K_1(A) \to K_1(B) \oplus K_1(C) \to K_1(D) \to \\ \to K_0(A) \to K_0(B) \oplus K_0(C) \to K_0(D). \end{aligned}$$

Relative groups. Let I be an ideal in a ring A. Then we have the cartesian square of rings

$$\begin{array}{ccc} A \times_{A/I} A & \xrightarrow{\pi_1} & A \\ {\scriptstyle \pi_2}\downarrow & & \downarrow \\ A & \longrightarrow & A/I. \end{array}$$

Following Milnor [262], we define the relative K-groups by

$$K_i(A, I) = \mathrm{Ker}\left((\pi_2)_* : K_i(A \times_{A/I} A) \to K_i(A)\right), \quad i = 0, 1, 2.$$

The 9-term Mayer–Vietoris exact sequence from Theorem 9.3 then yields the following exact sequence

$$\begin{aligned} K_2(A, I) \xrightarrow{(\pi_1)_*} K_2(A) \longrightarrow K_2(A/I) \longrightarrow \\ \longrightarrow K_1(A, I) \xrightarrow{(\pi_1)_*} K_1(A) \longrightarrow K_1(A/I) \longrightarrow \\ \longrightarrow K_0(A, I) \xrightarrow{(\pi_1)_*} K_0(A) \longrightarrow K_0(A/I). \end{aligned} \tag{9.3}$$

The relative group $K_1(A, I)$ can alternatively be defined by

$$K_1(A, I) = \mathrm{GL}(A, I)/\,\mathrm{E}(A, I) \tag{9.4}$$

where $\mathrm{GL}(A, I) \subset \mathrm{GL}(A)$ is the subgroup of the matrices congruent to the identity matrix modulo I, and $\mathrm{E}(A, I) \subset \mathrm{E}(A)$ is the minimal normal subgroup containing the matrices of the form e_{ij}^a, $i \neq j$, $a \in I$. In this way the relative K_1-group is defined in [19, Ch. V, §2].

We also have the relative groups

$$\mathrm{SL}(A, I) = \mathrm{SL}(A) \cap \mathrm{GL}(A, I), \qquad SK_1(A, I) = \mathrm{SL}(A, I)/\,\mathrm{E}(A, I).$$

Remark 9.4. (a) It is known that $K_0(A, I)$ does not depend on the ambient ring, i. e., K_0 satisfies the *excision property*: for a commutative diagram of the form

$$\begin{array}{ccc} I & \hookrightarrow & A \\ \downarrow & & \downarrow{\scriptstyle f} \\ J & \hookrightarrow & B \end{array}$$

where $f : A \to B$ is a ring homomorphism mapping the ideal $I \subset A$ bijectively onto the ideal $J \subset B$, the resulting homomorphism $K_0(A, I) \to K_0(B, J)$ is an isomorphism [19, Ch. IX, Ex. 6.5].

(b) Swan [346] has shown that, firstly, already K_1 fails to satisfy excision and, secondly, there is no functor $K_3 : \mathsf{Rings} \to \mathsf{AbGroups}$ that would extend the 9-term exact sequence in Theorem 9.3 to the corresponding 12-term exact sequence. The nontrivial elements in the relative groups constructed by Swan will play an important role later in this chapter.

(c) The higher relative groups $K_i(A, I)$ are defined as the homotopy groups of the homotopy fiber of the map $\mathrm{BGL}(A)^+ \to \mathrm{BGL}(A/I)^+$ between certain K-theoretical spaces, whose homotopy groups, introduced by Quillen [295], are the higher K-groups of A and A/I.

Since higher K-groups do not satisfy excision, one could ask for which nonunital rings I the relative groups $K_i(A, I)$ are independent of the ring A containing I as an ideal.This so-called *excision problem* was solved by Suslin and Wodzicki [343].

The fundamental theorem and a converse result. The K_1-analogue of Theorem 8.36 was established by Bass, Heller, and Swan [23]: $K_1(R) = K_1(R[X])$ for all regular rings R. This was extended to *all* higher K-groups at once by Quillen in his fundamental work [296]. This is the so-called *K-homotopy invariance* of regular rings. Below we state the general result, known as the *fundamental theorem of K-theory*, containing the homotopy invariance of regular rings.

Let R be a ring (not necessarily regular) and $\mathsf{F} : \mathsf{Rings} \to \mathsf{AbGroups}$ be a functor. Then we have the group homomorphisms

$$\begin{array}{ccc} \mathsf{F}(R) & \longrightarrow & \mathsf{F}(R[X]) \\ \downarrow & & \downarrow \\ \mathsf{F}(R[X^{-1}]) & \longrightarrow & \mathsf{F}(R[X,X^{-1}]) \end{array}$$

induced by the natural ring inclusions. The group $\mathsf{F}(R)$ functorially splits off from the other three groups in this diagram, and we can consider the decomposition $\mathsf{F}(R[X]) = \mathsf{F}(R) \oplus N\mathsf{F}(R)$.

Theorem 9.5 (Fundamental theorem of K-theory). *Let R be a ring and $i \in \mathbb{Z}$.*

(a) The maps $K_i(R[X]) \to K_i(R[X,X^{-1}])$ and $K_i(R[X^{-1}]) \to K_i(R[X,X^{-1}])$ are split monomorphisms. Moreover, we have the functorial splitting

$$K_i(R[X,X^{-1}]) = K_i(R) \oplus NK_i^+(R) \oplus NK_i^-(R) \oplus K_{i-1}(R)$$

where $K_i(R)\oplus NK_i^+(R)$ is the isomorphic image of $K_i(R[X])$ and $K_i(R)\oplus NK_i^-(R)$ is that of $K_i(R[X^{-1}])$.

(b) If R is regular, then $NK_i(R) = 0$, and so $K_i(R) = K_i(R[X])$.

Remark 9.6. (a) Theorem 9.5 is stated for all integer indices i, while the group $K_i(R)$ was defined above only for $i = 0,1,2$. In this book we will not work out higher K-groups in detail, although they will be mentioned occasionally. The groups $K_i(R)$ are higher homotopy groups of certain huge CW-complexes $\mathrm{BQ}\mathbb{P}(R)$, built up from the category of projective R-modules, see [296]. That the alternative definition of higher K-groups, mentioned in Remark 9.4(c), coincides with this one is also due to Quillen ; see Grayson [149]. The proof of Theorem 9.5(a) is spread over [296] and [149]. See also Remark 10.4(c).

(b) Theorem 9.5 holds for schemes that are not necessarily affine.

(c) Theorem 9.5(a) is actually the definition of the groups K_i for $i < 0$. Namely, one applies the following formula iteratively, starting with $i = 0$:

$$K_{i-1}(R) = \operatorname{Coker}\Big(K_i(R[X]) \oplus K_i(R[X^{-1}]) \to K_i(R[X,X^{-1}])\Big);$$

see [19, Ch. XII, §7].

(d) Since $K_{i-1}(R)$ is a natural direct summand of $K_i(R[X,X^{-1}])$, the higher K-groups can be considered as invariants of rings that are finer than the lower K-groups. In particular, it is natural to expect that higher analogues of the results in Chapter 8 are more difficult to obtain, and as we will see in this chapter, this is indeed the case.

(e) The question to what extent the K-homotopy invariance characterizes regular rings, i. e., whether the inverse of Theorem 9.5(b) is true, dates back to the 1970s. In [97] Cortiñas, Haesemeyer, and Weibel prove the following converse of Theorem 9.5(b), confirming a conjecture of Vorst [368]. Let R be a localization of an affine algebra over a field of characteristic 0 and $i \geq \dim R + 1$. Then

$$K_i(R) = K_i(R[\mathbb{Z}_+^r]) \text{ for all } r \geq 0 \quad \Longrightarrow \quad R \text{ is regular.}$$

(f) If R is a localization of an affine algebra over a *large* field of characteristic 0 then the implication in (e) can be strengthened as follows:

$$NK_i(R) = 0 \quad \Longrightarrow \quad R \text{ is regular}.$$

This follows from Cortiñas et al. [95] which answers Bass' question [20] (for the class of rings mentioned) whether $NK_i(R) = 0$ implies $K_i(R) = K_i(R[\mathbb{Z}_+^r])$, $r \in \mathbb{N}$. Here a field is called "large" if it has infinite transcendence degree over its prime subfield.

(g) In the special case when R is an affine algebra over a large field of characteristic 0, there is an alternative very short argument for Bass' question [171]. It is based on Murthy and Pedrini [270] and the action of big Witt vectors on NK-groups (to be discussed in Section 10.F).

9.B The nontriviality of $SK_1(R[M])$

We will now show that for a seminormal simplicial affine monoid M and a regular ring R one has $K_1(R) = K_1(R[M])$ if and only if M is free. Together with some variations, this theorem will be derived from a more precise result on SK_1.

In the following we will frequently use the notation $\Phi(M)$ and $M|W$ introduced on p. 290.

The main result. In Section 8.I the homotopy invariance of K_0 (Theorem 8.36) has been generalized as far as possible to a statement about monoid rings (Theorem 8.37). In view of Theorem 9.5 it is natural to expect that

$$K_i(R) = K_i(R[M]), \quad i \geq 0,$$

for all regular rings R and all positive seminormal monoids M. It is clear that we have to require the positivity of M: for nonpositive monoids the equation $K_i(R) = K_i(R[M])$ fails already for $i = 1$ because $\mathrm{U}(R)$ is a proper direct summand of $\mathrm{U}(R[M])$.

However, we will show below that the group $K_i(R[M])/K_i(R)$, $i \geq 1$, dramatically fails to be trivial for essentially all nonfree affine positive monoids. In other words, a monoid generalization of the fundamental theorem 9.5(b) is impossible for higher K-groups.

The main result of this section is

Theorem 9.7. *Let R be a regular ring and M be an affine simplicial monoid M. Then*

(a) $K_1(R[M]) = K_1(R[M][X])$ if and only if M is free;
(b) if M is seminormal, then $K_1(R) = K_1(R[M])$ if and only if M is free;
(c) if $\Omega_R \neq 0$ then $K_1(R) = K_1(R[M])$ if and only if M is free.

In part (c) Ω_R denotes the module of absolute differentials of R (see [253]).

Combinatorial preparations. We need some auxiliary results about affine monoids. The first concerns rank 2 affine monoids.

Lemma 9.8. *Let M be a nonfree affine positive normal monoid of rank 2. Then M is isomorphic to a submonoid $L \subset \mathbb{Z}_+^2$ which is integrally closed in $\mathbb{Z}_+^2$ and satisfies the conditions $(0,1) \in L$, $(1,1) \in L_*$, $(1,0) \notin L$.*

Proof. We identify $\operatorname{gp}(M)$ with $\mathbb{Z}^2$. Each of the two extreme integral generators of M can be extended to a basis of $\mathbb{Z}^2$, and therefore we can assume that $x = (1,0)$ is one of them. Moreover we can assume that M is contained in the upper halfplane. Since M is nonfree, its second extreme integral generator (m,n) has $n > 1$, and, moreover, m and n are coprime. We choose $k \in \mathbb{Z}$ such that

$$k < \frac{m}{n} < k+1$$

and set $z = (k,1)$. Then $\{x,z\}$ is a basis of $\mathbb{Z}^2$. After a change of bases keeping x fixed and sending z to $(0,1)$ we have reached the desired embedding. □

The proof of the next lemma is similar to that of Lemma 3.34.

Lemma 9.9. *Let M be an affine positive normal monoid of rank r and x be an extreme generator of M. Then there exists an affine normal submonoid $L \subset M$ of rank $r-1$ such that $L_* \subset M_*$ and $M[-x] = \mathbb{Z}x + L$.*

Proof. By Proposition 2.32 we have $M[-x] \cong \mathbb{Z} \oplus M_0$ for some affine positive normal monoid of rank $r-1$. Moreover, the isomorphism maps $\mathbb{Z}x$ to $\mathbb{Z}$. Taking the isomorphic image of M_0 in $M[-x]$ we can assume that $M_0 \subset M[-x]$. By Proposition 2.17(e) there is a free intermediate monoid F with $M_0 \subset F \subset \operatorname{gp}(M_0)$ such that $\operatorname{gp}(F) = \operatorname{gp}(M_0)$. Let $F = \mathbb{Z}_+x_1 + \cdots + \mathbb{Z}_+x_{r-1}$.

If c is a sufficiently large natural number then the monoid

$$L = M \cap \big(\mathbb{Z}_+(x_1+cx) + \cdots + \mathbb{Z}_+(x_{r-1}+cx)\big)$$

satisfies the desired condition. In fact, if c is big enough, then elementary geometric considerations show that

$$\operatorname{int}(\mathbb{R}_+M) \cap \operatorname{int}(\mathbb{R}_+L) = \operatorname{int}(\mathbb{R}_+M) \cap \operatorname{int}\Big(\sum \mathbb{R}_+(x_i+cx)\Big) \neq \emptyset.$$

This implies $\operatorname{rank} L = \dim \operatorname{aff}(L) = r-1$.

On the other hand, since $\operatorname{gp}(F) = \mathbb{Z}x + \mathbb{Z}(x_1+cx) + \cdots + \mathbb{Z}(x_{r-1}+cx) = \operatorname{gp}(M)$ and M is normal, we also have that L is integrally closed in $\mathbb{Z}_+(x_1+cx) + \cdots + \mathbb{Z}_+(x_{r-1}+cx)$. By Corollary 2.25 we have $\operatorname{gp}(L) = \mathbb{Z}(x_1+cx) + \cdots + \mathbb{Z}(x_{r-1}+cx)$. In particular, $\operatorname{gp}(M) = \mathbb{Z}x + \operatorname{gp}(L)$.

Finally, the equality of the two normal affine monoids $M[-x]$ and $\mathbb{Z}x + L$ follows since their groups of differences coincide as well as the cones they span in $\mathbb{R}M$. □

Finally we formulate a very simple criterion for the freeness of simplicial affine monoids:

Lemma 9.10. *Let M be a simplicial affine monoid and let $v \in \Phi(M)$ be a vertex. Then M is free if the following conditions hold:*

(a) the extreme submonoids $M' \subset M$ of rank 2 with $v \in \Phi(M')$ are free;
(b) there exists a free submonoid $F \subset M$ such that $\operatorname{gp}(F) = \operatorname{gp}(M)$, v is a vertex of $\Phi(F)$, and the cones spanned by the polytopes $\Phi(M)$ and $\Phi(F)$ at their vertex v are the same.

Proof. Let $t_1, \dots, t_r$ denote the extreme integral generators of M, labeled in such a way that $v \in \mathbb{R}_+ t_1$, and consider the submonoid $L = \mathbb{Z}_+ t_1 + \cdots + \mathbb{Z}_+ t_r$. We claim that $L = \bar{M}$. Then $L = M$ since $L \subset M$.

Since $\Phi(L) = \Phi(\bar{M})$ and L and $\bar{M}$ are normal, we only need to prove that $\operatorname{gp}(L) = \operatorname{gp}(M)$ $(= \operatorname{gp}(\bar{M}))$. Let $F = \mathbb{Z}_+ s_1 + \cdots + \mathbb{Z}_+ s_r$ with $v \in \mathbb{R}_+ s_1$. Then $s_1 = t_1$. Moreover the hypotheses imply $s_i \in \mathbb{Z}_+ t_1 + \mathbb{Z}_+ t_i$ for $i = 2, \dots, r$. In particular, $\operatorname{gp}(M) = \operatorname{gp}(F) \subset \operatorname{gp}(L)$. □

Swan's elements in $K_2(R[X,Y]/(XY))$. For a ring A and elements $u, v \in A$ such that $1 + uv \in \mathrm{U}(A)$, Dennis and Stein [333] define an element $\langle u, v\rangle \in K_2(A)$. In the special case when $A = R[X,Y]/(XY)$ for some ring R and u and v are the residue classes of X and Y, the element $\langle u, v\rangle$ coincides with $[x_{12}^u, x_{21}^v] \in K_2(A)$, used by Swan in [346]. Here the x_{ij}^* denote the standard generators of the Steinberg group (Section 9.A). The inclusion $[x_{12}^u, x_{21}^v] \in K_2(A)$ follows from the equation $[e_{12}^u, e_{21}^v] = 1$ and Theorem 9.2.

Lemma 9.11.

(a) For every ring A and all elements $u \in A$, one has $\langle u, 0\rangle = \langle 0, u\rangle = 0$.
(b) For $A = R[X,Y]/(XY)$ and u and v as above, one has $\langle u^a, v^b\rangle = 0$ for all natural numbers a and b with $\max(a,b) \geq 2$.

The claims (a) and (b) follow immediately from the general relations between the Dennis–Stein elements; see Dennis and Krusemeyer [109].

The following theorem is Corollary 4.8 in [109], stated there for a bigger class of coefficient rings R.

Theorem 9.12. *Let R be a regular ring and $R[u,v] = R[X,Y]/(XY)$ as above. Then $K_2(R[u,v]) = K_2(R) \oplus \langle u, v\rangle R$ where $\langle u, v\rangle R$ is the rank 1 free R-module over the basis $\{\langle u, v\rangle\}$.*

Remark 9.13. The R-module structure on $K_2(R[u,v])/K_2(R) \cong \langle u, v\rangle R$ is defined by $r \cdot \langle u, v\rangle = \langle ru, v\rangle$. More generally, for any graded ring $A = A_0 \oplus A_1 \oplus \cdots$ and all indices i the relative groups $K_i(A)/K_i(A_0)$ are modules over the *big Witt vectors* $\operatorname{Witt}(A_0)$. The module structures are given by the *Bloch–Stienstra–Weibel* operations on nil-K-theory [34], [335], [374]; see Section 10.F.

As usual, we make the identification $R[\mathbb{Z}_+^2] = R[X,Y]$. Let $R[u,v]$ be as above and consider the commutative diagram

$$\begin{array}{ccc} \mathrm{St}(R[X,Y]) & \xrightarrow{\psi} & \mathrm{E}(R[X,Y]) \\ \varphi \downarrow & & \downarrow \vartheta \\ \mathrm{St}(R[u,v]) & \longrightarrow & \mathrm{E}(R[u,v]). \end{array} \tag{9.5}$$

The key technical fact used in the proof of Theorem 9.7 is contained in the following:

Proposition 9.14. *For every convex neighborhood W of the point $\Phi(XY) \in \Phi(\mathbb{Z}_+^2)$ there exists a preimage $z \in \mathrm{St}(R[X,Y])$ of $\langle u,v\rangle$ such that*

$$\psi(z) \in \mathrm{E}\big(R[X,Y]\big) \cap \mathrm{SL}\big(R[\mathbb{Z}_+^2|W]\big)$$

and, simultaneously, $\psi(z)$ reduces to the unit element of $\mathrm{SL}(R)$ *modulo (XY).*

Proof. Consider the following system of φ-preimages of $\langle u,v\rangle$ in $\mathrm{St}(R[X,Y])$:

$$z_1 = [x_{12}(X), x_{21}(Y)],$$
$$z_{i+1} = x_{12}(X^{i+1}Y^i)z_i x_{21}(-X^iY^{i+1}), \quad i \geq 1.$$

(In these formula we have used functional notation for Steinberg symbols instead of the usual exponential one.) That the z_i are preimages of $\langle u,v\rangle$ follows by induction on i if one observes that $x_{12}(X^{i+1}Y^i)$ and $x_{21}(-X^iY^{i+1})$ map to the identity element.

We now show that the images $A_i = \psi(z_i)$ are given by the following matrices:

$$\psi(z_i) = \begin{pmatrix} f_i & -X^{i+1}Y^i \\ X^iY^{i+1} & 1-XY \end{pmatrix} \in \mathrm{SL}_2(R[X,Y]),$$

where $f_i = \sum_{j=0}^{2i}(XY)^j$. Indeed,

$$\psi(z_1) = \begin{pmatrix} 1+XY+X^2Y^2 & -X^2Y \\ XY^2 & 1-XY \end{pmatrix}$$

and

$$\psi(z_i) = \begin{pmatrix} f_{i-1}+X^{2i-1}Y^{2i-1}+X^{2i}Y^{2i} & -X^{i+1}Y^i \\ X^iY^{i+1} & 1-XY \end{pmatrix}, \quad i \geq 2,$$

so induction on i applies.

Let W_i denote the subsegment $[\Phi(X^{i+1}Y^i), \Phi(X^iY^{i+1})] \subset \Phi(\mathbb{Z}_+^2)$. It is clear that W_i is a neighborhood of $\Phi(XY)$ in $\Phi(\mathbb{Z}_+^2)$, $W_1 \supset W_2 \supset W_3 \supset \cdots$, and $\bigcap_i W_i = \Phi(XY)$. Since $\psi(z_i) \in \mathrm{SL}(R[\mathbb{Z}_+^2|W_i])$ and every $\psi(z_i)$ reduces to the identity matrix modulo (XY), we are done. □

The case of rank 2 seminormal monoids. For seminormal rank 2 monoids we can prove a stronger version of Theorem 9.7. We need it in the proof of Proposition 9.16.

Proposition 9.15. *Let R be a regular ring and M be a nonfree affine positive seminormal monoid of rank 2. Then $SK_1(R)$ is a proper subgroup of the image of $SK_1(R[M_*])$ in $SK_1(R[M])$.*

Proof. We use the notation introduced in Proposition 9.14 and the commutative diagram (9.5).

(1) First we consider the case in which M is normal. By Lemma 9.8 we can assume that $M \subset \mathbb{Z}_+^2$, $Y \in M$, $XY \in M_*$, and $X \notin M$. In this situation $\Phi(M_*)$ is a convex neighborhood of $\Phi(XY)$ in $\Phi(\mathbb{Z}_*^2)$. By Proposition 9.14 there is a φ-preimage $z \in \mathrm{St}(R[X,Y])$ of $\langle u,v\rangle \in K_2(R[u,v])$ such that

$$\psi(z) \in \mathrm{E}(R[X,Y]) \cap \mathrm{SL}(R[M_*]).$$

Let ζ denote the class of $\psi(z)$ in $SK_1(R[M])$. We claim that $\zeta \notin SK_1(R)$. Assume to the contrary that $\zeta \in SK_1(R)$.

Observe that

$$\psi(z) \in \mathrm{SL}\big(R[X,Y],(XY)\big),$$

equivalently, $\vartheta(\psi(z))$ is the identity matrix since $\varphi(z) \in K_2(R[u,v])$. Therefore the assumption on ζ implies $\zeta = 0$ in $SK_1(R[M])$, or, in other words, $\psi(z) \in \mathrm{E}(R[M])$.

Let s be a preimage of $\psi(z)$ in $\mathrm{St}(R[M])$, s_1 be the image of s in $\mathrm{St}(R[X,Y])$, and s_2 be the image of s_1 in $\mathrm{St}(R[u,v])$. Since both s_1 and z map to $\psi(z) \in E(R[X,Y])$ we have $s_1 z^{-1} \in K_2(R[X,Y])$. By Theorem 9.5(b) $K_2(R[X,Y]) = K_2(R)$. Therefore, $s_2\langle u,v\rangle^{-1} \in K_2(R) \subset K_2(R[u,v])$. Dividing s by this element of $K_2(R)$, we can assume that $s_2 = \langle u,v\rangle$. In other words, s also maps to $\langle u,v\rangle$.

On the other hand, the image of $R[M]$ in $R[u,v]$ under the composite map $R[M] \to R[X,Y] \to R[u,v]$ is $R[v]$ because of our special embedding of M into $\mathbb{Z}_+^2$. We get $\langle u,v\rangle \in \mathrm{St}(R[v]) \subset \mathrm{St}(R[u,v])$. But then the commutative diagram

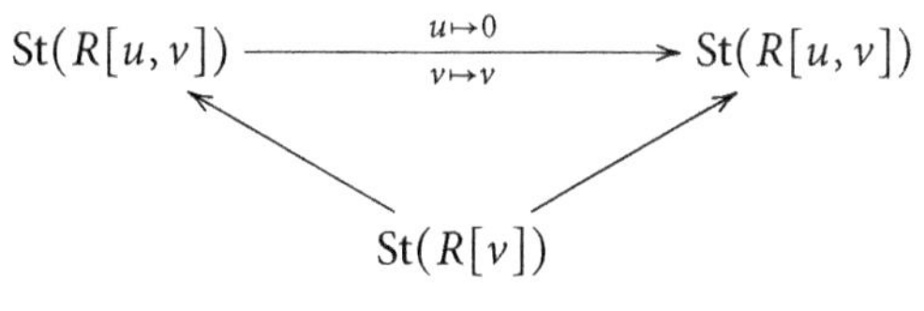

implies $\langle u,v\rangle = \langle 0,v\rangle = 0$ (Lemma 9.11(a))—in contradiction with Theorem 9.12.

(2) Now we treat the case in which M is seminormal and its normalization $\bar{M}$ is *not* free. By Proposition 2.40 we have $(\bar{M})_* = M_*$. Consider the commutative diagram

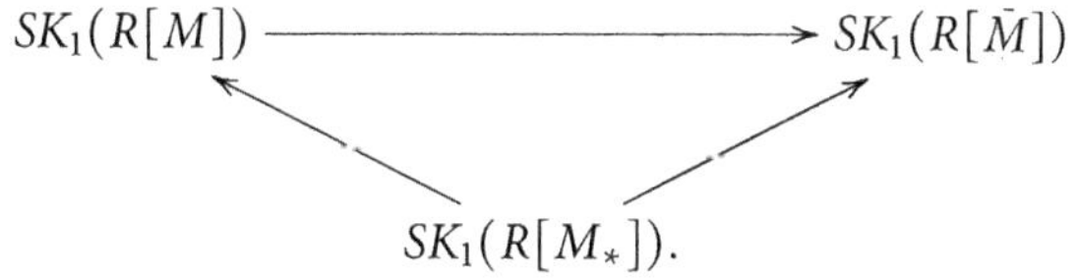

By the previous case there exists $\zeta \in SK_1(R[M_*])$, whose image in $SK_1(R[\bar{M}])$ does not belong to $SK_1(R)$. Then the diagram above implies that the image of ζ in $SK_1(R[M])$ does not belong to $SK_1(R)$.

(3) It remains to consider the case when M is seminormal and $\bar{M} = \mathbb{Z}_+^2$. Then Proposition 2.40 implies that there exist natural numbers a and b satisfying the following conditions:

(i) $\max(a, b) \geq 2$;
(ii) $M = \mathbb{Z}_+^2 \setminus \{X^i : a \nmid i\} \cup \{Y^j : b \nmid j\}$.

Then, as in case (1), we can find a preimage $z \in \mathrm{St}(R[X, Y])$ of $\langle u, v\rangle \in K_2(R[u, v])$ such that

$$\psi(z) \in \mathrm{E}(R[X, Y]) \cap \mathrm{SL}\big(R[X, Y], (XY)\big) \cap \mathrm{SL}(R[M_*]).$$

Let ζ denote the corresponding element of $SK_1(R[M])$. We claim that $\zeta \notin SK_1(R)$.

Assume to the contrary that $\zeta \in SK_1(R)$. Then, as in (1), $\psi(z) \in \mathrm{E}(R[M])$. Let s be a lifting of $\psi(z)$ in $\mathrm{St}(R[M])$. Consider the commutative diagram of rings

$$\begin{array}{ccc} R[M] & \longrightarrow & R[X, Y] \\ \downarrow & & \downarrow \\ R[u^a, v^b] & \longrightarrow & R[u, v] \end{array}$$

where the vertical maps represent the reduction modulo $\mathrm{int}(\mathbb{Z}_+^2)R$. Let s_0 denote the image of s in $\mathrm{St}(R[u^a, v^b])$, s_1 denote the image of s in $\mathrm{St}(R[X, Y])$, and s_2 denote the image of s_1 in $\mathrm{St}(R[u, v])$. Both s_1 and z map to $\psi(z) \in \mathrm{E}(R[M])$. Therefore, $s_1 z^{-1} \in K_2(R[X, Y]) = K_2(R)$. In particular, $s_2\langle u, v\rangle^{-1} \in K_2(R)$. Modifying s as in (1), we can assume $s_1 = z$ and $s_2 = \langle u, v\rangle$. On the other hand, s_2 is the image of s_0 under the map $\mathrm{St}(R[u^a, v^b]) \to \mathrm{St}(R[u, v])$ and, simultaneously, the image of s_0 in $\mathrm{E}(R[u^a, v^b])$ is 1. Therefore, $s_0 \in K_2(R[u^a, v^b])$. By Theorem 9.12, applied to the ring $R[u^a, v^b]$ ($\cong R[u, v]$), we have $s_0 = p \oplus r\langle u^a, v^b\rangle$ for some $p \in K_2(R)$ and $r \in R$. (Here we have used the notation $\langle u^a, v^b\rangle = \left[x_{12}^{u^a}, x_{21}^{v^b}\right]$.) By Lemma 9.11(b) s_0 maps to p under the map $K_2(R[u^a, v^b]) \to K_2(R[u, v])$. Thus we obtain $\langle u, v\rangle \in K_2(R)$, a contradiction to Theorem 9.12. □

The case of free extreme submonoids. Now we deal with the case in which there is no immediate inductive argument.

Proposition 9.16. *Let R be regular ring and M be a nonfree, affine, simplicial, seminormal monoid. Assume that for every facet $\Delta \subset \Phi(M)$ the extreme submonoid $M|\Delta \subset M$ is free. Then $SK_1(R) \neq SK_1(R[M])$.*

We need an auxiliary result.

Lemma 9.17. *Let R be a regular ring and M be an affine positive seminormal monoid all of whose proper extreme submonoids are free. Then the following conditions are equivalent:*

(a) $SK_1(R) \neq SK_1(R[M])$;
(b) $SK_1(R) \neq \mathrm{Im}\big(SK_1(R[M_*]) \to SK_1(R[M])\big)$.

Proof. First we prove the equality

$$SK_1(R) = SK_1\big(R[M]/(R\,\mathrm{int}(M))\big). \tag{9.6}$$

The proof is based on the ideas in Section 8.D, making use of the Mayer–Vietoris sequence (9.2).

As in the proof of Lemma 8.18, we let $M_1, \dots, M_n$ be the extreme submonoids, labeled in such a way that $M_i \supset M_j$ implies $i \leq j$. In particular, $M_1 = M$ and $M_n = 0$. In the proof of Lemma 8.18 we have constructed Milnor squares

$$\begin{array}{ccc} R[(M_i)_*] & \longrightarrow & A_{i-1} \\ \downarrow & & \downarrow \\ R & \longrightarrow & A_i \end{array}, \tag{9.7}$$

$i = 1, 2, \dots, n$, where $A_0 = R[M]$, $A_1 = R[M]/(R\,\mathrm{int}(M))$, and $A_{n-1} = R$. Observe that in these squares the map $R[(M_i)_*] \to A_{i-1}$ factors through the subring $R[M_i] \subset A_{i-1}$ for every index $i, 1 \leq i \leq n$.

In our situation the rings $R[M_i]$, $i = 2, \dots, n$, are polynomial rings. So by the Bass–Heller–Swan theorem (Theorem 9.5 for $i = 1$) we have $SK_1(R) = SK_1(R[M_i])$, $i \geq 2$. In particular, the Mayer–Vietoris exact sequences, associated to the squares (9.7), yield the *embeddings*

$$SK_1(A_1) \to SK_1(A_2) \to \cdots \to SK_1(A_{n-1}) = SK_1(R).$$

But $SK_1(R)$ is naturally a subgroup of $SK_1(A_1)$ and, hence, (9.6) follows.

Now we turn to the proof of the equivalence (a) $\iff$ (b). Clearly, we only need to show the implication (a) $\implies$ (b). Assume to the contrary that $SK_1(R) = \mathrm{Im}\big(SK_1(R[M_*]) \to SK_1(R[M])\big)$. Then the Mayer–Vietoris sequence, associated to (9.7) for $i = 1$, shows that we have an embedding

$$SK_1(R[M]) = SK_1(A_0) \to SK_1(A_1) = SK_1\big(R[M]/(R\,\mathrm{int}(M))\big).$$

In view of (9.6) this implies $SK_1(R) = SK_1(R[M])$, contradicting (a). □

Proof of Proposition 9.16. We use induction on $r = \mathrm{rank}\, M$. For $r = 1$ there is nothing to prove since a rank 1 seminormal monoid is free. The case $r = 2$ follows from Proposition 9.15. So we can assume $r \geq 3$ and that Proposition 9.16 has been proved for monoids of rank $r - 1$. Assume to the contrary $SK_1(R) = SK_1(R[M])$.

(1) Let $v_1, \dots, v_r$ be the vertices of $\Phi(M)$ and let t_i denote the generator of the monoid $M|v_i \cong \mathbb{Z}_+$.

By Lemma 9.9 there exists a submonoid $L \subset \bar{M}$ satisfying the following conditions (in multiplicative notation):

- L is integrally closed in $\bar{M}$;
- $L_* \subset \bar{M}_* = M_*$ (Proposition 2.40);
- $(\bar{M}|v_1)^{-1}\bar{M} = \mathrm{gp}(\bar{M}|v_1) \oplus L \; (\cong \mathbb{Z} \oplus L)$.

We have the sequence of four injective monoid homomorphisms followed by the projection onto L:

$$L_* \to M_* \to M \to \bar{M} \to \mathrm{gp}(M|v_1) \oplus L \to L.$$

Notice that the composite map coincides with the embedding $L_* \to L$ and, moreover, the image of M_* in L is L_*. In particular, there is the commutative diagram of monoids

$$\begin{array}{ccc} M_* & \longrightarrow & L_* \\ \downarrow & & \downarrow \\ M & \longrightarrow & L \end{array} \tag{9.8}$$

whose vertical arrows are the inclusion maps.

(2) Let L' be the image of M in L. It is an affine simplicial monoid for which $\Phi(L') = \Phi(L)$. By (9.8) $L_* \subset L'$.

Consider a facet $\delta \subset \Phi(L)$. Let us show that the extreme submonoid $L'|\delta \subset L'$ is free. In view of the inclusion $L_* \subset L'$ and Proposition 2.40, this would imply that L' is seminormal.

Let $\Delta \subset \Phi(M)$ be the uniquely determined facet that contains δ and v_1. There exists j, $2 \le j \le r$, such that $\Delta = \mathrm{conv}(v_i : i \ne j)$. By the condition on M the monoid $M|\Delta$ is free. It is generated by $\{t_i \; : \; i \ne j\}$. But then $L'|\delta$ is generated by the images of t_i, $i \ne 1, j$. Since $L'|\delta$ is generated by $\operatorname{rank} L'|\delta$ elements, the images of t_i, $i \ne 1, j$, must generate the monoid $L'|\delta$ freely.

Next we show that $L' = L$. Suppose $L' \ne L$. Then, in view of the equalities $\Phi(L') = \Phi(L)$ and $\mathrm{gp}(L') = \mathrm{gp}(L)$, the monoid L' cannot be normal. In particular, L' is not free. So L' satisfies all conditions in Proposition 9.16. By the induction hypothesis on rank we have $SK_1(R) \ne SK_1(R[L'])$. By Lemma 9.17 this is equivalent to $SK_1(R) \ne \mathrm{Im}\,(SK_1(R[M_*]) \to SK_1(R[M]))$.

On the other hand, the commutative diagram

$$\begin{array}{ccc} SK_1(R[L_*]) & \longrightarrow & SK_1(R[L']) \\ & \searrow \quad \nearrow & \\ & SK_1(R[M]) & \end{array}$$

in conjunction with the assumption $SK_1(R) = SK_1(R[M])$ implies that $SK_1(R) = \mathrm{Im}\big(SK_1(R[L_*]) \to SK_1(R[L'])\big)$. This contradiction shows that $L' = L$ and, moreover, L is free by induction on rank.

(3) Here we prove that M is free, in contradiction with the hypothesis that M is not free. Then our assumption $SK_1(R) = SK_1(R[M])$ has been shown to be wrong.

Let s_i denote the generator of $\bar{M}|v_i \cong \mathbb{Z}_+$, $i = 1, \dots, r$. For every index i we have $t_i = s_i^{a_i}$ for some $a_i \in \mathbb{N}$. First we show that $a_i = 1$ for all i.

By (2) L is a free monoid. Let τ_i, $2 \le i \le r-1$, denote the free generators of L so that $\Phi(\tau_i) \in \operatorname{conv}(\Phi(t_1), \Phi(t_i))$. The equality $\operatorname{gp}(M) = \operatorname{gp}(\bar{M}|v_1) \oplus \operatorname{gp}(L)$ and the freeness of L imply that s_1 and τ_i generate $\operatorname{gp}(\bar{M}|\Phi(s_1, s_i))$ for every $i = 2, \dots, r$. In particular, there exist representations

$$s_i = \tau_i^{x_i} s_1^{-y_i}, \quad x_i \in \mathbb{N}, \quad y_i \in \mathbb{Z}_+, \quad i = 2, \dots, r.$$

The image of $t_i = s_i^{a_i}$ in L under the map $M \to L$ is $\tau_i^{a_i x_i}$ for $i \ge 2$, and it is 1 for $i = 1$. By (2) we know that these images generate the free monoid L. In particular, $a_i x_i = 1$, and so $a_i = 1$ for $i = 2, \dots, r$, as claimed.

Applying all the arguments above to any other vertex $v_i \in \Phi(M)$ different from v_1 we get $t_1 = s_1$ as well.

Now we are in the situation of Lemma 9.10 with respect to the monoid M, the vertex $v_1 \in \Phi(M)$, and the free submonoid of M generated by t_1 and L. Hence the desired freeness of M. □

The general case. Before proving Theorem 9.7 we explain the relationship with the module of absolute differentials.

Lemma 9.18. *For every commutative ring R there exists a surjective homomorphism $K_2(R[X]/(X^2))/K_2(R) \to \Omega_R$.*

Proof. We only indicate the argument, as it involves techniques beyond the scope of our book.

It was shown by van der Kallen, Maazen, and Stienstra [361] that for a commutative ring A and any nilpotent ideal I the relative group $K_2(A, I)$ has a presentation by Dennis–Stein symbols [333]. In the special case $A = R[X]/(X^2)$ and $I = R\varepsilon$, where ε is the residue class of X modulo (X^2), if one factors out the extra relation $\langle r\varepsilon, s\varepsilon \rangle = 0$, $r, s \in R$, the Dennis–Stein presentation becomes precisely the definition of Ω_R. So there is always a surjection $K_2(R[X]/(X^2))/K_2(R) \to \Omega_R$ whose kernel is generated by Dennis–Stein symbols $\langle r\varepsilon, s\varepsilon \rangle$. □

Remark 9.19. The map $K_2(R[X]/(X^2))/K_2(R) \to \Omega_R$ is the first glimpse of the *Chern class maps* from higher K-theory to Hochschild and eventually to cyclic homology groups (Karoubi [218], Loday [243]). Deep facts from this theory (such as *Goodwillie's theorem* [144] and Cortiñas' proof of the *KABI conjecture* [93]) play a crucial role in [168] which presents a higher analogue of Theorem 8.4; see Section 9.C, in particular Theorem 9.33.

Lemma 9.20. *Let R be a regular ring and X a variable. Then there is a surjective homomorphism $SK_1(R[X^2, X^3])/SK_1(R) \to \Omega_R$.*

In the special case when R is a field, this has been proved by Krusemeyer [234]. However, his argument implies the general case, as we now show.

Proof. Being regular, R is a reduced ring. In particular, $\mathrm{U}(R) = \mathrm{U}(R[X])$. It then follows easily that $K_1(R[X], (X^2)) = SK_1(R[X], (X^2))$. Thus by (9.3) we have the exact sequence

$$K_2(R[X]) \to K_2(R[X]/(X^2)) \to SK_1(R[X],(X^2)) \to K_1(R[X]).$$

By Theorem 9.5(b) $K_2(R) = K_2(R[X])$ and $K_1(R) = K_1(R[X])$. On the other hand $SK_1\big(R[X],(X^2)\big) \to K_1(R)$ is the zero map. So we get

$$SK_1\big(R[X],(X^2)\big) = K_2\big(R[X],(X^2)\big)/K_2(R).$$

Now we are done by Lemma 9.18 once we notice that $SK_1(R[X^2,X^3])/SK_1(R)$ maps naturally onto $SK_1\big(R[X],(X^2)\big)$—an immediate consequence of the definition of relative K_1-groups in (9.4). □

Proof of Theorem 9.7. First observe that the equations for K_1 in the theorem hold if M is free; this follows from Theorem 9.5(b). Therefore, we only need to prove the following three implications (restated here for the reader's convenience):

(a) if $K_1(R[M]) = K_1(R[M][X])$, then M is free;
(b) if M is seminormal and $K_1(R) = K_1(R[M])$, then M is free;
(c) if $\Omega_R \neq 0$ and $K_1(R) = K_1(R[M])$, then M is free.

Let us first deduce (a) from (b). Assume $K_1(R[M]) = K_1(R[M][X])$. For an arbitrary ring A the following implication is known (Vorst [368]):

$$K_i(A) = K_i(A[X]) \implies K_{i-1}(A) = K_{i-1}(A[X]), \quad i \in \mathbb{N}.$$

On the other hand if $K_0(A) = K_0(A[X])$ then $\operatorname{Pic}(A) = \operatorname{Pic}(A[X])$: exactly the same argument as in the proof of the implication (i) $\implies$ (ii) in Theorem 8.37 applies. Therefore, Theorem 4.74 and the equality $K_1(R[M]) = K_1(R[M][X])$ imply that the ring $R[M]$ is seminormal. By Theorem 4.76 M is seminormal. But we also have $K_1(R) = K_1(R[M])$, because the monoid ring $R[M]$ admits a grading with $(R[M])_0 = R$, and Lemma 8.39 applies. Thus we are in the situation of (b). In particular, M is free.

Now we prove (b). The case $r = \operatorname{rank} M = 1$ is trivial because already seminormality guarantees that $M \cong \mathbb{Z}_+$. Suppose $r \geq 2$. Let $N \subset M$ be a proper extreme submonoid. Since the inclusion map $R[N] \subset R[M]$ splits the face projection $R[M] \to R[N]$, the equality $K_1(R) = K_1(R[M])$ implies $K_1(R) = K_1(R[N])$. Because of the functorial splitting $K_1 = \mathrm{U} \oplus SK_1$, we get $SK_1(R) = SK_1(R[N])$. By the induction hypothesis on rank we conclude that N is free. That is, all proper extreme submonoids of M are free. By Proposition 9.16 M is itself free.

To prove (c) we first notice that, as mentioned above, $K_1(R) = K_1(R[M])$ implies $SK_1(R) = SK_1(R[M])$.

Consider the case $r = \operatorname{rank} M = 1$. The monoid operation will be written additively. Without loss of generality we can assume $M \subset \mathbb{Z}_+$ and $\operatorname{gp}(M) = \mathbb{Z}$. By Proposition 2.33 there exists $m \in M$ such that $m + \mathbb{Z}_+ \subset M$. We want to show that $M = \mathbb{Z}_+$. Assume to the contrary that $M \neq \mathbb{Z}_+$. Then $M \subset \{0,2,3,\dots\}$. Both $R[M]$ and $R[X^2,X^3]$ contain the ideal $I = (X^m, X^{m+1}, X^{m+2}, \dots)$. (Here we make the identification $R[\mathbb{Z}_+] = R[X]$.) Since $R[X^2,X^3]/I$ is a nilpotent extension of R, any element of $SK_1(R[X^2,X^3])$ reduces modulo I to an element of $SK_1(R)$; see Lemma 9.1. Therefore, every element of $SK_1(R[X^2,X^3])$ has a representative in $\mathrm{SL}(R[X^2,X^3],I) \subset$

$\mathrm{SL}(R[M])$. In particular, the homomorphism $SK_1(R[M]) \to SK_1(R[X^2, X^3])$ is surjective. We are done because the target of this map is larger than $SK_1(R)$ by Lemma 9.20.

Assume $r \geq 2$, $SK_1(R) = SK_1(R[M])$, and that (c) is proved for affine simplicial monoids of rank $r-1$. As in the proof of (b), all proper extreme submonoids of M are free. By Proposition 2.40, $M|\Delta = \mathrm{sn}(M)|\Delta$ for any proper face $\Delta \subset \Phi(M)$. (Recall, $\mathrm{sn}(M)$ is the seminormalization of M, p. 66.)

We claim that $SK_1(R) = SK_1(R[\mathrm{sn}(M)])$. By Proposition 2.33 there exists an element $m \in \mathrm{int}(M)$ such that $m + \bar{M} \subset M$ (writing the monoid operation additively). Let $I = m\bar{M}R$ (switching back to multiplicative notation). Then I is an ideal of both $R[M]$ and $R[\mathrm{sn}(M)]$. Moreover, we have

$$(R[M]/I)_{\mathrm{red}} = (R[\mathrm{sn}(M)]/I)_{\mathrm{red}} = R[F]/(\mathrm{int}(F))$$

for the free submonoid F of M generated by $M|v_1 \cup \cdots \cup M|v_r$ where $v_1, \ldots, v_r$ are the vertices of the simplex $\Phi(M)$. Equation (9.6) in the proof of Lemma 9.17 implies $SK_1(R) = SK_1\big(R[F]/(\mathrm{int}(F))\big)$. Then the same argument we have used in the case $r = 1$ shows that the homomorphism $SK_1(R[M]) \to SK_1(R[\mathrm{sn}(M)])$ is surjective. Hence the equality $SK_1(R) = SK_1(R[\mathrm{sn}(M)])$.

Proposition 9.16 implies that $\mathrm{sn}(M)$ is free. Now the equations $\mathrm{sn}(M)|\Delta = F|\Delta$, Δ running through the proper faces of $\Phi(M)$, imply that $\mathrm{sn}(M) = F$. We are done because $F \subset M \subset \mathrm{sn}(M) \subset F$. □

Remark 9.21. (a) It follows from the $\mathrm{Witt}(R)$-module structure on the quotient group $K_1(R[M])/K_1(R)$, mentioned in Remark 9.13 (for general graded rings) that if R contains $\mathbb{Q}$ then the group $K_1(R[M])/K_1(R)$ is not even finitely generated once it is nontrivial. We will make an essential use of this fact in Section 10.F.

(b) For a natural number c and a monoid M we have the monoidal Frobenius endomorphism $c_* : R[M] \to R[M]$, $m \mapsto m^c$, introduced on p. 240. It gives rise to a group endomorphism $K_1(R[M])/K_1(R) \to K_1(R[M])/K_1(R)$, which we again denote by c_*. It follows easily from Lemma 9.11(b) that all nonzero elements of $K_1(R[M])/K_1(R)$, produced in the proof of Theorem 9.7, are trivialized by an endomorphism of type c_* for $c > 1$. We do not know whether the endomorphisms c_*, $c > 1$, are all zero maps for a regular ring R and a simplicial monoid M. A "nilpotent" version for the unstable groups is given by Theorem 9.35(b) below.

(c) Let R be a regular ring. The proof of Theorem 9.7 actually produces many nonsimplicial affine positive monoids M for which $K_1(R[M])/K_1(R) \neq 0$; for instance see [161, Theorem 9.1(b)]. However, we do not know whether Theorem 9.7 can be generalized to all affine positive monoids.

(d) It is interesting to remark that in the special case of affine normal positive monoids it is enough to consider only rank 3 monoids:

$$K_1(R[M])/K_1(R) \neq 0 \text{ for all nonfree such monoids } M \text{ of rank} \leq 3$$
$$\iff K_1(R[M])/K_1(R) \neq 0 \text{ for all nonfree such monoids } M.$$

The proof of this equivalence is based on the fact that polytopes in dimension ≥ 3 that are both simple and simplicial must be simplices. In more detail, assume M is an affine normal positive nonfree monoid. If M is simplicial, then Theorem 9.7 applies. If M is not simplicial, then either there exists a nonsimplicial extreme submonoid $N \subset M$ or there is a nonsimple vertex $v \in \Phi(M)$ and we have a nonsimplicial monoid $L \subset M$ as in Lemma 9.9. In the first case $R[N]$ is an R-retract of $R[M]$ and in the second case $R[L]$ is an R-retract of $R[M]$, so that the induction on rank goes through.

Examples and applications. We illustrate the main theorem by the discussion of some special cases.

Example 9.22. (*Krusemeyer's ring*) The argument in the proof of Lemma 9.18 shows that in the special case when $\Omega_R = 0$ and the Dennis–Stein symbols are trivial there are nonfree monoids M with $K_1(R) = K_1(R[M])$. In particular, we recover Krusemeyer's example: $\Bbbk^\times = K_1(\Bbbk) = K_1(\Bbbk[X^2, X^3])$ for any algebraic extension field $\Bbbk$ of $\mathbb{Q}$. Notice, though, $K_1(\Bbbk[X^2, X^3]) \neq K_1(\Bbbk[X^2, X^3][Y])$ by Theorem 9.7(a).

Example 9.23. (*Srinivas' element*) Let R be a regular ring and consider the algebra $R[X^2, XY, Y^2]$. It is isomorphic to the monomial algebra $R[U, UV, UV^2]$ (via $X^2 \mapsto U$, $XY \mapsto UV$, $Y^2 \mapsto UV^2$). The embedding of the multiplicative monoid of the monomials of the second algebra into the free monoid of all monomials in U and V satisfies the condition in Lemma 9.8. Let $\alpha \in \mathrm{SL}_2(R[U, UV, UV^2])$ be the same matrix as $\psi(z_1)$ in the proof of Lemma 9.14. Then step (1) in the proof of Proposition 9.15 shows that $[\psi(z_1)] \in SK_1(R[U, UV, UV^2]) \setminus SK_1(R)$. Let $\alpha \in \mathrm{SL}_2(R[U, UV, UV^2])$ be the matrix obtained from $\psi(z_1)$ by permuting rows and columns. Then $[\alpha] = [\psi(z_1)] \in SK_1(R[U, UV, UV^2]) \setminus SK_1(R)$.

For a ring A and comaximal elements $a, b \in A$ one defines the *Mennicke symbol* $[a, b] \in SK_1(A)$ as follows [19, Ch. VI, §1]. Choose $c, d \in A$ so that $ad - bc = 1$. Let

$$[a, b] = \text{class of} \begin{pmatrix} a & b \\ c & d \end{pmatrix} \in SK_1(A).$$

It can be shown that $[a, b]$ is independent of the choice of c and d [19].

Returning to our example we obtain that

$$[1 - UV, UV^2] \in SK_1(R[U, UV, UV^2]) \setminus SK_1(R),$$

or equivalently

$$[1 - XY, Y^2] \in SK_1(R[X^2, XY, Y^2]) \setminus SK_1(R).$$

In the particular case $R = \mathbb{C}$ we recover Srinivas' example [322].

Example 9.24. (*Toric deformation*) Let $\Bbbk$ be a field and $<$ be a positive monomial order on the polynomial ring $\Bbbk[X_1, \dots, X_n]$ (Section 7.A). Let $A \subset \Bbbk[X_1, \dots, X_n]$ be an affine subring. If the initial algebra $\mathrm{in}(A) = \Bbbk\{\mathrm{in}(f) : f \in A\} \subset \Bbbk[X_1, \dots, X_n]$ is finitely generated over $\Bbbk$, then many ring theoretic properties of $\mathrm{in}_<(A)$ transfer to A; see Remark 7.5. However, this is not the case for the property "all projective modules are free." The following example is taken from [165].

Let $A = \Bbbk[X, Y, Z^2, Z^3 - XYZ]$ and $<$ be the lexicographic term order induced by $Z < Y < X$. Then $SK_0(A) \neq 0$ (i. e., there are projective A-modules which are not even stably free), while $\operatorname{in}(A)$ is finitely generated and all projective $\operatorname{in}(A)$-modules are free.

To prove that $SK_0(A) \neq 0$ observe that $(Z^2 - XY)\Bbbk[X, Y, Z] \subset A$, yielding the Milnor square

$$\begin{array}{ccc} A & \hookrightarrow & \Bbbk[X,Y,Z] \\ \downarrow & & \downarrow \\ \Bbbk[U,V] & \longrightarrow & \Bbbk[S^2, ST, T^2] \end{array}$$

where U, V, S, T denote variables, and $X \mapsto S^2$, $Y \mapsto T^2$, $Z \mapsto ST$, $U \mapsto S^2$, $V \mapsto T^2$. The Mayer–Vietoris sequence, associated to the diagram above, implies

$$SK_0(A) = SK_1(\Bbbk[S^2, ST, T^2]) \neq 0,$$

the inequality being a consequence of Theorem 9.7.

Evidently $\Bbbk[X, Y, Z^2, XYZ] \subset \operatorname{in}(A)$, and equality follows since both rings have the same Hilbert function (Proposition 7.1). Moreover, $\Bbbk[X, Y, Z^2, XYZ]$ is a seminormal monoid $\Bbbk$-algebra: in view of the representation

$$\Bbbk[X, Y, Z^2, XYZ] = \Bbbk[X, Y] + \Bbbk[X, Z^2] + \Bbbk[Y, Z^2] + \Bbbk[\operatorname{int}(\mathbb{Z}^3_+)]$$

the seminormality of $\operatorname{in}(A)$ follows from Proposition 2.40. So by Theorem 8.4 projective modules over $\operatorname{in}(A)$ are free.

Example 9.25. (Pic *can be smaller than* $\mathrm{U}(K_0)$) Let A be as in Example 9.24 above. Then the six-term U-Pic Mayer–Vietoris sequence, i. e., the lower rows in (8.12) and (9.2), associated to the Milnor square in Example 9.24, implies $\operatorname{Pic}(A) = 0$.

Above we have shown that $SK_0(A) \neq 0$. In particular, $\tilde{K}_0(A) \neq 0$. Now the desired nontrivial elements in $\mathrm{U}(K_0(A))$ are given by $1 + x \in K_0(A)$ where $x \in \tilde{K}_0(A)$, $x \neq 0$. In fact, it is known that $\tilde{K}_0(B)$ is a nilpotent ideal in the ring $K_0(B)$ for any commutative ring B [19, Ch. IX, Prop. 4.6].

9.C Further results: a survey

Nonstable results. The Quillen–Suslin theorem 8.5 contains more information than $K_0(\Bbbk[X_1, \dots, X_n]) = \mathbb{Z}$: it says that projective modules over $\Bbbk[X_1, \dots, X_n]$ are free, and not just stably free. Can one strengthen the equality $K_1(\Bbbk) = K_1(\Bbbk[X_1, \dots, X_n])$ in the same way, namely to the assertion $\mathrm{SL}_r(\Bbbk[X_1, \dots, X_n]) = \mathrm{E}_r(\Bbbk[X_1, \dots, X_n])$? The following theorem of Suslin [342] shows that this is indeed the case, at least if r is large enough:

Theorem 9.26. *Let R be a noetherian ring of finite Krull dimension* $\dim R$. *Then the natural map*

$$\mathrm{GL}_r(R[X_1,\dots,X_n])/\mathrm{E}_r(R[X_1,\dots,X_n]) \to K_1(R[X_1,\dots,X_n])$$

is surjective for $r \geq \max(\dim R + 1, 2)$ *and bijective for* $r \geq \max(\dim R + 2, 3)$.

Remark 9.27. (a) For $r \geq 3$ the source of the map in Theorem 9.26 is not merely the set of cosets: suslin [342] has shown that $\mathrm{E}_r(A)$ is a normal subgroup of $\mathrm{GL}_r(A)$ for any ring A whenever $r \geq 3$.

(b) In the special case $n = 0$ Theorem 9.26 is essentially equivalent to the Bass–Vaserstein surjective K_1-stabilization estimate [19], [362] for noetherian rings.

Combining Theorems 9.5(b) and 9.26, we get

Corollary 9.28. *For every field* $\Bbbk$ *and all* $n \in \mathbb{N}$ *we have*

$$\mathrm{SL}_n(\Bbbk[X_1,\dots,X_r]) = \mathrm{E}_r(\Bbbk[X_1,\dots,X_n]), \quad r \geq 3.$$

That the results above are sharp follows from Cohn's observation [87]: for every field $\Bbbk$

$$\begin{pmatrix} 1+XY & X^2 \\ -Y^2 & 1-XY \end{pmatrix} \notin \mathrm{E}_2(\Bbbk[X,Y]).$$

See [237, Ch. I, §8] for Park's elementary proof of this result. It follows from Example 9.23 that the matrix above does not belong to the group $\mathrm{E}_n(\Bbbk[X^2, XY, Y^2])$, no matter how large n is.

The proof of Theorem 9.26 in [342] is a K_1-analogue of Quillen's solution [297] to the Serre problem, sketched in Section 8.C. (See [237, Ch. VI] for a very lucid exposition of Suslin's proof.) It involves several results on invertible matrices over polynomial rings that are of independent interest, like matrix analogues of Quillen's local-global principle 8.10 and of the affine Murthy–Horrocks theorem 8.12. The proof is constructive and, like that in [297], can be converted into an algorithm for factoring elements of $\mathrm{SL}_r(\Bbbk[X_1,\dots,X_n])$ into elementary matrices; see Park and Woodburn [281]. Such algorithms have applications in *signal processing* [281].

Suslin's techniques can be generalized to monoid rings. This opens up a possibility for results like Theorems 9.29 and 9.35 below.

Elementary actions on unimodular rows. To state the next result we introduce a special class of cones. A cone $C \subset \mathbb{R}^r$ is called of *simplicial growth* if there is a sequence of cones

$$\{0\} = C_0 \subset C_1 \subset \cdots \subset C_k = C$$

such that the closures of the sets $C_i \setminus C_{i-1}$, $i = 1,\dots,k$, are simplicial cones. It is easily seen that all cones of dimension ≤ 3 are of simplicial growth (Figure 9.1), while starting from dimension 4 this is no longer the case. For instance, the 4-cone $C \subset \mathbb{R}^4$ over an octahedron is not of simplicial growth, for otherwise at least one of the vertices of the octahedron would be simple.

For a commutative ring A an r-row $(a_1,\dots,a_r) \in A^r$ is *unimodular* if $Aa_1 + \cdots + Aa_r = A$. The set of unimodular r-rows over A is denoted by $\mathrm{Um}_r(A)$. The general

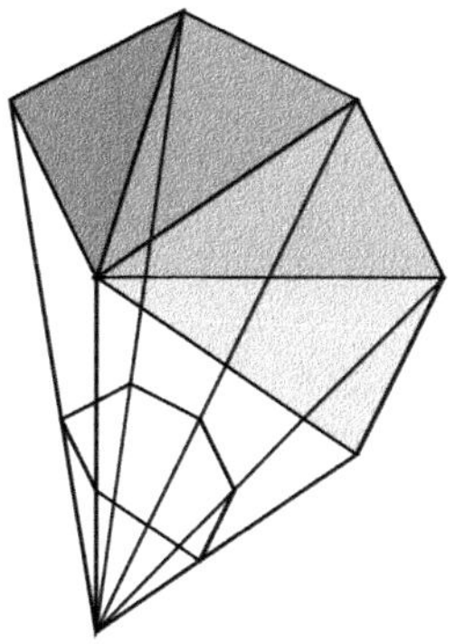

Fig. 9.1. 3-cones are of simplicial growth

linear group $\mathrm{GL}_r(A)$ acts on $\mathrm{Um}_r(A)$ by right multiplication, and we can restrict the action to the subgroup $\mathrm{E}_r(A)$.

The following extension of the surjectivity part of Theorem 9.26 has been obtained in [158], [159].

Theorem 9.29. *Let R be a noetherian ring of finite Krull dimension and $M \subset \mathbb{Z}^d$ be an affine positive monoid for which the cone $C(M) \subset \mathbb{R}^d$ is of simplicial growth. Then the group of elementary matrices $\mathrm{E}_r(R[M])$ acts transitively on the set of unimodular rows $\mathrm{Um}_r(R[M])$ whenever $r \geq \max(\dim R + 2,\ 3)$.*

Remark 9.30. (a) Theorem 9.29 is nontrivial already in the special case of simplicial affine monoids.

(b) It follows from Theorem 9.29 that the map

$$\mathrm{GL}_r(R[M])/\,\mathrm{E}_r(R[M]) \to K_1(R[M])$$

is surjective for $r \geq \max(\dim R + 1, 2)$. In the special case $M = \mathbb{Z}_+^r$ we recover the surjectivity part of Theorem 9.26. This may indicate that general monoid rings and polynomial rings have similar stabilizations for all higher K-groups. However, currently not even the validity of Theorem 9.29 for general monoids is known. For more information on this topic see [164].

Elementary actions and subintegral extensions. Concluding this section we give an application of Theorem 9.29 which goes beyond the class of monoid rings. It is a K_1-stabilization analogue of Swan's theorem on projective modules over subintegral extensions (Theorem 8.46).

Corollary 9.31. *Let $A \subset B$ be a subintegral extension and $\bar{a} = (a_1, \dots, a_r) \in \mathrm{Um}_r(A)$ for some $r \geq 3$. Then we have the equivalence*

$$\bar{a} \sim_{\mathrm{E}_r(A)} (1, 0, \dots, 0) \iff \bar{a} \sim_{\mathrm{E}_r(B)} (1, 0, \dots, 0).$$

An auxiliary result we need is the following *Milnor patching for unimodular rows.*

Lemma 9.32. *Let*

$$\begin{array}{ccc} A & \xrightarrow{\alpha} & B \\ {\scriptstyle\beta}\downarrow & & \downarrow{\scriptstyle\delta} \\ C & \xrightarrow[\gamma]{} & D \end{array}$$

be a Milnor square. Assume the image $\bar{d} \in \mathrm{Um}_r(D)$ *of* $\bar{b} \in \mathrm{Um}_r(B)$ *is in the* $\mathrm{E}_r(D)$*-orbit of* $(1, 0, \ldots, 0)$. *Then there is an element* $\bar{a} \in \mathrm{Um}_r(A)$ *such that* $\alpha(\bar{a}) \sim_{\mathrm{E}_r(B)} \bar{b}$.

Proof. Let $\varepsilon_D \in \mathrm{E}_r(D)$ be such that $\bar{d}\varepsilon_D = (1, 0, \ldots 0)$.

(1) γ is surjective. In this situation $\mathrm{E}_r(C) \to \mathrm{E}_r(D)$ is a surjective homomorphism. Let $\varepsilon_C \in \mathrm{E}_r(C)$ be a preimage of ε_D^{-1}. Then the first row of ε_C, say $\bar{c} \in C^r$, maps to $\bar{d}$ under γ and is unimodular. Because the diagram is cartesian, there exists a (unique) row $\bar{a} = (a_1, \ldots, a_r) \in A^r$ mapping respectively to $\bar{b}$ and $\bar{c}$. We only need to prove that $\bar{a} \in \mathrm{Um}_r(A)$. Since in our situation α is also surjective, the ideal of A generated by the components of $\bar{a}$ contains an element of the from $1 + x$ for some $x \in \mathrm{Ker}(\alpha)$. If we knew that this ideal would contain also $x + y$ for some $y \in \mathrm{Ker}(\beta)$, then the desired unimodularity would follow because $1 + x - x\big((1 + x) - (x + y)\big) = 1 + xy = 1$. (Here we use that $\mathrm{Ker}(\alpha) \cap \mathrm{Ker}(\beta) = 0$.) Now the existence of such element y is established as follows. There are $z_1, \ldots, z_r \in C$ such that

$$\big(\beta(x)z_1\big)c_1 + \cdots + \big(\beta(x)z_r\big)c_r = \beta(x)$$

where $\bar{c} = (c_1, \ldots, c_r)$. Since

$$\beta(x)z_1, \ldots, \beta(x)z_r \in \mathrm{Ker}(\gamma),$$

there are elements $a'_1, \ldots, a'_r \in A$ mapping to $\beta(x)z_1, \ldots, \beta(x)z_r$, respectively. Then $a_1a'_1 + \cdots + a_ra'_1 = x + y$ for some $y \in \mathrm{Ker}(\beta)$.

(2) δ is surjective. In this situation ε_D has a preimage $\varepsilon_B \in \mathrm{E}_r(B)$. The elements $\bar{b}\varepsilon_B$ and $(1, 0, \ldots, 0) \in \mathrm{Um}_r(C)$ have the same image in $\mathrm{Um}_r(D)$. Then the argument in (1) shows that $\bar{b}\varepsilon_B$ has a preimage in $\mathrm{Um}_r(A)$, and this completes the proof. □

Proof of Corollary 9.31. We use Theorem 9.29 with $R = \mathbb{Z}$: for an affine simplicial monoid M the group of elementary matrices $\mathrm{E}_r(\mathbb{Z}[M])$ acts transitively on $\mathrm{Um}_r(\mathbb{Z}[M])$ if $r \geq 3$.

By induction it is enough to treat the case of an elementary subintegral extension $B = A[x]$ with $x^2, x^3 \in A$, $x \notin A$. Let $S = A \cup \{x\}$ and consider the polynomial algebra $\mathbb{Z}[t_s : s \in S]$. The ring homomorphism

$$g : \mathbb{Z}[t_s : s \in S] \to B, \qquad t_s \mapsto s, \quad s \in S,$$

is surjective and we have the Milnor square

$$\begin{array}{ccc} D & \longrightarrow & A \\ \downarrow & & \downarrow \\ \mathbb{Z}[t_s : s \in S] & \xrightarrow[g]{} & B. \end{array}$$

Since $A \to B$ is injective, D can be identified with the preimage of A in $\mathbb{Z}[t_s : s \in S]$. By Lemma 9.32 it is enough to show that $\mathrm{E}_r(D)$ acts transitively on $\mathrm{Um}_r(D)$ for $r \geq 3$.

Let N be the multiplicative monoid consisting of 1 and those monomials in the indeterminates t_s which are divisible by t_x^2. Then N is a filtered union of affine simplicial monoids. In fact, for a finite subset $T \subset \{t_s : s \in A\}$ let M_T be the monoid of those monomials that involve only indeterminates from $T \cup \{t_x\}$ and, in addition, are divisible by t_x^2. In the following we identify monomials in M_T with their exponent vectors in $\mathbb{Z}^{n+1}$, $n = \#T$.

Let M_{Tk}, $k \in \mathbb{N}$, be the integral closure in M_T of the submonoid of M_T generated by the monomials t_x^2 and $t_x^2 t_s^k$, $s \in T$. Then the M_{Tk} form an ascending chain of affine simplicial monoids and their union is M_T. (We leave the easy proof of this claim to the reader.) Thus M_T is the filtered union of simplicial submonoids. Now N, too, is the filtered union of affine simplicial monoids because it is the filtered union of the M_T. In particular, $\mathrm{E}_r(\mathbb{Z}[N])$ acts transitively on $\mathrm{Um}_r(\mathbb{Z}[N])$ for $r \geq 3$.

The image of the ideal $I = \mathbb{Z}(N \setminus \{1\}) \subset \mathbb{Z}[t_s : s \in S]$ under g is contained in A since any natural number ≥ 2 is a nonnegative integral combination of 2 and 3. Therefore, I is an ideal of the subring $D \subset \mathbb{Z}[t_s : s \in S]$ as well. Consider the Milnor square

$$\begin{array}{ccc} \mathbb{Z}[N] & \longrightarrow & D \\ \downarrow & & \downarrow \\ \mathbb{Z} & \longrightarrow & D/I. \end{array}$$

If the action of $\mathrm{E}_r(D/I)$ on $\mathrm{Um}_r(D/I)$ were transitive, then, in view of the transitivity of $\mathrm{E}_r(\mathbb{Z}[N])$ on $\mathrm{Um}_r(\mathbb{Z}[N])$, Milnor patching would imply the desired transitivity of $\mathrm{E}_r(D)$ on $\mathrm{Um}_r(D)$. But $(D/I)_{\mathrm{red}} = \mathbb{Z}[t_s : s \in A]$, and the property we are interested in is invariant modulo nilpotents; see Exercise 9.7. So it is enough to apply Theorem 9.29 again, this time to $\mathbb{Z}[t_s : s \in A]$. □

Nilpotence of higher K-theory of monoid rings. Let c be a natural number, F : $\mathsf{Rings} \to \mathsf{AbGroups}$ a functor, R a ring, and M a monoid.

The group endomorphism $\mathsf{F}(R[M]) \to \mathsf{F}(R[M])$, induced by the monoidal Frobenius homomorphism $c_* : R[M] \to R[M]$ (see p. 240), will be denoted by the same symbol c_*.

We say that the multiplicative monoid $\mathbb{N}$ of natural numbers acts *nilpotently* on $\mathsf{F}(R[M])$ if for every natural number $c \geq 2$ and every element $x \in \mathsf{F}(R[M])$ we have

$$(c^j)_*(x) \in \mathsf{F}(R), \qquad j \gg \infty \quad (\text{potentially depending on } c \text{ and } x),$$

where $\mathsf{F}(R)$ is thought of as a subgroup of $\mathsf{F}(R[M])$ in a natural way. Equivalently, $\mathbb{N}$ acts nilpotently on $\mathsf{F}(R[M])$ if the induced action of the multiplicative monoid $\mathbb{N}$ on the quotient group $\mathsf{F}(R[M])/\mathsf{F}(R)$ is nilpotent.

In the special case when the functor F commutes with filtered direct limits—*as is the case for the functors K_i, $i \geq 0$*—there is another equivalent definition of the nilpotence above: $\mathbb{N}$ acts nilpotently on $\mathsf{F}(R[M])$ if and only if

$$\mathsf{F}(R) = \mathsf{F}\left(R\left[M^{c^{-\infty}}\right]\right)$$

where

$$M^{c^{-\infty}} = \varinjlim \left(M \xrightarrow{c_*} M \xrightarrow{c_*} \cdots\right),$$

writing the monoid operation multiplicatively. In other words, $M^{c^{-\infty}}$ is the c-divisible hull of M mentioned in Exercise 8.8.

This leads us to the following class of monoids: let c be a natural number; a monoid M is called *c-divisible* if for any element $m \in M$ there exists $n \in M$ such that $n^c = m$.

As soon as $c \geq 2$, (1) no c-divisible monoid save the trivial one is finitely generated, and (2) all c-divisible monoids are seminormal (prove this!).

Now we are going to discuss the relevance of monoidal Frobenius homomorphisms for K-theory of monoid rings, already indicated by the fact that c-divisible monoids are seminormal for $c \geq 2$.

Somewhat unexpectedly, Theorem 8.37(a) is equivalent to the nilpotence of the multiplicative action of $\mathbb{N}$ on the Grothendieck group. More precisely, the following equivalence is shown in [166] for an *arbitrary* ring R:

$$\mathbb{N} \text{ acts nilpotently on } K_0(R[M]) \text{ for all monoids } M \iff K_0(R) = K_0(R[M]) \text{ for all seminormal monoids.}$$

(See Exercise 9.4 for the nontrivial implication $\Longrightarrow$.) One should remark that, in view of Theorem 9.7 and the results discussed in this subsection, no higher (or even K_1-) version of this equivalence is possible.

It is therefore natural to interpret the potential nilpotence of the action of $\mathbb{N}$ on the higher groups $K_i(R[M])$, R a regular ring, as a higher stable version of Theorem 8.4. It is clear that for higher K-groups one has to restrict oneself to positive monoids. Otherwise the expected nilpotence fails already for the functorial direct summand $\mathrm{U}(R[M]) \subset K_1(R[M])$.

Even more is true: quillen's theorem 9.5(b) is an easy consequence of the nilpotence of higher K-groups of monoid rings: for every $r \in \mathbb{N}$ one has

$$\mathbb{N} \text{ acts nilpotently on } K_i(R[\mathbb{Z}_+^r]) \iff K_i(R) = K_i(R[\mathbb{Z}_+^r]).$$

In fact, assume $\mathbb{N}$ acts nilpotently on $K_i(R[\mathbb{Z}_+^r])$ for some $i, r \in \mathbb{N}$. Let c be a natural number ≥ 2. The monoidal Frobenius $c_* : R[\mathbb{Z}_+^r] \to R[\mathbb{Z}_+^r]$ makes $R[\mathbb{Z}_+^r]$ a free module over itself of rank c^r. Then using the *transfer maps* in K-theory one concludes that for every element $x \in K_i(R[\mathbb{Z}_+^r])/K_i(R)$ we have $c^j \cdot x = 0$, $j \gg 0$. Choosing another natural number $d \geq 2$, coprime to c, we get $d^j \cdot x = 0$ for $j \gg 0$. In view of the condition $\gcd(c, d) = 1$ the element x must be 0.

Free monoids are very special in that there is no way to use the trick above for nonfree (even affine) monoids—a consequence of Theorem 9.7. Moreover, as shown in [57], for an affine monoid M and a natural number $c \geq 2$ the monoidal Frobenius $c_* : R[M] \to R[M]$ makes $R[M]$ an $R[M]$-module of finite projective dimension[1] if and only if $M \cong \mathbb{Z}_+^s$ for some $s \in \mathbb{Z}_+$ (or $c = 1$).

[1] This is what is actually needed to make the transfer maps work, as in the case of free monoids.

The observations listed above have motivated the following theorem, first proved in [168] in the critical case when the coefficient ring is a field and then extended to a bigger class of rings in [170].

Theorem 9.33. *Let R be a regular ring containing $\mathbb{Q}$. Let $i, c \in \mathbb{N}$, $c \geq 2$, and M be a positive monoid. Then the multiplicative monoid $\mathbb{N}$ acts nilpotently on $K_i(R[M])$.*

Remark 9.34. The argument spread over [166] and [168] is a higher K-theoretical version of the proof of Theorem 8.4. However, one faces a number of issues nonexisting for projective modules and K_0. One notable example is that in a shrinking process, similar to the one in Section 8.G, the choice of the polytopes Γ_i and the points v_i must be made much more carefully [166, §6].

Based on the recent advances in K-theory of singular varieties, afforded in the papers [94], [97], [175], Cortiñas et al. gave an alternative proof of the crucial case of Theorem 9.33 when R is a characteristic 0 field [96]. Although the new proof is shorter, the two approaches are "cousins" in the word of the authors of [96] in that both have a common ancestor— Cortiñas verification of the KABI conjecture [93].

For the Bass–Whitehead and Milnor groups, we can do much better by not only dropping the condition $\mathbb{Q} \subset R$, but also by proving nilpotence for the unstable groups [169]. This makes the result a direct K_i-analogue ($i = 1, 2$) of Theorem 8.4.

Theorem 9.35. *Let M be a positive monoid, $c \geq 2$ a natural number, R a regular ring, and $\Bbbk$ a field. Then:*

(a) *$\mathbb{N}$ acts nilpotently on $K_{2,r}(R[M])$ for $r \geq \max(5, \dim R + 3)$.*
(b) *For every matrix $A \in \mathrm{GL}_r(R[M])$, $r \geq \max(3, \dim R + 2)$, we have $(c^j)_*(A) = \alpha\beta$ with $\alpha \in \mathrm{E}_r(R[M])$, $\beta \in \mathrm{GL}_r(R)$, $j \gg 0$.*
(c) *There is an algorithm which for any matrix $A = \mathrm{SL}_r(\Bbbk[M])$, $r \geq 3$, finds an integer number $j_A \geq 0$ and a factorization of the form*

$$(c^{j_A})_*(A) = \prod_k e_{p_k q_k}(\lambda_k), \qquad e_{p_k q_k}(\lambda_k) \in \mathrm{E}_r(\Bbbk[M]).$$

In the theorem we have used the following notation: for a ring A the group $K_{2,r}(A)$ is the kernel of the homomorphism $\mathrm{St}_r(A) \to \mathrm{E}_r(A)$ where $\mathrm{St}_r(A)$ is generated by the Steinberg symbols x_{ij}^{λ}, $1 \leq i, j \leq r$, subject to the usual Steinberg relations.

Exercises

9.1. Show that for a ring R and a matrix $\alpha \in \mathrm{GL}(R)$ one has

$$\left[\begin{pmatrix} \alpha & 0 & \cdots & 0 \\ 0 & \alpha & \cdots & 0 \\ \vdots & \vdots & \ddots & \vdots \\ 0 & 0 & \cdots & \alpha \end{pmatrix}\right] - [\alpha^m] = m[\alpha] \in K_1(R).$$

9.2. Give an example of a commutative noncancellative monoid M_1 and a commutative nontorsionfree monoid M_2 such that such that $SK_1(R) = SK_1(R[M_1]) = SK_1(R[M_2])$ for every regular ring R.

9.3. Let Γ be a rational conical complex whose cones $C_c \subset \mathbb{R}^{d_c}$ are simplicial. Let the monoids of the monoidal complex $\mathbb{M}$ given by $M_c = C_c \cap \mathbb{Z}^{d_c}$.
Show that the following conditions are equivalent for every regular ring R:
(a) $K_1(R[\mathbb{M}]) = K_1(R[\mathbb{M}][X])$;
(b) $K_1(R) = K_1(R[\mathbb{M}])$;
(c) every cone C_c, $c \in \Gamma$, is unimodular.

9.4. Let (F, G) be an admissible pair of functors in the sense of Exercise 8.7 and $c, d \geq 2$ be coprime natural numbers.
(a) Show the following implications

$$\mathsf{F}(R) = \mathsf{F}(R[M]) \text{ for all positive } c\text{-divisible and } d\text{-divisible monoids } M \implies \mathsf{F}(R) = \mathsf{F}(R[M]) \text{ for all positive monoids.}$$

Hint: use the same approach as suggested in Exercise 8.8.
(b) Show the same implication as in (a) without the restriction to positive monoids.

9.5. For a regular ring R give a detailed proof of the equivalence

$$K_1(R[M])/K_1(R) \neq 0 \text{ for all affine normal positive nonfree monoids } M \iff K_1(R[M])/K_1(R) \neq 0 \text{ for all monoids } M \text{ of rank } \leq 3 \text{ in the same class,}$$

which has been mentioned in Remark 9.21.

9.6. Prove (without using Theorem 9.35) that when $M \subset \mathbb{Q}^d$ is a c-divisible monoid for some $c \geq 2$ and $\mathbb{R}_+ M \subset \mathbb{R}^d$ is a simplicial cone then $K_1(R) = K_1(R[M])$ for a regular ring R.
Hint: apply claim (7) in the hint for Exercise 8.8 to Milnor squares similar to the ones in the proof of Lemma 8.18.

9.7. For a natural number $r \geq 2$ and a ring R show the equivalence

$$(a_1, a_2, \dots, a_n) \sim_{\mathrm{E}_r(R)} (1, 0, \dots, 0) \iff (\bar{a}_1, \bar{a}_2, \dots, \bar{a}_n) \sim_{\mathrm{E}_r(R/\mathrm{J}(R))} (\bar{1}, \bar{0}, \dots, \bar{0}).$$

($\mathrm{J}(R)$ is the Jacobson radical of R.)

9.8. For a natural number $r \geq 2$, a ring R, an affine monoid M, and a proper ideal $I \subset M$ show the implication

$$\mathrm{E}_r(R[M]) \text{ acts transitively on } \mathrm{Um}_r(R[M]) \implies \mathrm{E}_r(R[M]/RI) \text{ acts transitively on } \mathrm{Um}_r(R[M]/RI).$$

Hint: use the same strategy as in the proof of Theorem 8.48. This is possible in view of Lemma 9.32 and Exercise 9.7.

9.9. Let R be a noetherian ring of finite Krull dimension, $r \geq \max(\dim R + 2, 3)$ a natural number, M an affine positive monoid, and $I \subset M$ a nonempty proper ideal. Suppose that every proper face of $\mathbb{R}_+ M$ is simplicial if rank $M > 4$. Using Theorem 9.29, prove that $\mathrm{E}_r(R[M]/RI)$ acts transitively on $\mathrm{Um}_r(R[M]/RI)$.

Notes

Milnor patching and Mayer–Vietoris sequences show that projective modules and invertible matrices, as well as their stable counterparts, are intrinsically linked to each other. These results and the fundamental theorem for K_0 and K_1 constitute an essential part of classical K-theory. Other important topics in the field include the congruence subgroup problem, stabilization for projective modules and invertible matrices, K-theory of integer group rings (motivated from topology), and number theoretical applications. Bass' book [19] offered the most comprehensive description of the state of art by the end of the 1960s, although essential progress took place in some of these directions shortly after its publication. Milnor's K_2 is introduced in [262] while many technical aspects of the concept had been known since Steinberg's work [334] on universal central extensions of Chevalley groups. Then there was a flurry of definitions of higher K-theory (by Swan, Gersten, Volodin, and others), culminating in Quillen's decisive work [295], [296]. Weibel wrote a well-documented survey [376] of the history and development of algebraic K-theory before 1980.

The dichotomy between the K_0 and K_1 behavior of affine monoid rings is caused by the absence of the excision property for K_1 and K_2, as shown in this chapter. Exercise 9.6 indicates that the multiplicative action of $\mathbb{N}$ on the lower K-groups of $R[M]$ compensates the lack of excision with respect to the ideals of interior monomials. This was extended to higher K-groups in [160], using results from [343], and this served as a precursor to Theorem 9.33—a nilpotent analogue of Theorem 8.4. A similar variant of Theorem 8.48 was proved in [160]. However, the actual computation of higher K-groups of monomial quotients is a hard problem, even for rings of the form $R[X]/(X^n)$; see Hesselholt [187].

Although a direct extension of Theorem 8.4 to higher K-groups is impossible, there is another conjectural indirect analogue. It is discussed in [164] and phrased in terms of stabilizations of Volodin's higher K-groups of monoid rings. For K_1 and K_2 this is partially achieved in Theorems 9.35 and 9.29. However, the situation is complicated by the fact that already for polynomial rings the necessary technique is available only for K_1 (Suslin [342]) and K_2 (Tulenbaev [359]).

10

Varieties

In the first nine chapters we investigated affine monoid algebras with combinatorial, ring-theoretic, and K-theoretic methods. The class of objects will now be considerably extended. Historically, the subject drew inspirations from various fields, such as algebraic geometry, topology, representation theory —all of them relevant in Chapter 10.[1]

Affine varieties serve as the building blocks of general algebraic varieties, and the algebraic varieties built from the spectra of affine monoid algebras are toric varieties.

Several invariants of algebraic varieties can be considered as natural generalizations of their affine specializations that we have encountered already. But also new constructions, in particular Chow groups and Chow rings, will enter the scene. As in the other chapters, we will provide short introductions and surveys of the relevant theories.

Section 10.A serves as an overview of that part of algebraic geometry that is particularly important for us: vector bundles, Cartier and Weil divisors, homological and cohomological Grothendieck groups.

Toric varieties are introduced in Section 10.B. We give a brief sketch of their construction from rational fans and describe the divisors on projective toric varieties since the ample Cartier divisors correspond to lattice polytopes—objects of primary interest in the previous chapters. We will use this connection for the computation of the automorphism group of a projective toric variety from the corresponding ring-theoretic counterpart in Section 5.D. The overview includes the description of the homogeneous coordinate ring of a toric variety, following Cox' work. It is a broad generalization of the realization of normal affine monoid rings as rings of invariants in Section 5.B.

Chow groups beyond divisors will be introduced in Section 10.C, together with their computation in the toric case. This section is a natural extension of the description of divisor class groups in Section 4.F.

[1] One obvious omission from the list is arithmetic, which does not play any essential role in our treatment.

W. Bruns, J. Gubeladze, *Polytopes, Rings, and K-Thoery*, Springer Monographs in Mathematics,
DOI 10.1007/b105283_10,

Section 10.D is again a digression into the general theory, surveying intersection theory, Chow rings, Chern classes, and the Grothendieck–Riemann–Roch theorem together with its consequences for the computation of Grothendieck groups. An application of the Hirzebruch–Riemann–Roch theorem to lattice point counting builds a bridge to the toric combinatorics studied in Chapter 6.

Chow rings of toric varieties are considered in Section 10.E. We discuss the operational Chow ring of complete toric varieties, following the works of Fulton, MacPherson, Sottile, and Sturmfels, and its relationship to McMullen's polytope algebra. For simplicial toric varieties the Chow ring has been computed by Jurkiewicz and Danilov. Its concrete combinatorial description leads to Stanley's proof of McMullen's g-theorem that we encountered already in Section 6.E. The computations of Chow cohomology and Grothendieck groups are interwoven via the Grothendieck–Riemann–Roch theorem (for singular varieties)—a recurring theme in this chapter. The K-theoretical computations become rather explicit for complete smooth toric varieties in Section 10.E, culminating in a polytopal description of the Grothendieck ring (and its consequence for higher K-theory).

The last two sections are in a sense a continuation of Chapters 8 and 9. In Chapter 8 it has been proved that the Grothendieck K_0 is the smallest possible for seminormal affine monoid algebras. In Section 10.F we will see that K_0 becomes an extremely complicated object already one step outside the affine or smooth world. The source of such complexity of K_0 is the nontriviality of the SK_1-groups of affine normal monoid rings proved in Chapter 9.

The final section, Section 10.G, deals with the triviality of projective modules or vector bundles in the equivariant setting where an additional structure is given by a group operation, and simultaneously the trivializing isomorphisms are required to respect this operation. We will prove the theorem of Masuda, Moser–Jauslin, and Petrie, solving the equivariant Serre problem for the action of diagonalizable groups on affine spaces. A key step is the application of Theorem 8.4.

In contrast to the previous chapters, we have not provided separate notes. They are dispersed in numerous comments throughout the text.

Throughout this chapter we assume that $\Bbbk$ is an algebraically closed field and all varieties are defined over $\Bbbk$. Exceptions from this convention will be explicitly mentioned.

10.A Vector bundles, coherent sheaves, and Grothendieck groups

This section gives a survey of the basic objects of algebraic geometry, related to K-theory. The theory of Grothendieck groups will not be developed for general schemes or of exact categories. Instead we restrict ourselves right away to varieties over an algebraically closed field. The reader should consult:

(a) Fulton [134] and Hartshorne [178] for algebraic geometry, and
(b) Friedlander and Weibel [128, Lect. I,III,V], Quillen [296, §§ 7,8], and Swan [352, Sect. 1–4] for K-theoretic aspects.

These sources provide the details that we cannot include, as well as a more general treatment.

Varieties, sheaves, and vector bundles. A *variety* is an integral separated scheme of finite type over $\Bbbk$. The varieties $\mathcal{V}$ form the category $\mathsf{Var}(\Bbbk)$. The structure sheaf of $\mathcal{V}$ is denoted by $\mathcal{O}_{\mathcal{V}}$, and $\mathcal{O}_{\mathcal{V}}(X)$ is the ring of sections of $\mathcal{O}_{\mathcal{V}}$ over an open subset $X \subset \mathcal{V}$. For simplicity we write $\mathcal{O}(\mathcal{V})$ for $\mathcal{O}_{\mathcal{V}}(\mathcal{V})$. A *subvariety* is a closed integral subscheme and an *open subset* is open in the Zariski topology.

We let $\mathrm{R}(\mathcal{V})$ be the field of rational functions on $\mathcal{V}$. The *local ring of $\mathcal{V}$ along a subvariety* $\mathcal{U} \subset \mathcal{V}$ is denoted by $\mathcal{O}_{\mathcal{U},\mathcal{V}}$. It is defined as the subring of $\mathrm{R}(\mathcal{V})$ consisting of all rational functions regular on a nonempty open subset of $\mathcal{U}$. Alternatively, it can be defined as follows. Let $X \subset \mathcal{V}$ be an affine open subset meeting $\mathcal{U}$. Then $X \cap \mathcal{U}$ is the zero set of a prime ideal $\mathfrak{p} \subset \mathcal{O}(X)$, and $\mathcal{O}_{\mathcal{U},\mathcal{V}} = \mathcal{O}(X)_{\mathfrak{p}}$.

The category of sheaves of $\mathcal{O}_{\mathcal{V}}$-modules will be denoted by $\mathsf{Mod}(\mathcal{O}_{\mathcal{V}})$. It is an abelian category with enough injectives. We write $\Gamma(X, \mathcal{F})$ for the $\mathcal{O}_{\mathcal{V}}(X)$-module of global sections of $\mathcal{F} \in \mathsf{Mod}(\mathcal{O}_{\mathcal{V}})$ over an open subset $X \subset \mathcal{V}$. The cohomology groups are the right derived functors of $\Gamma(\mathcal{V}, -) : \mathsf{Mod}(\mathcal{O}_{\mathcal{V}}) \to \mathsf{AbGroups}$ (for their existence injective resolutions are needed). The full subcategory of $\mathsf{Mod}(\mathcal{V})$ formed by the *coherent sheaves* is denoted by $\mathsf{Coh}(V)$.

A morphism of varieties $f : \mathcal{U} \to \mathcal{V}$ induces a pair of adjoint functors:

(a) the *direct image functor* $f_* : \mathsf{Mod}(\mathcal{O}_{\mathcal{U}}) \to \mathsf{Mod}(\mathcal{O}_{\mathcal{V}})$, and
(b) the *inverse image functor* $f^* : \mathsf{Mod}(\mathcal{O}_{\mathcal{V}}) \to \mathsf{Mod}(\mathcal{O}_{\mathcal{U}})$.

In the affine case these are scalar restriction and extension along the underlying ring homomorphism.

An *algebraic vector bundle of rank n* on a variety $\mathcal{V}$ is a pair, given by a variety $\mathcal{E}$ and a morphism $\mathrm{p} : \mathcal{E} \to \mathcal{V}$ that admits an affine open cover $\mathcal{V} = \bigcup_{i=1}^{m} \mathcal{U}_i$ and isomorphisms

$$\sigma_i : \mathrm{p}^{-1}(\mathcal{U}_i) \to \mathcal{U}_i \times \Bbbk^n, \qquad i = 1, \ldots, m,$$

such that

(a) the diagram

$$\begin{array}{ccc} \mathrm{p}^{-1}(\mathcal{U}_i) & \xrightarrow{\sigma_i} & \mathcal{U}_i \times \Bbbk^n \\ & {\scriptstyle \mathrm{p}}\searrow \quad \swarrow{\scriptstyle \mathrm{pr}_{\mathcal{U}_i}} & \\ & \mathcal{U}_i & \end{array}$$

commutes; here $\mathrm{pr}_{\mathcal{U}_i}$ is the projection onto the first component;
(b) the induced map $\sigma_j \sigma_i^{-1} : (\mathcal{U}_i \cap \mathcal{U}_j) \times \Bbbk^n \to (\mathcal{U}_i \cap \mathcal{U}_j) \times \Bbbk^n$ is $\Bbbk$-linear on the fibers for all i and j.

We will usually suppress the adjective "algebraic" when we deal with vector bundles. One calls $\mathcal{V}$ the *base variety*, $\mathcal{E}$ the *total variety*, and p the *projection* of $\mathcal{E}$.

The restriction of a vector bundle $\mathrm{p} : \mathcal{E} \to \mathcal{V}$ to an open set or subvariety $U \subset \mathcal{V}$ is denoted by $\mathcal{E}|_U$. More precisely, $\mathcal{E}|_U$ is the vector bundle $\mathrm{p}|_{\mathrm{p}^{-1}(U)} : \mathrm{p}^{-1}(U) \to U$. A *section* of $\mathcal{E}$ over U is a morphism $\sigma : U \to \mathcal{E}$ such that $\mathrm{p}\,\sigma = \mathbf{1}_U$.

Vector bundles of rank 1 are called *line bundles*. Their isomorphism classes form the *Picard group* of $\mathcal{V}$, denoted by $\mathrm{Pic}(\mathcal{V})$. The multiplication is induced by the tensor product.

A *homomorphism* $f : \mathscr{E}_1 \to \mathscr{E}_2$ between two vector bundles on $\mathscr{V}$ is a morphism of varieties that is compatible with the projections and is $\Bbbk$-linear fiberwise. The set of homomorphisms $f : \mathscr{E}_1 \to \mathscr{E}_2$ is denoted by $\mathsf{Hom}(\mathscr{E}_1, \mathscr{E}_2)$; it is itself a vector bundle in a natural way. A vector bundle isomorphic to $\mathrm{pr}_{\mathscr{V}} : \mathscr{V} \times \Bbbk^n \to \mathscr{V}$ is called *trivial.*

The category of vector bundles on $\mathscr{V}$ is denoted by $\mathsf{Vect}(\mathscr{V})$. It is a fundamental fact that the functor assigning to every vector bundle $\mathrm{p} : \mathscr{E} \to \mathscr{V}$ the sheaf of sections

$$X \mapsto \{\text{sections } \sigma : X \to \mathscr{E}\}, \qquad X \subset \mathscr{V} \text{ open},$$

establishes an equivalence between $\mathsf{Vect}(\mathscr{V})$ and the full subcategory of $\mathsf{Coh}(\mathscr{V})$ consisting of locally free sheaves on $\mathscr{V}$. Here a coherent sheaf $\mathscr{F}$ on $\mathscr{V}$ is called *locally free* if there is an affine open cover $\mathscr{V} = \bigcup_i \mathscr{U}_i$ such that $\Gamma(\mathscr{U}_i, \mathscr{F})$ is a free $\mathscr{O}_{\mathscr{V}}(\mathscr{U}_i)$-module. In view of the equivalence we will often identify vector bundles and locally free sheaves on $\mathscr{V}$.

When $\mathscr{V}$ is an affine variety, the functor of global sections $\mathscr{E} \mapsto \Gamma(\mathscr{V}, \mathscr{E})$ is an equivalence between $\mathsf{Vect}(\mathscr{V})$ and the category $\mathbb{P}(\mathscr{O}(\mathscr{V}))$ of finitely generated projective modules over $\mathscr{O}(\mathscr{V})$. In particular, $\operatorname{Pic}(\mathscr{V}) = \operatorname{Pic}(\mathscr{O}(\mathscr{V}))$.

Remark 10.1. (a) A similar statement on the equivalence of vector bundles and projective modules is valid in the topological, differentiable, and analytic categories (Rosenberg [306, 1.6], Swan [344]).

(b) It is natural to expect that (algebraic) vector bundles on affine spaces behave like their topological counterparts on contractible spaces, and the Quillen–Suslin theorem 8.5 fulfills this expectation: vector bundles on affine spaces are trivial.

If $\mathscr{F}$ is the coherent sheaf associated to a vector bundle $\mathscr{E} \to \mathscr{V}$ then the *dual* vector bundle $\mathscr{E}^*$ is defined by the coherent sheaf $\mathsf{Hom}_{\mathscr{O}_{\mathscr{V}}}(\mathscr{F}, \mathscr{O}_{\mathscr{V}})$. For a line bundle $\mathscr{L}$ on $\mathscr{V}$ we have $\mathscr{L}^{-1} = \mathscr{L}^*$.

The *direct sum* of vector bundles $\mathrm{p}_i : \mathscr{E}_i \to \mathscr{V}$, $i = 1, 2$, corresponds to the direct sum of the associated coherent sheaves. Other familiar constructions are the tensor product $\mathscr{E}_1 \otimes \mathscr{E}_2$, the exterior powers $\bigwedge^k \mathscr{E}_1$, the vector bundle $\mathsf{Hom}(\mathscr{E}_1, \mathscr{E}_2)$, and the kernel bundle $\operatorname{Ker}(f)$, provided $f : \mathscr{E}_1 \to \mathscr{E}_2$ is a fiberwise surjective homomorphism.

The *tangent vector bundle* $\mathrm{T}_{\mathscr{V}}$ of a *smooth* variety $\mathscr{V}$ corresponds to the dual $\Omega^*_{\mathscr{V}/\Bbbk}$ of the locally free sheaf $\Omega_{\mathscr{V}/\Bbbk}$ of $\Bbbk$-differentials: for an affine open subset $X \subset \mathscr{V}$ with affine coordinate ring A we have the natural isomorphism

$$\Gamma(X, \Omega_{\mathscr{V}/\Bbbk}) \cong I(A)/I^2(A), \quad I(A) = \operatorname{Ker}(A \otimes_{\Bbbk} A \to A) \quad a \otimes a' \mapsto aa'.$$

Cartier and Weil divisors. A *Cartier divisor* D is given by an open cover $\mathscr{V} = \bigcup_\alpha X_\alpha$ and functions $f_\alpha \in \mathrm{R}(X_\alpha)^\times = \mathrm{R}(\mathscr{V})^\times$ such that $f_\alpha f_\beta^{-1}$ is a regular nowhere vanishing function on $X_\alpha \cap X_\beta$ for all α, β. Cartier divisors D and D' given by open covers $\bigcup_\alpha X_\alpha$, $\bigcup_\beta Y_\beta$ and families of rational functions (f_α), (g_β) are equal if $f_\alpha g_\beta^{-1}$ is a nowhere vanishing regular function on $X_\alpha \cap Y_\beta$ for all α, β. We denote the set of Cartier divisors by $\operatorname{CDiv}(\mathscr{V})$. A function $f \in \mathrm{R}(\mathscr{V})^\times$ gives rise to a *principal Cartier divisor*, denoted by $\operatorname{div}(f)$.

Let Cartier divisors D and D' be given by open covers $\bigcup_\alpha X_\alpha$, $\bigcup_\beta Y_\beta$, and rational functions f_α, g_β. Assigning to (D, D') the Cartier divisor defined by the cover $\bigcup_{\alpha,\beta}(X_\alpha \cap Y_\beta)$ and the rational functions $f_\alpha g_\beta$, makes $\mathrm{CDiv}(\mathcal{V})$ an abelian group, whose quotient by the subgroup of principal Cartier divisors is isomorphic to $\mathrm{Pic}(\mathcal{V})$, as we explain now.

The line bundle that corresponds to the Cartier divisor D, given by the cover $\bigcup_\alpha X_\alpha$ and $f_\alpha \in \mathrm{R}(X_\alpha)^\times$, is obtained as follows. Over the intersection $X_\alpha \cap X_\beta$ we glue the trivial line bundles $\mathcal{L}_\alpha = X_\alpha \times \Bbbk$ and $\mathcal{L}_\beta = X_\beta \times \Bbbk$ via the isomorphism $\mathcal{L}_\alpha|_{X_\beta} \to \mathcal{L}_\beta|_{X_\alpha}$ which on the fiber over a point $\xi \in X_\alpha \cap X_\beta$ is the multiplication by $(f_\alpha f_\beta^{-1})(\xi)$. This patching process produces a line bundle on $\mathcal{V}$ whose isomorphism class depends only on D. Conversely, to a line bundle $\mathcal{L}$ on $\mathcal{V}$ we associate the divisor D defined as follows: we find an affine open cover $\mathcal{V} = \bigcup_\alpha X_\alpha$ and isomorphisms $\varphi_\alpha : \mathcal{L}|_{X_\alpha} \to X_\alpha \times \Bbbk$. This yields a cocycle $g_{\alpha\beta} = \varphi_\alpha \circ \varphi_\beta^{-1}|_{X_\alpha \cap X_\beta}$. It can be written as $g_{\alpha\beta} = f_\alpha f_\beta^{-1}$ with rational functions f_α, f_β since the sheaf of rational functions is constant (see [178, II.6.15]). Then the pairs (X_α, f_α) give rise to a Cartier divisor on $\mathcal{V}$. We denote the line bundle associated to a Cartier divisor D by $\mathcal{L}(D)$.

If $\mathcal{V}$ is normal and $\mathcal{U} \subset \mathcal{V}$ is a codimension one subvariety, then $\mathcal{O}_{\mathcal{U},\mathcal{V}}$ is a discrete valuation domain with fraction field $\mathrm{R}(\mathcal{V})$. Thus we have the group homomorphism

$$\mathrm{ord}_{\mathcal{U}} : \mathrm{R}(\mathcal{V})^\times \to \mathbb{Z}, \quad rs^{-1} \mapsto \mathrm{ord}_{\mathcal{U}}(r) - \mathrm{ord}_{\mathcal{U}}(s), \quad r, s \in \mathcal{O}_{\mathcal{U},\mathcal{V}}, \quad r, s \neq 0.$$

Now assume $\mathcal{V}$ is a general variety and $\mathcal{U} \subset \mathcal{V}$ is again a codimension 1 subvariety. The following definition gives rise to a group homomorphism, coinciding with the one above in the normal case:

$$\mathrm{ord}_{\mathcal{U}} : \mathrm{R}(\mathcal{V})^\times \to \mathbb{Z}, \quad rs^{-1} \mapsto \lambda_A(A/rA) - \lambda_A(A/sA),$$
$$A = \mathcal{O}_{\mathcal{U},\mathcal{V}}, \; r, s \in A, \quad r, s \neq 0,$$

$\lambda(-)$ denoting the length of the artinian ring under consideration.

Let $\mathcal{V}$ be a variety and $d = \dim \mathcal{V}$. The group $Z_{d-1}(\mathcal{V})$ of *Weil divisors* on $\mathcal{V}$ is the abelian group freely generated by the $(d-1)$-dimensional subvarieties $\mathcal{U}$ of $\mathcal{V}$.

For a rational function $r \in \mathrm{R}(\mathcal{V})^\times$ we define the Weil divisor

$$[\mathrm{div}(r)] = \sum_{\substack{\mathcal{U} \subset \mathcal{V} \\ \mathrm{codim}\, \mathcal{U} = 1}} \mathrm{ord}_{\mathcal{U}}(r)[\mathcal{U}] \in Z_{d-1}(\mathcal{V}).$$

The sum on the right makes sense because $\mathrm{ord}_{\mathcal{U}}(r) = 0$ for all but finitely many $\mathcal{U}$.

Two Weil divisors are *rationally equivalent* if their difference is in the subgroup generated by the divisors $[\mathrm{div}(f)]$ of the rational functions. The corresponding quotient group of $Z_{d-1}(\mathcal{V})$ is called the *Chow group of $(d-1)$-cycles*, denoted by $\mathrm{A}_{d-1}(\mathcal{V})$.

Let D be a Cartier divisor given by the family (f_α). For a codimension 1 subvariety $\mathcal{W} \subset \mathcal{V}$ set

$$\mathrm{ord}_{\mathcal{W}}(D) = \mathrm{ord}_{\mathcal{W}}(f_\alpha)$$

where $\operatorname{ord}_{\mathscr{W}}$ is the order function on $\mathrm{R}(\mathscr{V})^{\times}$ corresponding to $\mathscr{W}$, as defined above. Then $\operatorname{ord}_{\mathscr{W}}(D)$ does not depend on the chosen family (f_α), and we have the associated Weil divisor

$$[D] = \sum_{\substack{\mathscr{W}\subset\mathscr{V}\\ \operatorname{codim}\mathscr{W}=1}} \operatorname{ord}_{\mathscr{W}}(D)[\mathscr{W}].$$

(In this sum all but finitely many summands are 0.)

Summarizing the discussion above, we get the following commutative diagram of abelian groups:

$$\begin{array}{ccc} \mathrm{CDiv}(\mathscr{V}) & \longrightarrow & Z_{d-1}(\mathscr{V}) \\ \downarrow & & \downarrow \\ \mathrm{Pic}(\mathscr{V}) & \longrightarrow & \mathrm{A}_{d-1}(\mathscr{V}). \end{array}$$

The vertical arrows are the natural epimorphism to the quotient groups by principal divisors and rational equivalence. The horizontal arrows are embeddings if $\mathscr{V}$ is normal, and they are isomorphisms if $\mathscr{V}$ is locally factorial, for example, if it is smooth.

Suppose $\mathscr{V} = \operatorname{Spec} A$ is a normal affine variety. Then Weil divisors on $\mathscr{V}$ correspond bijectively to fractional divisorial ideals of A so that $Z_{d-1}(\mathscr{V}) = \operatorname{Div}(A)$, and $\mathrm{A}_{d-1}(\mathscr{V}) = \operatorname{Cl}(A)$.

Cohomological and homological Grothendieck groups. The *Grothendieck group of coherent sheaves* on a variety $\mathscr{V}$, denoted $G_0(\mathscr{V})$, is defined by generators $[\mathscr{F}]$, the isomorphism classes of coherent sheaves on $\mathscr{V}$, subject to the relations $[\mathscr{F}] = [\mathscr{F}'] + [\mathscr{F}'']$ for every exact sequence

$$0 \to \mathscr{F}' \to \mathscr{F} \to \mathscr{F}'' \to 0.$$

The *Grothendieck group of vector bundles* on a variety $\mathscr{V}$, denoted by $K_0(\mathscr{V})$, is defined by generators $[\mathscr{E}]$, the isomorphism classes of vector bundles on $\mathscr{V}$, subject to the relations $[\mathscr{E}] = [\mathscr{E}'] + [\mathscr{E}'']$ for every exact sequence

$$0 \to \mathscr{E}' \to \mathscr{E} \to \mathscr{E}'' \to 0.$$

In the affine case every short exact sequence of vector bundles splits. Therefore, $K_0(\mathscr{V}) = K_0(A)$ for an affine variety $\mathscr{V} = \operatorname{Spec} A$, where $K_0(A)$ is the Grothendieck group of projective modules introduced in Section 8.I.

However, unlike in the affine case, short exact sequences in $\mathsf{Vect}(\mathscr{V})$ usually do not split. A simple example is given by the Koszul complex on the projective line; see p. 363.

The embedding of the category of locally free sheaves on $\mathscr{V}$ into that of coherent sheaves induces the *Cartan homomorphism*

$$K_0(\mathscr{V}) \to G_0(\mathscr{V}).$$

This homomorphism is in general neither injective nor surjective; for toric varieties this topic will be discussed in Section 10.F. But there is one classical situation when it is an isomorphism:

Theorem 10.2. *For a smooth variety the Cartan homomorphism is an isomorphism.*

The proof is based on the basic fact that for a smooth variety $\mathscr{V}$ every coherent sheaf $\mathscr{F} \in \mathsf{Coh}(\mathscr{V})$ admits a resolution

$$0 \to \mathscr{P}_n \to \cdots \to \mathscr{P}_1 \to \mathscr{P}_0 \to \mathscr{F} \to 0$$

by locally free sheaves $\mathscr{P}_i$ [178, Exercise III.6.9]. The inverse map $G_0(\mathscr{V}) \to K_0(\mathscr{V})$ is given by $[\mathscr{F}] \mapsto \sum_{i=0}^{n}(-1)^i[\mathscr{P}_i]$.

Remark 10.3. The property of a variety $\mathscr{V}$ that every sheaf $\mathscr{F} \in \mathsf{Coh}(\mathscr{V})$ is a surjective image of a vector bundle $\mathscr{E} \in \mathsf{Vect}(\mathscr{V})$ is called the *resolution property*. It is equivalent to the existence of (not necessarily finite) resolutions of coherent sheaves by vector bundles. Classically, quasiprojective as well as locally factorial, or even $\mathbb{Q}$-factorial varieties have the resolution property (Illusie [29, pp. 160–221]). A variety is called $\mathbb{Q}$-*factorial* if every Weil divisor is $\mathbb{Q}$-Cartier; see Exercise 10.4 for the latter notion. Whether all varieties have the resolution property is an open problem (Totaro [355]). Even for toric varieties the answer is not known (compare p. 367).

Next we list a few general facts on the functorial properties of the two Grothendieck groups:

(a) The tensor product of vector bundles induces a commutative ring structure on $K_0(\mathscr{V})$. This follows from the exactness of tensor product with a vector bundle.

(b) For the same reason, $G_0(\mathscr{V})$ is naturally a $K_0(\mathscr{V})$-module.

(c) As in the affine case (Section 8.I), $\operatorname{Pic}(\mathscr{V})$ is a subgroup of the unit group of the ring $K_0(\mathscr{V})$.

(d) The inverse image functor $f^* : \mathsf{Vect}(\mathscr{V}) \to \mathsf{Vect}(\mathscr{U})$, associated to a morphism of varieties $f : \mathscr{U} \to \mathscr{V}$, induces a ring homomorphism $f^* : K_0(\mathscr{V}) \to K_0(\mathscr{U})$.

(e) The inverse image of a coherent sheaf under a morphism of varieties is again coherent. The direct image usually fails coherence unless the morphism is proper. (Note that finite morphisms, especially closed embeddings, and projective morphisms are proper [178, II.4.8, II.4.9].) The direct image functor between the categories of coherent sheaves is exact for finite morphisms while for the exactness of the inverse image one needs the morphism to be flat.

(f) It follows from (e) that a finite morphism $f : \mathscr{U} \to \mathscr{V}$ induces a group homomorphism $f_* : G_0(\mathscr{U}) \to G_0(\mathscr{V})$.

(g) Moreover, a proper morphism $f : \mathscr{U} \to \mathscr{V}$ induces a group homomorphism $G_0(\mathscr{U}) \to G_0(\mathscr{V})$ [134, 15.1] by the assignment

$$[\mathscr{F}] \mapsto \sum_{i \geq 0}(-1)^i[R^i f_* \mathscr{F}], \qquad \mathscr{F} \in \mathsf{Coh}(\mathscr{U}), \tag{10.1}$$

where $R^i f_* \mathscr{F}$ is the ith higher direct image of $\mathscr{F}$, defined as the ith derived functor of f_*. (Since f is proper, $R^i f_* \mathscr{F}$ is coherent.) The homomorphism $G_0(\mathscr{U}) \to G_0(\mathscr{V})$ will also be denoted by f_*.

(h) All definitions and constructions introduced above extend in a straightforward way from varieties to: (i) closed subsets of varieties, (ii) open subsets of closed subsets of varieties, called *locally closed subvarieties*, and (iii) finite unions of locally closed subvarieties, called *constructible sets*.

The extended theory enjoys the same functorial properties as in the irreducible case. Let $Y \subset X$ be a closed embedding of constructible sets and $U = X \setminus Y$. Then we have the *localization exact sequence*

$$G_0(Y) \xrightarrow{i_*} G_0(X) \xrightarrow{j^*} G_0(U) \to 0, \qquad (10.2)$$

where $i : Y \to X$ and $j : U \to X$ are the inclusion maps.

(i) The Grothendieck group $G_0(\mathcal{V})$ has a natural filtration $F_* G_0(\mathcal{V})$ defined as follows [134, 16.1.5]: $F_k G_0(\mathcal{V})$ is the subgroup of $G_0(\mathcal{V})$ generated by the sheaves $\mathcal{F}$ whose support (i. e., the set of points $x \in \mathcal{V}$ where the stalks of $\mathcal{F}$ are nonzero) has dimension $\leq k$. Equivalently, $F_k G_0(\mathcal{V})$ is generated by the classes $[\mathcal{O}_{\mathcal{U}}]$ where $\mathcal{U}$ runs over subvarieties of $\mathcal{V}$ of dimension $\leq k$. Using the localization sequence, one reduces this claim to the affine case, in which the equivalence results from the following general fact: every finitely generated module M over a noetherian ring A admits a system of short exact sequences

$$0 \to L_i \to M_i \to M_{i+1} \to 0, \quad i = 0, \dots, t,$$

where the L_i are cyclic, $M_0 = M$, $M_{t+1} = 0$, and $\operatorname{Supp} M_i \subset \operatorname{Supp} M$ for all i.

(j) For a proper morphism $f : \mathcal{U} \to \mathcal{V}$ and elements $\alpha \in K_0(\mathcal{V})$ and $\beta \in G_0(\mathcal{U})$ we have the *projection formula*

$$f_*(f^*(\alpha) \cdot \beta) = \alpha \cdot f_*(\beta)$$

where the product is given by the K_0-module structure on G_0.

(k) For every variety $\mathcal{V}$ the projection map $\mathrm{pr}_{\mathcal{V}} : \mathcal{V} \times \mathbb{A}^1_{\Bbbk} \to \mathcal{V}$ is flat and, thus, we have the canonical homomorphism $(\mathrm{pr}_{\mathcal{V}})^* : G_0(\mathcal{V}) \to G_0(\mathcal{V} \times \mathbb{A}^1_{\Bbbk})$. It is always an isomorphism. Therefore, by Theorem 10.2 we get $K_0(\mathcal{V}) = K_0(\mathcal{V} \times \mathbb{A}^1_{\Bbbk})$ for a smooth variety $\mathcal{V}$.

(l) Although we do not introduce higher K-theory, it is worth mentioning that all properties listed above extend to the groups $K_i(\mathcal{V})$, $i \geq 0$.

Because of their similarities with functorial properties of singular homology and cohomology of topological spaces, the groups $K_0(\mathcal{V})$ and $G_0(\mathcal{V})$ are often called the *cohomological* and *homological Grothendieck group*, respectively.

Projective spaces. A complete description of vector bundles on a variety is in general a hopeless task. They are extremely complicated already for such classical varieties as projective spaces $\mathbb{P}^d_{\Bbbk}$, $d \geq 2$ (Hartshorne [179], Okonek, Schneider, and Spindler [279]).

By its very definition, $\mathbb{P}^d_{\Bbbk}$ comes with a vector bundle $\mathcal{O}(-1)$: the fiber over each point x is the line through the origin of $\mathbb{A}^{d+1}_{\Bbbk}$ that is represented by x. The trivial bundle is $\mathcal{O} = \mathcal{O}(0)$. For $n > 0$ one defines $\mathcal{O}(-n)$ as the n-fold tensor product $\mathcal{O}(-1)^n$ whereas $\mathcal{O}(n)$ is the dual of $\mathcal{O}(-n)$.

Over the projective line $\mathbb{P}^1_{\Bbbk}$ vector bundles are completely classified by Grothendieck's theorem [155]. A vector bundle on $\mathbb{P}^1_{\Bbbk}$ is given by a triple (P^+, P^-, ϑ) where $P^+ \in \mathbb{P}(\Bbbk[X])$, $P^- \in \mathbb{P}(\Bbbk[X^{-1}])$, and $\vartheta : P^+_X \to P^-_{X^{-1}}$ is a $\Bbbk[X, X^{-1}]$-isomorphism. The bundle $\mathcal{O}(n)$, $n \in \mathbb{Z}$, is then given by $(\Bbbk[X], \Bbbk[X^{-1}], \cdot X^{-n})$. Grothendieck's theorem says that every vector bundle on $\mathbb{P}^1_{\Bbbk}$ of rank r is of the form

$$\mathcal{O}(n_1) \oplus \mathcal{O}(n_2) \oplus \cdots \oplus \mathcal{O}(n_r), \quad r \in \mathbb{Z}_+,\ n_1, \ldots, n_r \in \mathbb{Z};$$

moreover, the numbers $n_1, \ldots, n_r$ are unique up to permutation. An elementary proof of this result has been given by Hazewinkel and Martin [181].

In particular, the group $K_0(\mathbb{P}^1_{\Bbbk})$ is generated by the classes $[\mathcal{O}(n)]$. Actually, the classes $[\mathcal{O}]$ and $[\mathcal{O}(1)]$ suffice. In fact, we have the Koszul complex

$$0 \to \mathcal{O} \to \mathcal{O}(1) \oplus \mathcal{O}(1) \to \mathcal{O}(2) \to 0, \tag{10.3}$$

defined by the commutative diagram of $\Bbbk[X, X^{-1}]$-linear maps

$$\begin{array}{ccccccc}
0 \to \Bbbk[X, X^{-1}] & \xrightarrow{(X,-1)} & \Bbbk[X, X^{-1}] \oplus \Bbbk[X, X^{-1}] & \xrightarrow{\binom{1}{X}} & \Bbbk[X, X^{-1}] \to 0 \\
\downarrow{\scriptstyle 1} & & \downarrow{\scriptstyle X^{-1}\oplus X^{-1}} & & \downarrow{\scriptstyle X^{-2}} \\
0 \to \Bbbk[X, X^{-1}] & \xrightarrow{(1,-X^{-1})} & \Bbbk[X, X^{-1}] \oplus \Bbbk[X, X^{-1}] & \xrightarrow{\binom{X^{-1}}{1}} & \Bbbk[X, X^{-1}] \to 0.
\end{array}$$

The sequence (10.3) implies the equation $[\mathcal{O}(2)] = 2[\mathcal{O}(1)] - [\mathcal{O}]$ in the Grothendieck group $K_0(\mathbb{P}^1_{\Bbbk})$. Tensoring with $\mathcal{O}(n-2)$ we get

$$[\mathcal{O}(n)] = 2[\mathcal{O}(n-1)] - [\mathcal{O}(n-2)], \quad n \in \mathbb{Z}.$$

Hence the claim on the generation by $[\mathcal{O}]$ and $[\mathcal{O}(1)]$. As it turns out, the two classes even generate $K_0(\mathbb{P}^1_{\Bbbk})$ freely. More generally, Berthelot's theorem [29, p. 365] says

$$K_0(\mathbb{P}^d_{\Bbbk}) = \bigoplus_{i=0}^{d} \mathbb{Z}[\mathcal{O}(i)]. \tag{10.4}$$

Remark 10.4. (a) Equation (10.4) in particular shows that the passage from vector bundles to the Grothendieck group may collapse much of the nontriviality. Formula (10.4) extends to the much wider class of smooth toric varieties introduced in Section 10.B; see Corollary 10.29 and Theorem 10.46.

(b) However, in Section 10.F we will show that for a certain class of singular toric varieties, closely resembling projective spaces, the K_0-groups can be huge. This observation explains the lack of a satisfactory description of $K_0(\mathcal{V})$ for most varieties. On the contrary, the rank of the G_0-groups of general toric varieties can be described in fairly simple combinatorial terms, as shown by Theorem 10.22 and Corollary 10.27.

(c) Quillen has extended Berthelot's formula to all higher K-groups in [296, §8, Th 2.1]: for any noetherian separated scheme X one has

$$K_i(\mathbb{P}_X^d) = K_i(X)^{d+1}. \tag{10.5}$$

(Actually, Quillen proves the formula for an arbitrary projective bundle on X, similar to the one considered for Chow groups in Section 10.D below.) The fundamental theorem of K-theory (Theorem 9.5(a)) is based on the special case of this formula for $d = 1$: first one derives the localization sequence in higher K-theory, and then the fundamental theorem follows from it, see Grayson's exposition [149] for the details. Quillen's work has inspired many deep insights, sometimes beyond the original framework of his Q-construction. A notable example is Thomason [354], which develops a new localization technique for singular varieties. We will need (a small part of) it in Section 10.F.

10.B Toric varieties

In the previous chapters we have studied various algebraic and combinatorial aspects of affine monoid rings. Their spectra are the building blocks of toric varieties: these are constructed by gluing the spectra of normal toric varieties along a combinatorial pattern supplied by a fan.

Two other approaches to toric varieties will be surveyed, namely via torus equivariant actions and via quotients of torus actions. In the affine case we have met both these approaches in Chapter 5.

We have discussed lattice polytopes and affine monoids in a real vector space, usually identified with $\mathbb{R}^d$. In working with toric varieties one has to consider fans as the primary combinatorial objects, for example normal fans of lattice polytopes. Therefore it is useful and customary to consider the fans as objects in the "primary" space and polytopes and affine monoids in the space dual to it. Henceforth we will follow this convention (with some explicitly marked exceptions).

Let V be an $\mathbb{R}$-vector space of dimension d, and $L \subset V$ a lattice of rank d. Furthermore, let $\mathscr{F}$ be a fan in V, consisting of pointed cones that are rational with respect to L. Then the dual cone $C^* \subset V^*$ of a cone $C \in \mathscr{F}$ is rational with respect to the dual lattice $L^* \subset V^*$. By Proposition 1.19, C^* is pointed if and only if $\dim C = d$.

Let $C_1 \subset C_2$ be cones in $\mathscr{F}$. Then the affine normal monoid $M_1 = C_1^* \cap L^*$ is a localization of $M_2 = C_2^* \cap L^*$—the discrete analogue of Exercise 1.13. Therefore, $\operatorname{Spec} \Bbbk[M_1] \subset \operatorname{Spec} \Bbbk[M_2]$ is an open subset. In particular, if $C_1 \subset C_2, C_3$ are cones in $\mathscr{F}$, then we can patch together the affine varieties $\operatorname{Spec} \Bbbk[M_2]$ and $\operatorname{Spec}(\Bbbk[M_3])$ along the common open subset $\operatorname{Spec} \Bbbk[M_1]$, $M_i = C_i^* \cap L^*$, $i = 1, 2, 3$. After performing all gluing operations encoded in the fan $\mathscr{F}$, we have produced a variety $\mathscr{V}(\mathscr{F})$. It is called the *toric variety* corresponding to $\mathscr{F}$.

It follows directly from the construction that $\mathscr{V}(\mathscr{F})$ is covered by the *standard affine charts*

$$\mathscr{V}(\mathscr{F}) = \bigcup_{C \in \max \mathscr{F}} \operatorname{Spec} \Bbbk[C^* \cap L^*].$$

Furthermore, the torus $\mathbb{T}^d = \operatorname{Spec} \Bbbk[L^*]$ is an open subset of every standard affine chart. It is called the *embedded torus* of $\mathscr{V}(\mathscr{F})$ and naturally identified with the torus $\operatorname{Hom}_{\mathbb{Z}}(L^*, \Bbbk^\times)$.

Example 10.5. (a) If $\mathscr{F}$ is the fan of all faces of a cone $C \subset V$ then

$$\mathscr{V}(\mathscr{F}) = \operatorname{Spec} \mathbb{k}[C^* \cap L^*].$$

(b) If $e_1, \ldots, e_d \in V$ form a basis of V and $\mathscr{F}$ is the fan with set of facets

$$\{\mathbb{R}_+ e_1 + \cdots + \mathbb{R}_+ e_d\} \cup \{\mathbb{R}_+(-e_1 - \cdots - e_d) + \sum_{j \neq i} \mathbb{R}_+ e_j : i = 1, \ldots, d\},$$

then

$$\mathscr{V}(\mathscr{F}) = \mathbb{P}^d_{\mathbb{k}},$$

and the standard open cover coincides with the one arising from the representation $\mathbb{P}^d_{\mathbb{k}} = \operatorname{Proj} \mathbb{k}[T_0, \ldots, T_d]$. General projective toric varieties will be discussed below.

(c) If $\mathscr{F} = \{0\}$, then $\mathscr{V}(\mathscr{F}) = \mathbb{T}^d$.

The embedded torus $\mathbb{T}^d$ acts algebraically on the standard affine charts, and these actions are compatible with the gluing procedure. Therefore $\mathbb{T}^d$ acts algebraically on $\mathscr{V}(\mathscr{F})$. In other words, the resulting map $\mathbb{T}^d \times \mathscr{V}(\mathscr{F}) \to \mathscr{V}(\mathscr{F})$ is a morphism of varieties. (In the affine case the induced action of $\mathbb{T}^d$ on the affine coordinate ring is rational.)

As it turns out, the existence of such an action is the defining property of toric varieties.

Theorem 10.6. *A variety $\mathscr{V}$ is isomorphic to a toric variety $\mathscr{V}(\mathscr{F})$ if and only if $\mathscr{V}$ is normal and contains an open torus $\mathbb{T} \subset \mathscr{V}$ whose group structure extends to an algebraic action of $\mathbb{T}$ on $\mathscr{V}$.*

The necessity of the condition is obvious from the definition of toric variety. The essential argument for sufficiency is a theorem of Sumihiro [340, Lemma 8, Cor. 2]: $\mathscr{V}$ is the union of open affine $\mathbb{T}$-stable subsets. Since varieties are quasicompact, $\mathscr{V}$ is covered by a finite number of such subsets, and because of irreducibility each of them intersects $\mathbb{T}$ nontrivially, and therefore contains the $\mathbb{T}$-orbit $\mathbb{T}$. Starting from the affine case dealt with in Exercise 5.17, one now constructs the fan $\mathscr{F}$ for $\mathscr{V}$ by reversing the construction of $\mathscr{V}(\mathscr{F})$ from $\mathscr{F}$.

If a variety $\mathscr{V}$ is isomorphic to a toric variety, it often has many embedded tori. In fact, one can use the automorphisms $\alpha \in \operatorname{Aut}(\mathscr{V})$ to duplicate $\mathbb{T}^d$.

With two alternative characterizations of toric varieties at our disposal, we can investigate to what extent the geometric properties of these varieties can be read from the combinatorial data. This has been the leitmotif of the interaction between convex and algebraic geometries since the early 1970s, starting with the pioneering works of Demazure [107] and Kempf et al. [221]. Standard references are Ewald [117], Fulton [133], and Oda [275] (the latter allows fans with an infinite number of cones). For more recent developments see Cox [100]. We would also like to direct the reader's attention to the upcoming book by Cox, Little, and Schenck [101].

Below we review some basic aspects of this interaction. For detailed proofs the reader is referred to [133, Sect. 2,3] and [275, Ch. 1,2].

(a) The category of toric varieties and torus equivariant morphisms is equivalent to the category of rational fans and their maps. Here a morphism between toric varieties $\varphi : \mathcal{V}(\mathcal{F}) \to \mathcal{V}(\mathcal{F}')$ with embedded tori $\mathbb{T}$ and $\mathbb{T}'$ is called *equivariant* if $\varphi(\mathbb{T}) \subset \mathbb{T}'$ and $\varphi(tx) = \varphi(t)\varphi(x)$ for all $t \in \mathbb{T}$ and $x \in \mathcal{V}(\mathcal{F})$.

A *morphism* $\mathcal{F} \to \mathcal{F}'$ of fans, living correspondingly in V and V', means an integral linear map $f : V \to V'$ such that every cone $C \in \mathcal{F}$ is mapped to a cone $C' \in \mathcal{F}'$.

(b) An equivariant morphism of toric varieties $\mathcal{V}(\mathcal{F}) \to \mathcal{V}(\mathcal{F}')$ is proper and birational if the corresponding map $L \to L'$ is an isomorphism and, after the identification of V with V' along the isomorphism, $\mathcal{F}$ becomes a subdivision of $\mathcal{F}'$.

(c) A toric variety is smooth if and only if each cone of its fan is unimodular. This follows from Proposition 4.45.

(d) It follows from (b) and (c) that every toric variety has an equivariant resolution of singularities; see Theorem 2.72.

(e) The toric variety $\mathcal{V}(\mathcal{F})$ is complete (projective) if and only if the fan $\mathcal{F}$ is complete (projective). The projective case will be discussed below; see Proposition 10.10.

Complete and, in particular, projective toric varieties have equivariant resolutions of singularities by projective toric varieties, as follows from Corollary 2.73.

(f) The set of orbits of the action of $\mathbb{T}^d$ on $\mathcal{V}(\mathcal{F})$ is in bijective correspondence with $\mathcal{F}$. The orbit τ_C corresponding to $C \in \mathcal{F}$ is isomorphic to the quotient torus

$$\mathbb{T}_C = \operatorname{Hom}_{\mathbb{Z}}(L^* \cap (C^*)_0, \Bbbk^\times)$$

of the embedded torus $\mathbb{T}^d = \operatorname{Hom}_{\mathbb{Z}}(L^*, \Bbbk^\times)$ where $(C^*)_0$ is the maximal linear subspace of C^* (see Proposition 1.18).

(g) The closure in $\mathcal{V}(\mathcal{F})$ of an orbit τ_C is given by

$$\overline{\tau}_C = \bigcup_{C' \supset C} \tau_{C'}.$$

(h) One has

$$\operatorname{Spec} \Bbbk[C^* \cap L^*] = \bigcup_{C' \supset C} \tau_{C'}$$

and τ_C is the only closed orbit in $\operatorname{Spec} \Bbbk[C^* \cap L^*]$.

(i) For fans $\mathcal{F}_1$ and $\mathcal{F}_2$ in $\mathbb{R}^{d_1}$ and $\mathbb{R}^{d_2}$ we have

$$\mathcal{V}(\mathcal{F}_1) \times \mathcal{V}(\mathcal{F}_2) = \mathcal{V}(\mathcal{F}_1 \times \mathcal{F}_2),$$

where

$$\mathcal{F}_1 \times \mathcal{F}_2 = \{C_1 \times C_2 : C_1 \in \mathcal{F}_1,\ C_2 \in \mathcal{F}_2\}$$

is the product fan in $\mathbb{R}^{d_1+d_2}$ (Exercise 10.2).

(j) A toric variety $\mathcal{V}(\mathcal{F})$ is called *nondegenerate* if it is not the product of a lower-dimensional toric variety and a subtorus of the embedded torus. It follows that $\mathcal{V}(\mathcal{F})$

is nondegenerate if and only if $\mathcal{F}$ is a full-dimensional fan, i. e., $|\mathcal{F}|$ spans V as a vector space.

(k) A toric variety $\mathcal{V}(\mathcal{F})$ is called *simplicial* if $\mathcal{F}$ is a *simplicial fan*, i. e., consists of simplicial cones. A toric variety is simplicial if and only if its standard affine charts are of type $\operatorname{Spec} \Bbbk[M]$ for simplicial affine normal monoids M.

A general variety is called an *orbifold* (or *V-manifold*) if it is covered by affine open sets that are quotients of smooth varieties by linear actions of finite groups. Among the toric varieties the simplicial ones are orbifolds (Exercise 10.12).

(l) Simplicial toric varieties have the resolution property because they are $\mathbb{Q}$-factorial (Exercise 10.4); compare Remark 10.3. A direct proof has been given by Morelli [265, Prop. 1]. However, it is still not known whether all toric varieties have the resolution property; see Payne [285].

Weil and Cartier divisors on toric varieties. In the following we will have to use certain subsets of a fan, namely its e-skeleton $\mathcal{F}^{(e)}$ for some $e \le \dim \mathcal{F}$ and the (smaller) set $\mathcal{F}^{[e]}$ consisting only of the faces of dimension e. In particular, $\mathcal{F}^{[1]}$ denotes the set of rays of $\mathcal{F}$.

Let $C \subset V$ be a rational cone, and $\mathcal{F}$ the fan of its faces. The face of C^* dual to an extreme ray $\rho \subset C$ is a facet. In view of Theorem 4.54 and Lemma 5.10(d) we have a natural group isomorphism

$$Z_{n-1}^{\mathbb{T}}(\operatorname{Spec} \Bbbk[C^* \cap L^*]) = \operatorname{Div}^{\mathbb{T}}(\Bbbk[C^* \cap L^*]) \cong \bigoplus_{\rho \in \mathcal{F}[1]} \mathbb{Z},$$

where $\mathbb{T} = \mathbb{T}^d$ and $Z_{n-1}^{\mathbb{T}}(\operatorname{Spec} \Bbbk[C^* \cap L^*])$ refers to the group of $\mathbb{T}$-stable Weil divisors, i. e., the elements of Z_{n-1} that are invariant under the induced action of $\mathbb{T}$.

As to Cartier divisors, the equality $\operatorname{Pic}(\Bbbk[C^* \cap L^*]) = 0$ (Theorem 8.4) implies that the $\mathbb{T}$-stable Cartier divisors on $\operatorname{Spec} \Bbbk[C^* \cap L^*]$ are given by rational $\mathbb{T}$-eigenfunctions, i. e., are of the form $\operatorname{div}(m)$ for a Laurent monomial $m \in L^*$. Therefore this divisor can be identified with the integral linear map $m \in L^* \subset V^*$.

Let $\mathcal{F}$ be a rational fan in V. Given cones $C_1, C_2 \in \mathcal{F}$, the $\mathbb{T}$-stable Cartier divisors $\operatorname{div}(m_i) \in \operatorname{Div}(\Bbbk[C_i^* \cap L^*])$, $i = 1, 2$, agree on

$$\operatorname{Spec} \Bbbk[C_1^* \cap L^*] \cap \operatorname{Spec} \Bbbk[C_2^* \cap L^*]$$

if and only if the corresponding maps m_1 and m_2 coincide on $C_1 \cap C_2$. These remarks lead to a combinatorial description of the $\mathbb{T}$-stable Cartier and Weil divisors.

For a fan $\mathcal{F}$ in V we let $\operatorname{PL}_{\mathbb{Z}}(\mathcal{F})$ denote the set of the functions $\varphi : |\mathcal{F}| \to \mathbb{R}$ which are integral linear on every cone in $C \in \mathcal{F}$.

Lemma 10.7. *Let $\mathcal{F}$ be a fan in V with n rays, and $\mathbb{T} \subset \mathcal{V}(\mathcal{F})$ be the embedded torus. Then $\operatorname{Div}^{\mathbb{T}}(\mathcal{V}(\mathcal{F})) \cong \operatorname{PL}_{\mathbb{Z}}(\mathcal{F}^{(1)}) \cong \mathbb{Z}^n$ and $\operatorname{CDiv}^{\mathbb{T}}(\mathcal{V}(\mathcal{F})) \cong \operatorname{PL}_{\mathbb{Z}}(\mathcal{F})$.*

For a ray $\rho \in \mathcal{F}^{[1]}$ let D_ρ denote the $\mathbb{T}$-stable Weil divisor on $\mathcal{V}(\mathcal{F})$ corresponding to the piecewise linear function on $|\mathcal{F}^{(1)}|$ whose restriction to $\rho \cap L$ is the isomorphism $\rho \cap L \cong \mathbb{Z}_+$ and which vanishes on all other rays. The complement of the union of the (zero sets of the) divisors D_ρ in $\mathcal{V}$ is the embedded torus.

An $\mathbb{T}$-*equivariant* line bundle on $\mathscr{V} = \mathscr{V}(\mathscr{F})$ is a line bundle $\mathscr{L}$ together with an algebraic action of the embedded torus $\mathbb{T}$ that is compatible with the projection map $\mathscr{L} \to \mathscr{V}$ and is linear on the fibers.

Lemma 10.8. *Every Weil divisor D on a toric variety $\mathscr{V}(\mathscr{F})$ is equivalent to a $\mathbb{T}$-stable Weil divisor. Consequently, every line bundle on $\mathscr{V}(\mathscr{F})$ is isomorphic to a $\mathbb{T}$-equivariant line bundle.*

Proof. The restriction of D to $\mathbb{T}$ is a principal Cartier divisor since $\mathrm{Cl}(\Bbbk[L^*]) = 0$. Let f be a rational function on $\mathscr{V}(\mathscr{F})$ such that $\operatorname{div}(f) \in Z_{d-1}(\mathbb{T})$ equals $D|_{\mathbb{T}}$. Since $\mathscr{V}(\mathscr{F}) \setminus \bigcup_i D_\rho = \mathbb{T}$, we see that the Weil divisor $\mathscr{O}(\mathscr{L}) - \operatorname{div}(f)$ is supported on $\bigcup_\rho D_\rho$. In particular, it is a $\mathbb{T}$-stable Weil divisor. But D and $\mathscr{O}(\mathscr{L})$ represent the same rational equivalence class of Weil divisors.

The second claim of the lemma follows because the correspondence between Cartier divisors and line bundles respects $\mathbb{T}$-actions. □

If $\mathscr{F}$ is full-dimensional, then the evaluation map $L^* \to \mathrm{PL}_{\mathbb{Z}}(\mathscr{F})$ is injective. In view of Lemmas 10.7 and 10.8 we therefore obtain a combinatorial description of the Picard group and the Chow group of Weil divisors of (nondegenerate) toric varieties.

Theorem 10.9. *Let $\mathscr{F}$ be a full-dimensional fan in V. Then we have the following diagram with exact horizontal sequences and injective vertical maps:*

$$\begin{array}{ccccccccc} 0 & \longrightarrow & L^* & \longrightarrow & \mathrm{PL}_{\mathbb{Z}}(\mathscr{F}) & \longrightarrow & \mathrm{Pic}(\mathscr{V}(\mathscr{F})) & \longrightarrow & 0 \\ & & \| & & \downarrow & & \downarrow & & \\ 0 & \longrightarrow & L^* & \longrightarrow & \mathrm{PL}_{\mathbb{Z}}(\mathscr{F}^{(1)}) & \longrightarrow & \mathrm{A}_{d-1}(\mathscr{V}(\mathscr{F})) & \longrightarrow & 0. \end{array} \tag{10.6}$$

The reader should compare the proof of Theorem 4.60.

Projective toric varieties. Let $P \subset V^*$ be a lattice polytope. Then the polytopal algebra $\Bbbk[P]$, graded in the standard way

$$\Bbbk[P] = \Bbbk \oplus A_1 \oplus \cdots, \qquad A_1 = \sum_{x \in \operatorname{lat}(P)} \Bbbk(x,1),$$

defines a projective variety

$$\operatorname{Proj} \Bbbk[P] \subset \mathbb{P}_{\Bbbk}^{n-1}, \quad n = \#\operatorname{lat}(P).$$

For a vertex v of P let M_v be the monoid generated by the vectors $x - v$, $x \in \operatorname{lat}(P)$. Note that $\Bbbk[M_v]$ is the dehomogenization of $\Bbbk[P]$ with respect to $(v,1) \in \Bbbk[P]$. Because these elements generate an ideal primary to the maximal irrelevant ideal, the variety $\operatorname{Proj} \Bbbk[P]$ is covered by the affine open subsets

$$\operatorname{Spec} \Bbbk[M_v], \quad v \in \operatorname{vert}(P).$$

In particular, Proj $\Bbbk[P]$ is normal if and only if M_v is a normal monoid for every vertex v of P. Moreover, this variety contains the torus Spec $\Bbbk[L^*]$ as an open subset if and only if L^* is generated by the differences $x - y$, $x, y \in \operatorname{lat}(P)$.

This motivates the following terminology, already introduced in Exercise 2.23: a lattice polytope $P \subset V^*$ is *very ample* if all monoids M_v, $v \in \operatorname{vert}(P)$, are normal and L^* is generated by the differences $x - y$, $x, y \in \operatorname{lat}(P)$. In other words, P is very ample if and only if $\operatorname{Hilb}(\mathbb{R}_+(P - v)) \subset P - v$ for all vertices v of P.

It follows directly that for a very ample lattice polytope $P \subset V^*$ one has

$$\operatorname{Proj} \Bbbk[P] = \mathscr{V}(\mathscr{N}(P)), \tag{10.7}$$

where $\mathscr{N}(P) \subset V$ is the normal fan of the polytope P (see Proposition 1.67).

Integrally closed polytopes are very ample, but the converse is not true (Exercises 2.23 and 2.24).

The attribute "very ample" is taken from the terminology introduced for line bundles $\mathscr{L}$ on a variety $\mathscr{V}$: one calls $\mathscr{L}$ *very ample* if there exist global sections $s_0, \dots, s_n$ of $\mathscr{L}$ such that the assignment $x \mapsto \Bbbk(s_0(x), \dots, s_n(x))$ embeds $\mathscr{V}$ into $\mathbb{P}^n_{\Bbbk}$ as an open subvariety of a closed subvariety. In the cases we are interested in, one calls $\mathscr{L}$ *ample* if $\mathscr{L}^k$ is very ample for some $k > 0$. Moreover, $\mathscr{L}$ is *generated by global sections* if for each $x \in \mathscr{V}$ there exists a global section s such that $s(x) \neq 0$.

Proposition 10.10. *The toric variety $\mathscr{V}(\mathscr{F})$ is projective if and only if $\mathscr{F}$ is a projective fan.*

The implication $\Longleftarrow$ has essentially been proved already. Indeed, given a projective fan $\mathscr{F}$, we find a lattice polytope P with $\mathscr{F} = \mathscr{N}(P)$. Replacing P by a multiple, we may assume that P is very ample (or even integrally closed); see Corollary 2.57. But as just seen, $\mathscr{V} \cong \operatorname{Proj} \Bbbk[P]$ is a projective toric variety then. For the converse, one starts from a very ample line bundle $\mathscr{L}$ on $\mathscr{V}$ supplied by a closed embedding of $\mathscr{V}$ into a projective space $\mathbb{P}^n_{\Bbbk}$. It is isomorphic to a $\mathbb{T}$-equivariant line bundle $\mathscr{L}$. The $\mathbb{T}$-equivariance implies that the homogeneous coordinate ring $R = \bigoplus_{k=0}^{\infty} \Gamma(\mathscr{V}, \mathscr{L}^k)$ is an affine monoid ring contained in $\Bbbk[L^* \oplus \mathbb{Z}_+]$. Then R is the integral closure of the algebra $\Bbbk[P]$, where P is the lattice polytope in V^* spanned by the height 1 lattice points of M. The very ampleness of $\mathscr{L}$ implies that P is very ample, and one has $\mathscr{F} \cong \mathscr{N}(P)$, $\mathscr{V} \cong \operatorname{Proj} R \cong \operatorname{Proj} \Bbbk[P] \subset \mathbb{P}(\Gamma(\mathscr{V}, \mathscr{L}))$.

The correspondence between very ample line bundles on projective toric varieties and very ample lattice polytope can be based on a more general construction. Already Lemma 10.7 allows us to associate a (possibly empty or unbounded) rational polyhedron with an equivariant line bundle $\mathscr{L}$ on an arbitrary toric variety $\mathscr{V}(\mathscr{F})$, namely

$$P(\mathscr{L}) = \bigcap_{\rho \in \mathscr{F}[1]} \{x \in V^* : \langle \rho, x \rangle \geq -\sigma(\rho_1)\}$$

where $\sigma \in \operatorname{PL}_{\mathbb{Z}}(\mathscr{F})$ corresponds to $\mathscr{L}$, ρ_1 generates the monoid $\rho \cap L$, and $\langle -, - \rangle$ is the natural pairing $V \times V^* \to \mathbb{R}$. The statements in the following theorems can be generalized with varying hypotheses; see [133, 3.4]. For simplicity we restrict ourselves to the projective case.

Theorem 10.11. *Let $\mathscr{F}$ be a projective fan, and $\mathscr{L}, \mathscr{L}'$ be two $\mathbb{T}$-equivariant line bundles on $\mathscr{V} = \mathscr{V}(\mathscr{F})$.*

(a) $P(\mathscr{L})$ is a (possibly empty) rational polytope.

(b) The space $\Gamma(\mathscr{V}, \mathscr{L})$ of global sections of $\mathscr{L}$ is isomorphic to the vector space $\bigoplus_{x \in \mathrm{lat}(P)} \Bbbk x$ where x is considered as an element of $\Bbbk[L^*]$ and $P = P(\mathscr{L})$. In particular, $\dim \Gamma(\mathscr{V}, \mathscr{L}) = \# \mathrm{lat}(P)$.

(c) The following are equivalent:

(i) $\mathscr{L}$ is generated by global sections;

(ii) $P(\mathscr{L})$ is a nonempty lattice polytope whose normal fan is refined by $\mathscr{F}$;

(iii) the function $\psi_{\mathscr{L}} \in \mathrm{PL}_{\mathbb{Z}}(\mathscr{F})$ corresponding to $\mathscr{L}$ is concave.

(d) $\mathscr{L}$ is ample if and only if $\mathscr{N}(P(\mathscr{L})) = \mathscr{F}$, equivalently, if $-\psi_{\mathscr{L}}$ is a support function of the regular conical subdivision $\mathscr{F}$ of V.

(e) Suppose $\mathscr{L}$ and $\mathscr{L}'$ are generated by global sections. Then

(i) $\mathscr{L} \cong \mathscr{L}'$ if and only if $P(\mathscr{L})$ and $P(\mathscr{L}')$ differ by an integral translation;

(ii) $P(\mathscr{L}) + P(\mathscr{L}') = P(\mathscr{L} \otimes \mathscr{L}')$.

Let P be a full-dimensional lattice polytope and $\mathscr{F} = \mathscr{N}(P)$. Then P defines a concave function $\psi \in \mathrm{PL}_{\mathbb{Z}}(\mathscr{F})$, and therefore a line bundle $\mathscr{L}(P)$ on $\mathscr{V}(\mathscr{F})$. Clearly $P = P(\mathscr{L}(P))$ and $\mathscr{L} = \mathscr{L}(P(\mathscr{L}))$ under suitable hypotheses.

Theorem 10.11 allows us to relate the enumerative combinatorics of lattice polytopes to projective algebraic geometry:

Corollary 10.12. *Let P be a full-dimensional lattice polytope in V^* and set $\mathscr{V} = \mathscr{V}(\mathscr{N}(P))$. Then $\bigoplus_{k=0}^{\infty} \Gamma(\mathscr{V}, \mathscr{L}(P)^k)$ is the integral closure of $\Bbbk[P]$ in $\Bbbk[L^*]$ and*

$$E_P(t) = \sum_{k=0}^{\infty} \dim_{\Bbbk} \Gamma\big(\mathscr{V}, \mathscr{L}(P)^k\big). \tag{10.8}$$

The corollary follows from Theorem 10.11(b) and (e) and the observation that the kth graded component of the integral closure of $\Bbbk[P]$ is $\bigoplus_{x \in \mathrm{lat}(kP)} \Bbbk x$.

Theorem 10.11 gives a combinatorial description of the space of global sections $\Gamma(\mathscr{V}, \mathscr{L}) = H^0(\mathscr{V}, \mathscr{L})$ of a line bundle $\mathscr{L}$ on a projective toric variety $\mathscr{V}$. As it turns out, this captures all of the cohomological information on $\mathscr{V}$ that an ample line bundle can provide:

Theorem 10.13. *Let $\mathscr{V}$ be a toric variety and $\mathscr{L}$ a line bundle on $\mathscr{V}$. Assume $\mathscr{V}$ is complete and $\mathscr{L}$ is generated by global sections. Then the higher cohomologies of $\mathscr{V}$ with coefficients in $\mathscr{L}$ vanish:*

$$H^q(\mathscr{V}, \mathscr{L}) = 0, \qquad q > 0.$$

For the proof one computes the cohomology in question as Čech cohomology, with respect to the standard affine covering of $\mathscr{V}$ by affine toric varieties; see [133, Section 3.5], [275, Section 2.2]. (For the general background on the relationship between cohomologies with coefficients in a (quasi)coherent sheaf and Čech cohomology see [178, III.4].) The vanishing of higher cohomology of ample line bundles can also be derived from Frobenius splitting; see Brion and Kumar [44, 1.2.8 and 1.3.E.(6)].

Remark 10.14. Similar in spirit to Theorem 10.13 is the following fact. For a fan $\mathscr{G}$ subdividing another fan $\mathscr{F}$, the proper map $f : \mathscr{V}(\mathscr{G}) \to \mathscr{V}(\mathscr{F})$ one has

$$f_*(\mathscr{O}_{\mathscr{V}(\mathscr{G})}) = \mathscr{O}_{\mathscr{V}(\mathscr{F})} \qquad \text{and} \qquad R^i f_*(\mathscr{O}_{\mathscr{V}(\mathscr{G})}) = 0 \quad \text{for} \quad i > 0.$$

See Danilov [102, Proposition 8.5.1] for details. Applied to equivariant resolution of toric singularities, this means that any toric variety has rational singularities.

That the Minkowski sum of two very ample polytopes with the same normal fan is also very ample is quite obvious, by arguments of combinatorics or algebraic geometry. However, the analogous claim for integrally closed polytopes seems to be open. Related to this remark, we know by Exercise 2.24 that very ample polytopes are in general not normal. But for one important special class, normality is still an open question. Call a lattice polytope $P \subset V^*$ *smooth* if the projective variety $\mathscr{V}(\mathscr{N}(P))$ is smooth, equivalently, if the corner cones $C_x = \mathbb{R}_+(P - x) \subset V^*$, $x \in \operatorname{vert}(P)$, are unimodular. (With all the information above at hand, it is not difficult to prove Demazure's theorem stating that every ample divisor on a smooth projective variety is very ample.)

The following two questions have attracted interest of researchers in recent years: (a) (Oda) Are all smooth polytopes normal? (b) (Bøgvad) Let P be a smooth lattice polytope, and let I_P be the toric ideal defining $\Bbbk[P]$. Is I_P generated by quadrics, or is $\Bbbk[P]$ even a Koszul algebra?

Application to the automorphism group. Now we discuss an application of the polytopal interpretation of projective toric varieties and very ample line bundles to the automorphism groups of these varieties. Namely, we will show that Theorem 5.45 on the structure of the graded automorphisms of a polytopal monoid ring implies a complete description of the automorphism group of a projective toric variety. The link is via the existence of "fully symmetric" polytopes. In the next subsection we indicate an alternative approach, based on the concept of the homogeneous coordinate ring of a toric variety.

Let $P \subset V^*$ be a lattice polytope. We may assume $\dim P = \dim V = d$ and that L^* is the affine lattice generated by the lattice points of P. As in Section 5.D, we let $\Gamma_{\Bbbk}(P)$ denote the group of graded $\Bbbk$-algebra automorphisms of $\Bbbk[P]$.

In the dual space $V = V^{**}$ of V^*, a column vector $v \in \operatorname{Col}(P)$ corresponds to an integral affine hyperplane H intersecting exactly one of the rays in $\mathscr{N}(P)$ (this is the condition $\operatorname{ht}_G(v) \geq 0$ for $G \neq F$) and such that there is no lattice point strictly between H and the parallel of H through 0 (this is the condition $\operatorname{ht}_F(v) = -1$). This shows that the column vectors correspond to Demazure's roots of unimodular fans [107].

It follows from formula (10.7) that for a very ample polytope $P \subset V^*$ the standard affine charts of $\operatorname{Proj} \Bbbk[P]$ are given by $\operatorname{Spec} \Bbbk[C_x^* \cap L^*]$, $x \in \operatorname{vert}(P)$, where $C_x^* = \mathbb{R}_+(P - x) \subset V^*$ for a vertex $x \in \operatorname{vert}(P)$. (Recall, $-^0$ refers to the connected component of an algebraic group, Section 5.A.)

Lemma 10.15. *Let $\mathscr{V}(\mathscr{F})$ be a projective toric variety and $\mathscr{L}(P), \mathscr{L}(Q) \in \operatorname{Pic}(\mathscr{V})$ be very ample line bundles. Then the quotient groups $\Gamma_{\Bbbk}(P)^0/\Bbbk^\times$ and $\Gamma_{\Bbbk}(Q)^0/\Bbbk^\times$ are naturally isomorphic.*

Proof. Consider the canonical antihomomorphisms $\Gamma_{\Bbbk}(P)^0 \to \operatorname{Aut}\mathcal{V}(\mathcal{F})$ and $\Gamma_{\Bbbk}(Q)^0 \to \operatorname{Aut}\mathcal{V}(\mathcal{F})$. Let $\bar{\Gamma}_{\Bbbk}(P)$ and $\bar{\Gamma}_{\Bbbk}(Q)$ denote the images.

We have $\mathcal{N}(P) = \mathcal{N}(Q)$. Therefore, $\operatorname{Col}(P) = \operatorname{Col}(Q)$ (Remark 5.38). Choose a column vector v (for both polytopes) and an element $\lambda \in \Bbbk$ and let $e_v^\lambda(P) \in \Gamma_{\Bbbk}(P)$ and $e_v^\lambda(Q) \in \Gamma_{\Bbbk}(Q)$ be the corresponding elementary automorphisms. We claim that their images ε_P and ε_Q in $\operatorname{Aut}\mathcal{V}(\mathcal{F})$ coincide. To this end we choose vertices $x \in \operatorname{vert}(F)$ and $y \in \operatorname{vert}(G)$ of the base facets $F = P_v$ and $G = Q_v$ such that $C_x = C_y$. The restrictions of ε_P and ε_Q are both given by the formula

$$z \mapsto (1 + \lambda v)^{\operatorname{ht}_F(z)} z.$$

(Both F and G define the same support hyperplane of $C_x = C_y$.) Since ε_P and ε_Q coincide on an open subset of $\mathcal{V}(\mathcal{F})$ they must be equal.

It is also clear that the natural tori $\mathbb{T}^{d+1} \subset \Gamma_{\Bbbk}(P)^0$ and $\mathbb{T}^{d+1} \subset \Gamma_{\Bbbk}(Q)^0$ have the same image in $\operatorname{Aut}\mathcal{V}(\mathcal{F})$. By Theorem 5.47(a) we arrive at the equation $\bar{\Gamma}(P) = \bar{\Gamma}(Q)$. It only remains to notice that $\Bbbk^\times = \operatorname{Ker}(\Gamma_{\Bbbk}(P)^0 \to \bar{\Gamma}_{\Bbbk}(P)) = \operatorname{Ker}(\Gamma_{\Bbbk}(Q)^0 \to \bar{\Gamma}_{\Bbbk}(Q))$. □

Example 10.16. Lemma 10.15 cannot be improved. For example, let P be the unit 1-simplex Δ_1 and $Q = 2P$. Then $\mathbb{C}[P] = \mathbb{C}[X_1, X_2]$, and $\mathbb{C}[Q] = \mathbb{C}[X_1^2, X_1X_2, X_2^2]$ is its second Veronese subring. Both polytopes have the same symmetries and column vectors, and moreover the torus action on $\mathbb{C}[P]$ is induced by that on $\mathbb{C}[Q]$. Therefore the natural map $\Gamma_{\mathbb{C}}(P) \to \Gamma_{\mathbb{C}}(Q)$ is surjective; in fact, $\Gamma_{\mathbb{C}}(P) = \mathrm{GL}_2(\mathbb{C})$ and $\Gamma_{\mathbb{C}}(Q) = \mathrm{GL}_2(\mathbb{C})/\{\pm 1\}$. If there were an isomorphism between these groups, then $\mathrm{SL}_2(\mathbb{C})$ and $\mathrm{SL}_2(\mathbb{C})/\{\pm 1\}$ would also be isomorphic. This can be easily excluded by inspecting the list of finite subgroups of $\mathrm{SL}_2(\mathbb{C})$.

For a lattice polytope P we denote the group opposite to $\Gamma_{\Bbbk}(P)^0/\Bbbk^\times$ by $\hat{\Gamma}_{\Bbbk}(P)$, the projective toric variety $\operatorname{Proj}\Bbbk[P]$ by $\mathcal{V}$, and the symmetry group of a rational fan $\mathcal{F}$ in V by $\Sigma(\mathcal{F})$. (Thus $\Sigma(\mathcal{F})$ is the subgroup of $\mathrm{GL}_d(\mathbb{Z}) = \operatorname{Aut}_{\mathbb{Z}}(L)$ that leaves $\mathcal{F}$ invariant.) Furthermore we consider $\hat{\Gamma}_{\Bbbk}(P)$ as a subgroup of $\operatorname{Aut}(\mathcal{V})$ in a natural way.

Theorem 10.17. *For all very ample lattice d-polytopes $P \subset V^*$ the automorphism group* $\operatorname{Aut}(\mathcal{V})$ *is generated by* $\hat{\Gamma}_{\Bbbk}(P)$ *and* $\Sigma(\mathcal{N}(P))$. *Moreover,* $\operatorname{Aut}(\mathcal{V})^0 = \hat{\Gamma}_{\Bbbk}(P)$, $\dim \hat{\Gamma}_{\Bbbk}(P) = \#\operatorname{Col}(P) + d$, *and the embedded torus $\mathbb{T}^d \subset \mathcal{V}$ is a maximal closed torus of* $\operatorname{Aut}(\mathcal{V})$.

Proof. Assume for the moment that $\mathcal{L}(P) \in \operatorname{Pic}(\mathcal{V})$ is preserved by every element $\alpha \in \operatorname{Aut}(\mathcal{V})$, that is $\alpha^*\mathcal{L}(P) \cong \mathcal{L}(P)$.

By Corollary 10.12 we have $\Bbbk[\bar{M}(P)] = \bigoplus_{i \geq 0} \Gamma(X, \mathcal{L}(P)^i)$. Since $\mathcal{L}(P)$ is invariant under $\operatorname{Aut}(\mathcal{V})$, any automorphism of $\mathcal{V}$ lifts to a graded automorphism of $\Bbbk[\bar{M}(P)]$, that is, to an element of $\Gamma_{\Bbbk}(P)$ (see the argument in Hartshorne [178, Example 7.1.1, p. 151]). In other words, the natural antihomomorphism $\Gamma_{\Bbbk}(P) \to \operatorname{Aut}(\mathcal{V})$ is surjective. Now Theorem 5.47 gives the desired result once we notice that the symmetry group of P (as a lattice polytope) is mapped to $\Sigma(\mathcal{F})$.

Therefore, and in view of Lemma 10.15, the proof is complete once we have shown that there is a very ample polytope Q such that $\mathcal{N}(P) = \mathcal{N}(Q)$ and the isomorphism class of $\mathcal{L}(Q)$ is invariant under $\mathrm{Aut}(\mathcal{V})$.

For any automorphism $\alpha \in \mathrm{Aut}(\mathcal{V})$ we have

$$\dim_{\Bbbk} \Gamma(\mathcal{V}, \alpha^* \mathcal{L}) = \dim_{\Bbbk} \Gamma(\mathcal{V}, \mathcal{L}) = \# \operatorname{lat}(P).$$

(For the second equality see Theorem 10.11.) Easy arguments ensure that the number of equivalence classes (up to parallel translation) of very ample lattice d-polytopes $Q \subset V^*$ with $\mathcal{N}(Q) = \mathcal{N}(P)$ and $\# \operatorname{lat}(Q) = \# \operatorname{lat}(P)$ is finite. Then Theorem 10.11 implies that the set of isomorphism classes of very ample line bundles to which $\mathcal{L}(P)$ is mapped by the automorphisms of $\mathcal{V}$ is finite. But then every element $\alpha \in \mathrm{Aut}(\mathcal{V})$ must permute this finite set. Let $\mathcal{L}(Q_i) \in \mathrm{Pic}(\mathcal{V})$, $i = 1, \dots, t$, be representatives of these classes. Then

$$\alpha^* \big(\mathcal{L}(Q_1) \otimes \cdots \otimes \mathcal{L}(Q_t)\big) \cong \mathcal{L}(Q_1) \otimes \cdots \otimes \mathcal{L}(Q_t)$$

for every $\alpha \in \mathrm{Aut}(\mathcal{V})$. Since $\mathcal{L}(Q_1) \otimes \cdots \otimes \mathcal{L}(Q_t) = \mathcal{L}(Q_1 + \cdots + Q_t)$ (Theorem 10.11) the Minkowski sum $Q_1 + \cdots + Q_t$ is the desired fully symmetric polytope. □

The homogeneous coordinate ring of a toric variety. We conclude the section by an overview of an alternative approach to toric varieties following Cox [99]. It leads to a description of the automorphism group of a complete simplicial toric variety. When the variety is in addition projective, the result is essentially equivalent to the simplicial case of Theorem 10.17.

The guiding idea is to extend to general toric varieties the classical relationship between $\mathsf{Coh}(\mathbb{P}^d_{\Bbbk})$, graded modules over $\Bbbk[T_0, T_1, \dots, T_d]$, $\mathrm{Aut}(\mathbb{P}^d_{\Bbbk})$ and $\mathrm{PGL}_{d+1}(\Bbbk) = \mathrm{GL}_{d+1}(\Bbbk)/\Bbbk^*$. The homogeneous coordinate ring construction has been a recurring topic. See Audin [12], Musson [271], and the discussion following [99, Th. 2.1].

From now on we assume $\operatorname{char} \Bbbk = 0$. This is related to Mumford's theory of moduli spaces [266] used in [99], and the role of character groups in Cox's argument.

Throughout this subsection we let $\mathcal{F}$ be a full-dimensional fan in V, $\mathbb{T} \subset \mathcal{V}(\mathcal{F})$ the embedded torus, and n the number of rays in $\mathcal{F}$. It follows from Theorem 10.9 that $\mathrm{A}_{d-1}(\mathcal{V}(\mathcal{F}))$ is a finitely generated abelian group of rank $n - d$. Therefore,

$$G = \operatorname{Hom}_{\mathbb{Z}}(A_{d-1}(\mathcal{V}(\mathcal{F})), \Bbbk^\times)$$

is a direct product of $(\Bbbk^\times)^{n-d}$ and a finite group. By Theorem 5.3 it is diagonalizable. In particular, $G^0 \cong \mathbb{T}^{n-d}$.

The polynomial ring $S = \Bbbk[\mathrm{T}_\rho : \rho \in \mathcal{F}^{[1]}]$ carries the following $\mathrm{A}_{d-1}(\mathcal{V}(\mathcal{F}))$-grading, where the Chow group $\mathrm{A}_{d-1}(\mathcal{V}(\mathcal{F}))$ of $(d-1)$-cycles is viewed as the quotient $\big(\bigoplus_{\mathcal{F}^{[1]}} \mathbb{Z}\big)/L^*$ (Theorem 10.9):

$$S = \bigoplus_{A_{d-1}(\mathcal{V}(\mathcal{F}))} S_\alpha,$$

$$S_\alpha = \bigoplus_{[a_1, \dots, a_n] = \alpha} \Bbbk T^{a_1} \cdots T^{a_n}, \qquad \alpha \in \mathrm{A}_{d-1}(\mathcal{V}(\mathcal{F})).$$

In the case of an affine toric variety, this is exactly the class ring constructed in Theorem 4.62.

Applying $\operatorname{Hom}_{\mathbb{Z}}(-, \Bbbk^\times)$ to the bottom row of the diagram (10.6) we get the exact sequence (in which we use multiplicative notation):

$$1 \to G \to (\Bbbk^\times)^n \to \mathbb{T} \to 1. \tag{10.9}$$

The torus $(\Bbbk^\times)^n$ acts on the affine space

$$\mathbb{A}^n_{\Bbbk} = \{(t_\rho : \rho \in \mathscr{F}^{[1]})\} = \operatorname{Spec} S$$

in the natural way, and the induced action of the subgroup G is given by the formula

$$g \cdot \xi = (g([D_\rho])t_\rho), \qquad g \in G, \quad \xi = (t_\rho), \quad \rho \in \mathscr{F}^{[1]}.$$

It follows that this action is compatible with the $A_{d-1}(\mathscr{V}(\mathscr{F}))$-grading of S: for any element $\alpha \in A_{d-1}(\mathscr{V}(\mathscr{F}))$, viewed as a character on G via the isomorphism $A_{d-1}(\mathscr{V}(\mathscr{F})) \cong \operatorname{Hom}_{\mathbb{Z}}(G, \Bbbk^\times)$, the homogeneous summand S_α is the α-weight space of the corresponding rational representation $G \to \operatorname{Aut}_{\Bbbk}(S)$. (Here we have used that $\operatorname{char} \Bbbk = 0$; compare Theorem 5.17 in the affine case.)

The construction of S and the action of G on S only depend on $\mathscr{F}^{(1)}$. Now we define a union of certain coordinate subspaces in $\mathbb{A}^n_{\Bbbk}$ that captures the full structure of $\mathscr{F}$. For each cone $\sigma \in \mathscr{F}$ consider the monomial

$$T^{\hat{\sigma}} = \prod_{\rho \notin \sigma} T_\rho,$$

and set

$$\mathscr{Z} = \{\xi \in \mathbb{A}^n_{\Bbbk} : T^{\hat{\sigma}}(\xi) = 0, \ \sigma \in \mathscr{F}\} = \bigcap_{\sigma \in \mathscr{F}} \{(t_\rho) \mid t_\rho = 0 \text{ for some } \rho \notin \sigma\}.$$

If $\mathscr{F}$ is a simplicial fan, then the ideal generated by the monomials $T^{\hat{\sigma}}$ is just the ideal $I(\mathscr{F})$ where $\mathscr{F}$ is viewed as an abstract simplicial complex; see Proposition 4.38. In the general case the ideal is given by Proposition 7.12(d).

The first observation is that the complement $\mathbb{A}^n_{\Bbbk} \setminus \mathscr{Z}$ is invariant under the action of G, and even under the action of $\mathbb{T}^n$. Therefore it makes sense to consider the quotient by this action.

Let Γ be an algebraic group, and $\mathscr{V}$ a variety. One calls $\mathscr{V}$ a Γ-*variety* if Γ acts as a group on $\mathscr{V}$ so that the associated map $\Gamma \times \mathscr{V}$ is a morphism of varieties. A morphism $f : \mathscr{V} \to \mathscr{W}$ is a *categorical quotient* if (i) it is Γ-equivariant with respect to the trivial action of Γ on $\mathscr{W}$ and (ii) every Γ-equivariant morphism $\mathscr{V} \to \mathscr{W}'$ factors through f. The categorical quotient is unique up to isomorphism. A much subtler notion is that of *geometric quotient*: for Γ and $\mathscr{V}$ as above, an equivariant morphism $f : \mathscr{V} \to \mathscr{W}$ with respect to the trivial action of Γ on $\mathscr{W}$ is a geometric quotient if the following conditions are satisfied [266, 0.1]:

(1) f is surjective and the preimages of the closed points of $\mathscr{W}$ are precisely the orbits of the action of Γ on the closed points of $\mathscr{V}$;

(2) $f_*(\mathcal{O}_{\mathcal{V}})^\Gamma = \mathcal{O}_{\mathcal{W}}$; in other words the structure sheaf of $\mathcal{W}$ is the subsheaf of $f_*\mathcal{O}_{\mathcal{V}}$ consisting of the Γ-invariant functions;
(3) f is submersive, i. e., a subset of $\mathcal{W}$ is open if and only if its preimage in $\mathcal{V}$ is open.

For reductive groups Γ categorical quotients always exist. Geometric quotients are also categorical quotients, but the existence of geometric quotients is a much more rare phenomenon [266].

The main results of [99] are summarized in the next theorem. (For a subgroup B of a group A the normalizer and the centralizer of B in A are denoted by $\mathrm{N}_A(B)$ and $\mathrm{C}_A(B)$.)

Theorem 10.18. *With the notation introduced above, we have the following:*

(a) $\mathcal{V}(\mathcal{F})$ *is the categorical quotient of* $\mathbb{A}^n_{\Bbbk} \setminus \mathcal{Z}$ *by* G.
(b) $\mathcal{V}(\mathcal{F})$ *is the geometric quotient of* $\mathbb{A}^n_{\Bbbk} \setminus \mathcal{Z}$ *by* G *if and only if* $\mathcal{F}$ *is simplicial.*
(c) There is a natural exact functor from the category of $\mathrm{A}_{d-1}(\mathcal{V}(\mathcal{F}))$*-graded* S*-modules to* $\mathcal{O}_{\mathcal{V}(\mathcal{F})}$*-modules, which maps finitely generated modules to coherent sheaves.*
(d) All coherent sheaves on $\mathcal{V}(\mathcal{F})$ *arise this way (Mustaţă [272]).*

Suppose now that $\mathcal{F}$ *is simplicial and complete.*

(e) $\mathrm{N}_{\mathrm{Aut}_{\Bbbk}(S)}(G)$ *is a linear algebraic group and* $\mathrm{N}_{\mathrm{Aut}_{\Bbbk}(S)}(G)^0 = \mathrm{C}_{\mathrm{Aut}_{\Bbbk}(S)}(G)$.
(f) $\mathrm{C}_{\mathrm{Aut}_{\Bbbk}(S)}(G)$ *is naturally (anti)isomorphic to the group of* $\mathrm{A}_{d-1}(\mathcal{V}(\mathcal{F}))$*-graded* $\Bbbk$*-automorphisms of* S.
(g) We have an exact sequence

$$1 \to G \to \mathrm{N}_{\mathrm{Aut}_{\Bbbk}(S)}(G) \to \mathrm{Aut}(\mathcal{V}(\mathcal{F})) \to 1.$$

For affine toric varieties, part (a) of the theorem is Corollary 5.19. In the simplicial case part (d) was proved in [99].

Finally, for a complete simplicial fan $\mathcal{F}$ a precise analysis of the group $\mathrm{C}_{\mathrm{Aut}_{\Bbbk}(S)}(G)$ (made possible by the fact that S is a polynomial ring) is carried out in [99]. For simplicial projective toric varieties the resulting description of the automorphism group is essentially the same as the combination of Theorems 5.45, 5.47, and 10.17 for this class of varieties.

10.C Chow groups of toric varieties

Divisor class groups classify codimension 1 subvarieties up to rational equivalence. Chow groups generalize their construction to subvarieties of higher codimension. In the first part of the section we develop the general notions and discuss the functorial behavior of Chow groups. In the second part we compute the Chow groups of toric varieties.

Chow groups of varieties. For proofs of the claims in this subsection see [134, Ch. 1,2]. Let $\mathcal{V}$ be a variety and $\mathcal{I} \subset \mathcal{O}_{\mathcal{V}}$ be a sheaf of ideals. It defines a subscheme $\mathcal{Z}$ of $\mathcal{V}$.

Let $\mathscr{U}$ be an irreducible component of $\mathscr{X}$. We choose an affine open subset $X \subset \mathscr{V}$ meeting $\mathscr{U}$. Then the *multiplicity* μ of $\mathscr{U}$ in $\mathscr{X}$ is the length of the artinian quotient ring $\mathscr{O}_{\mathscr{U},\mathscr{V}}/\mathscr{I}(X)\mathscr{O}_{\mathscr{U},\mathscr{V}}$.

A *k-cycle* on $\mathscr{V}$ is a finite formal sum $\sum n_i[\mathscr{V}_i]$ where the n_i are integers and the $\mathscr{V}_i$ are k-dimensional subvarieties. In other words, the *group* $Z_k(\mathscr{V})$ *of k-cycles* on $\mathscr{V}$ is the free abelian group on the k-dimensional subvarieties of $\mathscr{V}$. For $k = d - 1$ we recover the definition of the group of Weil divisors.

One calls k-cycles in $Z_k(\mathscr{V})$ *rationally equivalent* if their difference is in the subgroup generated by the divisors $[\operatorname{div}(f)]$ of the rational functions on the $(k+1)$-dimensional subvarieties $\mathscr{W} \subset \mathscr{V}$.

The corresponding quotient group of $Z_k(\mathscr{V})$ is called the *Chow group of k-cycles*, denoted by $A_k(\mathscr{V})$. We set

$$A_*(\mathscr{V}) = \bigoplus_{i=0}^{n} A_k(\mathscr{V}), \qquad n = \dim \mathscr{V}.$$

By convention, $A_k(\mathscr{V}) = 0$ for $k < 0$ and $k > \dim \mathscr{V}$. Again, for $k = d - 1$ we have just repeated the definition of $A_{d-1}(\mathscr{V})$.

We now list several properties of Chow groups and constructions related to them.

Covariant functoriality for proper morphisms. Chow groups are covariant functors with respect to proper morphisms. More precisely, such a morphism $f : \mathscr{V} \to \mathscr{V}'$ gives rise to the following functorial homomorphism which respects rational equivalence:

$$Z_k(\mathscr{V}) \to Z_k(\mathscr{V}'), \qquad [\mathscr{W}] \mapsto \deg(\mathscr{W}/f(\mathscr{W})) \cdot [f(\mathscr{W})],$$
$$\mathscr{W} \subset \mathscr{V} \text{ a } k\text{-dimensional subvariety,}$$

where

$$\deg(\mathscr{W}/f(\mathscr{W})) = \begin{cases} [R(\mathscr{W}) : R(f(\mathscr{W}))], & \dim \mathscr{W} = \dim f(\mathscr{W}), \\ 0, & \text{else.} \end{cases}$$

So we have the induced group homomorphism $f_* : A_k(\mathscr{V}) \to A_k(\mathscr{V}')$. (Properness implies that $f(\mathscr{W})$ is a subvariety of $\mathscr{V}'$.)

Degree. For a complete variety $\mathscr{V}$ the functoriality of Chow groups with respect to proper morphisms yields the *degree homomorphism*

$$\deg : A_0(\mathscr{V}) \to \mathbb{Z}$$

given by

$$\alpha \mapsto \sum_P n_P \quad \text{for} \quad \alpha = \sum_P n_P P \in A_0(\mathscr{V}).$$

In fact, deg is a well-defined map simply because $\deg(\alpha) = p_*(\alpha)$ for the structural morphism $p : \mathscr{V} \to \operatorname{Spec} \Bbbk$.

Contravariant functoriality for flat morphisms. Chow groups are contravariant functors with respect to flat morphisms. Every flat morphism $f : \mathscr{V} \to \mathscr{V}'$ of varieties has constant relative dimension $n = \dim \mathscr{V} - \dim \mathscr{V}'$, i. e., for every subvariety $\mathscr{W}' \subset \mathscr{V}'$ all irreducible components of $f^{-1}(\mathscr{W}')$ have dimension $n + \dim \mathscr{W}'$ [178, III.9.6]. Assume $\mathscr{W}' \subset \mathscr{V}'$ is a k-dimensional subvariety and $\mathscr{V}_1, \dots, \mathscr{V}_t$ are the irreducible components of the scheme-theoretic preimage $f^{-1}(\mathscr{W}')$. Let μ_i be the multiplicity of $\mathscr{V}_i$ in $f^{-1}(\mathscr{W}')$. Then the assignment

$$[\mathscr{W}'] \mapsto \sum_{i=1}^{t} \mu_i [\mathscr{V}_i]$$

gives rise to a group homomorphism $Z_k(\mathscr{V}') \to Z_{k+n}(\mathscr{V})$, which respects rational equivalence. Hence the group homomorphism

$$f^* : \mathrm{A}_k(\mathscr{V}') \to \mathrm{A}_{k+n}(\mathscr{V}).$$

The functoriality under flat maps permits an alternative definition of rational equivalence, which actually explains the terminology. Consider the projection map $\mathrm{pr} = \mathrm{pr}_{\mathbb{P}^1_{\Bbbk}} : \mathscr{V} \times \mathbb{P}^1_{\Bbbk} \to \mathbb{P}^1_{\Bbbk}$. A cycle $\alpha \in Z_k(\mathscr{V})$ is rationally equivalent to 0 if and only if there exist $(k+1)$-dimensional subvarieties $\mathscr{W}_1, \dots, \mathscr{W}_t \in \mathscr{V} \times \mathbb{P}^1_{\Bbbk}$ that project dominantly on $\mathbb{P}^1_{\Bbbk}$ and

$$\alpha = \sum_{i=1}^{t} [\mathscr{W}_i(0) - \mathscr{W}_i(\infty)],$$

where for a point $z \in \mathbb{P}^1_{\Bbbk}$ and a subvariety $\mathscr{W} \subset \mathscr{V} \times \mathbb{P}^1_{\Bbbk}$ the image of $\mathrm{pr}^{-1}(z) \cap \mathscr{W}$ under the projection $\mathscr{V} \times \mathbb{P}^1_{\Bbbk} \to \mathscr{V}$ is denoted by $\mathscr{W}(z)$. We have used that the restriction of pr to each $\mathscr{W}_i$ is flat [178, III.9.7].

Homotopy property. Let $\mathrm{p} : \mathscr{E} \to \mathscr{V}$ be a vector bundle of rank n. Then p is flat and the induced map $\mathrm{p}^* : \mathrm{A}_k(\mathscr{V}) \to \mathrm{A}_{k+n}(\mathscr{E})$ is an isomorphism for every k.

If we relax "vector bundle" to *affine fibration* (i. e., $\mathscr{E} \to \mathscr{V}$ is a locally trivial fibration whose fibers are of type $\mathbb{A}^n_{\Bbbk}$) then the maps $\mathrm{p}^* : \mathrm{A}_k(\mathscr{V}) \to \mathrm{A}_{k+n}(\mathscr{E})$ are still surjective. Surjectivity holds even for locally trivial fibrations whose fibers are copies of an open subset of $\mathbb{A}^n_{\Bbbk}$.

Localization sequence. There is a localization sequence for Chow groups, analogous to that for the homological Grothendieck group; see (10.2). More precisely, first one observes that all definitions and constructions introduced above extend in a straightforward way from varieties to constructible sets. If $i : Y \to X$ is a closed embedding in this category, $U = X \setminus Y$, and $j : U \to X$ is the inclusion map, then we have the *localization exact sequence*

$$\mathrm{A}_k(Y) \xrightarrow{i_*} \mathrm{A}_k(X) \xrightarrow{j^*} \mathrm{A}_k(U) \longrightarrow 0. \tag{10.10}$$

Filtrations and graded objects. The Chow group $\mathrm{A}_*(\mathscr{V})$ of an arbitrary variety $\mathscr{V}$ has the filtration $F_* \mathrm{A}_*(\mathscr{V})$, defined by

$$F_k \mathrm{A}_*(\mathscr{V}) = \bigoplus_{i \le k} \mathrm{A}_i(\mathscr{V}).$$

Also the homological Grothendieck group of $\mathscr{V}$ has a natural filtration $F_*\mathrm{G}_0(\mathscr{V})$; see property (i) of G_0, p. 362. This property also implies that the homomorphism

$$Z_k(\mathscr{V}) \to F_k\mathrm{G}_0(\mathscr{V})/F_{k-1}\mathrm{G}_0(\mathscr{V}), \qquad [\mathscr{W}] \mapsto [\mathscr{O}_{\mathscr{W}}],$$

is surjective. It respects rational equivalence; see [134, Ch. 15]. Therefore, we have the induced surjective group homomorphism

$$\varphi : \mathrm{A}_*(\mathscr{V}) = \mathrm{gr}\,\mathrm{A}_*(\mathscr{V}) \to \mathrm{gr}\,\mathrm{G}_0(\mathscr{V}). \tag{10.11}$$

Toric varieties. Our exposition is based on the article [136] by Fulton et al. In particular, we give a complete description of the Chow groups of a toric variety. The results discussed in this subsection are proved in [136] for varieties with an action of a connected solvable linear group. These include spherical varieties. Totaro has considered an even larger class varieties; see Remark 10.36.

Recall that the set of cones in $\mathscr{F}$ of dimension k is denoted by $\mathscr{F}^{[k]}$. Often it is more convenient to work with codimension, and therefore we introduce the set $\mathscr{F}_{[k]}$ of cones of codimension k.

The starting point is the following observation, for which we have developed all technical details:

Proposition 10.19. *Let $\mathscr{F}$ be a fan in V and $k \in \{1,\dots,d\}$, $d = \dim V$.*

(a) $\mathrm{A}_k(\mathscr{V}(\mathscr{F}))$ is generated by the classes of the orbit closures $\overline{\tau}_C$, $C \in \mathscr{F}_{[k]}$. Equivalently, $\mathrm{A}_k(\mathscr{V}(\mathscr{F}))$ is generated by the classes of $\mathbb{T}$-stable k-cycles.

(b) $\mathrm{G}_0(\mathscr{V}(\mathscr{F}))$ is generated by the classes of the ideal sheaves $\mathscr{I}(\overline{\tau}_C)$, $C \in \mathscr{F}$, on $\mathscr{V}(\mathscr{F})$, corresponding to the $\mathbb{T}$-stable subvarieties $\overline{\tau}_C \subset \mathscr{V}(\mathscr{F})$.

Proof. (a) It follows from the description of the orbits of the embedded torus and their closures, given in Section 10.B, that we have the filtration by closed subsets

$$\mathscr{V}(\mathscr{F}) = \mathscr{V}_d \supset \mathscr{V}_{d-1} \supset \cdots \supset \mathscr{V}_{-1} = \varnothing, \qquad \mathscr{V}_i = \bigcup_{\operatorname{codim} C \le i} \overline{\tau}_C = \bigcup_{\operatorname{codim} C \le i} \tau_C.$$

The successive differences decompose into disjoint unions of $\mathbb{T}$-orbits:

$$\mathscr{V}_i \setminus \mathscr{V}_{i-1} = \bigcup_{C \in \mathscr{F}_{[i]}} \tau_C, \qquad i = 0,\dots,d.$$

The localization sequence (10.10) specializes to the exact sequences

$$\mathrm{A}_k(\mathscr{V}_{i-1}) \to \mathrm{A}_k(\mathscr{V}_i) \to \bigoplus_{C \in \mathscr{F}_{[i]}} \mathrm{A}_k(\tau_C) \to 0, \qquad i = 0,\dots,d. \tag{10.12}$$

Here we have used the obvious equality $\mathrm{A}_k(X) = \bigoplus_{i=1}^m \mathrm{A}_k(X_i)$ where $X = \bigcup_{i=1}^m X_i$ is a disjoint union of closed subsets. Notice that in this situation we have similar equations $\mathrm{G}_0(X) = \bigoplus_{i=1}^m \mathrm{G}_0(X_i)$ and $K_0(X) = \bigoplus_{i=1}^m K_0(X_i)$.

Each of the orbits τ_C, $C \in \mathscr{F}_{[i]}$, is an open torus in the affine space $\mathbb{A}^i$ (over $\Bbbk$). So we have the localization sequence

$$A_k(\mathbb{A}^i \setminus \tau_C) \to A_k(\mathbb{A}^i) \to A_k(\tau_C) \to 0.$$

The homotopy invariance of Chow groups implies

$$A_k(\mathbb{A}^i) = \begin{cases} \mathbb{Z}\mathbb{A}^i \cong \mathbb{Z}, & k = i, \\ 0, & \text{else.} \end{cases}$$

For reasons of dimension, $A_i(\mathbb{A}^i \setminus \tau_C) = 0$. Therefore,

$$A_k(\tau_C) = \begin{cases} \mathbb{Z}\tau_C \cong \mathbb{Z}, & k = i, \\ 0, & \text{else.} \end{cases}$$

Finally, the homomorphism (induced by restriction) $A_k(\mathscr{V}_i) \to A_k(\tau_C)$ maps $\overline{\tau}_C$ to τ_C. Therefore, ascending induction with respect to i shows that for every index i the group $A_k(\mathscr{V}_i)$ is generated by the classes of the orbit closures $\overline{\tau}_C$, $C \in \mathscr{F}_{[k]}$.

(b) One uses the localization sequence (10.2) and essentially the same argument as in (a), based on the fact that $G_0(\mathbb{A}^i) = G_0(\tau_C) = \mathbb{Z}$. □

Remark 10.20. (a) In Proposition 10.19(b) it is enough to consider the ideal sheaves $\mathscr{I}(\overline{\tau}_C)$, $C \in \mathscr{F}^{[1]}$ corresponding to the $\mathbb{T}$-stable codimension 1 subvarieties $\overline{\tau}_C \subset \mathscr{V}(\mathscr{F})$; see Brion and Vergne [45, Corollary 1.2]. This follows from the fact that every coherent sheaf on a toric variety has a finite resolution by finite direct sums of such ideal sheaves [45, Theorem 1.1]. As a consequence, the Grothendieck group of a smooth toric variety is generated by the classes of line bundles.

(b) It is also shown in [45, Corollary 2.1] that the rational Cartan homomorphism $K_0(\mathscr{V}(\mathscr{F}))_{\mathbb{Q}} \to G_0(\mathscr{V}(\mathscr{F}))_{\mathbb{Q}}$ is surjective whenever $\mathscr{F}$ is simplicial.

(c) For weighted projective spaces, which are special cases of projective simplcial toric varieties, Chow and G_0 groups have been determined by Al Amrani [2]. For the standard projective spaces these computations specialize to Berthelot's theorem (10.4).

The subgroup of $\mathbb{T}$-stable k-cycles

$$Z_k^{\mathbb{T}}(\mathscr{V}(\mathscr{F})) \subset Z_k(\mathscr{V}(\mathscr{F}))$$

contains the subgroup of divisors defined by $\mathbb{T}$-eigenfunctions on $\mathbb{T}$-stable $(k+1)$-dimensional subvarieties. Here a $\mathbb{T}$-eigenfunction f is a rational function for which there exists a character $\chi = \chi_f \in X(\mathbb{T})$ such that

$$t \cdot f = \chi(t) f, \quad t \in \mathbb{T},$$

(with respect to the action of $\mathbb{T}$ on the field $R(\mathscr{V}(\mathscr{F}))$). Denote the corresponding quotient of $Z_k^{\mathbb{T}}(\mathscr{V}(\mathscr{F}))$ by $A_k^{\mathbb{T}}(\mathscr{V}(\mathscr{F}))$.

Theorem 10.21. *The natural homomorphism* $A_k^{\mathbb{T}}(\mathscr{V}(\mathscr{F})) \to A_k(\mathscr{V}(\mathscr{F}))$ *is an isomorphism.*

In the case of Weil divisors this theorem is represented by the lower row in diagram (10.6). In the general case, after Proposition 10.12 only the injectivity of the homomorphism $A_k^{\mathbb{T}}(\mathscr{V}(\mathscr{F})) \to A_k(\mathscr{V}(\mathscr{F}))$ need be shown. The proof Theorem 10.21 in [136] is based on previous work of Hirschowitz [192].

Theorem 10.21 and the explicit characterization of $\mathbb{T}$-stable Cartier divisors on toric varieties (Lemma 10.7) eventually lead to the following explicit description of the Chow group $A_*(\mathscr{V}(\mathscr{F}))$:

Theorem 10.22. *The group $A_k(\mathscr{V}(\mathscr{F}))$ is generated by the $\mathbb{T}$-stable k-cycles*

$$\overline{\tau}_C, \quad C \in \mathscr{F}_{[k]},$$

subject to the relations

$$\sum_{C_1} n_{C_1,C}(u) \cdot \overline{\tau}_{C_1} = 0,$$

where

(1) u runs over the elements in L^* such that $\xi(u) = 0$ for all $\xi \in C$,
(2) C_1 runs over the cones in $\mathscr{F}_{[k-1]}$ containing C,
(3) $n_{C_1,C}$ is an element of $C_1 \cap L$ representing a generator of the quotient group

$$\frac{\mathbb{R}C_1 \cap L}{\mathbb{R}C \cap L} \cong \mathbb{Z}.$$

Using Theorem 10.22 one can show that the Chow group A_* of a complete smooth toric variety is a free abelian group (Exercise 10.6).

10.D Intersection theory

In this section we give a general overview of classical intersection theory. We include all necessary definitions and constructions, referring the reader for proofs to the definitive account by Fulton [134] and the short, but very readable exposition [132] by the same author. A classical source is Grothendieck et al. [29]. The exposition culminates in the Grothendieck–Riemann–Roch theorem for smooth varieties. We then work out the rational isomorphism between the Chow and Grothendieck groups, a consequence of this theorem. We include a brief synopsis of intersection theory on singular varieties as developed in [134] and Fulton and MacPherson [135], and explain lattice point counting via the Hirzebruch–Riemann–Roch theorem.

As in previous sections, our exposition is not scheme-theoretic in full generality—it is essentially restricted to varieties.

Gysin maps and intersection products. Proofs of the claims used in this subsection are found in [134, Ch. 6,8].

The construction of the Gysin homomorphisms uses *deformation to the normal cone*. Let $\mathscr{W}' \subset \mathscr{W}$ be a subscheme, given by a sheaf of ideals $\mathscr{I}$. The *normal cone of $\mathscr{W}'$ in $\mathscr{W}$*, denoted by $C_{\mathscr{W}'}(\mathscr{W})$, is the scheme with the affine charts

$$C_{X\cap\mathscr{W}'}(X) = \operatorname{Spec}\left(\bigoplus_{j=0}^{\infty}\mathscr{I}(X)^j/\mathscr{I}(X)^{j+1}\right), \quad X \subset \mathscr{W} \text{ an affine open subset.}$$

The normal cone comes with the projection $\mathrm{p} : C_{\mathscr{W}'}(\mathscr{W}) \to \mathscr{W}'$, locally given by the graded ring homomorphism

$$\mathscr{O}_{\mathscr{W}}(X)/\mathscr{I}(X) = \mathscr{I}(X)^0/\mathscr{I}(X)^1 \to \bigoplus_{j=0}^{\infty}\mathscr{I}(X)^j/\mathscr{I}(X)^{j+1}.$$

Let $\mathscr{U}$ be a codimension r subscheme of $\mathscr{W}$ defined by a sheaf of ideals $\mathscr{I}$. The closed embedding of $\mathscr{U}$ is *regular* if every point of $\mathscr{U}$ has an affine neighborhood X such that the ideal $\mathscr{I}(X) \subset \mathscr{O}(X)$ is generated by a regular sequence. Then $\mathrm{p} : C_{\mathscr{U}}(\mathscr{W}) \to \mathscr{U}$ is a vector bundle of rank r, called the *normal bundle to $\mathscr{U}$ in $\mathscr{W}$* and denoted by $\mathrm{N}_{\mathscr{U}}(\mathscr{W})$. Locally, the associated graded ring $\bigoplus_{j=0}^{\infty}\mathscr{I}(X)^j/\mathscr{I}(X)^{j+1}$ is a polynomial algebra over $\mathscr{I}(X)^0/\mathscr{I}(X)^1$ since $\mathscr{I}(X)$ is generated by a regular sequence [72, 1.1.8]; the $\mathbb{k}$-linear structure on the fibers of $\mathrm{N}_{\mathscr{U}}(\mathscr{W})$ comes from the the graded structure of the rings of the affine charts above.

Later in this section, in the context of rational isomorphism between Chow and Grothendieck groups, we will need the following fact on tangent and normal bundles: for a closed regular embedding of smooth varieties $\iota : \mathscr{U} \subset \mathscr{V}$ there is a short exact sequence

$$0 \to \mathrm{T}_{\mathscr{U}} \to \iota^*(\mathrm{T}_{\mathscr{V}}) \to \mathrm{N}_{\mathscr{U}}(\mathscr{V}) \to 0 \tag{10.13}$$

of vector bundles on $\mathscr{U}$; see [178, II.8].

We continue with the concept of intersection product on a smooth variety. Suppose $f : \mathscr{V} \to \mathscr{V}'$ is a regular embedding of varieties of codimension r. We identify $\mathscr{V}$ with its image in $\mathscr{V}'$, given by the sheaf of ideals $\mathscr{I}_{\mathscr{V}}$, and construct *Gysin homomorphism* $f^{\bullet} : \mathrm{A}_k(\mathscr{V}') \to \mathrm{A}_{k-r}(\mathscr{V})$ as follows.

Let $\mathscr{W} \subset \mathscr{V}'$ be a k-dimensional subvariety. Consider the sheaf of ideals $\mathscr{J} = \iota^*(\mathscr{I}_{\mathscr{V}})$ on $\mathscr{W}$ where $\iota : \mathscr{W} \to \mathscr{V}'$ is the inclusion map. We have the closed subscheme

$$C_{\mathscr{W}\cap\mathscr{V}}(\mathscr{W}) \hookrightarrow \mathrm{N}_{\mathscr{V}}(\mathscr{V}')$$

resulting from the surjective graded ring homomorphisms

$$\bigoplus_{j=0}^{\infty}\mathscr{I}_{\mathscr{V}}(X)^j/\mathscr{I}_{\mathscr{V}}(X)^{j+1} \to \bigoplus_{j=0}^{\infty}\mathscr{J}(\mathscr{W}\cap X)^j/\mathscr{J}(\mathscr{W}\cap X)^{j+1},$$

$$X \subset \mathscr{V}' \text{ an open affine subset meeting } \mathscr{W}.$$

Every irreducible component of the cone $C_{\mathscr{W}\cap\delta(\mathscr{V})}(\mathscr{W})$ is of dimension k. See [134, B.6.6] for this essential point. Denote these components by $Z_1, \ldots, Z_s$ and consider the class

$$\sum_{i-1}^{s}\mu_i Z_i \in \mathrm{A}_k(\mathrm{N}_{\mathscr{V}}(\mathscr{V}')),$$

where μ_i is the multiplicity of Z_i in $C_{\mathscr{W}\cap\mathscr{V}}(\mathscr{W})$.

By the homotopy property of Chow groups, we have $\mathrm{A}_k(\mathrm{N}_{\mathcal{V}}(\mathcal{V}')) = \mathrm{A}_{k-n}(\mathcal{V})$. Therefore the class above gives rise to an element of $\mathrm{A}_{k-n}(\mathcal{V})$, which we denote by $\mathcal{V} \cdot_f \mathcal{W}$. The crucial fact is that $\mathcal{V} \cdot_f \mathcal{W}$ only depends on the class $\mathcal{W} \in \mathrm{A}_k(\mathcal{V}')$.

The Gysin map $f^\bullet : \mathrm{A}_k(\mathcal{V}') \to \mathrm{A}_{k-r}(\mathcal{V})$ is defined by the assignment

$$\sum n_i \mathcal{W}_i \mapsto \sum_i n_i (\mathcal{V} \cdot_f \mathcal{W}_i).$$

Observe that there is a (functorial) *external product* operation on Chow groups: for varieties $\mathcal{U}$ and $\mathcal{W}$ the assignment

$$(\mathcal{U}', \mathcal{W}') \mapsto \mathcal{U}' \times \mathcal{W}', \qquad \mathcal{U}' \subset \mathcal{U},\ \mathcal{W}' \subset \mathcal{W} \text{ subvarieties},$$

gives rise to a homomorphisms $Z_k(\mathcal{U}) \times Z_l(\mathcal{W}) \to Z_{k+l}(\mathcal{U} \times \mathcal{W})$ which respects rational equivalence. So we have the resulting external product maps

$$\mathrm{A}_k(\mathcal{U}) \times \mathrm{A}_l(\mathcal{W}) \to \mathrm{A}_{k+l}(\mathcal{U} \times \mathcal{W}).$$

For a smooth variety $\mathcal{V}$ we will use the notation

$$\mathrm{CH}^k(\mathcal{V}) = \mathrm{A}_{n-k}(\mathcal{V}).$$

The intersection product on $\mathcal{V}$ is defined by the following formula:

$$\mathrm{CH}^k(\mathcal{V}) \times \mathrm{CH}^l(\mathcal{V}) \to \mathrm{CH}^{k+l}(\mathcal{V}), \qquad (\mathcal{W}_1, \mathcal{W}_2) \mapsto \delta^\bullet(\mathcal{W}_1 \times \mathcal{W}_2)$$

where $\delta : \mathcal{V} \to \mathcal{V} \times \mathcal{V}$ is the diagonal embedding. It is important that δ is a regular embedding, a consequence of smoothness.

The intersection product turns $\mathrm{CH}^*(\mathcal{V}) = \bigoplus_{i=0}^n \mathrm{CH}^k(\mathcal{V})$ into a commutative graded ring, with $\mathcal{V}$ as the identity element. It is called the *Chow ring* of the smooth variety $\mathcal{V}$.

The Chow ring construction defines a contravariant functor from the category of smooth varieties to the category of abelian groups. Let $f : \mathcal{V} \to \mathcal{V}'$ be a morphism of smooth varieties. Then the graph embedding $\gamma_f : \mathcal{V} \to \mathcal{V} \times \mathcal{V}'$, $\xi \mapsto (\xi, f(\xi))$, is a regular embedding. We define $f^* : \mathrm{CH}^*(\mathcal{V}') \to \mathrm{CH}^*(\mathcal{V})$ by

$$f^*(\alpha) = \gamma_f^\bullet(\mathcal{V} \times \alpha).$$

When f is flat, the ring homomorphism f^* coincides with the inverse image homomorphism discussed above.

Suppose $f : \mathcal{V} \to \mathcal{V}'$ is a proper morphism of smooth varieties. Then the homomorphism $f_* : \mathrm{CH}^*(\mathcal{V}) \to \mathrm{CH}^*(\mathcal{V}')$ is a $\mathrm{CH}^*(\mathcal{V}')$-module homomorphism where $\mathrm{CH}^*(\mathcal{V})$ is made a $\mathrm{CH}^*(\mathcal{V}')$-module via f^*. This property is called *projection formula*.

Chern classes. For proofs of the claims below see [134, Ch. 3]. Let $\mathcal{L}$ be a line bundle on an n-dimensional variety $\mathcal{V}$ and $\mathcal{U} \subset \mathcal{V}$ be a subvariety of dimension k. Choose a Cartier divisor D on $\mathscr{U}$ with the property

$$\mathscr{L}|_{\mathscr{U}} \cong \mathscr{L}(D).$$

The assignment

$$\mathscr{U} \mapsto D$$

gives rise to well-defined group homomorphisms

$$c_1(\mathscr{L}) \cap - : \mathrm{A}_k(\mathscr{V}) \to \mathrm{A}_{k-1}(\mathscr{V}), \quad k = 1, \ldots, n,$$

called the *first Chern class of* $\mathscr{L}$.

In the special case when $\mathscr{V}$ is smooth and $\mathscr{L} = \mathscr{L}(D)$ for some Cartier divisor D on $\mathscr{V}$, we have compatibility with the intersection product:

$$c_1(\mathscr{L}) \cap \alpha = D\alpha, \quad \alpha \in \mathrm{A}_k(\mathscr{V}), \quad k = 1, \ldots, n. \tag{10.14}$$

Next we define higher Segre classes for vector bundles of arbitrary rank. Let $\mathscr{E} \to \mathscr{V}$ be a vector bundle. Projectivization of the fibers leads to the corresponding projective bundle

$$\mathrm{p} : \mathbb{P}(\mathscr{E}) \to \mathscr{V},$$

i. e., p is locally isomorphic to $\mathrm{pr}_{\mathscr{U}} : \mathbb{P}^{r-1}_{\mathbb{k}} \times \mathscr{U} \to \mathscr{U}$. It is both projective (in particular, proper) and flat. The canonical line bundle on $\mathbb{P}(\mathscr{E})$ is denoted by $\mathscr{L}_{\mathscr{E}}$. For an element $\alpha \in \mathrm{A}_k(\mathscr{V})$, $k = 1, \ldots, n$, the assignment

$$\alpha \mapsto \mathrm{p}_* \left(c_1(\mathscr{L}_{\mathscr{E}})^{r-1+i} \cap \mathrm{p}^* \alpha \right)$$

gives rise to a group homomorphism

$$s_i(\mathscr{E}) \cap : \mathrm{A}_k(\mathscr{V}) \to \mathrm{A}_{k-i}(\mathscr{V}),$$

called the ith *Segre class of* $\mathscr{E}$. Here the exponentiation of the first Chern class means its iterative application.

The Segre classes commute with each other—a consequence of the commutativity of the first Chern classes. Moreover, we have $s_0(\mathscr{E}) = \mathbf{1}$ (the identity map of $\mathrm{A}_*(\mathscr{V})$) and $s_i(\mathscr{E}) = 0$ for $i < 0$.

The higher *Chern classes* $c_i(\mathscr{E})$ of a vector bundle $\mathscr{E} \to \mathscr{V}$ are defined by the equation of formal power series

$$1 + c_1(\mathscr{E})T + c_2(\mathscr{E})T^2 + \cdots = \left(1 + s_1(\mathscr{E})T + s_2(\mathscr{E})T^2 + \cdots\right)^{-1}.$$

One sets $c_0(\mathscr{E}) = \mathbf{1}$. The Chern classes enjoy the following basic properties.

Commutativity. For vector bundles $\mathscr{E}_1$ and $\mathscr{E}_2$ on $\mathscr{V}$ and an element $\alpha \in \mathrm{A}_*(\mathscr{V})$ we have

$$c_i(\mathscr{E}_1) \cap (c_j(\mathscr{E}_2) \cap \alpha) = c_j(\mathscr{E}_2) \cap (c_i(\mathscr{E}_1) \cap \alpha).$$

Projection formula for Chern classes. For a proper morphism $f : \mathscr{U} \to \mathscr{V}$, an element $\alpha \in \mathrm{A}_*(\mathscr{U})$, and a vector bundle $\mathscr{E} \to \mathscr{V}$ we have the projection formula

$$f_*\big(c_i(f^*(\mathscr{E}))\cap\alpha\big)=c_i(\mathscr{E})\cap f_*(\alpha).$$

Inverse image formula. Let $\mathscr{E}\to\mathscr{V}$ be a vector bundle and $f:\mathscr{V}'\to\mathscr{V}$ be a flat morphism of varieties. Then for every cycle α on $\mathscr{V}$ we have

$$c_i(f^*(\mathscr{E}))\cap f^*(\alpha)=f^*\big(c_i(\mathscr{E})\cap\alpha\big). \tag{10.15}$$

Vanishing. For a vector bundle $\mathscr{E}\to\mathscr{V}$ of rank r we have $c_i(\mathscr{E})=0$ whenever $i>$ rank $\mathscr{E}$ or $i\ge 1$ and $\mathscr{E}$ is a trivial bundle.

The ring of Chern classes. We see that the Chern classes of all possible vector bundles on a (not necessarily smooth) variety $\mathscr{V}$ of dimension n generate a graded commutative ring with identity, concentrated in degrees $0,\dots,n$, whose multiplication is the composition of maps between Chow groups. It is a subring of the endomorphism ring $\mathrm{Hom}_{\mathbb{Z}}\big(\mathrm{A}_*(\mathscr{V}),\mathrm{A}_*(\mathscr{V})\big)$. We will denote this ring by

$$\tilde{\mathrm{A}}^*(\mathscr{V}).$$

It coincides with the image of the "operational" Chern classes of all possible vector bundles on $\mathscr{V}$ under the tautological map from the operational Chow cohomology of Fulton and MacPherson $\mathrm{A}^*(\mathscr{V})\to\mathrm{Hom}_{\mathbb{Z}}(\mathrm{A}_*(\mathscr{V}),\mathrm{A}_*(\mathscr{V}))$, discussed later in this section. With respect to flat maps, $\tilde{\mathrm{A}}^*$ is a contravariant functor to the category of commutative rings; see (10.15).

Whitney sum formula. For any exact sequence of vector bundles on $\mathscr{V}$

$$0\to\mathscr{E}'\to\mathscr{E}\to\mathscr{E}''\to 0$$

we have

$$c_i(\mathscr{E})=\sum_{j+k=i}c_j(\mathscr{E}')c_k(\mathscr{E}'').$$

Projective bundle formula. Let $\mathscr{V}$ be a variety, $\mathscr{E}\to\mathscr{V}$ be a vector bundle of rank r, and $\mathrm{p}:\mathbb{P}(\mathscr{E})\to\mathscr{V}$ be the canonical map. Then

$$\bigoplus_{i=0}^{r-1}\mathrm{A}_*(\mathscr{V})\cong\mathrm{A}_*(\mathbb{P}(\mathscr{E}))$$

where the isomorphism is given by

$$(\alpha_0,\dots,\alpha_{r-1})\mapsto \mathrm{p}^*(\alpha_0)+c_1(\mathscr{L}_{\mathscr{E}})\cap\mathrm{p}^*(\alpha_1)+\dots+c_1(\mathscr{L}_{\mathscr{E}})^{r-1}\cap\mathrm{p}^*(\alpha_{r-1}).$$

In particular, the embedding of the zeroth summand is p^*.

Now we explain that for a smooth variety $\mathscr{V}$ the ring $\tilde{\mathrm{A}}^*(\mathscr{V})$ is actually a subring of $\mathrm{CH}^*(\mathscr{V})$. To this end we introduce Chern roots.

Let $\mathscr{V}$ be an arbitrary variety and $\mathscr{E}\to\mathscr{V}$ be a vector bundle of rank r. Suppose first that $\mathscr{E}$ has a filtration by vector bundles

$$\mathscr{E} = \mathscr{E}_r \supset \mathscr{E}_{r-1} \supset \cdots \supset \mathscr{E}_1 \supset \mathscr{E}_0 = 0 \tag{10.16}$$

whose successive quotients are line bundles $\mathscr{L}_i = \mathscr{E}_i/\mathscr{E}_{i-1}$. Then the *Chern roots* of $\mathscr{E}$ are defined by

$$x_i(\mathscr{E}) = c_1(\mathscr{L}_i), \quad i = 1, \ldots, r.$$

Second, let $\mathscr{E} \to \mathscr{V}$ be a general vector bundle of rank r. The *splitting construction* yields a variety $\mathscr{V}'$ and a flat map $f : \mathscr{V}' \to \mathscr{V}$ satisfying the following conditions:

(a) $f^* : \mathrm{A}_*(\mathscr{V}) \to \mathrm{A}_*(\mathscr{V}')$ is an injective homomorphism;
(b) the vector bundle $f^*(\mathscr{E})$ has a filtration of the form (10.16);
(c) if $\mathscr{V}$ is smooth, then $\mathscr{V}'$ is smooth, too.

One constructs $\mathscr{V}'$ by induction with respect to r as follows. Consider the variety $\mathscr{W} = \mathbb{P}(\mathscr{E})$, the projective bundle of $\mathscr{E}$, and the quotient vector bundle

$$\mathscr{E}' = \mathrm{p}^*(\mathscr{E})/\mathscr{L}_{\mathscr{E}} \to \mathscr{W}, \quad \mathrm{rank}(\mathscr{E}') = r - 1.$$

Here $\mathrm{p} : \mathscr{W} \to \mathscr{V}$ is the canonical projective map (which is also flat) and $\mathscr{L}_{\mathscr{E}}$ is the canonical line bundle on $\mathscr{W}$ which is naturally a subbundle of $\mathrm{p}^*(\mathscr{E})$. The induction assumption applies to the vector bundle $\mathscr{E}' \to \mathscr{W}$, producing a flat morphism $g : \mathscr{V}' \to \mathscr{W}$ and an appropriate filtration of the vector bundle $g^*(\mathscr{E}') \in \mathsf{Vect}(\mathscr{V}')$. So we can take $f = \mathrm{p}\, g$. The condition (a) above follows from the injectivity of p^*, stated in the projective bundle formula above. The validity of (b) and (c) is clear.

Finally we set

$$x_i(\mathscr{E}) = x_i(f^*(\mathscr{E})) \in \tilde{\mathrm{A}}^*(\mathscr{V}').$$

Now the relationship between Chern classes and Chern roots is given by

$$1 + c_1(\mathscr{E})T + c_2(\mathscr{E})T^2 + \cdots = \prod_{i=1}^{r}(1 + x_i(\mathscr{E})T), \tag{10.17}$$

an equation in $\tilde{\mathrm{A}}^*(\mathscr{V}')$. In other words, Chern classes are elementary symmetric polynomials of the corresponding Chern roots.

If $\mathscr{V}$ is smooth of dimension n and $\mathscr{E} \to \mathscr{V}$ is a vector bundle with a filtration as in (10.16), then the Chern roots $x_i(\mathscr{E})$ belong to $\mathrm{CH}^1(\mathscr{V})$ because of (10.14). Therefore, by (10.17), the Chern classes $c_i(\mathscr{E})$ belong to the ring $\mathrm{CH}^*(\mathscr{V})$. Now assume that $\mathscr{E}$ is an arbitrary vector bundle on $\mathscr{V}$. Then we find $\mathscr{V}'$ and a flat map $f : \mathscr{V}' \to \mathscr{V}$ as in the splitting principle. The Chern roots $x_i(f^*(\mathscr{E}))$ belong to $\mathrm{CH}^1(\mathscr{V}')$. We express $c_i(\mathscr{E})$ in terms of the Chern roots, obtaining $c_i(\mathscr{E})$ as an element of $\mathrm{CH}^i(\mathscr{V}')$. Consider the commutative diagram

$$\begin{array}{ccc} \mathrm{CH}^*(\mathscr{V}) & \overset{f^*}{\hookrightarrow} & \mathrm{CH}^*(\mathscr{V}') \\ {\scriptstyle c_i(\mathscr{E})}\downarrow & & \downarrow{\scriptstyle c_i(\mathscr{E})} \\ \mathrm{CH}^*(\mathscr{V}) & \overset{f^*}{\hookrightarrow} & \mathrm{CH}^*(\mathscr{V}'). \end{array}$$

Multiplication by $c_i(\mathscr{E}) \in \mathrm{CH}^*(\mathscr{V}')$ leaves the subring $\mathrm{CH}^*(\mathscr{V})$ invariant, therefore $c_i(\mathscr{E})$ must belong to $\mathrm{CH}^*(\mathscr{V})$.

Grothendieck–Riemann–Roch theorem. In this subsection all varieties are assumed to be smooth. For proofs of the claims we refer the reader to [134, Ch. 15].

Often we will have to pass from an abelian group A to the $\mathbb{Q}$-vector space $\mathbb{Q} \otimes A$, for which we use the shorthand notation $A_{\mathbb{Q}}$.

The *Chern character* $\mathrm{ch}(\mathscr{E})$ of a vector bundle $\mathscr{E} \to \mathscr{V}$ of rank r is given by

$$\mathrm{ch}(\mathscr{E}) = \sum_{k=0}^{\infty} \frac{p_k(\mathscr{E})}{k!} \in \mathrm{CH}^*(\mathscr{V})_{\mathbb{Q}}, \qquad p_k(\mathscr{E}) = x_1(\mathscr{E})^k + \cdots + x_r(\mathscr{E})^k.$$

In particular, for a line bundle $\mathscr{L}$ and its corresponding Cartier divisor D on $\mathscr{V}$ we have:

$$\begin{aligned} \mathrm{ch}(\mathscr{L}) &= 1 + c_1(\mathscr{L}) + \frac{c_1(\mathscr{L})^2}{2!} + \cdots + \frac{c_1(\mathscr{L})^n}{n!} \\ &= 1 + D + \frac{D^2}{2!} + \cdots + \frac{D^n}{n!}, \qquad n = \dim \mathscr{V}. \end{aligned} \tag{10.18}$$

A determinantal expression for $\mathrm{ch}(\mathscr{E})$ is

$$p_k(\mathscr{E}) = \det \begin{pmatrix} c_1(\mathscr{E}) & 1 & 0 & \cdots & 0 \\ 2c_2(\mathscr{E}) & c_1(\mathscr{E}) & 1 & \cdots & 0 \\ \vdots & \vdots & \ddots & \ddots & \vdots \\ (k-1)c_{k-1}(\mathscr{E}) & (k-2)c_{k-2}(\mathscr{E}) & (k-3)c_{k-3}(\mathscr{E}) & \cdots & 1 \\ kc_k(\mathscr{E}) & (k-1)c_{k-1}(\mathscr{E}) & (k-2)c_{k-2}(\mathscr{E}) & \cdots & c_1(\mathscr{E}) \end{pmatrix}.$$

The first terms in the explicit expansion of $\mathrm{ch}(\mathscr{E})$ are

$$\begin{aligned} \mathrm{ch}(\mathscr{E}) = \mathrm{rank}(\mathscr{E}) + c_1(\mathscr{E}) &+ \frac{1}{2}(c_1(\mathscr{E})^2 - 2c_2(\mathscr{E})) \\ &+ \frac{1}{6}(c_1(\mathscr{E})^3 - 3c_1(\mathscr{E})c_2(\mathscr{E}) + 3c_3(\mathscr{E})) + \cdots. \end{aligned} \tag{10.19}$$

Because higher Chern classes vanish for a trivial vector bundle $\mathscr{E} \to \mathscr{V}$, we have

$$\mathrm{ch}(\mathscr{E}) = \mathrm{rank}(\mathscr{E}) = \mathrm{rank}(\mathscr{E}) \cdot \mathscr{V} \in \mathrm{CH}^0(\mathscr{V}). \tag{10.20}$$

The Whitney sum formula for Chern classes implies

$$\mathrm{ch}(\mathscr{E}) = \mathrm{ch}(\mathscr{E}') + \mathrm{ch}(\mathscr{E}'')$$

for any short exact sequence of vector bundles $0 \to \mathscr{E}' \to \mathscr{E} \to \mathscr{E}'' \to 0$. Moreover, for vector bundles $\mathscr{E}_1$ and $\mathscr{E}_2$ one has

$$\mathrm{ch}(\mathscr{E}_1 \otimes \mathscr{E}_2) = \mathrm{ch}(\mathscr{E}_1)\,\mathrm{ch}(\mathscr{E}_2).$$

In particular, one gets a ring homomorphism, also called *Chern character*,

$$\mathrm{ch}_{\mathcal{V}} : K_0(\mathcal{V}) \to \mathrm{CH}^*(\mathcal{V})_{\mathbb{Q}}. \tag{10.21}$$

Consequently there are two covariant functors from the category of smooth varieties and proper morphisms to AbGroups: $\mathrm{CH}^*_{\mathbb{Q}}$ and $K_0 = G_0$ (Theorem 10.2). As it turns out, the Chern character *fails* to be a natural transformation. The Grothendieck–Riemann–Roch theorem says that this failure is measured precisely by the Todd class.

The *Todd class* of a vector bundles $\mathcal{E} \to \mathcal{V}$ is defined in terms of its Chern roots:

$$\mathrm{td}(\mathcal{E}) = \prod_{i=1}^{r} \frac{x_i(\mathcal{E})}{1 - \exp(-x_i(\mathcal{E}))} \in \tilde{\mathrm{A}}^*(\mathcal{V}).$$

The Todd class expands as follows:

$$\mathrm{td}(\mathcal{E}) = 1 + \frac{1}{2}c_1(\mathcal{E}) + \frac{1}{12}\big(c_1(\mathcal{E})^2 + c_2(\mathcal{E})\big) + \frac{1}{24}c_1(\mathcal{E})c_2(\mathcal{E}) + \cdots. \tag{10.22}$$

It is multiplicative:

$$\mathrm{td}(\mathcal{E}) = \mathrm{td}(\mathcal{E}') \cdot \mathrm{td}(\mathcal{E}'') \tag{10.23}$$

for a short exact sequence of vector bundles $0 \to \mathcal{E}' \to \mathcal{E} \to \mathcal{E}'' \to 0$.

Since all Chern classes are nilpotent elements, the Todd class of any vector bundle is a unit of the ring $\mathrm{CH}^*(\mathcal{V})$.

Theorem 10.23 (Grothendieck–Riemann–Roch). *Let $f : \mathcal{V} \to \mathcal{V}'$ be a proper map of smooth varieties. Then for every $\alpha \in K_0(\mathcal{V})$ we have*

$$\mathrm{ch}_{\mathcal{V}'}\big(f_*(\alpha)\big) \cdot \mathrm{td}(\mathrm{T}_{\mathcal{V}'}) = f_*\big(\mathrm{ch}_{\mathcal{V}}(\alpha) \cdot \mathrm{td}(\mathrm{T}_{\mathcal{V}})\big).$$

Equivalently, the assignment $[\mathcal{E}] \mapsto \mathrm{ch}_{\mathcal{V}}(\mathcal{E}) \cdot \mathrm{td}(\mathrm{T}_{\mathcal{V}})$ gives rise to a natural transformation $K_0 \to \mathrm{CH}^*_{\mathbb{Q}}$ of covariant functors from the category of smooth varieties and proper morphisms to the category of abelian groups.

Remark 10.24. Both K_0 and CH^* are contravariant functors from the category of smooth varieties to Rings. In contrast to its covariant behavior, the Chern character is a natural transformation from K_0 to $\mathrm{CH}^*_{\mathbb{Q}}$.

When $\mathcal{V}$ is a complete variety and $f : \mathcal{V} \to \operatorname{Spec} \mathbb{k}$ is the structural map, Theorem 10.23 specializes to the following classical theorem:

Theorem 10.25 (Hirzebruch–Riemann–Roch). *Let $\mathcal{E}$ be a vector bundle on a complete variety $\mathcal{V}$. Then*

$$\chi(\mathcal{V}, \mathcal{E}) = \int_{\mathcal{V}} \mathrm{ch}(\mathcal{E}) \cdot \mathrm{td}(\mathrm{T}_{\mathcal{V}}).$$

In the theorem

$$\chi(\mathcal{V}, \mathcal{E}) = \sum_{i \geq 0} (-1)^i \dim_{\mathbb{k}} H^i(\mathcal{V}, \mathcal{E})$$

denotes the *Euler characteristic* of the bundle $\mathcal{E}$, and

$$\int_{\mathcal{V}} : \mathrm{CH}^*(\mathcal{V}) \to \mathbb{Z}$$

is the group homomorphism, defined by

$$\int_{\mathscr{V}} \alpha = \begin{cases} \deg(\alpha), & \alpha \text{ is a 0-cycle,} \\ 0, & \text{else.} \end{cases}$$

Theorem 10.25 is derived from Theorem 10.23 as follows. Let $f : \mathscr{V} \to \operatorname{Spec} \Bbbk$ be the structure map. Then by [178, III.8.5] we have an isomorphism of $\Bbbk$-vector spaces

$$H^i(\mathscr{V}, \mathscr{E}) \cong R^i f_*(\mathscr{E}).$$

Therefore, the definition of the group homomorphism

$$f_* : G_0(\mathscr{V}) \to G_0(\operatorname{Spec} \Bbbk) = \mathbb{Z} \quad \big(\subset \mathbb{Q} = \mathrm{CH}^*(\operatorname{Spec} \Bbbk)_{\mathbb{Q}} \big)$$

in (10.1) yields the natural identification $f_*(\mathscr{E}) = \chi(\mathscr{V}, \mathscr{E})$. Thus Theorem 10.23 implies the following equalities in $\mathrm{CH}^*(\operatorname{Spec} \Bbbk)_{\mathbb{Q}}$:

$$\chi(\mathscr{V}, \mathscr{E}) = \mathrm{ch}_{\operatorname{Spec} \Bbbk}\big(f_*(\mathscr{E}) \cdot \mathrm{td}(\mathrm{T}_{\operatorname{Spec} \Bbbk})\big) = f_*\big(\mathrm{ch}_{\mathscr{V}}(\mathscr{E}) \cdot \mathrm{td}(\mathrm{T}_{\mathscr{V}})\big) = \int_{\mathscr{V}} \mathrm{ch}(\mathscr{E}) \cdot \mathrm{td}(\mathrm{T}_{\mathscr{V}}).$$

Counting lattice points in polytopes revisited. Theorem 10.25 offers a cohomological approach to the classic theme of counting lattice points in lattice polytopes. This has been and still remains a very active direction of research. We only explain the link, not really delving into this vast topic.

Let $\mathscr{L}$ be a line bundle on a smooth projective toric variety $\mathscr{V}$. Assume $\mathscr{L}$ is generated by global sections and $P = P(\mathscr{L}) \subset V^*$ is the corresponding polytope (in the notation of Theorem 10.11). Then Theorems 10.11(b) and 10.13, imply

$$\chi(\mathscr{V}, \mathscr{L}) = \# \operatorname{lat}(P).$$

On the other hand, equation (10.18) implies

$$\int_{\mathscr{V}} \mathrm{ch}(\mathscr{L}) \cdot \mathrm{td}(\mathrm{T}_{\mathscr{V}}) = \sum_{k=0}^{n} \frac{1}{k!} \deg\big(D^k \cdot \mathrm{td}_k(\mathscr{V})\big),$$

where D is the divisor on $\mathscr{V}$, corresponding to the line bundle $\mathscr{L}$, and $\mathrm{td}_k(\mathscr{V})$ is the degree k summand of the Todd class $\mathrm{td}(\mathrm{T}_{\mathscr{V}})$. By Theorem 10.25 we get

$$\# \operatorname{lat}(P) = \sum_{k=0}^{n} \frac{1}{k!} \deg\big(D^k \cdot \mathrm{td}_k(\mathscr{V})\big). \tag{10.24}$$

To count the lattice points in an arbitrary polytope one can use equivariant resolutions of toric singularities. Namely, for an arbitrary full-dimensional polytope $P \subset V^*$ we consider an equivariant projective resolution of singularities $f : \mathscr{V}(\mathscr{G}) \to \mathscr{V}(\mathscr{N}(P))$. Let $\mathscr{L} = \mathscr{L}(P)$ be the ample line bundle on $\mathscr{V}(\mathscr{N}(P))$, corresponding to P; see the discussion immediately after Theorem 10.11. Then

$$\dim_{\Bbbk} \Gamma\big(\mathscr{V}(\mathscr{N}(P)), \mathscr{L}\big) = \dim_{\Bbbk} \Gamma\big(\mathscr{V}(\mathscr{G}), f^* \mathscr{L}\big)$$

and, as a consequence, one obtains an expression of $\#\operatorname{lat}(P)$ by applying (10.24) to the variety $\mathscr{V}(\mathscr{G})$ and the line bundle $f^*\mathscr{L}$ on it. Of course, it is critical that $f^*\mathscr{L} \in \operatorname{Pic}(\mathscr{V}(\mathscr{G}))$ is generated by global sections, although it usually fails to be very ample—even if P is a very ample polytope.

Based on this approach, one can derive Ehrhart's reciprocity for lattice polytopes (Theorem 6.51(d)) and Pick's formula (Exercise 6.17); see Danilov [102, §11]. In Exercise 10.7 we indicate how the cohomological approach leads to the existence of Ehrhart polynomials and their multivariate analogues.

Alternatively, one can attack the lattice point counting problem in general (lattice) polytopes via an extension of the Hirzebruch–Riemann–Roch Theorem for general varieties; see the formulas (10.27) and (10.28) below.

Rational isomorphism between Grothendieck and Chow groups. It is an important consequence of Theorem 10.23 that Grothendieck and Chow groups coincide rationally.

Theorem 10.26. *Let $\mathscr{V}$ be a smooth variety.*

(a) For every subvariety $\mathscr{U} \subset \mathscr{V}$ of dimension k there exists $\alpha \in F_{k-1}\operatorname{A}_*(\mathscr{U})$ such that

$$\operatorname{ch}_{\mathscr{V}}\big(\iota_*([\mathscr{O}_{\mathscr{U}}])\big) = \iota_*(\mathscr{U} + \alpha)$$

where $\iota : \mathscr{U} \to \mathscr{V}$ is the inclusion map and $\iota_*([\mathscr{O}_{\mathscr{U}}]) \in \operatorname{G}_0(\mathscr{V}) = K_0(\mathscr{V})$.

(b) We have the ring isomorphism

$$(\operatorname{ch}_{\mathscr{V}})_{\mathbb{Q}} : K_0(\mathscr{V})_{\mathbb{Q}} \to \operatorname{CH}^*(\mathscr{V})_{\mathbb{Q}}, \qquad [\mathscr{E}] \mapsto \operatorname{ch}_{\mathscr{V}}(\mathscr{E}).$$

(c) The group isomorphism

$$(\operatorname{ch}_{\mathscr{V}})_{\mathbb{Q}} \cdot \operatorname{td}(\operatorname{T}_{\mathscr{V}}) : K_0(\mathscr{V})_{\mathbb{Q}} \to \operatorname{CH}^*(\mathscr{V})_{\mathbb{Q}}, \qquad [\mathscr{E}] \mapsto \operatorname{ch}_{\mathscr{V}}(\mathscr{E})\operatorname{td}(\operatorname{T}_{\mathscr{V}}),$$

is natural in $\mathscr{V}$ with respect to proper morphisms of smooth varieties.

Proof. (a) There is a closed subset $S \subsetneq \mathscr{U}$ such that $\mathscr{U} \setminus S$ is smooth and regularly embedded in $\mathscr{V} \setminus S$. In the localization sequence

$$\operatorname{A}_*(S) \to \operatorname{A}_*(\mathscr{V}) \to \operatorname{A}_*(\mathscr{V} \setminus S) \to 0$$

the image of $\operatorname{A}_*(S)$ is in $\iota_*(F_{k-1}\operatorname{A}_*(\mathscr{U}))$. So there is no loss of generality in assuming that $S = \varnothing$, i. e., $\mathscr{U}$ is smooth and regularly embedded in $\mathscr{V}$. In particular, the corresponding short exact sequence (10.13) applies. Then the multiplicative property of the Todd class (10.23) yields

$$\operatorname{td}\big(\iota^*(\operatorname{T}_{\mathscr{V}})\big)^{-1}\operatorname{td}(\operatorname{T}_{\mathscr{U}}) = \operatorname{td}\big(\operatorname{N}_{\mathscr{U}}(\mathscr{V})\big)^{-1}. \tag{10.25}$$

By Theorem 10.23

$$\operatorname{ch}_{\mathscr{V}}\big(\iota_*([\mathscr{O}_{\mathscr{U}}])\big) = \operatorname{td}(\operatorname{T}_{\mathscr{V}})^{-1}\iota_*\big(\operatorname{td}(\operatorname{T}_{\mathscr{U}})\operatorname{ch}_{\mathscr{U}}([\mathscr{O}_{\mathscr{U}}])\big).$$

In view of the projection formula for Chern classes, this equation can be written as

$$\operatorname{ch}_{\mathscr{V}}\big(\iota_*([\mathscr{O}_{\mathscr{U}}])\big) = \iota_*\big(\iota^*(\operatorname{td}(\mathrm{T}_{\mathscr{V}})^{-1})\operatorname{td}(\mathrm{T}_{\mathscr{U}})\operatorname{ch}_{\mathscr{U}}([\mathscr{O}_{\mathscr{U}}])\big). \tag{10.26}$$

It follows from the inverse image formula for Chern classes that ι^* and Chern classes commute. Consequently, ι^* and td also commute. Since ι^* is a ring homomorphism, it respects inverses. These observations show

$$\iota^*(\operatorname{td}(\mathrm{T}_{\mathscr{V}})^{-1}) = \operatorname{td}\big(\iota^*(\mathrm{T}_{\mathscr{V}})\big)^{-1}.$$

In particular, by (10.20) and (10.25), equation (10.26) amounts to

$$\operatorname{ch}_{\mathscr{V}}(\iota_*([\mathscr{O}_{\mathscr{U}}])) = \iota_*\big(\operatorname{td}(\mathrm{N}_{\mathscr{U}}(\mathscr{V}))^{-1}\cdot\mathscr{U}\big).$$

Now the claim follows from the expansion (10.22) for the Todd class and the fact that $c_i(\beta) \in \mathrm{CH}_{k-i}(\mathscr{V})$ for all $\beta \in \mathrm{CH}_k(\mathscr{V})$.

(b) By (a) the Chern character maps $F_k G_0(\mathscr{V})$ to $F_k\,\mathrm{A}_*(\mathscr{V})_{\mathbb{Q}}$. So we have the induced homomorphism

$$\psi : \operatorname{gr} G_0(\mathscr{V}) \to \operatorname{gr}\mathrm{A}_*(\mathscr{V})_{\mathbb{Q}} = \mathrm{CH}^*(\mathscr{V})_{\mathbb{Q}}.$$

Again by (a), the composite $\psi\varphi$ is the canonical map $\mathrm{CH}^*(\mathscr{V}) \to \mathrm{CH}^*(\mathscr{V})_{\mathbb{Q}}$, where φ is as in (10.11). In view of the surjectivity of φ both maps $\varphi_{\mathbb{Q}}$ and $\psi_{\mathbb{Q}}$ are isomorphisms.

(c) follows from Theorem 10.23, part (b), and the fact that $\operatorname{td}(\mathrm{T}_{\mathscr{V}})$ is a unit in $\mathrm{CH}^*(\mathscr{V})$. □

General varieties. Motivated by the analogy with singular cohomology and the Chow ring construction in the smooth case, Fulton and MacPherson have introduced the so-called *operational* A^**-theory* for schemes, in particular for not necessarily smooth varieties [134, Ch. 17], [135]. (There is another variant of A^*-theory, see Baum, Fulton, and MacPherson [25].)

An element c of the operational group $\mathrm{A}^i(\mathscr{V})$ of a variety $\mathscr{V}$ is a collection of homomorphisms

$$c_{\mathscr{W}} : \mathrm{A}_k(\mathscr{W}) \to \mathrm{A}_{k-i}(\mathscr{W}) \quad (\text{written } c_{\mathscr{W}}(\alpha) = c \cap \alpha)$$

for every morphism $\mathscr{W} \to \mathscr{V}$ and all $k \geq i$ having the expected properties, like the projection formula, etc. The ring structure on $\mathrm{A}^*(\mathscr{V}) = \bigoplus \mathrm{A}^i(\mathscr{V})$ is introduced by composing homomorphisms, written as a cup product $c \cup c'$. When $\mathscr{V}$ is smooth, we have $\mathrm{A}^*(\mathscr{V}) = \mathrm{CH}^*(\mathscr{V})$.

One defines Chern classes of the vector bundles on $\mathscr{V}$ as certain elements of the ring $\mathrm{A}^*(\mathscr{V})$. Observe, the Chern classes we introduced before are actually the images of the latter under the map $\mathrm{A}^*(\mathscr{V}) \to \operatorname{Hom}_{\mathbb{Z}}(\mathrm{A}_*(\mathscr{V})\,\mathrm{A}_*(\mathscr{V}))$, resulting from the identity map $\mathscr{W} = \mathscr{V}$.

The commutativity of the ring $\mathrm{A}^*(\mathscr{V})$ relies on resolution of singularities—so it is known in characteristic 0, but not in general. For general complete toric varieties the operational Chow ring will be discussed below.

It is proved in [25] and [134, Ch. 18] that for every variety $\mathscr{V}$ there is a homomorphism

$$\tau_{\mathscr{V}} : G_0(\mathscr{V}) \to A_*(\mathscr{V})_{\mathbb{Q}}$$

satisfying the following conditions:

(1) *Grothendieck–Riemann–Roch for singular varieties.* For every proper morphism $f : \mathscr{V} \to \mathscr{W}$ the square

$$\begin{array}{ccc} G_0(\mathscr{V}) & \xrightarrow{\tau_{\mathscr{V}}} & A_*(\mathscr{V})_{\mathbb{Q}} \\ {\scriptstyle f_*}\downarrow & & \downarrow{\scriptstyle f_*} \\ G_0(\mathscr{W}) & \xrightarrow{\tau_{\mathscr{W}}} & A_*(\mathscr{W})_{\mathbb{Q}} \end{array}$$

is commutative.

(2) For all $\alpha \in K_0(\mathscr{V})$ and $\beta \in G_0(\mathscr{V})$,

$$\tau_{\mathscr{V}}(\alpha \otimes \beta) = \operatorname{ch}(\alpha) \cap \tau_{\mathscr{V}}(\beta).$$

(The Chern character on the right makes sense because vector bundles on $\mathscr{V}$ have Chern classes in $A^*(\mathscr{V})$.)

(3) For the closed embedding of a subvariety $\iota : \mathscr{U} \to \mathscr{V}$ one has

$$\tau_{\mathscr{V}}(\iota_*[\mathscr{O}_{\mathscr{U}}]) = \mathscr{U} + \text{lower terms}.$$

(4) If $\mathscr{V}$ is smooth and $\alpha \in K_0(\mathscr{V})$, then $\tau_{\mathscr{V}}(\alpha) = \operatorname{ch}(\alpha)\operatorname{td}(T_{\mathscr{V}})$.

For $\mathscr{V}$ smooth, (3) is equivalent to Theorem 10.26(a). The equivalence follows from the Todd class expansion (10.22) and (4).

Finally, exactly the same reasoning as in Theorem 10.26(b) and (c) implies

Corollary 10.27. *For every variety $\mathscr{V}$,*

$$(\tau_{\mathscr{V}})_{\mathbb{Q}} : G_0(\mathscr{V})_{\mathbb{Q}} \to A_*(\mathscr{V})_{\mathbb{Q}}$$

is a group isomorphism.

We record one more consequence of the Grothendieck–Riemann–Roch theorem for singular varieties: for a complete variety $\mathscr{V}$ and a vector bundle $\mathscr{E}$ on it,

$$\chi(\mathscr{V}, \mathscr{E}) = \int_{\mathscr{V}} \operatorname{ch}(\mathscr{E}) \cap \operatorname{Td}(\mathscr{V}), \tag{10.27}$$

where $\operatorname{Td}(\mathscr{V}) = \tau_{\mathscr{V}}(\mathscr{O}_{\mathscr{V}}) \in A_*(\mathscr{V})$; see [134, Sect. 18.3]. Then, similarly to (10.24), the lattice points in a full-dimensional L^*-polytope $P \subset V^*$ can be counted by

$$\#\operatorname{lat}(P) = \sum_{k=0}^{n} \frac{1}{k!} \deg\left(c_1(\mathscr{L}(P)) \cap \operatorname{Td}_k(\mathscr{V})\right). \tag{10.28}$$

This formula is the main motivation behind the works aiming at explicit expressions for the Todd classes of singular toric varieties: brion and Vergne [45], [46], Morelli [264], Pommersheim [290], [291], and Pukhlikov and Khovanskiĭ [294], to list just a few.

Cycle map. Now we discuss the cycle map from Chow groups to Borel–Moore homology. For more information on Borel–Moore homology in general see Iversen [210], and Danilov [103] in the context of algebraic varieties.

In this subsection the ground field is assumed to be $\mathbb{C}$, and for a variety $\mathscr{V}$ the associated complex analytic space is denoted by $\mathscr{V}(\mathbb{C})$.

For any locally compact space X one defines integral *Borel–Moore homology* $H_i^{\mathrm{BM}}(X,\mathbb{Z})$ as the homology of the chain complex of locally finite singular chains, i. e., instead of the chain group of finite $\mathbb{Z}$-linear combinations of singular simplices, one allows infinite linear combinations such that every point has a neighborhood meeting only finitely many of the singular simplices. In particular, for compact spaces, Borel–Moore homology coincides with singular homology: $H_i(X,\mathbb{Z}) = H_i^{\mathrm{BM}}(X,\mathbb{Z})$.

For a space of the form $Z = X \smallsetminus Y$, where X is a finite CW-complex and Y is any closed subcomplex, Borel–Moore homology can be defined in terms of (relative) singular homology:

$$H_i^{\mathrm{BM}}(Z,\mathbb{Z}) = \varprojlim H_i(Z, Z \smallsetminus K, \mathbb{Z}),$$

where K runs over all compact subsets of Z. This includes spaces of the form $\mathscr{V}(\mathbb{C})$: every algebraic variety is a Zariski open subset of a compact variety, and every compact algebraic variety can be triangulated into finitely many simplices (Hironaka [191]).

For rational coefficients (or over any field of coefficients), the duality between singular homology and cohomology has its Borel–Moore analogue:

$$H_c^i(X,\mathbb{Q}) = \mathrm{Hom}_{\mathbb{Q}}(H_i^{\mathrm{BM}}(X,\mathbb{Q}),\mathbb{Q}),$$

where on the left we have cohomology with compact support.

When $\mathscr{V}$ is a complete smooth variety the Poincaré–Lefschetz duality states

$$H_i^{\mathrm{BM}}(\mathscr{V}(\mathbb{C}),\mathbb{Z}) = H_i(\mathscr{V}(\mathbb{C}),\mathbb{Z}) = H^{2n-i}(\mathscr{V}(\mathbb{C}),\mathbb{Z}), \qquad n = \dim \mathscr{V}.$$

Borel–Moore homology is obviously a covariant functor on varieties and proper maps. Moreover, since every subvariety $\mathscr{U} \subset \mathscr{V}$ can be triangulated [191] (in general, into infinitely many singular simplices), we obtain a homomorphism

$$Z_i(\mathscr{V}) \to H_{2i}^{\mathrm{BM}}(\mathscr{V}(\mathbb{C}),\mathbb{Z}).$$

Then cycles, rationally equivalent to 0, are mapped to locally finite chains, homologous to 0 [134, Ch. 19]. Consequently, one obtains the *cycle map*

$$\mathrm{cl}_{\mathscr{V}} : \mathrm{A}_i(\mathscr{V}) \to H_{2i}^{\mathrm{BM}}(\mathscr{V}(\mathbb{C}),\mathbb{Z}),$$

natural with respect to proper morphisms. The cycle map has further good properties, such as compatibility with Chern classes of vector bundles; see [134, Ch. 19].

For complete smooth varieties the cycle map takes the shape

$$\mathrm{cl}_{\mathscr{V}}^* : \mathrm{CH}^*(\mathscr{V}) \to H^{2*}(\mathscr{V},\mathbb{Z}).$$

It is a homomorphism of graded rings, natural (contravariantly) with respect to all morphisms of smooth varieties [134, Ch. 19].

There are classes of varieties (not necessarily smooth) for which $\mathrm{cl}_{\mathcal{V}}$ is an isomorphism. For instance, using the localization sequence for Chow groups and its analogue for Borel–Moore homology, one shows that $\mathrm{cl}_{\mathcal{V}}$ is an isomorphism for the varieties $\mathcal{V}$ admitting a *cellular decomposition:*

$$\mathcal{V} = \mathcal{V}_m \supset \mathcal{V}_{m-1} \supset \cdots \supset \mathcal{V}_0 \supset \mathcal{V}_{-1} = \varnothing,$$

where for each j the complement $\mathcal{V}_j \setminus \mathcal{V}_{j-1}$ is a disjoint union of affine spaces $\mathbb{A}^{n_{jk}}$.

However, $\mathrm{cl}_{\mathcal{V}}$ in general fails to be an isomorphism, even rationally and even for smooth projective curves. For a smooth projective curve $\mathcal{V}$ of genus g we have:

(a) $\mathrm{A}_1(\mathcal{V}) = \mathbb{Z}\mathcal{V} \cong \mathbb{Z} \cong H_2(\mathcal{V}, \mathbb{Z})$;
(b) the map $\mathrm{cl}_{\mathcal{V}} : \mathrm{A}_0(\mathcal{V}) \to H_0(\mathcal{V}, \mathbb{Z})$ is the degree map whose kernel is the Jacobian of $\mathcal{V}$, a compact torus of dimension $2g$;
(c) $\mathrm{Coker}(\mathrm{cl}_{\mathcal{V}}) = H_1(\mathcal{V}, \mathbb{Z}) \cong \mathbb{Z}^{2g}$.

Moving beyond the smooth category, even the existence of a "contravariant cycle map" from the operational Chow group $\mathrm{cl}^*_{\mathcal{V}} : \mathrm{A}^*(\mathcal{V}) \to H^{2*}(\mathcal{V}, \mathbb{Z})$ with good properties is problematic: totaro has shown that there is no such a map $\mathrm{A}^1(\mathcal{V})_{\mathbb{Q}} \to H^2(\mathcal{V}, \mathbb{Q})$ even for the class of linear varieties, a natural extension of varieties with cellular decomposition that contains toric varieties; see Remark 10.36(f).

However, as we will see in the next section, the cycle map is an isomorphism for complete smooth toric varieties, and it is a rational isomorphism for complete simplicial toric varieties.

10.E Chow cohomology of toric varieties

The general theory of the last section will now be applied to toric varieties. We start with the combinatorial description of the operational Chow groups for general complete toric varieties developed by Fulton and Sturmfels. The strongest result, due Jurkiewicz and Danilov, hold for complete smooth toric varieties. After an extension to simplicial complete toric varieties (but with rational coefficients) it forms the basis of Stanley's proof of McMullen's g-theorem. Finally we will discuss the connection with McMullen's polytope algebra and Morelli's computation of the Grothendieck group of a smooth complete toric variety.

Complete toric varieties. The results are drawn from Fulton et al. [136] and Fulton and Sturmfels [137]. In particular, we give a combinatorial description of the operational Chow cohomology groups for all complete toric varieties. Accordingly, in this subsection $\mathscr{F}$ is always assumed to be a complete fan in V.

Recall that the ring structure of the operational Chow cohomology A^* is denoted by cup product, and the A^*-module structure on the Chow groups A_* is denoted by cap product.

The following theorem is the crucial fact that lets us interpret Chow cohomology in combinatorial terms. Actually, it holds under the much more general hypothesis that the variety admits an action by a connected solvable linear group with finitely many orbits; see [136, Th. 3]. It can be further generalized to linear varieties (see Remark 10.36(b)).

Theorem 10.28. *For all k we have the isomorphism*

$$A^k(\mathscr{V}(\mathscr{F})) \to \operatorname{Hom}_{\mathbb{Z}}(A_k(\mathscr{V}(\mathscr{F})), \mathbb{Z}), \quad c \mapsto [\alpha \mapsto \deg(c \cap \alpha)].$$

Corollary 10.29. *Let $\mathscr{F}$ be a unimodular fan (so that $\mathscr{V}(\mathscr{F})$ is smooth). Then the groups $\mathrm{CH}^*(\mathscr{V}(\mathscr{F}))$ and $K_0(\mathscr{V}(\mathscr{F})) \cong G_0(\mathscr{V}(\mathscr{F}))$ are free abelian groups of rank $\#\max\mathscr{F}$.*

Proof. Here we show that $\mathrm{CH}^*(\mathscr{V}(\mathscr{F}))$ and $G_0(\mathscr{V}(\mathscr{F}))$ are isomorphic free abelian groups. The claim on the rank then follows from the last equation in Theorem 10.37 below.

First of all, by Theorem 10.2, we have $G_0(\mathscr{V}(\mathscr{F})) = K_0(\mathscr{V}(\mathscr{F}))$. The Chow group $\mathrm{CH}^*(\mathscr{V}(\mathscr{F}))$ is finitely generated by Proposition 10.19 and torsionfree by Theorem 10.28.

In view of Theorem 10.26(b) (or (c)) we only need show that the group $G_0(\mathscr{V}(\mathscr{F}))$ is torsionfree. As in the proof of Theorem 10.26, we have two group homomorphisms,

$$\varphi : \mathrm{CH}^*(\mathscr{V}(\mathscr{F})) \to \operatorname{gr} G_0(\mathscr{V}(\mathscr{F})) \quad \text{and} \quad \psi : \operatorname{gr} G_0(\mathscr{V}(\mathscr{F})) \to \mathrm{CH}^*(\mathscr{V}(\mathscr{F})),$$

such that φ is surjective and the composite $\psi\varphi : \mathrm{CH}^*(\mathscr{V}(\mathscr{F})) \to \mathrm{CH}^*(\mathscr{V}(\mathscr{F}))_{\mathbb{Q}}$ is the canonical embedding. In particular, $\operatorname{gr} G_0(\mathscr{V}(\mathscr{F}))$ is a torsionfree group and, since the filtration $F_* G_0(\mathscr{V}(\mathscr{F}))$ is exhaustive, $G_0(\mathscr{V}(\mathscr{F}))$ is torsionfree, too. □

Remark 10.30. Vezzosi and Vistoli [363, Cor. 6.10(ii)] show that for a unimodular but not necessarily complete fan $\mathscr{F}$, for which the subset

$$D = \bigcap_{C \in \mathscr{F}} \bigcup_{\substack{C' \in \mathscr{F} \\ C \subset C'}} (C' + \mathbb{R}C) \subset V$$

is full-dimensional, the group $K_0(\mathscr{V}(\mathscr{F}))$ is free of rank equal to $\#\max\mathscr{F}$. Notice that completeness of $\mathscr{F}$ is sufficient for the full-dimensionality of D, but it is not necessary.

The freeness of the Grothendieck group of smooth projective toric models of tori over a not necessarily algebraically closed field was previously derived by Merkurjev and Panin [259, Sect. 5].

Generators and relations for the Chow ring $\mathrm{CH}^*(\mathscr{V}(\mathscr{F}))$ will be given in Theorem 10.37.

Theorems 10.22 and 10.28 identify the Chow cohomology classes of $\mathscr{V}(\mathscr{F})$ quite explicitly with certain functions on cones in $\mathscr{F}$. A *weight of codimension k on $\mathscr{F}$* is a function $c : \mathscr{F}_{[k]} \to \mathbb{Z}$. It is called a *Minkowski weight* if the following condition is satisfied for every $C \in \mathscr{F}_{[k+1]}$:

$$\sum_{C_1} n_{C_1,C}(u) \cdot c(C_1) = 0, \tag{10.29}$$

where u, C_1, and $n_{C_1,C}$ are as in Theorem 10.22. Using also the combinatorial description of $\mathrm{Pic}(\mathscr{V}(\mathscr{F}))$, given in Theorem 10.9, one derives the following:

Corollary 10.31.

(a) As an abelian group, $\mathrm{A}^k(\mathscr{V}(\mathscr{F}))$ *is isomorphic to the group of Minkowski weights on* $\mathscr{F}$ *of codimension* k.
(b) The first Chern class yields an isomorphism $\mathrm{Pic}(\mathscr{V}(\mathscr{F})) \to \mathrm{A}^1(\mathscr{V}(\mathscr{F}))$.

Remark 10.32. Brion has shown the analogue of Corollary 10.31(b) for arbitrary complete spherical varieties. The argument in the projective case appears in Brion [42].

Another consequence of the explicit description of the Chow groups of toric varieties in Theorem 10.22 is the following

Corollary 10.33. *If* $\mathscr{V}$ *and* $\mathscr{W}$ *are toric varieties, then the homomorphism*

$$\mathrm{A}_*(\mathscr{V}) \otimes \mathrm{A}_*(\mathscr{W}) \to \mathrm{A}_*(\mathscr{V} \times \mathscr{W}), \qquad \mathscr{V}' \otimes \mathscr{W}' \mapsto \mathscr{V}' \times \mathscr{W}',$$
$$\mathscr{V}' \subset \mathscr{V},\ \mathscr{W}' \subset \mathscr{W} \text{ subvarieties,}$$

is an isomorphism.

Corollary 10.33 gives a possibility to introduce the numbers

$$m_{C,C_1,C_2} \in \mathbb{Z}$$

for any triple of cones $C, C_1, C_2 \in \mathscr{F}$, satisfying the conditions

$$C \subset C_1 \cap C_2 \quad \text{and} \quad \operatorname{codim} C_1 + \operatorname{codim} C_2 = \operatorname{codim} C.$$

Namely, consider the diagonal embedding

$$\delta : \mathscr{V}(\mathscr{F}) \to \mathscr{V}(\mathscr{F}) \times \mathscr{V}(\mathscr{F}) = \mathscr{V}(\mathscr{F} \times \mathscr{F}).$$

Then $\delta(\overline{\tau}_C) \subset \overline{\tau}_C \times \overline{\tau}_C$ and, therefore, there are integers m_{C,C_1,C_2}, indexed as above, such that the following equation holds true in $\mathrm{A}_{\operatorname{codim} C}(\mathscr{V}(\mathscr{F}))$:

$$\delta(\overline{\tau}_C) = \sum_{C_1,C_2} m_{C,C_1,C_2} \cdot \left(\overline{\tau}_{C_1} \times \overline{\tau}_{C_2}\right). \tag{10.30}$$

Observe that the numbers m_{C,C_1,C_2} are *not* uniquely determined by the cones C, C_1, C_2. Set $m_{C,C_1,C_2} = 0$ if $C \not\subset C_1 \cap C_2$.

In the next theorem, which gives a complete description of the intersection theory for complete toric varieties, the Chow cohomology classes are viewed as the corresponding Minkowski weights (Corollary 10.31). Like Theorem 10.28 it can be generalized to a larger class of varieties.

Theorem 10.34. *Let $c \in A^p(\mathscr{V}(\mathscr{F}))$, $\tilde{c} \in A^q(\mathscr{V}(\mathscr{F}))$, and $C \in \mathscr{F}_{[k]}$. Then for all choices of the numbers m_{C,C_1,C_2} satisfying* (10.30) *the following equations hold:*

$$c \cap \overline{\tau}_C = \sum_{\text{codim } C_1 + \text{codim } C_2 = \text{codim } C} m_{C,C_1,C_2} \cdot c(C_1) \cdot \overline{\tau}_{C_2}$$

and

$$(c \cup \tilde{c})(C) = \sum_{\text{codim } C_1 + \text{codim } C_2 = \text{codim } C} m_{C,C_1,C_2} \cdot c(C_1) \cdot \tilde{c}(C_2).$$

The next result provides an effective combinatorial algorithm, phrased in terms of certain displacements in the fan $\mathscr{F}$, for finding a particular system of integers m_{C,C_1,C_2}, satisfying (10.30).

Theorem 10.35. *For a cone $C \in \mathscr{F}$ and a generic element $v \in L$ we have*

$$\delta(\overline{\tau}_C) = \sum_{C_1,C_2} m_{C,C_1,C_2}(\overline{\tau}_{C_1} \times \overline{\tau}_{C_2})$$

where

(1) $m_{C,C_1,C_2} = \#[L : \text{gp}(C_1 \cap L) + \text{gp}(C_2 \cap L)]$,
(2) $C_1 \cap (C_2 + v) \neq \varnothing$,
(3) $\text{codim } C_1 + \text{codim } C_2 = \text{codim } C$,
(4) $C \subset C_1 \cap C_2$.

In the case $C = 0$ of Theorem 10.35 an element $v \in L$ is called *generic* if the existence of exactly one element $x \in V$ with $(v,0)+(x,x) \in C_1 \times C_2$ implies $\dim C_1 + \dim C_2 = d$. (In general, the existence of exactly one such element $x \in V$ implies $\dim C_1 + \dim C_2 \leq d$.) For $C \neq 0$ this condition must be evaluated after all objects have been replaced by their homomorphic images modulo $\mathbb{R}C$.

The work [137] contains also explicit combinatorial formulas for (i) the action of $A^*(\mathscr{V}(\mathscr{F}))$ on $A_*(\mathscr{V}(\mathscr{F}'))$ via a morphism of complete fans $\mathscr{F}' \to \mathscr{F}$ and (ii) the corresponding contravariant homomorphism $A^*(\mathscr{V}(\mathscr{F})) \to A^*(\mathscr{V}(\mathscr{F}'))$.

The equivariant Chow cohomology of toric varieties was computed by Payne [283]. Chow rings of toric varieties defined by atomic lattices were studied by Feichtner and Yuzvinsky [121].

Remark 10.36. Totaro [356] describes Chow groups for *linear varieties*, which are substantially more general than the class of toric varieties. These varieties are defined inductively according to the following rules: (i) an affine space is a linear variety; (ii) the complement of a linear variety embedded in an affine space is also a linear variety; (iii) a variety that can be stratified into finitely many linear subvarieties is linear itself. In particular, if a variety admits a solvable group action with finitely many orbits then it is a linear variety because the orbits are of the form $\mathbb{T}^a \times \mathbb{A}^b$. So the results in [356] extend those in [136]. The following is a brief summary of the former:

(a) the groups $A_i(\mathscr{V})$ are described in terms of generators and relations when $\mathscr{V}$ can be stratified into finitely many strata of the form $\mathbb{T}^a \times \mathbb{A}^b$.

(b) The analogue of Theorem 10.28 holds true for complete linear varieties:

$$A^i(\mathscr{V}) = \operatorname{Hom}_{\mathbb{Z}}(A_i(\mathscr{V}), \mathbb{Z}).$$

(c) For any linear variety $\mathscr{V}$ over $\mathbb{C}$ the cycle map is a rational isomorphism between the Chow groups $A_i(\mathscr{V})_{\mathbb{Q}}$ and the smallest nonzero subspace of Borel–Moore homology $H_{2i}^{\mathrm{BM}}(\mathscr{V}(\mathbb{C}), \mathbb{Q})$ with respect to the Deligne's weight filtration.

(d) For any variety $\mathscr{V}$ over $\mathbb{C}$, which is stratified into finitely many strata of the form $\mathbb{T}^a \times \mathbb{A}^b$, there is an explicit chain complex whose homology computes the weight-graded pieces of the Borel–Moore homology $HB_*^{\mathrm{BM}}(\mathscr{V}(\mathbb{C}), \mathbb{Q})$.

(e) For a complete toric variety $\mathscr{V}$ over $\mathbb{C}$ there is a natural isomorphism

$$A^i(\mathscr{V})_{\mathbb{Q}} \to H^{2i}(\mathscr{V}(\mathbb{C}), \mathbb{Q}) \cap F^i H^{2i}(\mathscr{V}(\mathbb{C}), \mathbb{C}),$$

where F^i refers to the Hodge filtration.

(f) Unlike the toric case in (e) above, there is no "good" functorial homomorphism $A^1(\mathscr{V})_{\mathbb{Q}} \to H^2(\mathscr{V}(\mathbb{C}), \mathbb{Q})$ for general normal projective linear varieties over $\mathbb{C}$.

Related results on higher K- and Bloch's higher Chow groups of linear varieties have been obtained by Joshua [213].

Simplicial complete toric varieties. In this subsection we assume that $\mathscr{F}$ is a simplicial complete unimodular fan in V. The lattice of reference in V is again denoted by L.

First we consider the case in which $\mathscr{F}$ is even unimodular. The representation of the Chow ring given below is due to Jurkiewicz (in the projective case) [214] and Danilov [102]. We will discuss its proof below when the g-theorem 10.41 will be derived from it.

Theorem 10.37. *Let $\mathscr{F}$ be a complete unimodular fan. For simplicity denote the cones $C \in \mathscr{F}^{[1]}$ by the generators ρ of the monoids $C \cap L$, and choose an indeterminate X_ρ for each ρ. Define the following ideals in the polynomial ring $\mathbb{Z}[X_\rho : \rho \in \mathscr{F}^{[1]}]$:*

$$I_{\mathscr{F}} = (X_{\rho_1} \cdots X_{\rho_u} : \mathbb{R}_+\rho_1 + \cdots + \mathbb{R}_+\rho_u \notin \mathscr{F}),$$
$$J_{\mathscr{F}} = (\ell_\sigma : \sigma \in L^*), \qquad \ell_\sigma = \sum_{\rho \in \mathscr{F}^{[1]}} \sigma(\rho) X_\rho.$$

Then the assignment $X_\rho \mapsto \overline{\tau}_\rho$ induces a ring isomorphism

$$\mathbb{Z}[X_\rho : \rho \in \mathscr{F}^{[1]}]/(I_{\mathscr{F}} + J_{\mathscr{F}}) \cong \mathrm{CH}^*(\mathscr{V}(\mathscr{F})).$$

In particular, $\mathrm{CH}^*(\mathscr{V}(\mathscr{F}))$ is generated by the classes of the $\mathbb{T}$-stable divisors $\overline{\tau}_\rho$, $\rho \in \mathscr{F}^{[1]}$. Moreover,

$$\operatorname{rank} \mathrm{CH}^k(\mathscr{V}(\mathscr{F})) = \sum_{i=k}^{d} (-1)^{i-k} \binom{i}{k} \# \mathscr{F}^{[d-i]}$$

and, consequently,

$$\operatorname{rank} \mathrm{CH}^*(\mathscr{V}(\mathscr{F})) = \# \max \mathscr{F}.$$

Next we exhibit an explicit free basis of the group $\mathrm{CH}^*(\mathcal{V}(\mathcal{F}))$ when $\mathcal{F}$ is additionally assumed to be projective. The result is proved in [102]. We present the version of Fulton [133, Sect. 5.2].

Fix a function $f : V \to \mathbb{R}$ that belongs to $\mathrm{SF}(\mathcal{F}, V)$. Choose a point $\xi \in V^*$ in general position so that the numbers

$$f_C(\xi), \quad C \in \max \mathcal{F} = \mathcal{F}^{[d]}$$

are all different, where $f_C : V \to \mathbb{R}$ is the linear extension of $f|_C$. We introduce the linear order on $\max \mathcal{F}$ by

$$C_1 < C_2 \iff f_{C_1}(\xi) < f_{C_2}(\xi).$$

Call a facet D of a cone $C \in \max \mathcal{F}$ *positive* if $C < C_1$ for the cone $C_1 \in \max \mathcal{F}$, different from C and containing D. Finally, for a cone $C \in \max \mathcal{F}$ denote by $\Gamma(C)$ the intersection of all positive facets of C. (This construction amounts to a shelling of the fan $\mathcal{F}$; see Exercise 10.10.)

Theorem 10.38. *The classes of the* $\mathbb{T}$*-stable cycles* $\overline{\tau}_{\Gamma(C)}$, $C \in \max \mathcal{F}$, *form a free basis of the Chow group* $\mathrm{CH}^*(\mathcal{V}(\mathcal{F}))$.

The construction of this explicit basis plays a crucial role in the proof of Theorem 10.37; see Remark 10.40(b). The proof of Corollary 10.29 and Theorem 10.38 imply that $G_0(\mathcal{V}(\mathcal{F}))$ is generated by the classes of the coherent sheaves that correspond to equivariant closed subsets.

Fulton [133] observes that all one needs for the validity of Theorem 10.38 under the assumptions of completeness and smoothness is the existence of a linear order on $\max \mathcal{F}$ satisfying the following condition ($\Gamma(-)$ as above):

$$C_1, C_2 \in \max \mathcal{F}, \quad \Gamma(C_1) \subset C_2 \quad \Longrightarrow \quad C_1 \le C_2.$$

There are nonprojective complete (smooth) fans $\mathcal{F}$ admitting such orders on $\max \mathcal{F}$, but this is not always the case.

Now we relax the hypothesis on $\mathcal{F}$ and assume it is a simplicial complete fan. On the variety $\mathcal{V} = \mathcal{V}(\mathcal{F})$ the intersection product can be defined if one allows rational coefficients. The main reason is that every Weil divisor has a Cartier multiple (Exercise 10.4). Setting $A^k(\mathcal{V})_{\mathbb{Q}} = A_{d-k}(\mathcal{V})_{\mathbb{Q}}$, $d = \dim \mathcal{F}$, one obtains a ring structure on

$$\mathrm{A}^*(\mathcal{V})_{\mathbb{Q}} = \bigoplus_{k=0}^{d} A^k(\mathcal{V})_{\mathbb{Q}}$$

whose multiplication table satisfies the equation

$$\overline{\tau}(C)\overline{\tau}(D) = \begin{cases} \frac{\mu(C)\mu(D)}{\mu(C+D)} \overline{\tau}_{C+D}, & C + D \in \mathcal{F}, \ \dim C + \dim D = \dim(C+D), \\ 0, & C + D \notin \mathcal{F}; \end{cases} \tag{10.31}$$

see [133, Sect. 5.1]. In the smooth case all coefficients are equal to 1.

Remark 10.39. The ring structure on $\mathrm{A}^*(\mathscr{V})_{\mathbb{Q}}$ indicated above above is naturally isomorphic to the rational operational Chow cohomology of $\mathscr{V}$ and so the notation is consistent with the previous use of $\mathrm{A}^*(\mathscr{V})$. More generally, on a variety $\mathscr{W}$ which is locally a quotient of a smooth variety by a finite group the maps

$$\mathrm{A}^k(\mathscr{W})_{\mathbb{Q}} \to \mathrm{A}_{d-k}(\mathscr{W})_{\mathbb{Q}}, \quad c \mapsto c \cap \mathscr{W}, \quad k = 0, \dots, d, \quad d = \dim \mathscr{W},$$

are isomorphisms; see Vistoli [365]. That simplicial toric varieties are locally quotients by finite (abelian) groups is the content of Exercise 10.12.

An alternate way to recognize the rational operational Chow cohomology under the product formula (10.31) is via Theorems 10.28 and 10.34.

Choose an indeterminate X_ρ for each $\rho \in \mathscr{F}^{[1]}$, and consider the ideals $I_{\mathscr{F}}$ and $J_{\mathscr{F}}$ as in Theorem 10.37. We also use the convention on ρ in the theorem. Then the assignment $X_\rho \mapsto \overline{\tau}(\rho)$, $\rho \in \mathscr{F}^{[1]}$, induces a homomorphism

$$\varphi : \mathbb{Q}[X_\rho : \rho \in \mathscr{F}^{[1]}] \to \mathrm{A}^*(\mathscr{V})_{\mathbb{Q}}.$$

It results immediately from (10.31) that φ is surjective, and (10.31) also shows that $I_{\mathscr{F}} \subset \operatorname{Ker} \varphi$. This holds for $J_{\mathscr{F}}$, too, since $\sum_{\rho \in \mathscr{F}^{[1]}} \sigma(\rho)\overline{\tau}_\rho$ is the principal divisor of $\sigma \in L^*$ considered as a rational function (compare Theorem 10.9).

By its definition

$$R_{\mathscr{F}} = \mathbb{Q}[X_\rho : \rho \in \mathscr{F}^{[1]}]/I_{\mathscr{F}}.$$

is the Stanley–Reisner ring of the $(d-1)$-dimensional simplicial complex $\Delta_{\mathscr{F}}$ formed by the sets $\operatorname{conv}(\rho_1, \dots, \rho_u)$, $\mathbb{R}_+\rho_1 + \cdots + \mathbb{R}_+\rho_u \in \mathscr{F}$. Using standard properties of Stanley–Reisner rings, one shows that $R_{\mathscr{F}}$ is a Gorenstein ring and $J_{\mathscr{F}} R_{\mathscr{F}}$ is generated by a linear system of parameters in $R_{\mathscr{F}}$. Therefore the Hilbert series of $S_{\mathscr{F}} = R_{\mathscr{F}}/J_{\mathscr{F}} R_{\mathscr{F}}$ is

$$H_{S_{\mathscr{F}}}(t) = 1 + h_1 t + \cdots + h_d t^d, \qquad h_k = \sum_{i=k}^{d} (-1)^{i-k} \binom{i}{k} \# \mathscr{F}^{[d-i]}.$$

See Exercise 10.9 for the details.

As observed above, one has a surjective epimorphism $\psi : S_{\mathscr{F}} \to \mathrm{A}^*(\mathscr{V}(\mathscr{F}))$ induced by φ. We claim that

$$S_{\mathscr{F}} \cong \mathrm{A}^*(\mathscr{V})_{\mathbb{Q}}. \tag{10.32}$$

Since ψ is surjective, it suffices that both vector spaces have the same dimension. The dimensions do not depend on $\Bbbk$, as follows from the construction of $S_{\mathscr{F}}$ on the one hand and from Theorems 10.22 and 10.28 on the other. Therefore we can choose $\Bbbk = \mathbb{C}$.

For the proof of (10.32) and the application to the g-theorem below, the singular cohomology groups $H^{2k}(\mathscr{V}, \mathbb{Q})$ are brought into play via the cycle map $\mathrm{cl}_{\mathscr{V}}$, and it suffices to show that the composite

$$S_{\mathscr{F}} \xrightarrow{\varphi} \mathrm{A}^*(\mathscr{V})_{\mathbb{Q}} \xrightarrow{\mathrm{cl}_{\mathscr{V}}} \bigoplus_{k=0}^{d} H^{2k}(\mathscr{V}, \mathbb{Q}) \tag{10.33}$$

is an isomorphism of graded $\mathbb{Q}$-algebras that doubles degrees. For the surjectivity of $(\mathrm{cl}_{\mathcal{V}})_{\mathbb{Q}}$ we refer the reader to [102]. Both ends have the same dimension over $\mathbb{Q}$: for $S_{\mathcal{F}}$ this is the multiplicity of $R_{\mathcal{F}}$, equal to the number of d-dimensional cones in $\mathcal{F}$. This is also the dimension of $H^*(\mathcal{V}, \mathbb{Q})$; see [102] and also the following remark.

Remark 10.40. (a) The isomorphism (10.32) generalizes Theorem 10.37 to simplicial complete toric varieties. In the smooth case, all arguments work over $\mathbb{Z}$ and prove Theorem 10.37. The reader should compare [102, §10] or [133, Sect. 5].

(b) The basis of $\mathrm{CH}^*(\mathcal{V})$ in Theorem 10.38 is used to show the surjectivity of the map $(\mathrm{cl}_{\mathcal{V}})_{\mathbb{Q}}$ (or $\mathrm{cl}_{\mathcal{V}}$ itself in the smooth case), after a reduction to the smooth projective case via a projective resolution of singularities (compare p. 366).

(c) Rational singular cohomology of complete simplicial toric varieties vanishes in odd degrees as shown in [102, 12.11]. Completeness alone is not sufficient [102, 12.12].

The interpretation of the h-vector of the fan $\mathcal{F}$ as the Hilbert series of the cohomology ring $H^*(\mathcal{V}, \mathbb{Q})$ has a beautiful combinatorial application observed by Stanley [328]. It proves the g-theorem previously conjectured by McMullen [254]:

Theorem 10.41. *Let P be a simplicial d-polytope with h-vector $(h_0 = 1, h_1, \dots, h_d)$. Then the vector*

$$g = (1, h_1 - h_0, \dots, h_u - h_{u-1}), \qquad u = \lfloor d/2 \rfloor,$$

is a Macaulay sequence.

We sketch the proof. Perturbing the vertices of P by small displacements, one can assume that P is rational, and, after a translation, that it contains the origin in its interior. Now choose $\mathcal{F}$ as the fan whose cones are spanned by the faces of P.

As observed above, the h-vector of P represents the Hilbert function of the cohomology ring $H^*(\mathcal{V}, \mathbb{Q})$, $\mathcal{V} = \mathcal{V}(\mathcal{F})$ (in twice the degrees). The fan $\mathcal{F}$ is projective (see Proposition 1.67), and therefore $\mathcal{V}$ is a projective variety. For such a variety the intersection cohomology $IH(\mathcal{V})$ (of middle perversity) is a module over the cohomology ring (Goresky and MacPherson [146, Sect. 1.4 and 2.4]). It was proved by Beilinson, Bernstein, and Deligne that intersection cohomology [27, 5.4.10] satisfies the *strong Lefschetz theorem*: let $\omega \in H^2(\mathcal{V}, \mathbb{Q})$ be the class of a generic hyperplane section; then the multiplication by ω^j is an isomorphism $IH^{d-j}(\mathcal{V}, \mathbb{Q}) \cong IH^{d+j}(\mathcal{V}, \mathbb{Q})$ for all j.

Since $\mathcal{F}$ is simplicial, the standard affine charts of $\mathcal{V}$ are quotients of affine spaces by finite abelian groups (see Exercise 10.12), and therefore $\mathcal{V}$ is an orbifold (see page 367). In this case, intersection cohomology and singular cohomology are isomorphic, and for orbifolds they are isomorphic [146, Sect. 6.4]. Therefore multiplication by ω is injective on $H^{2j}(\mathcal{V}, \mathbb{Q})$ for $j < d/2$ and surjective for $j \geq d/2$. It follows that the vector g represents the Hilbert function of $H^*(\mathcal{V}, \mathbb{Q})/\omega H^*(\mathcal{V}, \mathbb{Q})$, an algebra generated in degree 1 by (10.33).

Remark 10.42. One can assign a (generalized) h-vector to an arbitrary polytope (Stanley [331]). However, the h-vector is difficult to interpret in the nonsimplicial case. If the polytope is rational, it corresponds to a toric variety as above, and the entries of the h-vector give the dimensions of the intersection cohomologies also in the general case (see Fieseler [122] for a proof). Therefore the strong Lefschetz theorem implies unimodality of the h-vector [331, 3.2] also for nonsimplicial polytopes. See Remark 10.45 for a further extension.

The polytope algebra. When $\mathscr{F}$ is a projective fan, the ring structure of $\mathrm{A}^*(\mathscr{V}(\mathscr{F}))$ is related to McMullen's polytope algebra [256], [257], [258]. (Recall, the fan $\mathscr{F}$ lives in an $\mathbb{R}$-vector space V in which we have fixed a lattice L.) For a deeper study of the polytope algebra we recommend Brion's article [43].

The rational version of McMullen's *polytope algebra* Π has generators $[P]$ for every rational polytope P in the dual space V^*, where the rational structure is defined by $\mathbb{Q}L^* \subset V^*$. The generators satisfy the relations

(1) $[P \cup Q] + [P \cap Q] = [P] + [Q]$ whenever $P \cup Q$ is a polytope;
(2) $[P + x] = [P]$ for every $x \in \mathbb{Q}L^*$;
(3) $[P] \cdot [Q] = [P + Q]$.

The multiplicative unit of this ring is the class of a rational point.

Let $P \subset V^*$ be rational polytope. Denote by $\Pi(P)$ the subalgebra of Π, generated by the classes $[Q]$ for which there exist a positive rational number λ and a rational polytope R with $P = \lambda Q + R$; such a polytope Q is called a *Minkowski summand* of P. Thus, if P and P' are two L^*-polytopes with $\mathscr{N}(P) = \mathscr{N}(P')$ then $\Pi(P) = \Pi(P')$ (Exercise 10.11). This observation allows us to introduce the following subring of Π:

$$\Pi(\mathscr{F}) = \Pi(P), \quad P \text{ an } L^*\text{-polytope with } \mathscr{N}(P) = \mathscr{F}.$$

Recall from Theorem 10.11 an ample line bundle $\mathscr{L}$ on $\mathscr{V}(\mathscr{N}(P))$ defines an L^*-polytope $P(\mathscr{L})$ with $\mathscr{N}(P(\mathscr{L})) = \mathscr{N}(P)$; if $\mathscr{L}$ is only generated by its global sections then $\mathscr{L}$ defines a an L^*-polytope $P(\mathscr{L})$ such that $\mathscr{F}$ is a subdivision of $\mathscr{N}(P(\mathscr{L}))$. It follows that $[P(\mathscr{L})] \in \Pi(\mathscr{F})$ when $\mathscr{L}$ is generated by its global sections (Exercise 10.11).

Any Cartier divisor D on $\mathscr{V}(\mathscr{F})$ can be viewed as a line bundle on $\mathscr{V}(\mathscr{F})$ and, simultaneously, as an element of $\mathrm{A}^1(\mathscr{V}(\mathscr{F}))$ (Corollary 10.31(b)). We put

$$\exp(D) = \sum_{r=0}^{d} \frac{D^r}{r!} \in \mathrm{A}^*(\mathscr{V}(\mathscr{F}))_{\mathbb{Q}}, \qquad d = \dim_{\mathbb{R}} V,$$

where the rth powers on the right are understood in the sense of the Chow cohomology ring structure. When $\mathscr{F}$ is unimodular, this is the Chern character D, as introduced in equation (10.18).

Theorem 10.43.

(a) There exists an injective ring homomorphism

$$\vartheta : \Pi(\mathscr{F}) \to \mathrm{A}^*(\mathscr{V}(\mathscr{F}))_{\mathbb{Q}}$$

such that $\vartheta([P(D)]) = \exp(D)$ *for any ample Cartier divisor D on $\mathscr{V}(\mathscr{F})$. The image of ϑ equals the subring of $\mathrm{A}^*(\mathscr{V}(\mathscr{F}))_{\mathbb{Q}}$, generated by the rational Picard group* $\operatorname{Pic}(\mathscr{V}(\mathscr{F}))_{\mathbb{Q}} = \mathrm{A}^1(\mathscr{V}(\mathscr{F}))_{\mathbb{Q}}$.

(b) *We have a ring isomorphism* $\varinjlim A^*(\mathscr{V}(\mathscr{G})) = \Pi$ *where the direct limit is considered with respect to subdivisions of complete fans in* V, *contravariantly inducing injective Chow cohomology ring homomorphisms.*

Observe that in Theorem 10.43(b) we can restrict ourselves to projective, even projective unimodular fans $\mathscr{G}$ in V. In fact, every complete fan has a projective unimodular triangulation (Corollary 2.73).

Next we present McMullen's polytopal formula [256] for computing the product of two elements in $\Pi(\mathscr{F})$. It serves as the motivation (for the terminology as well) for the product rule in the Chow cohomology ring of a complete toric variety, given in Theorems 10.34 and 10.35. Simultaneously, this formula can be deduced from these theorems, with use of Theorem 10.43(a).

First we fix a positive definite inner product $\langle -,- \rangle$ on V^*. In particular, the elements of V are thought of as the mappings $\langle v,- \rangle : V^* \to \mathbb{R}$, $v \in V^*$. For any affine subspace $H \subset V^*$ we have the associated volume function vol_H and the notion of a vector in V^* perpendicular to H.

For a polytope $P \subset V^*$ a *k-weight* is a function ω from the set of k-faces of P to $\mathbb{R}$ such that for every $(k+1)$-dimensional face $G \subset P$ we have

$$\sum_{\substack{F \subset G \\ \dim F = k}} \omega(F) \cdot \nu_{F,G} = 0,$$

where $\nu_{F,G} \in V^*$ denotes the unit vector, parallel to $\mathrm{aff}(G) \subset V$ and perpendicular to $\mathrm{aff}(F)$. Denote by $\Omega_k(P)$ the real vector space of k-weights on P and

$$\Omega(P) = \bigoplus_{k=0}^{\dim P} \Omega_k(P).$$

Let $Q, P \subset V^*$ be polytopes with $[Q] \in \Pi(\mathscr{N}(P))$. Then to a k-face $F \subset P$ there corresponds a unique face $F' \subset Q$ of dimension $\leq k$. When $\dim F' = k$, then F and F' are parallel. Define

$$\omega(F) = \begin{cases} \mathrm{vol}_{\mathrm{aff}(F')}(F'), & \dim F' = k, \\ 0, & \text{else.} \end{cases}$$

In view of Minkowski's reconstruction theorem [154, p. 332], this gives rise to an element $\omega \in \Omega_k(P)$. We have the resulting group homomorphism $\Pi(\mathscr{N}(P)) \to \Omega(P)$.

Let $P, Q \subset V^*$ be polytopes with $\dim(P+Q) = \dim P + \dim Q$. Then there exists a unique real number $\alpha_{F,G}$ only depending on $\mathrm{aff}(P)$ and $\mathrm{aff}(Q)$, such that

$$\mathrm{vol}_{\mathrm{aff}(P+Q)} = \alpha_{F,G} \, \mathrm{vol}_{\mathrm{aff}(P)} \, \mathrm{vol}_{\mathrm{aff}(Q)} .$$

Here vol denotes the euclidean volume forms in the respective affine subspaces of V^*.

A face H of $2P$ can be dissected into a union of sums $F + G$ of faces P such that $\dim F + \dim G = \dim H$. A way to construct such a subdivision is to use a *mixed decomposition*. Assume that $\dim P = \dim V^*$, and fix a sufficiently general element

$v \in V = V^{**}$. Then the faces of the bottom (in the sense of Remark 1.58) of the polytope

$$\operatorname{conv}\{(p+p', v(p)) \ : \ p, p' \in P\} \subset V^* \oplus \mathbb{R},$$

project onto polytopes of the form $F+G$, where $F, G \subset P$ are faces and $\dim(F+G) = \dim F + \dim G$. Moreover, these projections define a regular subdivision of $2P$, the mixed subdivision of $2P$ corresponding to v; it is denoted by $\Delta_v(2P)$. For further information in mixed subdivisions see Sturmfels [337].

Part (b) of the following theorem is McMullen's description of the cup product in $\mathrm{A}^*(\mathscr{F})$ for $\mathscr{F} = \mathscr{N}(P)$ (in the notation of [137]).

Theorem 10.44.

(a) The map $\Pi(\mathscr{N}(P)) \to \Omega(P)$ *is an injective homomorphism of rational vector spaces.*

(b) Assume $v \in V$ *is a sufficiently general element. If* $x_1, x_2 \in \Pi(\mathscr{N}(P))$ *are given by the weights* $\omega_1, \omega_2 \in \Omega(P)$ *then the weight corresponding to* x_1x_2 *is given by*

$$\omega(H) = \sum_{F+G\in\Delta_v(P)} \alpha_{F,G}\cdot\omega_1(F)\cdot\omega_2(G), \qquad H \in \mathscr{N}(P)^{[p+q]},$$

where the sum is over all faces F, G *of* H *with* $\dim F + \dim G = \dim H$ *and* $F+G \in \Delta_v(2P)$.

McMullen formulates the product rule in [257, p. 426] and proves it in full generality in [258]. It can also be derived from Theorem 10.34; see [137].

Remark 10.45. (a) For simple polytopes McMullen proves that $\Pi(\mathscr{N}(P)) \cong \Omega(P)$. Moreover, he proves the g-theorem [257, Th. 7.2] and the strong Lefschetz theorem [257, Th. 13.1] for $\Pi(\mathscr{N}(P))$.

(b) McMullen's arguments do not require that the polytope P be rational. Recently Karu [219] has generalized the strong Lefschetz theorem to (the intersection cohomology of) arbitrary polytopes.

We conclude the subsection by a polytopal description of the ring $K_0(\mathscr{V}(\mathscr{F}))$, due to Morelli [265]. It represents an extension to higher rank bundles (in the smooth complete case) of the polytopal description of the Picard group of a toric variety, given in Section 10.B. Simultaneously, it should be viewed as an integral version of the polytope algebra, in which $\mathscr{F}$ is not necessarily projective, but smooth.

Let $\mathsf{Pol}(L^*)$ denote the ring generated by symbols $[P]$ for the L^*-polytopes $P \subset V^*$, subject to the relations:

(1) $[P\cup Q] + [P\cap Q] = [P] + [Q]$ whenever $P\cup Q$ and $P\cap Q$ is are L^*-polytopes;
(2) $[P] = [x+P]$ whenever $x \in L^*$;
(3) $[P]\cdot[Q] = [P+Q]$.

Let $\mathsf{Con}(L^*)$ be the group generated by symbols $[C]$ for the cones $C \subset V^*$ that are rational with respect to the lattice L^* but not necessarily pointed, subject to the relations

$$[C_1] + [C_2] = [C_1 \cup C_2] + [C_1 \cap C_2]$$

whenever $C_1 \cup C_2$ is a rational cone. Denote by

$$\mathrm{Con}(L^*, \mathscr{F}) \subset \mathrm{Con}(L^*)$$

the subgroup generated by the dual cones $C^* \subset V^*$, $C \in \mathscr{F}$.

Finally, let $\mathrm{Pol}(L^*, \mathscr{F}) \subset \mathrm{Pol}(L^*)$ be the subset of those elements $[P]$ for which

$$[\mathbb{R}_+(P - x)] \in \mathrm{Con}(L^*, \mathscr{F}) \text{ for every point } x \in \mathrm{lat}(P).$$

Morelli gives a polytopal description of $K_0(\mathscr{V}(\mathscr{F}))$ as follows:

Theorem 10.46. *Let $\mathscr{F}$ be a complete unimodular fan. Then $\mathrm{Pol}(L^*, \mathscr{F}) \subset \mathrm{Pol}(L^*)$ is a subring, isomorphic to the ring $K_0(\mathscr{V}(\mathscr{F}))$.*

Actually, in [265] one even obtains a polytopal description of the λ-operations in the Grothendieck ring, defined in terms of the exterior powers of vector bundles.

Remark 10.47. (a) It is shown in [363, Cor. 6.10(iii)] that for a smooth toric variety $\mathscr{V}(\mathscr{F})$, whose fan $\mathscr{F}$ satisfies the condition mentioned in Remark 10.30, the natural map

$$K_*(\Bbbk) \otimes K_0(\mathscr{V}) \to K_*(\mathscr{V})$$

is an isomorphism of rings with respect to the product operations in higher K-theory: let $\mathscr{W}$ be a variety; then its K-theory $K_*(\mathscr{W}) = \bigoplus_{i=0}^{\infty} K_i(\mathscr{W})$ carries a graded-commutative ring structure, extending the ring structure on $K_0(\mathscr{W})$; this follows from Waldhausen's work [370]. So Theorem 10.46 also offers a polytopal view on higher K-theory of a smooth complete toric variety.

(b) Because of Theorem 10.26(b), the two rings, described respectively in Theorems 10.37 and 10.46, are rationally isomorphic.

The K-theoretical results for toric varieties in this section, drawn from [259], [265], and [363], are all derived by first considering equivariant K-groups. (This also applies to [45].) Au, Huang, and Walker show [11] that the higher equivariant K-theory of an arbitrary affine toric variety is completely determined by the K-theory of the ground field. In this book we do not develop equivariant theories. However, in the last section we completely characterize equivariant vector bundles in one important case—when the base variety is a representation space of a torus.

10.F Toric varieties with huge Grothendieck group

In this section $\Bbbk$ will be an algebraically closed field of characteristic 0.

For some time it has been an open question whether the Cartan homomorphism $K_0(\mathscr{V})_{\mathbb{Q}} \to G_0(\mathscr{V})_{\mathbb{Q}}$ is an isomorphism for a simplicial toric variety $\mathscr{V}$ (Brion and

Vergne [45], Cox [100, §7.12]). In view of Theorems 8.4 and 10.2 this question is interesting only for nonaffine singular toric varieties, and one might think that Corollary 10.27 hints at a positive answer in general. It has even been conjectured that the rational Cartan homomorphism is an isomorphism for an arbitrary quasiprojective orbifold.

In this section we will show that the Cartan homomorphism is usually far from being a rational isomorphism, even for projective simplicial toric varieties in dimensions ≥ 3. The source of nontrivial vector bundles on such varieties is the nonvanishing of the SK_1-groups of their standard affine charts—a consequence of Theorem 9.7: nontrivial invertible matrices will be used to glue trivial vector bundles on toric affine charts in such a way that the effect of the twisting survives even in the Grothendieck group K_0.

Our strategy is as follows. By Proposition 10.19(b) the group $G_0(\mathscr{V}(\mathscr{F}))_{\mathbb{Q}}$ is finitely generated for an arbitrary fan $\mathscr{F}$. Corollary 10.27 and Theorem 10.22 tell us how to compute the rank of $G_0(\mathscr{V}(\mathscr{F}))_{\mathbb{Q}}$ in terms of $\mathscr{F}$, and the map $K_0(\mathscr{V})_{\mathbb{Q}} \to G_0(\mathscr{V})_{\mathbb{Q}}$ is surjective whenever $\mathscr{F}$ is simplicial (Remark 10.20(b)). Yet Theorem 10.56 will provide many examples of projective simplicial toric varieties $\mathscr{V}(\mathscr{F})$ with rank $K_0(\mathscr{V}(\mathscr{F})) \geq \dim_{\mathbb{Q}} \Bbbk$.

***K*-theoretical background.** We will employ two independent K-theoretical techniques: the action of big Witt vectors on nil-K-theory (in the general graded situation) and Thomason's Mayer–Vietoris sequence for singular varieties. Below we survey the basic definitions and properties to be used later on. We confine ourselves to the functors K_0 and K_1; however, see Remarks 10.48(b) and 10.51.

Let Λ be a (commutative) ring. The additive group of the ring of *big Witt vectors* $\operatorname{Witt}(\Lambda)$ is the multiplicative group $1 + T\Lambda[[T]]$, T an indeterminate. The decreasing filtration of subgroups $I_m(\Lambda) \subset 1 + T^m\Lambda[[T]]$ makes $\operatorname{Witt}(\Lambda)$ a topological group, and every element $\omega(T) \in \operatorname{Witt}(\Lambda)$ has a unique convergent expansion $\omega(T) = \prod_{n\geq 1}(1 - r_nT^n)$, $r_n \in \Lambda$. When $\omega(T) \in I_m(\Lambda)$, then the expansion is of the form $\omega(T) = \prod_{n\geq m}(1-r_nT^n)$. The multiplicative structure on $\operatorname{Witt}(\Lambda)$ is the unique continuous extension of the pairing

$$(1 - rT^m) * (1 - sT^n) = (1 - r^{n/d}s^{m/d}T^{mn/d})^d, \quad r, s \in R, \quad d = \gcd(m, n)$$

to $\operatorname{Witt}(\Lambda)$.

When $\mathbb{Q} \subset \Lambda$, the *ghost isomorphism* between the multiplicative and additive groups

$$-T \cdot \frac{d}{dT}(\log) : 1 + T\Lambda[[T]] \to T\Lambda[[T]], \quad \alpha \mapsto \frac{-T}{\alpha} \cdot \frac{d\alpha}{dt}$$

is actually a ring isomorphism $\operatorname{Witt}(\Lambda) \to \prod_1^\infty \Lambda$ where the right hand side is viewed as $T\Lambda[[T]]$ under the assignment $(r_1, r_2 \dots) \mapsto r_1T + r_2T^2 + \cdots$. In the following we will use that the diagonal embedding $\Lambda \to \prod_1^\infty \Lambda = \operatorname{Witt}(\Lambda)$ is a ring homomorphism.

Bloch [34] and, in a systematic way, Stienstra [335] defined a $\operatorname{Witt}(\Lambda)$-module structure on $NK_i(\Lambda)$. For the module structure it is enough to specify the action of the elements $1 - rT^m \in \mathrm{Witt}(\Lambda)$, $r \in \Lambda$, on $NK_i(\Lambda)$ because it can be shown that any element of $NK_i(\Lambda)$ is annihilated by the ideal $I_m(\Lambda)$ for some m [335, §8]; in other words, $NK_i(\Lambda)$ is a *continuous* $\mathrm{Witt}(\Lambda)$-module.

The action of $1 - rT^m \in \mathrm{Witt}(\Lambda)$, $r \in \Lambda$, $m \in \mathbb{N}$, on $NK_i(\Lambda)$ is the effect of the composite of the upper row in the commutative diagram

$$
\begin{array}{ccccccc}
0 & & 0 & & 0 & & 0 \\
\downarrow & & \downarrow & & \downarrow & & \downarrow \\
NK_i(\Lambda) & \longrightarrow & NK_i(\Lambda) & \longrightarrow & NK_i(\Lambda) & \longrightarrow & NK_i(\Lambda) \\
\downarrow & & \downarrow & & \downarrow & & \downarrow \\
K_i(\Lambda[T]) & \xrightarrow{m^*} & K_i(\Lambda[T]) & \xrightarrow{r_*} & K_i(\Lambda[T]) & \xrightarrow{m_*} & K_i(\Lambda[T]) \\
\downarrow & & \downarrow & & \downarrow & & \downarrow \\
K_i(\Lambda) & \xrightarrow{m\cdot -} & K_i(\Lambda) & \xrightarrow{1} & K_i(\Lambda) & \xrightarrow{1} & K_i(\Lambda) \\
\downarrow & & \downarrow & & \downarrow & & \downarrow \\
0 & & 0 & & 0 & & 0
\end{array}
\tag{10.34}
$$

where:

(1) m_* corresponds to scalar extension through the endomorphism $\Lambda[T] \to \Lambda[T]$, $T \mapsto T^m$, of Λ-algebras;

(2) m^* corresponds to *scalar restriction* through the same endomorphism $\Lambda[T] \to \Lambda[T]$;

(3) r_* corresponds to scalar extension through the Λ-algebra endomorphism $\Lambda[T] \to \Lambda[T]$, $T \mapsto rT$;[2]

(4) $m \cdot -$ is multiplication by m.

We have to discuss the map m^* and, in particular, to explain why it induces multiplication by m on $K_i(\Lambda)$. The endomorphism $\Lambda[T] \to \Lambda[T]$, $T \mapsto T^m$, makes $\Lambda[T]$ a free rank m-module over itself. This yields the scalar restriction functor $\mathbf{m}^* : \mathbb{P}(\Lambda[T]) \to \mathbb{P}(\Lambda[T])$, which is exact (and multiplies rank by m). Hence the homomorphism m^* for $i = 0$. Now consider a matrix $\alpha \in \mathrm{GL}_n(\Lambda[T])$. It can be viewed as an automorphism of $\Lambda[T]^n$. Via the functor $\mathbf{m}^*$ we get an automorphism of $\Lambda[T]^{mn}$. Upon fixing a $\Lambda[T]$-basis, the latter automorphism gives rise to a matrix $\alpha^* \in \mathrm{GL}_{mn}(\Lambda[T])$. Since various bases correspond to conjugate matrices, the element $[\alpha^*] \in K_1(\Lambda[T])$ is well-defined. Hence the homomorphism m^* for $i = 1$. Observe that for an extended projective module $P = \Lambda[T] \otimes_\Lambda P_0$, $P_0 \in \mathbb{P}(\Lambda)$, and a matrix $\alpha \in \mathrm{GL}(\Lambda)$ one has $\mathbf{m}^*(P) \cong P^m$ and

---

[2] Observe that our placing of $*$ is dual to that in Section 10.A, because here we work with rings rather than with the dual category of affine schemes.

$$m^*([\alpha]) = \left[\begin{pmatrix} \alpha & 0 & \cdots & 0 \\ 0 & \alpha & \cdots & 0 \\ \vdots & \vdots & \ddots & \vdots \\ 0 & 0 & \cdots & \alpha \end{pmatrix}\right] = [\alpha^m] = m[\alpha] \in K_1(\Lambda[T]).$$

(See Exercise 9.1.) This shows that the restriction of m^* to $K_i(\Lambda)$, $i = 0, 1$, is the multiplication by m.

Weibel [374] has generalized the Witt(Λ)-module structure to the graded situation so that for a graded ring $A = \bigoplus_{k\geq 0} A_k$ and a subring $\Lambda \subset A_0$ there is a *functorial* continuous Witt(Λ)-module structure on the *relative groups* $K_i(A, A^+)$, where $A^+ = \bigoplus_{k>0} A_k$. (We restrict ourselves to the commutative case; Weibel considers the general situation in which Λ is in the center of A). Without giving a full-blown definition of relative K-groups, we just mention that in the graded situation the groups $K_i(A, A^+)$ are naturally isomorphic to the quotient groups $K_i(A)/K_i(A_0)$ from the canonical splitting $K_i(A) = \big(K_i(A)/K_i(A_0)\big) \oplus K_i(A)$, induced by the augmentation $A \to A_0$.

Weibel's Witt(Λ)-module structure on $NK_i(A, A^+)$ is the restriction of that on $NK_i(A)$ under the embedding $NK_i(A, A^+) \to NK_i(A)$ induced by the *graded* homomorphism (from the Swan–Weibel homotopy trick, Lemma 8.39)

$$\begin{aligned} \mathrm{w} : A_0 \oplus A_1 \oplus A_2 \oplus \cdots &\to A[T] = A + TA + T^2 A + \cdots, \\ a_0 \oplus a_1 \oplus a_2 \oplus \cdots &\mapsto a_0 + a_1 T + a_2 T^2 + \cdots, \end{aligned}$$

where the Witt(Λ)-module structure on $NK_i(A)$ corresponds to scalar restriction through the functorial map Witt$(\Lambda) \to$ Witt(A). That $NK_i(A, A^+) \to NK_i(A)$ is an embedding follows from the fact that w splits the (nongraded!) augmentation $A[T] \to A$, $T \mapsto 1$, in particular $\mathrm{w}_* : K_i(A) \to K_i(A[T])$ is a split monomorphism.

Remark 10.48. (a) The group endomorphism $V_m : NK_i(\Lambda) \to NK_i(\Lambda)$, induced by the Λ-algebra endomorphism $\Lambda[T] \to \Lambda[T]$, $T \mapsto T^m$, is called the *mth Verschiebung*. In [335] it is defined in terms of the *category of nilpotent endomorphisms* Nil(Λ) and later, in [335, Theorem 4.7], identified as the map m^*.

(b) We have the exact functors (using self-explanatory notation)

$$\mathbb{P}(\Lambda[T]) \xrightarrow{\mathbf{m}^*} \mathbb{P}(\Lambda[T]) \xrightarrow{\mathbf{r}_*} \mathbb{P}(\Lambda[T]) \xrightarrow{\mathbf{m}_*} \mathbb{P}(\Lambda[T]),$$

and, as an immediate consequence of Quillen's definition of higher K-theory [296], Witt(Λ) acts on all groups $NK_i(\Lambda)$, $i \geq 0$.

The big Witt vectors' action implies

Proposition 10.49. *For graded* $\Bbbk$*-algebras* $A = A_0 \oplus A_1 \oplus \cdots$ *and* $B = B_0 \oplus B_1 \oplus \cdots$ *the natural map* $K_i(A)/K_i(A_0) \to K_i(B)/K_i(B_0)$ *is a homomorphism of* $\Bbbk$*-vector spaces.*

The second K-theoretical tool to be applied is the Mayer–Vietoris sequence for singular varieties proved by Thomason [354, Theorem 8.1] using his localization technique. (The original scheme theoretic version is much more general.)

Theorem 10.50. *Let U, V be open subsets of some $\Bbbk$-variety. Then there is a natural long exact sequence*

$$\cdots \to K_i(U \cup V) \to K_i(U) \oplus K_i(V) \to K_i(U \cap V) \to K_{i-1}(U \cup V) \to \cdots.$$

Remark 10.51. (a) The indices i in Theorem 10.50 can be arbitrary integers, but we are only interested in the segment of the long exact sequence that involves K_1 and K_0. We have not formally introduced the K_1-group of a variety. However, only the case of an affine variety will come up in the proof of Proposition 10.53. So the theory of Bass–Whitehead groups developed in Section 9.A suffices for our purposes.

(b) The K-groups in Theorem 10.50 are actually those of Waldhausen, associated to the appropriate categories of perfect complexes on the given varieties. However, these K-groups coincide with Quillen's K-groups for quasiprojective varieties [354, Sect. 3], in particular, for the varieties of type $\mathscr{V}(\mathfrak{C})$ in Proposition 10.53 below.

Basic configurations. Suppose M is a positive normal affine monoid, $\operatorname{rank} M = r$, and $\operatorname{gp}(M) = \mathbb{Z}^r$. In particular, $\operatorname{gp}(\mathbb{Z} \oplus M) = \mathbb{Z}^{r+1}$. *For simplicity of notation we assume $n = r + 1$ throughout the section.* By Proposition 2.17 there is a basis $x_1, \dots, x_r$ of $\operatorname{gp}(M)$ such that $\mathbb{R}_+ M \subset \mathbb{R}_+ x_1 + \cdots + \mathbb{R}_+ x_r$.

Put $e = (1, 0) \in \mathbb{Z} \oplus M$ and consider the sequence of submonoids $M = M_0, M_1, M_2, \dots$ defined by

$$M_i = \big(\mathbb{R} \oplus \mathbb{R}_+ M\big) \cap \big(\mathbb{Z}(x_1 - ie) + \cdots + \mathbb{Z}(x_r - ie)\big), \qquad i \in \mathbb{Z}_+.$$

In this definition and in the following we consider $\mathbb{R}M$ as a subspace of $\mathbb{R} \oplus \mathbb{R}_+ M$ via the embedding $y \mapsto (0, y)$.

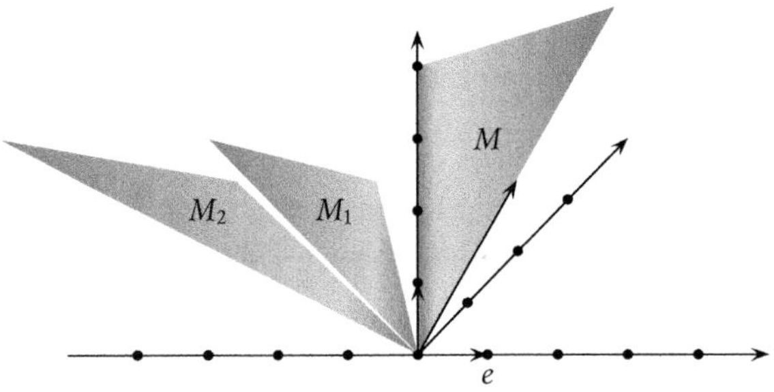

Fig. 10.1. The monoids M_i

Lemma 10.52. *In the notation above we have:*

(a) the following direct sum representations

$$\mathbb{Z} \oplus M = \mathbb{Z}e + M_0 = \mathbb{Z}e + M_1 = \mathbb{Z}e + M_2 = \cdots;$$

(b) inclusions

$$\mathbb{Z}_+ e + M_0 \subset \mathbb{Z}_+ e + M_1 \subset \cdots$$

leading to the equality

$$M_\infty = \bigcup_{i=0}^{\infty}(\mathbb{Z}_+e + M_i) = (\mathbb{Z} \oplus M) \setminus \{-ie : i \in \mathbb{Z}_+,\ i > 1\};$$

(c) *isomorphisms* $\alpha_i : M_i \to M_{i+1}$, $i \in \mathbb{Z}_+$, *making the following diagrams commute:*

$$\begin{array}{ccc} \mathbb{Z}_+e + M_i & \hookrightarrow & \mathbb{Z}_+e + M_{i+1} \\ \downarrow{\scriptstyle 1+\alpha_i} & & \downarrow{\scriptstyle 1+\alpha_{i+1}} \\ \mathbb{Z}_+e + M_{i+1} & \hookrightarrow & \mathbb{Z}_+e + M_{i+2}. \end{array}$$

The lemma is easily proved. The isomorphisms α_i, for instance, are the restrictions of the automorphisms $\mathbb{Z}^n \to \mathbb{Z}^n$, $e \mapsto e$, $x_j \mapsto x_j - e$, $j = 1, \dots, r$.

Choose affine normal submonoids $N_+ \subset \mathbb{Z}_+e + M$ and $N_- \subset \mathbb{Z}_-e + M_1$ so that the following conditions hold:

(i) $e \in N_+$, $-e \in N_-$,
(ii) $N_- \cap M_1 = \{0\}$,
(iii) $\mathbb{Z}e + N_+ = \mathbb{Z}e + N_- = \mathbb{Z}e + M$.

In order to construct N_+ and N_-, one fixes two rational cones $C_+ \subset \mathbb{R}_+e + \mathbb{R}_+M$ and $C_- \subset \mathbb{R}_-e + \mathbb{R}_+M_1$ so that C_+ is bounded by the facets of $\mathbb{R}_+e + \mathbb{R}_+M$, containing e, and one more hyperplane through the origin; C_- is chosen similarly with respect to $\mathbb{R}_-e + \mathbb{R}_+M_1$ under the following additional requirement: the facet of C_- that does not contain $-e$ intersects $\mathbb{R}_+M_1$ only at the origin. Then $N_+ = C_+ \cap \mathbb{Z}^n$ and $N_- = C_- \cap \mathbb{Z}^n$.

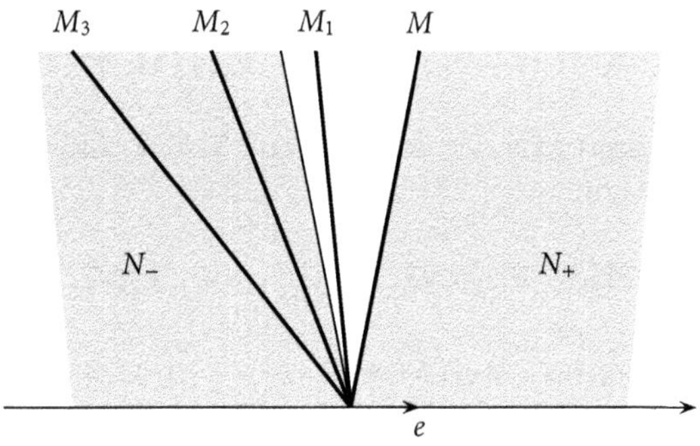

Fig. 10.2. Cross-section of a basic configuration

A triple of type $\mathfrak{C} = (M, N_+, N_-)$ will be called a *basic configuration.*

The noncomplete case. Now we are in position to produce toric varieties with huge Grothendieck group, starting with a noncomplete case. It constitutes the K-theoretical core of the general case.

Let $\mathfrak{C}$ be a basic configuration. By $\mathscr{V}(\mathfrak{C})$ we denote the scheme obtained by gluing $\operatorname{Spec} \Bbbk[N_+]$ and $\operatorname{Spec} \Bbbk[N_-]$ along their common open subscheme $\operatorname{Spec} \Bbbk[\mathbb{Z} \oplus M]$. This is a toric variety whose fan has two maximal n-dimensional cones—the dual cones $(\mathbb{R}_+N_+)^*, (\mathbb{R}_+N_-)^* \subset (\mathbb{R}^n)^*$. (Contrary to the conventions in Section 10.B the monoids live in "original" space, and the fan in its dual.) The quasiprojectivity of $\mathscr{V}(\mathfrak{C})$ is easily shown; see Lemma 10.55(a) below.

Proposition 10.53. *Let $\mathfrak{C} = (M, N_+, N_-)$ be a basic configuration in which the monoid M is simplicial and nonfree. Then* $\operatorname{rank} K_0(\mathscr{V}(\mathfrak{C})) \geq \dim_{\mathbb{Q}} \Bbbk$.

Proof. We will use multiplicative notation for the monoid operation with $T = e$, where e is chosen as in the definition of basic configuration. We simplify notation by writing $\Bbbk[M, T]$ for $\Bbbk[M][T]$ and $\Bbbk[M, T, T^{-1}]$ for $\Bbbk[M][T, T^{-1}]$.

Theorem 10.50 yields the exact sequence

$$K_1(\Bbbk[N_+]) \oplus K_1(\Bbbk[N_-]) \xrightarrow{\kappa} K_1(\Bbbk[M, T, T^{-1}]) \to K_0(\mathscr{V}(\mathfrak{C}))$$

and, hence, it is enough to show that $\operatorname{rank} \operatorname{Coker} \kappa \geq \dim_{\mathbb{Q}} \Bbbk$.

The map $K_1(\Bbbk[M, T, T^{-1}]) \to K_0(\mathscr{V}(\mathfrak{C}))$ corresponds to gluing trivial vector bundles on $\operatorname{Spec} \Bbbk[N_+]$ and $\operatorname{Spec} \Bbbk[N_-]$ of same rank, using invertible linear maps over the common part $\operatorname{Spec} \Bbbk[M, T, T^{-1}]$. That nontrivial gluings are possible follows from the nontriviality of $K_1(\Bbbk[M, T, T^{-1}])$.

On the other hand, $\Bbbk[M, T, T^{-1}] = \Bbbk[M_1, T, T^{-1}]$ by Lemma 10.52(a) and

$$K_1(\Bbbk[M_1, T, T^{-1}]) = K_1(\Bbbk[M_1]) \oplus NK_1^+(\Bbbk[M_1]) \oplus NK_1^-(\Bbbk[M_1]) \oplus K_0(\Bbbk[M_1])$$

by the fundamental theorem 9.5(a). Here

$$K_1(\Bbbk[M_1]) \oplus NK_1^+(\Bbbk[M_1]) = K_1(\Bbbk[M_1, T]),$$
$$K_1(\Bbbk[M_1]) \oplus NK_1^-(\Bbbk[M_1]) = K_1(\Bbbk[M_1, T^{-1}]).$$

Because of the inclusions $\Bbbk[N_+] \subset \Bbbk[M, T]$ and $\Bbbk[N_-] \subset \Bbbk + T^{-1}\Bbbk[M_1, T^{-1}]$ the cokernel of κ surjects onto

$$\operatorname{Coker}\Big(K_1(\Bbbk[M, T]) \oplus NK_1^-(\Bbbk[M_1]) \to K_1(\Bbbk[M_1, T, T^{-1}])\Big).$$

But the latter group contains $\operatorname{Coker}\big(K_1(\Bbbk[M, T]) \to K_1(\Bbbk[M_1, T])\big)$ as a direct summand because, first, the homomorphism $\Bbbk[M, T] \to \Bbbk[M_1, T, T^{-1}]$ factors through $\Bbbk[M_1, T] \to \Bbbk[M_1, T, T^{-1}]$ and, second, the groups $K_1(\Bbbk[M_1, T])$ and $NK_1^-(\Bbbk[M_1])$ inside $K_1(\Bbbk[M_1, T, T^{-1}])$ are direct summands with 0 intersection. It is therefore enough to show the inequality

$$\begin{aligned}
&\operatorname{rank} \operatorname{Coker}\Big(K_1(\Bbbk[M, T]) \to K_1(\Bbbk[M_1, T])\Big) \\
&\qquad = \operatorname{rank} \operatorname{Coker}\Big(K_1(\Bbbk[M, T])/\Bbbk^\times \to K_1(\Bbbk[M_1, T])/\Bbbk^\times\Big) \geq \dim_{\mathbb{Q}} \Bbbk.
\end{aligned}$$

(We view the multiplicative group $\Bbbk^\times$ as a natural direct summand of the K_1-groups.) Fix a grading $\Bbbk[M_1, T] = \Bbbk \oplus A_1 \oplus A_2 \oplus \cdots$ so that all elements of M_1 as well as the variable T are homogeneous (Proposition 2.17(f)). This grading restricts to a grading on $\Bbbk[M, T]$.

By Proposition 10.49 the homomorphism

$$K_1(\Bbbk[M,T])/\Bbbk^\times \to K_1(\Bbbk[M_1,T])/\Bbbk^\times$$

is a homomorphism of $\Bbbk$-vector spaces. Thus our claim finally reduces to the nonsurjectivity of this homomorphism or, equivalently, to the nonsurjectivity of

$$K_1(\Bbbk[M,T]) \xrightarrow{\lambda} K_1(\Bbbk[M_1,T]).$$

Assume λ is surjective. There is a representation $\Bbbk[M_\infty] = \bigcup_i \Bbbk[M_i, T]$ as a filtered union with M_∞ as in Lemma 10.52(b). In particular,

$$K_1(\Bbbk[M_\infty]) = \varinjlim\Bigl(K_1(\Bbbk[M,T]) \to \cdots \to K_1(\Bbbk[M_i,T]) \to \cdots\Bigr).$$

By Lemma 10.52(c) the mapping $K_1(\Bbbk[M_i,T]) \to K_1(\Bbbk[M_{i+1},T])$ coincides with λ for every index i, up to an isomorphic transformation. Therefore, by the surjectivity assumption all these mappings are surjective. In particular, the limit map $K_1(\Bbbk[M,T]) \to K_1(\Bbbk[M_\infty])$ is also surjective. But it is injective as well because so is the composite map $K_1(\Bbbk[M,T]) \to K_1(\Bbbk[M_\infty]) \to K_1(\Bbbk[M,T,T^{-1}])$ (Theorem 9.5(a)).

The diagram

$$\begin{array}{ccc} \Bbbk[M_\infty] & \hookrightarrow & \Bbbk[M,T,T^{-1}] \\ \downarrow & & \downarrow \\ \Bbbk[T] & \hookrightarrow & \Bbbk[T,T^{-1}] \end{array}$$

is a Cartesian square by Lemma 10.52(b). Here the vertical maps are defined by $m \mapsto 0 \in \Bbbk$ for all nontrivial elements $m \in M$. In view of the equation $K_1(\Bbbk[M,T]) = K_1(\Bbbk[M_\infty])$, the associated Milnor Mayer–Vietoris sequence (the middle row in the diagram (9.2)) reads as

$$K_1(\Bbbk[M,T]) \to K_1(\Bbbk[T]) \oplus K_1(\Bbbk[M,T,T^{-1}]) \to K_1(\Bbbk[T,T^{-1}]).$$

(Note, just the surjectivity of $K_1(\Bbbk[M,T]) \to K_1(\Bbbk[M_\infty])$ is needed for this sequence.) Therefore, $NK_1(\Bbbk[M]) = 0$ by the fundamental theorem, i. e., $K_1(\Bbbk[M]) = K_1(\Bbbk[M,T^{-1}])$. This is a contradiction with Theorem 9.7(a). □

Proposition 10.53 already gives many examples of simplicial toric varieties in dimensions ≥ 3 for which the Cartan homomorphism is not an isomorphism over the rationals. But we proceed further to projective examples.

The projective case. The localization technique of Thomason, used already in the previous subsection, is the key to the construction of relevant projective completions of the varieties in Proposition 10.53.

Lemma 10.54. *Let $\mathcal{F}$ be a fan in the dual space $(\mathbb{R}^n)^*$ and $\mathfrak{C} = (M, N_+, N_-)$ be a basic configuration in which the monoid M is simplicial and nonfree. Suppose the dual cones $(\mathbb{R}_+N_+)^*, (\mathbb{R}_+N_-)^* \subset (\mathbb{R}^n)^*$ are maximal cones of $\mathcal{F}$ and that all other maximal cones of $\mathcal{F}$ are unimodular. Then* rank $K_0(\mathcal{V}(\mathcal{F})) \geq \dim_{\mathbb{Q}} \Bbbk$.

Proof. The variety $\mathcal{V}(\mathcal{F})$ has the open cover $\mathcal{V}(\mathfrak{C}) \cup \mathcal{A}_1 \cup \dots \cup \mathcal{A}_s$, where the $\mathcal{A}_j$ are the smooth affine toric varieties that correspond to the unimodular facets of $\mathcal{F}$. For each $j = 1, \dots, s$ Theorem 10.50 yields the exact sequence

$$K_0\big(\mathcal{V}(\mathfrak{C}) \cup \mathcal{A}_1 \cup \dots \cup \mathcal{A}_j\big) \to K_0\big(\mathcal{V}(\mathfrak{C}) \cup \mathcal{A}_1 \cup \dots \cup \mathcal{A}_{j-1}\big) \oplus K_0(\mathcal{A}_j) \to K_0(\mathcal{U}_j),$$

where $\mathcal{U}_j = \big(\mathcal{V}(\mathfrak{C}) \cup \mathcal{A}_1 \cup \dots \cup \mathcal{A}_{j-1}\big) \cap \mathcal{A}_j$. Since $\mathcal{U}_j$ is open in the smooth variety $\mathcal{A}_j$ and $K_0(\mathcal{A}_j) = \mathbb{Z}$ we have $K_0(\mathcal{U}_j) = \mathbb{Z}$, $j = 1, \dots, s$, as follows from the exact sequence (10.2) (and Theorem 10.2). Therefore, induction on $j = 1, \dots, s$ shows that rank $K_0(\mathcal{V}(\mathcal{F})) \geq \dim_{\mathbb{Q}} \Bbbk$. (Here one needs $\dim_{\mathbb{Q}} \Bbbk = \infty$.) □

The first part of the next lemma says that the variety $\mathcal{V}(\mathfrak{C})$ can be compactified to a projective simplicial toric variety. However, the difficulty in applying Lemma 10.54 to the latter is that its affine toric charts, different from Spec $\Bbbk[N_+]$ and Spec $\Bbbk[N_-]$, are in general not smooth. We overcome this difficulty by resolving the corresponding toric singularities without affecting Spec $\Bbbk[N_+]$ and Spec $\Bbbk[N_-]$.

Call a basic configuration $\mathfrak{C} = (M, N_+, N_-)$ *admissible* if all facets of $(\mathbb{R}_+N_+)^*$ and $(\mathbb{R}_+N_-)^*$ different from their common facet are unimodular.

Lemma 10.55. *Suppose $\mathfrak{C} = (M, N_+, N_-)$ is a basic configuration.*

(a) Then there exists an equivariant open embedding of $\mathcal{V}(\mathfrak{C})$ into a projective toric variety $\mathcal{V}$.

(b) If $\mathfrak{C}$ is admissible, then $\mathcal{V}$ can be chosen in such a way that its standard affine charts not contained in $\mathcal{V}(\mathfrak{C})$ are smooth.

Proof. (a) Consider the intersection $\Delta(\mathfrak{C}) = (e + \mathbb{R}_+N_-) \cap \mathbb{R}_+N_+$. This is a rational n-polytope with a pair of corners, spanning respectively the cones $\mathbb{R}_+N_+$ and $e + \mathbb{R}_+N_-$. Let $\mathcal{F}$ be the normal fan of $\Delta(\mathfrak{C})$ and set $\mathcal{V} = \mathcal{V}(\mathcal{F})$.

(b) We now remove from $\mathcal{F}$ (as a set of cones) the cones $(\mathbb{R}_+N_+)^*$ and $(\mathbb{R}_+N_-)^*$ and their maximal common face. The remaining set of cones is a fan $\mathcal{F}'$ whose support is the complement of the interior of the union $(\mathbb{R}_+N_+)^* \cup (\mathbb{R}_+N_-)^*$. The fan $\mathcal{F}'$ has a unimodular regular subdivision $\mathcal{F}''$ according to Theorem 2.72. But since the common faces of $\mathcal{F}'$ with $(\mathbb{R}_+N_+)^*$ and $(\mathbb{R}_+N_-)^*$ are unimodular by hypothesis, they have not been subdivided in the sequence of stellar subdivisions leading from $\mathcal{F}'$ to $\mathcal{F}''$. Therefore we can insert the three removed faces of $\mathcal{F}$ into $\mathcal{F}''$, obtaining a fan $\mathcal{G}$. The resulting subdivision $\mathcal{G}$ of $\mathcal{F}$ is regular since it has been obtained from $\mathcal{F}$ by successive stellar subdivisions—one applies the conical version of Lemma 1.65. Finally we choose $\mathcal{V} = \mathcal{V}(\mathcal{G})$. Since all facets of $\mathcal{G}$ apart from $(\mathbb{R}_+N_+)^*$ and $(\mathbb{R}_+N_-)^*$ are unimodular cones, we are done. □

Now we are ready to prove the next theorem:

Theorem 10.56. *For each $n \geq 3$ there are projective simplicial toric varieties $\mathscr{V}$ of dimension n for which* rank $K_0(\mathscr{V}) \geq \dim_{\mathbb{Q}} \Bbbk$.

Proof. In view of Lemmas 10.54 and 10.55 we only need to show the existence of admissible configurations $\mathfrak{C} = (M, N_+, N_-)$ in which M, N_+ and N_- are simplicial and M is not free.

There are simplicial rational n-cones whose facets, with exactly one exception, are unimodular (Exercise 10.14). Fix such a cone C in the dual space $(\mathbb{R}^n)^*$, rational with respect to the dual lattice $(\mathbb{Z}^n)^*$. Let $F \subset C$ be the nonunimodular facet. The dual cone C^* is a rational simplicial cone. Let l be the extreme ray of C^* that corresponds to F under the duality. Put $N_+ = C^* \cap \mathbb{Z}^n$ and let e denote the extreme integral generator of C^* in l. By Proposition 2.32 we have $\mathbb{Z}_- e + N_+ = \mathbb{Z}e + M'$ for some rank r simplicial normal monoid M'. Moreover, $(\mathbb{R}e + \mathbb{R}_+ M')^* = F$.

By Proposition 2.17 we can find a basis $x_1, \dots, x_r$ of $\operatorname{gp}(M')$ such that $\mathbb{R}_+ M' \subset \mathbb{R}_+ x_1 + \cdots + \mathbb{R}_+ x_r$. If $c \in \mathbb{N}$ is large enough, then the monoid $M = \mathbb{Z}^n \cap \big(\mathbb{R}_+(x_1 - ce) + \cdots + \mathbb{R}_+(x_r - ce)\big)$ satisfies the conditions $\mathbb{R}_+ N_+ \subset \mathbb{R}_+ e + \mathbb{R}_+ M$ (equivalently, $N_+ \subset \mathbb{Z}_+ e + M$) and $\mathbb{R}_+ N_+ \cap \mathbb{R}_+ M = 0$. We fix such c and define the monoid M_1 to be $\mathbb{Z}^n \cap \big(\mathbb{R}_+(x_1 - (c+1)e) + \cdots + \mathbb{R}_+(x_r - (c+1)e)\big)$. To construct N_- we make the identification of the monoids $\mathbb{Z}_+ e + M$ and $\mathbb{Z}_- e + M_1$ along the isomorphism induced by $e \mapsto -e$, $x_j \mapsto x_j - e$, $j = 1, \dots, r$. Then N_- is by definition the corresponding copy of N_+. It follows from the construction that (M, N_+, N_-) is an admissible basic configuration. The nonfreeness of M results from the nonunimodularity of F. □

Note that all steps in the proof of Theorem 10.56 are constructive. The reader may practice in constructing an explicit 3-dimensional normal polytopes P with $K_0(\operatorname{Proj} \Bbbk[P])$ huge—this is Exercise 10.15.

Remark 10.57. Theorem 10.56 was originally proved in [167]. Observe that the main result remains true under the weaker condition that $\Bbbk$ is not necessarily algebraically closed, but $\dim_{\mathbb{Q}} \Bbbk = \infty$. Later an alternative argument was offered by Cortiñas et al. [96]: using the techniques of [97], one constructs examples of noncomplete simplicial toric varieties with huge Grothendieck groups, similar to the one in Proposition 10.53; the passage to the projective case in [96] is the same as above.

10.G The equivariant Serre problem for abelian groups

In this section G denotes an algebraic group over $\Bbbk$. Let $\mathscr{V}$ be a G-variety. A *G-equivariant vector bundle*, or just *G-vector bundle*, on $\mathscr{V}$ is an algebraic vector bundle $\mathrm{p} : \mathscr{E} \to \mathscr{V}$ together with an action of G on $\mathscr{E}$ for which the projection p is equivariant and G acts linearly on the fibers.

An *equivariant homomorphism*, or just *homomorphism* of G-vector bundles on $\mathscr{V}$ is a homomorphism of vector bundles that respects the G-actions.

As for bundles without group actions, if $\mathscr{E}_1$ and $\mathscr{E}_2$ are G-vector bundles, then so are $\mathscr{E}_1 \oplus \mathscr{E}_2$, $\mathscr{E}_1 \otimes \mathscr{E}_2$, $\bigwedge^k \mathscr{E}_1$, and $\operatorname{Ker}(\mathbf{f})$ provided $\mathbf{f} : \mathscr{E}_1 \to \mathscr{E}_2$ is an equivariant fiberwise surjective homomorphism. Moreover, $\mathsf{Hom}(\mathscr{E}_1, \mathscr{E}_2)$ is a G-vector bundle with respect to the following G-action:

$$g(\mathbf{f})(\xi) = g^{-1}(\mathbf{f}(g(\xi))), \quad g \in G, \quad \mathbf{f} \in \mathsf{Hom}(\mathscr{E}_1, \mathscr{E}_2), \quad \xi \in \mathscr{E}_1.$$

A G-vector bundle over $\mathscr{V}$ is called *trivial* if it is isomorphic to a bundle of the form $\operatorname{pr}_{\mathscr{V}} : \mathscr{V} \times W \to \mathscr{V}$ for some *rational G-module* W, i. e., a *finite-dimensional* $\Bbbk$-vector space with a rational G-representation $G \to \operatorname{GL}(W)$. For brevity we will usually suppress the attribute "rational". The trivial G-vector bundle on $\mathscr{V}$ associated to a G-module W will be denoted by $\mathbb{W}_{\mathscr{V}}$.

From now on we make the blanket assumption that G is a *linearly reductive* group: every G-module splits into a direct sum of simple G-modules.

The following theorem of Masuda, Moser–Jauslin, and Petrie [251] solves the *equivariant Serre problem for abelian groups.*

Theorem 10.58. *Suppose all G-modules are sums of* 1*-dimensional G-modules. Then every G-vector bundle on a G-module, viewed as a G-variety, is trivial.*

Remark 10.59. (a) The group G in Theorem 10.58 is necessarily diagonalizable and, hence, commutative. On the other hand, every linearly reductive abelian group is diagonalizable (Exercise 10.16). However, the commutativity of G cannot be dropped. In fact, Knop [224] has shown that for every connected noncommutative group G there exists a G-module with a nontrivial G-vector bundle.

(b) Below we will see and use that the theorem can be translated into the language of projective modules over $\Bbbk[X_1, \ldots, X_n]$ with a suitable G-action, by passing from a vector bundle to the module of its global sections. In the extreme case when $G = 1$, Theorem 10.58 is nothing but the Quillen–Suslin Theorem 8.5. On the other extreme, when G is the embedded torus $\mathbb{T} \subset V$ (and, more generally, if $\mathbb{T}$ has a unique minimal closed orbit) Theorem 10.58 is fairly easy to prove; see Exercise 10.18.

(c) Part of the motivation for Theorem 10.58 comes from the well-known *linearization problem* for an algebraic action of a reductive group G on an affine space. Namely, all nonlinearizable actions found so far are derived from nontrivial G-vector bundles on representation spaces. But nobody so far has been able to construct a nonlinearizable action when G is commutative. So Theorem 10.58 may be viewed as an indication that such actions are extremely rare for commutative G. See Kraft [228, Section 6] for a survey. This expectation is further supported by the result of Koras and Russell [227] stating that all $\mathbb{C}^{\times}$-actions on $\mathbb{C}^3$ are linearizable.

(d) Masuda [249] extended Theorem 10.58 by showing that all G-vector bundles on any G-equivariant affine toric variety are trivial. The picture is far more complicated for nonaffine toric varieties $\mathscr{V}(\mathscr{F})$, even when the group G is the embedded torus $\mathbb{T} \subset \mathscr{V}(\mathscr{F})$. A brief overview of results on $\mathbb{T}$-vector bundles: on $\mathbb{P}^d_{\Bbbk}$ all bundles of rank $< d$ split into direct sums of (equivariant) line bundles (Kaneyama [216]); for a general toric variety $\mathscr{V}(\mathscr{F})$ Klyachko [222] describes $\mathbb{T}$-vector bundles in terms of certain vector spaces with compatible filtrations; moduli spaces of $\mathbb{T}$-vector bundles on $\mathscr{V}(\mathscr{F})$ are studied by Payne [284]. Also see Masuda's survey [250].

(e) Kraft and Schwarz [229] have given another, more geometric proof of Theorem 10.58.

The proof of Theorem 10.58 presented below follows [251] closely, with the only difference that we work out various steps in more detail, including the auxiliary results drawn from other sources.

A-G-modules. Let $\mathcal{V} = \operatorname{Spec} A$ be an affine G-variety. Then A, a finitely generated $\Bbbk$-domain, is a filtered union of G-modules [36, 1.9].

The fundamental equivalence of the category of vector bundles on an affine variety and projective modules over the coordinate ring via the functor of global sections (see Section 10.A), extends to the equivariant setting. In other words, the functor of (algebraic) global sections establishes an equivalence between the category of G-vector bundles on $\mathcal{V}$ and that of finitely generated projective A-G-modules:

(1) an *A-G-module* means an A-module M that is a filtered union of G-modules such that the following condition holds:

$$g(am) = g(a)\cdot g(m), \qquad g \in G,\ a \in A,\ m \in M,$$

the G-action on A corresponding to that on $\mathcal{V}$;

(2) if an A-G-module M is finitely generated as an A-module, then we call it a *finitely generated A-G-module*, and if it is a projective A-module, then it is a *projective A-G-module*. (This notion of projectivity is justified by Lemma 10.63 below.)

In particular, A itself is a projective A-G-module. As in the previous chapters, we assume projectivity includes finite generation.

The category $\mathsf{Mod}(A\text{-}G)$ of A-G-modules and their morphisms (i. e., G-equivariant A-module homomorphisms) is abelian and closed under tensor products over A. Moreover, if $M, N \in \mathsf{Mod}(A\text{-}G)$ and M is finitely generated, then the G-action on $\operatorname{Hom}_A(M,N)$, defined by

$$(gf)(m) = g(f(g^{-1}m)), \quad g \in G,\ f \in \operatorname{Hom}_A(M,N),\ m \in M,$$

makes $\operatorname{Hom}_A(M,N)$ an A-G-module. The equality

$$\operatorname{Hom}_{A\text{-}G}(M,N) = \operatorname{Hom}_A(M,N)^G \tag{10.35}$$

then follows immediately from the definition of the G-structure on $\operatorname{Hom}_A(M,N)$. The tensor product $M \otimes_A N$ of A-G-modules is an A-G–module under the diagonal action given by $g(m \otimes n) = g(m) \otimes g(n)$.

Lemma 10.60. *Let V and W be G-modules and $\mathcal{V} = \operatorname{Spec} A$ be an affine G-variety. Then we have the natural isomorphisms of A-G-modules*

$$\begin{aligned}\mathsf{Hom}(\mathbb{V}_{\mathcal{V}}, \mathbb{W}_{\mathcal{V}}) &\cong \operatorname{Hom}_{\Bbbk[G]}(V, A \otimes_{\Bbbk} W)\\ &\cong (A \otimes_{\Bbbk} V^* \otimes_{\Bbbk} W)^G \cong \mathsf{Mor}_G(\mathcal{V}, \operatorname{Hom}_{\Bbbk}(V,W)),\end{aligned}$$

where $\operatorname{Hom}_{\Bbbk}(V,W)$ is the vector space of $\Bbbk$-linear maps $V \to W$ carrying the G-module structure defined above and $\mathsf{Mor}_G(\mathcal{V}, \operatorname{Hom}_{\Bbbk}(V,W))$ is the set G-equivariant morphisms of affine G-varieties.

Proof. Using the functor of global sections, the first isomorphism is shown as follows:

$$\begin{aligned}\mathsf{Hom}(\mathbb{V}_{\mathscr{V}}, \mathbb{W}_{\mathscr{V}}) &\cong \mathrm{Hom}_{A\text{-}G}(\Gamma(\mathscr{V}, \mathbb{V}_{\mathscr{V}}), \Gamma(\mathscr{V}, \mathbb{W}_{\mathscr{V}}))\\ &= \mathrm{Hom}_{A\text{-}G}(A \otimes_{\Bbbk} V, A \otimes_{\Bbbk} W) = \mathrm{Hom}_{\Bbbk[G]}(1 \otimes_{\Bbbk} V, A \otimes_{\Bbbk} W)\\ &= \mathrm{Hom}_{\Bbbk[G]}(V, A \otimes_{\Bbbk} W).\end{aligned}$$

The second isomorphism follows from the standard adjunction between the functors Hom and $\otimes$ for $\Bbbk$-vector spaces:

$$\mathrm{Hom}_{\Bbbk[G]}(V, A \otimes_{\Bbbk} W) = \mathrm{Hom}_{\Bbbk}(V, A \otimes_{\Bbbk} W)^G \cong (A \otimes_{\Bbbk} V^* \otimes_{\Bbbk} W)^G.$$

The G-action on $V^* = \mathrm{Hom}_{\Bbbk}(V, \Bbbk)$ is given by

$$g(f(x)) = f(g^{-1}(x)), \quad g \in G,\ f \in V^*, x \in V,$$

since G acts trivially on $\Bbbk$.

Finally, instead of the third isomorphism, we describe directly the isomorphism

$$\mathsf{Hom}(\mathbb{V}_{\mathscr{V}}, \mathbb{W}_{\mathscr{V}}) \cong \mathsf{Mor}_G(\mathscr{V}, \mathrm{Hom}_{\Bbbk}(V, W)).$$

Namely, let $\Phi \in \mathsf{Hom}(\mathbb{V}_{\mathscr{V}}, \mathbb{W}_{\mathscr{V}})$; then

$$\Phi \mapsto \varphi, \qquad \Phi(x, v) = \big(x, \varphi(x)(v)\big), \quad x \in \mathscr{V},\ v \in V.$$

That this is in fact an isomorphism of A-G-modules is an easy exercise. □

We record a consequence of Lemma 10.60 that will be useful in the proof of Theorem 10.58:

Lemma 10.61. *Let $\mathscr{V} = \operatorname{Spec} A$ be an affine G-variety. Suppose V and W are 1-dimensional G-modules, and let χ be the character of G on $V \otimes W^*$. Then $\mathsf{Hom}(\mathbb{V}_{\mathscr{V}}, \mathbb{W}_{\mathscr{V}})$ is naturally identified with the G-submodule of A formed by the semi-invariants of weight χ.*

One simply uses that the opposite of the weight of $V \otimes W^*$ is the weight of $(V \otimes W^*)^* = V^* \otimes W$, and $A \otimes_{\Bbbk} V^* \otimes_{\Bbbk} W \cong A$ as $\Bbbk$-vector spaces.

Stably trivial equivariant vector bundles. The first major step in our argument is provided by the Bass–Haboush result [22] that equivariant vector bundles on G-modules are *stably trivial*:

Theorem 10.62. *Every G-vector bundle $\mathscr{E}$ over a G-module V is stably trivial, i. e., there exist G-modules F and S such that $\mathscr{E} \oplus \mathbb{S}_V \cong \mathbb{F}_V \oplus \mathbb{S}_V$.*

Because of its importance for our goals, we discuss how Theorem 10.62, not formulated explicitly in [22], follows from the equivariant homotopy invariance in loc. cit.

Sketch of proof of Theorem 10.62. Let V be a G-module and $A = \mathrm{Sym}_{\Bbbk}(V)$ its symmetric algebra, equipped with the induced G-action. Then A is a polynomial (in particular, regular) $\Bbbk$-algebra and by [22, Theorem 2.3] one has the natural isomorphism $K_0(\Bbbk\text{-}G) \to K_0(A\text{-}G)$ for the corresponding Grothendieck groups of G-equivariant projective modules.[3] (When $G = 1$ one recovers the familiar equality $K_0(\Bbbk[X_1, \ldots, X_n]) = \mathbb{Z}$, $n = \dim V$.)

Since G is linearly reductive, every short exact sequence of projective A-G-modules splits by 10.63 below. Then a standard argument on K_0 (see, for instance, [19, VII, 1.3(b)]) ensures that for any projective A-G-module P there are a G-module F and a projective A-G-module Q such that

$$P \oplus Q \cong (A \otimes_{\Bbbk} F) \oplus Q$$

as A-G-modules.

Next we show that $Q \oplus R \cong A \otimes_{\Bbbk} S$ as A-G-modules for some G-module S and some A-G-module R. By the very definition of A-G-modules and the finite generation of Q we can choose a G-invariant finite-dimensional $\Bbbk$-vector subspace $S \subset Q$ that contains an A-generating subset of Q. Then we have a surjective A-G-homomorphism $A \otimes_{\Bbbk} S \to Q$. It splits in the category of A-modules and, by Lemma 10.63, also in the category of A-G-modules. So we can choose $R = \mathrm{Ker}\big(A \otimes_{\Bbbk} S \to Q\big)$. (This last step is [21, Corollary 4.2(c)].)

Now Theorem 10.62 follows from the equivalence of G-equivariant vector bundles on V and projective A-G-modules via the functor of global sections. □

The following splitting lemma is due to Bass and Haboush [21, (4.1)].

Lemma 10.63. *A short exact sequence of A-G-modules $0 \to M' \to M \to M'' \to 0$ with M'' a finitely generated module splits in the category of A-modules if and only if it splits in the category of A-G-modules.*

Proof. Assume

$$0 \longrightarrow M' \longrightarrow M \xrightarrow{\alpha} M'' \longrightarrow 0$$

splits A-linearly. Then the A-linear map

$$\mathrm{Hom}_A(M'', M) \xrightarrow{\alpha\circ-} \mathrm{Hom}_A(M'', M'')$$

is surjective. Since M'' is a finitely generated A-module, we actually have a surjective homomorphism of A-G-modules. Because G is linearly reductive, it is also a surjective homomorphism of direct sums of simple G-modules. Therefore, and in view of (10.35), taking G-invariants yields a surjective map

$$\mathrm{Hom}_{A\text{-}G}(M'', M) \longrightarrow \mathrm{Hom}_{A\text{-}G}(M'', M'') \ .$$

The desired splitting homomorphism $M'' \to M$ of A-G-modules is obtained by lifting $\mathbf{1}_{M''}$ to $\mathrm{Hom}_{A\text{-}G}(M'', M)$. □

[3] The ring $K_0(\Bbbk\text{-}G)$ with the multiplicative structure corresponding to tensor products is known as the *character ring* of G, usually denoted by $\mathrm{R}(G)$.

For an affine G-variety $\mathcal{V} = \operatorname{Spec} A$ and G-modules S and F let $\mathsf{Vect}(\mathcal{V}, F, S)$ denote the set of isomorphism classes of G-vector bundles $\mathscr{E}$ on $\mathcal{V}$ such that $\mathscr{E} \oplus \mathbb{S}_{\mathcal{V}} \cong \mathbb{F}_{\mathcal{V}} \oplus \mathbb{S}_{\mathcal{V}}$.

Theorem 10.62 shows that Theorem 10.58 can be reformulated as the equality

$$\mathsf{Vect}(V, F, S) = \{ [\mathbb{F}_V] \}, \tag{10.36}$$

where V is a G-module, provided that all simple G-modules are 1-dimensional. Here $[-]$ refers to the isomorphism class.

Returning to the case of a general affine G-variety $\mathcal{V} = \operatorname{Spec} A$, for G-vector bundles $\mathscr{E}_1$ and $\mathscr{E}_2$ on $\mathcal{V}$ we denote the automorphism group of the G-vector bundle $\mathscr{E}_1$ by $\mathsf{Aut}(\mathscr{E}_1)$, and the set of all homomorphisms between $\mathscr{E}_1$ and $\mathscr{E}_2$ that are surjective on fibers by $\mathsf{Sur}(\mathscr{E}_1, \mathscr{E}_2)$.

The group $\mathsf{Aut}(\mathscr{E}_1)$ acts on the set $\mathsf{Sur}(\mathscr{E}_1, \mathscr{E}_2)$ by multiplication from the left. The corresponding orbit set will be denoted by $\mathsf{Sur}(\mathscr{E}_1, \mathscr{E}_2)/\mathsf{Aut}(\mathscr{E}_1)$. Thus the A-G-homomorphisms, corresponding to the elements of $\mathsf{Sur}(\mathscr{E}_1, \mathscr{E}_2)$, are the surjective maps in $\operatorname{Hom}_{A\text{-}G}(\Gamma(\mathcal{V}, \mathscr{E}_1), \Gamma(\mathcal{V}, \mathscr{E}_2))$.

We need the following lemma [252, Corollary 1.2]:

Lemma 10.64. *In the notation above, assume* $\mathbf{f}, \mathbf{g} \in \mathsf{Sur}(\mathscr{E}_1, \mathscr{E}_2)$. *Then we have the equivalence*

$$\operatorname{Ker}(\mathbf{f}) \cong \operatorname{Ker}(\mathbf{g}) \iff [\mathbf{f}] = [\mathbf{g}] \text{ in } \mathsf{Sur}(\mathscr{E}_1, \mathscr{E}_2)/\mathsf{Aut}(\mathscr{E}_1).$$

In particular, for arbitrary G-modules S and F the map

$$\mathsf{Sur}(\mathbb{S}_{\mathcal{V}} \oplus \mathbb{F}_{\mathcal{V}}, \mathbb{S}_{\mathcal{V}})/\mathsf{Aut}(\mathbb{S}_{\mathcal{V}} \oplus \mathbb{F}_{\mathcal{V}}) \to \mathsf{Vect}(\mathcal{V}, F, S), \qquad [\mathbf{f}] \mapsto [\operatorname{Ker}(\mathbf{f})]$$

is bijective.

Proof. The implication $\Longleftarrow$ is straightforward.

The other implication amounts to showing that for any isomorphism $\vartheta : \operatorname{Ker}(\mathbf{f}) \to \operatorname{Ker}(\mathbf{g})$ there is an element $\Theta \in \mathsf{Aut}(\mathscr{E}_1)$ making the following diagram commute:

$$\begin{array}{ccccccccc} 0 & \longrightarrow & \operatorname{Ker}(\mathbf{f}) & \longrightarrow & \mathscr{E}_1 & \xrightarrow{\mathbf{f}} & \mathscr{E}_2 & \longrightarrow & 0 \\ & & \downarrow{\scriptstyle \vartheta} & & \downarrow{\scriptstyle \Theta} & & \downarrow{\scriptstyle \mathbf{1}} & & \\ 0 & \longrightarrow & \operatorname{Ker}(\mathbf{g}) & \longrightarrow & \mathscr{E}_1 & \xrightarrow{\mathbf{g}} & \mathscr{E}_2 & \longrightarrow & 0. \end{array}$$

By Lemma 10.63 and the fact that $\Gamma(\mathcal{V}, \mathscr{E}_i)$, $i = 1, 2$, is a projective A-module, the diagram above is equivalent to a diagram of A-G-modules of the form

$$\begin{array}{ccccccccc} 0 & \longrightarrow & P' & \longrightarrow & P & \longrightarrow & P'' & \longrightarrow & 0 \\ & & \downarrow{\scriptstyle \wr} & & \downarrow & & \downarrow{\scriptstyle \mathbf{1}} & & \\ 0 & \longrightarrow & P' & \longrightarrow & P & \longrightarrow & P'' & \longrightarrow & 0 \end{array}$$

whose rows are A-G-split exact sequences. Hence the existence of the middle map is obvious. □

Matrix interpretation. We continue to assume that $\mathscr{V}$ is a general affine G-variety. Let F and S be G-modules. An element of $\mathsf{Sur}(\mathbb{F}_{\mathscr{V}} \oplus \mathbb{S}_{\mathscr{V}}, \mathbb{S}_{\mathscr{V}})$ can be uniquely represented by a 1×2-matrix (Φ, T) with $\Phi \in \mathsf{Hom}(\mathbb{F}_{\mathscr{V}}, \mathbb{S}_{\mathscr{V}})$ and $T \in \mathsf{Hom}(\mathbb{S}_{\mathscr{V}}, \mathbb{S}_{\mathscr{V}})$.

Similarly, any automorphism $\mathbf{f} \in \mathsf{Aut}(\mathbb{F}_{\mathscr{V}} \oplus \mathbb{S}_{\mathscr{V}})$ can be interpreted as an invertible 2×2-matrix

$$\begin{pmatrix} \Theta_{11} & \Theta_{12} \\ \Theta_{21} & \Theta_{22} \end{pmatrix}, \qquad \begin{array}{ll} \Theta_{11} \in \mathsf{Hom}(\mathbb{F}_{\mathscr{V}}, \mathbb{F}_{\mathscr{V}}), & \Theta_{12} \in \mathsf{Hom}(\mathbb{S}_{\mathscr{V}}, \mathbb{F}_{\mathscr{V}}), \\ \Theta_{21} \in \mathsf{Hom}(\mathbb{F}_{\mathscr{V}}, \mathbb{S}_{\mathscr{V}}), & \Theta_{22} \in \mathsf{Hom}(\mathbb{S}_{\mathscr{V}}, \mathbb{S}_{\mathscr{V}}). \end{array}$$

Now we can rewrite Theorem 10.58 in terms of matrices. Let V, F, S be G-modules. Since $\mathbb{F}_V \in \mathrm{Ker}((0,1)) \in \mathsf{Sur}(\mathbb{F}_V \oplus \mathbb{S}_V, \mathbb{S}_V)$, equation (10.36) and Lemma 10.64 imply that Theorem 10.58 is equivalent to

Theorem 10.65. *Let V, F, S be G-modules such that V and S are direct sums of* 1-*dimensional G-modules. Then every matrix $(\Phi, T) \in \mathsf{Sur}(\mathbb{F}_V \oplus \mathbb{S}_V, \mathbb{S}_V)$ can be completed to an invertible* 2×2-*matrix in* $\mathsf{Aut}(\mathbb{F}_V \oplus \mathbb{S}_V)$.

In this form Theorem 10.58 will be proved. Later on the following technical result, stated for a general affine G-variety $\mathscr{V}$, will be useful:

Lemma 10.66. *Suppose F'' and S are G-modules such that every composition $\mathbb{S}_{\mathscr{V}} \to \mathbb{F}''_{\mathscr{V}} \to \mathbb{S}_{\mathscr{V}}$ is zero. Then for every G-module F', the map*

$$\mathsf{Vect}(\mathscr{V}, F', S) \to \mathsf{Vect}(\mathscr{V}, F' \oplus F'', S), \qquad [\mathscr{E}] \mapsto [\mathscr{E} \oplus \mathbb{F}''_{\mathscr{V}}],$$

is surjective.

Proof. Let $F = F' \oplus F''$. By Lemma 10.64 any element of $\mathsf{Vect}(\mathscr{V}, F, S)$ is of the form

$$[\mathrm{Ker}(\mathbf{f})], \qquad \mathbf{f} = (\Phi, T) \in \mathsf{Sur}(\mathbb{F}_{\mathscr{V}} \oplus \mathbb{S}_{\mathscr{V}}, \mathbb{S}_{\mathscr{V}}).$$

Assume

$$\Phi = (\Phi', \Phi''), \qquad \Phi' \in \mathsf{Hom}(\mathbb{F}'_{\mathscr{V}}, \mathbb{S}_{\mathscr{V}}), \quad \Phi'' \in \mathsf{Hom}(\mathbb{F}''_{\mathscr{V}}, \mathbb{S}_{\mathscr{V}}).$$

Again by Lemma 10.64, we want to prove the existence of a 3×3-matrix

$$\Theta \in \mathsf{Aut}(\mathbb{F}_{\mathscr{V}} \oplus \mathbb{S}_{\mathscr{V}}) = \mathsf{Aut}(\mathbb{F}'_{\mathscr{V}} \oplus \mathbb{F}''_{\mathscr{V}} \oplus \mathbb{S}_{\mathscr{V}})$$

such that

$$(\Phi', \Phi'', T) = (\Phi', 0, T)\Theta. \tag{10.37}$$

Since $(\Phi, T) : \mathbb{F}_{\mathscr{V}} \oplus \mathbb{S}_{\mathscr{V}} \to \mathbb{S}_{\mathscr{V}}$ is fiberwise surjective, Lemma 10.63 implies the existence of a splitting

$$\begin{pmatrix} \Psi \\ Y \end{pmatrix} : \mathbb{S}_{\mathscr{V}} \to \mathbb{F}_{\mathscr{V}} \oplus \mathbb{S}_{\mathscr{V}}, \qquad \Psi \in \mathsf{Hom}(\mathbb{S}_{\mathscr{V}}, \mathbb{F}_{\mathscr{V}}), \quad Y \in \mathsf{Hom}(\mathbb{S}_{\mathscr{V}}, \mathbb{S}_{\mathscr{V}}).$$

In particular,

$$\Phi\Psi + TY = \mathbf{1}_{\mathbb{S}_{\mathcal{V}}}.$$

Assume $\Psi = (\Psi', \Psi'')$ for some $\Psi' \in \mathrm{Hom}(\mathbb{S}_{\mathcal{V}}, \mathbb{F}'_{\mathcal{V}})$ and $\Psi'' \in \mathrm{Hom}(\mathbb{S}_{\mathcal{V}}, \mathbb{F}''_{\mathcal{V}})$. Since $\Phi''\Psi'' = 0$ by hypothesis, the equation $\Phi\Psi + TY = \mathbf{1}_{\mathbb{S}_{\mathcal{V}}}$ reduces to $\Phi'\Psi' + TY = \mathbf{1}_{\mathbb{S}_{\mathcal{V}}}$.

Finally, the matrix

$$\Theta = \begin{pmatrix} 1 & \Psi'\Phi'' & 0 \\ 0 & 1 & 0 \\ 0 & Y\Phi'' & 1 \end{pmatrix} \in \mathrm{Aut}(\mathbb{F}'_{\mathcal{V}} \oplus \mathbb{F}''_{\mathcal{V}} \oplus \mathbb{S}_{\mathcal{V}})$$

satisfies (10.37). (Why does this matrix represent an automorphism?) □

***Proof of Theorem* 10.65.** Several lemmas will be used in the course of the proof. First we fix some notation. Let V, F, S be G-modules such that V and S split into direct sums of 1-dimensional G-modules. For simplicity we will always skip the subscript $-_V$ in denoting trivial bundles on V: $\mathbb{F} = \mathbb{F}_V$, etc.

Set $n = \dim V$ and let $(\Phi, T) \in \mathrm{Sur}(\mathbb{F} \oplus \mathbb{S}, \mathbb{S})$. We want to show that there is a matrix $\Theta \in \mathrm{Aut}(\mathbb{F} \oplus \mathbb{S})$ such that

$$(\Phi, T) = (0, 1)\Theta. \tag{10.38}$$

The proof uses induction on n, the case $n = 0$ being vacuously true.

Suppose the existence of Θ as (10.38) has been shown for G-vector bundles on representation spaces of dimension $n - 1$.

By hypothesis, V is a sum of 1-dimensional G-modules. Therefore V^* also decomposes into a direct sum of 1-dimensional G-modules. Hence

$$\mathscr{O}(V) = \Bbbk[x_1, \ldots, x_n],$$

where $x_1, \ldots, x_n$ are semi-invariants of the action of G on V^*.

The first reduction in Theorem 10.65 is

Lemma 10.67. *We may assume that S is o1-dimensional and $F = \bigoplus_{i=1}^m F_i$ is a sum of 1-dimensional G-modules such that $\mathrm{Hom}(\mathbb{S}, \mathbb{F}_i) \neq 0$ for $i = 1, \ldots, m$.*

Proof. Let $\mathbb{S} = \mathbb{S}_1 \oplus \cdots \oplus \mathbb{S}_k$ be a splitting into 1-dimensional trivial G-vector bundles. If each of these direct summands can be canceled on both sides of isomorphisms, then

$$\mathscr{E} \oplus \mathbb{S}_1 \oplus \cdots \oplus \mathbb{S}_k \cong \mathbb{F} \oplus \mathbb{S}_1 \oplus \cdots \oplus \mathbb{S}_k \quad \Longrightarrow \quad \mathscr{E} \cong \mathbb{F}.$$

Therefore it is enough to consider the case $\dim_{\Bbbk} S = 1$.

By Lemma 10.66 we may further assume that $F = \bigoplus_{i=1}^m F_i$ where the F_i are simple G-modules for which

$$\mathrm{Hom}(\mathbb{F}_i, \mathbb{S}) \neq 0, \quad \mathrm{Hom}(\mathbb{S}, \mathbb{F}_i) \neq 0, \quad i = 1, \ldots, m. \tag{10.39}$$

By Lemma 10.60

$$\mathsf{Hom}(\mathbb{F}_i, \mathbb{S}) \cong \mathrm{Hom}_{\Bbbk[G]}(F_i, \Bbbk[x_1, \ldots, x_n] \otimes_{\Bbbk} S), \quad i = 1, \ldots, m.$$

Using the decomposition into 1-dimensional G-modules,

$$\Bbbk[x_1, \ldots, x_n] = \bigoplus_{a=(a_1,\ldots,a_n)\in\mathbb{Z}_+^n} \Bbbk x_1^{a_1} \cdots x_n^{a_n} = \bigoplus_{\mathbb{Z}_+^n} T_a,$$

and the fact that F_i is a finitely generated $\Bbbk[G]$-module, we get

$$\mathrm{Hom}_{\Bbbk[G]}(F_i, \Bbbk[x_1, \ldots, x_n] \otimes_{\Bbbk} S) \cong \bigoplus_{\mathbb{Z}_+^n} \mathrm{Hom}_{\Bbbk[G]}(F_i, T_a \otimes_{\Bbbk} S).$$

In particular, $\mathrm{Hom}_{\Bbbk[G]}(F_i, T_a \otimes_{\Bbbk} S) \neq 0$ for at least one a. But F_i is simple and $\dim_{\Bbbk} T_a \otimes_{\Bbbk} S = 1$. So $F_i \cong T_a \otimes_{\Bbbk} S$. □

In the notation introduced above we have

Corollary 10.68.
(a) *The elements of* $\mathsf{Hom}(\mathbb{F}_i, \mathbb{S})$ *and* $\mathsf{Hom}(\mathbb{S}, \mathbb{F}_i)$ *are represented by semi-invariants in* $\Bbbk[x_1, \ldots, x_n]$, *and the elements of* $\mathsf{Hom}(\mathbb{S}, \mathbb{S})$ *(such as T) are represented by invariants, i. e., elements of* $\Bbbk[x_1, \ldots, x_n]^G$.
(b) *Moreover, if* $(\Phi, T) = (\Phi_1, \ldots, \Phi_m, T)$ *for some* $\Phi_i \in \mathsf{Hom}(\mathbb{F}_i, \mathbb{S})$ *then the* $m+1$ *polynomials in* $\Bbbk[x_1, \ldots, x_n]$, *corresponding to the homomorphisms* $\Phi_1, \ldots, \Phi_m, T$, *have no common zero.*

In the following we do not distinguish notationally between the homomorphisms from $\mathsf{Hom}(\mathbb{F}_i, \mathbb{S})$, $\mathsf{Hom}(\mathbb{S}, \mathbb{F}_i)$, or $\mathsf{Hom}(\mathbb{S}, \mathbb{S})$ and the corresponding semi-invariants in $\Bbbk[x_1, \ldots, x_n]$. We also let Φ_i $(i = 1, \ldots, m)$ denote the homomorphisms as in the corollary above. Thus, in view of Lemma 10.60,

$$(\Phi, T)(z, a_1, \ldots, a_m, b) = (z, \Phi_1(z)a_1 + \cdots + \Phi_m(z)a_m + T(z)b),$$
$$z \in V, \ a_1 \in F_1, \ldots, a_m \in F_m, \ b \in S. \quad (10.40)$$

Proof of Corollary 10.68. The first part of the corollary follows directly from Lemma 10.60.

As for the second part, it is a consequence of the fiberwise surjectivity condition on (Φ, T): if there were a common zero $z \in V$ of the polynomials $\Phi_1, \ldots, \Phi_m, T$, then by the formula (10.40) above the homomorphism (Φ, T) would induce the zero map on the fiber over z. □

Now we make the second reduction in Theorem 10.65. Denote the zero set of x_i by V_i. It is a G-submodule of V of dimension $n-1$. Consider $Z = \bigcup_{i=1}^n V_i$—the union of the coordinate hyperplanes in V.

For a (Zariski) closed G-invariant subset $\mathscr{Y} \subset V$ we let

$$(\Phi, T)|_{\mathscr{Y}} \in \mathsf{Sur}(\mathbb{F}|_{\mathscr{Y}} \oplus \mathbb{S}|_{\mathscr{Y}}, \mathbb{S}_{\mathscr{Y}}) = \mathsf{Sur}((\mathbb{F} \oplus \mathbb{S})|_{\mathscr{Y}}, \mathbb{S}_{\mathscr{Y}})$$

denote the corresponding objects obtained by restriction to $\mathscr{Z} \subset \mathscr{V}$. In other words, we apply the inverse image functor $\mathsf{Vect}(V) \to \mathsf{Vect}(\mathscr{Z})$ corresponding to the natural inclusion $\mathscr{Z} \to V$. We also use similar notation $\Theta|_{\mathscr{Z}}$ for $\Theta \in \mathsf{Aut}(\mathbb{F} \oplus \mathbb{S})$.

Observe that neither of the maps

$$\mathsf{Sur}(\mathbb{F} \oplus \mathbb{S}, \mathbb{S}) \to \mathsf{Sur}(\mathbb{F}|_{\mathscr{Z}} \oplus \mathbb{S}|_{\mathscr{Z}}, \mathbb{S}_{\mathscr{Z}}), \quad \mathsf{Aut}(\mathbb{F} \oplus \mathbb{S}) \to \mathsf{Aut}(\mathbb{F}|_{\mathscr{Z}} \oplus \mathbb{S}|_{\mathscr{Z}})$$

(the latter being a group homomorphism) is necessarily surjective. If the embedding $\mathscr{Z} \subset V$ has a right G-equivariant inverse then, of course, these maps are surjective. This happens, for instance, if $\mathscr{Z} = V_i$ for which the desired right inverse is provided by the corresponding coordinate projection.

Lemma 10.69. *In addition to the condition in Lemma* 10.67 *we can assume without loss of generality that* $(\Phi, T)|_Z = (0,1)$.

The argument below uses ideas very similar to those in Lemmas 8.18 and 9.17, where we showed that essential information on projective modules and invertible matrices comes from the interior monoids.

Proof. For an induction on $k = 1, \dots, n$ suppose we already have

$$(\Phi, T)|_{V_j} = (0,1), \quad j = 1, \dots, k-1. \tag{10.41}$$

The induction assumption on n implies that there exists an automorphism $\Theta_k \in \mathsf{Aut}(\mathbb{F}|_{V_k} \oplus \mathbb{S}|_{V_k})$ such that

$$(\Phi, T)|_{V_k} = (0,1)\Theta_k. \tag{10.42}$$

Let $\pi : V \to V_k$ be the coordinate projection. Applying the inverse image functor π^*, we get an automorphism $\pi^*(\Theta_k) \in \mathsf{Aut}(\mathbb{F} \oplus \mathbb{S})$ whose restriction to $\mathbb{F}|_{V_k} \oplus \mathbb{S}|_{V_k}$ is the same Θ_k. Because of the equations (10.41) and (10.42) and the fact that π projects V_j to $V_j \cap V_k$, we arrive at the conclusion that the restriction of $\pi^*(\Theta_k)$ on $V_1, \dots, V_{k-1}$ has the form

$$\pi^*(\Theta_k)|_{V_j} = \begin{pmatrix} * & * \\ 0 & 1 \end{pmatrix}, \quad j = 1, \dots, k-1. \tag{10.43}$$

Now the equations (10.41)–(10.43) imply

$$((\Phi, T)\pi^*(\Theta_k)^{-1})|_{V_j} = (0,1), \qquad j = 1, \dots, k.$$

Therefore, replacing the surjection (Φ, T) by the matrix $(\Phi, T)\pi^*(\Theta_k)^{-1}$ in the same $\mathsf{Aut}(\mathbb{F} \oplus \mathbb{S})$-orbit, we may assume

$$(\Phi, T)|_{V_j} = (0,1), \qquad j = 1, \dots, k.$$

For $k = n$ the desired conclusion follows.

Finally, we have not altered the vector bundles $\mathbb{F}$ and $\mathbb{S}$. Consequently, the validity of the conditions in Lemma 10.67 is not affected. □

From now on we assume that G-modules S and F and the homomorphism $(\Phi, T) : \mathbb{F} \oplus \mathbb{S} \to \mathbb{S}$ satisfy the conditions mentioned in Lemmas 10.67, 10.69, and, consequently, Corollary 10.68.

Let $\Pi = \prod_{j=1}^{n} x_j$. By Lemma 10.69 the polynomials Φ_i are divisible by Π and $T \equiv 1 \mod \Pi$. Since the semi-invariants of a given weight are $\Bbbk$-linear combinations of monomials of the same weight, the condition $\mathrm{Hom}(\mathbb{S}, \mathbb{F}_i) \neq 0$ (Lemma 10.67) implies the existence of a nonzero monomial $\Psi_i \in \mathrm{Hom}(\mathbb{S}, \mathbb{F}_i)$ for $i = 1, \ldots, m$.

Consider the vector

$$(\Phi', T) = (\Phi_1\Psi_1, \ldots, \Phi_m\Psi_m, T)$$

with entries from $B = \Bbbk[x_1, \ldots, x_n]^G$. By Theorem 5.14, B is an affine normal monomial $\Bbbk$-subalgebra of $\Bbbk[x_1, \ldots, x_n]$. Let $\mathcal{V} = \operatorname{Spec} B$ be the corresponding affine toric variety. Observe that $\Phi_1\Psi_1, \ldots, \Phi_m\Psi_m, T \in B$.

Lemma 10.70. *There is a matrix $\Theta \in \mathrm{GL}_{m+1}(B)$ with last row $(\Phi_1\Psi_1, \ldots, \Phi_m\Psi_m, T)$.*

Proof. The polynomials $\Phi_1\Psi_1, \ldots, \Phi_m\Psi_m$ and T do not have a common zero because this is true for the components of (Φ, T), $T \equiv 1 \mod \Pi$, and the Ψ_i are all monomials. In particular, the homomorphism

$$\begin{aligned} \mathbf{f} : \mathcal{V} \times \Bbbk^{m+1} &\to \mathcal{V} \times \Bbbk, \\ (x, a_1, \ldots, a_m, b) &\mapsto (x, (\Phi_1\Psi_1)(x)a_1 + \cdots + (\Phi_m\Psi_m)(x)a_m + T(x)b), \\ & \qquad x \in \mathcal{V},\ a_1, \ldots, a_m, b \in \Bbbk, \end{aligned}$$

of trivial vector bundles on $\mathcal{V}$ is fiberwise surjective. Then we have the vector bundle $\mathrm{Ker}(\mathbf{f}) \to \mathcal{V}$. By Theorem 8.4 it is a trivial bundle. Thus, by Lemma 10.64 (stated for a trivial group action) and our matrix theoretical interpretation, there is a matrix $\Theta \in \mathrm{GL}_{m+1}(B)$ such that

$$(\Phi_1\Psi_1, \ldots, \Phi_m\Psi_m, T) = (0, \ldots, 0, 1)\Theta. \qquad \square$$

Lemma 10.71. *For every natural number N there is a matrix $\Theta \in \mathrm{GL}_{m+1}(B)$ such that:*

(a) the last row of Θ is $(\Phi_1\Psi_1, \ldots, \Phi_m\Psi_m, T)$;

(b) the off-diagonal entries of Θ outside the last row and the last column are divisible by Π^N.

Proof. Owing to the fact that $\Pi\Bbbk[x_1, \ldots, x_n] \cap B \subset B$ is a radical monomial ideal, the lemma follows from the more general Lemma 10.72 below. $\square$

Lemma 10.72. *Let R be a ring, $m \in \mathbb{Z}_+$, M a positive affine monoid, and $I \subset R[M]$ be a radical monomial ideal. Suppose the matrix $\alpha = (a_{pq}) \in \mathrm{GL}_{m+1}(R[M])$ satisfies the condition*

$$(a_{m+11}, \ldots, a_{m+1m}, a_{m+1m+1}) \equiv (0, \ldots, 0, 1) \mod I.$$

Then for any natural number N there exists a matrix $\beta = (b_{pq}) \in \mathrm{GL}_{m+1}(R[M])$ such that

(1) $(a_{m+11},\dots,a_{m+1m},a_{m+1m+1}) = (b_{m+11},\dots,b_{m+1m},b_{m+1m+1})$,
(2)

$$\beta \equiv \begin{pmatrix} * & 0 & \dots & 0 & * \\ 0 & * & \dots & 0 & * \\ \vdots & \vdots & \ddots & \vdots & \vdots \\ 0 & 0 & \dots & * & * \\ a_{m+1m} & a_{m+12} & \dots & a_{m+1m} & a_{m+1m+1} \end{pmatrix} \mod I^N.$$

Proof. By subtracting multiples of the last row from the first m rows we can assume that

$$\begin{pmatrix} a_{1m+1} \\ \vdots \\ a_{mm+1} \\ a_{m+1m+1} \end{pmatrix} \equiv \begin{pmatrix} 0 \\ \vdots \\ 0 \\ a_{m+1m+1} \end{pmatrix} \mod I,$$

that is

$$\bar{\alpha} = \begin{pmatrix} \gamma & 0 \\ 0 & 1 \end{pmatrix} \quad \text{in} \quad \mathrm{GL}_m(R[M]/I).$$

Let $M_1,\dots,M_r \subset M$ be the facet submonoids not meeting I (see Proposition 2.36(a)). The natural inclusions $R[M_i] \to R[M]$ induce embeddings $R[M_i] \to R[M]/I$. We will identify the rings $R[M_i]$ with their images in $R[M]/I$. The embedding $R[M_i] \to R[M]$ splits the surjective homomorphism

$$\pi_i : R[M]/I \to R[M_i], \quad \operatorname{Ker}\pi_i = R(M \setminus M_i)/I.$$

Consider the following recursively defined sequence of matrices:

$$\gamma_1 = \pi_1(\gamma),$$
$$\gamma_i = \pi_i(\gamma_{i-1}^{-1}\cdots\gamma_1^{-1}\gamma), \quad i = 2,\dots,r.$$

The matrices γ_i are defined over $R[M_i] \subset R[M]$ and, therefore, can be viewed as elements of $\mathrm{GL}_m(R[M])$. Hence the matrix

$$\beta' = \begin{pmatrix} \gamma_1\cdots\gamma_r & 0 \\ 0 & 1 \end{pmatrix}^{-1} \cdot \alpha \in \mathrm{GL}_{m+1}(R[M])$$

reduces to the identity matrix modulo I. Consequently, β' reduces modulo I^N to a matrix of the form

$$\delta = \begin{pmatrix} 1+b_{11} & b_{12} & \dots & b_{1m} & b_{1m+1} \\ b_{21} & 1+b_{22} & \dots & b_{2m} & b_{2m+1} \\ \vdots & \vdots & \vdots & \vdots & \vdots \\ b_{m+11} & b_{m+12} & \dots & b_{m+1m} & 1+b_{1m+1} \end{pmatrix} \in \mathrm{GL}_{m+1}\big(R[M]/I^N\big)$$

where the elements

$$b_{pq} \in \mathrm{GL}_{m+1}\big(R[M]/I^N\big), \quad 1 \le p,\ q \le m+1,$$

are all nilpotent.

Then we transform δ by elementary row operations as follows. First subtract the $(1+b_{11})^{-1}b_{p1}$-multiples of the 2nd, 3rd,..., mth row. Then, in the new matrix, we subtract the corresponding multiples of the second row from the 1st, 3rd,..., mth row, and so on (never touching the $(m+1)$st row). After m steps, each involving $m-1$ row transformations, we will have reached a matrix in $\mathrm{GL}_{m+1}(R[M]/I^N)$ whose upperleft $m \times m$-corner is a diagonal matrix. All these elementary transformations lift to elementary row transformations of β', none of which affects the last row of β'. The corresponding transformed matrix β has the desired properties. □

Now we finish the proof of Theorem 10.65. Choose a natural number N large enough so that the following divisibility condition holds

$$\Psi_i | \Pi^N, \quad i = 1, \dots, m.$$

Let $\Theta = (\vartheta_{pq})$ be an $(m+1) \times (m+1)$ matrix as in Lemma 10.71. Then the matrix

$$\Theta' = \begin{pmatrix} \Psi_1 & 0 & \dots & 0 & 0 \\ 0 & \Psi_2 & \dots & 0 & 0 \\ \vdots & \vdots & \ddots & \vdots & \vdots \\ 0 & 0 & \dots & \Psi_m & 0 \\ 0 & 0 & \dots & 0 & 1 \end{pmatrix} \Theta \begin{pmatrix} \Psi_1^{-1} & 0 & \dots & 0 & 0 \\ 0 & \Psi_2^{-1} & \dots & 0 & 0 \\ \vdots & \vdots & \ddots & \vdots & \vdots \\ 0 & 0 & \dots & \Psi_m^{-1} & 0 \\ 0 & 0 & \dots & 0 & 1 \end{pmatrix}$$

has the form

$$\begin{pmatrix} \vartheta'_{11} & \vartheta'_{12} & \dots & \vartheta_{1m+1}\Psi_1 \\ \vartheta'_{21} & \vartheta'_{22} & \dots & \vartheta_{2m+1}\Psi_2 \\ \vdots & \vdots & \vdots & \vdots \\ \vartheta'_{m1} & \vartheta'_{m2} & \dots & \vartheta_{mm+1}\Psi_m \\ \Phi_1 & \Phi_2 & \dots & T \end{pmatrix}$$

where $\vartheta'_{pq} = \Psi_p^{-1}\vartheta_{pq}\Psi_q$ for $1 \le p, q \le m$. In particular, all entries of Θ' belong to $\Bbbk[x_1, \dots, x_n]$ and the whole matrix defines a homomorphism $\mathbb{F} \oplus \mathbb{S} \to \mathbb{F} \oplus \mathbb{S}$. The equation $\det(\Theta) = \det(\Theta')$ shows that Θ is an invertible matrix and we are done because

$$(\Phi_1, \dots, \Phi_m, T) = (0, \dots, 0, 1)\Theta'.$$

Exercises

10.1. Prove that a morphism between toric varieties $\varphi : \mathscr{V}(\mathscr{F}) \to \mathscr{V}(\mathscr{F}')$ with embedded tori $\mathbb{T}$ and $\mathbb{T}'$ is equivariant if it fits in a commutative diagram of the form

$$\begin{array}{ccc} \mathscr{V}(\mathscr{F}) & \xrightarrow{\ \varphi\ } & \mathscr{V}(\mathscr{F}') \\ \uparrow & & \uparrow \\ \mathbb{T} & \xrightarrow{\ \varphi_0\ } & \mathbb{T}' \end{array}$$

with φ_0 a homomorphism of linear groups.

10.2. Prove the statement about products of fans in property (i) of toric varieties on p. 366.

10.3. Show that every finitely generated abelian group G can be realized as the Picard group of a smooth toric variety (necessarily noncomplete if G has torsion).

10.4. A Weil divisor D is called $\mathbb{Q}$-*Cartier* if there exists $k \in \mathbb{N}$ such that kD is a Cartier divisor. Prove that the following are equivalent for a toric variety $\mathscr{V}$:

(a) $\mathscr{V}$ is simplicial;

(b) every Weil divisor on $\mathscr{V}$ is $\mathbb{Q}$-Cartier.

Moreover, give a uniform value of k in this case.

10.5. In this problem on a "bordism polytope" of an equivariant resolution of toric singularities, we have exchanged the roles of $\mathbb{R}^n$ and its dual $(\mathbb{R}^n)^*$.

Let $P \subset \mathbb{R}^n$ be a lattice polytope. Show that there exists a lattice polytope $\tilde{P} \subset \mathbb{R}^{n+1}$, containing P as a facet and having another facet Q such that:

(1) Q is parallel to P;

(2) the corner cones $\mathbb{R}_+(\tilde{P} - v)$, $v \in \operatorname{vert}(Q)$, are unimodular (in particular, Q is a smooth polytope);

(3) $\mathscr{N}(Q)$ is a (unimodular) triangulation of $\mathscr{N}(P)$.

Hint: let $\mathscr{N}(P) = \mathscr{F}_1, \dots, \mathscr{F}_k$ be a sequence of fans in $(\mathbb{R}^n)^*$ such that $\mathscr{F}_j$ is obtained from $\mathscr{F}_{j-1}$ by a stellar subdivision with respect to some vector. Then there are natural numbers $c_2, \dots, c_n$ and lattice n-polytopes $P = P_1, P_2, \dots, P_k = Q \subset \mathbb{R}^n$ such that:

(i) P_j is obtained by "cutting off" a codimension ≥ 2 face from $c_j P_{j-1}$;

(ii) $\mathscr{F}_j = \mathscr{N}(P_j)$;

(iii) P_k is smooth.

Consider the sequence of polytopes in $\mathbb{R}^{n+1}$:

$$\begin{aligned}
\tilde{P}_1 &= (P_1, 1),\\
\tilde{P}_2 &= \operatorname{conv}\big(\tilde{P}_1, c_2\tilde{P}_1, (P_2, c_2 + 1)\big),\\
\tilde{P}_3 &= \operatorname{conv}\big(\tilde{P}_1, c_3\tilde{P}_2, (P_3, c_3(c_2 + 1) + 1)\big),\\
&\ \vdots\\
\tilde{P}_k &= \operatorname{conv}\big(\tilde{P}_1, c_k\tilde{P}_{k-1}, (P_k, c_k(c_{k-1}\cdots(c_2 + 1) + \cdots + 1) + 1)\big).
\end{aligned}$$

If $c_2, \dots, c_k \gg 0$ then $\tilde{P} = \tilde{P}_k$ has the desired properties.

10.6. Using Theorem 10.22, prove that the Chow group A_* of a complete smooth toric variety is a free abelian group.

10.7. Let V be an n-dimensional $\mathbb{R}$-vector space and $L \subset V$ a lattice.

(a) Using Riemann–Roch, give a new proof of the existence of the Ehrhart polynomial $q_P(t) \in \mathbb{Q}[t]$ of an L^*-polytope $P \subset V^*$.

Hint: the discussion, following (10.24), shows that the Ehrhart function $E(P, v)$ can be replaced by the function

$$\chi(\mathscr{V}(\mathscr{G}), \mathscr{L}^v) = \sum_{k=0}^{n} \frac{1}{k!} \deg\big((vD)^k \cdot \operatorname{td}_k(\mathscr{V}(\mathscr{G}))\big) = \sum_{k=0}^{n} a_k v^k,$$

$$a_k = \frac{1}{k!} \deg\big(D^k \cdot \operatorname{td}_k(\mathscr{V}(\mathscr{F}))\big),$$

where $\mathscr{F}$ is a unimodular projective fan, $\mathscr{L}$ is a line bundle on $\mathscr{V}(\mathscr{F})$ generated by global sections, and D is the corresponding Cartier divisor.

(b) Show the existence of the following multivariate version of Ehrhart polynomials. Let $P_1, \dots, P_m \subset V^*$ be full-dimensional L^*-polytopes. Then there exists a polynomial $q_{P_1,\dots,P_m} \in \mathbb{Q}[t_1, \dots, t_m]$ of total degree n such that

$$\#(L^* \cap (\nu_1 P_1 + \cdots + \nu_m P_m)) = q_{P_1,\dots,P_m}(\nu_1, \dots, \nu_m), \quad \nu_1, \dots, \nu_m \in \mathbb{Z}_+.$$

(See [102, 11.12.1] and the references given there.)

Hint: the same approach as in (a), together with Theorem 10.11(e), leads to an expression of the form

$$\#(L^* \cap (\nu_1 P_1 + \cdots + \nu_m P_m)) = \sum_{k=0}^{n} \frac{1}{k!} \deg\left((\nu_1 D_1 + \cdots + \nu_m D_m)^k \cdot \mathrm{td}_k(\mathscr{V}(\mathscr{F}))\right),$$

where $\mathscr{F}$ is a projective unimodular fan, subdividing each of the normal fans $\mathscr{N}(P_1), \dots, \mathscr{N}(P_m)$ and the D_j are appropriate Cartier divisors in $\mathscr{V}(\mathscr{F})$.

10.8. (a) Let vol be the volume function on the dual space V^*, normalized with respect to L^*. For a full-dimensional L^*-polytope $P \subset V^*$ and an equivariant projective resolution of singularities $f : \mathscr{V}(\mathscr{F}) \to \mathscr{V}(\mathscr{N}(P))$. show that

$$\mathrm{vol}(P) = \frac{D^n}{n!}, \qquad n = \dim V,$$

where D is the Cartier divisor on $\mathscr{V}(\mathscr{F})$, corresponding to the line bundle $f^*\mathscr{L}(P)$.

Hint: use Exercise 10.7(a), $\lim_{\nu\to\infty} E(P,\nu)/\nu^n = \mathrm{vol}(P)$, and $\mathrm{td}_0(\mathscr{V}(\mathscr{F})) = 1$. Also compare Theorem 6.54.

(b) Show the following multivariate version of the formula in (a): for full-dimensional L^*-polytopes $P_1, \dots, P_m \subset V^*$ and equivariant projective resolutions of singularities $f_j : \mathscr{V}(\mathscr{F}) \to \mathscr{V}(\mathscr{N}(P_j))$, $j = 1, \dots, m$, we have

$$\mathrm{vol}(\nu_1 P_1 + \cdots + \nu_m P_m) = \frac{(\nu_1 D_1 + \cdots + \nu_m D_m)^n}{n!}, \qquad \nu_1, \dots, \nu_m \in \mathbb{Z}_+,$$

where D_j is the Cartier divisor on $\mathscr{V}(\mathscr{F})$, corresponding to the line bundle $f_j^*\mathscr{L}(P_j) \in \mathrm{Pic}(\mathscr{V}(\mathscr{F}))$, $j = 1, \dots, m$.

Remarks. (1) This exercise explains why for full-dimensional L^*-polytopes $P_1, \dots, P_n \subset V^*$ the coefficient of $\nu_1 \cdots \nu_n$ in $(\nu_1 D_1 + \cdots + \nu_n D_n)^n$, $n = \dim V^*$, is called the *mixed volume of* $P_1, \dots, P_n$.

(2) In view of (10.28), one can relax the unimodularity and projectivity conditions on $\mathscr{F}$ (but keeping the completeness)—the result is then phrased in terms of the operational Chow cohomology.

10.9. We use the notation introduced for the proofs of Theorems 10.37 and 10.41. Recall, $R_{\mathscr{F}} = \mathbb{Q}[X_\rho : \rho \in \mathscr{F}^{[1]}]/I_{\mathscr{F}}$ is the Stanley–Reisner ring of the $(d-1)$-dimensional simplicial complex $\Delta_{\mathscr{F}}$ formed by the sets $\mathrm{conv}(\rho_1, \dots, \rho_u)$, $\mathbb{R}_+\rho_1 + \cdots + \mathbb{R}_+\rho_u \in \mathscr{F}$.

(a) Use [72, p. 213] to prove that the Hilbert series of $R_{\mathscr{F}}$ is

$$H_{R_{\mathscr{F}}}(t) = \frac{1 + h_1 t + \cdots + h_d t^d}{(1-t)^d}, \qquad h_k = \sum_{i=0}^{k} (-1)^{k-i} \binom{d-i}{k-i} \#\mathscr{F}^{[i]}.$$

(b) Suppose $\mathscr{F}$ is complete. Show that $\Delta_{\mathscr{F}}$ is homeomorphic to the unit sphere in V, and therefore $R_{\mathscr{F}}$ is Gorenstein [72, 5.6.5]. It follows that the h-vector $(1, h_1, \dots, h_d)$ is palindromic.

(c) Prove: the ideal $J_{\mathscr{F}}$ generated by the linear forms ℓ_σ is in fact generated by d elements $\ell_1, \dots, \ell_d$ corresponding to a basis $\sigma_1, \dots \sigma_d$ of L^*.

(d) Show next that $S_{\mathscr{F}} = R_{\mathscr{F}}/J_{\mathscr{F}}R_{\mathscr{F}}$ is a finite-dimensional $\mathbb{Q}$-vector space generated by the residue classes of the squarefree monomials. ([72, 5.1.16] gives a hint, or look up [102].)

(e) Conclude from (c) and (d) that $H_{S_{\mathscr{F}}}(t) = 1 + h_1 t + \dots + h_d t^d$.

(f) Finally, use that the h-vector is palindromic in order to derive the formula in Theorem 10.37.

10.10. Study the notion of shelling of a polytopal or conical complex in [381] or [72] and relate Theorem 10.38 to it. In particular, compare the theorem to [72, 5.1.16(c)].

10.11. Let P be a polytope with normal fan $\mathscr{F}$. Show that the polytope Q is a Minkowski summand of P if and only if $\mathscr{F}$ subdivides $\mathscr{N}(Q)$.

10.12. Let $\mathscr{V}$ be a variety over an algebraically closed field $\Bbbk$ of characteristic 0. Show that $\mathscr{V}$ is a simplicial toric variety if and only if locally $\mathscr{V}$ is a quotient of an affine space by a linear action of a finite abelian group.

Hint: for the "if" part use Corollary 5.19; for the "only if" part use Theorem 5.3.

10.13. (a) Let $\Bbbk \subset \Bbbk'$ be an algebraic extension of characteristic 0 fields, not necessarily algebraically closed, and R a $\Bbbk$-algebra. Show that the maps $NK_i(R) \to NK_i(\Bbbk' \otimes_{\Bbbk} R)$, $i = 0, 1,$are injective.

(b) Show the K_1 part of [270, Th. 1.3] that for a ring R and a monic polynomial $f \in R[X]$ the map $K_1(R) \to K_1(R[X]_f)$ is injective. (For the K_0 part see Exercise 8.10.)

(c) Show that the injectivity in (a) remains true for an arbitrary extension $\Bbbk \subset \Bbbk'$ of characteristic 0 fields.

Hint for (a): use the big Witt vectors' action, transfer maps, and the fact that K-groups commute with filtered inductive limits.

10.14. Construct examples of rational simplicial cones $C \subset \mathbb{R}^n$, $n \geq 3$, all of whose facets, except exactly one, are unimodular.

10.15. Construct an explicit normal lattice simple 3-polytope P whose projective toric variety over $\mathbb{C}$ fails the rational isomorphism property for the Cartan homomorphism.

10.16. Show that every abelian linearly reductive group G is diagonalizable.

Hint: G^0 is a torus [36, p. 125].

10.17. Explain why the finite generation of the A-module M is important for the definition of an A-G-module structure on $\operatorname{Hom}_A(M, N)$ (p. 415).

10.18. Let $\mathscr{V}$ be an affine $\Bbbk$-variety with coordinate ring A and an action of a torus $\mathbb{T}$. The decomposition into weight spaces makes A a graded ring with grading group $\mathbb{Z}^r = X(\mathbb{T})$, $r = \dim \mathbb{T}$. Moreover, any A-$\mathbb{T}$-module is naturally graded over A.

(a) Show that a $\mathbb{T}$-equivariant vector bundle on $\mathscr{V}$ is trivial if and only if its module of global sections M is free with a homogeneous basis.

(b) Prove that the set of closed orbits of $\mathbb{T}$ has a unique minimal element if and only if A is *local.

Hint: consider the prime ideal of functions vanishing on the unique minimal closed orbit.

(c) Show that a $\mathbb{T}$-equivariant bundle on $\mathscr{V}$ is trivial if $\mathbb{T}$ has a unique minimal closed orbit. (Compare Remark 6.16.)

(d) Prove that the hypothesis of (c) is satisfied if $\mathscr{V}$ is toric with embedded torus $\mathbb{T}$. Give other examples, too.

References

1. S. Aguzzoli and D. Mundici. An algorithmic desingularization of 3-dimensional toric varieties. *Tôhoku Math. J.* **46** (1994), 557–572.
2. A. Al Amrani. Groupe de Chow et K-théorie cohérente des espaces projectifs tordus. In *Proceedings of Research Symposium on K-Theory and its Applications (Ibadan, 1987)*, vol. 2, 1989, pp. 579–602.
3. H. Anand, V.C. Dumir, and H. Gupta. A combinatorial distribution problem. *Duke Math. J.* **33** (1966), 757–769.
4. D.D. Anderson and D.F. Anderson. Divisorial ideals and invertible ideals in a graded integral domain. *J. Algebra* **76** (1982), 549–569.
5. D.F. Anderson. Projective modules over subrings of $k[x, y]$ generated by monomials. *Pac. J. Math.* **79** (1978), 5–17.
6. D.F. Anderson. The Picard group of a monoid domain. II. *Arch. Math.* **55** (1990), 143–145.
7. D.J. Anick. A counterexample to a conjecture of Serre. *Ann. Math.* **115** (1982), 1–33.
8. D.J. Anick. On the homology of associative algebras. *Trans. Amer. Math. Soc.* **296** (1986), 641–659.
9. C.A. Athanasiadis. Ehrhart polynomials, simplicial polytopes, magic squares and a conjecture of Stanley. *J. Reine Angew. Math.* **583** (2005), 163–174.
10. M.F. Atiyah and I.G. Macdonald. *Introduction to commutative algebra*. Addison-Wesley, Reading MA–Menlo Park, CA–London–Don Mills , Ont., 1969.
11. S. Au, M. Huang, and M.E. Walker. The equivariant K-theory of toric varieties. *Preprint*.
12. M. Audin. *The topology of torus actions on symplectic manifolds*. Birkhäuser, Boston, 1991.
13. J. Backelin and R. Fröberg. Koszul algebras, Veronese subrings, and rings with linear resolutions. *Rev. Roumaine Pure Appl.* **30** (1985), 85–97.
14. C. Baeţica. Cohen-Macaulay classes which are not conic. *Commun. Algebra* **32** (2004), 1183–1188.
15. I. Bárány and A. Vershik. On the number of convex lattice polytopes. *Geom. Funct. Anal.* **2** (1992), 381–393.
16. V. Barucci, D.E. Dobbs, and M. Fontana. *Maximality properties in numerical semigroups and applications to one-dimensional analytically irreducible local domains*. Mem. Amer. Math. Soc. **598**, 1997.
17. A. Barvinok. *A course in convexity*. American Mathematical Society, Providence, RI, 2002.
18. H. Bass. Big projective modules are free. *Ill. J. Math.* **7** (1963), 24–31.

19. H. Bass. *Algebraic K-Theory*. W.A. Benjamin, Inc., New York–Amsterdam, 1968.
20. H. Bass. Some problems in "classical" algebraic K-theory. In *Algebraic K-theory, II: "Classical" algebraic K-theory and connections with arithmetic*, Lect. Notes Math. **342**. Springer, Berlin–Heidelberg–New York, 1973, pp. 3–73.
21. H. Bass and W. Haboush. Linearizing certain reductive group actions. *Trans. Amer. Math. Soc.* **292** (1985), 463–482.
22. H. Bass and W. Haboush. Some equivariant K-theory of affine algebraic group actions. *Commun. Algebra* **15** (1987), 181–217.
23. H. Bass, A. Heller, and R.G. Swan. The Whitehead group of a polynomial extension. *Publ. Math. I.H.E.S.* **22** (1964), 61–79.
24. V. Batyrev. Dual polyhedra and mirror symmetry for Calabi–Yau hypersurfaces in toric varieties. *J. Algebr. Geom.* **3** (1994), 493–535.
25. P. Baum, W. Fulton, and R. MacPherson. Riemann-Roch for singular varieties. *Publ. Math. I.H.E.S.* **45** (1975), 101–145.
26. M. Beck and S. Robbins. *Computing the continuous discretely*. Springer, Berlin–Heidelberg–New York, 2006.
27. A.A. Beĭlinson, J. Bernstein, and P. Deligne. Faisceaux pervers. In *Analysis and topology on singular spaces, I*, Astérisque **100**. 1982, pp. 5–171.
28. F. Berchtold and J. Hausen. Demushkin's theorem in codimension one. *Math. Z.* **244** (2003), 697–703.
29. P. Berthelot, A. Grothendieck, and L. Illusie, Eds. *Théorie des intersections et théorème de Riemann-Roch*. Lect. Notes Math. **225**, Springer, Berlin–Heidelberg–New York, 1971. Séminaire de Géométrie Algébrique du Bois–Marie 1966–1967 (SGA 6).
30. S.M. Bhatwadekar and R.A. Rao. On a question of Quillen. *Trans. Amer. Math. Soc.* **279** (1983), 801–810.
31. A.M. Bigatti, R. La Scala, and L. Robbiano. Computing toric ideals. *J. Symb. Comp.* **27** (1999), 351–365.
32. L.J. Billera and C.W. Lee. A proof of the sufficiency of McMullen's conditions for f-vectors of simplicial convex polytopes. *J. Combinatorial Theory Ser. A* **31** (1981), 237–255.
33. A. Björner, M. Las Vergnas, B. Sturmfels, N. White, and G.M. Ziegler. *Oriented matroids*, second ed. Cambridge University Press, Cambridge, 1999.
34. S. Bloch. Some formulas pertaining to the K-theory of commutative groupschemes. *J. Algebra* **53** (1978), 304–326.
35. S. Blum. Base-sortable matroids and Koszulness of semigroup rings. *Eur. J. Comb.* **22** (2001), 937–951.
36. A. Borel. *Linear algebraic groups*. Springer, Berlin–Heidelberg–New York, 1991.
37. N. Bourbaki. *Algèbre commutative, Chap. I–IX*. Hermann, Masson, Paris, 1961–1983.
38. N. Bourbaki. *Lie groups and Lie algebras, Ch. 4–6*. Springer, Berlin–Heidelberg–New York, 2002.
39. J.-F. Boutot. Singularités rationelles et quotients par les groupes réductifs. *Invent. Math.* **88** (1987), 65–68.
40. C. Bouvier and G. Gonzalez–Sprinberg. Système générateurs minimal, diviseurs essentiels et G-désingularizations de variétés toriques. *Tôhoku Math. J.* **46** (1994), 125–149.
41. J.W. Brewer, D.L. Costa, and K. McCrimmon. Seminormality and root closure in polynomial rings and algebraic curves. *J. Algebra* **58** (1979), 217–226.
42. M. Brion. Groupe de Picard et nombres caractéristiques des variétés sphériques. *Duke Math. J.* **58** (1989), 397–424.
43. M. Brion. The structure of the polytope algebra. *Tôhoku Math. J.* **49** (1997), 1–32.
44. M. Brion and S. Kumar. *Frobenius splitting methods in geometry and representation theory*. Birkhäuser, Boston, 2005.

45. M. Brion and M. Vergne. An equivariant Riemann-Roch theorem for complete, simplicial toric varieties. *J. Reine Angew. Math.* **482** (1997), 67–92.
46. M. Brion and M. Vergne. Lattice points in simple polytopes. *J. Amer. Math. Soc.* **10** (1997), 371–392.
47. M.P. Brodmann and R.Y. Sharp. *Local cohomology: an algebraic introduction with geometric applications.* Cambridge University Press, Cambridge, 1998.
48. M. Brun, W. Bruns, and T. Römer. Cohomology of partially ordered sets and local cohomology of section rings. *Adv. Math.* **208** (2007), 210–235.
49. M. Brun and T. Römer. Subdivisions of toric complexes. *J. Algebr. Comb.* **21** (2005), 423–448.
50. M. Brun and T. Römer. On algebras associated to partially ordered sets. *Math. Scand.* **103** (2008), 1–17.
51. W. Bruns. Conic divisor classes over a normal monoid algebra. In *Commutative algebra and algebraic geometry*, S. Ghorpade, H. Srinivasan, and J. Verma, Eds., Contemp. Math. **390**, Amer. Math. Soc., 2005, pp. 63–71.
52. W. Bruns. Commutative algebra arising from the Anand–Dumir–Gupta conjectures. In *Commutative Algebra and Combinatorics*, W. Bruns, Ed., Ramanujan Math. Soc. Lect. Notes Ser. **4** (Part II). Ramanujan Math. Soc., 2007, pp. 1–38.
53. W. Bruns. On the integral Carathéodory property. *Experiment. Math.* **16** (2007), 359–365.
54. W. Bruns and A. Conca. Gröbner bases and determinantal ideals. In *Commutative algebra, singularities and computer algebra*, J. Herzog and V. Vuletescu, Eds., NATO Sci. Ser. II, Math. Phys. Chem. **115**, Kluwer, 2003, pp. 9–66.
55. W. Bruns and A. Conca. Gröbner bases, initial ideals and initial algebras. In *Homological methods in commutative algebra*, L.L. Avramov et al., Eds., IPM Proceedings **II**, IPM Teheran, 2004, pp. 17–34.
56. W. Bruns, A. Conca, and T. Römer. Koszul homology and the N_p property of Veronese subalgebras. *In preparation.*
57. W. Bruns and J. Gubeladze. A regularity criterion for semigroup rings. *Georg. Math. J.* **6** (1999), 259–262.
58. W. Bruns and J. Gubeladze. Normality and covering properties of affine semigroups. *J. Reine Angew. Math.* **510** (1999), 151–178.
59. W. Bruns and J. Gubeladze. Rectangular simplicial semigroups. In *Commutative algebra, algebraic geometry, and computational methods*, D. Eisenbud, Ed., Springer Singapore, 1999, pp. 201–214.
60. W. Bruns and J. Gubeladze. Polyhedral algebras, arrangements of toric varieties, and their groups. In *Computational Commutative Algebra and Combinatorics*, T. Hibi, Ed., Advanced Studies in Pure Mathematics **33**, Math. Soc. of Japan, 2002, pp. 1–51.
61. W. Bruns and J. Gubeladze. Polyhedral K_2. *Manuscripta Math.* **109** (2002), 367–404.
62. W. Bruns and J. Gubeladze. Polytopal linear algebra. *Beitr. Algebra Geom.* **43** (2002), 479–500.
63. W. Bruns and J. Gubeladze. Polytopal linear retractions. *Trans. Amer. Math. Soc.* **354** (2002), 179–203.
64. W. Bruns and J. Gubeladze. Semigroup algebras and discrete geometry. In *Geometry of toric varieties*, L. Bonavero and M. Brion, Eds., Sémin. Congr. **6**, Soc. Math. de France, 2002, pp. 43–127.
65. W. Bruns and J. Gubeladze. Unimodular covers of multiples of polytopes. *Doc. Math., J. DMV* **7** (2002), 463–480.
66. W. Bruns and J. Gubeladze. Divisorial linear algebra of normal semigroup rings. *Algebr. Represent. Theory* **6** (2003), 139–168.

67. W. Bruns and J. Gubeladze. Higher polyhedral K-groups. *J. Pure Appl. Algebra* **184** (2003), 175–228.
68. W. Bruns, J. Gubeladze, M. Henk, A. Martin, and R. Weismantel. A counterexample to an integer analogue of Carathéodory's theorem. *J. Reine Angew. Math.* **510** (1999), 179–185.
69. W. Bruns, J. Gubeladze, and N.V. Trung. Normal polytopes, triangulations, and Koszul algebras. *J. Reine Angew. Math.* **485** (1997), 123–160.
70. W. Bruns, J. Gubeladze, and N.V. Trung. Problems and algorithms for affine semigroups. *Semigroup Forum* **64** (2002), 180–212.
71. W. Bruns and J. Herzog. Semigroup rings and simplicial complexes. *J. Pure Appl. Algebra* **122** (1997), 185–205.
72. W. Bruns and J. Herzog. *Cohen–Macaulay rings*, revised ed. Cambridge University Press, Cambridge, 1998.
73. W. Bruns, J. Herzog, and U. Vetter. Syzygies and walks. In *Commutative algebra*, A. Simis, N.V. Trung, and G. Valla, Eds., World Scientific, 1994, pp. 36–57.
74. W. Bruns and B. Ichim. Normaliz 2.0. Computing normalizations of affine semigroups. Available from http://www.math.uos.de/normaliz.
75. W. Bruns and B. Ichim. On the coefficients of Hilbert quasipolynomials. *Proc. Amer. Math. Soc.* **135** (2007), 1305–1308.
76. W. Bruns, R. Koch, and T. Römer. Gröbner bases and Betti numbers of monoidal complexes. *Michigan Math. J.* **57** (2008), 71–91.
77. W. Bruns, P. Li, and T. Römer. On seminormal monoid rings. *J. Algebra* **302** (2006), 361–386.
78. W. Bruns and T. Römer. h-vectors of Gorenstein polytopes. *J. Combinatorial Theory Ser. A* **114** (2007), 65–76.
79. W. Bruns, W.V. Vasconcelos, and R.H. Villarreal. Degree bounds in monomial subrings. *Ill. J. Math.* **41** (1997), 341–353.
80. W. Bruns and U. Vetter. *Determinantal rings*. Lect. Notes Math. **1327**, Springer, Berlin–Heidelberg–New York, 1988.
81. A. Campillo, J. Grabowski, and G. Müller. Derivation algebras of toric varieties. *Compositio Math.* **116** (1999), 119–132.
82. W. Castryck and J. Voight. On nondegeneracy of curves. *Preprint*.
83. G. Caviglia. The pinched Veronese is Koszul. *J. Algebr. Comb.* (to appear).
84. L.G. Chouinard, II. Krull semigroups and divisor class groups. *Canad. J. Math.* **33** (1981), 1459–1468.
85. L.G. Chouinard, II. Projective modules over Krull semigroup rings. *Michigan Math. J.* **29** (1982), 143–148.
86. CoCoATeam. CoCoA: a system for doing computations in commutative algebra. Available at http://cocoa.dima.unige.it.
87. P.M. Cohn. The similarity reduction of matrices over a skew field. *Math. Z.* **132** (1973), 151–163.
88. A. Conca, J. Herzog, and G. Valla. SAGBI bases with applications to blow-up algebras. *J. Reine Angew. Math.* **474** (1996), 113–138.
89. A. Conca, S. Hoşten, and R.R. Thomas. Nice initial complexes of some classical ideals. In *Algebraic and geometric combinatorics*, C.A. Athanasiadis, Ed., Contemp. Math. **423**. 2006, pp. 11–42.
90. A. Conca, N.V. Trung, and G. Valla. Koszul property for points in projective spaces. *Math. Scand.* **89** (2001), 201–216.
91. W. Cook, J. Fonlupt, and A. Schrijver. An integer analogue of Carathéodory's theorem. *J. Combinatorial Theory Ser. B* **40** (1986), 63–70.

92. T. Coquand. On seminormality. *J. Algebra* **305**, 1 (2006), 577–584.
93. G. Cortiñas. The obstruction to excision in K-theory and in cyclic homology. *Invent. Math.* **164** (2006), 143–173.
94. G. Cortiñas, C. Haesemeyer, M. Schlichting, and C.A. Weibel. Cyclic homology, *cdh*-cohomology and negative K-theory. *Ann. Math.* **167** (2008), 549–573.
95. G. Cortiñas, C. Haesemeyer, M.E. Walker, and C.A. Weibel. Bass' *NK*-groups and *cdh*-fibrant Hochschild homology. *Preprint.*
96. G. Cortiñas, C. Haesemeyer, M.E. Walker, and C.A. Weibel. The K-theory of toric varieties. *Trans. Amer. Math. Soc.* (to appear).
97. G. Cortiñas, C. Haesemeyer, and C.A. Weibel. K-regularity, *cdh*-fibrant Hochschild homology, and a conjecture of Vorst. *J. Amer. Math. Soc.* **21** (2008), 547–561.
98. D.L. Costa. Seminormality and projective modules. In *Paul Dubreil and Marie-Paule Malliavin Algebra Seminar*, Lect. Notes Math. **924**. Springer, 1982, pp. 400–412.
99. D.A. Cox. The homogeneous coordinate ring of a toric variety. *J. Algebr. Geom.* **4** (1995), 17–50.
100. D.A. Cox. Update on toric geometry. In *Geometry of toric varieties*, L. Bonavero and M. Brion, Eds., Sémin. Congr. **6**. Soc. Math. de France, 2002, pp. 1–41.
101. D.A. Cox, J. Little, and H. Schenck. *Toric varieties*. In preparation.
102. V.I. Danilov. The geometry of toric varieties. *Russian Math. Surveys* **33** (1978), 97–154.
103. V.I. Danilov. Cohomology of algebraic varieties. In *Current problems in mathematics. Fundamental directions, Vol. 35 (Russian)*, Itogi Nauki i Tekhniki. 1989, pp. 5–130, 272.
104. B.H. Dayton. The Picard group of a reduced G-algebra. *J. Pure Appl. Algebra* **59** (1989), 237–253.
105. C. De Concini, D. Eisenbud, and C. Procesi. *Hodge Algebras*. Astérisque **91**, 1982.
106. J.A. De Loera, J. Rambau, and F. Santos. *Triangulations of point sets: Applications, Structures, Algorithms*. In preparation.
107. M. Demazure. Sous-groupes algébriques de rang maximum du groupe de Cremona. *Ann. Sci. Ec. Norm. Sup. (4)* **3** (1970), 507–588.
108. A. Demushkin. Combinatoric invariance of toric singularities. *Mosc. Univ. Math. Bull.* **37** (1982), 104–111.
109. R.K. Dennis and M. Krusemeyer. $K_2(A[X,Y]/(XY))$, a problem of Swan and related computations. *J. Pure Appl. Algebra* **15** (1979), 125–148.
110. X. Dong. Canonical modules of semigroup rings and a conjecture of Reiner. *Discrete Comput. Geom.* **27** (2002), 85–97.
111. E. Ehrhart. *Polynômes arithméthiques et méthode des polyédres en combinatoire*. Birkhäuser, Boston, 1977.
112. D. Eisenbud. *Commutative algebra with a view toward algebraic geometry*. Springer, Berlin–Heidelberg–New York, 1995.
113. D. Eisenbud, A. Reeves, and B. Totaro. Initial ideals, Veronese subrings, and rates of algebras. *Adv. Math.* **109** (1994), 168–187.
114. D. Eisenbud and B. Sturmfels. Binomial ideals. *Duke Math. J.* **84** (1996), 1–45.
115. E.B. Elliott. On linear homogeneous diophantine equations. *Quart. J. Math.* **34** (1902), 348–377.
116. K. Eto. Multiplicity and Hilbert–Kunz multiplicity of monoid rings. *Tokyo J. Math.* **25** (2002), 241–245.
117. G. Ewald. *Combinatorial convexity and algebraic geometry*. Springer, Berlin–Heidelberg–New York, 1996.
118. G. Ewald and A. Schmeinck. Representation of the Hirzebruch–Kleinschmidt varieties by quadrics. *Beitr. Algebra Geom.* **34** (1993), 151–156.

119. G. Ewald and U. Wessels. On the ampleness of invertible sheaves in complete projective toric varieties. *Result. Math.* **19** (1991), 275–278.
120. M. Farber, J.-C. Hausman, and D. Schuetz. On the conjecture of Kevin Walker. *Preprint.*
121. E.M. Feichtner and S. Yuzvinsky. Chow rings of toric varieties defined by atomic lattices. *Invent. Math.* **155** (2004), 515–536.
122. K.-H. Fieseler. Rational intersection cohomology of projective toric varieties. *J. Reine Angew. Math.* **413** (1991), 88–98.
123. D.R. Finston and S. Maubach. The automorphism group of certain factorial threefolds and a cancellation problem. *Isr. J. Math.* **163** (2008), 369–381.
124. R.T. Firla and G.M. Ziegler. Hilbert bases, unimodular triangulations, and binary covers of rational polyhedral cones. *Discrete Comp. Geom.* **21** (1999), 205–216.
125. N. Fitchas and A. Galligo. Nullstellensatz effectif et conjecture de Serre (théorème de Quillen-Suslin) pour le calcul formel. *Math. Nachr.* **149** (1990), 231–253.
126. R.M. Fossum. *The divisor class group of a Krull domain.* Springer, Berlin–Heidelberg–New York, 1973.
127. G. Freudenburg and P. Russell. Open problems in affine algebraic geometry. In *Affine algebraic geometry*, J. Gutierrez, Ed., Contemp. Math. **369**. 2005, pp. 1–30.
128. E.M. Friedlander and C.A. Weibel. An overview of algebraic K-theory. In *Algebraic K-theory and its applications*, H. Bass et al., Eds. World Sci. Publ., 1999, pp. 1–119.
129. F.G. Frobenius. *Gesammelte Abhandlungen, Band 3.* Springer, Berlin–Heidelberg–New York, 1968.
130. R. Fröberg. Determination of a class of Poincaré series. *Math. Scand.* **37** (1975), 29–39.
131. R. Fröberg. Koszul algebras. In *Advances in commutative ring theory*, D.E. Dobbs et al., Eds., Lect. Notes Pure Appl. Math **205**, M. Dekker, 1999, pp. 337–350.
132. W. Fulton. *Introduction to intersection theory in algebraic geometry.* CBMS Regional Conference Series in Mathematics **54**, Amer. Math. Soc., Providence, RI, 1984.
133. W. Fulton. *Introduction to toric varieties.* Princeton University Press, Princeton, NJ, 1993.
134. W. Fulton. *Intersection theory*, second ed. Springer, Berlin–Heidelberg–New York, 1998.
135. W. Fulton and R. MacPherson. *Categorical framework for the study of singular spaces.* Mem. Amer. Math. Soc. **243**, 1981.
136. W. Fulton, R. MacPherson, F. Sottile, and B. Sturmfels. Intersection theory on spherical varieties. *J. Algebr. Geom.* **4** (1995), 181–193.
137. W. Fulton and B. Sturmfels. Intersection theory on toric varieties. *Topology* **36** (1997), 335–353.
138. V. Gasharov, I. Peeva, and V. Welker. Rationality for generic toric rings. *Math. Z.* **233** (2000), 93–102.
139. E. Gawrilow and M. Joswig. polymake: a framework for analyzing convex polytopes. In *Polytopes — Combinatorics and Computation*, G. Kalai and G.M. Ziegler, Eds. Birkhäuser, 2000, pp. 43–74.
140. I.M. Gelfand, M.M. Kapranov, and A.V. Zelevinsky. *Discriminants, resultants, and multidimensional determinants.* Birkhäuser, Boston, 1994.
141. A. Geroldinger and F. Halter–Koch. *Non-unique factorizations.* Pure and Applied Mathematics (Boca Raton) **278**, Chapman & Hall/CRC, Boca Raton, FL, 2006.
142. R. Gilmer. *Commutative semigroup rings.* University of Chicago Press, Chicago–London, 1984.
143. P.D. González Pérez and G. Gonzalez–Sprinberg. Analytical invariants of quasi-ordinary hypersurface singularities associated to divisorial valuations. *Kodai Math. J.* **27** (2004), 164–173.

144. T.G. GOODWILLIE. Relative algebraic K-theory and cyclic homology. *Ann. Math.* **124** (1986), 347–402.
145. P. GORDAN. Über die Auflösung linearer Gleichungen mit reellen Coefficienten. *Math. Ann.* **6** (1873), 23–28.
146. M. GORESKY AND R. MACPHERSON. Intersection homology theory. *Topology* **19** (1980), 135–165.
147. S. GOTO, N. SUZUKI, AND K.-I. WATANABE. On affine semigroup rings. *Japan J. Math.* **2** (1976), 1–12.
148. S. GOTO AND K.-I. WATANABE. On graded rings, II ($\mathbb{Z}^n$-graded rings). *Tokyo J. Math.* **1** (1978), 237–261.
149. D.R. GRAYSON. Higher algebraic K-theory: II. In *Algebr. K-Theory*, Lect. Notes Math. **551**. Springer, 1976, pp. 217–240.
150. D.R. GRAYSON AND M.E. STILLMAN. Macaulay 2, a software system for research in algebraic geometry. Available at http://www.math.uiuc.edu/Macaulay2/.
151. M. GREEN AND R. LAZARSFELD. On the projective normality of complete linear series on an algebraic curve. *Invent. Math.* **83** (1986), 73–90.
152. G.-M. GREUEL, G. PFISTER, AND H. SCHÖNEMANN. Singular. A computer algebra system for polynomial computations. Available from http://www.singular.uni-kl.de/.
153. P.A. GRILLET. *Commutative semigroups.* Kluwer, Dordrecht, 2001.
154. B. GRÜNBAUM. *Convex polytopes*, second ed. Springer, Berlin–Heidelberg–New York, 2003.
155. A. GROTHENDIECK. Sur la classification des fibrés holomorphes sur la sphère de Riemann. *Amer. J. Math.* **79** (1957), 121–138.
156. J. GUBELADZE. *Serre's generalized problem for affine rings, generated by monomials.* Tbilisi State University Press, 1982.
157. J. GUBELADZE. Anderson's conjecture and the maximal monoid class over which projective modules are free. *Math. USSR-Sb.* **63** (1989), 165–180.
158. J. GUBELADZE. The elementary action on unimodular rows over a monoid ring. *J. Algebra* **148** (1992), 135–161.
159. J. GUBELADZE. The elementary action on unimodular rows over a monoid ring II. *J. Algebra* **155** (1993), 171–194.
160. J. GUBELADZE. Geometric and algebraic representations of commutative cancellative monoids. *Proc. A. Razmadze Math. Inst., Georg. Acad. Sci.* **113** (1995), 31–81.
161. J. GUBELADZE. Nontriviality of $SK_1(R[M])$. *J. Pure Appl. Algebra* **104** (1995), 169–190.
162. J. GUBELADZE. The isomorphism problem for monoid rings of rank 2 monoids. *J. Pure Appl. Algebra* **105** (1995), 17–51.
163. J. GUBELADZE. The isomorphism problem for commutative monoid rings. *J. Pure Appl. Algebra* **129** (1998), 35–65.
164. J. GUBELADZE. K-theory of affine toric varieties. *Homology Homotopy Appl.* **1** (1999), 135–145.
165. J. GUBELADZE. Toric degenerations and vector bundles. *Proc. Amer. Math. Soc.* **127** (1999), 3493–3494.
166. J. GUBELADZE. Higher K-theory of toric varieties. *K-Theory* **28** (2003), 285–327.
167. J. GUBELADZE. Toric varieties with huge Grothendieck group. *Adv. Math.* **186** (2004), 117–124.
168. J. GUBELADZE. The nilpotence conjecture in K-theory of toric varieties. *Invent. Math.* **160** (2005), 173–216.
169. J. GUBELADZE. The Steinberg group of a monoid ring, nilpotence, and algorithms. *J. Algebra* **307** (2007), 461–496.

170. J. Gubeladze. Global coefficient ring in the nilpotence conjecture. *Proc. Amer. Math. Soc.* **136** (2008), 499–503.
171. J. Gubeladze. On Bass' question for finitely generated algebras over large fields. *Bull. London Math. Soc.* (to appear).
172. C. Haase and I.V. Melnikov. The reflexive dimension of a lattice polytope. *Ann. Comb.* **10** (2006), 211–217.
173. C. Haase and B. Nill. Lattices generated by skeletons of reflexive polytopes. *J. Combinatorial Theory Ser. A* **115** (2008), 340–344.
174. C. Haase and G.M. Ziegler. On the maximal width of empty lattice simplices. *Eur. J. Comb.* **21** (2000), 111–119.
175. C. Haesemeyer. Descent properties of homotopy K-theory. *Duke Math. J.* **125** (2004), 589–620.
176. E. Hamann. The R-invariance of $R[X]$. *J. Algebra* **35** (1975), 1–16.
177. E. Hamann. A postscript on steadfast rings and R-algebras in one variable. *J. Algebra* **83** (1983), 490–501.
178. R. Hartshorne. *Algebraic geometry*. Springer, Berlin–Heidelberg–New York, 1977.
179. R. Hartshorne. Algebraic vector bundles on projective spaces: a problem list. *Topology* **18** (1979), 117–128.
180. M. Hashimoto. Determinantal ideals without minimal free resolutions. *Nagoya Math. J.* **118** (1990), 203–216.
181. M. Hazewinkel and C.F. Martin. A short elementary proof of Grothendieck's theorem on algebraic vectorbundles over the projective line. *J. Pure Appl. Algebra* **25** (1982), 207–211.
182. M. Hellus, L.T. Hoa, and J. Stückrad. Gröbner bases of simplicial toric ideals. *Preprint*.
183. M. Hering, H. Schenck, and G.G. Smith. Syzygies, multigraded regularity and toric varieties. *Compositio Math.* **142** (2006), 1499–1506.
184. J. Herzog and T. Hibi. Discrete polymatroids. *J. Algebr. Comb.* **16** (2002), 239–268.
185. J. Herzog and T. Hibi. Distributive lattices, bipartite graphs and Alexander duality. *J. Algebr. Comb.* **22** (2005), 289–302.
186. J. Herzog, V. Reiner, and V. Welker. The Koszul property in affine semigroup rings. *Pac. J. Math.* **186** (1998), 39–65.
187. L. Hesselholt. K-theory of truncated polynomial algebras. In *Handbook of K-theory. Vol. 1, 2.* Springer, Berlin, 2005, pp. 71–110.
188. T. Hibi. *Algebraic combinatorics on convex polytopes.* Carslaw Publications, Glebe, 1992.
189. D. Hilbert. Über die Theorie der algebraischen Formen. *Math. Ann.* **36** (1890), 473–534.
190. D. Hilbert. *Theory of algebraic invariants. Lectures. Transl. by R.C. Laubenbacher, ed. and with an introduction by Bernd Sturmfels.* Cambridge University Press, Cambridge, 1993.
191. H. Hironaka. Triangulations of algebraic sets. In *Algebraic geometry*, Proc. Sympos. Pure Math. **29**. Amer. Math. Soc., 1975, pp. 165–185.
192. A. Hirschowitz. Le groupe de Chow équivariant. (Equivariant Chow group). *C.R. Acad. Sc. Paris Sér. I* **298** (1984), 87–89.
193. L.T. Hoa. The Gorenstein property depends upon characteristic for affine semigroup rings. *Arch. Math.* **56** (1991), 228–235.
194. M. Hochster. Nonuniqueness of coefficient rings in a polynomial ring. *Proc. Amer. Math. Soc.* **34** (1972), 81–82.
195. M. Hochster. Rings of invariants of tori, Cohen–Macaulay rings generated by monomials, and polytopes. *Ann. Math.* **96** (1972), 318–337.
196. M. Hochster and C. Huneke. Applications of the existence of big Cohen–Macaulay algebras. *Adv. Math.* **113** (1995), 45–117.

197. M. Hochster and J.L. Roberts. Rings of invariants of reductive groups acting on regular rings are Cohen–Macaulay. *Adv. Math.* **13** (1974), 115–175.
198. M. Hochster and J.L. Roberts. The purity of the Frobenius and local cohomology. *Adv. Math.* **21** (1976), 117–172.
199. G. Horrocks. Projective modules over an extension of a local ring. *Proc. Lond. Math. Soc., III. Ser.* **14** (1964), 714–718.
200. S. Hoşten. A survey of toric initial ideals. In *Commutative Algebra and Combinatorics*, W. Bruns, Ed., Ramanujan Math. Soc. Lect. Notes Ser. **4** (Part II). Ramanujan Math. Soc., 2007, pp. 39–90.
201. S. Hoşten and B. Sturmfels. GRIN: an implementation of Gröbner bases for integer programming. In *Integer programming and combinatorial optimization*, Lect. Notes Comput. Sci. **920**. Springer, 1995, pp. 267–276.
202. J.E. Humphreys. *Linear algebraic groups*. Springer, Berlin–Heidelberg–New York, 1975.
203. C. Huneke. *Tight closure and its applications*. CBMS Regional Conference Series in Mathematics **88**, American Mathematical Society, Providence, RI, 1996.
204. C. Huneke, M.A. McDermott, and P. Monsky. Hilbert-Kunz functions for normal rings. *Math. Res. Lett.* **11** (2004), 539–546.
205. C. Huneke and I. Swanson. *Integral closure of ideals, rings, and modules*. London Math. Soc. Lecture Notes Series **336**, Cambridge University Press, Cambridge, 2006.
206. B. Ichim and T. Römer. On toric face rings. *J. Pure Appl. Algebra* **210** (2007), 249–266.
207. B. Ichim and T. Römer. On canonical modules of toric face rings. *Nagoya Math. J.* (to appear).
208. F. Ischebeck and R.A. Rao. *Ideals and reality*. Springer, Berlin–Heidelberg–New York, 2005.
209. M.-N. Ishida. The local cohomology groups of an affine semigroup ring. In *Algebraic geometry and commutative algebra in honor of Masayoshi Nagata, Vol. I*, H. Hijikata et al., Eds., Konikuniya, 1987, pp. 141–153.
210. B. Iversen. *Cohomology of sheaves*. Springer, Berlin–Heidelberg–New York, 1986.
211. H. Jarchow. *Locally convex spaces*. Teubner, Stuttgart, 1981.
212. A.N. Jensen. CaTS. A software system for toric state polytopes. Available at http://www.soopadoopa.dk/anders/cats/cats.html.
213. R. Joshua. Algebraic K-theory and higher Chow groups of linear varieties. *Math. Proc. Cambridge Philos. Soc.* **130**, 1 (2001), 37–60.
214. J. Jurkiewicz. Chow ring of projective non-singular torus embedding. *Colloq. Math.* **43** (1980), 261–270.
215. S. Kaliman, M. Koras, L. Makar–Limanov, and P. Russell. $\mathbb{C}^*$-actions on $\mathbb{C}^3$ are linearizable. *Electron. Res. Announc. Amer. Math. Soc.* **3** (1997), 63–71.
216. T. Kaneyama. Torus-equivariant vector bundles on projective spaces. *Nagoya Math. J.* **111** (1988), 25–40.
217. J.M. Kantor and K.S. Sarkaria. On primitive subdivisions of tetrahedra. *Pac. J. Math.* **211** (2003), 123–155.
218. M. Karoubi. *Homologie cyclique et K-théorie*. Astérisque **149**, 1987.
219. K. Karu. Hard Lefschetz theorem for nonrational polytopes. *Invent. Math.* **157** (2004), 419–447.
220. G. Kempf. The Hochster–Roberts theorem of invariant theory. *Michigan Math. J.* **26** (1979), 19–32.
221. G. Kempf, F. Knudsen, D. Mumford, and B. Saint–Donat. *Toroidal embeddings I*. Lect. Notes Math. **339**, Springer, Berlin–Heidelberg–New York, 1973.

222. A. Klyachko. Vector bundles, linear representations, and spectral problems. In *Proceedings of the International Congress of Mathematicians, Vol. II (Beijing, 2002)*, Higher Ed. Press, 2002, pp. 599–613.
223. F. Knop. Die Cohen–Macaulay–Eigenschaft von Invariantenringen. Available from http://www.math.rutgers.edu/~knop/papers/CM.html.
224. F. Knop. Nichtlinearisierbare Operationen halbeinfacher Gruppen auf affinen Räumen. *Invent. Math.* **105** (1991), 217–220.
225. R.J. Koelman. A criterion for the ideal of a projectively embedded toric surface to be generated by quadrics. *Beitr. Algebra Geom.* **34** (1993), 57–62.
226. R.J. Koelman. Generators for the ideal of a projectively embedded toric surface. *Tôhoku Math. J.* **45** (1993), 385–392.
227. M. Koras and P. Russell. $\mathbf{C}^*$-actions on $\mathbf{C}^3$: the smooth locus of the quotient is not of hyperbolic type. *J. Algebr. Geom.* **8** (1999), 603–694.
228. H. Kraft. Challenging problems on affine n-space. In *Séminaire Bourbaki, Vol. 1994/95*, Astérisque **237**. 1996, pp. 295–317.
229. H. Kraft and G. Schwarz. Finite automorphisms of affine N-space. In *Automorphisms of affine spaces.*, A. v. d. Essen, Ed., Kluwer, 1995, pp. 55–66.
230. M. Kreuzer and L. Robbiano. *Computational commutative algebra. I.* Springer, Berlin–Heidelberg–New York, 2000.
231. M. Kreuzer and H. Skarke. On the classification of reflexive polyhedra. *Commun. Math. Phys.* **185** (1997), 495–508.
232. M. Kreuzer and H. Skarke. Classification of reflexive polyhedra in three dimensions. *Adv. Theor. Math. Phys.* **2** (1998), 853–871.
233. M. Kreuzer and H. Skarke. Complete classification of reflexive polyhedra in four dimensions. *Adv. Theor. Math. Phys.* **4** (2000), 1209–1230.
234. M. Krusemeyer. Fundamental groups, algebraic K-theory, and a problem of Abhyankar. *Invent. Math.* **19** (1973), 15–47.
235. E. Kunz. *Introduction to commutative algebra and algebraic geometry.* Birkhäuser, Boston, 1985.
236. J.C. Lagarias and G.M. Ziegler. Private communication.
237. T.Y. Lam. *Serre's problem on projective modules.* Springer, Berlin–Heidelberg–New York, 2005.
238. S. Lang. *Algebra.* Springer, Berlin–Heidelberg–New York, 2002.
239. R.C. Laubenbacher and C.J. Woodburn. An algorithm for the Quillen-Suslin theorem for monoid rings. *J. Pure Appl. Algebra* **117/118** (1997), 395–429.
240. O. Laudal and A. Sletjøe. Betti numbers of monoid algebras. Applications to 2-dimensional torus embeddings. *Math. Scand.* **56** (1985), 145–162.
241. H. Lindel. On the Bass–Quillen conjecture concerning projective modules over polynomial rings. *Invent. Math.* **65** (1981), 319–323.
242. J. Liu, L.E. Trotter, and G.M. Ziegler. On the height of the minimal Hilbert basis. *Result. Math.* **23** (1993), 374–376.
243. J.-L. Loday. *Cyclic Homology*, second ed. Springer, Berlin–Heidelberg–New York, 1998.
244. A. Logar and B. Sturmfels. Algorithms for the Quillen-Suslin theorem. *J. Algebra* **145** (1992), 231–239.
245. D. Maclagan, S. Hoşten, and B.Sturmfels. Supernormal vector configurations. *J. Algebr. Comb.* **19** (2004), 297–313.
246. D. Maclagan and R.R. Thomas. Computational algebra and combinatorics of toric ideals. In *Commutative Algebra and Combinatorics*, Ramanujan Math. Soc. Lect. Notes Ser. **4** (Part I). Ramanujan Math. Soc., 2007, pp. 1–106.

247. P.A. MacMahon. *Combinatory analysis. Vol. I, II.* Chelsea Publishing Company., 1960. Originally published by Cambridge University Press, 1915, 1916.
248. S. Mandal. *Projective modules and complete intersections.* Lect. Notes Math. **1672**, Springer, Berlin–Heidelberg–New York, 1997.
249. M. Masuda. Equivariant algebraic vector bundles over affine toric varieties. In *Transformation group theory.* Korea Adv. Inst. Sci. Tech., 1997, pp. 76–84.
250. M. Masuda. Equivariant algebraic vector bundles over representations—a survey. In *Current trends in transformation groups*, K-Monogr. Math. **7**, Kluwer Acad. Publ., 2002, pp. 25–36.
251. M. Masuda, L. Moser-Jauslin, and T. Petrie. The equivariant Serre problem for abelian groups. *Topology* **35** (1996), 329–334.
252. M. Masuda and T. Petrie. Stably trivial equivariant algebraic vector bundles. *J. Amer. Math. Soc.* **8** (1995), 687–714.
253. H. Matsumura. *Commutative ring theory.* Cambridge University Press, Cambridge, 1989.
254. P. McMullen. The numbers of faces of simplicial polytopes. *Isr. J. Math.* **9** (1971), 559–570.
255. P. McMullen. Lattice invariant valuations on rational polytopes. *Arch. Math.* **31** (1978), 509–516.
256. P. McMullen. The polytope algebra. *Adv. Math.* **78** (1989), 76–130.
257. P. McMullen. On simple polytopes. *Invent. Math.* **113** (1993), 419–444.
258. P. McMullen. Separation in the polytope algebra. *Beitr. Algebra Geom.* **34** (1993), 15–30.
259. A.S. Merkurjev and I.A. Panin. K-theory of algebraic tori and toric varieties. *K-Theory* **12** (1997), 101–143.
260. E. Miller and V. Reiner. Reciprocal domains and Cohen-Macaulay d-complexes in $\mathbb{R}^d$. *Electron. J. Combin.* **11** (2004/06), Note 1, 9 pp. (electronic).
261. E. Miller and B. Sturmfels. *Combinatorial commutative algebra.* Springer, Berlin–Heidelberg–New York, 2005.
262. J. Milnor. *Introduction to algebraic K-theory.* Princeton University Press, Princeton, NJ, 1971.
263. M. Miyanishi and T. Sugie. Affine surfaces containing cylinderlike open sets. *J. Math. Kyoto Univ.* **20** (1980), 11–42.
264. R. Morelli. Pick's theorem and the Todd class of a toric variety. *Adv. Math.* **100** (1993), 183–231.
265. R. Morelli. The K-theory of a toric variety. *Adv. Math.* **100** (1993), 154–182.
266. D. Mumford, J. Fogarty, and F. Kirwan. *Geometric invariant theory*, third ed. Springer, Berlin–Heidelberg–New York, 1994.
267. J.R. Munkres. *Topology: a first course.* Prentice–Hall Inc., Englewood Cliffs, NJ, 1975.
268. M.P. Murthy. Projective $A[X]$-modules. *Proc. London Math. Soc.* **41** (1966), 453–456.
269. M.P. Murthy. A survey of obstruction theory for projective modules of top rank. In *Algebra, K-theory, groups, and education*, Contemp. Math. **243**. 1999, pp. 153–174.
270. M.P. Murthy and C. Pedrini. K_0 and K_1 of polynomial rings. In *Algebraic K-theory, II*, Lect. Notes Math. **342**. Springer, 1973, pp. 109–121.
271. I.M. Musson. Differential operators on toric varieties. *J. Pure Appl. Algebra* **95** (1994), 303–315.
272. M. Mustaţă. Vanishing theorems on toric varieties. *Tôhoku Math. J.* **54** (2002), 451–470.
273. M. Mustaţă and S. Payne. Ehrhart polynomials and stringy Betti numbers. *Math. Ann.* **333** (2005), 787–795.
274. M. Noether. Paul Gordan. *Math. Ann.* **75** (1914), 1–41.
275. T. Oda. *Convex bodies and algebraic geometry (An introduction to the theory of toric varieties).* Springer, Berlin–Heidelberg–New York, 1988.

276. S. Ogata. On projective toric varieties whose defining ideals have minimal generators of the highest degree. *Ann. Inst. Fourier* **53** (2003), 2243–2255.
277. S. Ogata. On higher syzygies of projective toric varieties. *Math. J. Okayama Univ.* **48** (2006), 47–55.
278. H. Ohsugi and T. Hibi. Toric ideals generalized by quadratic binomials. *J. Algebra* **218** (1999), 509–527.
279. C. Okonek, M. Schneider, and H. Spindler. *Vector bundles on complex projective spaces*. Birkhäuser, Boston, 1980.
280. E. O'Shea and R.R. Thomas. Toric initial ideals of Δ-normal configurations: Cohen-Macaulayness and degree bounds. *J. Algebr. Comb.* **21** (2005), 247–268.
281. H. Park and C.J. Woodburn. An algorithmic proof of Suslin's stability theorem for polynomial rings. *J. Algebra* **178** (1995), 277–298.
282. S. Payne. Lattice polytopes cut out by root systems and the Koszul property. *Preprint*.
283. S. Payne. Equivariant Chow cohomology of toric varieties. *Math. Res. Lett.* **13** (2006), 29–41.
284. S. Payne. Moduli of toric vector bundles. *Compositio Math.* (to appear).
285. S. Payne. Toric vector bundles, branched covers of fans, and the resolution property. *J. Algebr. Geom.* (to appear).
286. I. Peeva. Infinite free resolutions over toric rings. In *Syzygies and Hilbert Functions*, I. Peeva, Ed., Lect. Notes Pure Appl. Math **254**, Chapman&Hall/CRC, 2007, pp. 233–247.
287. I. Peeva, V. Reiner, and B. Sturmfels. How to shell a monoid. *Math. Ann.* **310** (1998), 379–393.
288. I. Peeva and B. Sturmfels. Syzygies of codimension 2 lattice ideals. *Math. Z.* **229** (1998), 163–194.
289. A. Polishchuk and L. Positselski. *Quadratic algebras*. Amer. Math. Soc., Providence, RI, 2005.
290. J.E. Pommersheim. Toric varieties, lattice points and Dedekind sums. *Math. Ann.* **295** (1993), 1–24.
291. J.E. Pommersheim and H. Thomas. Cycles representing the Todd class of a toric variety. *J. Amer. Math. Soc.* **17** (2004), 983–994.
292. D. Popescu. General Néron desingularizations. *Nagoya Math. J.* **100** (1985), 97–126.
293. S.B. Priddy. Koszul resolutions. *Trans. Amer. Math. Soc.* **152** (1970), 39–60.
294. A.V. Pukhlikov and A.G. Khovanskiĭ. The Riemann-Roch theorem for integrals and sums of quasipolynomials on virtual polytopes. *Algebra i Analiz* **4** (1992), 188–216.
295. D. Quillen. On the cohomology and K-theory of the general linear groups over a finite field. *Ann. Math.* **96** (1972), 552–586.
296. D. Quillen. Higher algebraic K-theory: I. In *Algebr. K-Theory I*, Lect. Notes Math. **341**, Springer, 1973, pp. 85–147.
297. D. Quillen. Projective modules over polynomial rings. *Invent. Math.* **36** (1976), 167–171.
298. D. Rees. *Lectures on the asymptotic theory of ideals*. LMS Lect. Note Ser. **113**, Cambridge University Press, Cambridge, 1988.
299. L. Reid and L.G. Roberts. Monomial subrings in arbitrary dimension. *J. Algebra* **236** (2001), 703–730.
300. L. Reid, L.G. Roberts, and M.A. Vitulli. Some results on normal homogeneous ideals. *Commun. Algebra* **31** (2003), 4485–4506.
301. L. Robbiano and M. Sweedler. Subalgebra bases. In *Commutative algebra*, W. Bruns and A. Simis, Eds., Lect. Notes Math. **1430**, Springer, 1990, pp. 61–87.
302. L.G. Roberts and B. Singh. Subintegrality, invertible modules and the Picard group. *Compositio Math.* **85** (1993), 249–279.

303. K.W. Roggenkamp. Group rings: units and the isomorphism problem. In *Group rings and class groups*, DMV Sem. **18**. Birkhäuser, Basel, 1992, pp. 1–152. With contributions by W. Kimmerle and A. Zimmermann.
304. J.-E. Roos. Commutative non Koszul algebras having a linear resolution of arbitrarily high order. Applications to torsion in loop space homology. *C.R. Acad. Sc. Paris Sér. I* **316** (1993), 113–128.
305. J.-E. Roos and B. Sturmfels. Un anneau toroidal à série de Poincaré–Betti irrationelle. (A toric ring with irrational Poincaré–Betti series). *C.R. Acad. Sc. Paris Sér. I* **326** (1998), 141–146.
306. J. Rosenberg. *Algebraic K-theory and its applications*. Springer, Berlin–Heidelberg–New York, 1994.
307. J. Rotman. *An introduction to homological algebra*. Academic Press, 1979.
308. D.E. Rush. Picard groups in abelian group rings. *J. Pure Appl. Algebra* **26** (1982), 101–114.
309. L. Sabalka. On rigidity and the isomorphism problem for tree braid groups. *Preprint*.
310. U. Schäfer and P. Schenzel. Dualizing complexes of affine semigroup rings. *Trans. Amer. Math. Soc.* **322** (1990), 561–582.
311. H. Schenck. Lattice polygons and Green's theorem. *Proc. Amer. Math. Soc.* **132** (2004), 3509–3512.
312. A. Schrijver. *Theory of linear and integer programming*. Wiley, Chichester, 1998.
313. A. Sebő. An introduction to empty lattice simplices. *Lect. Notes Comput. Sci.* **1610** (1999), 400–414.
314. A. Sebő. Hilbert bases, Carathéodory's theorem, and combinatorial optimization. In *Integer Programming and Combinatorial Optimization*, W.P.R. Kannan, Ed., University of Waterloo Press, Waterloo, 1990, pp. 431–456.
315. A. Seidenberg. A note on the dimension theory of rings. *Pac. J. Math.* **3** (1953), 505–512.
316. J.-P. Serre. Faisceaux algébriques cohérents. *Ann. Math.* **61** (1955), 197–278.
317. J.-P. Serre. Modules projectifs et espaces fibrés á fibre vectorielle. *Semin. Dubreil* **23** (1957-58).
318. J.-P. Serre. *Algèbre locale. Multiplicités*. Lect. Notes Math. **11**, Springer, Berlin–Heidelberg–New York, 1965.
319. C.S. Seshadri. Triviality of vector bundles over the affine space K^2. *Proc. Nat. Acad. Sci. U.S.A.* **44** (1958), 456–458.
320. I.P. Shestakov and U.U. Umirbaev. The tame and the wild automorphisms of polynomial rings in three variables. *J. Amer. Math. Soc.* **17** (2004), 197–227.
321. B. Singh. The Picard group and subintegrality in positive characteristic. *Compositio Math.* **95** (1995), 309–321.
322. V. Srinivas. K_1 of the cone over a curve. *J. Reine Angew. Math.* **381** (1987), 37–50.
323. R.P. Stanley. Linear homogeneous diophantine equations and magic labelings of graphs. *Duke Math. J.* **40** (1973), 607–632.
324. R.P. Stanley. Combinatorial reciprocity theorems. *Adv. Math.* **14** (1974), 194–253.
325. R.P. Stanley. Hilbert functions of graded algebras. *Adv. Math.* **28** (1978), 57–83.
326. R.P. Stanley. Combinatorics and invariant theory. In *Relations between combinatorics and other parts of mathematics*, Proc. Sympos. Pure Math. **XXXIV**. Amer. Math. Soc., 1979, pp. 345–355.
327. R.P. Stanley. Decompositions of rational convex polytopes. *Ann. Discrete Math.* **6** (1980).
328. R.P. Stanley. The number of faces of a simplicial convex polytope. *Adv. Math.* **35** (1980), 236–238.
329. R.P. Stanley. Linear diophantine equations and local cohomology. *Invent. Math.* **68** (1982), 175–193.

330. R.P. Stanley. *Enumerative combinatorics, Vol. I.* Wadsworth & Brooks/Cole, 1986.
331. R.P. Stanley. Generalized h-vectors, intersection cohomology of toric varieties, and related results. In *Commutative algebra and combinatorics*, Advanced Studies in Pure Mathematics **11**, 1987, pp. 187–213.
332. R.P. Stanley. *Combinatorics and commutative algebra*, second ed. Birkhäuser, Boston, 1996.
333. M.R. Stein and R.K. Dennis. K_2 of radical ideals and semi-local rings revisited. In *Algebr. K-Theory II*, Lect. Notes Math. **342**, Springer, 1973, pp. 281–303.
334. R. Steinberg. Générateurs, relations et revêtements de groupes algébriques. In *Centre Belge Rech. Math., Colloque Theor. Groupes Algebr.*, Bruxelles, 1962, pp. 113–127.
335. J. Stienstra. Operations in the higher K-theory of endomorphisms. In *Current trends in algebraic topology, Semin. London/Ont. 1981, CMS Conf. Proc. 2, 1*, Amer. Math. Soc., 1982, pp. 59–115.
336. B. Sturmfels. Gröbner bases of toric varieties. *Tôhoku Math. J.* **43** (1991), 249–261.
337. B. Sturmfels. On the Newton polytope of the resultant. *J. Algebr. Comb.* **3** (1994), 207–236.
338. B. Sturmfels. *Gröbner bases and convex polytopes.* Amer. Math. Soc., Providence, RI, 1996.
339. B. Sturmfels. Equations defining toric varieties. *Proc. Symp. Pure Math.* **62** (1997), 437–449.
340. H. Sumihiro. Equivariant completion. *J. Math. Kyoto Univ.* **14** (1974), 1–28.
341. A.A. Suslin. Projective modules over a polynomial ring are free. *Sov. Mat. Dokl.* **17** (1976), 1160–1164.
342. A.A. Suslin. On the structure of the special linear group over polynomial rings. *Math. USSR-Izv.* **11** (1977), 221–238.
343. A.A. Suslin and M. Wodzicki. Excision in algebraic K-theory. *Ann. Math.* **136** (1992), 51–122.
344. R.G. Swan. Vector bundles and projective modules. *Trans. Amer. Math. Soc.* **105** (1962), 264–277.
345. R.G. Swan. *Algebraic K-theory.* Lect. Notes Math. **76**, Springer, Berlin–Heidelberg–New York, 1968.
346. R.G. Swan. Excision in algebraic K-theory. *J. Pure Appl. Algebra* **1** (1971), 221–252.
347. R.G. Swan. On seminormality. *J. Algebra* **67** (1980), 210–229.
348. R.G. Swan. Projective modules over binary polyhedral groups. *J. Reine Angew. Math.* **342** (1983), 66–172.
349. R.G. Swan. Projective modules over Laurent polynomial rings. *Trans. Amer. Math. Soc.* **237** (1987), 111–120.
350. R.G. Swan. Vector bundles, projective modules and the K-theory of spheres. In *Algebraic topology and algebraic K-theory*, Ann. of Math. Stud. **113**. Princeton Univ. Press, Princeton, NJ, 1987, pp. 432–522.
351. R.G. Swan. Gubeladze's proof of Anderson's conjecture. *Contemp. Math.* **124** (1992), 215–250.
352. R.G. Swan. Higher algebraic K-theory. In *K-theory and algebraic geometry: connections with quadratic forms and division algebras*, B. Jacob et al., Eds., Proc. Sympos. Pure Math. **58**. Amer. Math. Soc., 1995, pp. 247–293.
353. R.G. Swan. Néron–Popescu desingularization. In *Lectures in algebra and geometry*, M.-C. Kang, Ed. International Press, 1998, pp. 135–192.
354. R.W. Thomason and T. Trobaugh. Higher algebraic K-theory of schemes and of derived categories. In *The Grothendieck Festschrift, Vol. III*. Birkhäuser, 1990, pp. 247–435.

355. B. Totaro. The resolution property for schemes and stacks. *J. Reine Angew. Math.* **577** (2004), 1–22.
356. B. Totaro. Chow groups, Chow cohomology, and linear varieties. *J. Algebr. Geom.* (to appear).
357. C. Traverso. Seminormality and Picard group. *Ann. Scuola Norm. Sup. Pisa (3)* **24** (1970), 585–595.
358. N.V. Trung and L.T. Hoa. Affine semigroups and Cohen–Macaulay rings generated by monomials. *Trans. Amer. Math. Soc.* **298** (1986), 145–167.
359. M.S. Tulenbaev. The Steinberg group of a polynomial ring. *Math. USSR-Sb.* **45** (1983), 139–154.
360. J.G. van der Corput. Konstruktion der Minimalbasen für spezielle Diophantische Systeme von linear-homogenen Gleichungen und Ungleichungen. *Proc. Royal Acad. Amsterdam* **34** (1931), 515–523.
361. W. van der Kallen, H. Maazen, and J. Stienstra. A presentation for some $K_2(n,R)$. *Bull. Amer. Math. Soc.* **81** (1975), 934–936.
362. L. Vaserstein. On stabilization of the general linear group over a ring. *Math. USSR-Sb.* **79** (1969), 405–424.
363. G. Vezzosi and A. Vistoli. Higher algebraic K-theory for actions of diagonalizable groups. *Invent. Math.* **153** (2003), 1–44.
364. R.H. Villarreal. *Monomial algebras.* Marcel Dekker, New York, 2001.
365. A. Vistoli. Alexander duality in intersection theory. *Compositio Math.* **70**, 3 (1989), 199–225.
366. M. von Thaden. Improved bounds for covers and triangulations of lattice polytopes. *In preparation.*
367. M. von Thaden. Unimodular covers and triangulations of lattice polytopes. Dissertation, Osnabrück, 2007.
368. T. Vorst. Localization of the K-theory of polynomial extensions. *Math. Ann.* **244** (1979), 33–53.
369. T. Vorst. The Serre problem for discrete Hodge algebras. *Math. Z.* **184** (1983), 425–433.
370. F. Waldhausen. Algebraic K-theory of generalized free products. I, II. *Ann. Math.* **108** (1978), 135–204.
371. K.-i. Watanabe. Hilbert–Kunz multiplicity of toric rings. *Proc. Inst. Nat. Sci., Nihon Univ.* **35** (2000), 173–177.
372. K.-i. Watanabe and K.-i. Yoshida. Hilbert–Kunz multiplicity and an inequality between multiplicity and colength. *J. Algebra* **230** (2000), 295–317.
373. W.C. Waterhouse. Automorphisms of $\det(x_{ij})$: the group scheme approach. *Adv. Math.* **65** (1987), 171–203.
374. C.A. Weibel. Module structures on the K-theory of graded rings. *J. Algebra* **105** (1987), 465–483.
375. C.A. Weibel. Pic is a contracted functor. *Invent. Math.* **103** (1991), 351–377.
376. C.A. Weibel. The development of algebraic K-theory before 1980. In *Algebra, K-theory, groups, and education*, Contemp. Math. **243**. 1999, pp. 211–238.
377. G.K. White. Lattice tetrahedra. *Canad. J. Math.* **16** (1964), 389–396.
378. N.L. White. The basis monomial ring of a matroid. *Adv. Math.* **24** (1977), 292–297.
379. S. Yuzvinsky. Cohen–Macaulay rings of sections. *Adv. Math.* **63** (1987), 172–195.
380. S. Yuzvinsky. Projective modules over rings of global sections of flasque sheaves. *J. Pure Appl. Algebra* **65** (1990), 191–197.
381. G.M. Ziegler. *Lectures on polytopes.* Springer, Berlin–Heidelberg–New York, 1998.

Notation

$\mathsf{Var}(\Bbbk)$	category of $\Bbbk$-varieties, 357
V_χ	weight space of V with respect to character χ, 168
$\mathsf{Vect}(\mathscr{V})$	category of vector bundles on variety $\mathscr{V}$, 358
$\mathsf{Vect}(\mathscr{V}, F, S)$	418
$\mathrm{vert}(P)$	set of vertices of P, 6
$\mathscr{V}(\mathscr{F})$	toric variety corresponding to fan $\mathscr{F}$, 364
vol_L	volume function defined by lattice L, 69
V^*	space of linear forms on V, 12
$\mathbb{V}_{\mathscr{V}}$	trivial G-bundle associated with G-module V, 414
w	Waterman map, 99
$\mathrm{Witt}(\Lambda)$	ring of Witt vectors over Λ, 405
$\lceil x \rceil$	$\min\{z \in \mathbb{Z} : z \geq x\}$, 52
$\lfloor x \rfloor$	$\max\{z \in \mathbb{Z} : z \leq x\}$, 52
$X(G)$	character group of algebraic group G, 167
$x_i(\mathscr{E})$	ith Chern root of vector bundle $\mathscr{E}$, 385
x_{ij}^a	Steinberg symbol, 329
$[x, y]$	line segment between x and y, 4
$\mathbb{Z}, \mathbb{Z}_+$	(nonnegative) integers
$Z_k^{\mathbb{T}}$	group of $\mathbb{T}$-stable k-cycles, 379
$Z_k^{\mathbb{T}}$	group of $\mathbb{T}$-stable k-cycles, 367
$Z_k(\mathscr{V})$	group of k-cycles on variety $\mathscr{V}$, 359, 376
$\mathbb{Z}(L, 1)$	69

Index